SOUTH-WESTERN

# MATH MATTERS BOOK 2

## An Integrated Approach

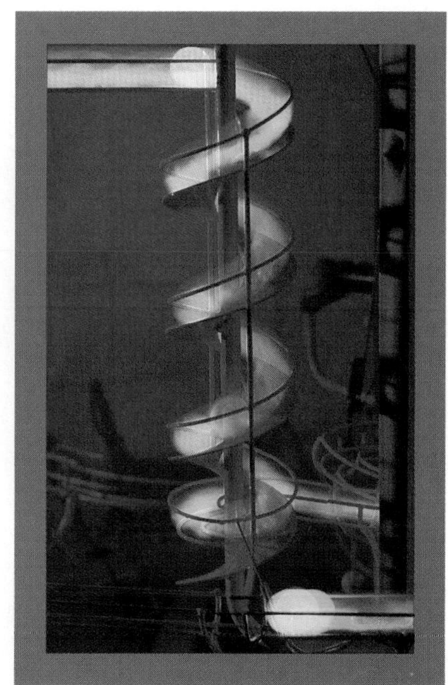


**CHICHA LYNCH**
*Capuchino High School*
*San Bruno, California*

**EUGENE OLMSTEAD**
*Elmira Free Academy*
*Elmira, New York*



SOUTH-WESTERN PUBLISHING CO.

# About the **Math Matters** Authors

## Chicha Lynch

Ms. Lynch is mathematics department head at Capuchino High School, San Bruno, California. As a specialist with the California Mathematics Project, Ms. Lynch helped her school district redesign courses to meet the state framework for Math A. Since 1988 she has served the state of California as a Mentor Teacher for Math A/B. Ms. Lynch is a graduate of the University of Florida and received the LaBoskey Award in 1988 from Stanford University for her contribution to its teacher education program. Most recently Ms. Lynch has been a research associate at the Far West Educational Laboratory, San Francisco, developing assessment for secondary mathematics teachers for certification by the National Board of Professional Teaching Standards. She was a state finalist in 1988 for the Presidential Award for Excellence in Mathematics Teaching.

## Eugene Olmstead

Mr. Olmstead is a mathematics teacher at Elmira Free Academy, Elmira, New York. He has had 20 years of public school experience teaching courses from general mathematics through calculus. Mr. Olmstead earned his B.S. in mathematics at State University College at Geneseo, New York, and received his M.S. in mathematics education at Elmira College, Elmira, New York. Mr. Olmstead has worked as a counselor at the Calculator and Computer Precalculus Institutes at The Ohio State University and frequently gives workshops on teaching mathematics using technology. He was a state finalist in 1991 for the Presidential Award for Excellence in Mathematics Teaching.

# Program Consultants

Bert K. Waits
Professor of Mathematics
The Ohio State University
Columbus, Ohio

Henry S. Kepner, Jr.
Professor of Mathematics
and Computer Education
University of Wisconsin-Milwaukee
Milwaukee, Wisconsin

Tommy Eads
Mathematics Teacher
North Lamar High School
Paris, Texas

Ann M. Farrell
Assistant Professor of
Mathematics and Statistics
Wright State University
Dayton, Ohio

# Reviewers

Rosemary Garmann
Mathematics Curriculum Consultant
Hamilton County Office of Education
Cincinnati, Ohio

Karen Nixon
Mathematics Teacher
Blue Valley High School
Stillwell, Kansas

Mary Alice Gembach
Mathematics Teacher
Powell Valley High School
Big Stone Gap, Virginia

Edward Okuniewski
Mathematics Teacher
Andover High School
Bloomfield Hills, Michigan

Gerald A. Haber
Assistant Principal
Winthrop Intermediate School
Brooklyn, New York

David M. Otte
Principal
St. Henry High School
Erlanger, Kentucky

Joan C. Lamborne
Supervisor of Mathematics
and Computers
Egg Harbor Township High School
Pleasantville, New Jersey

Linda K. Schoeff
Director, Tech Prep
Programs and Services
Department of Workforce
Development
Indianapolis, Indiana

Gordon Lewis
Supervising Director of Mathematics
District of Columbia Public Schools
Washington, D.C.

Jean Thiry
Mathematics Department Chair
Glenbard West High School
Glen Ellyn, Illinois

# Contents

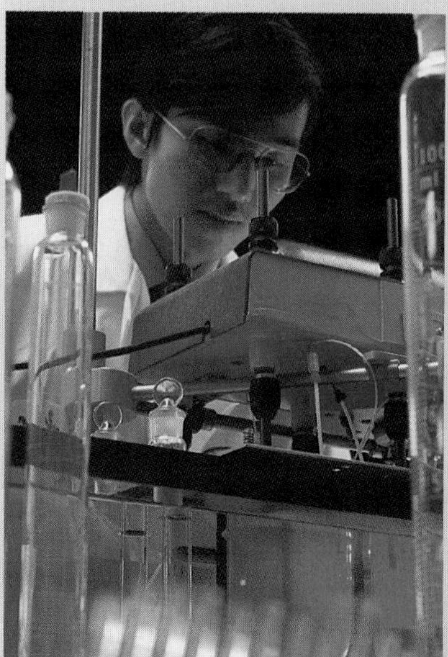

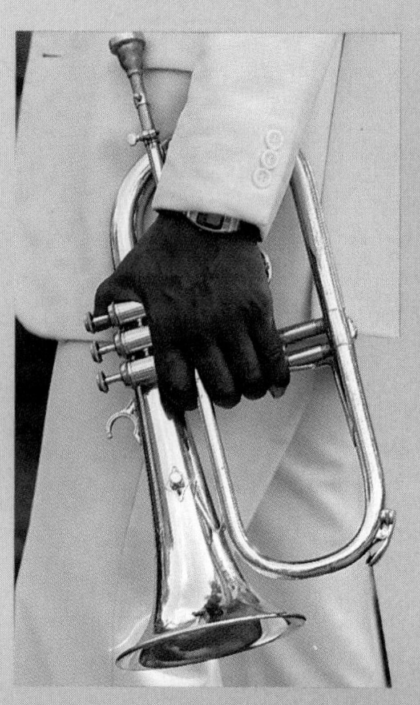

# Preface

Success stories can be found in all walks of life—in sports, in health services, in entertainment, in building and construction—in fact, in just about any career you can think of. Successful people, no matter what their field, are similar in two special ways: They set goals for themselves, and then they make and follow plans that help them to reach these goals. This is also true for your success as a student of mathematics. If you take some time now to learn how to use this textbook, it will greatly improve your chances for success in this course. Here are a few ideas that you may find worthwhile.

## GETTING STARTED WITH NOTEBOOKS AND LEARNING JOURNALS

Before you begin using *Math Matters,* you will find that keeping an organized notebook or learning journal is essential to successful learning throughout the school year. It is important that the organization of your notebook or learning journal reflect your particular needs and learning style. Begin by thinking about the kinds of information you will want to record and how you might easily get access to this information. Below are some ideas to help you organize your notebook or journal. Be sure, however, to listen carefully to what your teacher requires of you, since he or she may wish to add or change the kinds of information you record.

### Vocabulary List

For each unfamiliar word, definition, or property, write the correct definition as given in the text or Glossary. Also include an example if that will help. Then rewrite the definition and example in your own words. Review your vocabulary list periodically to make sure that you maintain a clear idea of what each term means.

### Notes and Questions

The contents of this part of your journal will depend somewhat on your teacher's method of instruction. This section should include any important problems, exercises, and solutions that your teacher puts on the chalkboard for discussion. It also might include highlights of key points that your teacher makes during a class. You should write any questions, examples, or exercises that need further explanation. You may want to mark your questions in some way—perhaps, with an asterisk (*) or a question mark in the margin. Then, when you get a satisfactory explanation or you are able to answer your own question, you can cross out or circle the symbol you used to indicate that you no longer need to be concerned about that problem.

### Summaries and Conclusions

In this section, you can reflect in your own words on what you have learned, which topics interest you the most, and which topics have been difficult. Summarizing and drawing conclusions is often the last step in mastering what you have learned. These skills are also helpful in clarifying areas for further study.

### Formulas, Properties, and Drawings

You may wish to put this section at the back of your notebook or learning journal so that you can quickly find what you need. Here you can record commonly used formulas, properties, and drawings that explain or summarize key information.

### Homework and Assignments

Have a special place in your notebook or learning journal where you can write the complete assignment and make a regular habit of recording all the assignments you receive. Doing these things will prevent last-minute "panic" phone calls to classmates, asking, "Do you know what the math homework is?" Be sure that you have some special way of indicating long-term projects so that the due day doesn't sneak up on you unexpectedly. Your teacher will tell you whether you should keep your homework in your notebook or turn it in each day.

## LEARN HOW TO USE THIS BOOK

Math Matters has a consistent organization and many recurring features. It is wise to get to know how the book is organized and what the purpose of each feature is. Knowing what to expect will allow you to focus efficiently on what you need to learn.

### Skills Preview

Each chapter begins with a Skills Preview that will help you to discover which topics of the chapter you may already know and which will require careful study. Don't be concerned by what you get wrong on the Skills Preview. By identifying what you already know, you will be able to spend more time on those topics that are new to you.

### Chapter Opener

The Chapter Opener helps to set the stage for the topics you are about to study. It tells you about the subject matter and theme of the chapter and presents some fascinating data and visual information for you to examine. Each opener requires that you use the data presented to make decisions and that you complete a group project or exploration tied to the mathematical content and theme of the chapter.

Every chapter is divided into a group of numbered sections, each of which is either four pages or two pages long. Every four-page section touches on a group of related math concepts or skills and follows a consistent organization.

### Explore and Explore/Working Together

Each four-page section begins with an activity that provides a stimulating introduction to the mathematical concepts or skills to be studied. These activities often involve hands-on investigations that lead to meaningful mathematical insights. Many of the Explore activities require you to work cooperatively with other students.

### Skills Development

This section explains the concepts and skills you are expected to master. Here you will find helpful discussions, along with essential definitions. The Skills Development presents the key points of the lesson through model Examples and worked-out Solutions for you to study and apply later on.

### Try These

These are questions and problems that you will likely discuss in your classroom. The items in Try These are similar to the ones presented in the Examples and Solutions of the Skills Development, so they are an excellent way for you to determine if you really have absorbed the key points of each lesson.

### Exercises

The Exercises provide numerous opportunities to practice and apply the mathematical concepts and skills that you have studied and to make connections with other areas of mathematics and with the use of mathematics in real-world situations and in other subject areas. Each section

contains three levels of Exercises, which provide increasingly challenging practice and problem solving. These items appear under the headings Practice/Solve Problems, Extend/Solve Problems, and Think Critically/Solve Problems.

Some four-page sections end with a special full-page feature called Problem Solving Applications. These pages make connections between the mathematics presented in the section and the practical application of those mathematical skills and concepts within a particular career or field.

**Problem Solving Strategies**
Each two-page Problem Solving Strategies section focuses on one important strategy. Each Problem Solving Strategies section has a brief introduction, model Problems and Solutions, and practice Problems. Problem Solving Strategies, introduced and applied throughout the book, include, for example, Make an Algebraic Model, Draw a Graph, Use a System of Equations, Explore with Simulations, and Solve a Simpler Problem. From time to time throughout the book there will be a special Choose a Strategy lesson that presents a variety of problems involving any of the strategies that have been presented up to that point in the book.

**Problem Solving Skills**
These are special two-page sections that introduce a skill essential to successful problem solving. Each Problem Solving Skills section has a brief introduction, model Problems and Solutions, and practice Problems. Problem Solving Skills, introduced and applied throughout the book, include Choose a Method of Computation, Identify a Misleading Use of Statistics, Interpret Graphs, and Make Predictions.

Within the sections of each chapter you will encounter a variety of special features. The following is a description of the various features and their corresponding icons.

**Check Understanding**
This feature helps you to evaluate your understanding of key concepts and skills by posing questions about model Examples or Exercises.

**Talk It Over**
This feature provides thought-provoking questions that will spark interesting group discussions between you and your classmates.

**Writing About Math**
In this feature you will be asked to demonstrate your knowledge of mathematical concepts and skills by writing about them in a variety of meaningful ways.

**Reading Math**
This feature introduces the vocabulary of mathematics and the special meanings that ordinary words sometimes have within the discipline.

**Connections**
This feature helps you to make connections between the mathematics you are learning and other areas of mathematics, other disciplines, and real-world problem solving situations.

**Math in the Workplace**
Here you will learn how the math you are studying is applied in the work world.

**Math: Who, Where, When**
In this feature you will learn about mathematical ideas from different cultures around the world and the contributions that people have made to the development of mathematics.

### Computer and Computer Tips

These features provide useful computer programs for solving math problems, along with helpful tips to make them work effectively and to adapt them to other uses.

### Calculator and Calculator Tips

These features explain how to solve problems using a calculator, as well as tips to improve your efficiency and accuracy when using a calculator.

### Problem Solving, Estimation, and Mental Math Tips

These features will help you to use a variety of strategies to solve problems that are presented in the text.

### Mixed Review

This feature presents a quick way to refresh and maintain your mastery of previously learned mathematical skills and concepts.

At the end of each chapter you will find several features to help you maintain your mastery of concepts and skills and to help you prepare for tests.

### Chapter Review

Each chapter ends with a two-page Chapter Review that highlights all the major rules and definitions in the chapter and provides additional examples and practice items. It is very helpful to look at the Chapter Review even before beginning a new chapter since the Review will help to focus your attention on the chapter's key concepts and skills.

### Chapter Test

The one-page Chapter Test is very much like the Skills Preview at the beginning of the chapter. Try comparing your answers on both these features to learn which concepts and skills you have mastered and which you should review more carefully.

### Cumulative Review

The Cumulative Review feature provides on-going opportunities to maintain math concepts and skills learned earlier in the course of study.

### Cumulative Test

The Cumulative Test feature provides additional opportunities to practice test-taking skills for concepts and skills taught throughout the year. The test items are in multiple-choice format similar to the type encountered on standardized tests.

### Data Files

This section provides fascinating data organized around major themes, such as Health and Fitness, Economics, Ecology, Useful Math References, and so on. The tables, charts, graphs, and other visual and verbal data provided in this section are needed to answer questions that appear from time to time under the heading Using Data.

### Glossary

This is an alphabetical listing of all the key terms introduced and defined in the text. Use the Glossary entries as you develop your vocabulary lists. Note that after each term there is a reference indicating the page of the chapter section where the term is first introduced.

### Selected Answers

This section provides answers to selected odd-number items of lessons and features.

# CHAPTER 1 SKILLS PREVIEW

Simplify each expression.

**1.** $36 - 4 \times 2$ **2.** $3 \times 6 - 9 + 5$ **3.** $\frac{1}{2} \times \frac{1}{2} + 2$

Evaluate each expression when $m = 12$.

**4.** $10 + m$ **5.** $2m - 3$ **6.** $\frac{1}{3}m + 11$

Complete.

**7.** ■ + 8 = 8 + 14 **8.** 3 × (4 × 9) = (3 × ■) × 9

**9.** 107 + ■ = 107 **10.** 5(■ + 3) = 5(5) + 5(3)

Find the next three terms in each pattern.

**11.** 3, 7, 11, 15, ■, ■, ■ **12.** 2, 3, 5, 8, 12, ■, ■, ■

**13.** 1, 3, 9, 27, ■, ■, ■

Copy and complete each function table.

**14.**

| $x$ | $x + 2$ |
|---|---|
| 1 | ■ |
| 2 | ■ |
| 5 | ■ |
| 8 | ■ |
| 10 | ■ |

**15.**

| $y$ | $2y + 1$ |
|---|---|
| 3 | ■ |
| 7 | ■ |
| 12 | ■ |
| 19 | ■ |
| 30 | ■ |

**16.**

| $z$ | $4z - 1$ |
|---|---|
| 2 | ■ |
| 5 | ■ |
| 9 | ■ |
| 12 | ■ |
| 20 | ■ |

Write each fraction or decimal as a percent.

**17.** $\frac{1}{2}$ **18.** 0.4 **19.** 0.36 **20.** $\frac{3}{4}$ **21.** 0.01

Estimate each answer by rounding.

**22.** 308 × 19 **23.** 7,831 − 2,259 **24.** 4,239 + 9,762 + 5,088

The bar graph shows the monthly sales of cars at a dealership.

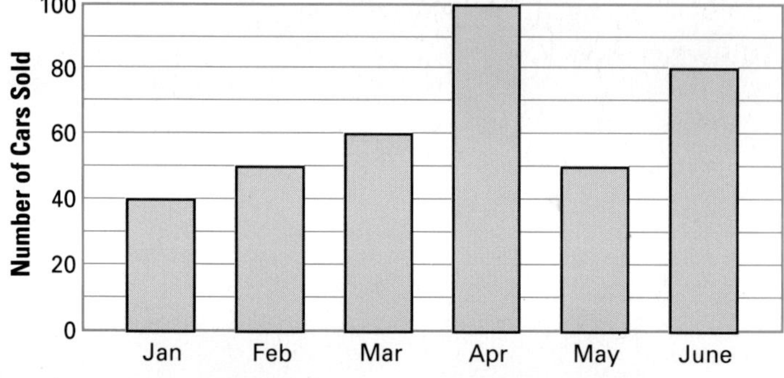


**25.** In which month were more cars sold, March or May?

**26.** How many cars in all were sold in January and February?

**27.** What is the increase in sales from May to June?

**28.** The January sales were __?__% of the April sales.

CHAPTER

# MATHEMATICS AND NUMBER SENSE

**THEME Immigration Waves**

You see large and small numbers of things all around you, but do you ever stop to think about what these numbers mean? Try to imagine, for example, what 1 million of something looks like. Now try imagining many times that number – say 23 million! What does 1 thousandth of an inch look like? How about 23 thousandths?

In this chapter, you will increase your understanding about our number system by working with large numbers and with fractional and decimal parts of numbers. You will discover number patterns and learn how to solve problems involving percents and estimation. You will also learn to obtain information by reading and analyzing several kinds of graphs.

Since the early 1600s, large numbers of people have come to live in the United States from other countries, forming waves of immigration. Between 1890 and 1939, more than 23 million people immigrated to the United States–the largest wave of immigration in the country's history. Over half of these people had to pass inspection at Ellis Island, in New York City, in order to enter the country. In September 1990, the Ellis Island Immigration Museum was dedicated to the memory of the contributions of these diverse groups of people.

## DECISION MAKING

### Using Data

Use the information in the table and in the circle graph to answer the following questions.

1. About how many more people immigrated to the United States between 1910 and 1914 than between 1915 and 1919?
2. After which year was there a sharp decrease in immigration?
3. About one half the total number of immigrants came from which three countries together?
4. Notice that, while 18.8% of immigrants came from Italy and 8.4% came from Canada, 14.7% came from "Other." Explain what is meant by "Other."

**IMMIGRATION TO THE UNITED STATES, 1890–1939**

**TOTAL NUMBER OF IMMIGRANTS**

| | |
|---|---|
| 1890–94 | 2,320,645 |
| 1895–99 | 1,373,649 |
| 1900–04 | 3,255,149 |
| 1905–09 | 4,947,239 |
| 1910–14 | 5,174,701 |
| 1915–19 | 1,172,679 |
| 1920–24 | 2,774,600 |
| 1925–29 | 1,520,910 |
| 1930–34 | 426,953 |
| 1935–39 | 272,422 |
| Total ⟶ | 23,238,947 |

**IMMIGRATION BY COUNTRY OF ORIGIN**

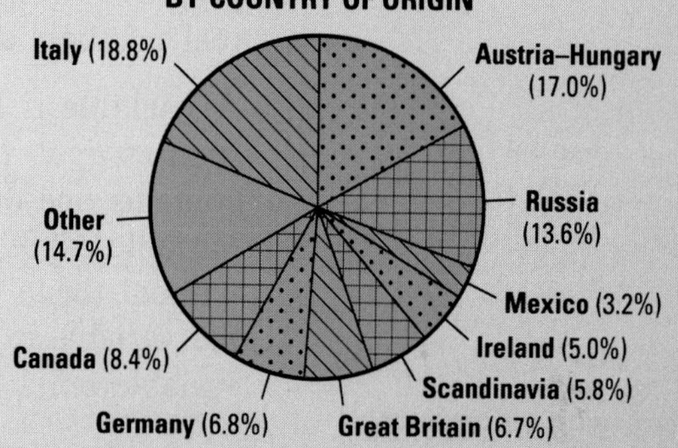


### Working Together

Immigration to the United States is ongoing. Research the latest immigration wave, which began about twenty years ago. Find out from which countries most people have come during this time and what living conditions were like in those countries. Try to find the approximate numbers of people who came from each country and where they have settled in the United States. Then look for patterns in their settlement. Read the world news and try to predict the population makeup of future waves of immigration. Explain your predictions. Discuss your findings.

If you and your family have come to this country during the last twenty years, you may wish to provide some of the data from your own experience.

# 1-1 Numbers, Variables, and Expressions

## EXPLORE

Determine where the symbols +, −, ×, and ( ) should be placed to make each sentence true.

**a.** 8   3   5 = 55

**b.** 5   7   6   5 = 65

**c.** 6   9   7 = 96

## SKILLS DEVELOPMENT

In your study of mathematics so far, you have worked with many kinds of numbers, such as 3, 0, −12, $-6\frac{1}{2}$, $0.\overline{45}$, $\pi$, and $\sqrt{2}$. Each of these numbers—integers (the positive and negative whole numbers and zero), fractions, and decimals—is called a *real number.* As you progress through this textbook, you will expand your knowledge of real numbers and the operations we perform with them.

A **variable** is a symbol used to represent a number. Usually, letters such as *a, x, F,* or *T* are used as variables. An expression that contains one or more variables is called a **variable expression.** Variable expressions that involve multiplication may be written with or without the × sign, with parentheses, or with the symbol •. Each of the variable expressions below represents the product of 6 and the variable *m*.

$6 \times m \qquad 6 \cdot m \qquad 6(m) \qquad 6m$

Similarly, the following expressions all represent division.

$n \div 6 \qquad 6\overline{)n} \qquad \frac{n}{6}$ ← fraction bar

To **evaluate** a variable expression, substitute a given number for each variable. The result is a **numerical expression** that may include operation symbols. The number represented by the numerical expression is the **value** of the variable expression for the value(s) of the variable(s) used. When you find the value of a numerical expression, you **simplify** the expression.

| 59 + *n* | Substitute 35 for *n*. → | 59 + 35 | Simplify. → | 94 |
|---|---|---|---|---|
| variable expression | | numerical expression | | value |

**MATH: WHO, WHERE, WHEN**

The idea of a variable—a letter that can be replaced by different numbers—is believed to have been developed by the French mathematician and lawyer François Viète (1540–1603). In a book written in 1591, Viète introduced what has become modern algebraic notation. Viète also developed practical methods for solving trigonometric problems and advanced equations.

## Example 1

Simplify each numerical expression.

a. $9.64 - 3.2$  b. $\frac{1}{2}(5 + 7)$

### Solution

a. $9.64 - 3.2 = 6.44$

b. $\frac{1}{2}(5 + 7) = \frac{1}{2}(12)$ ← Work inside parentheses first.
$= 6$ ◄

## Example 2

Evaluate each expression when $n = 1.8$.

a. $7 + n$  b. $10 - 3n$

### Solution

a. $7 + n = 7 + 1.8$ ← Substitute 1.8 for *n*.
$= 8.8$

b. $10 - 3n = 10 - 3(1.8)$ ← Evaluate the multiplication expression first.
$= 10 - 5.4$
$= 4.6$ ◄

A *mnemonic* is something that helps you remember. A mnemonic expression that is often used to remember the order of operations is "*P*lease *M*y *D*ear *A*unt *S*ally." The *P* stands for parentheses or other grouping symbols, the *M* for multiplication, the *D* for division, the *A* for addition, and the *S* for subtraction. Can you make up a different mnemonic expression for the order of operations?

The rules for the **order of operations** are:

► First, perform all calculations within parentheses and brackets.
► Next, multiply or divide in order, from left to right.
► Finally, add or subtract in order, from left to right.

## Example 3

Evaluate each expression when $x = \frac{3}{4}$ and $y = 20$.

a. $4x + 2y$  b. $\frac{xy}{5}$ ← *xy* means "*x* times *y*."

### Solution

a. $4x + 2y = 4\left(\frac{3}{4}\right) + 2(20)$
$= 3 + 40$
$= 43$

b. $\frac{xy}{5} = \frac{\frac{3}{4}(20)}{5}$ ← The fraction bar is also a grouping symbol. This could be written $(xy) \div 5$.
$= \frac{15}{5}$
$= 3$ ◄

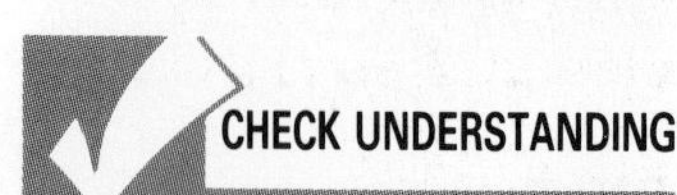


Write each expression in two other ways.

1. $24 \times 13$
2. $95 \div 5$
3. $7x$

You can write expressions to solve some word problems.

## Example 4

Mrs. Torres bought a loaf of bread for $1.69, 2 lb of ground beef at $2.39 a pound, and a carton of juice for $2.79. What was her total bill?

### Solution

Write a numerical expression for the problem. Then simplify.

$$1.69 + 2(2.39) + 2.79$$
$$= 1.69 + 4.78 + 2.79 = 9.26$$

Her total bill was $9.26. ◄

### Example 5

You multiply a number $k$ by 3 and then you add 6. Write an expression to represent the operations you performed.

**Solution**

$k$ ← Number you started with.

$3k$ ← Multiply $k$ by 3.

$3k + 6$ ← Add 6 to the product. 

## TRY THESE

Simplify each expression.

1. $8 - 2 \times 3$
2. $(15 + 9) \div 6$
3. $1.3 + (2.5 \div 1.25)$
4. $(14 - 4) \times 4 + 4$
5. $80 \div (3 + 7) - 1$
6. $\frac{1}{4}(3 \times 8) + 2 \times 12$

Evaluate each expression when $b = 25$.

7. $b + 4$
8. $\frac{1}{2}(b - 1)$
9. $10b$
10. $b + b$

Write an expression and solve.

11. Ana works 7 h a day Monday through Friday and 4 h on Saturday. Write an expression and find how many hours Ana works each week.
12. Divide a number $n$ by 4 and then add 10. Write an expression to represent these operations.

## EXERCISES

**PRACTICE/ SOLVE PROBLEMS**

Simplify each expression.

1. $4 \times 9 - 9 + 7$
2. $(20 + 8) \div 7 \times 6$
3. $1 \times 1 + 5$
4. $\frac{1}{2}\left(\frac{1}{2} + \frac{5}{8}\right)$
5. $6.9 - 3.2 \cdot (10 \div 5)$
6. $[10 + (2 \times 8)] \div 4$

Evaluate each expression when $w = 16$.

7. $30 - w$
8. $\frac{1}{4}w$
9. $w + 0$
10. $w \div 8$

11. $3w - 10$  12. $\frac{64}{w}$  13. $w + 2w$  14. $\frac{3}{4}w + 5$

Evaluate each expression when $c = 10$ and $d = 2.5$.

15. $c + d$  16. $c - 2d$  17. $c \div d$  18. $2cd$

19. $\frac{c}{2}(d + 1)$  20. $d(3c - 1)$  21. $(c + d)(c - d)$  22. $\frac{c + d}{c}$

23. Leon saved \$8 last week. He saved \$2 less than that amount this week. How much has Leon saved so far?

24. The difference between a number $y$ and 98 is multiplied by 11. Write an expression to represent this situation.

## EXTEND/ SOLVE PROBLEMS

Evaluate each expression when $n = 10.2$, $p = 6$, and $r = 0.8$.

25. $n \div p$  26. $p(n + r)$  27. $np$  28. $r \times p + n$

29. $\frac{n + p + r}{3p}$  30. $\frac{2}{3}n + \frac{1}{4}p$  31. $2n - 6r$  32. $8p \div 3r + n$

Write a word phrase for each expression.

33. $37.5 + n$  34. $5w - 72$

35. $2(c + t)$  36. $\frac{3}{4}b \div 10$

37. The distance from New York to Philadelphia is 88 mi. A bus driver makes 5 round trips each week. Write and simplify an expression to show the total number of miles the driver travels each week.

38. Write a variable expression that shows three different operations. Then write a word phrase for the expression.

## THINK CRITICALLY/ SOLVE PROBLEMS

39. Use exactly four 4s and any of the symbols +, −, ×, ÷, and ( ) to write expressions for each of the numbers from 1 to 5. Can you find more than one way for some of the numbers? If you wish, use the same rules and try to find expressions for numbers greater than 5. Compare your work with the work of others.

40. Simplify $\frac{6\frac{1}{2} - \frac{1}{2}}{3}$. Which operation did you perform first? Does this follow the rules for the order of operations? Explain why.

# 1-2 Properties and Mental Mathematics

**EXPLORE** Use graph paper for the activities below.

**a.** Draw a rectangle to model this expression: $8 \times (2 + 3)$. How many rows are in your rectangle? How many boxes in each row?

**b.** Draw two rectangles to model this expression: $(8 \times 2) + (8 \times 3)$. For each rectangle, tell how many rows are in the rectangle and how many boxes are in each row. Write and simplify a numerical expression for the total number of boxes in both rectangles.

**c.** What did you discover? What can you say about the expressions $8 \times (2 + 3)$ and $(8 \times 2) + (8 \times 3)$?

**SKILLS DEVELOPMENT** For some problems, you may be able to find the solution using mental math. The **properties of addition and multiplication** can help you. In the tables below, *a*, *b*, and *c* represent real numbers.

| ADDITION PROPERTIES | | |
|---|---|---|
| **Property** | **In Symbols** | **Example** |
| Commutative | $a + b = b + a$ | $38 + 26 = 26 + 38$ |
| Associative | $(a + b) + c = a + (b + c)$ | $\left(\frac{1}{4} + \frac{1}{3}\right) + \frac{1}{2} = \frac{1}{4} + \left(\frac{1}{3} + \frac{1}{2}\right)$ |
| Identity | $a + 0 = 0 + a = a$ | $3.7 + 0 = 0 + 3.7 = 3.7$ |

← Zero is the additive identity. The sum of 0 and any number is identical to the original number.

| MULTIPLICATION PROPERTIES | | |
|---|---|---|
| **Property** | **In Symbols** | **Example** |
| Commutative | $ab = ba$ | $12 \times 19 = 19 \times 12$ |
| Associative | $(ab)c = a(bc)$ | $\left(\frac{3}{4} \times \frac{1}{8}\right) \times \frac{5}{9} = \frac{3}{4} \times \left(\frac{1}{8} \times \frac{5}{9}\right)$ |
| Identity | $a \times 1 = 1 \times a = a$ | $29 \times 1 = 1 \times 29 = 29$ |
| Property of Zero | $a \times 0 = 0 \times a = 0$ | $64 \times 0 = 0 \times 64 = 0$ |

← One is the multiplicative identity. The product of 1 and any number is identical to the original number.

The **distributive property** relates addition and multiplication. It states that a factor outside parentheses can be used to multiply *each* term within the parentheses.

$a(b + c) = ab + ac \qquad 2(7.8 + 3.6) = (2 \times 7.8) + (2 \times 3.6)$

## Example 1

Complete. Tell which property is used.

**a.** $27 \times 5 = 5 \times ■$ **b.** $6.8 + ■ = 6.8$

**c.** $3\left(\frac{2}{3} + \frac{1}{6}\right) = \left(3 \times \frac{2}{3}\right) + \left(■ \times \frac{1}{6}\right)$

### Solution

**a.** Use the commutative property of multiplication.
$27 \times 5 = 5 \times 27$

**b.** Use the identity property of addition.
$6.8 + 0 = 6.8$

**c.** Use the distributive property.
$3\left(\frac{2}{3} + \frac{1}{6}\right) = \left(3 \times \frac{2}{3}\right) + \left(3 \times \frac{1}{6}\right)$ ◄

## Example 2

Evaluate each expression when $x = 3.5$, $y = 0$, and $z = 1$. Tell which property you used.

**a.** $x + y$ **b.** $xy$ **c.** $xz$ **d.** $\frac{y}{x}$ **e.** $x \div y$

### Solution

**a.** $x + y = 3.5 + 0 = 3.5$ ← additive identity

**b.** $xy = 3.5 \times 0 = 0$ ← multiplication property of zero

**c.** $xz = 3.5 \times 1 = 3.5$ ← multiplicative identity

**d.** $\frac{y}{x} = \frac{0}{3.5} = 0$ ← multiplication property of zero

**e.** $x \div y = 3.5 \div 0$
Think: If $3.5 \div 0$ equals some number $n$, then $n \times 0 = 3.5$, but the multiplication property of zero says that $n \times 0 = 0$. Therefore, division by 0 is undefined and there is no solution. ◄

**MENTAL MATH TIP**

To multiply by multiples of 10, 100, or 1,000, use a pattern.

Multiply 90 x 300.

9 x 3 = 27 Multiply the digits to the left of the zero in each factor.

90 x 300 Count the number of zeros in each factor.

3 zeros

27,000 Write the same number of zeros in the product.

90 x 300 = 27,000

Find each product mentally.

1. 20 x 60
2. 50 x 40
3. 70 x 900
4. 300 x 800
5. 6,000 x 5,000

Use the properties to group compatible numbers whose sum or product is easy to find mentally.

## Example 3

Use the addition properties to find the sum $(110 + 412) + 390$ mentally.

### Solution

$(110 + 412) + 390 = (412 + 110) + 390$ ← commutative property
$= 412 + (110 + 390)$ ← associative property
$= 412 + 500$
$= 912$ ◄

### Example 4

Use the multiplication properties to find each product mentally.

**a.** $15 \times (70 \times 4)$ **b.** $(0.2)[(9)(5)]$ **c.** $\frac{1}{3}(7 \cdot 27)$

**Solution**

**a.** $15 \times (70 \times 4) = 15 \times (4 \times 70)$
$= (15 \times 4) \times 70$
$= 60 \times 70 = 4{,}200$

**b.** $(0.2)[(9)(5)] = (0.2)[(5)(9)]$
$= [(0.2)(5)](9)$
$= 1(9) = 9$

**c.** $\frac{1}{3}(7 \cdot 27) = \frac{1}{3}(27 \cdot 7)$
$= (\frac{1}{3} \cdot 27)7$
$= 9 \cdot 7 = 63$ ◄

### Example 5

Use the distributive property to find each product.

**a.** $6 \times 73$ **b.** $12\left(5\frac{1}{4}\right)$

**Solution**

**a.** Think: $73 = 70 + 3$
$6 \times 73 = 6(70 + 3)$
$= (6 \times 70) + (6 \times 3)$
$= 420 + 18 = 438$

**b.** Think: $5\frac{1}{4} = 5 + \frac{1}{4}$
$12\left(5\frac{1}{4}\right) = 12\left(5 + \frac{1}{4}\right)$
$= 12(5) + 12\left(\frac{1}{4}\right)$
$= 60 + 3 = 63$ ◄

## TRY THESE

Complete. Tell which property you used.

1. $8 \times 7 = ■ \times 8$
2. $1.7 + ■ = 3.08 + 1.7$
3. $(4 \times 6) \times 9 = 4 \times (■ \times 9)$
4. $213 \times ■ = 213$
5. $729 \times ■ = 0$
6. $4\left(\frac{3}{4} + \frac{1}{8}\right) = 4(■) + 4\left(\frac{3}{4}\right)$
7. $0.6 + (1.3 + 0.2) = (■ + 1.3) + 0.2$
8. $■ \times 1 = 967$
9. $68(139) = 139(■)$
10. $388 + 1{,}065 = ■ + 388$
11. $8{,}297 = ■ \times 1$
12. $4.123 + ■ = 4.123$

Write the letter of the correct answer.

| | | | |
|---|---|---|---|
| 13. $18 + (39 + 42)$ | a. 89 | b. 99 | c. 111 |
| 14. $(7.4 + 1.3) + 2.6$ | a. 11.3 | b. 10.7 | c. 12.9 |
| 15. $2.5 \times (38 \times 4)$ | a. 380 | b. 3,800 | c. 38,000 |
| 16. $\frac{1}{5}(9 \cdot 45)$ | a. 45 | b. 90 | c. 81 |
| 17. $(25 \times 66) \times 40$ | a. 660 | b. 6,600 | c. 66,000 |

# EXERCISES

## PRACTICE/ SOLVE PROBLEMS

Complete.

1. ■ × (4×7) = (5×4) × 7
2. 73.9 × ■ = 0
3. $\left(\frac{1}{2} + \frac{3}{4}\right) + \frac{7}{16} = \frac{1}{2} + \left(■ + \frac{7}{16}\right)$
4. 10(0.07 + 4.9) = 10(0.07) + ■(4.9)

Evaluate each expression when $a = 24$, $b = 0$, $c = 1$.

5. $a + b$
6. $ac$
7. $ba$
8. $\frac{a}{c}$
9. $b \div a$
10. $c \div b$

Use the properties to find each sum or product mentally.

11. 20 × (93 × 5)
12. (5)[(89)(0.2)]
13. $15\left(4\frac{1}{3}\right)$
14. (4.6 + 7.8) + 5.4
15. Bellville received 3.8 in. of rain during April, 2.7 in. during May, and 3.2 in. during June. What was the total rainfall?

## EXTEND/ SOLVE PROBLEMS

Evaluate each expression when $p = 4$, $q = 9$, and $r = 25$.

16. $q \div r \times 0$
17. $p \times q \times r$
18. $(0.5)rp$
19. $p + r + 76$
20. $q + 0 + 131$
21. $(0.2)(q)(50)$

Use the properties to find each sum or product mentally.

22. 5.7 + (16.2 + 9.3)
23. (5 × 36.7) × 2
24. 8 × 36
25. $24 \times \left(9 \times \frac{3}{8}\right)$
26. $15\left(4\frac{2}{5}\right)$
27. (18 + 56) + (112 + 34)
28. Lou and Nina bought 250 flowers at a cost of \$55 and sold each flower for \$0.40. How much money did they make on the sale of the flowers?

## THINK CRITICALLY/ SOLVE PROBLEMS

29. Is there a commutative property for subtraction? Is there an associative property? Give examples to explain.
30. Is there a commutative property for division? Is there an associative property? Give examples.
31. Is multiplication distributive over subtraction; that is, does $a(b - c)$ equal $ab - ac$? Is multiplication distributive over multiplication; that is, does $a(bc)$ equal $(ab)(ac)$? Give examples.

# 1-3 Exploring Patterns

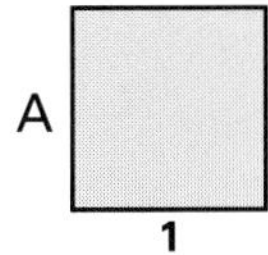


## EXPLORE

The search for patterns is an important problem solving tool in all areas of mathematics—arithmetic, algebra, geometry, trigonometry, and probability, to name a few. Patterns may involve just numbers or numbers and geometric figures. Look at the figures at the right. Figure A consists of 1 square. Figure B consists of 1 large square and 4 small squares. Figure C consists of 1 large square, 4 medium squares, and 9 small squares.

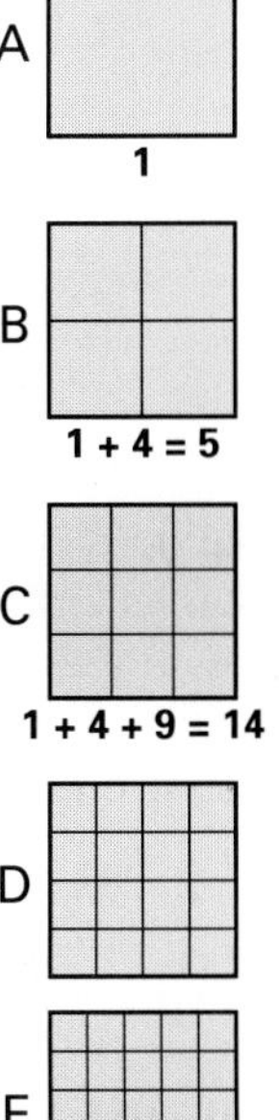


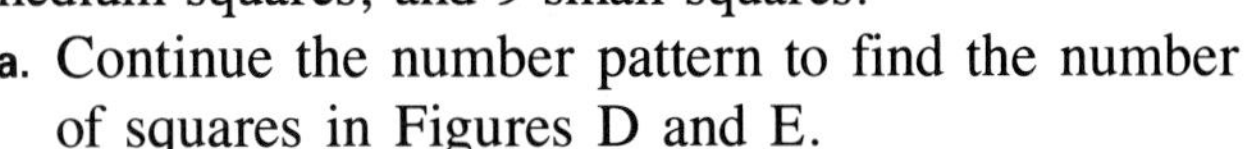

**a.** Continue the number pattern to find the number of squares in Figures D and E.

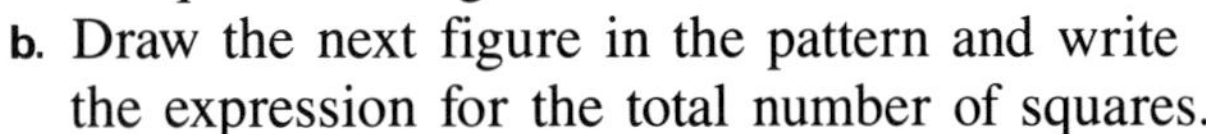

**b.** Draw the next figure in the pattern and write the expression for the total number of squares.

**c.** Without drawing, predict how many squares will be in each of the next two figures. Explain.

## SKILLS DEVELOPMENT

An arrangement of numbers according to a pattern is also called a **sequence.** Each number in a sequence is called a **term.** Thus, you can refer to the first term, the second term, and so on.

> **CALCULATOR**
>
> On many calculators, the key [=] or the constant key [con] can be used for repeated operations. For example, to find a pattern that results from adding 4 to a number, if you enter
>
> 2 [+] 4 [=] [=] [=] [=]
>
> you would see the pattern 2, 6, 14, 18.
>
> On some calculators, if you enter
>
> [×] 5 [con] [con] [con] [con]
>
> you would get 5, 25, 125, 625.
>
> Consult your calculator manual to find out how to work with repeated operations.

### Example 1

Identify each pattern and find the next three terms.

**a.** 1, 5, 9, 13, ■, ■, ■ **b.** 2, 4, 8, 16, ■, ■, ■

### Solution

**a.** Each number is 4 more than the preceding number. The pattern is *add 4.*

1, 5, 9, 13, 17, 21, 25

The next three terms are 17, 21, and 25.

**b.** Each number is 2 times the preceding number. The pattern is *multiply by 2.*

2, 4, 8, 16, 32, 64, 128

The next three terms are 32, 64, and 128. ◄

Some patterns involve more than one operation.

### Example 2

Identify the pattern and find the next three terms.

3, 8, 6, 11, 9, 14, ■, ■, ■

**Solution**

Examine the differences between the successive terms.
The pattern is *add 5, subtract 2.*

$$3,\ 8,\ 6,\ 11,\ 9,\ 14,\ 12,\ 17,\ 15$$
$$+5\quad -2\quad +5\quad -2\quad +5\quad -2\quad +5\quad -2$$

The next three terms are 12, 17, and 15. ◄

## Example 3

Identify the pattern and find the next three terms.

2, 5, 9, 14, 20, 27, ■, ■, ■

**Solution**

The second term is 3 more than the first term. The third term is 4 more than the second term. The fourth term is 5 more than the third term, and so on. The pattern is *add 3, add 4, add 5,* and so on.

$$2,\ 5,\ 9,\ 14,\ 20,\ 27,\ 35,\ 44,\ 54$$
$$+3\quad +4\quad +5\quad +6\quad +7\quad +8\quad +9\quad +10$$ ◄

**WRITING ABOUT MATH**

Make up some patterns of your own and exchange with a partner. Write descriptions of each other's patterns. Give the next three terms for each pattern.

## Example 4

A gardener plants 10 rows of roses in increasing numbers. The first row has 13 roses. The second row has 21. The third row has 29 roses, and the fourth has 37. If the pattern continues, how many roses are in the tenth row?

**Solution**

Make a table.

| Row | 1 | 2 | 3 | 4 | 5 | 6 | 7 | 8 | 9 | 10 |
|---|---|---|---|---|---|---|---|---|---|---|
| Number of Roses | 13 | 21 | 29 | 37 | ■ | ■ | ■ | ■ | ■ | ■ |

Notice the pattern *add 8.* Complete to the tenth row.

There are 85 roses in the tenth row of the garden. ◄

## TRY THESE

Identify each pattern and find the next three terms.

1. 6, 12, 18, 24, ■, ■, ■
2. 306, 302, 298, 294, ■, ■, ■
3. $1, \frac{1}{2}, \frac{1}{4}, \frac{1}{8},$ ■, ■, ■
4. 11, 14, 9, 12, 7, 10, ■, ■, ■
5. David is making a design. The first row has 1 star, the second row has 3 stars, the third row has 5 stars, and the fourth row has 7 stars. If the pattern continues, how many stars are in the eighth row?

# EXERCISES

## PRACTICE/ SOLVE PROBLEMS

Find the next three terms in each pattern.

1. 2.3, 2.5, 2.7, 2.9, ■, ■, ■
2. 2, 6, 18, 54, ■, ■, ■
3. 120, 105, 90, 75, ■, ■, ■
4. 8, 4, 11, 7, 14, 10, ■, ■, ■
5. 46, 37, 29, 22, ■, ■, ■
6. 14, 25, 38, 53, 70, ■, ■, ■
7. 380, 38, 3.8, 0.38, ■, ■, ■
8. 4, 12, 36, 108, 324, ■, ■, ■
9. A fund-raising group sent letters asking for donations. The first week, they sent 32 letters and received $350. The second week, they mailed 96 letters and received $1,050. The third week, 288 letters were sent, and they received $3,150. If the pattern continued, how many letters did they send out the seventh week and how much money was received?

## EXTEND/ SOLVE PROBLEMS

**MATH: WHO, WHERE, WHEN**

The triangular and square numbers in Exercises 10 and 11 are examples of **figurate numbers** that were recognized by the Greeks over 2,500 years ago. Another kind of figurate numbers are **pentagonal numbers.**

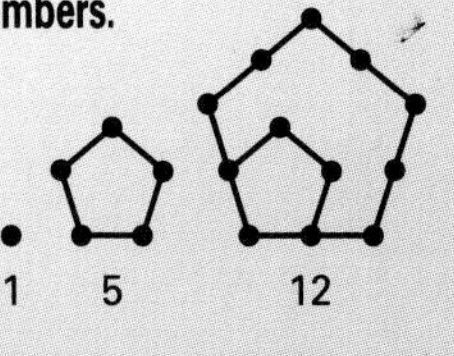


Can you find a number pattern?

10. These figures represent **triangular numbers.** Draw figures for the next three triangular numbers. Describe the pattern.

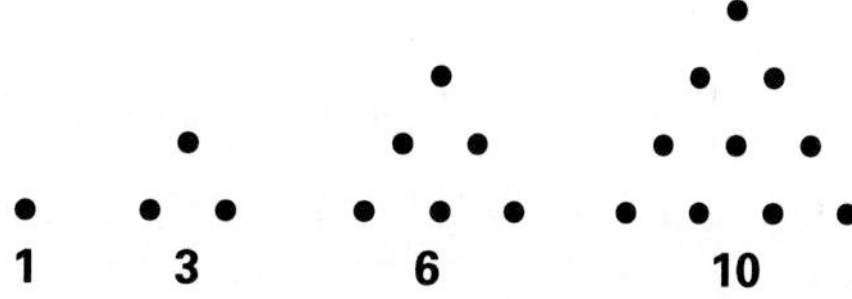


11. The figures below represent the **square numbers.** Draw figures for the next three square numbers. What number is represented by a $12 \times 12$ square?

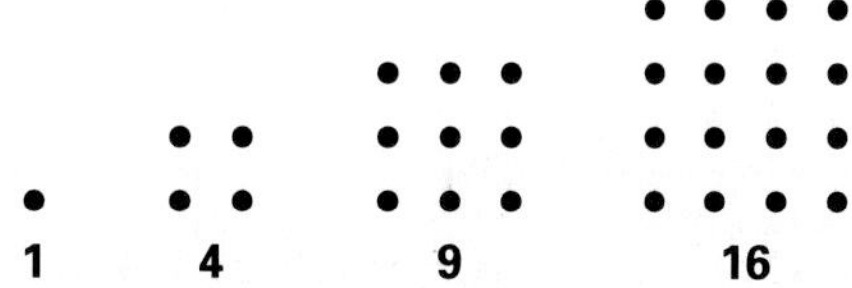


12. Use a calculator to multiply 12,345,679 by the first nine multiples of 9. What pattern do you observe?

## THINK CRITICALLY/ SOLVE PROBLEMS

13. Find two different ways to continue each pattern. Show as many terms as needed to establish the pattern.
    a. 2, 3, 4, 5, . . .
    b. 2, 6, 9, 27, . . .

14. Find the indicated term for each of the following sequences. Did you use a shortcut? Explain.
    a. The fiftieth term of 2, 4, 6, 8, . . .
    b. The hundredth term of 1, 4, 7, 10, . . .

# Problem Solving Applications:

- ► READ
- ► PLAN
- ► SOLVE
- ► ANSWER
- ► CHECK

## THE FIBONACCI SEQUENCE

One of the greatest mathematicians of the Middle Ages was Leonardo Fibonacci, born in Pisa, Italy. In 1202, Fibonacci wrote a book in which he proposed the following problem: A pair of rabbits one month old are too young to produce more rabbits, but suppose that each month from their second month on they produce a new pair. If each pair produces rabbits (and none of the rabbits die), how many pairs of rabbits will there be at the beginning of each month?

| | | |
|---|---|---|
| Beginning Month 1 (1 pair) | 1 |
| End Month 1 (1 pair) | 1 |
| End Month 2 (2 pairs) | $1 + 1 = 2$ |
| End Month 3 (3 pairs) | $1 + 2 = 3$ |
| End Month 4 (5 pairs) | $2 + 3 = 5$ |

The numbers in the solution of this problem form the **Fibonacci sequence.**

1, 1, 2, 3, 5, 8, 13, 21, . . . ← **Three dots mean that the sequence continues.**

The first two terms of the sequence are 1 and each succeeding term is the sum of the previous pair of terms. The numbers that reflect the Fibonacci sequence show up in a variety of real-world situations, including the number of petals on daisies and other flowers and the spirals on pine cones and nautilus shells. Moreover, the sequence itself has some amazing properties.

Try these experiments with the Fibonacci sequence.

1. Write the first fifteen terms of the Fibonacci sequence. (You will use these terms again below.)
2. Find the sum of the first *five* terms of the sequence. How does the sum compare to the *seventh* term of the sequence?
3. Find the sum of the first *eight* terms of the sequence. How does the sum compare to the *tenth* term of the sequence?
4. Try to predict the sum of the first *twelve* terms without adding.
5. Divide each of the first fifteen terms of the Fibonacci sequence by 4. The remainders for the first four terms are the terms themselves. Find the remainders for the rest of the terms and describe the pattern in the sequence of remainders.
6. Research and report on other applications of the Fibonacci sequence. For example, you might find out about different aspects of plant growth, or the arrangement of black and white keys on a piano.

# 1-4 Patterns and Functions

## EXPLORE

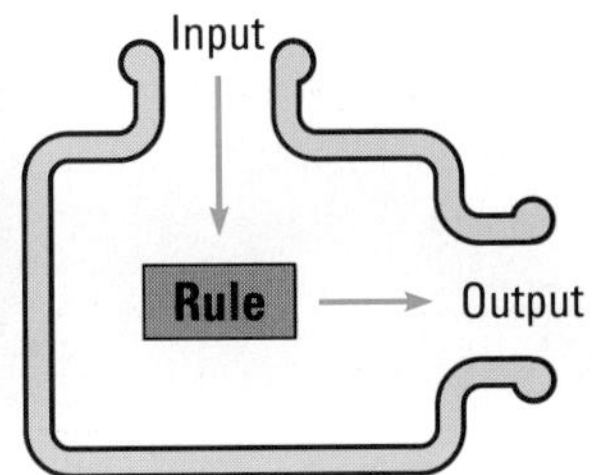


The machine shown at the left works as follows: when numbers are input, the machine processes them according to a specific rule and then gives the correct output number.

a. Suppose the rule is *add 5.* Copy and complete this input-output table.

| Input | 2 | 5 | 8 | 14 | 20 | 100 |
|---|---|---|---|---|---|---|
| Output | ■ | ■ | ■ | ■ | ■ | ■ |

b. For each value input, how many output values are there?

c. Look at the input-output table below. Can you determine the rule the machine was using this time?

| Input | 3 | 7 | 13 | 32 | 45 | 90 |
|---|---|---|---|---|---|---|
| Output | 1 | 5 | 11 | 30 | 43 | 88 |

## SKILLS DEVELOPMENT

READING MATH

Functions are sometimes written using **arrow notation:**

$x \rightarrow x + 5$

so

$1 \rightarrow 6$

In the Explore activity, the value of the output depended on the value of the input *and* the rule that was being used. To each value input, there corresponded one and only one output value. This type of correspondence is called a function. A **function** $f$ is a correspondence or relationship that pairs each member of a given set with *exactly* one member of another set. Often, both sets are sets of numbers, but they can also be sets of names or geometric figures.

The description of a function is called a **function rule.** Variable expressions are very useful for creating function rules. For example, if $x$ represents the value input, then the *add 5* rule becomes $x + 5$. A symbol like $f(x)$, $g(x)$, or $*(x)$ denotes the number assigned to $x$.

Read "$f$ of $x$ is equal to $x$ plus 5." → $f(x) = x + 5$

So,

$f(1) = 1 + 5 = 6 \qquad f(4) = 4 + 5 = 9$

You can also use the function rule to make a **function table,** like the one at the right.

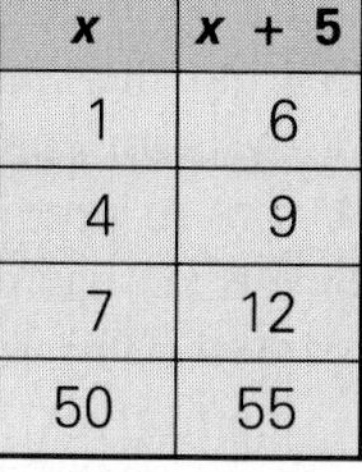

| $x$ | $x + 5$ |
|---|---|
| 1 | 6 |
| 4 | 9 |
| 7 | 12 |
| 50 | 55 |

### Example 1

Which of the following correspondences are functions?

a. Hamburger → \$2.75; Pizza → \$4.00; Chicken → \$2.75; Salad → \$2.50 (Hamburger, Pizza, Chicken, Salad; \$4.00, \$2.75, \$3.50, \$2.50)

b.

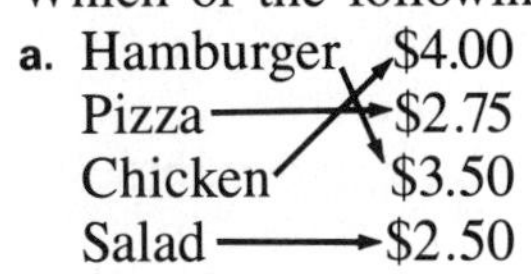


c.

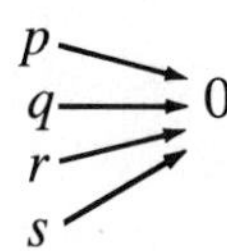

**Solution**

Correspondence **a** is a function. Correspondence **b** is not a function because each of the state names (input) corresponds to more than one city (output). Corespondence **c** is a function since each variable (input) corresponds to exactly one output. ◄

## Example 2

Copy and complete the function table at the right.

| $y$ | $y - \frac{1}{2}$ |
|---|---|
| 2 | ■ |
| $4\frac{3}{4}$ | ■ |
| 10 | ■ |
| $26\frac{1}{2}$ | ■ |
| $41\frac{1}{8}$ | ■ |

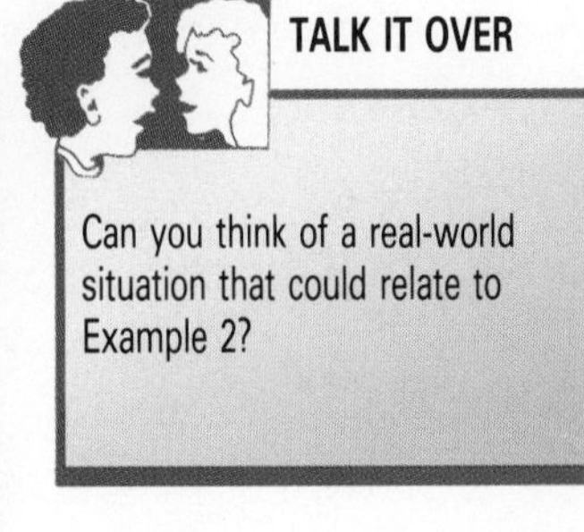


**Solution**

The function rule is $f(y) = y - \frac{1}{2}$. Substitute each value shown for $y$ in the expression $y - \frac{1}{2}$.

$f(2) = 2 - \frac{1}{2} = 1\frac{1}{2}$

$f\left(4\frac{3}{4}\right) = 4\frac{3}{4} - \frac{1}{2} = 4\frac{1}{4}$

$f(10) = 10 - \frac{1}{2} = 9\frac{1}{2}$

$f\left(26\frac{1}{2}\right) = 26\frac{1}{2} - \frac{1}{2} = 26$

$f\left(41\frac{1}{8}\right) = 41\frac{1}{8} - \frac{1}{2} = 40\frac{5}{8}$ ◄

Look for a pattern to help determine the function rule.

## Example 3

Write a function rule for the function table at the right.

| $x$ | ■ |
|---|---|
| 1 | 10 |
| 2 | 20 |
| 7 | 70 |
| 10 | 100 |
| 15 | 150 |

**Solution**

Think: What is the pattern? Each number in the second column is ten times the number in the first column. Check the pattern.

$10(1) = 10 \quad 10(2) = 20$

$10(7) = 70 \quad 10(10) = 100 \quad 10(15) = 150$

So, the function rule is $f(x) = 10x$. ◄

## Example 4

When lightning strikes, the gap between seeing the flash and hearing the thunder is a function of time (in seconds) and distance (in feet).

**a.** Copy and complete the function table using the function rule $f(t) = 1{,}100t$.

**b.** How far away from you is the lightning if you hear the thunder 6 s after you see the flash?

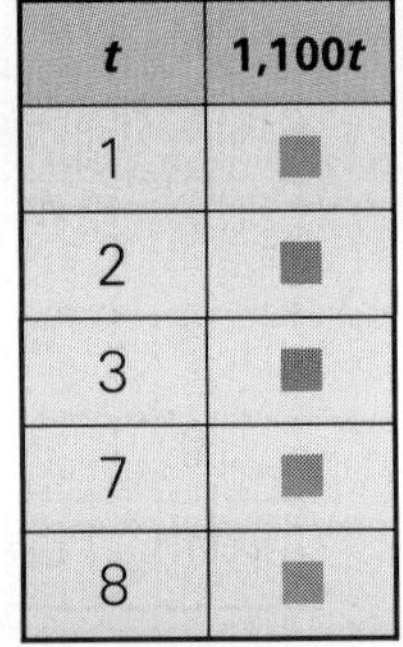

| $t$ | $1{,}100t$ |
|---|---|
| 1 | ■ |
| 2 | ■ |
| 3 | ■ |
| 7 | ■ |
| 8 | ■ |

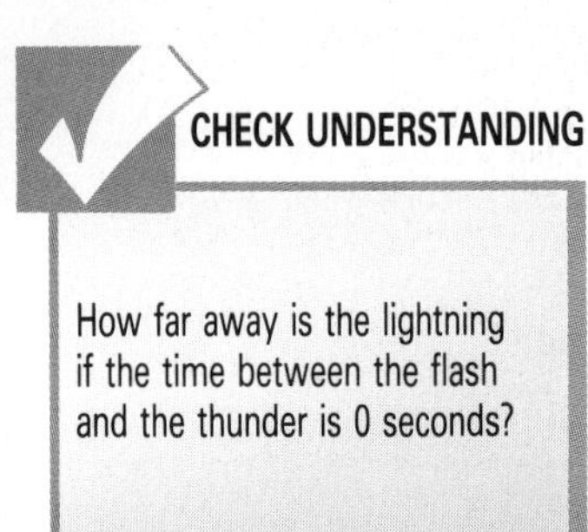


**Solution**

**a.** Substitute each value in the expression $f(t) = 1{,}100t$.

$f(1) = 1{,}100(1) = 1{,}100 \quad f(2) = 1{,}100(2) = 2{,}200$

$f(3) = 1{,}100(3) = 3{,}300 \quad f(7) = 1{,}100(7) = 7{,}700$

$f(8) = 1{,}100(8) = 8{,}800$

**b.** Find $f(6)$.

$f(6) = 1{,}100(6) = 6{,}600$

The lightning is 6,600 ft away. ◄

## TRY THESE

1. Which of the following correspondences are functions?

a. Ed → 135
Sue → 110
Ana → 118
Lee → 123

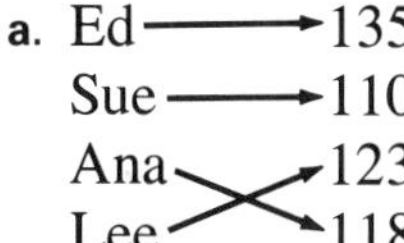

b. 1987 → *New York Times*
1988 → *Philadelphia Inquirer*
1989 → *Philadelphia Inquirer*
1990 → *Seattle Times*

c.

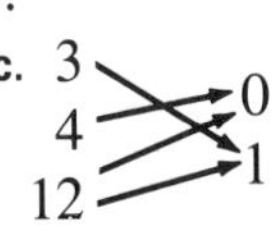


Copy and complete each function table.

2.

| $x$ | $x + 6$ |
|---|---|
| 0 | ■ |
| 1 | ■ |
| 4 | ■ |
| 10 | ■ |
| 18 | ■ |

3.

| $y$ | $0.1y$ |
|---|---|
| 2 | ■ |
| 7 | ■ |
| 10 | ■ |
| 35 | ■ |
| 100 | ■ |

4.

| $w$ | $3w - 1$ |
|---|---|
| $\frac{1}{2}$ | ■ |
| 3 | ■ |
| 8 | ■ |
| 15 | ■ |
| 21 | ■ |

Write each function using function notation.

5. $x$ corresponds to $2x - 4$

6. $y$ corresponds to $\frac{1}{3}y + 5$

7. For a car traveling at a constant rate of 45 mi/h, distance is a function of time: $d(t) = 45t$. Find $d(3)$. What does your answer mean?

## EXERCISES

### PRACTICE/ SOLVE PROBLEMS

Copy and complete each function table.

1.

| $m$ | $7m - 2$ |
|---|---|
| 1 | ■ |
| 4 | ■ |
| 6 | ■ |
| 10 | ■ |
| 21 | ■ |

2.

| $c$ | $4c + 3$ |
|---|---|
| 0 | ■ |
| 3 | ■ |
| 9 | ■ |
| 13 | ■ |
| 25 | ■ |

3.

| $d$ | $\frac{d}{5}$ |
|---|---|
| 0 | ■ |
| 1 | ■ |
| 5 | ■ |
| 20 | ■ |
| 45 | ■ |

Write the function rule for each function table.

4.

| $x$ | ■ |
|---|---|
| 1 | 4 |
| 2 | 5 |
| 3 | 6 |
| 6 | 9 |
| 9 | 12 |

5.

| $y$ | ■ |
|---|---|
| 2 | 1 |
| 8 | 4 |
| 15 | 7.5 |
| 20 | 10 |
| 33 | 16.5 |

6.

| $z$ | ■ |
|---|---|
| 4 | 10 |
| 6 | 14 |
| 8 | 18 |
| 10 | 22 |
| 12 | 26 |

7. Because the force of gravity is less on Mars than on Earth, a person would weigh less there. A person's weight on Mars can be given as a function of his or her weight on Earth ($w$) as follows: $M(w) = 0.4w$. How much would a 135-lb person weigh on Mars?

## EXTEND/ SOLVE PROBLEMS

A manufacturer determines that the total cost $C$ of producing gizmos depends on the number of gizmos and is given by the function $C(n) = 1.85n + 400$.

8. What is the total cost of producing 1 gizmo?

9. What is the total cost of producing each number of gizmos?
   a. 50 gizmos　　b. 1,000 gizmos

10. Postage for a first-class letter is a function of weight: 29¢ for the first ounce and 23¢ for each additional ounce. A mixed-number weight is rounded up to the next whole number.
    $p(w) = 29 + 23(w - 1)$
    a. Verify that $p(1) = 29$.
    b. Find the postage for each of the following letter weights.
    3 oz　$4\frac{1}{8}$ oz　$8\frac{1}{2}$ oz

## THINK CRITICALLY/ SOLVE PROBLEMS

11. Consider a new function called the star (*) function. The star correspondence for some numbers is shown below.
    *(5) = 17　*(9) = 29　*(12) = 38　*(20) = 62
    Can you star these numbers? 7, 19, 40

12. Another function is the check (✓) function. The check correspondence for some numbers is shown below.
    ✓(2) = 5　✓(3) = 10　✓(5) = 26　✓(7) = 50
    Can you check these numbers? 4, 8, 10

13. What function "undoes" $f(x) = x + 8$?

14. What function "undoes" $g(y) = 3y$?

# 1-5 Problem Solving Strategies:

## USE THE 5-STEP PLAN

Any immigrant who wants to seek employment in the United States must first apply for a work permit at an office of the United States Immigration and Naturalization Service (INS).

### PROBLEM

A photographer located near an INS office takes photos for work permits and passports. There is a sitting fee of \$8.75 and a charge of \$3.95 for each photo. One day, 19 people were photographed, each of whom bought 2 photos. How much did the photographer earn that day?

- READ
- PLAN
- SOLVE
- ANSWER
- CHECK

### SOLUTION

**5-STEP PROBLEM SOLVING PLAN**

1. READ . . . Read the entire problem slowly and carefully. Then go back and read the question again.

How much did the photographer earn that day?

2. PLAN . . . Decide what you can do to solve the problem.

Add to find the amount each person was charged, then multiply by the number of people photographed.

3. SOLVE . . . Put your plan into action.

\$8.75 + \$3.95 + \$3.95 = \$16.65 19 × \$16.65 = \$316.35

4. ANSWER . . . State your answer in terms of the original question.

The photographer earned \$316.35 that day.

5. CHECK . . . Determine whether or not your answer is *reasonable.*

Round each addend to the nearest dollar. Then add.
\$9 + \$4 + \$4 = \$17
Round the number of people photographed, 19, to 20. Think:
Since \$17 × 10 = \$170, then \$17 × 20 = \$340.
While \$340 is greater than \$316.35, the estimate is high because you rounded up, so the answer is reasonable.

**CONNECTIONS**

In the 1920s, the government passed laws restricting immigration from Europe. The Emergency Quota Act of 1921 limited the total number of immigrants to 350,000 a year. It also ruled that the number of people who could immigrate from any country would be limited to just 3% of the people of that nationality living in the United States in 1910. In 1924, another law lowered the 3% quota to 2%. The National Origins Act of 1929 further restricted the total number of immigrants per year to 150,000.

# PROBLEMS

Use the 5-step plan to solve these problems.

1. A driver of a truck carrying oranges made a sharp turn, spilling 2,604 oranges onto the highway. How many dozen oranges were spilled?
2. Maria earns $6.75 per hour at her after-school job. If she earned $108 one week, how many hours did she work?
3. Anya's round trip to and from school each day is 3.9 mi. If she attended school on 173 days one year, find the total number of miles she traveled.
4. At the start of a trip, the odometer reading in a car was 29,065.4. At the end of the trip, the reading was 29,337. What was the distance traveled? How much farther will the car travel before the odometer reads 30,000?
5. ***USING DATA*** Use the Data Index on pages 556–557 to find the number of people that immigrated to the U.S. in the 1980s. About as many people immigrated from the Philippines as immigrated from which three other countries together?

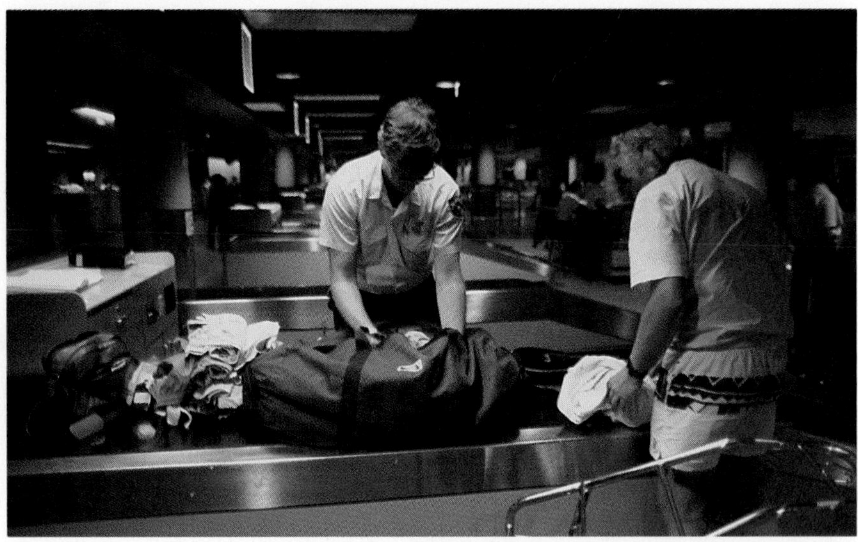

6. The distance between the bases in a major league baseball diamond is 27.43 m. How many kilometers did Hank Aaron run in making 755 home runs?
7. During one baseball season, the Cheetahs won 3 times as many games as the Lions and the Lions won $\frac{1}{4}$ as many games as the Bears. if the Bears won 3 games more than the Cheetahs, which team won 12 games? How many games did each of the other teams win?

**MIXED REVIEW**

Change the following measurements.

1. 2,500 m to kilometers
2. 10 c to quarts
3. 5 kg to grams

Complete.

4. 42 in. = ■ ft
5. 750 mL = ■ L
6. $2\frac{3}{4}$ lb = ■ oz
7. 4,500 cm = ■ m

Solve.

8. How much change would you receive if you paid for 3 scarves at $10.95 each with two twenty-dollar bills?

**PROBLEM SOLVING TIP**

Sometimes you have to change units of measure in order to solve a problem. What units must you change to solve Problem 6?

# 1-6 Solving Percent Problems

## EXPLORE

The paragraph below contains 100 letters. Use it to complete a table like the one below. Then answer the questions.

SOME IMMIGRANTS MAKE THE ADJUSTMENT TO AMERICAN LIFE QUICKLY; OTHERS NEVER COMPLETELY MAKE IT AT ALL AND MAY RETURN HOME.

| Letter | Number of Times | Fraction of 100 | Percent |
|---|---|---|---|
| A | | | |
| B | | | |
| C | | | |
| ⋮ | | | |

a. How many times does the letter C appear? What fraction of the total number of letters is this? What percent?

b. Which letter appears most often? What fraction of the total number of letters is this? What percent?

c. Which four letters account for 45% of the total?

d. What is the sum of the percents for all the letters?

## SKILLS DEVELOPMENT

Percent means "per hundred." The symbol 23% is read "twenty-three percent" and means 23 out of 100.

$$\text{percent form} \rightarrow 23\% = \underset{\uparrow \text{ fraction form}}{\frac{23}{100}} = 0.23 \leftarrow \text{decimal form}$$

CONNECTIONS

Recall that a **ratio** is the quotient of two numbers and is used to compare one number to another. The order of the numbers in a ratio is important. A percent is a special ratio in which a number is compared to 100 (that is, 100 is the second number in the ratio).

### Example 1

a. Write 8.5% as a decimal.

b. Write 46% as a fraction.

### Solution

a. $8.5\% = \frac{8.5}{100} = \frac{8.5 \times 10}{100 \times 10} = \frac{85}{1{,}000} = 0.085$

b. $46\% = \frac{46}{100} = \frac{23}{50}$ ← Write as a fraction with a denominator of 100 and simplify. ◄

### Example 2

a. Write 1.527 as a percent.

b. Write $\frac{1}{7}$ as a percent.

### Solution

a. $1.527 = \frac{1{,}527}{1{,}000} = \frac{1{,}527 \div 10}{1{,}000 \div 10} = \frac{152.7}{100} = 152.7\%$

b. Divide: $\frac{1}{7} \rightarrow 7\overline{)1.00}$ with quotient $0.14\frac{2}{7}$

$$0.14\frac{2}{7} = \frac{14\frac{2}{7}}{100} = 14\frac{2}{7}\%$$ ◄

Percent can be used to solve several types of problems.

**CHECK UNDERSTANDING**

How could you find what number is 60% of 85 by writing an equation using a fraction?

### Example 3

**a.** What number is 60% of 85? **b.** 91 is 140% of what number?

**Solution**

**a.** Find the percent of a number writing an equation using a decimal.

What number is 60% of 85?

$n = 0.60 \times 85$ Let $n$ represent the unknown number.

$n = 51$ So, 60% of 85 is 51.

**b.** To find a number when a percent of it is known, write an equation.

91 is 140% of what number?

$91 = 1.4 \times n$

$\frac{91}{1.4} = \frac{1.4}{1.4}n$ Divide.

$65 = n$ So, 91 is 140% of 65. ◄

**MENTAL MATH TIP**

A quick way to change a percent to a decimal is to divide by 100 by moving the decimal point 2 places to the left.

8.5% = 0.085

A quick way to change a decimal to a percent is to multiply by 100 by moving the decimal point 2 places to the right.

0.57 = 57%

### Example 4

A 4-oz serving of hot dogs contains about 2.2 oz of water. What percent of the hot dogs' weight is water?

**Solution**

Find what percent of 4 is 2.2.

$p \times 4 = 2.2$

$4p = 2.2$

$\frac{4p}{4} = \frac{2.2}{4}$

$p = 0.55$ so $p = 55\%$

About 55% of the weight of the hot dogs is water. ◄

## TRY THESE

Write each percent as a decimal *and* a fraction.

**1.** 57% **2.** 9% **3.** 8.2% **4.** 110% **5.** $\frac{7}{8}\%$

Write each fraction or decimal as a percent.

**6.** $\frac{12}{50}$ **7.** 0.13 **8.** 0.004 **9.** $\frac{9}{20}$ **10.** 0.365 **11.** $\frac{5}{8}$

Find the percent of each number.

**12.** 32% of 75 **13.** 0.5% of 66 **14.** 9% of 80 **15.** 125% of 24

**16.** The Recommended Daily Allowance (RDA) of calcium for teenagers is 1,200 mg. What percent of the RDA is 800 mg?

# EXERCISES

## PRACTICE/ SOLVE PROBLEMS

Write each fraction or decimal as a percent.

**1.** 0.79 **2.** $\frac{3}{4}$ **3.** $\frac{17}{20}$ **4.** 1.06 **5.** $\frac{5}{6}$ **6.** $0.02\frac{1}{3}$

Find the percent of each number.

**7.** $66\frac{2}{3}\%$ of 90 **8.** 15% of 68 **9.** 96% of 50 **10.** 110% of 70

Find each percent.

**11.** 16 is what percent of 25? **12.** 129 is what percent of 300?

Solve.

**13.** In 1980, there were 82,536 foreign-born persons living in Phoenix, Arizona. These people represented 5.5% of the city's population. To the nearest whole number, what was the population of Phoenix in 1980?

## EXTEND/ SOLVE PROBLEMS

Refer to the table and and use a calculator to solve the problems.

| AVERAGE STARTING SALARY | | |
|---|---|---|
| | **1985** | **1988** |
| Engineering | $26,880 | $29,856 |
| Accounting | 20,620 | 25,140 |
| Computing | 24,156 | 26,904 |

**14.** In 1988, the average starting salary for an accountant was what percent of the average starting salary for an engineer?

**15.** By how much did the salary for computing increase from 1985 to 1988? What percent of the 1985 salary does this increase represent? Round to the nearest tenth of a percent.

## THINK CRITICALLY/ SOLVE PROBLEMS

**16.** Owen sold 80% as many raffle tickets as Donna. Jorge sold 50% as many raffle tickets as Owen. Donna sold __?__% as many raffle tickets as Jorge.

**17.** A record store had a sale. All items were reduced by 10%. After the sale, prices were marked up 10% over the sale prices. How do prices after the sale differ from prices before the sale?

**18.** Fred has $20 and Freda has $25. Fred says he has 20% less than Freda. Freda says that he's wrong, and that she has 25% more than he does. Who is right? Explain.

- ► READ
- ► PLAN
- ► SOLVE
- ► ANSWER
- ► CHECK

# Problem Solving Applications:

## SALES TAX AND THE CONSUMER

Many state and local governments require consumers to pay a **sales tax** on items that they purchase. Sales tax is figured as a percent of the selling price of the item.

total cost = selling price + sales tax

Sales taxes are usually rounded up to the next cent.

Calculate the sales tax of 7% for each of the following items.

| | Item | Price |
|---|---|---|
| 1. | Hairdryer | \$19.95 |
| 2. | Tennis racket | \$52.50 |
| 3. | Shirt | \$14.75 |

4. The price of a pair of shoes is \$29.98. Calculate the total cost of the shoes if the sales tax is 6%.

5. A stereo system costs \$379.99. The sales tax is $8\frac{1}{4}$%.
   a. How much sales tax would be added to the price of the stereo?
   b. What would be the total cost of the stereo including tax?

6. Cesar bought the following items at a department store: hat—\$8.50, gloves—\$3.99, batteries—\$2.49, candy—\$4.50, jeans—\$19.95.

   The sales tax is 9%. Calculate Cesar's total bill.

**MENTAL MATH TIP**

You may find it helpful to memorize these equivalent percents, decimals, and fractions.

$12\frac{1}{2}\% = 0.125 = \frac{1}{8}$

$16\frac{2}{3}\% = 0.1\overline{6} = \frac{1}{6}$

$20\% = 0.2 = \frac{1}{5}$

$25\% = 0.25 = \frac{1}{4}$

$33\frac{1}{3}\% = 0.\overline{3} = \frac{1}{3}$

$37\frac{1}{2}\% = 0.375 = \frac{3}{8}$

$40\% = 0.4 = \frac{2}{5}$

$50\% = 0.5 = \frac{1}{2}$

$60\% = 0.6 = \frac{3}{5}$

$62\frac{1}{2}\% = 0.625 = \frac{5}{8}$

$66\frac{2}{3}\% = 0.\overline{6} = \frac{2}{3}$

$75\% = 0.75 = \frac{3}{4}$

$80\% = 0.8 = \frac{4}{5}$

$83\frac{1}{3}\% = 0.8\overline{3} = \frac{5}{6}$

***USING DATA*** Use the Data Index on pages 556–557 to find state sales tax rates.

7. How much state sales tax would be added to a bicycle selling for \$142.75 in Michigan?

8. Find the total cost of a computer selling for \$1,069 in Florida.

9. If the computer in Exercise 8 was selling for the same price in Louisiana, how much less would you pay for the computer?

10. Beth bought the following items.

| | | | |
|---|---|---|---|
| Watch | \$39.95 | Book | \$10.00 |
| Scarf | \$5.79 | Earrings | \$7.50 |

If Beth paid \$66.25 including sales tax, in which state did she make the purchase?

11. In many states, food items are not taxable. One day, Jack's total supermarket bill, including tax, was \$47.68. The amount of tax was \$1.42. If the tax rate was 6%, what amount of money was taxed? What amount of money was not taxed?

# 1-7 Using Estimation to Solve Problems

## EXPLORE/ WORKING TOGETHER

Work with a small group of students.

a. Without using pencil and paper or a calculator, each of you must pick *three* numbers from those shown in the chart to get a total as close to 1,000 as you can without going over.

b. Write your numbers on slips of paper. Then find an exact answer. Compare results with your group to determine who came the closest.

| 144 | 201 | 320 | 92 | 679 |
|---|---|---|---|---|
| 512 | 476 | 155 | 197 | 287 |

c. Discuss your strategies for picking the numbers. Did you estimate? What estimation method did you use? Why?

d. You can vary the activity by picking *four* numbers to get as close to 1,500 as you can.

## SKILLS DEVELOPMENT

When you solve problems, you must first decide whether you need an *estimate* or an *exact answer.* Problems that ask *about how much, about how many,* or *do you have enough* can often be answered just by using estimation. Even when problems require an exact answer, you can estimate *before* you start your calculations in order to get a "ballpark" figure or *after* you complete the work in order to check the reasonableness of your answer.

### Example 1

Decide whether an estimate or an exact answer is needed. Solve.

a. A caterer charges \$9.75 per person (tax and tip included) for buffet luncheons. You have \$400 to spend for food for a party for 38 people. Do you have enough money?

b. When you pay the bill for the party, you give the caterer \$400. How much change should you receive?

c. Suppose you were expecting 41 people, instead of 38. Do you have enough money?

### Solution

a. An estimate is sufficient if it shows that the actual cost will be less than \$400. Round \$9.75 up to \$10.

$$38 \times 10 = \$380$$

Since you used a per-person cost that was a little higher than the actual cost, you can be sure you have enough money.

b. You need an exact answer.

$$38 \times \$9.75 = \$370.50 \text{ and } \$400 - \$370.50 = \$29.50$$

**c.** If you estimate according to Solution **a,** you will find that

$$41 \times \$10 = \$410$$

Since one factor was rounded up, you cannot be sure whether you have enough money. Find the exact answer.

$$41 \times \$9.75 = \$399.75$$ ◄

You can use different estimation techniques depending on the degree of accuracy needed. **Rounding** is a quick way to estimate.

### Example 2

During the summer, Ramón Velez operates a snack bar at the beach. Last year he sold 947 canned soft drinks in June, 1,260 in July, and 1,488 in August. How many soft drinks should he order for the same period this year?

**Solution**

To make sure that he has enough soft drinks, Ramón decides to *overestimate*. He overestimates by rounding up the numbers of drinks sold and then adds to get the total.

| | | |
|---|---|---|
| 947 → | 1,000 | 1,000 |
| 1,260 → | 1,300 | 1,300 |
| 1,488 → | 1,500 | +1,500 |
| | | 3,800 |

Ramón estimates that he should order 3,800 soft drinks. ◄

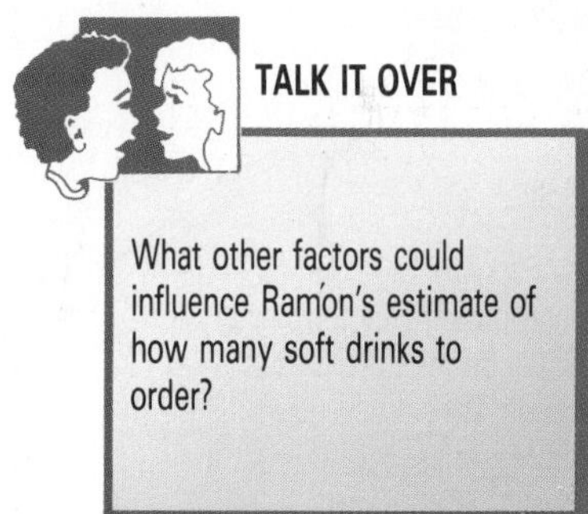


What other factors could influence Ramón's estimate of how many soft drinks to order?

Another useful method is **front-end estimation.**

### Example 3

Ramón wants to renovate his snack bar. His budget is $5,500. Can he afford new chairs and tables costing $2,620, a new awning for $439, a paint job costing $1,325, and hanging plants for $227?

**Solution**

Ramón uses front-end estimation with adjustment.

| Add the front-end digits. | | Then adjust the estimate. | | Think: |
|---|---|---|---|---|
| | $2,620 | | $2[620] | ← about 1,000 |
| | 439 | | [439] | |
| | 1,325 | | 1[325] | ← about 500 |
| | + 227 | | + [227] | |
| | $3,000 | | $4,500 | ← 3,000 + 1,000 + 500 |

Since his estimate is considerably less than his budget, Ramón decides he can afford all the items. ◄

**Compatible numbers,** or numbers with which you can compute mentally, are often used in estimating products and quotients.

MIXED REVIEW

Find each answer. Use rounding to estimate and compare to see if the answer is reasonable.

1. 8,245 + 796

2. 7 − 5.61

3. \$298 × 9

4. 366,960 ÷ 11

Complete.

5. 47.9 × ■ = 0

6. $5\left(\frac{3}{8} + \frac{1}{4}\right) = 5(■) + 5\left(\frac{1}{4}\right)$

7. ■ × 268 = 268

Solve.

8. Parkway High School set a fund-raising goal for charity of \$500. So far, \$309.85 has been raised. How much more money is needed to meet the goal?

## Example 4

Estimate.

**a.** 1,197 ÷ 29 **b.** $\frac{4}{5} \times 138$ **c.** 67% of 598

### Solution

**a.** 1,197 ÷ 29 **Think: 1,197 is close to 1,200 and 29 is close to 30.**
Mentally divide 1,200 by 30 to get an estimate of 40.

**b.** $\frac{4}{5} \times 138$ **Think: 138 is close to 150.**
Mentally find $\frac{1}{5} \times 150$, then mentally multiply by 4.
The estimate is 30 × 4 = 120.

**c.** 67% of 598 **Think: 67% is about $\frac{2}{3}$ and 598 is about 600.**
Mentally find $\frac{1}{3}$ of 600, then double the result.
The estimate is 2 × 200 = 400. ◄

## TRY THESE

Decide whether an estimate or an exact answer is needed. Solve.

1. A week-long cruise costs \$1,128.50 per person. Will \$5,000 cover the cost of the cruise for a family of 4?
2. A customer bought 2.5 lb of cashews at \$6.98 a pound. How much change should the customer get from a twenty-dollar bill?
3. Each guest at a barbecue will eat about $\frac{1}{2}$ of a chicken. How many chickens should be ordered for 56 guests?

Estimate each answer by using compatible numbers.

4. 2,368 ÷ 78
5. $\frac{5}{8} \times 314$
6. 72% of 773
7. 35.81 ÷ 5.9
8. A pharmacy had weekly sales of \$2,379, \$764, \$1,281, and \$1,788. Estimate the total sales for the four-week period. Use the front-end digits and adjust.

# EXERCISES

### PRACTICE/ SOLVE PROBLEMS

Choose the best estimate.

1. 207 × 18 **a.** 4,000 **b.** 4,400 **c.** 3,000
2. \$12.78 × 32 **a.** \$320 **b.** \$390 **c.** \$420
3. 28.6 × 3.7 **a.** 120 **b.** 102 **c.** 84
4. 26.80 ÷ 0.9 **a.** 2.8 **b.** 30 **c.** 3.0

5. $\frac{5}{6} \times 415$ a. 300 b. 350 c. 480

6. 38% of 7,169 a. 2,100 b. 2,400 c. 2,800

Estimate each sum. Use front-end digits and adjust.

7. $\begin{array}{r} 622 \\ 375 \\ +\ 409 \\ \hline \end{array}$ 8. $\begin{array}{r} 249 \\ 512 \\ 683 \\ +\ 1{,}805 \\ \hline \end{array}$ 9. $\begin{array}{r} 44{,}319 \\ 25{,}687 \\ +\ 9{,}692 \\ \hline \end{array}$ 10. $\begin{array}{r} 3.944 \\ 0.897 \\ 2.168 \\ +\ 1.315 \\ \hline \end{array}$

Decide whether an estimate or an exact answer is needed. Then solve.

11. Martha is a speed reader. On Monday she read 69 pages in one hour. If she continues to read at the same rate for one hour each day, can she finish her 367-page book on Friday?

12. The cost of sending a 1-lb overnight package is $21.50. Is $800 enough to cover the cost of sending 38 packages?

13. Tickets to a concert cost $23.75 each. How many tickets can Jon buy with $130?

14. Last week, Leon's average time for running a mile was 9.2 min. His goal has been to shorten his time by 10%. This week he ran a mile in 8.15 min. Did Leon meet his goal?

## EXTEND/ SOLVE PROBLEMS

15. Is the answer to 7,233 − 4,826 *less than* or *greater than* 3,000? Explain your reasoning.

16. Is 35,871 − 23,449 *less than* or *greater than* 12,000?

17. Tai is at the supermarket. She has collected the items listed at the right in her shopping cart, but she has only $15 to spend. What is the least costly item she can put back and still have enough money to pay for the other items?

| Item | Cost |
|---|---|
| Cheese | $2.29 |
| Fish | $5.80 |
| Chicken | $3.50 |
| Juice | $2.89 |
| Coffee | $3.95 |

## THINK CRITICALLY/ SOLVE PROBLEMS

Estimate to solve. Tell which amounts you overestimate and which you underestimate. Explain.

18. Gino is planning to drive to and from a crafts fair about 100 mi away. His gas tank is almost empty. Gas costs about $1.37 per gallon. His car travels about 24 mi/gal. How much money does Gino need for gas?

19. A photographer wants to be at the finish line for a bicycle race. The race course is 204 mi long and the faster racer is expected to average 24.5 mi/h. How long after the start of the race should the photographer arrive at the finish line?

# 1-8 Problem Solving Skills:
## CHOOSE A COMPUTATION METHOD

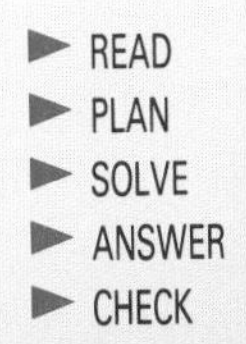


You can solve some math problems mentally. To solve others, you may choose to use paper and pencil or a calculator. In order to solve a problem accurately, first read it carefully, then ask yourself each of the following questions.

- Do I need an exact answer, or will an estimate do?
- Which number operation(s) do I need to use?
- Which computation method should I use to get an answer quickly and accurately—mental math, paper and pencil, or a calculator?

**COMPUTER**

If you have access to a computer, you will find that it can help you solve problems by providing you with still another method of computation.

This program finds the sum and difference of any two whole numbers.

```
10 PRINT "FIND THE SUM
   AND DIFFERENCE OF ANY
   TWO WHOLE NUMBERS."
20 PRINT
30 PRINT "ENTER THE
   GREATER OF TWO WHOLE
   NUMBERS.": INPUT A
40 PRINT "ENTER THE
   LESSER OF TWO WHOLE
   NUMBERS.": INPUT B
50 PRINT "THE SUM OF ";A;"
   AND ";B;" IS ";A + B;"."
60 PRINT "THE DIFFERENCE
   OF ";A;" AND ";B;" IS
   ";A - B;"."
```

### PROBLEM

Travis needs to replace seven of the large windows in his house. Each replacement window costs $142.50. He has saved $1,000 to pay for the entire job. Will Travis have enough money for all seven replacement windows?

### SOLUTION

Which computation method is appropriate?

**Mental Math**
Round $142.50 to $150
Think: 7 × $100 = $700, and 7 × $50 = $350
$700 + $350 = $1,050

Since $1,050 is more than $1,000, it appears that Travis *does not* have enough money for all seven of the replacement windows. But, since his estimates were high, he should use either pencil and paper or a calculator to get an exact answer.

**Paper and Pencil**

$$\begin{array}{r} \$142.50 \\ \times \quad 7 \\ \hline \$997.50 \end{array}$$

Think: Since $997.50 is less than $1,000, Travis *does* have enough money for seven windows.

**Calculator**
7 [×] 142.50 [=] 997.50 Think: Yes, Travis *does* have enough money for all seven windows.

# PROBLEMS

Indicate which computation method you would use to solve each problem. Then solve.

1. Adolfo ordered socks for $8.97, a belt for $11.95, and shirts for $48.85. What is the cost of his order?
2. There are 4 seedlings in a pack and 6 packs in a flat. How many seedlings are there in 12 flats?
3. Jo bought a cat carrier for $20.98, a litter box for $11.95, a bag of litter for $2.87, and 4 cans of cat food at $0.79 each. She has four ten-dollar bills with which to pay. Is this enough?
4. Fifteen cartons, each containing 4 dozen rose bushes, were delivered to Red Rose Nursery. The roses will be separated into smaller boxes with 8 bushes each. How many boxes will there be?

5. Curtis bought a pair of stereo speakers on which he paid a sales tax of 8.5%. His bill came to $382.50, including tax. If Curtis had bought the speakers in a nearby state where the sales tax is 6.5%, how much could he have saved on his purchase?
6. It takes Carlos 2 h to mow the lawn, while his younger brother, Julio, needs 4 h to do the same job. If each boy spent a total of 24 h mowing last year, how many times did Carlos mow the lawn? How many times did Julio mow it?
7. Jenna won a trip in a contest. She phoned two friends to tell them the news. Ten minutes later, each of these friends told two others. Ten minutes later, each of these four friends told two friends. In ten minutes more, these friends each phoned two friends. If the news continues to travel in this way, how many people will have been phoned during the ninth ten-minute period?

**TALK IT OVER**

There are two correct answers to Problem 5. Describe how you arrived at your answer. See if you can find the other answer. Compare your findings with those of your classmates.

# 1-9 Working with Graphs

**EXPLORE**

This **pictograph** shows the number of different kinds of recordings that were shipped by manufacturers to stores over a recent six-month period. Read the key at the bottom of the graph, then answer these questions.

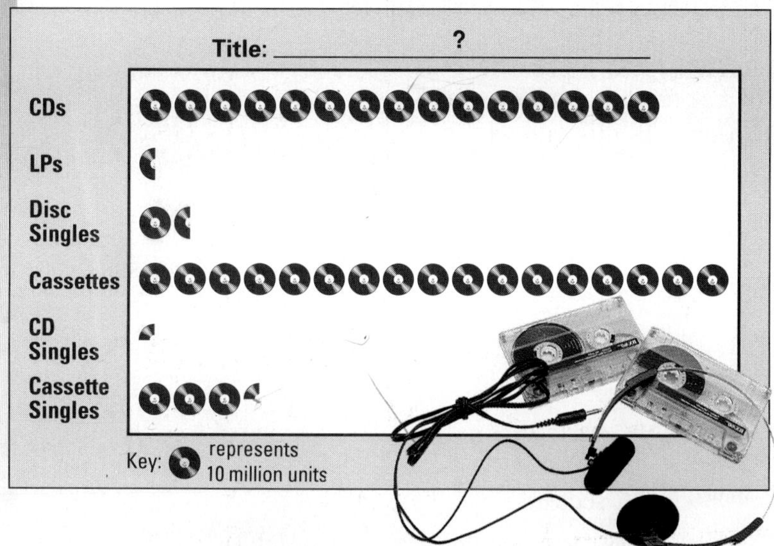


a. What is the value of the half symbol?

b. Which symbol represents 2.5 million units?

c. How many cassettes were shipped? How many cassette singles?

d. How many more CDs were shipped than CD singles?

e. What would be a good title for this pictograph?

**SKILLS DEVELOPMENT**

Many kinds of data can be organized and presented in a graph so that they are easy to compare visually.

Some data can be displayed in a **bar graph** with either horizontal or vertical bars. An appropriate scale for a bar graph is one marked in intervals that best reflect the data. To interpret the graph, follow the edge of each bar to the part of the scale with which it aligns to find its value. If the edge of a bar falls between marked intervals, you must estimate to determine the value of the bar.

### Example 1

Use the bar graph at the right.

a. In which year were about 3 million acres of forest land burned?

b. About how much forest land was burned in 1986?

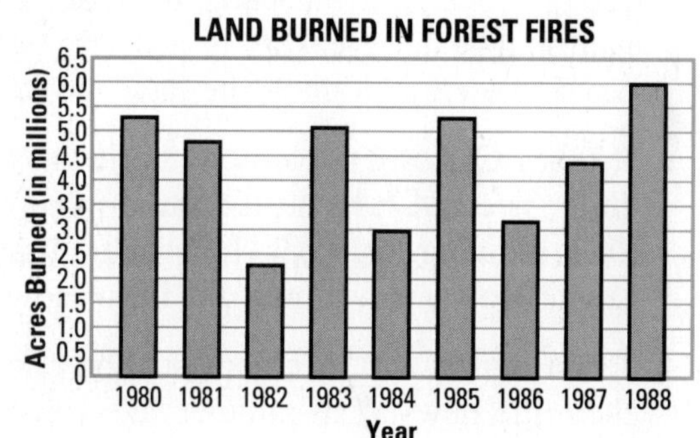


**Solution**

a. In 1984 about 3 million acres of land was burned.

b. About 3.2 million acres of land was burned in 1986. ◄

**Line graphs** show trends in data, or changes in data over periods of time. Read a line graph by locating the data points and relating them to the labeled points on the horizontal and vertical scales with which they align. The line segments connecting the points help you see the changes in the data.

### Example 2

Use the line graph at the right.

a. What was the population of the United States in 1920?

b. About how much did the population increase between 1960 and 1990?

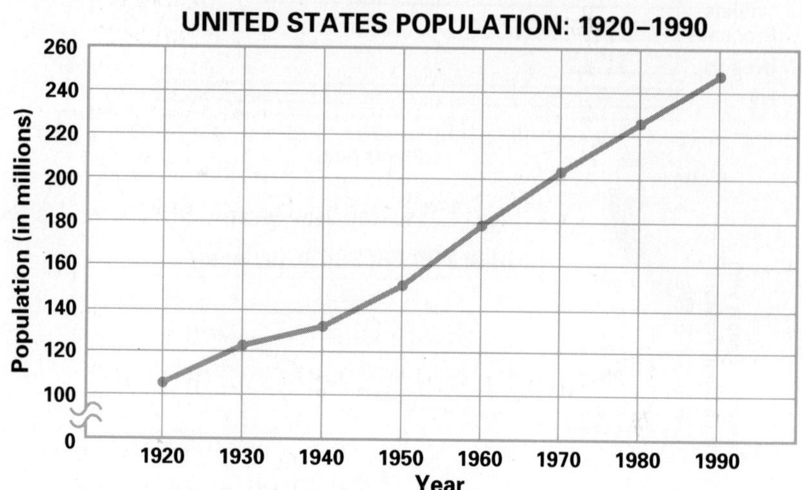

**Solution**

a. 106 million people

b. about 70 million people ◄

Data are displayed in a **circle graph** as individual parts of the whole circle. The parts are called sectors. Each sector represents a percent of the total amount of data. The total value of the sectors represents 100% of the data.

### Example 3

Use the circle graph at the right.

a. The percent spent on transportation is about the same as the percent spent on which other two categories together?

b. About how much money is spent on clothing per year?

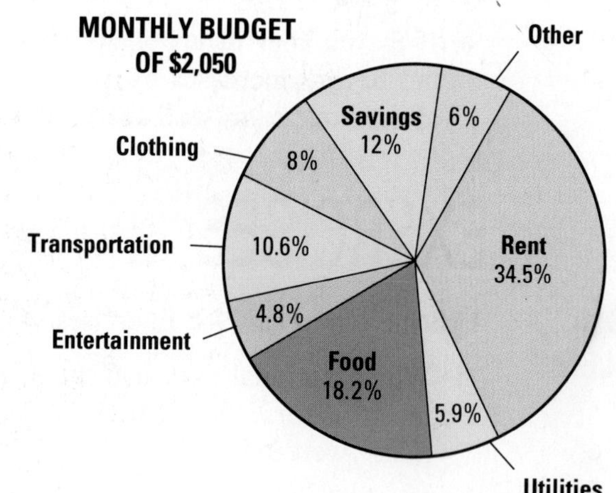


**Solution**

a. The percent spent on entertainment (4.8%) and utilities (5.4%) together is about equal to the percent spent on transportation (10.6%).

b. Find 8% of $2,050 to find the amount spent in one month, then multiply this amount by 12 to find how much is spent in one year, or 12 months.

$0.08 \times \$2,050 = \$164$

$12 \times \$164 = \$1,968$

$\$1,968 \approx \$2,000$

So, about $2,000 is spent on clothing per year. ◄

## TRY THESE

Use the bar graph for Exercises 1–2.

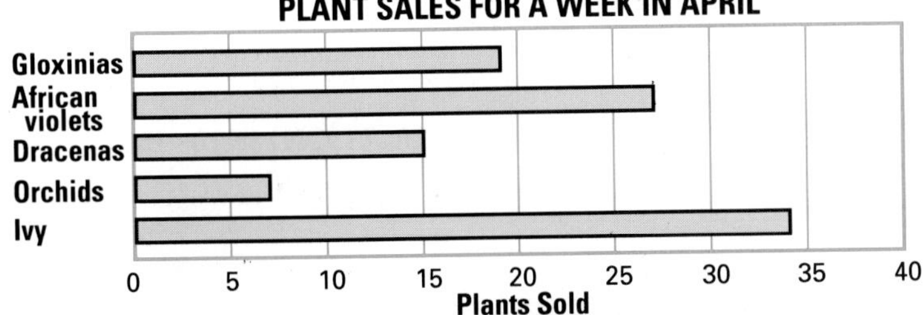


1. Fifteen of which kind of plant were sold?
2. About how many more African violets were sold than orchids?

Use the line graph for Exercises 3–4.

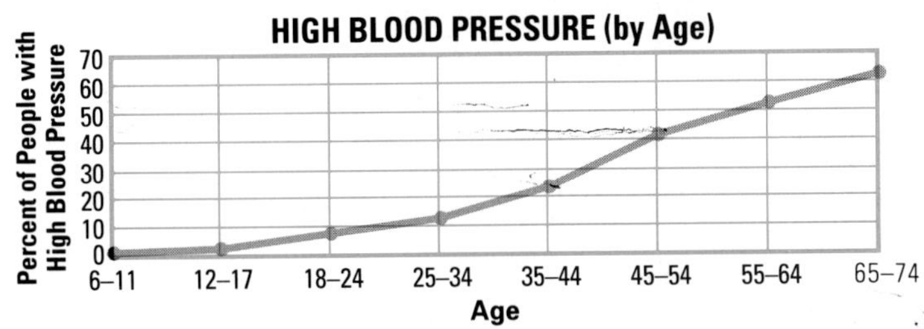


3. What percent of ages 45–54 have high blood pressure?
4. About how much greater is the percent of high blood pressure in people of ages 65–74 than in those of ages 35–44?

Use the circle graph for Exercises 5–6. Round answers to nearest **whole number**.

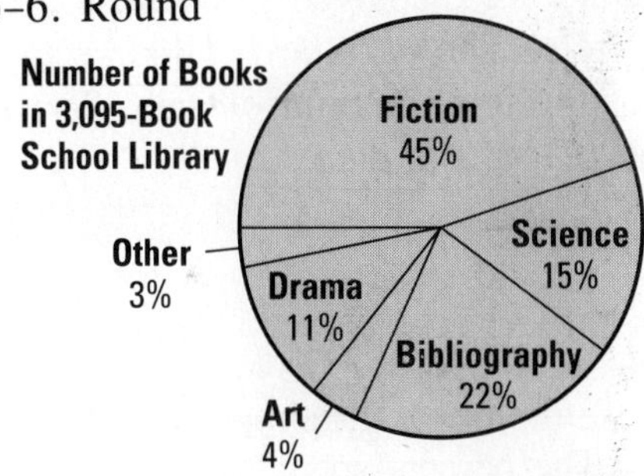


5. How many science books are there in the library?
6. If the number of art books in the library increases by 10%, about how many will there be?

# EXERCISES

## PRACTICE/ SOLVE PROBLEMS

Use the bar graph for Exercises 1–6.

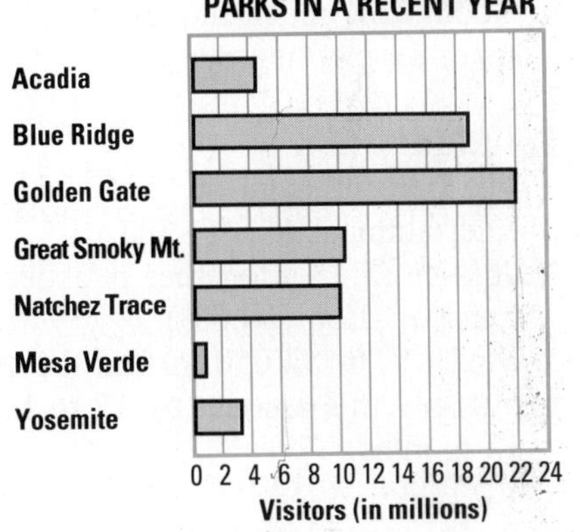


1. Which national park had the most visitors? About how many were there?
2. Which park had the fewest visitors? About how many were there?

About how many visited each?

3. Natchez Trace
4. Yosemite
5. Acadia
6. Blue Ridge had about six times as many visitors as did which other national park?

Use the circle graph for Exercises 7–9.

7. How much does Greg spend on movies each month?

8. How much will Greg spend on school lunches in a month?

9. Greg wants to buy a $15 ticket to a soccer game. How much will he need from his savings to add to the amount budgeted for sports events?

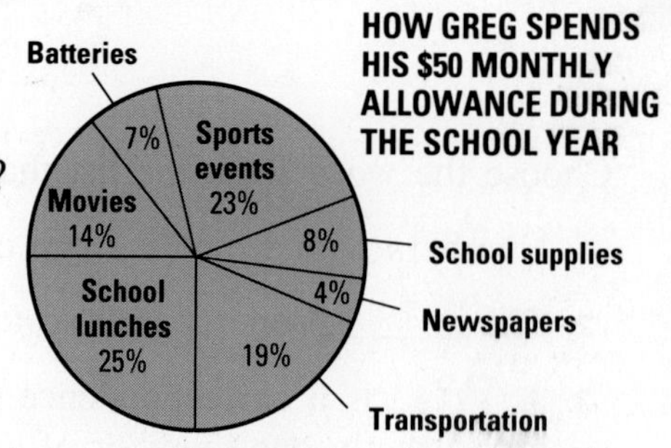


## EXTEND/ SOLVE PROBLEMS

Examine the set of data. Decide on the appropriate scales you can use to graph the data. Then construct the graph. Write a title.

10. Construct a bar graph for this immigration data. Start by deciding whether to draw the bars horizontally or vertically.

| NATIONS SENDING THE LARGEST PERCENTAGE OF TOTAL U.S. IMMIGRANTS, 1820–1987 | | | |
|---|---|---|---|
| Germany | 13% | Italy | 10% |
| United Kingdom | 9.4% | Mexico | 5% |
| Austria–Hungary | 6.0% | USSR | 6.3% |
| Canada | 7.9% | Ireland | 8.8% |

11. Construct a line graph for these plant-growth data.

| HEIGHT OF MARIGOLDS (from first week after seeding) | |
|---|---|
| Week | Height (in cm) |
| 1st | 0.5 |
| 2nd | 1.25 |
| 3rd | 2.00 |
| 4th | 3.25 |
| 5th | 5.75 |
| 6th | 7.50 |
| 7th | 10.00 |
| 8th | 12.50 |

## THINK CRITICALLY/ SOLVE PROBLEMS

**USING DATA** Use the Data Index on pages 556–557 to find the graph *Percent Distribution of U.S. Population Growth.* Use the graph for Exercises 12–15.

12. How is this graph like a line graph?

13. How is this graph like a bar graph?

14. Why is the bar for each decade the same height?

15. Explain why the values given in the first bar (1900–1910) are approximately the reverse of the values given in the last bar (1980–1990).

# CHAPTER 1 • REVIEW

Choose the word from the list that completes each statement.

1. Numbers with which we can compute mentally are ___?___ numbers.
2. The ___?___ property is used to rewrite $a(b + c)$ as $ab + ac$.
3. A ___?___ is a correspondence that pairs each member of a given set with exactly one member of another set.
4. Zero is the additive ___?___.

a. function
b. identity
c. distributive
d. compatible

## SECTION 1–1 NUMBERS, VARIABLES, AND EXPRESSIONS (pages 4–7)

► An expression that contains one or more variables is called a **variable expression.** To evaluate a variable expression, substitute a given number for each value.

Evaluate each expression when $x = 12$.

5. $x + 5$
6. $\frac{1}{3}x - 1$
7. $6x - 7$
8. $\frac{x}{2} + 10$
9. $(x)(x)$

## SECTION 1–2 PROPERTIES AND MENTAL MATHEMATICS (pages 8–11)

► We perform operations on numbers by using the **commutative, associative, distributive,** and **identity properties,** and the **property of zero for multiplication.**

Complete. Identify the property.

10. $43 \times$ ■ $= 43$
11. $(1.3 +$ ■$) + 0.9 = 1.3 + (4.8 + 0.9)$
12. $6(70 +$ ■$) = 6(70) + 6(8)$
13. ■ $+ 0 = 127$

## SECTION 1–3 EXPLORING PATTERNS (pages 12–15)

► An arrangement of numbers in a pattern is a **sequence.**

Find the next three terms in each pattern.

14. 95, 91, 87, 83, ■, ■, ■
15. $1, \frac{1}{3}, \frac{1}{9}, \frac{1}{27}$, ■, ■, ■

## SECTION 1–4 PATTERNS AND FUNCTIONS (pages 16–19)

► A **function** is a correspondence. The description of a function is called a **function rule.** Variable expressions are often used for function rules and for making **function tables.**

Make a function table for each of the following.

16. $f(x) = 3x + 2$; $x = 0, 1, 4, 7, 10$
17. $f(y) = \frac{1}{2}y - 1$; $y = 2, 6, 9, 14, 20$

## SECTIONS 1–5 AND 1–8 PROBLEM SOLVING (pages 20–21, 30–31)

► Decide which computation method is best for solving a given problem—mental math, pencil and paper, or calculator. Then, use your method to solve.

18. Kim bought a stapler for $5.00, file folders for $3.29, and a dictionary for $9.00. How much did she spend?

19. A theater group sold 486 tickets for the matinee show and 592 tickets for the evening show. The matinee tickets cost $12.95 each and the evening tickets cost $17.95 each. How much money did the group take in?

## SECTION 1–6 SOLVING PERCENT PROBLEMS (pages 22–25)

► Use an equation to find a percent of a number, to find what percent one number is of another, or to find a number when a percent of it is known.

20. Find 75% of 2,000.

21. Find 14% of 350.

22. What percent of 80 is 25?

23. 42 is 60% of what number?

## SECTION 1–7 USING ESTIMATION TO SOLVE PROBLEMS (pages 26–29)

► Sometimes you can solve a problem using estimation. Other times, when an exact answer is needed, estimation can help you decide if your solution is reasonable.

Decide whether an estimate or an exact answer is needed. Then solve.

24. Mr. Ramos must pay 23% of his gross income of $38,000 for taxes. Will $10,000 be enough to cover his tax bill?

25. It costs $0.87 to print each copy of a rules booklet. Can a club afford to print 150 copies if its budget is $135?

## SECTION 1–9 WORKING WITH GRAPHS (pages 32–35)

► A graph, such as a bar graph, a line graph, or a circle graph, may be an appropriate means of displaying certain data.

*USING DATA* Use the circle graph on page 3 to answer the following question.

26. Of the 23,238,947 people who immigrated to the United States between 1890 and 1939, how many came from Austria-Hungary?

# CHAPTER 1 ● TEST

Simplify each expression.

1. $20 + 8 \div 2$
2. $15 \div 3 + 7 - 8$
3. $1.5 \times 1.5 - 0.5$

Evaluate each expression when $c = 25$.

4. $2c - 15$
5. $\frac{1}{2}(c + 13)$
6. $c + c \times 4$

Complete.

7. $6(20 + 7) = 6(■) + 6(7)$
8. $51.9 \times ■ = 0$
9. $■ \times 83 = 83$
10. $\left(\frac{1}{2} + \frac{2}{3}\right) + \frac{7}{8} = \frac{1}{2} + \left(■ + \frac{7}{8}\right)$

Find the next three terms in each pattern.

11. 2, 3.2, 4.4, 5.6, ■, ■, ■
12. $\frac{1}{2}, \frac{1}{4}, \frac{1}{8}, \frac{1}{16}$, ■, ■, ■
13. 101, 100, 98, 95, 91, ■, ■, ■
14. 2, 4, 12, 48, ■, ■, ■

Copy and complete each function table.

15.

| $x$ | $4x + 4$ |
|---|---|
| 0 | ■ |
| 1 | ■ |
| 5 | ■ |
| 9 | ■ |
| 14 | ■ |

16.

| $w$ | $5w - 2$ |
|---|---|
| 2 | ■ |
| 7 | ■ |
| 12 | ■ |
| 15 | ■ |
| 21 | ■ |

17.

| $m$ | $\frac{1}{2}m + 3$ |
|---|---|
| 0 | ■ |
| 3 | ■ |
| 8 | ■ |
| 11 | ■ |
| 20 | ■ |

Write each fraction or decimal as a percent.

18. 0.07
19. $\frac{2}{3}$
20. 0.92
21. $\frac{7}{50}$
22. 1.6

Estimate each answer by using compatible numbers.

23. $3,590 \div 62$
24. 41% of 981
25. $\frac{3}{8} \times 476$

The line graph shows population data for a small town. Use the graph for Exercises 26–28.

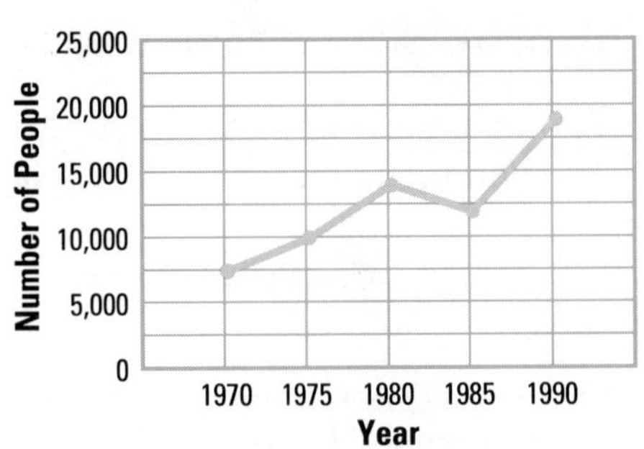


26. Did the population increase or decrease during the period 1975–1980?
27. During which five-year period did the greatest population increase take place?
28. Estimate the population of the town in 1985.

## CHAPTER 1 ● CUMULATIVE REVIEW

Simplify each expression.

1. $18 \cdot 2 \div 4$
2. $11 - 6 \div 3 \times 5$
3. $12 \cdot 5 + 6 \div 6$
4. $3 + 8 \cdot 9 + 4$

Complete.

5. $7(2 + 5) = (7 \times 2) + (\blacksquare \times 5)$
6. $7.3 \times \blacksquare = 7.3$
7. $\frac{2}{3} \times \blacksquare = 0$
8. $\blacksquare(9 - 5) = (2.5 \times 9) - (2.5 \times 5)$

Evaluate each expression when $s = 10$ and $t = 5.5$.

9. $3s + t$
10. $2st - 10$
11. $(s + t) - (s - t)$
12. $\frac{s}{2}(2t)$

Complete each function table.

13.

| $x$ | $2x - 2$ |
|---|---|
| 2 | ■ |
| 4 | ■ |
| 6 | ■ |
| 8 | ■ |
| 10 | ■ |

14.

| $m$ | $\frac{m}{3}$ |
|---|---|
| 0 | ■ |
| 1 | ■ |
| 3 | ■ |
| 12 | ■ |
| 18 | ■ |

Find the percent of each number.

15. 25% of 90
16. $33\frac{1}{3}$% of 50
17. $12\frac{1}{2}$% of 72

Find each percent to the nearest tenth.

18. 50 is what percent of 200?
19. 178 is what percent of 350?
20. 10 is what percent of 1,000?

The circle graph shows the number of calories allowed for each meal of a 2,500-calorie diet.

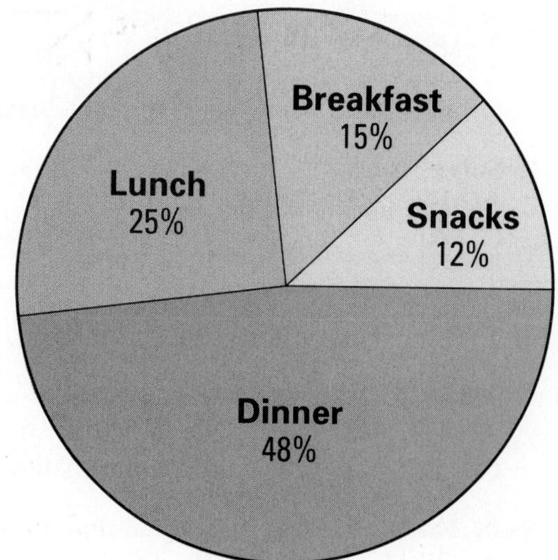


21. How many calories are allowed for lunch?
22. The number of calories allowed for breakfast and snacks together is about the same as the number allowed for which meal?
23. If someone on this diet had no snacks, but instead applied the calories allowed for snacks to dinner, how many calories would be allowed for dinner?

Simplify each expression.

1. $5 + 3 \times 10 \div 2$
   A. 40 B. 20
   C. 17.5 D. none of these

2. $5.3 - (4.8 \div 2)$
   A. 12.72 B. 0.25
   C. 2.9 D. none of these

3. $[48 - 4(17 - 9)] \div 8$
   A. 92.38 B. 44
   C. $-2$ D. none of these

4. Which expression has the value of 22 when $s = 3$ and $t = 5$?
   A. $4s - 2t$ B. $4s + 2t$
   C. $2st$ D. none of these

5. Solve. $78 + 25 + 62$
   A. 265 B. 155
   C. 165 D. none of these

6. Solve. $2.6 \times 28 \times 8.6$
   A. 62.6 B. 6.3
   C. 62,608 D. none of these

7. Solve. $6 \cdot 8 \cdot \frac{1}{3}$
   A. 144 B. 160
   C. 16 D. none of these

Find the next three terms in the pattern.

8. 4.6, 4.9, 5.2, __?__, __?__, __?__
   A. 4.9, 4.6, 4.4 B. 5.5, 5.7, 6.0
   C. 5.5, 5.8, 6.1 D. none of these

9. 6, 18, 9, __?__, __?__, __?__
   A. 27, 13.5, 40.5 B. 18, 9, 18
   C. 6, 9, 18 D. none of these

10. What is the function rule for this function table?

| y | ? |
|---|---|
| −2 | −7 |
| −1 | −3 |
| 0 | −1 |
| 1 | 2 |
| 2 | 5 |

   A. $f(y) = 3y - 1$
   B. $f(y) = 2y - 1$
   C. $f(y) = 3y + 1$
   D. $f(y) = 2y + 1$

11. 7 is what percent of 42?
   A. 6% B. $16\frac{2}{3}\%$
   C. 294% D. none of these

12. 14% of what number is 28?
   A. 0.5 B. 392
   C. 200 D. none of these

13. Jenna earned \$260. She banked \$60 and spent the rest. What percent of her earnings did she spend?
   A. 24% B. 15%
   C. 47.5% D. 76.9%

14. Estimate the answer by rounding. $80.95 \div 8.9$
   A. 10 B. 0.9
   C. 0.1 D. 9

The bar graph shows the most popular vacation destinations according to one travel agency.

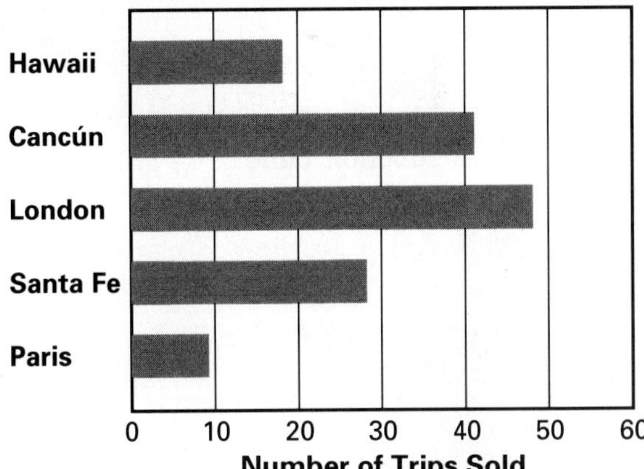


15. Which destination is the most popular?
   A. Cancún B. Paris
   C. Hawaii D. London

16. Which two destinations together are about as popular as Santa Fe?
   A. Hawaii and London
   B. Hawaii and Paris
   C. Cancún and London
   D. Paris and Cancún

# CHAPTER 2 SKILLS PREVIEW

Graph each set of numbers on a number line.

1. all real numbers less than $-1$
2. all real numbers greater than or equal to $-5$

Evaluate when $a = -6$.

3. $|-a|$ 4. $-|a|$ 5. $-(-a)$

Add or subtract.

6. $4 + (-3)$ 7. $-15 + (-3)$ 8. $61 + (-3)$ 9. $-19 + (-48)$

10. $4 - 6$ 11. $-2 - (-7)$ 12. $0 - 8$ 13. $-27 - 15$

Multiply or divide.

14. $2(-7)$ 15. $-6(-10)$ 16. $-14 \cdot 3$ 17. $-2 \cdot 6(-4)$

18. $-25 \div (-5)$ 19. $-98 \div 7$ 20. $70 \div (-14)$ 21. $-159 \div (53)$

Perform the indicated operation.

22. $7(-9) + 3.7$ 23. $\left(\frac{1}{6} - \frac{1}{3}\right)(-16)$ 24. $-14 \div (8.3 + 5.7)$

Evaluate each expression when $a = -9$ and $b = 0.25$.

25. $-4b$ 26. $\frac{a}{3}$ 27. $-(b - a)$ 28. $\frac{a}{2b}$

Simplify.

29. $3g - 4g$ 30. $-(-f + g)$ 31. $xy + 3xy$ 32. $3(ab - a) + 2(ab - b)$

Simplify.

33. $n^2 \cdot n^4$ 34. $(n^3)^3$ 35. $\frac{36w^6}{9w^2}$, $w \neq 0$ 36. $c \cdot c^2 \cdot c^4$

Evaluate each expression when $q = -2$ and $r = 3$.

37. $q^6 \cdot r^2$ 38. $q^{-5}$ 39. $(r^2)^{-2}$ 40. $q^2r^{-2}$

Write each number in scientific notation.

41. 305,000 42. 0.077 43. 0.00098

Solve.

44. The highest point in Colorado is Mt. Elbert at an altitude of 14,433 ft. The lowest point in Colorado is the Arkansas River at an altitude of 3,350 ft. Use an integer to describe the change from the highest point to the lowest point in Colorado.

CHAPTER 2

# REAL NUMBERS AND ALGEBRAIC EXPRESSIONS

THEME Geography

The word *algebra* comes from *al-jabr*–part of the title of a book written by a ninth-century Arabian mathematician named Mohammed ibn Musa al-Khowarizmi. Algebra is a special field of mathematics in which unknown numbers can be represented by letter symbols. Its rules are similar to those that you applied with whole numbers, decimals, and fractions.

In this chapter you will solve problems that involve algebraic symbols, integers, rational numbers, exponents, and scientific notation.

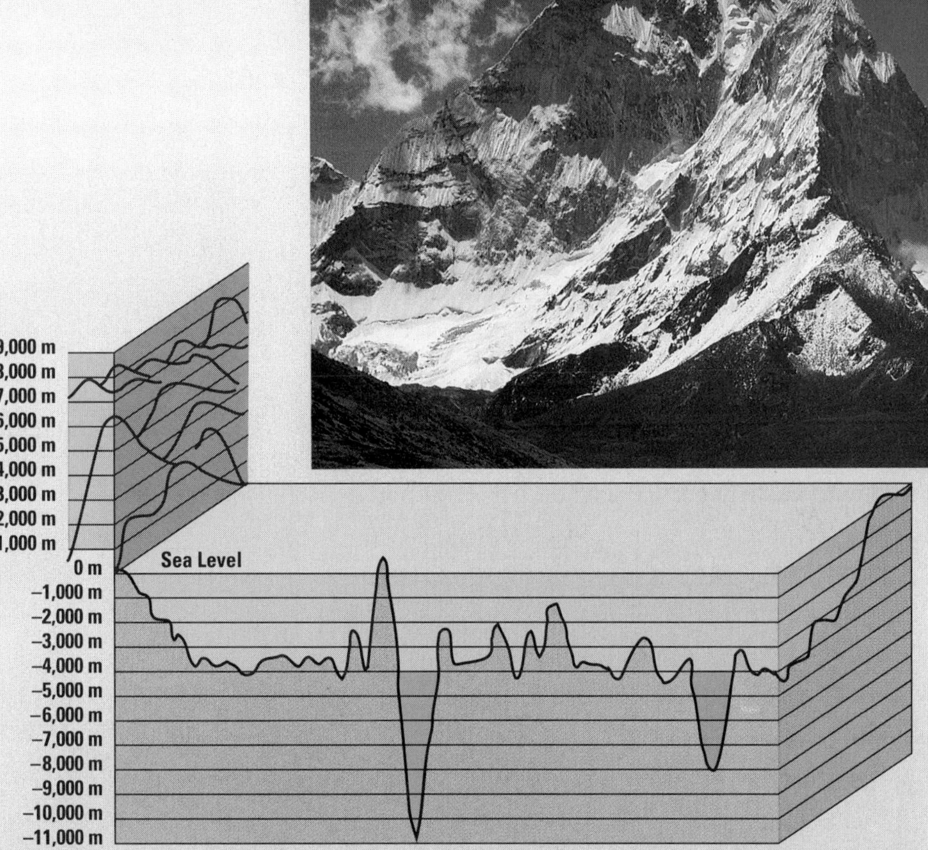

The level of ocean waters varies in different parts of the world and at different times. Thus, geographers use an average sea level as their point of reference for measuring height above the earth or depth below the earth.

| HEIGHT OR DEPTH OF VARIOUS LOCATIONS ON EARTH | |
|---|---|
| **Location** | **Height or Depth (Above or Below Sea Level)** |
| Mount Everest, Nepal/Tibet | +8,848 m |
| Mount McArthur, Canada | +4,344 m |
| Sea level | 0 m |
| Caspian Sea | −28 m |
| Death Valley, California | −86 m |
| Pacific Ocean (deepest descent by human) | −10,912 m |

## DECISION MAKING

### Using Data

Use the information in the chart above to answer the following.

1. How far below sea level is the Caspian Sea?
2. How much higher is Mount McArthur than Death Valley?
3. How much would Death Valley have to fall to be level with the deepest descent by a human?

### Working Together

Your group's task will be to find other locations in the world and their height or depth. For your research, use almanacs, encyclopedias, an atlas, or other sources of geographical information. Assemble the results of your research in a chart. Compare your group's chart with those made by other groups. Were some of the world's high and low spots selected by more than one group? Did your group find some places not found by any other group? Use the results of your research to answer the following question.

The world's deepest mine is a gold mine in South Africa. It is about 3,840 m deep. Suppose you wanted to describe or display this datum in a very dramatic way. What kinds of comparisons could you make to illustrate the extent of this depth? For example, this depth is about ten times the height of the Empire State Building.

# 2-1 The Real Number System

## EXPLORE

Draw a number line like the one below.

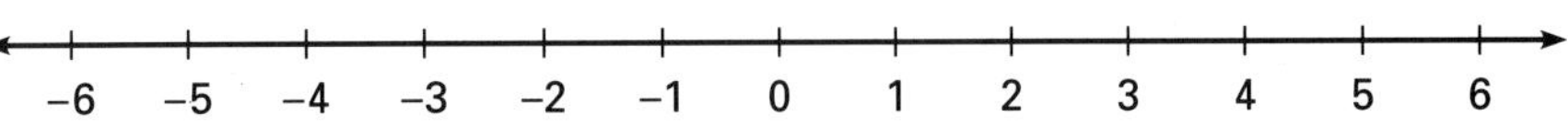


a. On the number line, locate a number greater than zero and a number less than zero.

b. What can you say about any number that corresponds to a point that is to the left of a given point on a number line?

c. Approximately where on the number line would you locate the point for $-2\frac{1}{3}$?

## SKILLS DEVELOPMENT

The set of **integers** consists of the **whole numbers** and their **opposites**. Opposites of whole numbers are the same distance from zero but in the opposite direction.

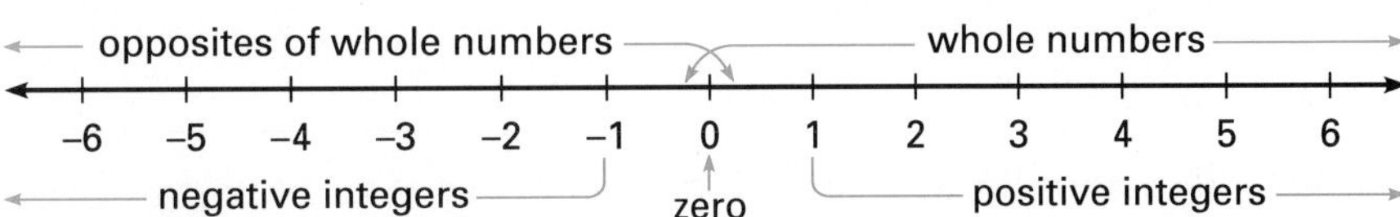


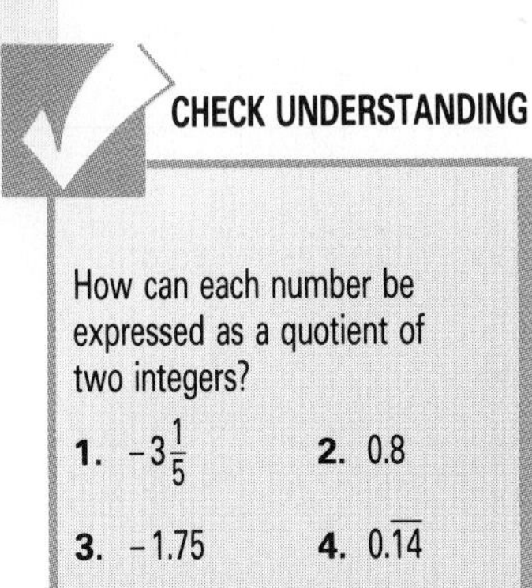

**CHECK UNDERSTANDING**

How can each number be expressed as a quotient of two integers?

1. $-3\frac{1}{5}$
2. 0.8
3. −1.75
4. $0.\overline{14}$

**Positive integers** are greater than zero. **Negative integers** are less than zero. Zero is neither positive nor negative and is its own opposite.

Numbers other than integers also can be shown on a number line. A **rational number** is a number that can be expressed as a ratio of two integers $a$ and $b$, where $b$ is not equal to zero. This is usually written $a/b$, $b \neq 0$. The symbol $\neq$ means "is not equal to."

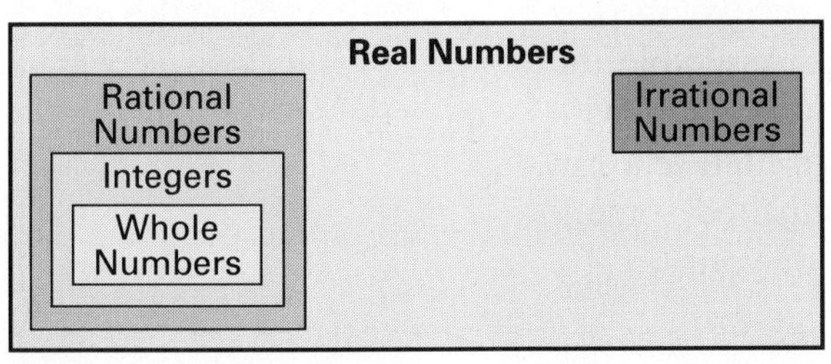


All rational numbers can be expressed as either **terminating decimals** or **repeating decimals**. Some numbers, such as π (pi) or the square root of 2, are nonterminating and nonrepeating decimals. Such numbers are called **irrational numbers**. Together, the sets of rational and irrational numbers make up the set of **real numbers**.

The number that corresponds to a point on a number line is called the **coordinate of the point**. Each real number corresponds to exactly one point on a number line. The point that corresponds to a number is called the **graph of the number** and is indicated on the number line by a solid dot. Each point on a number line corresponds to exactly one real number.

## Example 1

Graph this set of numbers on a number line.
$\{2.25, -1, -2\frac{3}{4}, -4\}$

**Solution**
Draw a number line. Use a solid dot to graph each number.

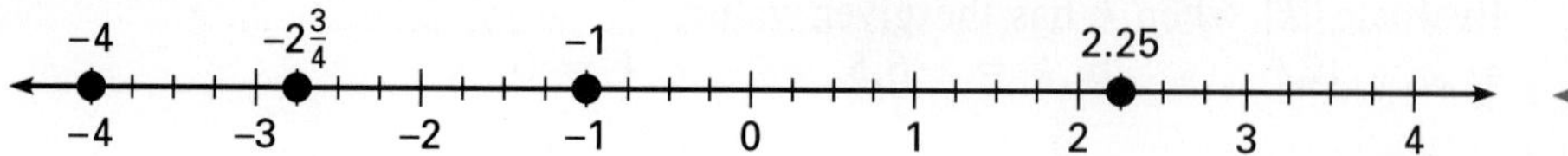


## Example 2

Use a number line to compare numbers. Replace each ● with <, >, or =.

**a.** $-3$ ● $1$ **b.** $2$ ● $0$ **c.** $\frac{1}{4}$ ● $-\frac{3}{4}$

**Solution**
Draw a number line and graph each number.

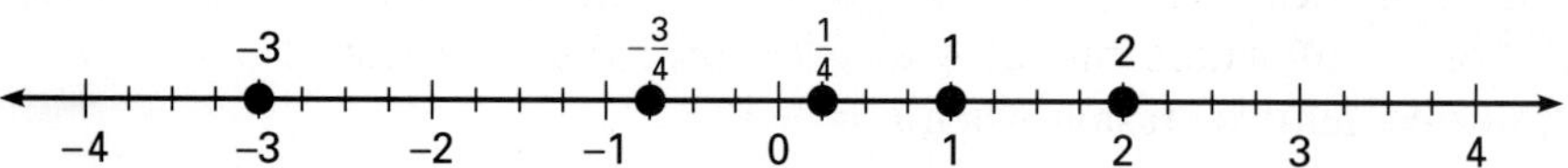


**a.** $-3$ is to the left of 1, so $-3 < 1$.
**b.** 2 is to the right of 0, so $2 > 0$.
**c.** $\frac{1}{4}$ is to the right of $-\frac{3}{4}$, so $\frac{1}{4} > -\frac{3}{4}$.

**READING MATH**

A **set** is a collection of objects, such as the set of integers or the set of whole numbers. The objects in the set are called **members,** or **elements,** of the set.

A set is indicated by braces { } enclosing the names of the set members. Inside the braces, the names of the set members are separated by commas. For example, the set of integers can be represented as {. . . , −3, −2, −1, 0, 1, 2, 3, . . .} where ". . ." means that the list of members continues without end in both directions.

You will learn more about sets and set notation in Chapter 13.

## Example 3

Graph each set of numbers on a number line.
**a.** the integers from $-3$ to 2
**b.** the real numbers from $-3$ to 2
**c.** all real numbers less than or equal to 2
**d.** all real numbers greater than $-1$

**Solution**

**a.** The set consists of $-3$, $-2$, $-1$, 0, 1, and 2. To graph the set, put a solid dot at each of these points on the number line.

−4 −3 −2 −1 0 1 2 3 4

**b.** The set consists of 2 and $-3$ and all the real numbers between. Graph the set by drawing solid dots at $-3$ and 2 and connecting the two points.

**c.** The set consists of 2 and all real numbers less than 2. Graph the set by drawing a solid arrow beginning at 2 and pointing to the left. To indicate that 2 is part of the set, draw a solid dot at 2.

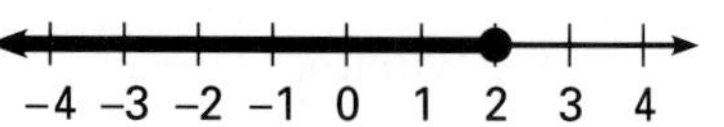


**d.** The set consists of all real numbers greater than $-1$. Graph the set by drawing a solid arrow beginning at $-1$ and pointing to the right. To indicate that $-1$ is not part of the set, draw an open circle at $-1$.

The distance a number is from zero on the number line is the **absolute value** of the number. Since opposite integers are the same distance from zero, opposite integers have the same absolute value. The absolute value of an integer $a$ is written as $|a|$.

### Example 4

Evaluate $|k|$ when $k$ has the given value.

**a.** $k = 103$ **b.** $k = -6.5$ **c.** $k = 0$

### Solution

**a.** Since $k$ is 103 units from zero, $|k| = 103$.
**b.** Since $k$ is 6.5 units from zero, $|k| = 6.5$.
**c.** Since $k$ is 0 units from zero, $|k| = 0$. ◀

When working with real numbers, the **opposite of the opposite** property can be useful. For every real number $n$, $-(-n) = n$.

### Example 5

Evaluate each expression when $t = -12$.

**a.** $-t$ **b.** $-(-t)$ **c.** $|-t|$

### Solution

**a.** Since $t = -12$, $-t = -(-12) = 12$.
**b.** Since, according to Solution **a**, $-t = 12$, then $-(-t) = -12$.
**c.** Since $-t = 12$, then $|-t| = |12| = 12$. ◀

## TRY THESE

Graph the given sets of numbers on a number line.

**1.** $\{2.5, 0.5, -1\frac{1}{2}, -2\}$ **2.** $\{2\frac{2}{3}, 1\frac{1}{3}, -1\frac{2}{3}, -2.\overline{3}\}$

Use a number line to compare numbers. Replace each ● with $<$, $>$, or $=$.

**3.** 2 ● $-2$ **4.** 0 ● 3 **5.** $-2\frac{2}{3}$ ● $1\frac{1}{3}$

Graph each set of numbers on a number line.

**6.** the integers from $-5$ to 1 **7.** all real numbers less than 4

Evaluate $|n|$ when $n$ has the given value.

**8.** $n = 79$ **9.** $n = 0$ **10.** $n = -212$

Evaluate each expression.

**11.** $-r$, when $r = 3$ **12.** $-|-k|$, when $k = -2$

# EXERCISES

## PRACTICE/ SOLVE PROBLEMS

Graph each set of numbers on a number line.

1. $\{-2\frac{1}{2}, -0.75, 0, 1\frac{1}{4}, 3\}$
2. the integers from $-7$ to $-2$
3. all real numbers greater than 6
4. all real numbers less than or equal to $-6$

Evaluate each expression.

5. $-a$, when $a = -4$
6. $-|-b|$, when $b = 6$
7. Which depth is closer to sea level, a depth of $-2{,}018$ ft or a depth of $-3{,}320$ ft?

## EXTEND/ SOLVE PROBLEMS

Replace each ● with <, >, or =.

8. $-4$ ● $|-4|$
9. $|16|$ ● $-|-20|$
10. $-|3|$ ● $3$

Write a rational number between the two given numbers.

11. $-3, -2$
12. $\frac{1}{2}, \frac{7}{8}$
13. $-\frac{1}{3}, -\frac{1}{10}$
14. $2\frac{1}{5}, 2\frac{3}{10}$

List or describe the numbers that are graphed on each number line.

15. 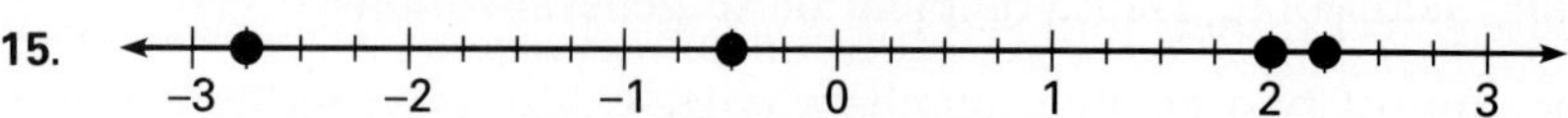


16. −6 −5 −4 −3 −2 −1 0 1 2 3 4 5 6

17. −6 −5 −4 −3 −2 −1 0 1 2 3 4 5 6

*USING DATA* Use the Data Index on pages 556–557 to answer the question.

18. Find the number of feet above sea level for the highest point on earth. Find the number of feet below sea level for the greatest known ocean depth. Compare to find which is closer to sea level.

## THINK CRITICALLY/ SOLVE PROBLEMS

Write *true* or *false* for each statement. For any false statement, give an example that demonstrates how you know the statement is false.

19. Between any two integers there is another integer.
20. Between any two rational numbers there is another rational number.
21. All real numbers are rational numbers.
22. All negative integers are rational numbers.
23. For any real number $n$, where $n < 0$, $|n| = n$.

# 2-2 Adding and Subtracting Integers

## EXPLORE

You can use a number line to add integers. To add a positive integer, move right. To add a negative integer, move left.

For example, to show $2 + (-5)$ on a number line, start at 0. Move right 2 units, then move left 5 units.

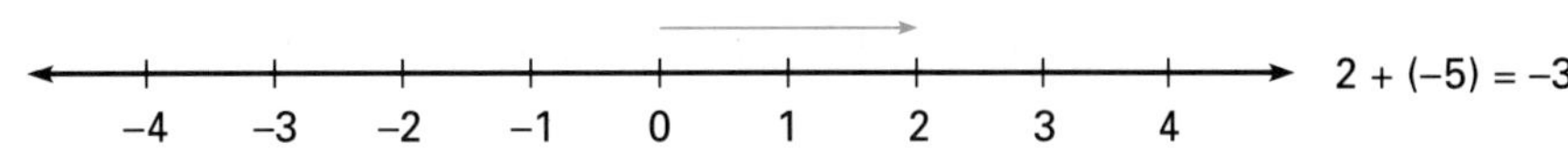


Use a number line from $-10$ to 10 to help you complete the chart at the right.

| First Move | Second Move | End | Number Sentence |
|---|---|---|---|
| 6 | 2 | ■ | $6 + 2 =$ ■ |
| −6 | −2 | ■ | $-6 + (-2) =$ ■ |
| 6 | −2 | ■ | $6 + (-2) =$ ■ |
| −6 | 2 | ■ | $-6 + 2 =$ ■ |

Try other additions. Then complete these generalizations.

- ► The sum of two positive numbers is __?__.
- ► The sum of two negative numbers is __?__.
- ► To find the sum of a positive and a negative number, you __?__.

## SKILLS DEVELOPMENT

Use your generalizations to add integers without a number line.

- ► If the integers have the **same** sign, add the absolute values and use the sign of the addends.
- ► If the integers have **different** signs, subtract the absolute values and use the sign of the addend with the greater absolute value.

### Example 1

**a.** $-5 + (-6)$ **b.** $12 + (-5)$

### Solution

**a.** $-5 + (-6)$ ← Signs are the same, so add the absolute values.
$5 + 6 = 11$
$-5 + (-6) = -11$ ← Use the sign of the addends.

**b.** $12 + (-5)$ ← Signs are different, so subtract absolute values.
$12 - 5 = 7$
$12 + (-5) = 7$ ← Use the sign of the addend with the larger absolute value. ◄

You can use the same methods, along with the commutative and associative properties, to add more than two integers.

### Example 2

a. $-14 + (-8) + (-10)$ b. $9 + (-23) + 6$

### Solution

a. $-14 + (-8) + (-10)$ ← Signs are the same, so add the absolute values.
$14 + 8 + 10 = 32$
$-14 + (-8) + (-10) = -32$ ← Use the sign of the addends.

b. $9 + (-23) + 6$ ← Signs are different, so use the commutative and associative properties and add integers with like signs.
$9 + [6 + (-23)]$
$(9 + 6) + (-23)$
$15 + (-23)$
$23 - 15 = 8$ ← Signs are different, so subtract absolute values.

$9 + (-23) + 6 = -8$ ← Since $23 > (9 + 6)$, the sum is negative. ◀

Suppose you are to find each of the sums at the right. What do you notice about the answers?

$-3 + 3$
$27 + (-27)$
$15{,}345 + (-15{,}345)$

Each addition involves a number and its opposite. These pairs of numbers are called **additive inverses**. Additive inverses are used to state the **addition property of opposites**.

► For every real number $a$, $a + (-a) = 0$.

Additive inverses are also used when you subtract integers.

$-15 + 7 = -8$ $-15 - (-7) = -8$ (same results; additive inverses)

$25 + (-39) = -14$ $25 - 39 = -14$ (same results; additive inverses)

Subtracting an integer is the same as adding its additive inverse.

### Example 3

a. $-18 - 9$ b. $33 - (-29)$ c. $-72 - (-59)$

### Solution

a. $-18 - 9 = -18 + (-9)$ ← Add the additive inverse of 9.
$= -27$ ← Add the integers.

b. $33 - (-29) = 33 + 29$ ← Add the additive inverse of $-29$.
$= 62$ ← Add the integers.

c. $-72 - (-59) = -72 + 59$ ← Add the additive inverse of $-59$.
$= -13$ ← Add the integers. ◀

**MIXED REVIEW**

Complete.

1. $\left(\frac{2}{3}+\blacksquare\right)+\frac{1}{5}=\frac{2}{3}+\left(\frac{1}{8}+\frac{1}{5}\right)$
2. $\blacksquare \times 783.5 = 783.5$
3. $\blacksquare \times \frac{3}{4} = 0$

Evaluate each expression when $m = 6$, $n = 8$, and $p = 10$.

4. $mn + p$ 5. $(p + n) \div m$

Find the next three terms.

6. 4.8, 6, 5.5, 6.7, 6.2, . . .
7. 4, 3, 6, 5, 10, . . .
8. 10, 22, 36, 52, 70, . . .
9. After adding 6 to a number $a$, you divide your result by 2. Write an expression to represent the operations you performed.
10. Park City received 4.2 cm of snow in December, 3.8 cm in January, and 5 cm in February. How much snow was that?

You add or subtract integers to evaluate variable expressions.

### Example 4

Evaluate each expression when $m = -17$ and $p = 21$.

**a.** $m + p$ **b.** $p - m$ **c.** $m - p$

### Solution

**a.** $m + p = -17 + 21 = 4$

**b.** $p - m = 21 - (-17) = 21 + 17 = 38$

**c.** $m - p = -17 - 21 = -17 + (-21) = -38$ ◄

## TRY THESE

Add.

1. $-9 + (-8)$ 2. $-7 + (-24)$ 3. $31 + (-49)$ 4. $-194 + 59$
5. $-7 + (-9) + (-4)$ 6. $-8 + 11 + (-5) + 6$ 7. $-70 + 89 + 43 + (-15)$

Subtract.

8. $-15 - 7$ 9. $90 - 23$ 10. $-38 - (-69)$ 11. $40 - (-53)$

Evaluate each expression when $a = -4$ and $b = 11$.

12. $a + b$ 13. $a - b$ 14. $-a - b$ 15. $b - a$
16. The average temperature in Norilsk, Russia, is $-11$°C. The average temperature in Timbuktu, Mali, is 29°C. What is the difference between the two temperatures?

## EXERCISES

### PRACTICE/ SOLVE PROBLEMS

Add.

1. $-5 + (-9)$ 2. $-29 + 7$ 3. $22 + (-22)$ 4. $-32 + (-13)$
5. $-4 + (-7) + (-8)$ 6. $-4 + (-9) + 4 + 12$ 7. $-51 + 70 + 23 + (-37)$

Subtract.

8. $-9 - 4$ 9. $13 - (-4)$ 10. $-100 - 15$ 11. $-135 - (-85)$

Evaluate each expression when $m = 17$ and $n = -23$.

12. $m - n$ 13. $m + n$ 14. $-m - n$ 15. $-n - m$

16. The average depth of the Pacific Ocean is 12,925 ft below sea level. The average depth of the Atlantic Ocean is 11,730 ft below sea level. Which ocean is deeper? How much deeper?

## EXTEND/ SOLVE PROBLEMS

Replace each ● with <, >, or =.

17. $-5 + (-9)$ ● $17 - 31$
18. $-8 + (-4)$ ● $17 + 10$
19. $-27 + (-14)$ ● $-50 - (-19)$
20. $62 - (-54)$ ● $18 - 29$

**USING DATA** Use the Data Index on pages 556–557 to find statistics on United States foreign trade. To find the United States *balance of trade* with a country, subtract the value of imports from exports.

21. What was the United States balance of trade with Norway in 1989?
22. What was the total value of United States exports to Ireland, France, and Turkey in 1980?
23. With which country did the United States have the most favorable trade in 1989?

**TALK IT OVER**

Could you answer Exercise 23 without doing any written calculations? Explain.

## THINK CRITICALLY/ SOLVE PROBLEMS

Integers can be modeled using integer chips. In the following models, [+] represents 1 and [−] represents −1. Match each model with the appropriate expression by writing the letter of the addition or subtraction each model represents. Write the sum or difference for each expression you choose.

24. 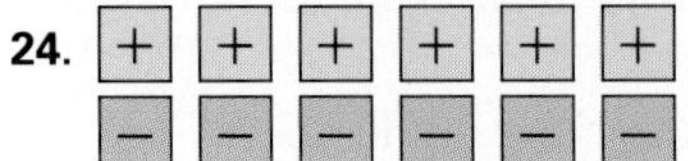

25. 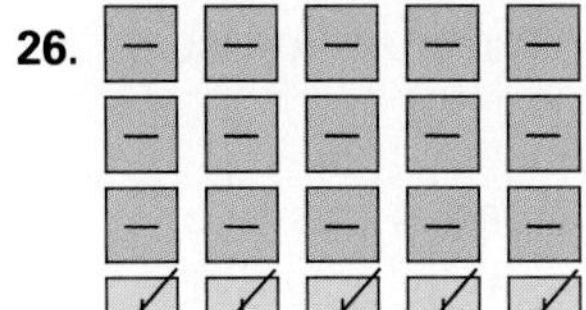

26. 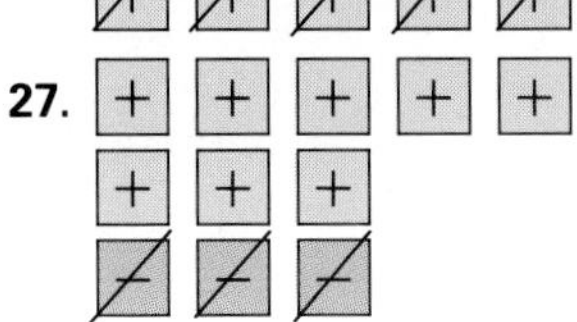

27. [+] [+] [+] [+] [+]
[+] [+] [+]
[+̸] [+̸] [+̸]

a. $-14 - (-8)$

b. $-10 - 5$

c. $6 + (-7)$

d. $5 - (-3)$

Draw a model and solve.

28. The temperature was $-2°F$ at dawn in Bangor, Maine. By noon the temperature had reached $7°F$. By how many degrees did the temperature change?

**WRITING ABOUT MATH**

Suppose that a friend of yours has trouble understanding the models in Exercises 26 and 27. Write a note to your friend that explains the models by relating them to the idea of the additive inverse.

# 2-3 Multiplying and Dividing Integers

## EXPLORE

Draw a number line like the one below.

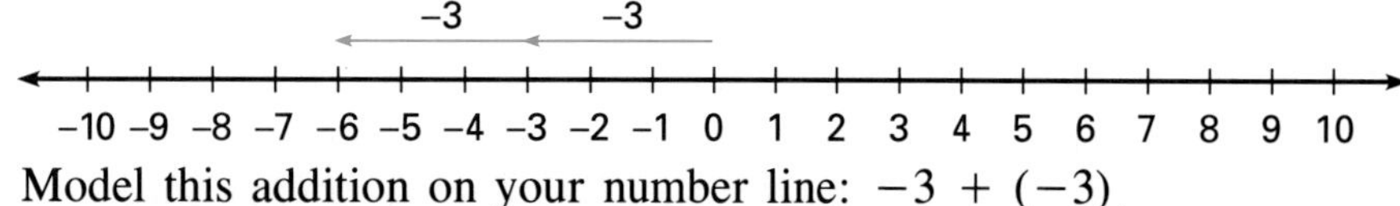


Model this addition on your number line: $-3 + (-3)$.

Complete. Addition sentence: $-3 + (-3) = ■$
Multiplication sentence: $2(-3) = ■$
Division sentence: $■ \div (-3) = 2$

Follow a pattern to complete the multiplication sentences below. Then complete the related division sentences.

| | | |
|---|---|---|
| Multiplication → | $2(-3) = ■$ | $1(-3) = ■$ |
| | $0(-3) = ■$ | $-1(-3) = ■$ |
| Division → | $■ \div (-3) = 2$ | $■ \div (-3) = 1$ |
| | $■ \div (-3) = 0$ | $■ \div (-3) = -1$ |

What patterns did you use to complete the sentences?

**a.** If two positive numbers are multiplied together, what is the sign of the product?
**b.** If two negative numbers are multiplied together, what is the sign of the product?
**c.** If a positive number and a negative number are multiplied together, what is the sign of the product?

## SKILLS DEVELOPMENT

In the Explore activity, you should have discovered how to decide whether a product is positive or negative.

► The product of two integers with the same signs is positive.
► The product of two integers with different signs is negative.

### Example 1

Find each product.
**a.** $(-4)(2)$ **b.** $(-6)(-7)$ **c.** $15(8)$

**Solution**

**a.** different signs
↓
negative product
$(-4)(2) = -8$

**b.** same signs
↓
positive product
$(-6)(-7) = 42$

**c.** same signs
↓
positive product
$15(8) = 120$ ◄

The associative and commutative properties of multiplication may be used to arrange factors in the order easiest for multiplying.

### Example 2

Find each product.

**a.** $(8)(-3)(5)$ **b.** $(4)(-6)(-2)$ **c.** $(-4)(-6)(5)(-25)$

**Solution**

**a.** $(8)(-3)(5) = [(8)(5)](-3)$ ← Use the commutative property to rearrange the factors.
$= (40)(-3)$ ← Multiply the first two factors.
$= -120$ ← Multiply the product by the third factor.

**b.** $(4)(-6)(-2) = [(4)(-2)](-6)$
$= (-8)(-6)$
$= 48$

**c.** $(-4)(-6)(5)(-25) = [(-4)(-25)][(-6)(5)]$ ← Use the commutative and associative properties.
$= (100)(-30)$ ← Multiply the two pairs of factors.
$= -3{,}000$ ← Multiply the two products. ◀

**MENTAL MATH TIP**

Look for products that are multiples of 10.

You know that multiplication and division are inverse operations.

$6 \times 2 = 12$, so $12 \div 2 = 6$ $10 \times -2 = -20$, so $-20 \div -2 = 10$

$-3 \times -5 = 15$, so $15 \div -5 = -3$ $-8 \times 4 = -32$, so $-32 \div 4 = -8$

From these examples, you should understand how to decide whether a quotient is positive or negative.

► The quotient of two integers with the same signs is positive.

► The quotient of two integers with different signs is negative.

Since multiplication and division are inverse operations, you can check division by multiplying.

### Example 3

Find each quotient. Then check by multiplying.

**a.** $-63 \div 7$ **b.** $-99 \div (-9)$ **c.** $484 \div (-4)$

**Solution**

**a.** Since the signs are different, the quotient is negative.
$-63 \div 7 = -9$ Check: $(-9)(7) = -63$

**b.** Since the signs are the same, the quotient is positive.
$-99 \div (-9) = 11$ Check: $(11)(-9) = -99$

**c.** Since the signs are different, the quotient is negative.
$484 \div (-4) = -121$ Check: $(-121)(-4) = 484$ ◀

You already know the identity property of 1.

► Multiplying any real number by 1 results in a product that is the same real number.

When working with integers, the **multiplication property of –1** is also very useful.

► Multiplying any real number by $-1$ results in a product that is the additive inverse of the real number.

$$-1(a) = -a, \text{ and } a(-1) = -a$$

Sometimes you will need to use this property along with the rules for multiplying and dividing integers to evaluate variable expressions.

## Example 4

Evaluate each expression when $r = 12$ and $s = -3$.

**a.** $rs$ **b.** $-rs$ **c.** $\frac{r}{s}$ **d.** $-\frac{r}{s}$

### Solution

**a.** $rs = (12)(-3) = -36$

**b.** $-rs = (-1)(rs)$ ← Use the multiplication property of –1.

$= (-1)(12)(-3)$

$= [(-1)(-3)](12)$ ← Use the commutative and associative properties.

$= (3)(12) = 36$

**c.** $\frac{r}{s} = \frac{12}{-3} = -4$

**d.** $-\frac{r}{s} = (-1)\left(\frac{r}{s}\right)$

$= (-1)\left(\frac{12}{-3}\right)$

$= (-1)(-4) = 4$ ◄

## TRY THESE

Find each product.

1. $(-7)(7)$
2. $6(-4)$
3. $(-3)(-15)$
4. $4(10)$
5. $(12)(-5)(2)$
6. $(7)(-10)(-3)$
7. $(20)(-7)(-5)(2)$

Find each quotient. Then check by multiplying.

8. $42 \div (-6)$
9. $-39 \div 13$
10. $-105 \div -5$
11. $633 \div -3$

Evaluate each expression when $a = 5$ and $b = -10$.

12. $ab$
13. $-ab$
14. $\frac{b}{a}$
15. $-\frac{b}{a}$
16. A business posts a balance of $-\$100$ for a 5-day work week. What was the average loss per day?

# EXERCISES

## PRACTICE/ SOLVE PROBLEMS

Find each product or quotient.

1. $7(-5)$ 2. $(-3)(-20)$ 3. $45 \div (-5)$ 4. $20(-60)$

5. $-82 \div (-2)$ 6. $(8)(-10)(-2)$ 7. $225 \div (-25)$

Evaluate each expression when $c = 4$ and $d = -20$.

8. $cd$ 9. $-cd$ 10. $\frac{c}{d}$ 11. $-\frac{c}{d}$

12. A deep-sea diver descended to a depth of 90 ft below sea level. The diver then ascended to the starting point in 6 equal stages. How many feet did the diver ascend per stage?

13. From 9 P.M. to 4 A.M. the temperature dropped 2°F per hour. What was the change in temperature during that time?

**CALCULATOR TIP**

Here is a shortcut you can use to multiply or divide integers in Exercises 14–22. Just enter their absolute values. Then determine the sign of the product or quotient.

## EXTEND/ SOLVE PROBLEMS

Find each product or quotient.

14. $-27(31)$ 15. $16(-37)$ 16. $-224 \div (-14)$

17. $-71(19)$ 18. $1{,}512 \div (-56)$ 19. $-1{,}344 \div 16$

20. $-83(-51)$ 21. $3{,}975 \div (-159)$ 22. $27(-17)(-83)$

What is the greatest integer that can be used to make each open sentence true?

23. $(-3) + (-10) < ■(-1)$ 24. $■(-1) > 72 \div 12$

25. During a 7-hour period one day in July in Sydney, Australia, the temperature dropped from a high of 60°F to a low of 46°F. If the change in temperature took place at a steady rate, how would you describe each hourly change during that period?

## THINK CRITICALLY/ SOLVE PROBLEMS

Replace each ● with +, −, ×, or ÷ to make a true sentence.

26. $-8$ ● $4$ ● $(-2) = -34$ 27. $-8$ ● $(-6)$ ● $14 = 0$

28. $21$ ● $(-7)$ ● $(-8) = -11$ 29. $49$ ● $(-9)$ ● $(-1) = 59$

Write *true* or *false* for each statement. For any false statement, give an example that demonstrates how you know the statement is false. For Exercises 30–33, $a$ and $b$ are integers and $b \neq 0$.

30. $-\frac{a}{b} = \frac{-a}{b}$ 31. $\frac{-a}{b} = -\frac{a}{-b}$

32. $-\frac{a}{-b} = -\frac{-a}{-b}$ 33. $-\frac{a}{b} \neq -\frac{-a}{-b}$

# 2-4 Problem Solving Skills:

## CHECK YOUR SOLUTION

- ► READ
- ► PLAN
- ► SOLVE
- ► ANSWER
- ► CHECK

After you solve any problem, you should check your solution. You can check your arithmetic by using an inverse operation. You should also look back at the original statement of the problem to make sure your answer makes sense.

### PROBLEM

The highest land mountain in the world is Mount Everest, which is on the border of Nepal and Tibet and has a height of 8,848 m above sea level. The tip of the highest sea mountain in the world is in the Tonga Trench and is 9,213 m lower than the tip of Mount Everest. How far below sea level is the tip of the sea mountain?

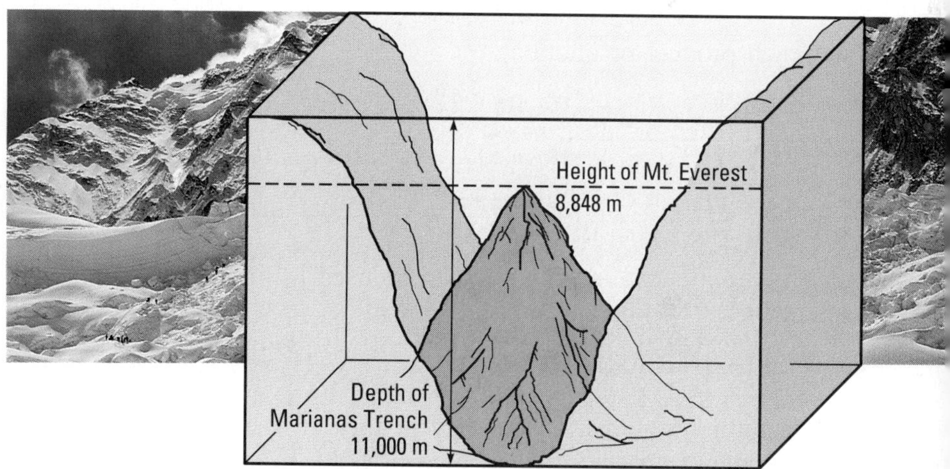


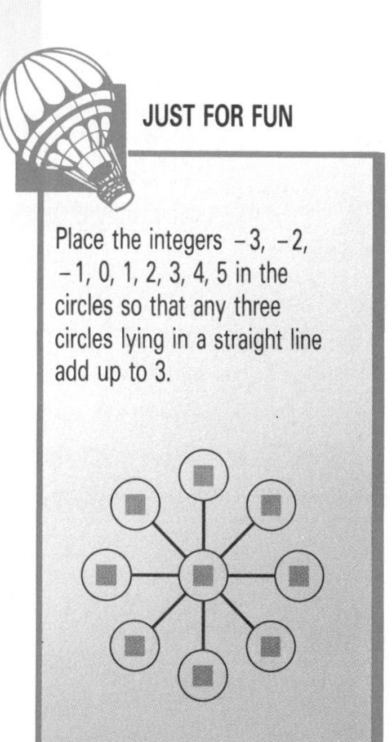


### SOLUTION

The height of Mount Everest is 8,848 m.
The tip of the highest sea mountain is 9,213 m lower.
Subtract 9,213 from 8,848.

$8,848 - 9,213 = -365$ ← **To check subtraction use the inverse operation, addition.**

$-365 + 9,213 = 8,848$

The tip of the highest sea mountain is at a depth of −365 m, or 365 m below sea level.

You can also check your solution by looking back at the original statement of the problem. The tip of Mount Everest is 9,213 m above the tip of the sea mountain, which is 365 m below sea level. So, −365 + 9,213 must give the height of Mount Everest, 8,848 m.

# PROBLEMS

1. In the countdown before the launch of a space shuttle, the time was displayed as −1 min 23 s. Assuming that the clock was accurate and in good working order, how much time had to pass before the display showed the time as 2 min?

2. A football team lost 6 yd on one play and gained 14 yd on the next play. What was the net loss or net gain for the two plays?

3. The opening price of a share of stock on Monday morning was \$105. The closing price on Friday was \$78. What was the price change during the week?

4. The shore of the Dead Sea in Israel is 1,286 ft below sea level. Death Valley in California is 282 ft below sea level. Which altitude is lower? How much lower?

5. The deepest well in the United States is a gas well in Washita County, Oklahoma, at a depth of about 6 mi. Assume that a well is to be drilled at a steady rate. To drill a well that is as deep as the one in Washita County in 25 months, about how many feet would have to be drilled each month?

6. Alfonso has a balance of \$221 in his checking account. If he writes a check for \$303, what will his new checking account balance be?

7. Andina wrote a check for \$25 to each of her six nieces and nephews. How did the total of checks change the balance in her checking account?

8. A business showed a loss of \$105 for the month of February and a loss of \$52 in the month of March. In which of the months did the business do better? How much better did it do?

9. In 1988 the United States balance of trade with Denmark was −\$786 million. In 1989 it was −\$484 million. What was the change in the United States balance of trade with Denmark from 1988 to 1989?

10. The highest temperature ever recorded in Huron, South Dakota, was 112°F. The lowest temperature recorded there was 151°F lower than the highest recorded temperature. What was the lowest recorded temperature?

**MATH: WHO, WHERE, WHEN**

Evelyn Boyd Granville is one of the first two African-American women to receive a doctorate in mathematics (Yale University, 1949), after having graduated summa cum laude from Smith College.

Granville spent sixteen years in government and industry working on the space program. Later she became a full professor of mathematics at California State University in Los Angeles. In addition to her college teaching, she has taught mathematics to teachers of elementary school and high school mathematics.

# 2-5 Operations with Rational Numbers

**EXPLORE/ WORKING TOGETHER**

Discuss with your group which of the following statements are *true* and which are *false*. For any false statement, give an example that demonstrates how you know the statement is false.

**a.** The sum of two integers is always an integer.
**b.** The difference of two whole numbers is always a whole number.
**c.** The product of two whole numbers is always a whole number.
**d.** The quotient of two integers is always an integer.

**SKILLS DEVELOPMENT**

When you add, subtract, multiply, or divide with rational numbers, you follow the same rules that you use when you perform those operations with integers.

### Example 1

Perform the indicated operations.

**a.** $2.5 + (-1.7)$ **b.** $-13 - \left(5\frac{1}{8}\right)$ **c.** $\left(-\frac{1}{4}\right)(5)$ **d.** $-27 \div (-5)$

**Solution**

**a.** Find the absolute values. $|2.5| = 2.5$ and $|-1.7| = 1.7$
Since the signs are different, subtract the absolute values.
$2.5 - 1.7 = 0.8$
The positive addend has the greater absolute value, so the sum is positive.
$2.5 + (-1.7) = 0.8$

**b.** Add the opposite of $5\frac{1}{8}$, which is $-5\frac{1}{8}$.
$-13 + \left(-5\frac{1}{8}\right) = -18\frac{1}{8}$
So $-13 - \left(5\frac{1}{8}\right) = -18\frac{1}{8}$

**c.** Since the signs are different, the product is negative.
$\left(-\frac{1}{4}\right)(5) = --1\frac{1}{4}$

**d.** Since the signs are the same, the quotient is positive.
$-27 \div (-5) = 5.4$ ◄

**CHECK UNDERSTANDING**

If there were no parentheses in Example **2b**, what would be the answer?

### Example 2

Simplify.

**a.** $2.5 \div 5 + (-0.2)$
**b.** $(-5.6 + 8) \div 8$
**c.** $1\frac{7}{10} - \left(-\frac{1}{10}\right)(-6)$
**d.** $\left(1\frac{1}{2} + 8\frac{1}{2}\right)2 + \left(-1\frac{7}{10}\right)$

**Solution**

**a.** $2.5 \div 5 + (-0.2)$
$= 0.5 + (-0.2) = 0.3$

**b.** $(-5.6 + 8) \div 8$
$= 2.4 \div 8 = 0.3$

c. $1\frac{7}{10} - \left(-\frac{1}{10}\right)(-6)$
$= 1\frac{7}{10} - \frac{6}{10}$
$= 1\frac{1}{10}$

d. $\left(1\frac{1}{2} + 8\frac{1}{2}\right)2 + \left(-1\frac{7}{10}\right)$
$= (10 \cdot 2) + \left(-1\frac{7}{10}\right)$
$= 20 + \left(-1\frac{7}{10}\right) = 18\frac{3}{10}$ ◄

Sometimes you will need to perform operations with rational numbers to evaluate expressions with variables.

### Example 3

Evaluate each expression when $q = -4$, $r = 0.5$, and $s = -0.25$.

a. $qs$ b. $qr + s$ c. $q(r \div s)$

**Solution**

a. $qs = (-4)(-0.25) = 1$
b. $qr + s = -4 \cdot 0.5 + (-0.25)$
$= -2 + (-0.25) = -2.25$
c. $q(r \div s) = -4[0.5 \div (-0.25)]$
$= -4[-2] = 8$ ◄

## TRY THESE

Perform the indicated operations.

1. $9\frac{4}{5} + \left(-6\frac{1}{10}\right)$ 2. $-8.1 - (-4.9)$ 3. $15\left(-7\frac{1}{3}\right)$ 4. $-42 \div 4$

Simplify.

5. $6 \div (-3) + 7.1$
6. $-4(-9) - (-30.5)$
7. $(7.6 - 1.6)(-0.2)$
8. $2[-7.7 \div (-99.5 + 98.5)]$

Evaluate each expression when $a = 6$, $b = 0.5$, and $c = 1.5$.

9. $ac$ 10. $(a + b)c$ 11. $a + bc$ 12. $ab \div c$

13. During an illness Martha's temperature was 101.5°F one day and 99.6°F the next day. What was the change in her temperature?

# EXERCISES

**PRACTICE/ SOLVE PROBLEMS**

Perform the indicated operations.

1. $3\frac{1}{3} + \left(-\frac{5}{6}\right)$ 2. $9.4(-2.8)$ 3. $-24.5 - 4.3$ 4. $8.5 + (-10.3)$
5. $-75 \div (-2.5)$ 6. $3\frac{1}{8} \div \left(-\frac{1}{8}\right)$ 7. $-\frac{3}{4} \cdot 2$ 8. $18\frac{1}{5} - \left(-\frac{1}{4}\right)$

## MIXED REVIEW

Write each as a percent.

**1.** 0.35 **2.** 0.5 **3.** 0.245

Find the percent of each.

**4.** 20% of 75

**5.** 1.5% of 60

**6.** 150% of 30

**7.** Recall that a person's weight on Mars can be given as a function of that person's weight on Earth ($w$): $M(w) = 0.4w$. How much would a 112-lb person weigh on Mars?

**9.** $4\frac{1}{5} + (3)(-6)$

**10.** $22 \div 1.1 - 4.8$

**11.** $7(-8) \div 16$

**12.** $-27 \div (-6) + \left(3\frac{1}{2}\right)$

**13.** $7.5 + 8.4 + (-4.3)$

**14.** $8\frac{1}{3} \div (-3) + 6$

Evaluate each expression when $b = -7$, $c = -2.5$, and $d = 2.5$.

**15.** $bd$ **16.** $b + cd$ **17.** $b(c + d)$ **18.** $(b + c) \div d$

**19.** Elena earns \$9.50 per hour for each hour of her regular 35-h work week. For each hour over 35 h, she earns $1\frac{1}{2}$ times that amount. How much will she earn in all if she works 45 h a week for 6 weeks?

**20.** The opening price of a stock on Monday morning was \$18.75. The following Friday afternoon the same stock had a closing price of \$10.25. Assuming that the value of the stock fell at a steady rate, how would you describe the daily change in the stock's value during that week?

**21.** Mark's checking account shows a balance of −\$10.50. Rebecca's checking account has a balance of −\$5.25. Whose balance is closer to 0?

## EXTEND/ SOLVE PROBLEMS

Perform the indicated operation. Express your answer as a decimal rounded to the nearest tenth.

**22.** $-1\frac{5}{8} \cdot 4.23$

**23.** $\frac{6}{7} \cdot \frac{1}{3}$

**24.** $\frac{2}{3}\left(-\frac{1}{3}\right)$

**25.** $-12\frac{3}{8} \div (-4.2)$

**26.** $4\frac{7}{16} - (-3.85)$

**27.** $\left(\frac{4}{15}\right)^2$

**USING DATA** Use the Data Index on pages 556–557 to find statistics on calories.

**28.** How many calories does a 100-lb person use playing tennis for 10 min?

**29.** Mary consumed 348 calories at lunch. If she weighs 100 lb, how long must she swim to use up those calories?

**30.** A 150-lb cross-country skier uses how many more calories in one hour than a 120-lb cross-country skier?

**31.** George weighs 150 lb. The food in his lunch provided him with 400 calories. If he swims for a half hour and plays tennis for a half hour, how many of those calories were not used?

## THINK CRITICALLY/ SOLVE PROBLEMS

**32.** You know that $\frac{1}{3} = 0.\overline{3}$ and $\frac{2}{3} = 0.\overline{6}$. What rational number do you think is named by $0.\overline{9}$? Explain your reasoning.

► READ
► PLAN
► SOLVE
► ANSWER
► CHECK

# Problem Solving Applications:

## WIND CHILL TEMPERATURE

Have you noticed that your face stings on a cold, windy day? The combination of cold and wind makes you feel colder than the actual temperature. For example, the chart shows that when the actual temperature is −5°F and the wind blows at 5 mi/h, the temperature on your face feels like −10°F. If you step into a wind of 20 mi/h, however, the temperature on your face feels like −46°F. The temperature that one would feel as a result of the wind speed can be called the **wind chill temperature**.

| Wind Speed (mi/h) | Actual Temperature (°F) | | | | | | | | | | | | |
|---|---|---|---|---|---|---|---|---|---|---|---|---|---|
| | **25** | **20** | **15** | **10** | **5** | **0** | **−5** | **−10** | **−15** | **−20** | **−25** | **−30** | **−35** |
| **5** | 21 | 19 | 12 | 7 | 0 | −5 | −10 | −15 | −21 | −26 | −31 | −36 | −42 |
| **10** | 10 | 3 | −3 | −9 | −15 | −22 | −27 | −34 | −40 | −46 | −52 | −58 | −64 |
| **15** | 2 | −5 | −11 | −18 | −25 | −31 | −38 | −45 | −51 | −58 | −65 | −72 | −78 |
| **20** | 3 | −10 | −17 | −24 | −31 | −39 | −46 | −53 | −60 | −67 | −74 | −81 | −88 |
| **25** | 7 | −15 | −22 | −29 | −36 | −44 | −51 | −59 | −66 | −74 | −81 | −88 | −96 |
| **30** | −10 | −18 | −25 | −33 | −41 | −49 | −56 | −64 | −71 | −79 | −86 | −93 | −101 |
| **35** | −12 | −20 | −27 | −35 | −43 | −52 | −58 | −67 | −74 | −82 | −89 | −97 | −105 |

1. What is the wind chill temperature when the actual temperature is 10°F and the wind speed is 15 mi/h?
2. When the wind is blowing at 25 mi/h and the wind chill temperature is −22°F, what is the actual temperature?

Which pair of conditions in Exercises 3–5 results in a colder wind chill temperature? The temperatures given are actual temperatures.

3. 30 mi/h, 15°F **or** 10 mi/h, 5°F
4. 35 mi/h, 0°F **or** 15 mi/h, −15°F
5. 10 mi/h, −25°F **or** 20 mi/h, −10°F

Suppose the actual temperature is 5°F.

6. What is the wind chill temperature with a wind speed of 15 mi/h?
7. If the actual temperature were to increase by 10°F without any change in wind speed, what would the new wind chill temperature be?

When Leslie started skiing one day, the actual temperature was 25°F and the wind speed was 10 mi/h.

8. What was the wind chill temperature?
9. Later, the actual temperature dropped 15°F, and the wind speed doubled. What was the new wind chill temperature?

# 2-6 Problem Solving Strategies: MAKE AN ALGEBRAIC MODEL

A **mathematical model** is a mathematical representation of an object or a situation. You probably are familiar with geometric models, such as scale drawings. In an **algebraic model**, you generally use a rule or a set of rules to describe the behavior of an object or a situation.

## PROBLEM

To park a car at the Acme Garage on a Saturday, the cost is \$4 for the first hour and \$0.50 for each additional hour. At the Park-Rite Garage, the cost is \$3.75 for the first hour and \$0.60 for each additional hour. Suppose that both garages are conveniently located.

**a.** Which garage is the better choice?

**b.** For each garage, write a function rule that represents the cost of parking your car there for $n$ hours. Then find the cost of parking your car at the garage for ten hours.

## SOLUTION

**a.** Make a table to compare the two choices.

| Number of Hours | Cost at Acme Garage | Cost at Park-Rite Garage |
|---|---|---|
| 1 | \$4.00 | \$3.75 |
| 2 | \$4.50 | \$4.35 |
| 3 | \$5.00 | \$4.95 |
| 4 | \$5.50 | \$5.55 |
| 5 | \$6.00 | \$6.15 |

If you need to park your car for three hours or less, Park-Rite Garage is the better choice. If you need to park your car for more than three hours, Acme Garage is the better choice.

**b.** To park for $n$ hours, you pay a fixed cost for the first hour and an additional charge per hour for the following $(n - 1)$ hours.

**Acme Garage**

$$\begin{aligned} f(n) &= 4 + 0.5(n - 1) \\ f(10) &= 4 + 0.5(10 - 1) \\ &= 4 + 0.5(9) \\ &= 8.5 \end{aligned}$$

**Park-Rite Garage**

$$\begin{aligned} f(n) &= 3.75 + 0.6(n - 1) \\ f(10) &= 3.75 + 0.6(10 - 1) \\ &= 3.75 + 0.6(9) \\ &= 9.15 \end{aligned}$$

To park your car for ten hours, it would cost you \$8.50 at the Acme Garage and \$9.15 at the Park-Rite Garage.

# PROBLEMS

1. Suppose you take a temporary job in another town. You plan to rent a compact car to drive back and forth between home and your job, a distance of 150 mi per day. The Drive-More Rent-a-Car Company charges \$30 a day plus \$0.10 per mile. The Miles Plus Rent-a-Car Company charges \$32 a day with unlimited free mileage.
   a. Which car rental company offers the better value?
   b. For each company, write a function rule that represents the cost of renting a compact car for *n* days. Then find the cost of renting a compact car for 9 days.

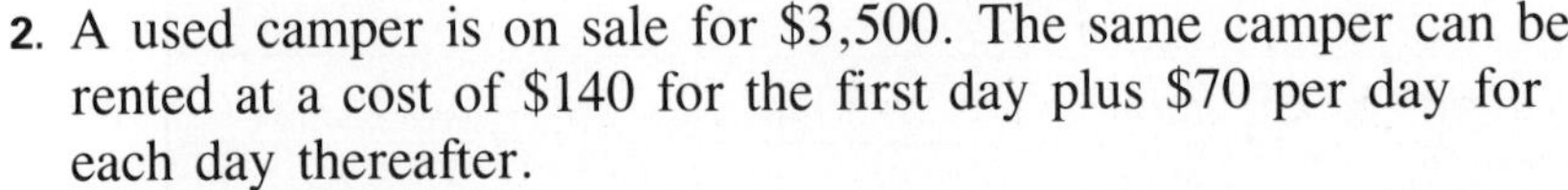

2. A used camper is on sale for \$3,500. The same camper can be rented at a cost of \$140 for the first day plus \$70 per day for each day thereafter.
   a. If you are planning a 60-day camping trip, compare the cost of buying the camper to the cost of renting it for that period of time.
   b. At what point does the camper become less expensive to buy than to rent?

3. A member of the See-More Video Club can rent any movie for \$3 as long as the movie is returned by 2 P.M. of the second day after it is taken out. If the movie is late being returned, however, the member is charged an additional \$2.50 per late day for a new release or an additional \$1.50 per late day for any other movie. The Movie Magic Video Club charges \$2.50 per movie as long as the movie is returned by 3 P.M. of the second day after it is taken out. If the movie is late being returned, the member is charged an additional \$2 for each day that the movie is overdue, regardless of whether the movie is a new release or not.
   a. At which video club would it cost you more to let a movie become overdue?
   b. For each video club, write a function rule that represents the cost of having a new release and an old classic movie overdue for *n* days. Then find the cost of having a new release and an old classic movie overdue for 5 days at each video club.

# 2-7 Simplifying Variable Expressions

## EXPLORE

Evaluate each variable expression.

| | | | Answers |
|---|---|---|---|
| a. | $x = 3.75$ | $x + x + x$ | |
| | | $3x$ | |
| b. | $t = -5.2$ | $5t - t$ | |
| | | $4t$ | |
| c. | $r = 10.51$ | $-r - r - 2r$ | |
| | | $-4r$ | |
| d. | $s = -0.25$ | $s + s - s + 2s$ | |
| | | $3s$ | |

e. The table shows two ways to find each answer. Which way is easier? Why?

## SKILLS DEVELOPMENT

The parts of a variable expression that are separated by addition or subtraction signs are called the **terms** of the expression.

one term → $2x$ two terms → $3x + 6y$

Terms that have identical variable parts are called **like terms**. Terms that have different variable parts are called **unlike terms**.

**like terms:** $2x$ and $4.5x$; $3st$ and $-\frac{1}{2}st$ **unlike terms:** $4a$ and $2.5b$; $7ab$ and $7ac$

You **simplify** a variable expression when you perform as many of the indicated operations as possible. Using the distributive property to simplify an expression that contains like terms is called **combining like terms**. Unlike terms cannot be combined.

PROBLEM SOLVING TIP

In simplifying Example **1b**, recall that $-n = -1 \cdot n$.

### Example 1

Simplify.

a. $3x - 2x$ b. $-9n - n + 3m$ c. $3ab - 2ac + 4ab$

### Solution

a. $3x - 2x = (3 - 2)x$ Use the distributive property.
$= 1x = x$

b. $-9n - n + 3m = (-9 - 1)n + 3m$ Use the distributive property.
$= [-9 + (-1)]n + 3m$ To subtract, add the opposite.
$= -10n + 3m$

c. $3ab - 2ac + 4ab = 3ab + 4ab - 2ac$ Rewrite using the commutative property.
$= (3 + 4)ab - 2ac$ Use the distributive property.
$= 7ab - 2ac$ ◄

The next example uses the distributive property in a different way.

### Example 2

Simplify.

a. $4(m + 7)$ b. $-8(4.5 - rs)$ c. $-(2x + y)$

### Solution

a. $4(m + 7) = 4m + 4(7)$ Use the distributive property.
$= 4m + 28$ Simplify 4(7).

b. $-8(4.5 - rs) = -8(4.5) - (-8)(rs)$ Use the distributive property.
$= -8(4.5) - (-8rs)$ To subtract, add the opposite.
$= -36 + 8rs$

c. $-(2x + y) = (-1)(2x + y)$ Apply the multiplication property of −1.
$= (-1)(2x) + (-1)y$ Use the distributive property.
$= -2x + (-y)$
$= -2x - y$ ◄

Show how to apply the multiplication property of −1 and the distributive property to obtain the property of the opposite of a sum.

Example **2c** is an illustration of the **property of the opposite of a sum**. For all real numbers $a$ and $b$, $-(a + b) = -a + (-b)$.

Think about how to simplify more complicated expressions.

### Example 3

Simplify.

a. $(4x - 1) + (3x - 5)$ b. $4(2p - 1) - (p + 5)$ c. $3(ab - a) + 3(b - a)$

### Solution

a. $(4x - 1) + (3x - 5) = 4x - 1 + 3x - 5$ Use the distributive property.
$= 7x - 6$ Combine like terms.

b. $4(2p - 1) - (p + 5) = 8p - 4 - (p + 5)$ Use the distributive property.
$= 8p - 4 - p + (-5)$ Use the property of the opposite of a sum.
$= 7p - 9$ Combine like terms.

c. $3(ab - a) + 3(b - a) = 3ab - 3a + 3b - 3a$ Use the distributive property.
$= 3ab - 6a + 3b$ Combine like terms. ◄

## TRY THESE

Simplify.

1. $5t + 2t$
2. $-3r + r$
3. $-6t - 6t$
4. $7xy - 2xy + xz$
5. $2(5 + s)$
6. $6(3.1 - ab)$

7. $4(g + gh)$

8. $-8(x - y)$

9. $(5a - 7) + (9a + 2)$

10. $6(a - 9) - (2a + 30)$

11. $7(cd + 8c) - (c + 5.2)$

12. $8(3x + y) - (-x + 21)$

# EXERCISES

## PRACTICE/ SOLVE PROBLEMS

Simplify.

1. $3a + 11a$
2. $-5m - 3m$
3. $16rs - 5rs$
4. $4vt + (-6vt) + 2t$
5. $3(ab - 2ab)$
6. $9(xy - 4x)$
7. $\frac{1}{2}(6a + 2ab - 2a)$
8. $-5x + 3(-x + 2y)$
9. $-(4a - 2) + 7a$
10. $10y + (3y - 9)$
11. $(8f + 8) - (f - 6)$
12. $15u + \frac{1}{3}(12u - 9n)$
13. $2(6x - 2y) + (x + y)$
14. $(a - 2) - 1.5(a + b)$
15. $6(p + 3q) - (7p + 4q)$
16. $(8m - 2n) - 3(-m + n)$
17. $-3(r + 2s) + 2(-3r - s)$
18. $4(ab - 5a) - (9a + ab)$
19. $7(wz + w) - 2(wz + z)$
20. $16(-c - cd) - (-cd + 12d)$

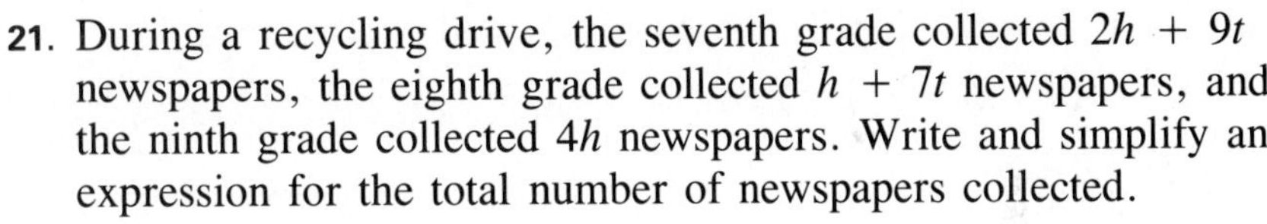

21. During a recycling drive, the seventh grade collected $2h + 9t$ newspapers, the eighth grade collected $h + 7t$ newspapers, and the ninth grade collected $4h$ newspapers. Write and simplify an expression for the total number of newspapers collected.

22. Jesse has a piece of ribbon that measured $3a + 2b$ inches before he used $a + 3b$ inches of it. Write and simplify an expression to show how many inches of ribbon Jesse has left.

## EXTEND/ SOLVE PROBLEMS

There are more than 150 different computer languages. BASIC (Beginner's All-purpose Symbolic Instruction Code) is a computer language that was developed at Dartmouth College in 1963 to help students learn computer programming. The following table shows some operation symbols used in BASIC.

| Operation | BASIC Symbol | Example |
|---|---|---|
| Addition | + | a + b |
| Subtraction | − | a − b |
| Multiplication | * | a * b |
| Division | / | a/b |

**COMPUTER**

Modify the program on page 30 so that it finds the product and quotient of two whole numbers.

Simplify each expression. For Exercises 23–31, write your answer using BASIC symbols.

23. $a + 5 + a$

24. $2b + 3 - b$

**25.** $-6a - 3b + 6b$

**26.** $7a - \frac{a}{2} + 5$

**27.** $(5r \div 2) + 2s + 10s$

**28.** $3.5(2a - 6) + 5$

**29.** $19 - 4(a - c)$

**30.** $(2a - 5) - (6 - a)$

**31.** $6\left(\frac{1}{3}mn - 2\right) - 2(mn + 1)$

**32.** What is the difference when you subtract $(4a + b) + (3a + 2b)$ from $6a + 8b + (-4b + 7a)$?

**33.** Find the sum of $4a - 3b$, $a - b$, and $a - 5b$.

**34.** Use the expressions in the box below. Match the expressions that are the same when simplified.

**a.** $5(a + b) - 6(a - b)$
**b.** $-5a + 6(5a - b)$
**c.** $2(12a + b) - (-a + 8b)$
**d.** $-29 + 3(a - b) + 6b$
**e.** $(a - 2a) + (5b + 6b)$
**f.** $(-15 + 3a) + (-14 + 3b)$

## THINK CRITICALLY/ SOLVE PROBLEMS

**35.** On Monday, Joe did $n$ situps, on Tuesday he did $n + 2$ situps, on Wednesday he did $n + 5$, and on Thursday he did $n + 9$ situps. If Joe exercises every day and the pattern continues, how many situps will he have done in all for the seven-day week?

**36.** Alicia and Frank each simplified this variable expression.

$$3(-4g + 6f) - 3(g + 2f)$$

| **Alicia's Work** | **Frank's Work** |
|---|---|
| $3(-4g + 6f) - 3(g + 2f)$ | $3(-4g + 6f) - 3(g + 2f)$ |
| $= -12g + 6f - 3g - 6f$ | $= 12g + 18f - 3g + 6f$ |
| $= -15g$ | $= -9g + 24f$ |

Explain where Alicia and Frank each made a mistake. Then simplify and write the variable expression correctly.

# 2-8 Properties of Exponents

## EXPLORE

Use the table at the right to help you answer the questions.

a. You know that $4 = 2^2$, which means $2 \times 2$. What does $8 = 2^3$ mean?

b. Copy and complete the following.
$4 \times 8 = 2^2 \times 2^3 = 2 \times 2 \times ? = 2^{■}$

c. What is $16 \times 64$? What is $8 \times 256$?

d. $(2^4)^2 = 2^4 \times 2^{■} = 2^{■}$
So, $(2^4)^2 = 2^{■ \times ■}$

e. When a number with an exponent is squared, what happens to the exponent?

f. $(2^2)^3 = 2^2 \times 2^{■} \times 2^{■} = 2^{■}$
So, $(2^2)^3 = 2^{■ \times ■}$

g. When a number with an exponent is cubed, what happens to the exponent?

h. What do you think happens to the exponent of a number when that number is raised to any power $n$?

| Number | Power of 2 |
|---|---|
| 4 | $2^2$ |
| 8 | $2^3$ |
| 16 | $2^4$ |
| 32 | $2^5$ |
| 64 | $2^6$ |
| 128 | $2^7$ |
| 256 | $2^8$ |
| 512 | $2^9$ |
| 1,024 | $2^{10}$ |
| 2,048 | $2^{11}$ |

## SKILLS DEVELOPMENT

A number written in **exponential form** has a base and an exponent.

$$\underbrace{a \times a \times a \times a \times a}_{\text{5 factors}} = a^5$$

exponent

base

The **base** tells what factor is being multiplied. The **exponent** tells how many equal factors there are. The expression above, $a^5$, is read as "$a$ to the fifth power." Any number raised to the first power is that number itself.

$n^1 = n$, so, $8^1 = 8$

### Example 1

Evaluate each expression. Let $x = 4$ and $y = -3$.

a. $x^2$ b. $y^3$ c. $xy^2$

### Solution

a. $x^2 = 4^2 = 4 \times 4 = 16$

b. $y^3 = (-3)^3 = (-3)(-3)(-3) = -27$

c. $xy^2 = (4)(-3)^2 = (4)(-3)(-3) = (4)(9) = 36$ ◄

**CHECK UNDERSTANDING**

If the expression in Example **1c** had been $(xy)^2$, how would you have evaluated it?

To find the product of $3^2 \times 3^4$ you could write out the factors for each term and then write the product using exponents.

$$3^2 \times 3^4 = \underbrace{\overbrace{3 \times 3}^{\text{2 factors}} \times \overbrace{3 \times 3 \times 3 \times 3}^{\text{4 factors}}}_{2 + 4 = 6 \text{ factors}} = 3^6 = 3^{2+4}$$

A shorter method is to use the **product rule.**

► To multiply numbers with the same base, add the exponents.
$a^m \times a^n = a^{m+n}$

Apply the property above to develop another important rule.
$(3^2)^4 = 3^2 \times 3^2 \times 3^2 \times 3^2 = 3^{2+2+2+2} = 3^8 = 3^{2\times4}$

This relationship is described by the **power rule.**

► To raise an exponential number to a power, multiply exponents.
$(a^m)^n = a^{mn}$

You can simplify a power of a product such as $(3y^2)^3$.

$$\begin{aligned}(3y^2)^3 &= (3y^2)(3y^2)(3y^2)\\ &= (3)(3)(3)(y^2)(y^2)(y^2)\\ &= (3^3)(y^2)^3 = 27y^6\end{aligned}$$

The result is described by the **power of a product rule.**

► To find the power of a product, find the power of each factor and multiply.
$(ab)^m = a^m b^m$

> **Properties of Exponents for Multiplication**
>
> For all real numbers $a$ and $b$, if $m$ and $n$ are integers then
> $a^m \cdot a^n = a^{m+n}$ $\quad (a^m)^n = a^{mn}$ $\quad (ab)^m = a^m b^m$

To simplify expressions with exponents, simplify the exponent first.
$4 \cdot 2^3 = 4 \cdot (2 \cdot 2 \cdot 2) = 32$

The order of operations extends to include exponents.

- Perform all calculations within parentheses and brackets.
- Do all calculations involving exponents.
- Multiply or divide in order, from left to right.
- Add or subtract in order, from left to right.

The order of operations can now be remembered by using the mnemonic "*P*lease *E*xcuse *M*y *D*ear *A*unt *S*ally," where *E* stands for *exponents.*

Simplify.
**1.** $3(2 + 5)^2$ **2.** $(10 - 2^3)^4$

## Example 2

Simplify.

**a.** $x^7 \cdot x^9$ **b.** $(r^4)^2$ **c.** $(u^3v)^4$

**Solution**

**a.** $x^7 \cdot x^9 = x^{7+9} = x^{16}$

**b.** $(r^4)^2 = r^{4\times2} = r^8$

**c.** $(u^3v)^4 = u^{3\times4}v^4 = u^{12}v^4$ ◄

Study the following examples.

$$3^4 \div 3^2 = \frac{3^4}{3^2} = \frac{3 \times 3 \times \cancel{3} \times \cancel{3}}{\cancel{3} \times \cancel{3}} = 3^2 = 3^{4-2} \qquad \left(\frac{2}{5}\right)^3 = \left(\frac{2}{5}\right)\left(\frac{2}{5}\right)\left(\frac{2}{5}\right) = \frac{2^3}{5^3}$$

These examples illustrate two properties of exponents for division. One property is the **quotient rule.**

► To divide numbers with the same base, subtract the exponents.
$a^m \div a^n = a^{m-n}$

Another property is the **power of a quotient rule.**

► To find the power of a quotient, find the power of each number and divide.
$\left(\frac{a}{b}\right)^m = \frac{a^m}{b^m}$

**Properties of Exponents for Division**

For all real numbers $a$ and $b$, if $m$ and $n$ are integers then

$$\frac{a^m}{a^n} = a^{m-n}, \text{ if } a \neq 0 \qquad \left(\frac{a}{b}\right)^m = \frac{a^m}{b^m}, \text{ if } b \neq 0$$

### Example 3

Simplify. **a.** $\frac{t^5}{t^2}$ **b.** $\left(\frac{d}{2}\right)^3$

### Solution

**a.** $\frac{t^5}{t^2} = t^{5-2}$
$= t^3$

**b.** $\left(\frac{d}{2}\right)^3 = \frac{d^3}{2^3}$
$= \frac{d^3}{8}$ ◄

## TRY THESE

Evaluate each expression when $x = 5$ and $y = -2$.

**1.** $y^2$ **2.** $x^4$ **3.** $x^2y$ **4.** $xy^3$

Simplify.

**5.** $c^4 \cdot c^7$ **6.** $(d^5)^2$ **7.** $(gh^2)^3$

**8.** $\frac{v^7}{v^3}$, $v \neq 0$ **9.** $\left(\frac{c}{3}\right)^3$ **10.** $z^5 \div z$, $z \neq 0$

**11.** Why was it necessary to say $v \neq 0$ in Exercise 8?

# EXERCISES

## PRACTICE/ SOLVE PROBLEMS

Evaluate each expression when $a = 3$ and $b = -4$.

**1.** $a^3$ **2.** $a^2 - b^2$ **3.** $b^4$ **4.** $\left(\frac{b}{7}\right)^2$

**5.** $4a^2b$ **6.** $\frac{a^8}{a^7}$, $a \neq 0$ **7.** $(2 + b)^3$ **8.** $(a^2 - 5)^2$

Use the properties of exponents to explain why $a^7 \cdot a = a^8$ and $\frac{a^7}{a} = a^6$.

Simplify.

**9.** $c^4 \cdot c^3$ **10.** $(x^3)^5$ **11.** $\frac{y^9}{y^4}$, $y \neq 0$ **12.** $g(g^5)$

**13.** $\frac{r^{10}}{r^9}$, $r \neq 0$ **14.** $e^{10} \cdot e^2$ **15.** $(z^4)^4$ **16.** $\left(\frac{1}{d}\right)^6$, $d \neq 0$

**17.** $\left(\frac{m}{4}\right)^3$ **18.** $\frac{15w^7}{3w^5}$, $w \neq 0$ **19.** $3(y^3)^4$ **20.** $(d^2)(d^3)(d^4)$

**21.** A piece of wire $n^5$ inches long is to be cut into pieces each $n^2$ in. long. How many pieces will there be?

**22.** For the piece of wire above, let $n = 5$. How long was the wire to begin with? How long will each piece be? How many pieces will there be?

Simplify. You may need to use more than one of the properties of exponents.

23. $a^3 \cdot a^7$
24. $\frac{c^{10}}{c^5}$, $c \neq 0$
25. $(x^2y)^3$
26. $m^2(m^3n^4)$
27. $\frac{f^2g^3}{fg}$, $f \neq 0$
28. $\frac{25z^8}{50z^6}$, $z \neq 0$
29. $t^{10} \div t^9$, $t \neq 0$
30. $\left(\frac{r}{s^2}\right)^5$, $s \neq 0$
31. $\left(\frac{x^8}{x^2}\right)^3$, $x \neq 0$
32. $(b^3 \cdot b^2)^4$
33. $\left(\frac{2}{w}\right)^4$, $w \neq 0$
34. $\frac{v^2(v^5)}{v^3}$, $v \neq 0$

35. Without multiplying, tell which of these numbers is the greatest.
$100^4$ $1{,}000^3$ $10{,}000^2$
Explain your reasoning.

36. According to data compiled by the United Nations, in 1975 the earth's population was 4 billion and doubling every 35 years. If the population continues to grow at the same rate, what will the population be in 2010? in 2045?

## EXTEND/ SOLVE PROBLEMS

### MATH: WHO, WHERE, WHEN

The Rhind Papyrus is a document that was found in the ruins of a small building in an Egyptian city.

The papyrus was a handbook of Egyptian mathematics containing exercises and practical examples. Mathematical historians have developed versions of these problems.

One problem deals with exponents and has been formulated as follows: "An estate consisted of seven houses; each house had seven cats, each cat ate seven mice, each mouse ate seven heads of wheat, and each head of wheat could yield seven hekat measures of grain. Houses, cats, mice, heads of wheat, and hekat measures of grain, how many of these in all were in the estate?" Try to solve this problem.

## THINK CRITICALLY/ SOLVE PROBLEMS

Write *true* or *false*. Give examples to support your answer.

37. If $a$ and $b$ are integers and $a < b$, then $a^2 < b^2$.

38. If $a$ and $b$ are integers and $a < b$, then $a^3 < b^3$.

Solve.

39. A certain type of bacteria doubles in number every 5 minutes. At 10 A.M. there were 16,384 bacteria. At what time had there been exactly half that amount?

# 2-9 Zero and Negative Exponents

**EXPLORE**

Use grid paper. Copy and label the figures at the right.

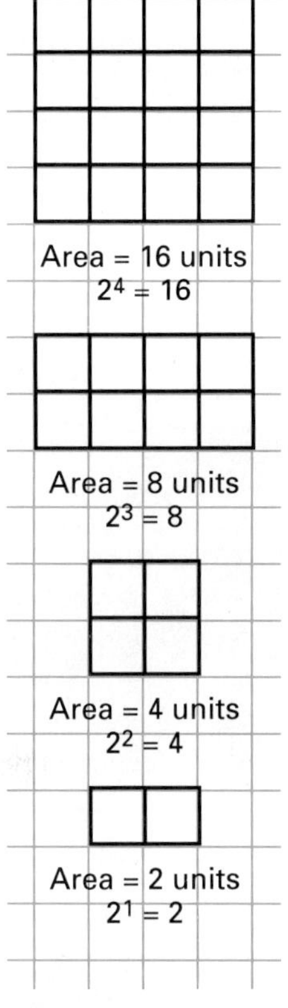


a. Describe the relationship between $2^4$ and $2^3$.

b. How are $2^3$ and $2^2$ related? What about $2^2$ and $2^1$? Write a statement summarizing the pattern.

c. Study the shape of each figure. What repeating pattern of shapes do you notice?

d. Follow the pattern. Draw the next figure. What is the area of the figure you drew? What power of 2 do you think this figure represents?

e. Draw the next three figures in the pattern. What is the area of each figure? What power of 2 do you think each figure represents?

f. Complete: $2^{-n} = \frac{■}{■}$

g. Use the values you found above for the powers of 2. Find each product by multiplying.

$2^3 \cdot 2^{-2} = ■ = 2^{■}$

$2^4 \cdot 2^{-3} = ■ = 2^{■}$

$2^2 \cdot 2^{-1} = ■ = 2^{■}$

h. Does $2^3 \cdot 2^{-2} = 2^{3+(-2)}$?

Does $2^4 \cdot 2^{-3} = 2^{4+(-3)}$?

Does $2^2 \cdot 2^{-1} = 2^{2+(-1)}$?

What property does this illustrate?

What can you conclude?

i. How would you find $\frac{2^4}{2^{-3}}$ and $\frac{2^3}{2^{-2}}$? What can you conclude?

**SKILLS DEVELOPMENT**

Any nonzero real number raised to the zero power has a value that is equal to 1.

► For every nonzero real number $a$, $a^0 = 1$.

Exponents can be negative numbers. The quotient property of exponents can help you understand the meaning of negative exponents.

$$\frac{4^3}{4^5} = \frac{\overset{1}{\cancel{4}} \cdot \overset{1}{\cancel{4}} \cdot \overset{1}{\cancel{4}}}{\underset{1}{\cancel{4}} \cdot \underset{1}{\cancel{4}} \cdot \underset{1}{\cancel{4}} \cdot 4 \cdot 4} = \frac{1}{4 \cdot 4} = \frac{1}{4^2}$$

If you use the quotient property, you obtain $\frac{4^3}{4^5} = 4^{3-5} = 4^{-2}$.
So, $4^{-2} = \frac{1}{4^2}$.

► For every nonzero real number $a$, if $n$ is an integer then $a^{-n} = \frac{1}{a^n}$. So, for any nonzero real number $a$, $a^{-n}$ is the reciprocal, or multiplicative inverse, of $a^n$.

### Example 1

Write each expression as a fraction.

**a.** $5^{-2}$ **b.** $(-2)^{-4}$

### Solution

**a.** $5^{-2} = \frac{1}{5^2} = \frac{1}{25}$ **b.** $(-2)^{-4} = \frac{1}{(-2)^4} = \frac{1}{16}$ ◄

The properties of exponents discussed in Section 2–8 can now be extended to include negative exponents.

### Example 2

Simplify each expression.

**a.** $x^7 \div x^{-3}$ **b.** $c^5 \cdot c^{-2}$ **c.** $(y^3)^{-4}$

### Solution

**a.** $x^7 \div x^{-3} = x^{7-(-3)} = x^{10}$

**b.** $c^5 \cdot c^{-2} = c^{5+(-2)} = c^3$

**c.** $(y^3)^{-4} = y^{3\times(-4)} = y^{-12}$ ◄

You can use negative exponents to evaluate variable expressions.

### Example 3

Evaluate each expression when $m = 2$ and $s = -3$.

**a.** $m^{-5}$ **b.** $(s^2)^{-3}$ **c.** $m^3s^{-2}$

### Solution

**a.** $m^{-5} = (2)^{-5} = \frac{1}{2^5} = \frac{1}{32}$

**b.** $(s^2)^{-3} = s^{2\times(-3)} = s^{-6} = (-3)^{-6} = \frac{1}{(-3)^6} = \frac{1}{729}$

**c.** $m^3s^{-2} = \frac{m^3}{s^2} = \frac{(2)^3}{(-3)^2} = \frac{8}{9}$ ◄

Scientists often deal with very large or very small numbers. In order to be able to write and to compute with such numbers more easily, the system of **scientific notation** was developed. A number written in scientific notation has two factors. The first factor is greater than or equal to 1 and less than 10. The second factor is a power of 10.

**MENTAL MATH TIP**

To multiply by $10^n$ when $n$ is a positive integer, move the decimal point $n$ places to the right.

$5.18 \times 10^7 = 51{,}800{,}000$ (7 places)

To multiply by $10^n$ when $n$ is a negative integer, move the decimal point $n$ places to the left.

$9.2 \times 10^{-4} = 0.00092$ (4 places)

## Example 4

**a.** Write 2,674,000 in scientific notation.
**b.** Write $6.3 \times 10^{-5}$ in standard form.

### Solution

**a.** $2{,}674{,}000 = 2.674 \times 1{,}000{,}000$
$= 2.674 \times 10^6$

**b.** $6.3 \times 10^{-5} = 6.3 \times 0.00001$
$= 0.000063$ ◄

## Example 5

The speed of light is $3.00 \times 10^5$ km/s. How far does light travel in 1 hour? Write the answer in scientific notation.

### Solution

Think: 1 h = 60 min = 60(60) s = 3,600 s
$3{,}600 = 3.6 \times 10^3$

To find the distance light travels in 1 hour, multiply.

$(3.00 \times 10^5)(3.6 \times 10^3) = (3.00 \times 3.6)(10^5 \times 10^3)$ — Use the commutative and associative properties of multiplication.
$= (10.8)(10^{5+3})$
$= 10.8 \times 10^8$
$= 1.08 \times 10^9$ ← Remember that the first factor must be less than 10.

Light travels at $1.08 \times 10^9$ km/h. ◄

## TRY THESE

**CALCULATOR TIP**

If your calculator has an EE or an EXP key, you can use it as a quick way to enter a number that is in scientific notation. For example, to enter $8 \times 10^5$, use this key sequence:

8 EXP 5

Write each expression as a fraction or an integer.

**1.** $(-4)^0$ **2.** $2^{-6}$ **3.** $3^{-5}$ **4.** $(-1)^{-3}$ **5.** $(-5)^{-3}$

Simplify.

**6.** $x^4 \div x^7$ **7.** $b^{-5} \cdot b^9$ **8.** $(w^5)^{-4}$ **9.** $d^3 \cdot d^{-8}$ **10.** $m^4 \div m^{-7}$

Evaluate each expression when $m = -4$ and $n = 5$.

**11.** $m^{-2}$ **12.** $n^{-4}$ **13.** $m^0n^{-2}$ **14.** $n^3 \times n^{-6}$

Write each number in scientific notation.

**15.** 29,000,000 **16.** 0.0039 **17.** 0.0000808

Write each number in standard form.

**18.** $7.6 \times 10^8$ **19.** $9.5 \times 10^{-3}$ **20.** $4.03 \times 10^{-6}$

**21.** The diameter of some white blood cells is about $8 \times 10^{-4}$ in. If 1 million of these cells were just touching in a straight line, how long would the line be? Write the answer in scientific notation.

# EXERCISES

## PRACTICE/ SOLVE PROBLEMS

Simplify.

1. $z^{-10} \div z^2$ 2. $y^{-4} \cdot y^{-6}$ 3. $(w^3)^{-3}$ 4. $a^8 \cdot a^{-8}$ 5. $x^5 \div x^8$

Evaluate each expression when $s = -1$ and $t = 4$.

6. $s^{-3}$ 7. $t^{-5}$ 8. $(st)^{-2}$ 9. $t^6 \times t^{-4}$

Write each number in scientific notation.

10. 8,450 11. 0.04658 12. 0.0000317

Write each number in standard form.

13. $2.01 \times 10^{-4}$ 14. $5.3 \times 10^8$ 15. $8.6 \times 10^{-6}$

Solve. Write each answer in scientific notation.

16. Mercury travels around the sun at a speed of approximately $1.72 \times 10^5$ km/h. At this rate, how far does Mercury travel in 30 days?
17. The wavelength of red light is about $6.5 \times 10^{-5}$ cm. Express this distance in meters.

## EXTEND/ SOLVE PROBLEMS

Write in order from least to greatest.

18. $2^3 \cdot 2^{-2}$; $(6 \cdot 3)^0$; $\left(\frac{1}{3}\right)^{-1}$
19. $(-4)^{-2}(-4)$; $(-5)^2(-5)$; $(-5)(-5)^{-2}$
20. $5^{-1}$; $3^{-2}$; $-(3^2)$; $36^0$; $-(10)^{-2}$

Solve. Write each answer in scientific notation.

21. The mass of a hydrogen atom is approximately $1.67 \times 10^{-24}$ g. What is the approximate mass of 10,000,000,000 hydrogen atoms?
22. The length of a microchip is $3 \times 10^{-7}$ mm. How many chips could be placed side by side on a 2.1-cm length of circuit board?

## THINK CRITICALLY/ SOLVE PROBLEMS

23. You know that $2^2 = 4$ and $2^{-2} = \frac{1}{4}$. What do you think is the value of $\frac{1}{2^{-2}}$? If $a$ is a nonzero real number and $n$ is an integer, what does $\frac{1}{a^{-n}}$ represent?

Use the results of Exercise 23. Rewrite each with positive exponents.

24. $\frac{1}{m^{-3}}$ 25. $\frac{x^{-2}}{y^{-2}}$ 26. $\frac{a^4}{b^{-3}}$ 27. $\frac{m^{-7}}{n^3}$

# CHAPTER 2 • REVIEW

Choose the word from the list that completes each statement.

1. Two different integers that are the same distance from 0 on the number line are __?__.
2. The point on a number line that corresponds to a number is called the __?__ of the number.
3. The distance a number is from 0 on the number line is the __?__ of the number.
4. The number that corresponds to a point on a number line is called the __?__ of the point.
5. The parts of a variable expression that are separated by addition and subtraction signs are its __?__.

a. absolute value
b. terms
c. graph
d. opposites
e. coordinate

## SECTION 2–1 THE REAL NUMBER SYSTEM (pages 44–47)

- Real numbers can be graphed on a number line.
- To show that a number is included in the set of numbers being graphed, put a solid dot at the point that corresponds to the number.
- To show that a number is not included in the set of numbers being graphed, put an open circle at the point that corresponds to the number.

Graph each set of numbers on a number line.

6. real numbers less than 1
7. real numbers greater than or equal to $-4$

## SECTION 2–2 ADDING AND SUBTRACTING INTEGERS (pages 48–51)

- To add integers with the same sign, add the absolute values of the integers. Give the sum the sign of the addends.
- To add integers with different signs, subtract the absolute values. Give the sum the sign of the addend(s) with the greater absolute value.
- To subtract an integer, add its opposite.

Add or subtract.

8. $-15 + (-6)$
9. $-57 + 29$
10. $33 - (-5)$
11. $-30 - 19$

## SECTION 2–3 MULTIPLYING AND DIVIDING INTEGERS (pages 52–55)

- The product or quotient of two integers with the same sign is positive.
- The product or quotient of two integers with different signs is negative.

Multiply or divide.

12. $-5(-16)$ 13. $4(-23)$ 14. $-55 \div (-5)$ 15. $86 \div (-2)$

## SECTIONS 2–4 AND 2–6 PROBLEM SOLVING (pages 56–57, 62–63)

► You can check your solution to a problem by using inverse operations.

**USING DATA** Solve using the table on page 43.

16. What is the difference in meters between the height of Mount Everest and the depth of the Caspian Sea?

## SECTION 2–5 OPERATIONS WITH RATIONAL NUMBERS (pages 58–61)

► Operations with rational numbers involve the same rules and properties that are used in performing those operations with integers.

Simplify.

17. $\frac{1}{2}\left(3\frac{1}{4} + 2\frac{1}{8}\right)$ 18. $(1.43 + 3.52) \div (-0.05)$ 19. $(15 - 18)(2.5 + 1.5)$

## SECTION 2–7 SIMPLIFYING VARIABLE EXPRESSIONS (pages 64–67)

► To simplify variable expressions, use the properties and combine like terms.

Simplify.

20. $6rs + rs - rt$ 21. $-2(mn + 3.2)$ 22. $-(c + 3d)$ 23. $2(xy + y) + 3(y - x)$

## SECTION 2–8 PROPERTIES OF EXPONENTS (pages 68–71)

► For all real numbers $a$ and $b$, if $m$ and $n$ are integers then

**Multiplication**

$a^m \cdot a^n = a^{m+n}$ $(a^m)^n = a^{mn}$

$(ab)^m = a^m b^m$

**Division**

$\frac{a^m}{a^n} = a^{m-n}$, if $a \neq 0$ $\left(\frac{a}{b}\right)^m = \frac{a^m}{b^m}$, if $b \neq 0$

Simplify.

24. $x^2 \cdot x^5$ 25. $(a^2)^4$ 26. $\frac{27a^3}{9a}$, $a \neq 0$ 27. $\left(\frac{27w^3}{9w}\right)^2$, $w \neq 0$

## SECTION 2–9 ZERO AND NEGATIVE EXPONENTS (pages 72–75)

► For every nonzero real number $a$, $a^0 = 1$ and $a^{-n} = \frac{1}{a^n}$.

Evaluate each expression when $r = -2$ and $s = 2$.

28. $r^0$ 29. $r^{-2}$ 30. $r^0 s^{-3}$ 31. $s^2 \times s^{-5}$

# CHAPTER 2 ● TEST

Graph each set of numbers on a number line.

1. all real numbers greater than $-5$
2. all real numbers less than or equal to $-2$

Evaluate when $r = -4$.

3. $|r|$
4. $-|-r|$
5. $-r$

Add or subtract.

6. $-6 + (-4)$
7. $-13 + 9$
8. $-17 + (-9)$
9. $35 + (-23)$
10. $-5 - 5$
11. $9 - (-7)$
12. $-26 - 25$
13. $55 - (-20)$

Multiply or divide.

14. $-6 \cdot 6$
15. $3(-8)$
16. $-44(-4)$
17. $3 \cdot 5(-7)$
18. $18 \div (-6)$
19. $-63 \div (-3)$
20. $-128 \div 4$
21. $-500 \div (-25)$

Perform the indicated operation.

22. $-2.1 + 3 \cdot 5.6$
23. $-\frac{1}{8} + \frac{5}{8} \div \frac{1}{2}$
24. $-2[7.5 + (-7.9)]$

Evaluate each expression when $m = -6$ and $n = \frac{3}{4}$.

25. $3m$
26. $-\frac{1}{2}n$
27. $-(m + n)$
28. $-n \div m$

Simplify.

29. $-(r + s)$
30. $ab - 5ab$
31. $-(3b + c)$
32. $5(rs + s) - 3(rs + r)$

Simplify.

33. $x^3 \cdot x^5$
34. $(n^2)^3$
35. $\frac{16b^6}{4b^2}$, $b \neq 0$
36. $(d^3)(d^5)$

Evaluate each expression when $a = 2$ and $b = -4$.

37. $b^0$
38. $a^{-3}$
39. $(a^3)^{-2}$
40. $b^2a^{-3}$

Write each number in scientific notation.

41. 0.00385
42. 9,380,000,000
43. 0.0543

Solve.

44. The highest point in Nevada is Boundary Peak with an altitude of 13,140 ft. The lowest point in Nevada is Mount Manchester at 479 ft. Use an integer to describe the change in altitude from the highest to the lowest point in Nevada.

Complete.

1. $4\left(\frac{1}{4} + \frac{2}{3}\right) = 4\left(\frac{1}{4}\right) + ■\left(\frac{2}{3}\right)$
2. $12.4 \times (-5.4) = -5.4 \times$ ■
3. $81 \times$ ■ $= 0$

Find the next three terms in each pattern.

4. 50, 46, 42, 38, ■, ■, ■
5. 650, 65, 6.5, 0.65, ■, ■, ■

Copy and complete the function table.

6.

| $x$ | $2x - 3$ |
|---|---|
| $\frac{1}{2}$ | ■ |
| 0 | ■ |
| 2 | ■ |
| 5 | ■ |
| 10 | ■ |

Find each percent to the nearest tenth.

7. What percent of 77 is 11?
8. 225 is what percent of 900?

The bar graph shows the maximum depth of the Great Lakes, in meters. Use the bar graph for Exercises 9–11.

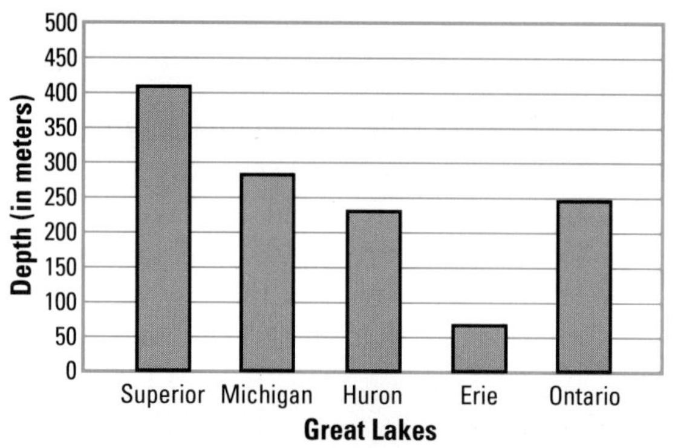


9. The maximum depth of Lake Erie is about one-fourth the maximum depth of which other lake?
10. How much deeper is Lake Superior than Lake Michigan?
11. The surface of Lake Huron is 176 m above sea level. About how much of Lake Huron is below sea level?

Add or subtract.

12. $13 + (-5)$
13. $-33 + (-19) + 5$
14. $-15 - 7$
15. $-16 - (-4)$

Multiply or divide.

16. $-4(-5)(-2)$
17. $-3\ (8)(-5)$
18. $150 \div (-15)$
19. $-444 \div 111$

Write in scientific notation.

20. 15,000,000
21. 0.00067

Evaluate when $c = 2.5$ and $d = -10$.

22. $\frac{d}{-c}$
23. $-(d)^2$

Perform the indicated operations.

24. $\frac{1}{2} + \left(-\frac{3}{4}\right)$
25. $3\frac{1}{8}\left(-\frac{2}{5}\right)$

Evaluate each expression when $a = 4$ and $b = -1$.

26. $(ab)^2$
27. $a^{-2}$

28. The lowest temperature recorded in Wyoming was −63°F in February 1933. The highest recorded temperature was in July 1900, when the temperature climbed to 177°F above the lowest recorded temperature. What was the highest recorded temperature?

1. Simplify. $20.6 + 8.2 \div 2 - 1$
A. 23.7 B. 13.4
C. 28.8 D. none of these

2. Simplify. $[48 - (18 \div 6)] \div 15$
A. $33\frac{1}{3}$ B. 3
C. 75 D. none of these

3. Peter swam 36 laps, which was 75% of his goal. What was his goal?
A. 72 laps B. 360 laps
C. 40 laps D. none of these

Choose the best estimate.

4. $52.7 \times 47.8$
A. 200 B. 250
C. 2,000 D. 2,500

5. 45% of 1,880
A. 1,000 B. 800
C. 640 D. 80

Use the circle graph for Exercises 6–7. It shows one family's monthly budget of $3,500.

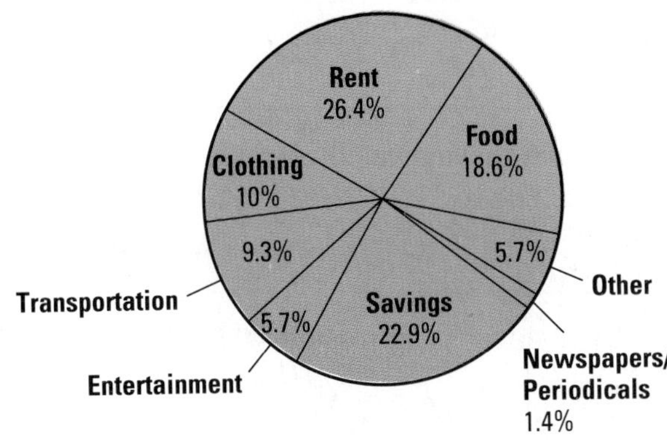


6. The percent budgeted for food is about equal to the percent budgeted for which other two categories together?
A. transportation and entertainment
B. clothing and transportation
C. clothing and entertainment
D. entertainment and other

7. About how much money is budgeted for entertainment per year?
A. about $200 B. about $2,200
C. about $2,400 D. about $240

8. Multiply. $-6(-8)$
A. 48 B. $-48$
C. 14 D. $-14$

9. Divide. $-250 \div 2.5$
A. $-0.1$ B. 0.1
C. $-100$ D. 100

10. Subtract. $-9.7 - 4.6$
A. $-13.1$ B. $-14.3$
C. $-5.1$ D. 5.1

11. Add. $-18 + (-17) + (-35)$
A. $-40$ B. $-50$
C. $-60$ D. $-70$

12. Simplify. $x \cdot y \cdot x \cdot y \cdot y$
A. $2x(3y)$ B. $x^2y^3$
C. $2xy^3$ D. $2x + 3y$

13. Write in scientific notation. 0.00748
A. $7.48 \times 10^{-3}$ B. $7.48 \times 10^{3}$
C. $7.48 \times 10^{-5}$ D. $7.48 \times 10^{5}$

14. Write in standard form. $6.6 \times 10^4$
A. 660 B. 6,600
C. 66,000 D. 660,000

15. Evaluate $x^{-3}$ when $x = -2$.
A. 8 B. $-8$ C. $\frac{1}{8}$ D. $-\frac{1}{8}$

16. Perform the indicated operations.
$-\left(\frac{1}{8} + \frac{1}{4}\right) \div (-3)$
A. $\frac{1}{8}$ B. $-\frac{1}{8}$ C. $\frac{3}{8}$ D. $-\frac{3}{8}$

17. In January, the average daily low temperature in Oslo, Norway, is −7°C. In Bangkok, Thailand, the average daily low temperature is 19°C in January. How much higher is the low temperature in Bangkok than in Oslo?
A. 12°C B. −12°C
C. 26°C D. −26°C

# CHAPTER 3 SKILLS PREVIEW

Write the symbol for each figure.

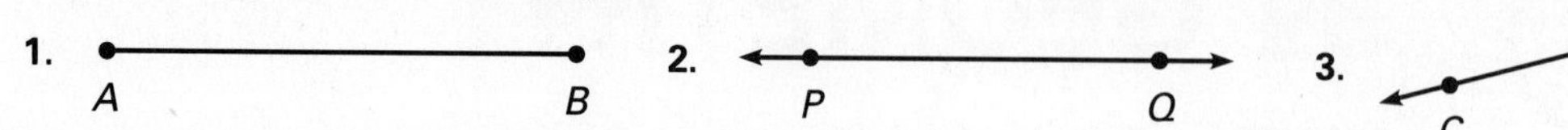


Refer to the figure to name the following.

4. all obtuse angles
5. a pair of complementary angles
6. two pairs of adjacent supplementary angles
7. a pair of perpendicular line segments

Refer to the figure at the right to name the following. Line $l \parallel$ line $m$.

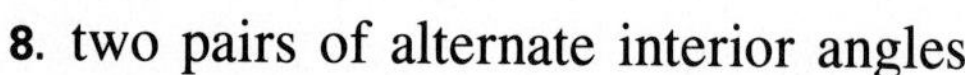

8. two pairs of alternate interior angles
9. all pairs of corresponding angles

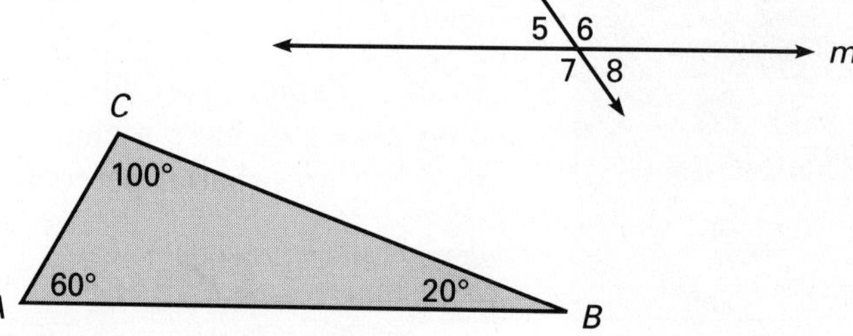


10. Name the longest and shortest sides of this triangle.
11. Write the converse of this statement: *If a figure is a triangle, then the figure is not a rectangle.*
12. Name the corresponding sides of these two triangles.

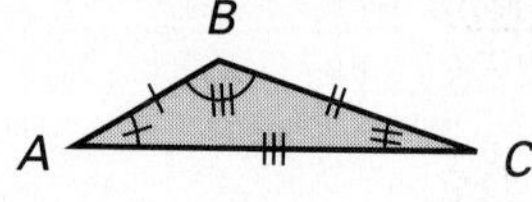


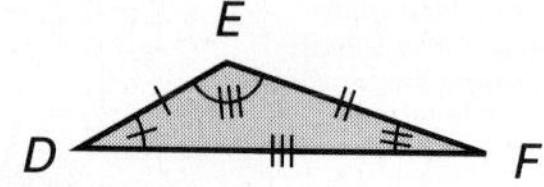


Refer to the circle to name the following.

13. a minor arc
14. a major arc
15. a central angle
16. an inscribed angle

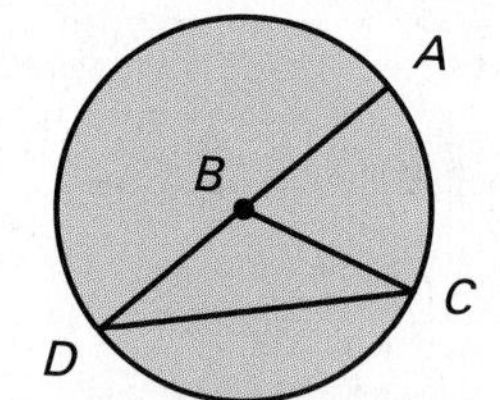


17. Find the approximate number of degrees in each central angle of a circle graph that could represent each category in the table shown at the right.

| MONTHLY EXPENSES | |
|---|---|
| Rent | 32% |
| Travel | 12% |
| Food | 36% |
| Clothing | 14% |
| Savings | 6% |

Refer to the figure to name the following.

18. a pair of intersecting edges
19. a pair of parallel edges
20. a pair of parallel faces
21. a pair of edges that are skew

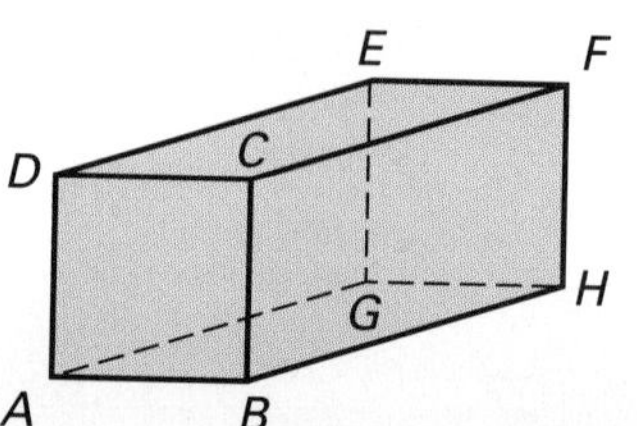

CHAPTER

# 3 REASONING IN GEOMETRY

**THEME Puzzles and Games**

Geometry is one area of mathematics that you actually see every day of your life. Each time you enter a room, you find geometric elements—the lines and planes that make up the walls, ceiling, and floor; the angles that make up the corners, and the two- and three-dimensional figures that make up just about everything else. In some special rooms, like the school gym, you may see circular shapes—the outline painted in the center of the basketball court, the rim of the net attached to a pole high off the ground, and the basketball itself.

In this chapter, you will work with and construct geometric figures. You also will have a chance to sharpen your skills in logical thinking and spatial perception.

*Trivial Pursuit, Clue,* and *Monopoly*

Many puzzles and games involve geometric shapes. Some puzzles require not only the use of common sense, but also the use of spatial sense. The best way to approach a puzzle is to first be sure you understand the challenge, then work in an orderly way to explore all possible ways of meeting it.

Look at the two sets of nested figures below. Into which group, A or B, does the figure in the middle belong? To solve the puzzle, follow this rule: *If a certain color or shape is inside another color or shape, the nested figure belongs in group A. Otherwise, it belongs in group B.*

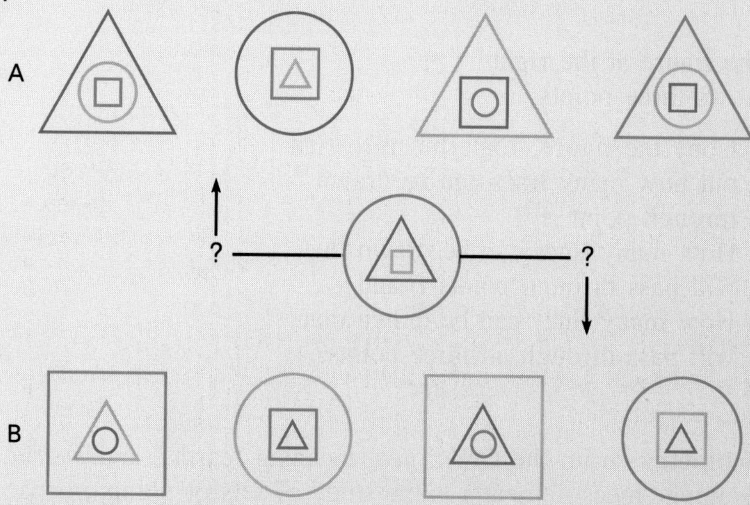


## DECISION MAKING

### Using Data

1. Identify the three shapes, colors, and sizes that make up each nested figure.
2. Can you determine how the figures in group A are alike? Can you determine how those in group B are alike?
3. How is the figure in the middle like the figures either in group A or in group B?
4. Name some board games you have played. How was the game board marked? Did geometric shapes appear on the board or on the playing pieces? Did those shapes affect the way in which the game was played?

### Working Together

Make up a puzzle of your own. You may use colors, shapes, numbers, letters, or even pictures. Do your own artwork or cut some from newspapers or magazines. Starting with a geometric format may help. Share your ideas and develop the promising ones together. (If you find this kind of puzzle making very easy, you might want to try a three-dimensional puzzle!)

Challenge others to solve your puzzle, and then try to solve theirs.

# 3-1 The Basic Elements of Geometry

## EXPLORE

The figure at the right shows three points.

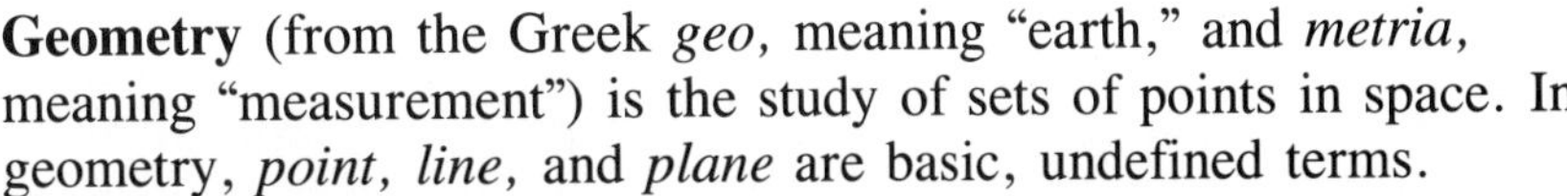


a. Copy the figure. Experiment to find out how many lines can be drawn through point $A$.

b. How many lines can be drawn that will pass through points $B$ and $C$?

c. How many lines can be drawn that will pass through all three points?

## SKILLS DEVELOPMENT

**MATH: WHO, WHERE, WHEN**

The Greek mathematician Euclid (c. 300 B.C.) wrote what has become the best known textbook in the world. The *Elements*, consisting of 13 books, has been used as a text in geometry for over 2,000 years. In it Euclid presents a careful, logical treatment of topics in plane and solid geometry.

**Geometry** (from the Greek *geo*, meaning "earth," and *metria*, meaning "measurement") is the study of sets of points in space. In geometry, *point, line,* and *plane* are basic, undefined terms.

A **point** is a location in space having no dimensions. A point is represented by a dot, which is named by a letter. Every other geometric figure is composed of sets of points. A **line** is a set of points that extends infinitely in two opposite directions. A **plane** is a flat surface that exists without end in all directions. A plane contains an infinite set of points and has no boundaries. **Space** is the set of all points.

point $P$ — line $TB$ ($\overleftrightarrow{TB}$), line $BT$ ($\overleftrightarrow{BT}$), or line $k$ — plane $M$

The following definitions follow from these basic terms.

**Collinear points** are points that lie on the same line.

**Noncollinear points** are points that do not lie on the same line.

**Coplanar points** are points that lie in the same plane.

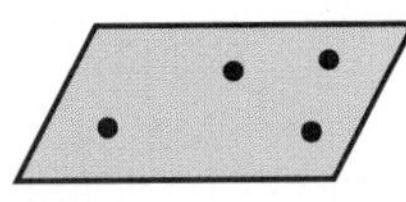

**Noncoplanar points** are points that do not lie in the same plane.

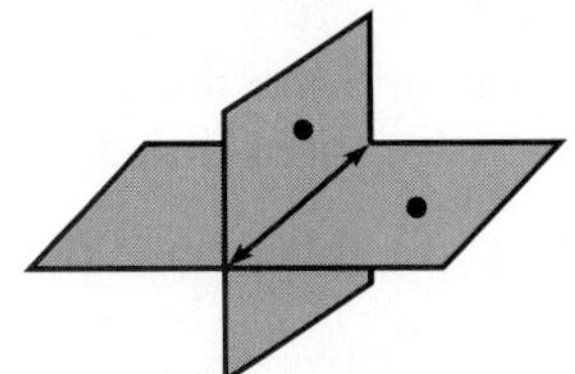

A **segment (line segment)** is a part of a line consisting of two **endpoints** and all points that lie between them.

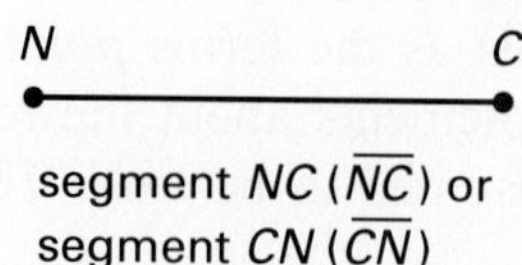


segment $NC$ ($\overline{NC}$) or segment $CN$ ($\overline{CN}$)

**Congruent line segments** have the same measure. The **midpoint of a segment** is the point that divides the segment into two congruent segments. The symbol ≅ means "is congruent to."

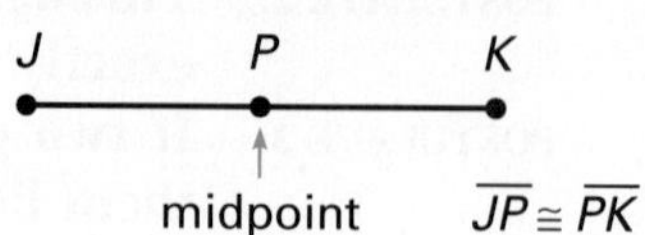


$\overline{JP} \cong \overline{PK}$

The **intersection** of two figures is the set of points that are in both figures. Two lines intersect in a point. A plane and a line intersect in a point.

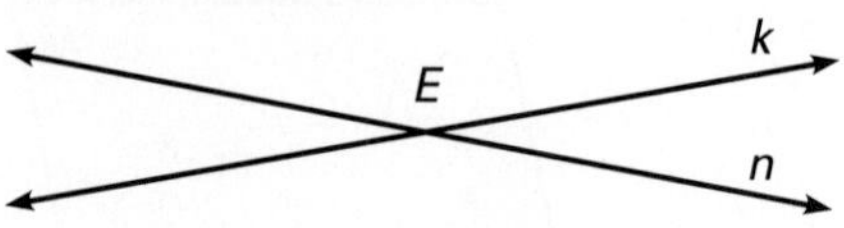


Line $k$ and line $n$ intersect in $E$.

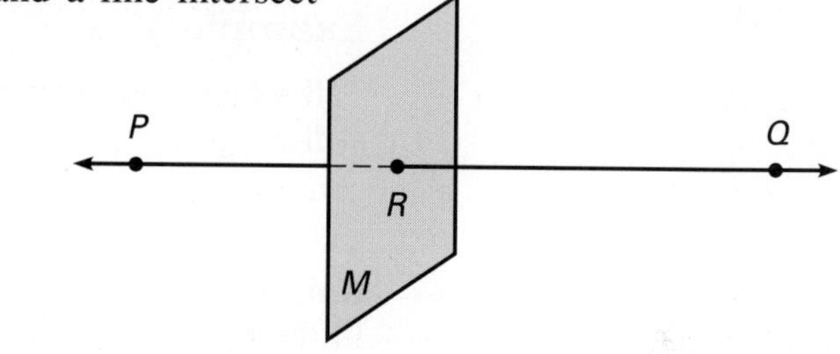


Line $PQ$ intersects plane $M$ in $R$.

A **bisector of a segment** is any line, segment, ray, or plane that intersects the segment at its midpoint.

A **ray** is part of a line that starts at one endpoint and extends without end in one direction.

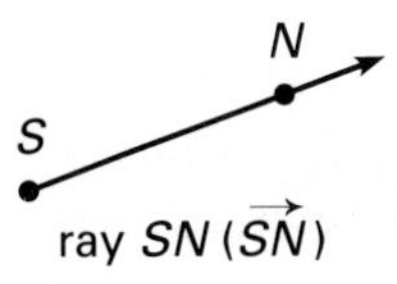


ray $SN$ ($\overrightarrow{SN}$)

Two rays, $\overrightarrow{BA}$ and $\overrightarrow{BC}$, are **opposite rays,** if $B$ is in $\overleftrightarrow{AC}$ and between $A$ and $C$.

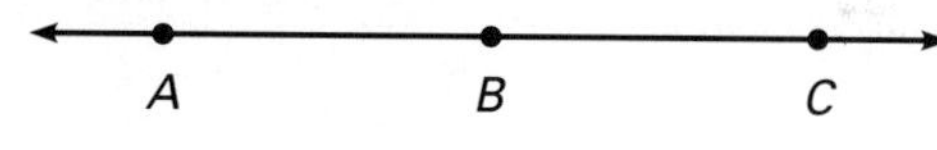


An **angle** is the figure formed by two rays that have a common endpoint. The endpoint is called the **vertex** (plural **vertices**) of the angle. Each ray forms a **side** of the angle.

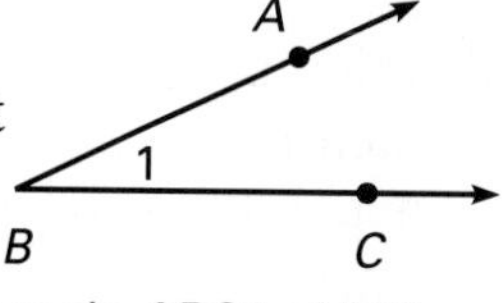


angle $ABC$ ($\angle ABC$) or
angle $CBA$ ($\angle CBA$) or
angle $B$ ($\angle B$) or angle 1 ($\angle 1$)

## Example 1

Name each figure and write the symbol for it.

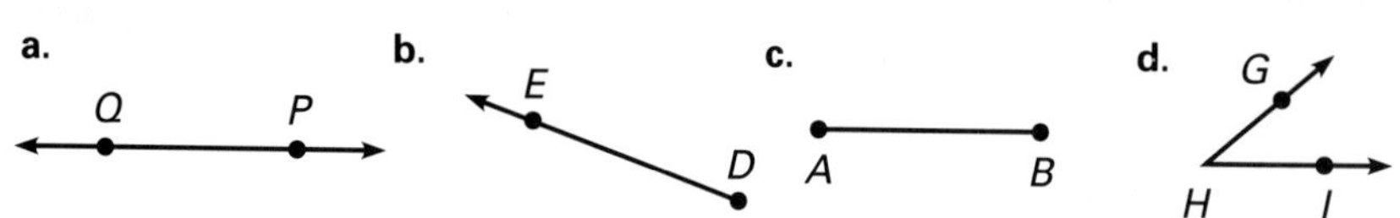


**Solution**

a. line $PQ$ ($\overleftrightarrow{PQ}$) or line $QP$ ($\overleftrightarrow{QP}$)

b. ray $DE$ ($\overrightarrow{DE}$)

c. segment $AB$ ($\overline{AB}$) or segment $BA$ ($\overline{BA}$)

d. angle $GHI$ ($\angle GHI$) or angle $IHG$ ($\angle IHG$) or angle $H$ ($\angle H$) ◄

**READING MATH**

In referring to "line segment $AB$," an overbar is written above the letters that name the segment.

$$\overline{AB} \perp \overline{CD}$$

In referring to the length, or measure, of the segment, no overbar is used.

$$AB = 5 \text{ cm}$$

Just as the terms *point, line,* and *plane* are undefined, there are statements about the relationships between these elements that are assumed to be true. These assumptions are called **postulates.**

POSTULATE 1: **Through any two points, there is exactly one line.**

POSTULATE 2: **Through any three noncollinear points, there is exactly one plane.**

POSTULATE 3: **If two points lie in a plane, then the line joining them lies in that plane.**

POSTULATE 4: **If two planes intersect, then their intersection is a line.**

### Example 2

Tell which postulate is illustrated in the figure at the right.

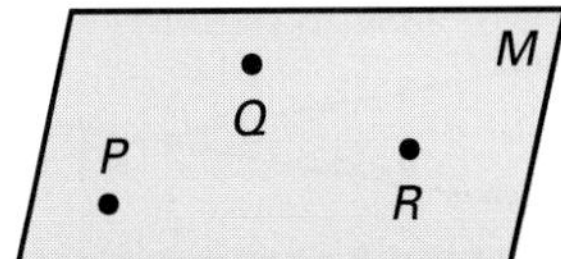


**Solution**
In the figure, the three noncollinear points *P*, *Q*, and *R* determine plane *M*. The figure illustrates Postulate 2. ◄

## TRY THESE

Name each figure and write the symbol for each figure.

1. R S

2.

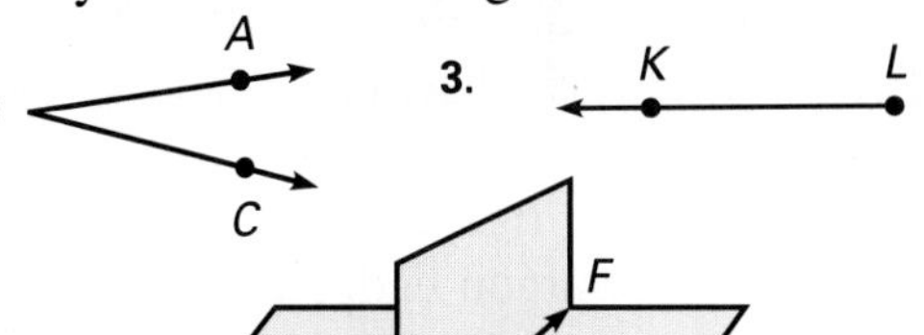


4. Tell which postulate is illustrated in the figure at the right.

F

E

## EXERCISES

**PRACTICE/ SOLVE PROBLEMS**

Write the symbol for each figure.

1.

2.

3.

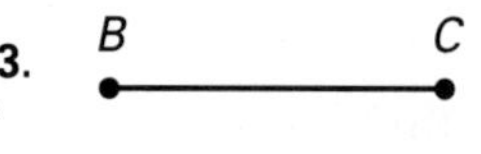


Draw and label a representation of each figure.

4. $\overleftrightarrow{DE}$ 5. $\overrightarrow{QR}$ 6. plane *N* 7. $\overline{ST}$ 8. $\angle KLM$

Identify the points in each figure as *collinear* or *noncollinear* and *coplanar* or *noncoplanar.*

9.

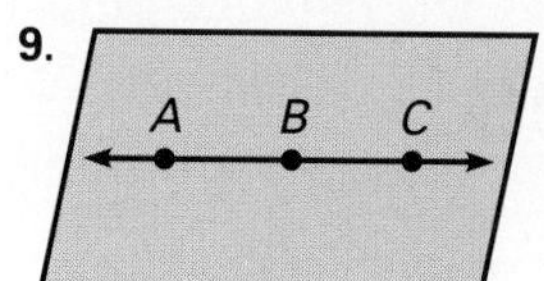


10.

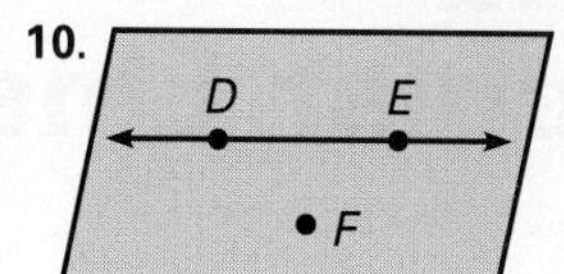


11.

## EXTEND/ SOLVE PROBLEMS

Choose the term at the right that matches each definition.

12. points that do not lie on the same line
13. part of a line consisting of two endpoints and all points that lie between them
14. any line, segment, ray, or plane that intersects a segment at its midpoint
15. points that lie in the same plane
16. figure formed by two rays having a common endpoint
17. segments that have the same measure

a. congruent line segments
b. coplanar points
c. noncollinear points
d. bisector of a segment
e. intersection of two figures
f. line segment
g. angle
h. collinear points

## THINK CRITICALLY/ SOLVE PROBLEMS

In the figure at the right, plane $M$ contains point $A$ and collinear points $P$, $Q$, and $R$. Point $E$ is not in plane $M$.

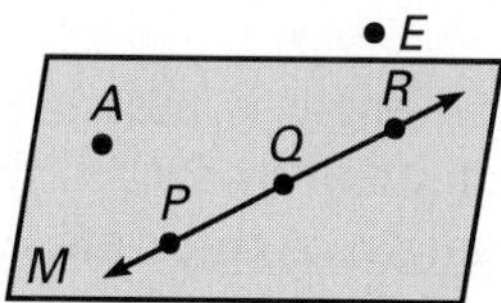

**TALK IT OVER**

What figure do you think the intersection of three planes might look like?

18. Which postulate tells you that exactly one plane contains $A$, $P$, and $Q$?
19. Which postulate tells you that exactly one line contains $A$ and $R$?
20. If $A$ and $P$ are in plane $M$, which postulate tells you that $\overleftrightarrow{AP}$ is in plane $M$?
21. Points $P$, $Q$, and $E$ determine a plane. Name the intersection of that plane and plane $M$.
22. Points $E$ and $Q$ determine a line. Name the intersection of that line and $\overleftrightarrow{PR}$ and of that line and plane $M$.
23. How are the intersection of a line and a plane and the intersection of two lines similar?

# 3-2 Angles and Perpendicular Lines

## EXPLORE

Use a straightedge and a compass to complete this construction.

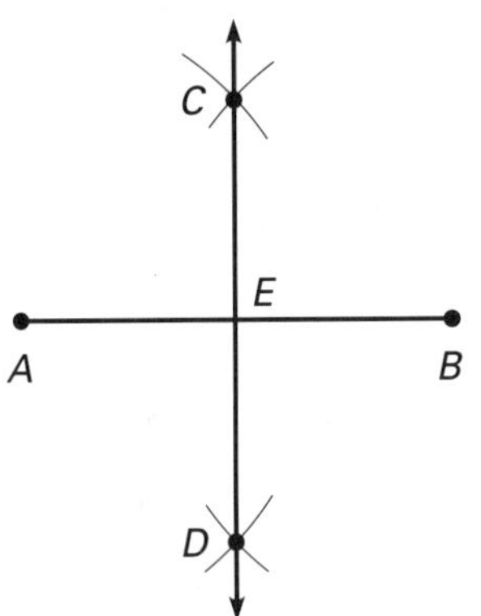


1. Draw a line segment. Label it $\overline{AB}$. Place the point of a compass at $A$. Open the compass to more than half the length of $\overline{AB}$. Draw one arc above and one arc below $\overline{AB}$.
2. With the compass point at $B$ and with the same compass setting as in Step 1, draw arcs above and below $\overline{AB}$. Label the points $C$ and $D$.
3. Use a straightedge to draw $\overleftrightarrow{CD}$. Label the point of intersection of $\overleftrightarrow{CD}$ and $\overline{AB}$ point $E$.

   a. Measure $\overline{AE}$ and $\overline{EB}$. What do you discover?
   b. Now measure $\angle CEA$ and $\angle CEB$. What do you discover?
   c. What can you conclude about $\overline{AB}$ and $\overleftrightarrow{CD}$?

## SKILLS DEVELOPMENT

The size of an angle is measured in a unit called the **degree.** The measure of an angle indicates the amount of openness between the sides of the angle. Angles are classified by their degree measure.

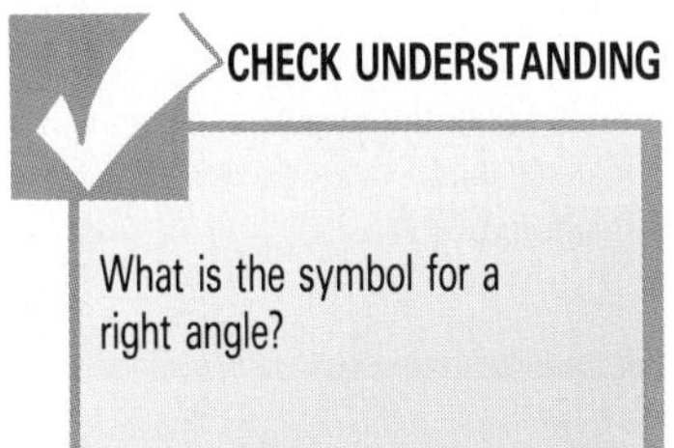


| Acute angle | Right angle | Obtuse angle | Straight angle |
|---|---|---|---|
| Angle measures less than 90°. | Angle measures exactly 90°. | Angle measures greater than 90°, but less than 180°. | Angle measures exactly 180°. |

Two angles whose measures have a sum of 90° are called **complementary angles.** Each angle is the *complement* of the other.

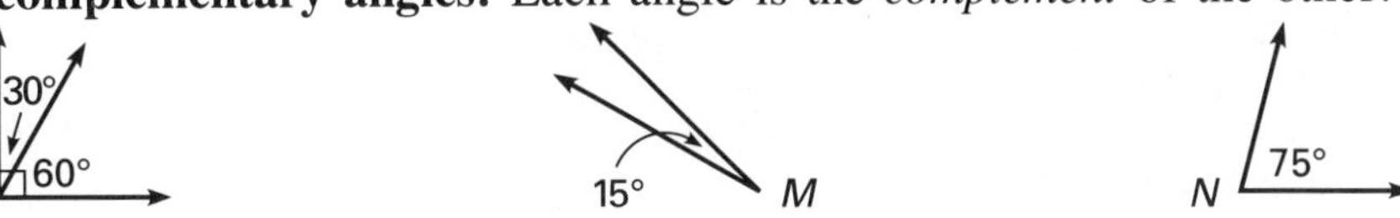

Two angles whose measures have a sum of 180° are called **supplementary angles.** Each angle is the *supplement* of the other.

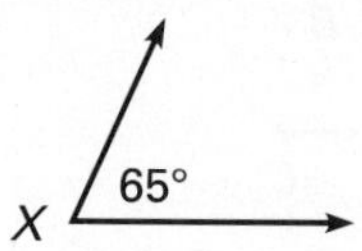


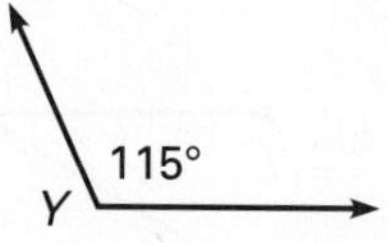


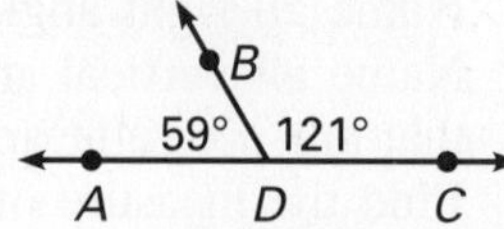


Two angles that have a common vertex and a common side, but no interior points in common, are called **adjacent angles.**

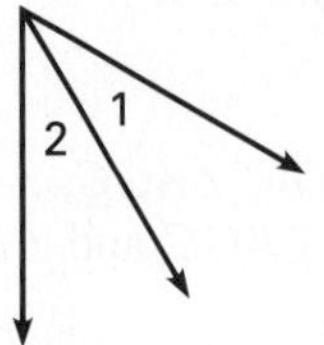


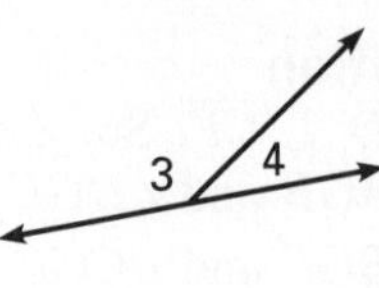


**Congruent angles** have the same measure.

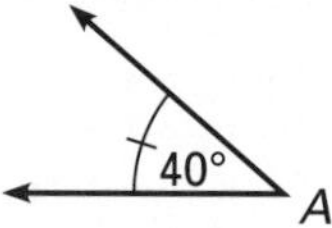


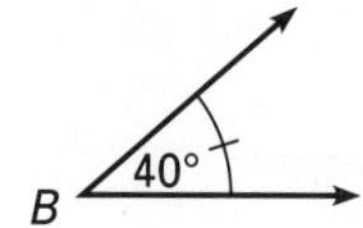


$\angle A \cong \angle B$

The **bisector of an angle** is a ray that divides the angle into two congruent adjacent angles.

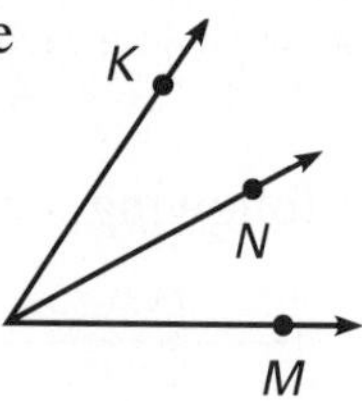


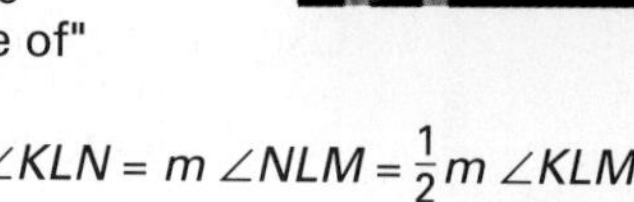

read "the measure of"

$m\angle KLN = m\angle NLM = \frac{1}{2} m\angle KLM$

**Perpendicular lines** are two lines that intersect to form equal adjacent right angles.

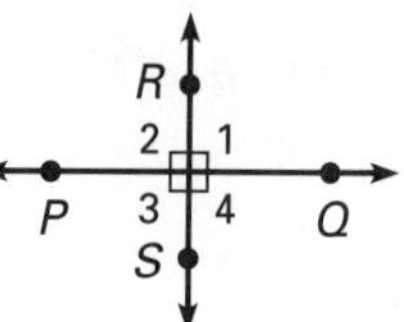


$\overleftrightarrow{PQ} \perp \overleftrightarrow{RS}$

$\angle 1 = \angle 2 = \angle 3 = \angle 4 = 90°$

TALK IT OVER

A line that is perpendicular to a segment and bisects that segment is called the **perpendicular bisector** of the segment. A segment has only one perpendicular bisector. In the Explore activity, name the line that is the perpendicular bisector of $\overline{AB}$.

When two lines intersect, the angles that are not adjacent to each other are called **vertical angles.** Vertical angles are congruent.

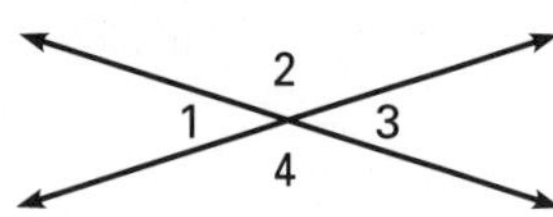


$\angle 1$ and $\angle 3$ are vertical angles.
$\angle 2$ and $\angle 4$ are vertical angles.

## Example 1

Use the figure to name the following.

**a.** all acute angles
**b.** two obtuse angles
**c.** a pair of adjacent complementary angles
**d.** two pairs of adjacent supplementary angles

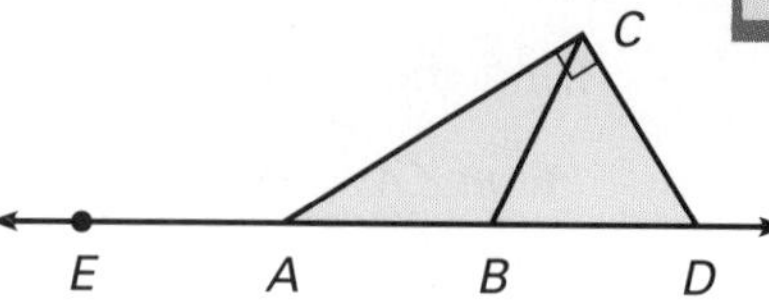


### Solution

**a.** $\angle CDB$, $\angle BCD$, $\angle DBC$, $\angle ACB$, and $\angle CAB$
**b.** $\angle CAE$ and $\angle CBA$
**c.** $\angle ACB$ and $\angle BCD$
**d.** $\angle EAC$ and $\angle CAB$; $\angle ABC$ and $\angle CBD$ ◄

**MIXED REVIEW**

Simplify.

**1.** (20 + 4) ÷ 4 x 2

**2.** [15 − (2 x 3)] ÷ 6

Evaluate each expression when $n$ = 2.5.

**3.** 8 + $n$ **4.** 15 − 3$n$

Write the function using function notation.

**5.** $x$ corresponds to 3$x$ − 4.

Find each product or quotient.

**6.** 8(−5) **7.** 36 ÷ (−6)

Simplify.

**8.** $8\left(3\frac{1}{8} - 2\frac{1}{4}\right)$

**9.** (8.2 + 7.1) − (9.6)

## Example 2

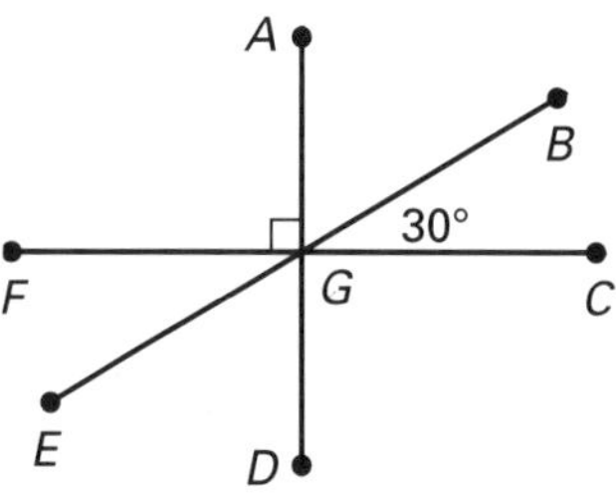


In the figure, $\overline{AD} \perp \overline{CF}$.

**a.** Name all right angles.
**b.** Name all vertical angles that are not right angles.
**c.** Find the measure of $\angle AGB$.
**d.** Find the measure of $\angle EGF$.

**Solution**

**a.** $\overline{AD} \perp \overline{CF}$. So, $\angle AGF$, $\angle AGC$, $\angle CGD$, and $\angle DGF$ are all right angles.
**b.** $\angle AGB$ and $\angle DGE$; $\angle BGC$ and $\angle EGF$ ; $\angle BGD$ and $\angle AGE$; $\angle BGF$ and $\angle CGE$
**c.** $\angle AGB$ and $\angle BGC$ are complementary angles.
So, $m\angle AGB = 90° - m\angle BGC$
$= 90° - 30° = 60°$
**d.** Vertical angles have equal measures.
So, $m\angle BGC = 30°$ and $m\angle EGF = 30°$. ◄

## TRY THESE

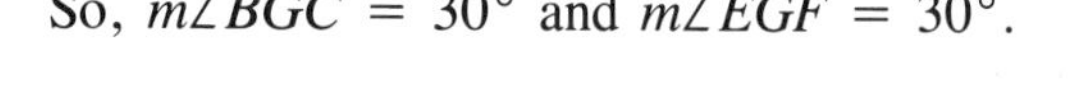

Use the figure to name the following.

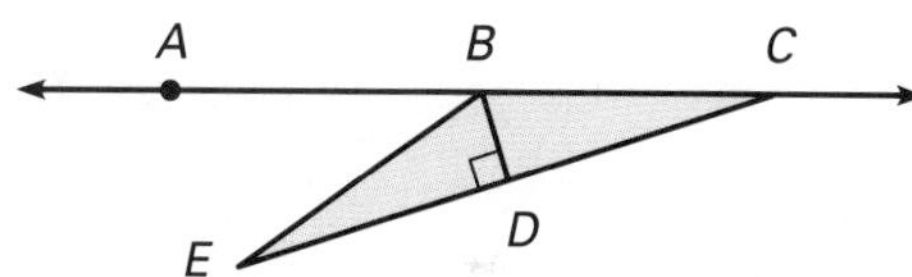

**1.** all acute angles

**2.** all obtuse angles
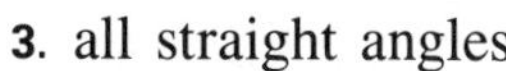
**3.** all straight angles

In the figure, $\overline{FG} \perp \overline{BE}$.

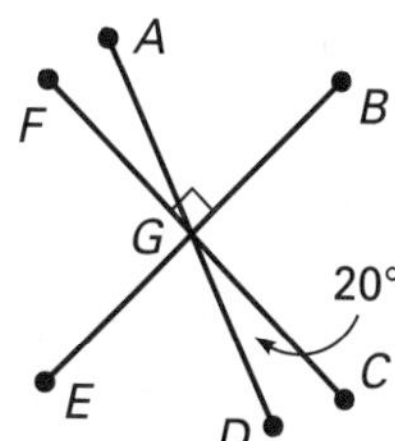


**4.** Name all right angles.
**5.** Name all pairs of vertical angles that are not right angles.
**6.** Find $m\angle AGB$.

# EXERCISES

**PRACTICE/ SOLVE PROBLEMS**

Find the measure of the complement and supplement of the angle.

**1.** $m\angle A = 14°$ **2.** $m\angle B = 45°$ **3.** $m\angle C = 30°$

Use the figure at the right for Exercises 4–10.

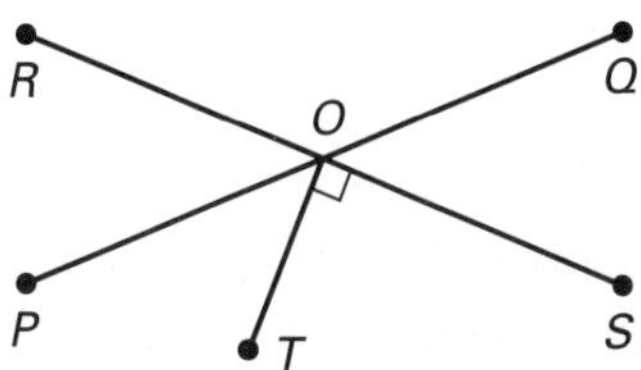


**4.** Name a pair of perpendicular segments.
**5.** Name two right angles.

6. Name two adjacent complementary angles.

7. Name all obtuse angles.

8. Name one straight angle.

9. Name two pairs of vertical angles.

10. If $O$ is the midpoint of $\overline{RS}$, then $\overline{OT}$ is the __?__ of $\overline{RS}$.

Use the diagram at the right for Exercises 11 and 12.

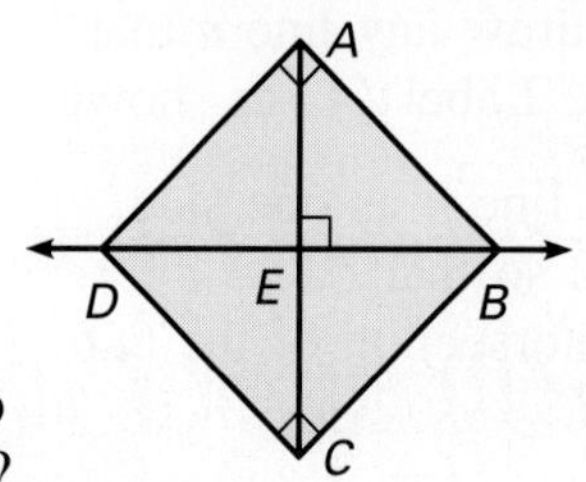


11. Name all pairs of perpendicular lines.

12. If $\overleftrightarrow{DB}$ bisects $\overline{AC}$, what two line segments are congruent?

## EXTEND/ SOLVE PROBLEMS

Shown at the right is the finished construction of a line perpendicular to $\overleftrightarrow{AB}$ from point $C$ not on $\overleftrightarrow{AB}$. Trace $\overleftrightarrow{AB}$ and $C$, and repeat the construction. Then complete the statements below outlining the steps of the construction.

13. *Step 1:* With the point of the compass at __?__, draw an arc intersecting $\overleftrightarrow{AB}$ at __?__ and __?__.

14. *Step 2:* With the point of the compass at $X$, open the compass more than half the length of __?__. Draw an arc below __?__.

15. *Step 3:* With the point of the compass at __?__ and with the same setting as in Step 2, draw an arc below $\overleftrightarrow{AB}$ that intersects the arc drawn in Step __?__. Label the point __?__.

16. *Step 4:* Draw __?__. __?__ ⊥ __?__.

## THINK CRITICALLY/ SOLVE PROBLEMS

17. The measure of $\angle A$ is the same as the measure of its complement. Find $m\angle A$.

18. $\angle C$ and $\angle D$ are supplementary angles with equal measures. Find the measure of each angle.

Classify each statement as *true* or *false*. Justify your answer.

19. Two vertical angles may also be adjacent.

20. The complement of an acute angle is an obtuse angle.

21. Two vertical angles may be complementary.

22. Show that the difference between the measures of the supplement and the complement of an angle is always 90°.

# 3-3 Parallel Lines

## EXPLORE

Trace line $m$ and point $P$ in the figure at the right. Then follow the steps.

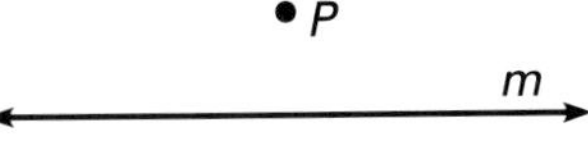


1. Through $P$ draw any line $n$ that intersects $m$. Label $\angle 1$, as shown.
2. At $P$, using line $n$ as one side, construct $\angle 2$ so that $\angle 2 \cong \angle 1$. Label the intersection of the two arcs point $X$.
3. Draw and label $\overleftrightarrow{PX}$. What seems to be true about $\overleftrightarrow{PX}$ and line $m$?

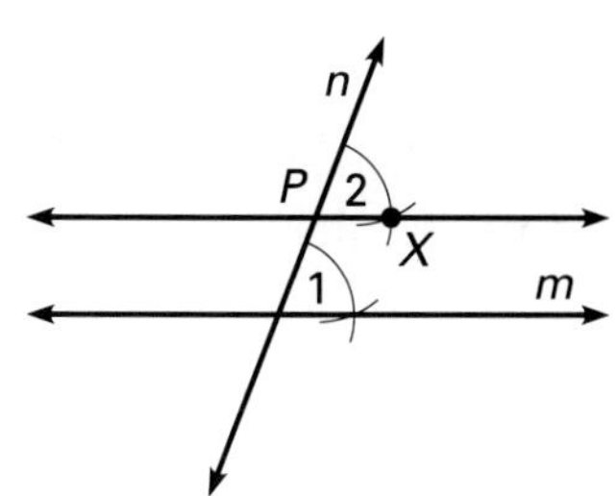


## SKILLS DEVELOPMENT

Coplanar lines may intersect in a point. Those coplanar lines that do not intersect are called **parallel lines.** In the figure, lines $k$ and $l$ intersect. Lines $m$ and $n$ are parallel lines.

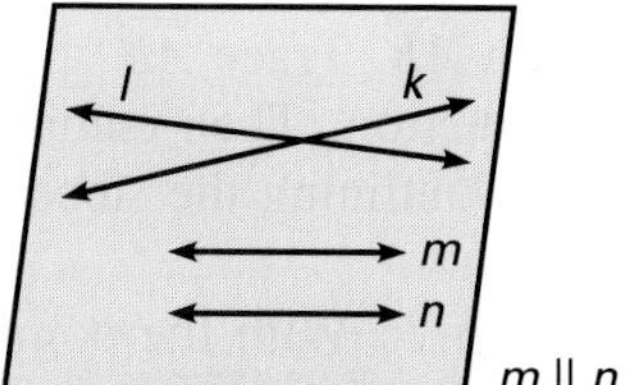


Planes that do not intersect are called parallel planes. Planes $P$ and $Q$ are **parallel planes.**

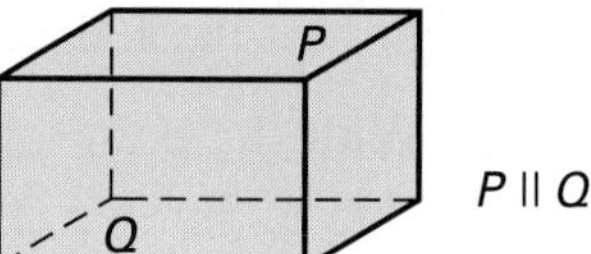


Noncoplanar lines that do not intersect and are not parallel are called **skew lines.**

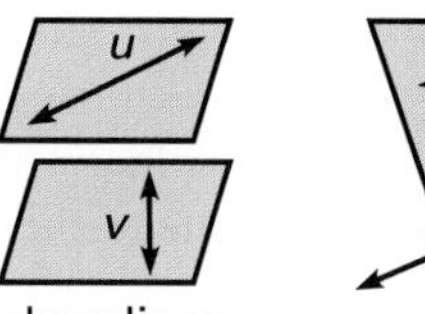


A **transversal** is a line that intersects each of two other coplanar lines in different points. The intersection of a transversal with two lines produces **interior** and **exterior angles.** In the figure at the right, transversal $t$ intersects lines $j$ and $k$.

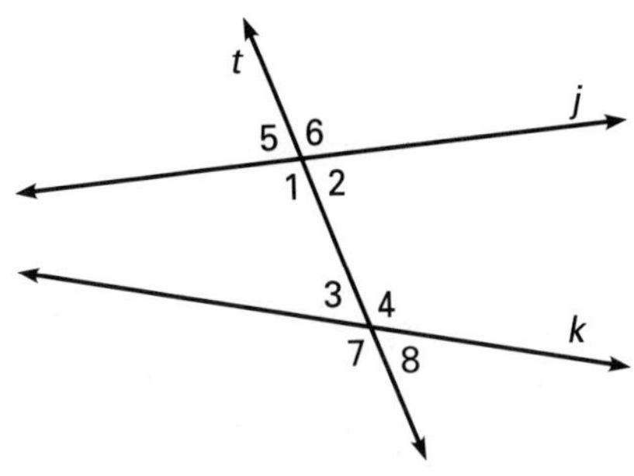

Two nonadjacent interior angles on opposite sides of the transversal are called **alternate interior angles.** The two pairs of alternate interior angles in the figure at the bottom of page 92 are ∠1 and ∠4, and ∠2 and ∠3.

Interior angles on the same side of a transversal are called **same-side interior angles.** The two pairs of same-side interior angles in the figure at the bottom of page 92 are ∠1 and ∠3, and ∠2 and ∠4.

Two nonadjacent exterior angles on opposite sides of the transversal are called **alternate exterior angles.** The two pairs of alternate exterior angles in the figure at the bottom of page 92 are ∠5 and ∠8, and ∠6 and ∠7.

Two angles in corresponding positions relative to two lines cut by a transversal are called **corresponding angles.** The four pairs of corresponding angles in the figure at the bottom of page 92 are ∠1 and ∠7; ∠2 and ∠8; ∠3 and ∠5; and ∠4 and ∠6.

## Example 1

Refer to the figure to name the following.

**a.** all pairs of alternate interior angles
**b.** all pairs of alternate exterior angles
**c.** all pairs of same-side interior angles
**d.** all pairs of corresponding angles

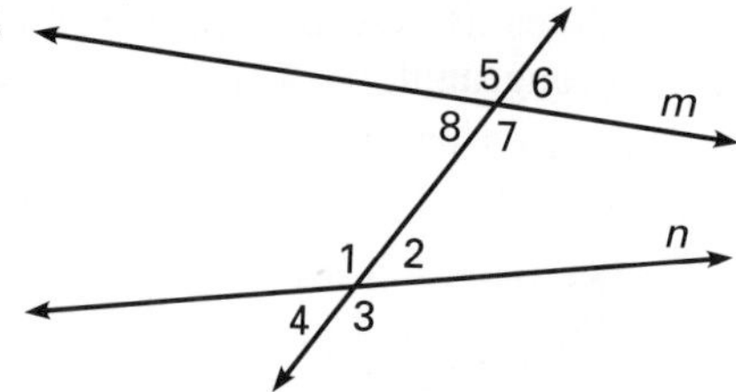


### Solution

**a.** alternate interior angles: ∠1 and ∠7, ∠2 and ∠8
**b.** alternate exterior angles: ∠5 and ∠3, ∠6 and ∠4
**c.** same-side interior angles: ∠2 and ∠7, ∠1 and ∠8
**d.** corresponding angles: ∠1 and ∠5, ∠2 and ∠6, ∠3 and ∠7, ∠4 and ∠8 ◄

When a transversal intersects two parallel lines, the angles formed have special relationships.

POSTULATE 5: **If two parallel lines are cut by a transversal, then corresponding angles are congruent.**

From this postulate, the following statements can be shown to be true.

5A: If two parallel lines are cut by a transversal, then alternate interior angles are congruent.

5B: If two parallel lines are cut by a transversal, then alternate exterior angles are congruent.

POSTULATE 6: **If two lines are cut by a transversal so that corresponding angles are congruent, then the lines are parallel.**

From this postulate, the following statements can be shown to be true.

**6A:** If two lines are cut by a transversal so that alternate interior angles are congruent, then the lines are parallel.

**6B:** If two lines are cut by a transversal so that alternate exterior angles are congruent, then the lines are parallel.

## Example 2

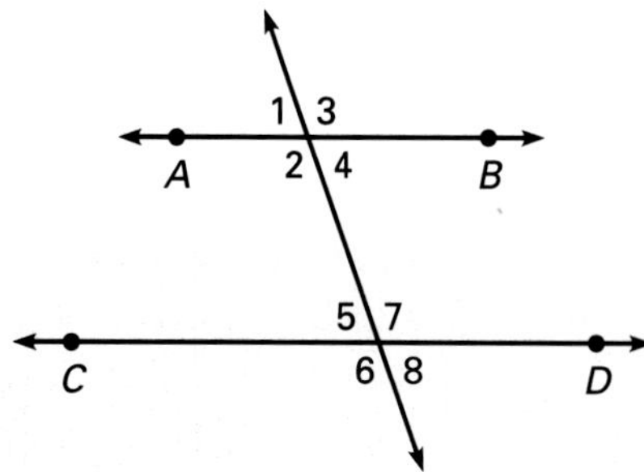


In the figure at the right, $\overleftrightarrow{AB} \parallel \overleftrightarrow{CD}$. Name the postulate or statement that gives the reason why each statement is true.

**a.** $\angle 2 \cong \angle 6$ **b.** $\angle 4 \cong \angle 5$
**c.** $\angle 1 \cong \angle 8$

**Solution**

**a.** $\angle 2$ and $\angle 6$ are corresponding angles. Postulate 5: If two parallel lines are cut by a transversal, then corresponding angles are congruent.

**b.** $\angle 4$ and $\angle 5$ are alternate interior angles. Statement 5A: If two parallel lines are cut by a transversal, then alternate interior angles are congruent.

**c.** $\angle 1$ and $\angle 8$ are alternate exterior angles. Statement 5B: If two parallel lines are cut by a transversal, then alternate exterior angles are congruent. ◄

## TRY THESE

Refer to the figure to name the following.

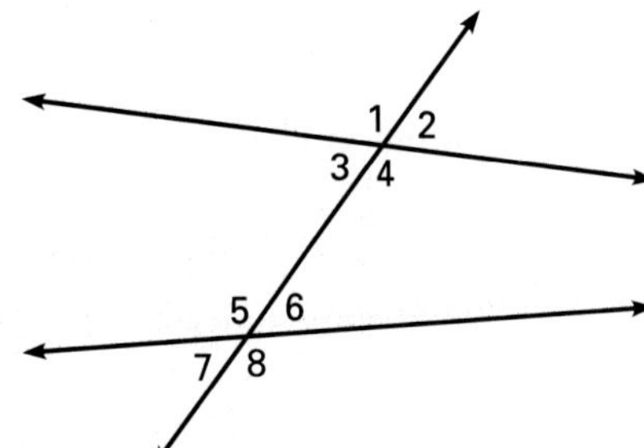


1. a pair of alternate interior angles
2. a pair of alternate exterior angles
3. a pair of same-side interior angles
4. all pairs of corresponding angles

In the figure at the right, $\overleftrightarrow{PQ} \parallel \overleftrightarrow{RS}$. Name the postulate or statement that gives the reason why each statement is true.

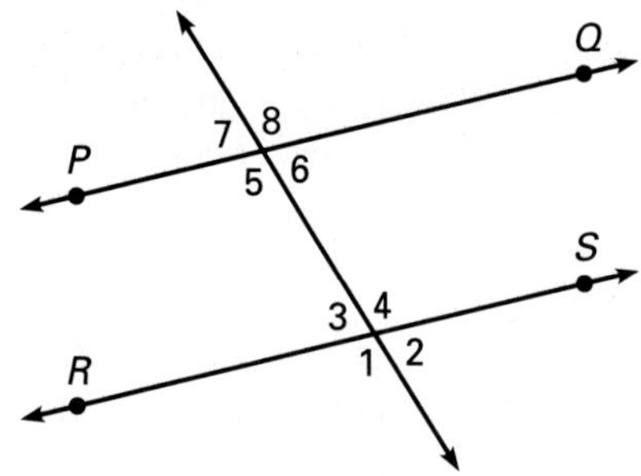


5. $\angle 5 \cong \angle 1$

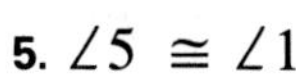

6. $\angle 8 \cong \angle 1$
7. $\angle 4 \cong \angle 5$

# EXERCISES

## PRACTICE/ SOLVE PROBLEMS

Refer to the figure to classify each pair of angles named.

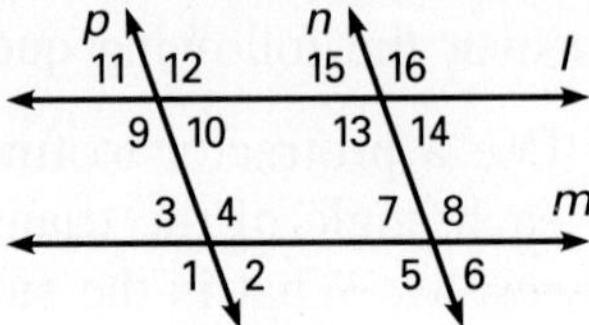


1. $\angle 1$ and $\angle 9$
2. $\angle 9$ and $\angle 3$
3. $\angle 1$ and $\angle 5$
4. $\angle 3$ and $\angle 10$
5. $\angle 15$ and $\angle 6$
6. $\angle 8$ and $\angle 13$
7. $\angle 5$ and $\angle 16$
8. $\angle 7$ and $\angle 13$

Find the measure of each angle. $\overleftrightarrow{PQ} \parallel \overleftrightarrow{RS}$.

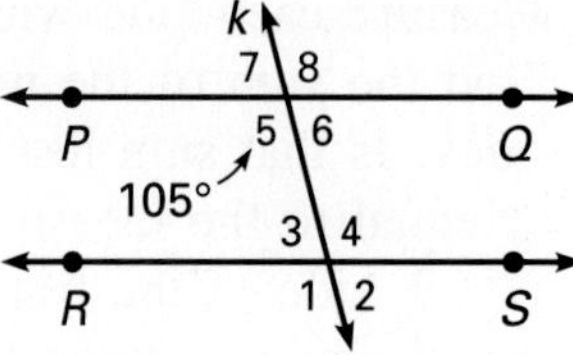


9. $\angle 1$
10. $\angle 2$
11. $\angle 3$
12. $\angle 4$
13. $\angle 6$
14. $\angle 7$
15. $\angle 8$

For each figure, write *P* if the lines cut by the transversal are parallel. Write *N* if they are not parallel.

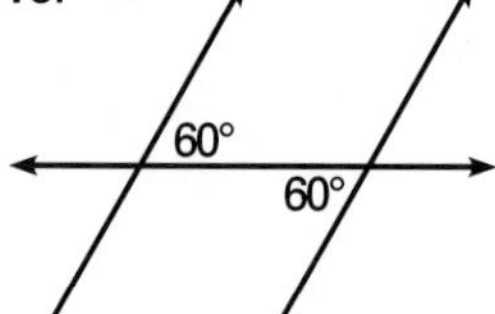


17.

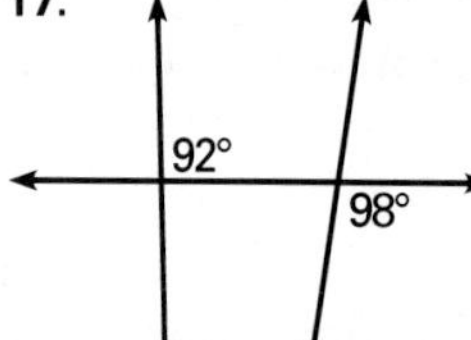


18.

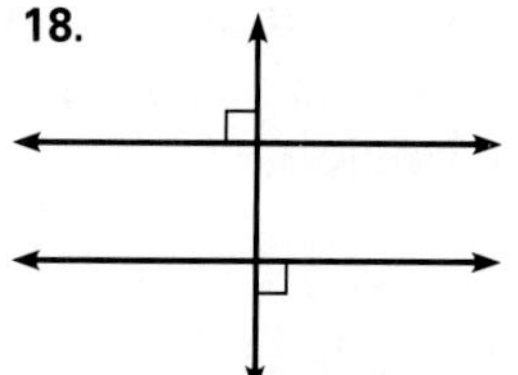

### CONNECTIONS

The following properties are true for any real numbers *a*, *b*, and *c*.

**Reflexive Property:**
$a = a$

**Symmetric Property:**
If $a = b$, then $b = a$.

**Transitive Property:**
If $a = b$ and $b = c$, then $a = c$.

**Substitution Property:**
If $a = b$, then $b$ may be substituted for $a$ in any equation or inequality.

## EXTEND/ SOLVE PROBLEMS

Find the value of the unknown measures.

19.

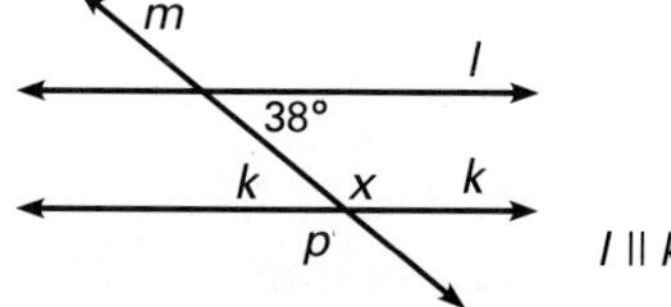


20.

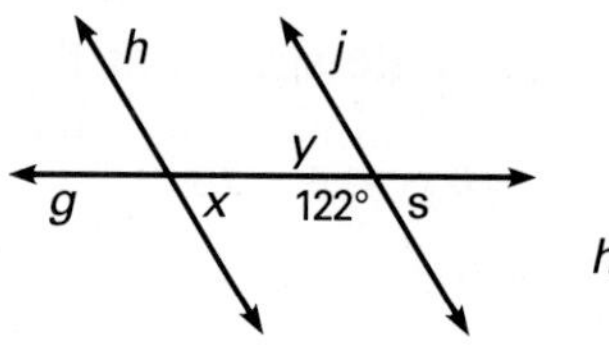


Classify each statement as *true* or *false*.

21. Two skew lines never intersect.
22. Two lines in the same plane are sometimes skew lines.
23. If a transversal is perpendicular to one of two parallel lines, then it is perpendicular to the other line.

## THINK CRITICALLY/ SOLVE PROBLEMS

24. In the figure at the right, $\overleftrightarrow{AB} \parallel \overleftrightarrow{CD}$. How could you show that alternate interior angles $\angle 1$ and $\angle 3$ are congruent, using Postulate 5 and other facts about angle measures?

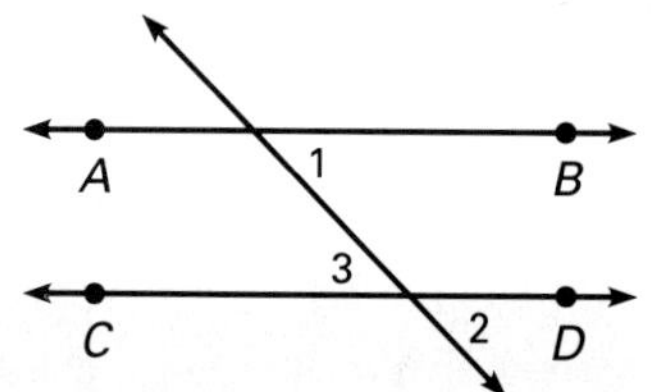

# 3-4 Triangles

**EXPLORE/ WORKING TOGETHER**

Trace the triangle shown at the right. Answer the following questions.

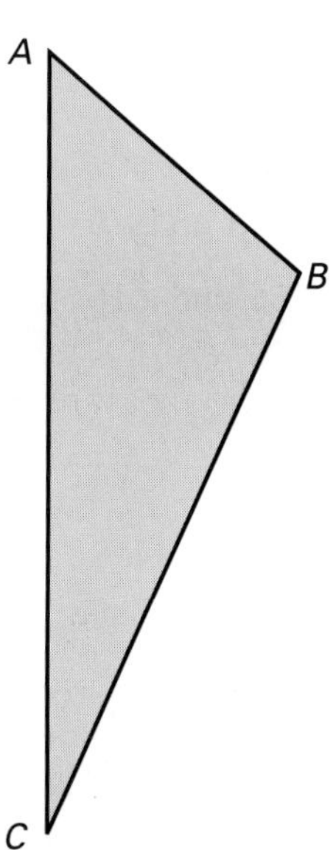


**a.** Use a protractor to find the measure of each angle of the triangle. Record your results. What is the sum of the three angles?

**b.** Measure each side with a metric ruler. Find the sum of the measures of any two sides. Is that sum less than, greater than, or equal to the measure of the third side?

**c.** Which side of the triangle is longest? Which angle is largest?

**d.** Which side of the triangle is shortest? Which angle is smallest?

**e.** Draw a triangle of your own. Have someone answer Exercises **a–d** for your triangle.

**SKILLS DEVELOPMENT**

A **triangle** is the closed plane figure formed by three line segments joining three noncollinear points. Each point is called a **vertex.** Each vertex names an **angle** of the triangle. Each of the line segments that join the vertices is called a side of the triangle.

For the triangle shown at the right,
Vertices: Points $P$, $Q$, and $R$
Sides: $\overline{PQ}$, $\overline{QR}$, $\overline{RP}$
Angles: $\angle P$, $\angle Q$, $\angle R$

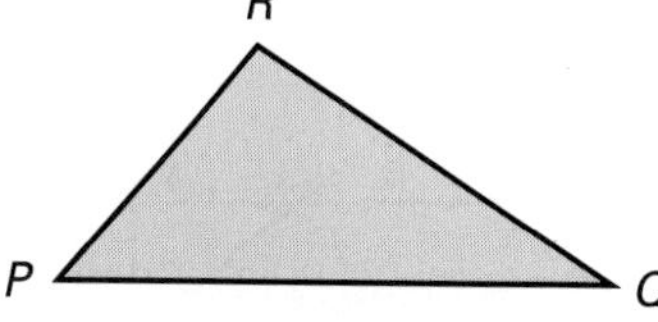

**READING MATH**

The word *scalene* comes from a Greek word meaning "uneven." Trace the origins and original meanings of these words: *isosceles, obtuse, parallel.*

A triangle is named by its vertices. The figure is named triangle $PQR$. This is symbolized as $\triangle PQR$.

Two sides of a triangle that have the same length are said to be **congruent sides.** Two angles of a triangle that have the same measure are said to be **congruent angles.** A triangle may be classified by its number of congruent sides.

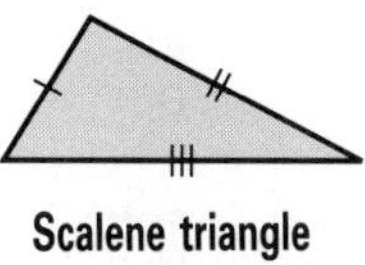

**Scalene triangle**
no congruent sides

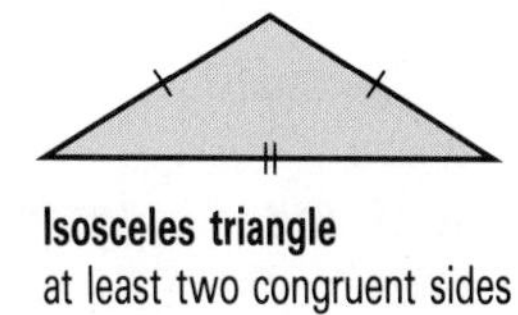

**Isosceles triangle**
at least two congruent sides

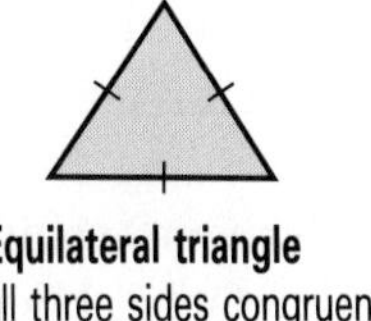

**Equilateral triangle**
all three sides congruent

A triangle may also be classified by its angles.

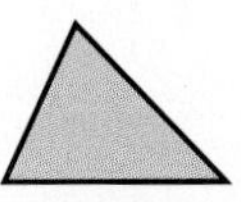

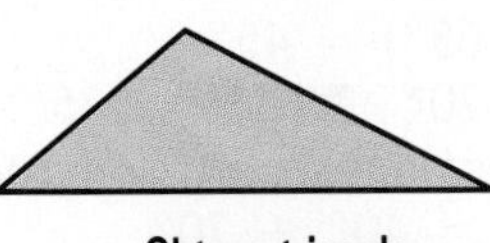

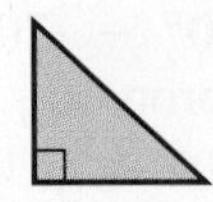

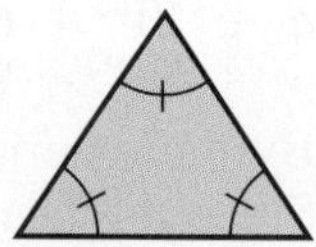

**Acute triangle**
three acute angles

**Obtuse triangle**
one obtuse angle

**Right triangle**
one right angle

**Equiangular triangle**
all three angles congruent

In the Explore activity, you observed several properties of triangles. These properties hold for all triangles.

- ► **The sum of the angles of a triangle is 180°.**
- ► **The sum of the lengths of any two sides of a triangle is greater than the length of the third, or remaining, side.**
- ► **In a triangle, the longest side is opposite the largest angle, and the shortest side is opposite the smallest angle.**

There are other important properties of triangles.

In the figure at the right, $\angle LNR$ is called an **exterior angle** of $\triangle LMN$. Notice that $m\angle LNR = m\angle L + m\angle M$. The figure illustrates the following property.

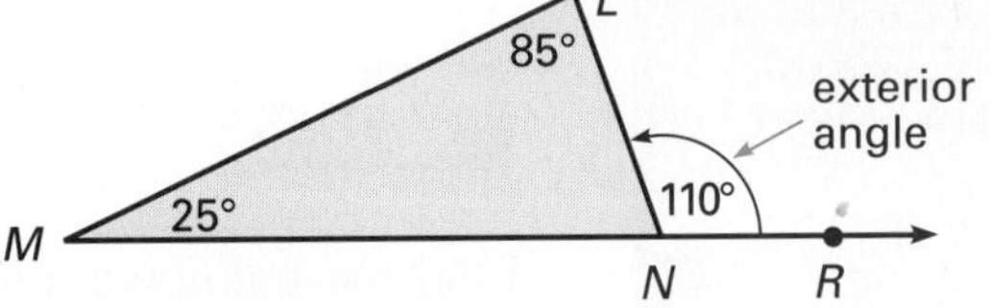


- ► **If one side of a triangle is extended, then the exterior angle so formed is equal to the sum of the two remote interior angles of the triangle.**

The following property is true for all isosceles triangles.

- ► **If two sides of a triangle are congruent, then the angles opposite those sides are congruent.**

Another way of stating this property is to say that *the base angles of an isosceles triangle are congruent*.

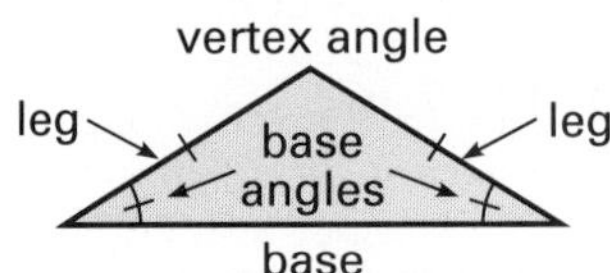


You can use these properties to determine angle measures of triangles. You can also determine whether it is possible to construct a triangle with sides of given lengths.

### Example 1

Find the values of $a$ and $b$.

**a**

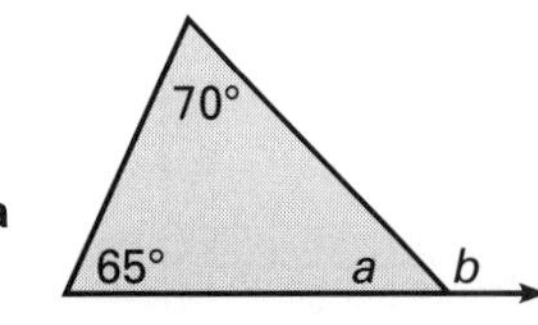


**b.**

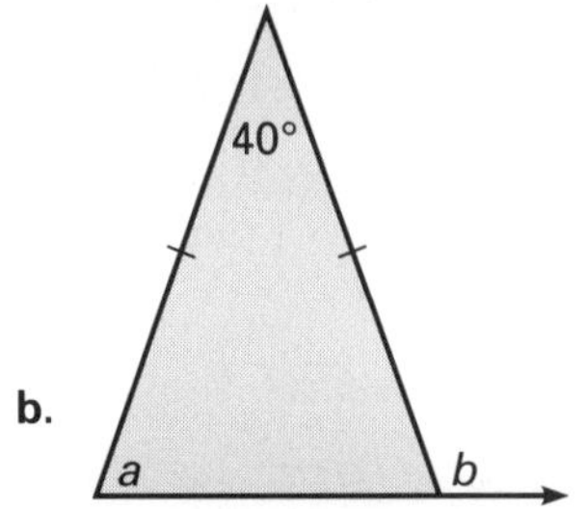

## Solution

a. $a + 70° + 65° = 180°$

$$a = 180° - (70° + 65°) = 45°$$

By the exterior angle property, $b = 70° + 65° = 135°$.

b. Since the triangle is isosceles, the base angles are congruent.

$$a + a + 40° = 180° \qquad b = 40° + 70°$$
$$2a = 180° - 40° \qquad b = 110° \quad ◄$$
$$2a = 140°$$
$$a = 70°$$

**COMPUTER**

This program tells you which three line segments of given measures can be used to form a triangle.

```
10 READ A, B, C
20 IF A + B <= C THEN 70
30 IF A + C <= B THEN 70
40 IF B + C <= A THEN 70
50 PRINT "THE SEGMENTS
   CAN FORM A TRIANGLE."
60 GOTO 10
70 PRINT "THE SEGMENTS
   CANNOT FORM A
   TRIANGLE."
80 GOTO 10
90 DATA 5, 6, 7, 13, 3, 5,
   12, 6, 8
```

Run the program, then modify line 90 to include your own data entries. Have a classmate run your program.

## Example 2

Tell whether it is possible to have a triangle with sides of these lengths.

a. 14, 9, 6 b. 8, 5, 3

## Solution

a. It is possible because $14 + 9 = 23$, $9 + 6 = 15$, and $14 + 6 = 20$. In each case, the sum of any two sides is greater than the third side.

b. It is not possible because $3 + 5 = 8$, which is not greater than the length of the remaining side, which is 8. ◄

## TRY THESE

Find the unknown angle measure in each figure.

1.

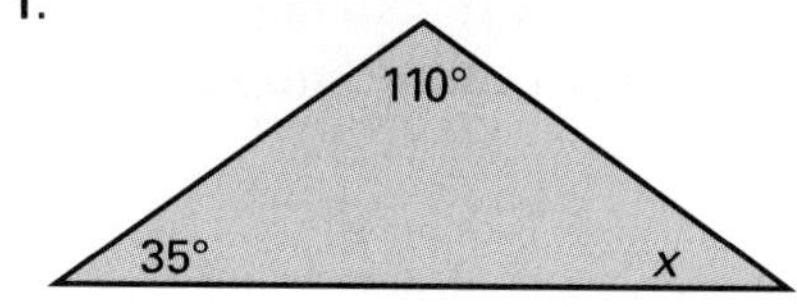


2.

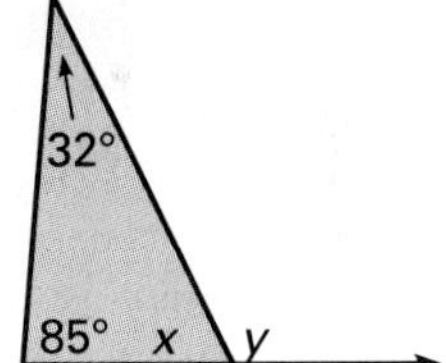


Tell whether it is possible to have a triangle with sides of the given lengths.

3. 12, 7, 6 4. 13, 7, 5 5. 11, 14, 16

# EXERCISES

## PRACTICE/ SOLVE PROBLEMS

In figure *ABCD*, classify each triangle by its sides.

1. $\triangle ABC$
2. $\triangle BDC$
3. $\triangle AEB$
4. $\triangle BCE$

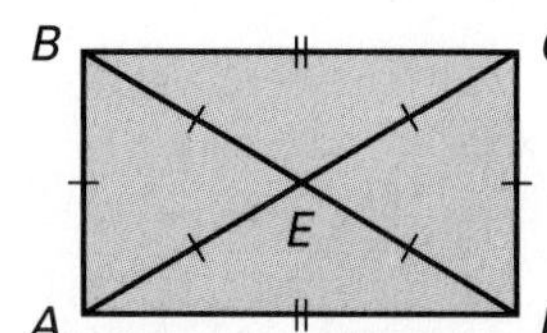

In the figure, $\overleftrightarrow{AB} \parallel \overleftrightarrow{CD}$ and $\overleftrightarrow{AD} \perp \overleftrightarrow{BC}$. Find the measure of each angle named.

**5.** $\angle 1$ **6.** $\angle 2$

**7.** $\angle 3$ **8.** $\angle 4$

**9.** $\angle 5$ **10.** $\angle 6$

**11.** $\angle 7$ **12.** $\angle 8$

**13.** $\angle 9$ **14.** $\angle 10$

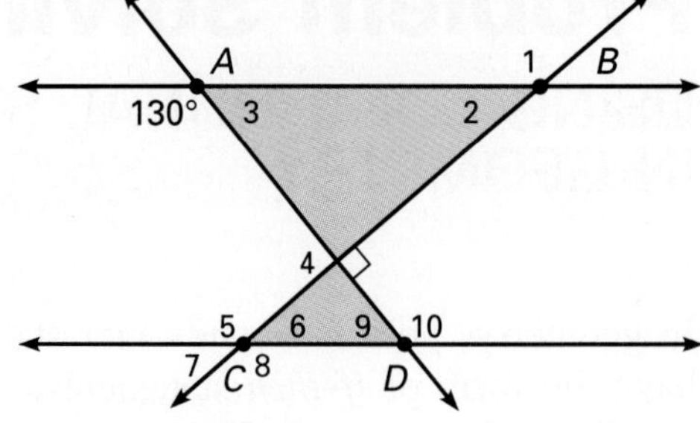


Find the unknown angle measure in each figure.

**15.**

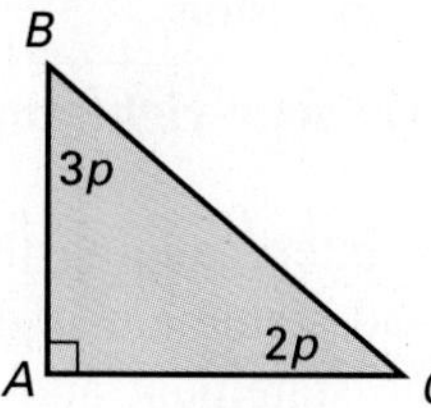


**16.**

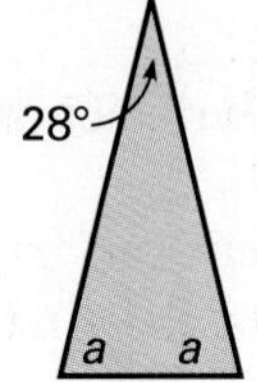


**17.**

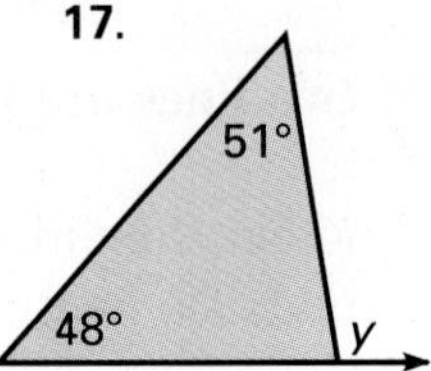


Complete each statement with *always, sometimes,* or *never.*

**18.** An isosceles triangle is __?__ scalene.

**19.** A right triangle is __?__ isosceles.

**20.** An exterior angle of a triangle is __?__ acute.

**21.** Isosceles triangles are __?__ acute.

**22.** In a right triangle, the acute angles are __?__ complementary.

## EXTEND/ SOLVE PROBLEMS

**TALK IT OVER**

How could you calculate the measure of each angle of an equilateral triangle? What is that measure?

In the figure, $m\angle 1 = x°$.

**23.** Write an expression for $m\angle 3$.

**24.** Write an expression for $m\angle 5$.

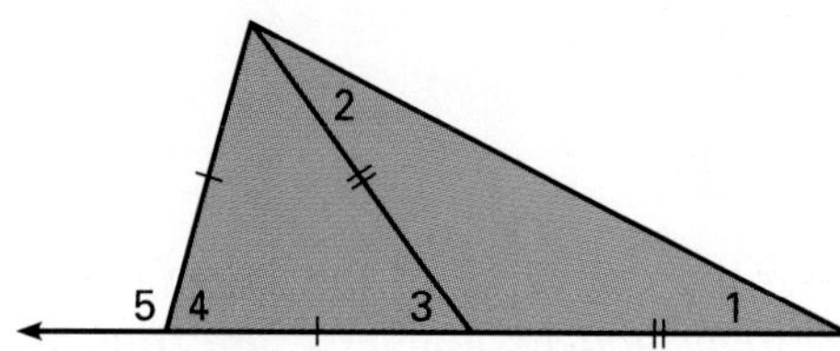


## THINK CRITICALLY/ SOLVE PROBLEMS

**25.** In the figure, line $m$ is drawn through point $C$ as parallel to $\overline{AB}$. How can you use the drawing to show that the sum of the angles of a triangle is equal to 180°?

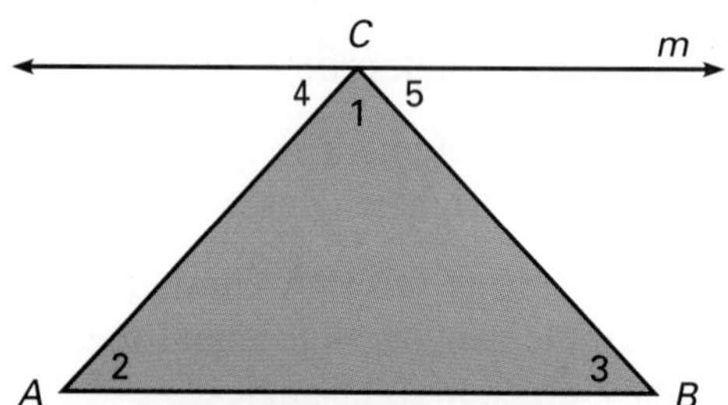

# 3-5 Problem Solving Skills:

## USING CONDITIONAL STATEMENTS IN GEOMETRY

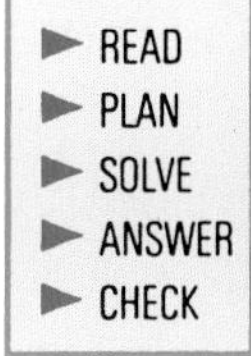


In geometry, postulates and other statements derived from them often have the form of *if–then* statements. An *if–then* statement is called a **conditional statement.** It has two parts, *hypothesis* and *conclusion*.

If **two lines are perpendicular** (hypothesis), then **the lines form four right angles.** (conclusion)

Another conditional statement can be formed by interchanging the hypothesis and conclusion of the first. The second statement is called the **converse** of the original. The converse of the statement above is as follows.

If **two lines form four right angles,** then **the lines are perpendicular.**

A conditional statement may be either true or false. The only way to show that a conditional is false is to find a **counterexample,** an instance that satisfies the hypothesis, but not the conclusion of the conditional statement. Just one counterexample proves that the conditional statement is false.

The fact that a conditional statement is true is no guarantee that its converse is true.

### PROBLEM

Write the converse of the statement. Then determine whether the statement and its converse are true or false. If false, give a counterexample.

If two angles are right angles, then the angles are congruent.

### SOLUTION

The converse of the statement is as follows.

If two angles are congruent, then the angles are right angles.

The original statement is true. Since a right angle has a measure of 90°, any two right angles will have equal measures.

The converse is false. A counterexample would be any two angles, for example, angles with equal measures of 30°. They are congruent, but they are not right angles.

**Managing Editor:** Eve Lewis
**Developmental Editor:** Judith A. Witt
**Marketing Manager:** Carol Ann Dana
**Mathematics Consultant:** William E. Royalty
**Coordinating Editor:** Patricia Matthews Boies
**Production Editor:** Martha G. Conway
**Cover Photo Photographer:** Wayne Sorce © June 1988
Audio Kinetic Sculptures by George Rhoads

**About the Math Matters Cover**

Artist George Rhoads' "audio kinetic sculptures" appear in shopping centers, malls, terminals, and museums throughout North America. The mechanisms in the sculptures are visible and easy to understand, but the mathematical design provides infinite variety. Crowds gather to watch the acrobatic activity and hear the clanging noises triggered by randomly propelled balls traveling down twisting skeletal ramps.



Authorized adaptation of a work first published by Nelson Canada,
A Division of Thomson Canada Limited,
1120 Birchmount Road,
Scarborough, Ontario
M1K 5G4



ISBN: 0-538-61112-X

Library of Congress Catalog Card Number: 91-62724

3 4 5 6 7 KI 98 97 96 95 94

Printed in the United States of America

Statements that do not have the form of conditional statements can sometimes be rephrased as *if–then* statements. This is possible as long as the two statements have the same meaning.

**PROBLEM SOLVING TIP**

The first thing to do in translating sentences into *if–then* statements is to determine which part you will use for the conclusion and which part for the hypothesis.

## PROBLEM

Rephrase this statement as an *if–then* statement.

All right angles have a measure of 90°.

## SOLUTION

The statement means that it is not possible for an angle to be a right angle and have a measure other than 90°. This is the same as the following statement.

If an angle is a right angle, then that angle has a measure of 90°.

# PROBLEMS

Write the converse of each statement. Then determine whether the statement and its converse are true or false. If false, give a counterexample.

1. If a triangle has two congruent sides, then it is isosceles.
2. If two supplementary angles are congruent, then the angles are right angles.
3. If two angles are vertical angles, then the angles are congruent.
4. If a figure is an equilateral triangle, then the figure is an isosceles triangle.
5. If two lines do not intersect, then the two lines are parallel.

Rephrase each statement as an *if–then* statement.

6. Two angles whose measures have a sum of 90° are complementary.
7. Collinear points lie on the same line.
8. Two distinct planes intersect in a line.
9. A rhombus has four congruent sides.
10. Two lines that are skew do not intersect.
11. Two coplanar lines intersect unless they are parallel.
12. A triangle is equilateral *only* if all three sides are congruent.

# 3-6 Congruent Triangles

## EXPLORE

You can construct a triangle, given three segments, provided that the sum of the measures of any two segments is greater than the measure of the third.

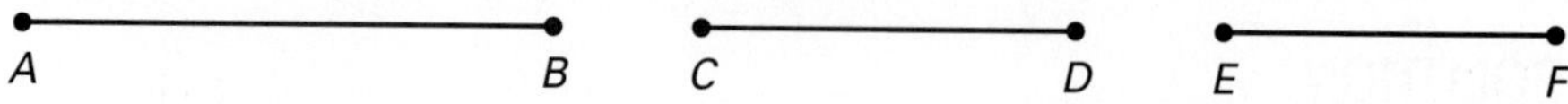


1. With a straightedge, draw line $m$.
2. Choose point $X$ on line $m$. Set a compass for the length of $\overline{AB}$. With $X$ as the center, draw an arc that intersects line $m$. Label that point $Y$.

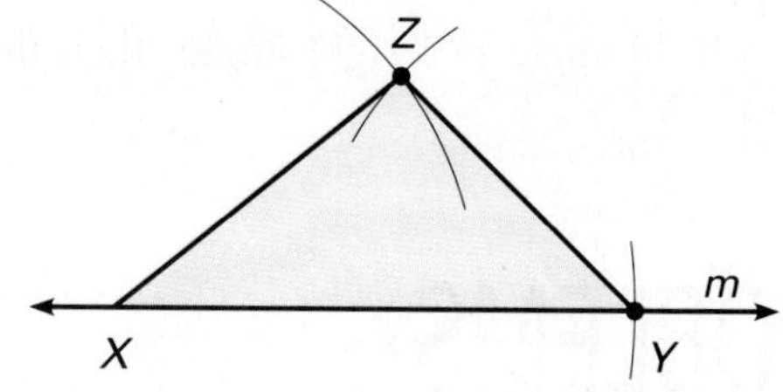


3. Set the compass for the length of $\overline{CD}$. With $X$ as the center, draw an arc of length equal to $\overline{CD}$. Do not let the arc intersect line $m$.
4. Set the compass for the length of $\overline{EF}$. With $Y$ as the center, draw an arc that intercepts the arc drawn in Step 3. Label that point $Z$.
5. Connect points $X$ and $Z$ and points $Z$ and $Y$.

Triangle $XYZ$ is the required triangle.

## SKILLS DEVELOPMENT

If you trace $\triangle KLM$ and slide it over $\triangle PQR$, the vertices of one triangle match, or correspond to, the vertices of the other and the angles and sides of one triangle match, or correspond to, the angles and sides of the other.

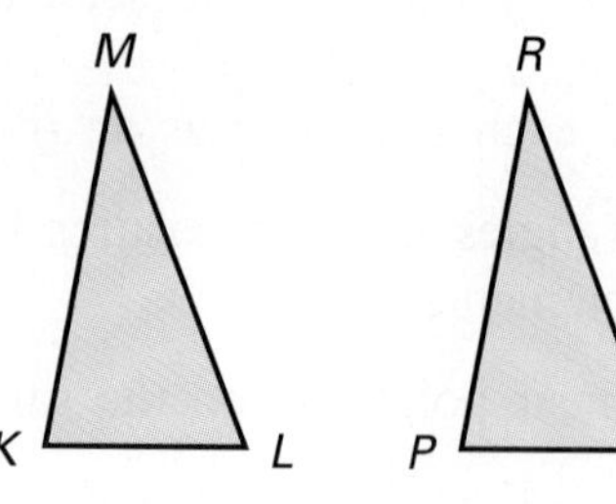


| Corresponding sides | Corresponding angles |
|---|---|
| $\overline{KL} \leftrightarrow \overline{PQ}$ | $\angle K \leftrightarrow \angle P$ |
| $\overline{LM} \leftrightarrow \overline{QR}$ | $\angle L \leftrightarrow \angle Q$ |
| $\overline{MK} \leftrightarrow \overline{RP}$ | $\angle M \leftrightarrow \angle R$ |

Corresponding parts have equal measures and therefore are congruent.

## Example 1

Name the congruent sides and angles of $\triangle KLM$ and $\triangle PQR$.

**Solution**

$$\overline{KL} \cong \overline{PQ} \qquad \angle K \cong \angle P$$
$$\overline{LM} \cong \overline{QR} \qquad \angle L \cong \angle Q$$
$$\overline{MK} \cong \overline{RP} \qquad \angle M \cong \angle R$$

Triangles *KLM* and *PQR* are congruent. That is,

$\triangle KLM \cong \triangle PQR$ ← Read "Triangle *KLM* is congruent to triangle *PQR*." ◄

Two triangles are **congruent** if their vertices can be matched so that corresponding parts of the triangles are congruent.

To show that two triangles are congruent, you need only show that at least three of the corresponding parts are congruent. The following congruence postulates state which corresponding parts must be congruent. (In the diagrams, an equal number of hatchmarks on any two corresponding parts indicates that the parts are congruent.)

**Side–Side–Side Postulate (SSS):**

If three sides of one triangle are congruent to three corresponding sides of another triangle, then the triangles are congruent.

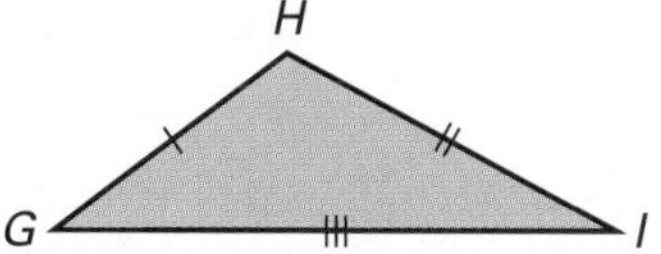


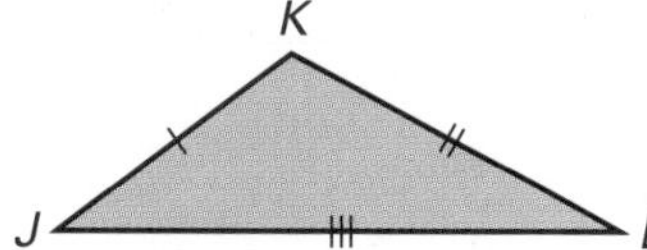


**Side–Angle–Side Postulate (SAS):**

If two sides and the included angle of one triangle are congruent to two corresponding sides and the included angle of another triangle, then the triangles are congruent.

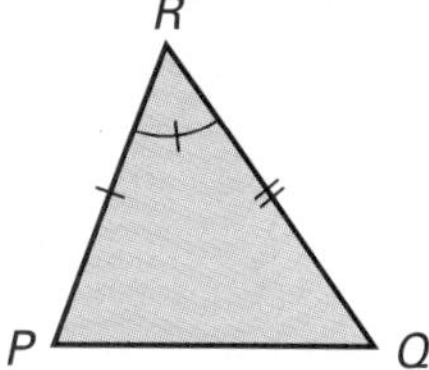


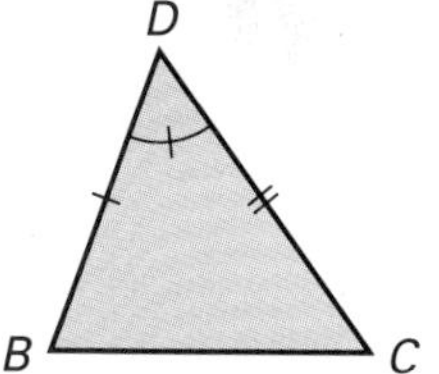


The angle referred to must be the angle that is formed by the two congruent sides. It is called the **included angle.** In $\triangle PQR$, $\angle R$ is *included* between $\overline{PR}$ and $\overline{RQ}$. In $\triangle BCD$, $\angle D$ is *included* between $\overline{BD}$ and $\overline{CD}$.

**Angle–Side–Angle Postulate (ASA):**

If two angles and the included side of one triangle are congruent to two corresponding angles and the included side of another triangle, then the triangles are congruent.

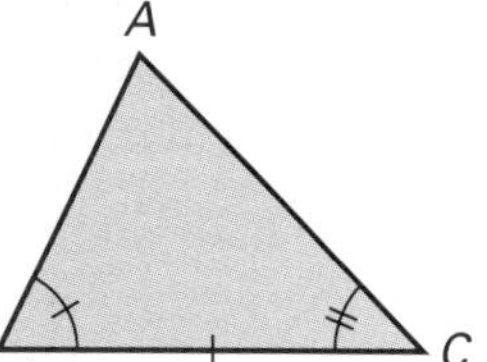


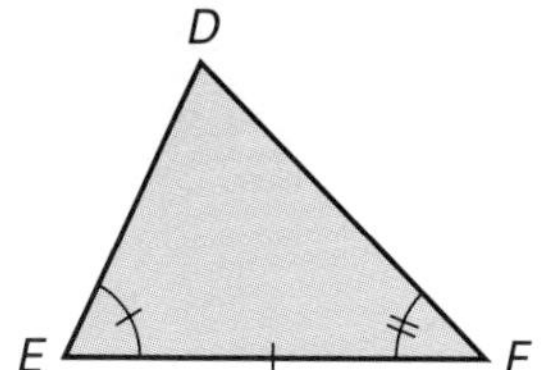


The side referred to must be the side included between the two corresponding congruent angles. It is called the **included side.** In $\triangle ABC$, $\overline{BC}$ is *included* between $\angle C$ and $\angle B$. In $\triangle DEF$, $\overline{EF}$ is *included* between $\angle E$ and $\angle F$.

## Example 2

If three angles of one triangle are congruent to three angles of another triangle, are the triangles congruent? Explain.

For each pair of triangles, state whether the pair is congruent by SSS, SAS, or ASA. If the pair is not congruent, write *not congruent*. Corresponding congruent parts are marked.

**a.** 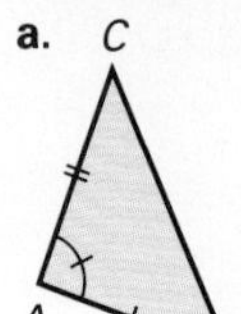

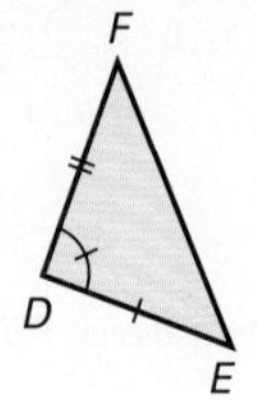


**b.** 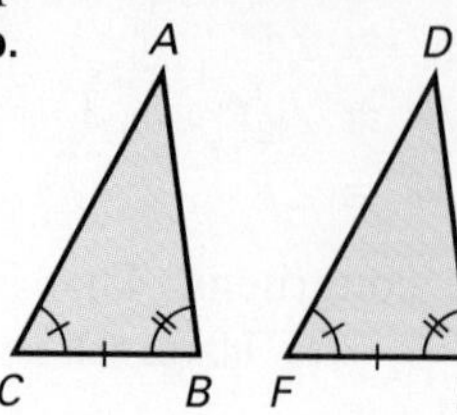


**c.** 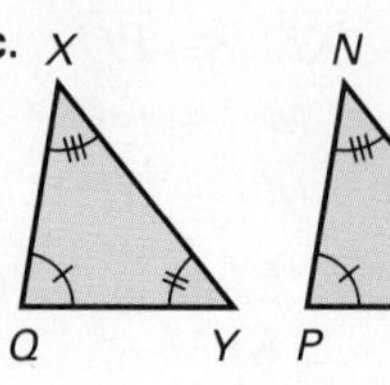


### Solution

**a.** $\triangle ABC \cong \triangle DEF$; SAS

**b.** $\triangle ABC \cong \triangle DEF$; ASA

**c.** not congruent ◄

## TRY THESE

1. In the figure at the right, $\triangle ABC \cong \triangle DEF$. Name the corresponding parts that are congruent.

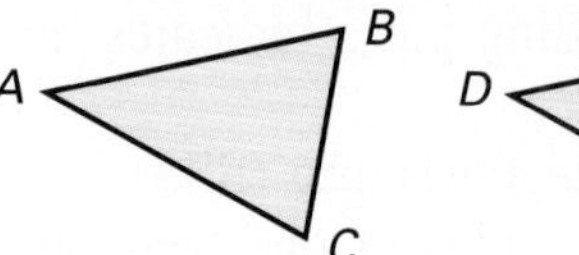

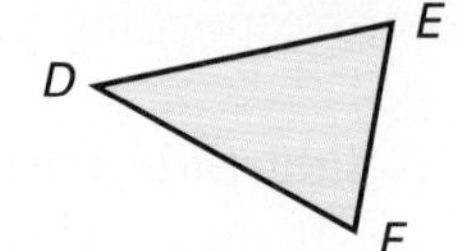


PROBLEM SOLVING TIP

How does the reflexive property help you to answer Exercise 3?

If each pair of angles is congruent, name the postulate that applies: SSS, SAS, or ASA. If the pair is not congruent, write *not congruent*.

2. 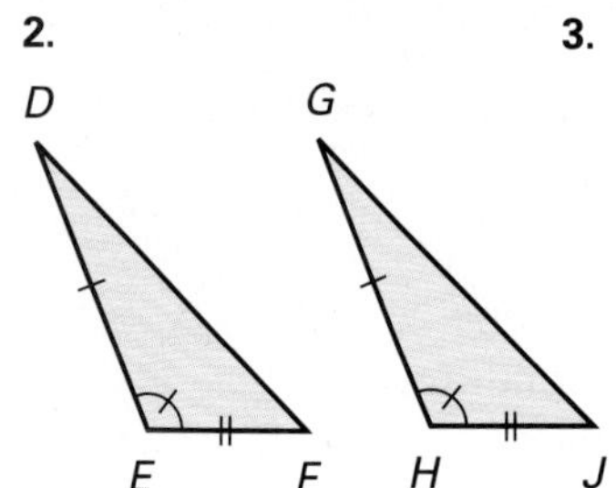


3. 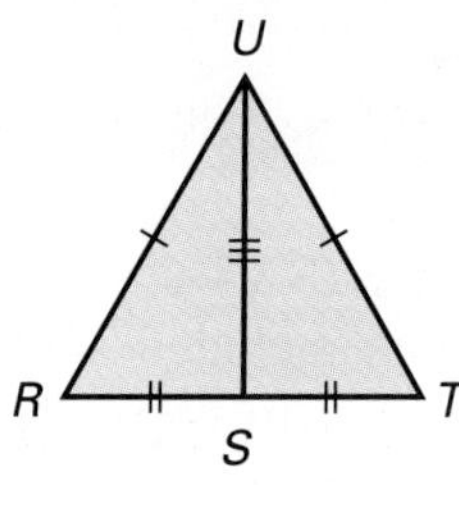


4. 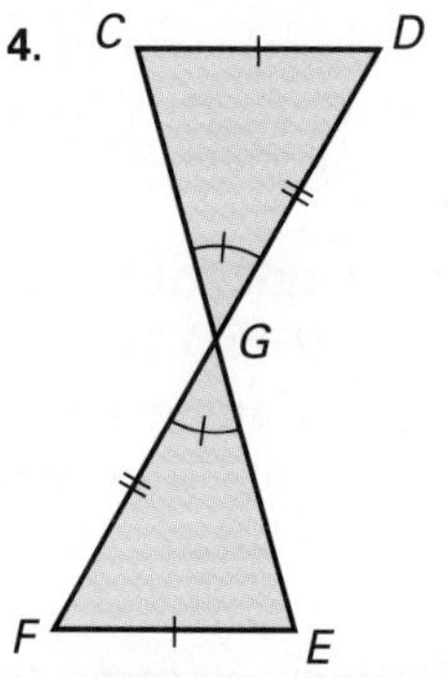


# EXERCISES

## PRACTICE/ SOLVE PROBLEMS

For Exercises 1–4, refer to the figure at the right.

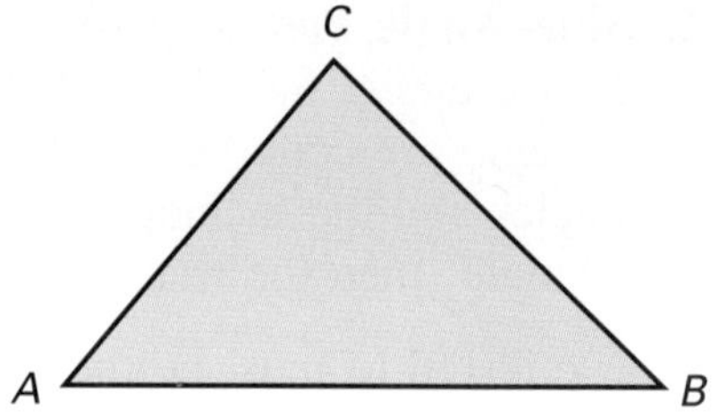


1. Which angle is included between $\overline{AB}$ and $\overline{BC}$?
2. Which angle is included between $\overline{AC}$ and $\overline{AB}$?
3. Which side is included between $\angle A$ and $\angle B$?
4. Which side is included between $\angle B$ and $\angle C$?

In Exercises 5–10, state whether the pair is congruent by SAS, ASA, or SSS.

5. 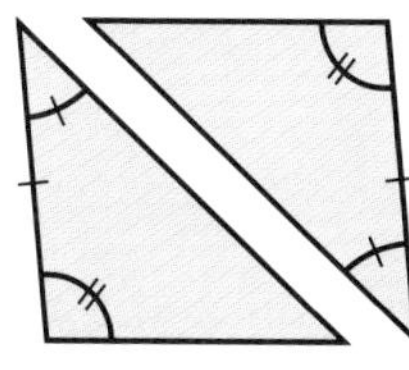

6. 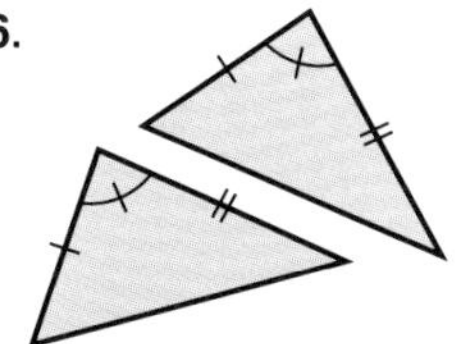

7. 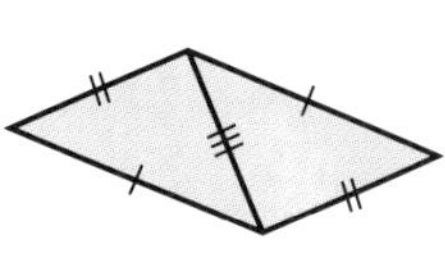

8. 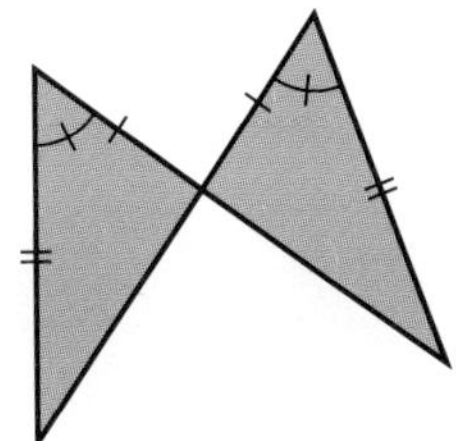

9. 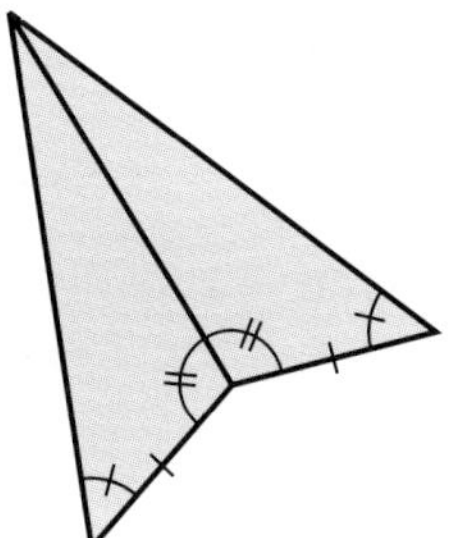

10. 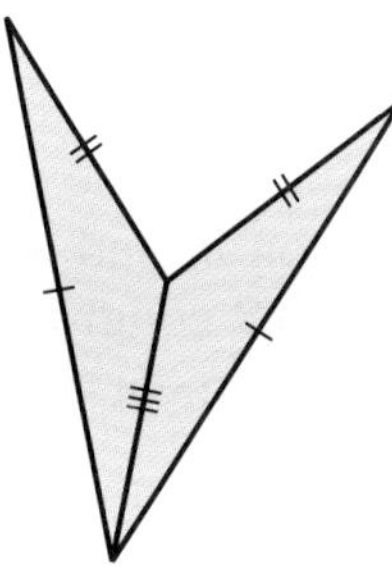

## EXTEND/ SOLVE PROBLEMS

Use the figure at the right for Exercises 11–14. Answer the questions in Exercises 11–13 with one of the following lettered statements.

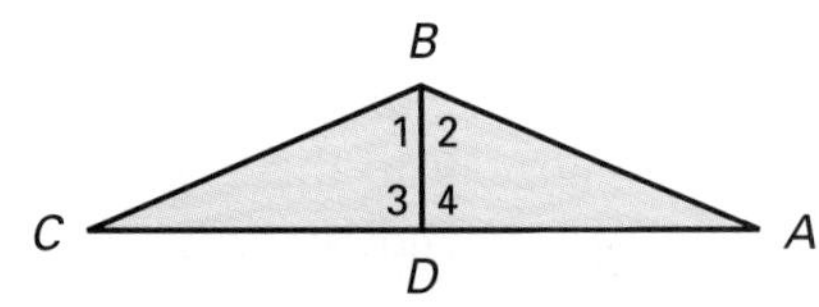


**a.** $\angle 1 \cong \angle 2$ **b.** $\angle 3 \cong \angle 4$ **c.** $\angle A \cong \angle C$ **d.** $\overline{AB} \cong \overline{CB}$ **e.** $\overline{AD} \cong \overline{DC}$

**11.** Suppose that $\overline{BD} \perp \overline{AC}$. Which statement would be true?

**12.** Suppose that $\overline{BD}$ bisects $\overline{AC}$. Which statement would be true?

**13.** Suppose that $\overline{BD}$ bisects $\angle ABC$. Which statement would be true?

**14.** Suppose that $\overline{BD} \perp \overline{AC}$ and bisects $\overline{AC}$. Could you show that $\triangle ABD \cong \triangle CBD$? Explain.

**15.** How can you show that the triangles in the figure at the right are congruent?

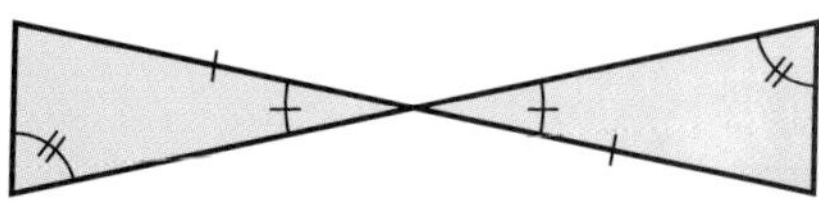

## THINK CRITICALLY/ SOLVE PROBLEMS

For Exercises 16–18, refer to the figure at the right. Triangle $ABC$ is isosceles. $\overline{AD} \perp \overline{BC}$ and $\overline{AD}$ bisects $\overline{BC}$.

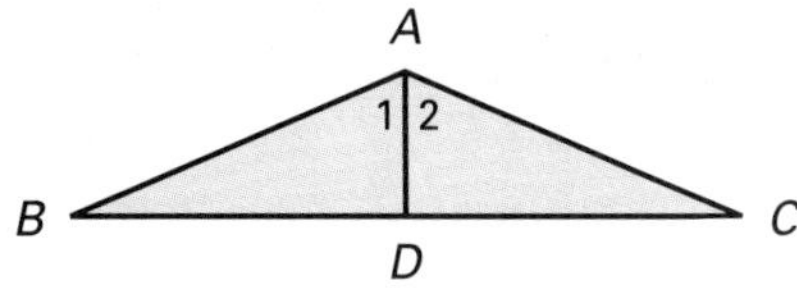


**16.** Which sides and angles are congruent? Explain.

**17.** How can you show that $\overline{AD}$ bisects $\angle A$?

**18.** Make a generalization about the perpendicular bisector drawn from the vertex to the base of an isosceles triangle.

# 3-7 Properties of Quadrilaterals

## EXPLORE

**TALK IT OVER**

How could you construct the parallelogram using a different set of steps?

Given two segments of different lengths, $a$ and $b$, you can construct a parallelogram. Follow these steps.

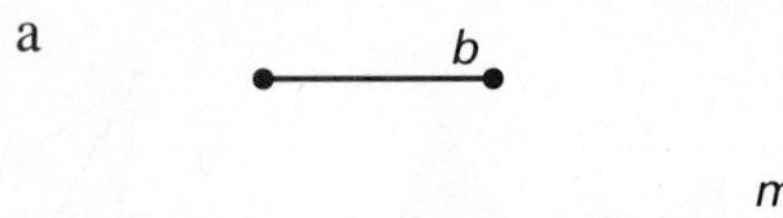


1. Draw line $m$. Choose point $A$ on $m$ and construct segment $AB$ congruent to the given segment of length $a$.
2. Through $A$, draw line $n$ intersecting line $m$ at $A$ at any convenient angle. Through $B$, construct line $p$ parallel to line $n$. (Hint: Construct an angle at $B$ that is congruent to the angle formed by $\overline{AB}$ and line $n$.)

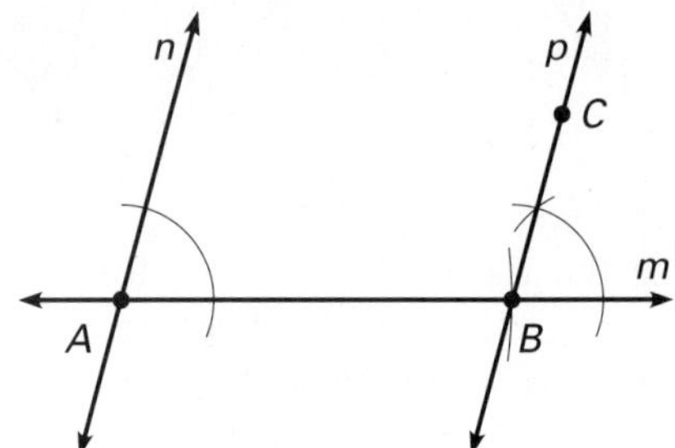


3. On $p$, construct segment $BC$ congruent to the given segment of length $b$.
4. On line $n$, construct segment $AD$ congruent to $\overline{BC}$. Draw the line connecting $C$ and $D$.

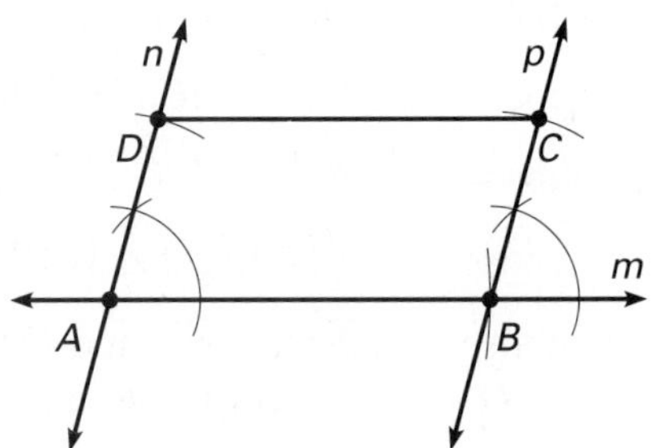


Figure $ABCD$ is the required parallelogram.

a. Which sides are parallel to each other?
b. Which sides are congruent? Explain.
c. Which angles are congruent? Explain.
d. What appears to be true about the measures of $\angle A$ and $\angle B$, $\angle B$ and $\angle C$, $\angle C$ and $\angle D$, and $\angle D$ and $\angle A$?
e. Based on your answer to Exercise **d**, what is the sum of the angles of a parallelogram?

## SKILLS DEVELOPMENT

A **quadrilateral** is a closed plane figure that has four sides. The relationships between different kinds of quadrilaterals are shown in the diagram at the top of page 107.

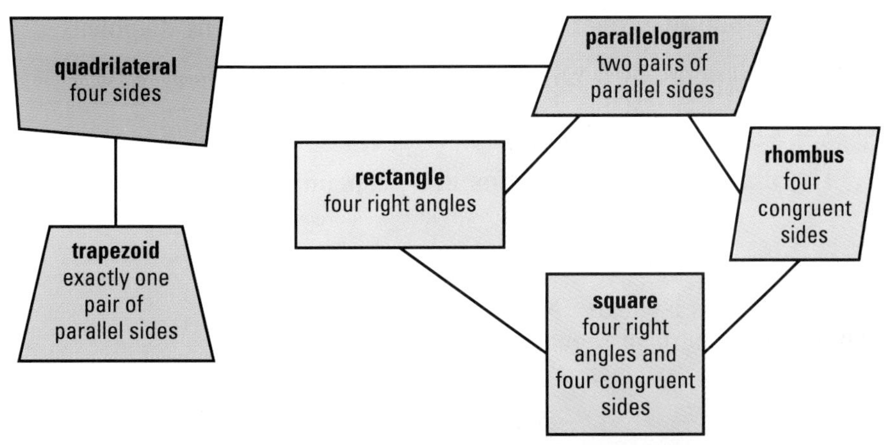


In a quadrilateral, such as parallelogram $ABCD$ at the right, the parts have special names.

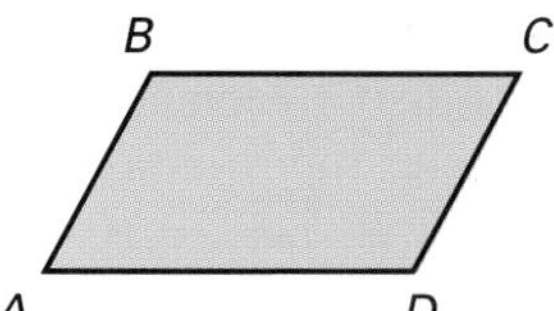


Angles such as $\angle A$ and $\angle C$ are **opposite angles.**
Angles such as $\angle A$ and $\angle B$ are **consecutive angles.**
Sides such as $\overline{AB}$ and $\overline{CD}$ are **opposite sides.**
Sides such as $\overline{AB}$ and $\overline{BC}$ are **consecutive sides.**

In the Explore activity, you discovered several important properties of parallelograms.

1. **The opposite sides of a parallelogram are congruent.**
2. **The opposite angles of a parallelogram are congruent.**
3. **The consecutive angles of a parallelogram are supplementary.**
4. **The sum of the angle measures of a parallelogram is 360°.**

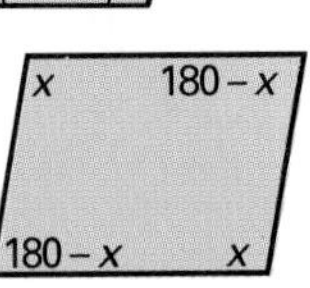


## Example 1

In $\square DEFG$, $m\angle D = 60°$. Find the measures of $\angle E$, $\angle F$, and $\angle G$. Give reasons for your answers.

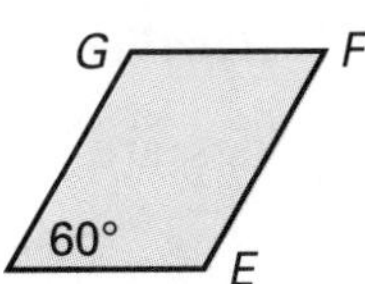


### Solution

$m\angle E + m\angle D = 180°$ — $\angle E$ and $\angle D$ are consecutive angles, so they are supplementary (Property 3).

$m\angle E + 60° = 180°$
$m\angle E = 180° - 60° = 120°$

$\angle F \cong \angle D$ — $\angle F$ and $\angle D$ are opposite angles, so they are congruent (Property 2).
$m\angle F = 60°$

$\angle E \cong \angle G$ — $\angle E$ and $\angle G$ are opposite angles, so they are also congruent. ◄
$m\angle G = 120°$

**MIXED REVIEW**

Complete each function table.

1.

| $x$ | $2x + 3$ |
|---|---|
| 1 | ■ |
| 3 | ■ |
| 5 | ■ |
| 6 | ■ |

Write each fraction or decimal as a percent.

2. $\frac{1}{5}$ 3. 0.07

Estimate each answer by rounding.

4. $341 \times 48$

5. $7{,}213 + 8{,}957 + 6{,}044$

Replace each ● with <, >, or =.

6. $-4 + (-8)$ ● $36 \div (-3)$

7. $9 - 3$ ● $42 \div (-7)$

Another group of properties of parallelograms concerns diagonals.

**5. The diagonals of a parallelogram bisect each other.**  

**6. The diagonals of a rectangle are congruent.**

**7. The diagonals of a rhombus are perpendicular.** 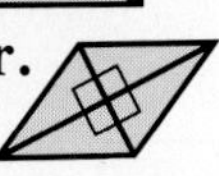

### Example 2

Figure *ABCD* is a square. Name all the pairs of congruent segments. Give reasons for your answers.

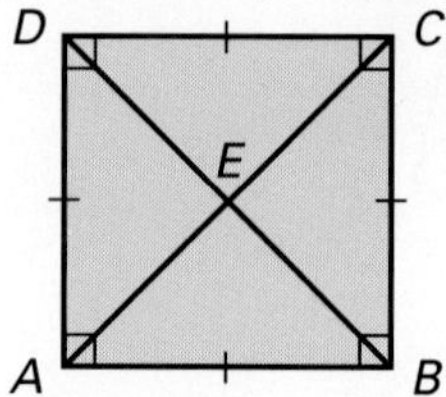


**Solution**

Since *ABCD* is a square, by definition $\overline{AB} \cong \overline{BC} \cong \overline{CD} \cong \overline{DA}$. Since a square is also a rectangle, by Property 6 $\overline{AC}$ and $\overline{BD}$ are congruent. Since the figure is a parallelogram, by Property 5 $\overline{AE} \cong \overline{CE}$ and $\overline{BE} \cong \overline{DE}$. ◄

## TRY THESE

1. In ▱*ABCD*, $m\angle A = 42°$. Find the measure of the other angles. Give reasons for your answers.

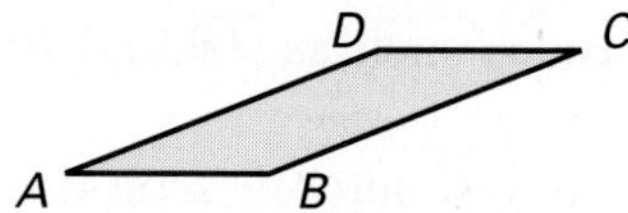


2. Name all the pairs of congruent segments in ▱*PQRS*. Give reasons for your answers.

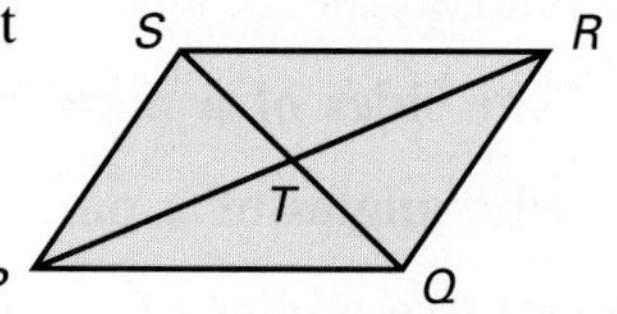


# EXERCISES

**PRACTICE/ SOLVE PROBLEMS**

Match the property named with all the quadrilaterals listed that have that property: *parallelogram, rectangle, rhombus, square.*

1. Opposite sides are congruent.
2. Adjacent sides are perpendicular.
3. Diagonals bisect each other.
4. Opposite angles are congruent.
5. Diagonals are congruent.
6. Diagonals are perpendicular.

Find the unknown angle measures in each parallelogram.

7. $m\angle D$
8. $m\angle B$
9. $m\angle C$

D, A 58°, C, B

10. $m\angle M$
11. $m\angle N$
12. $m\angle L$

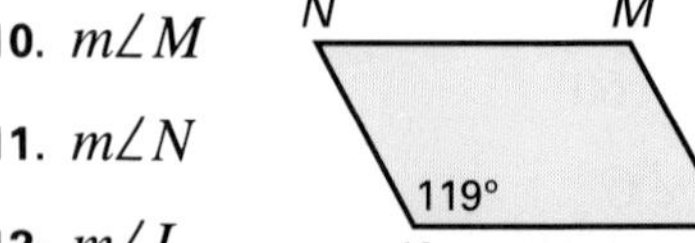

Use the properties of parallelograms to find the unknown measures for each parallelogram.

13. $m\ \overline{BC}$
14. $m\ \overline{CD}$
15. $m\angle B$
16. $m\angle C$
17. $m\angle D$

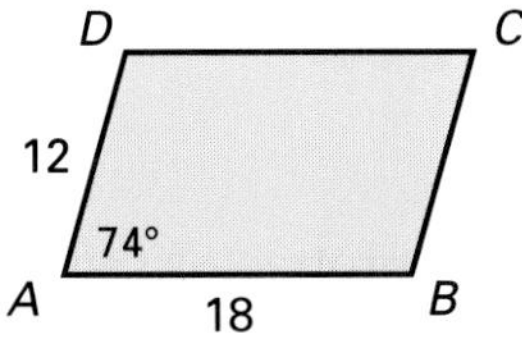


18. $m\ \overline{EF}$
19. $m\ \overline{FG}$
20. $m\ \overline{EG}$
21. $m\ \overline{HI}$
22. $m\ \overline{FH}$

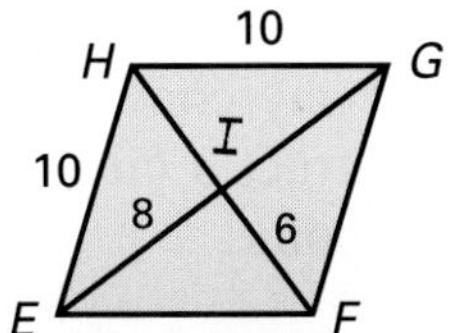


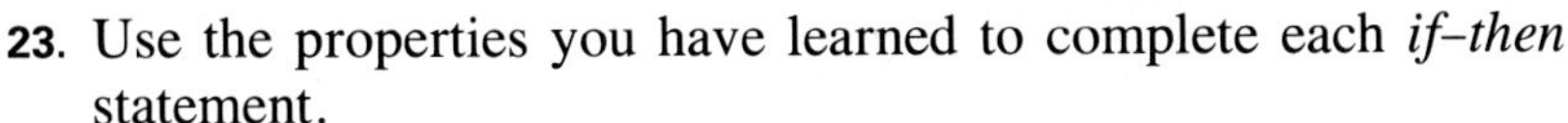

## EXTEND/ SOLVE PROBLEMS

23. Use the properties you have learned to complete each *if–then* statement.

    If a quadrilateral is a parallelogram, then __?__.
    If __?__, then the quadrilateral is a parallelogram.

24. Use the properties of parallelograms to show why a parallelogram having four angles of equal measure must be a rectangle. (Hint: Let $x$ = the measure of any angle.)

25. In the figure at the right, the adjustable ironing board is built so that when open, $\overline{AO} \cong \overline{OC} \cong \overline{BO} \cong \overline{OD}$. Thus, when opened in any of the adjustable positions, the ironing surface will be parallel to the floor. Explain why this is true.

## THINK CRITICALLY/ SOLVE PROBLEMS

An isosceles trapezoid is a trapezoid in which the nonparallel sides (or legs) are congruent. Isosceles trapczoids have special properties.

- **The base angles of an isosceles trapezoid are congruent.**
- **The diagonals of an isosceles trapezoid are congruent.**

26. Since the base angles of an isosceles trapezoid are congruent, how could you show that the diagonals of isosceles trapezoid *ABCD* are congruent? Use the figure at the right.

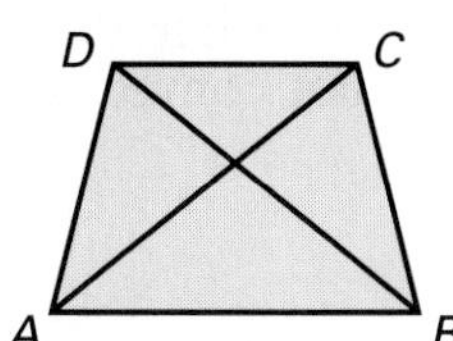

# 3-8 Polygons

## EXPLORE

**TALK IT OVER**

Suppose that you draw a radius from each point along the circle to the center.

1. How many triangles are there?
2. What is the length of each side of each triangle?
3. What kind of triangles are formed?
4. What is the measure of each angle of any triangle?

With a compass, a protractor, and a straightedge, you can construct a regular hexagon.

1. Set a compass at any convenient opening and construct a circle. Mark the center point.
2. With the same compass opening, set the compass point on the circle and mark off an arc along the circle. Then place the compass point on that mark and mark off another arc. Continue the process around the circle.
3. Use a straightedge to connect the points at which the arcs you made intersect the circle.

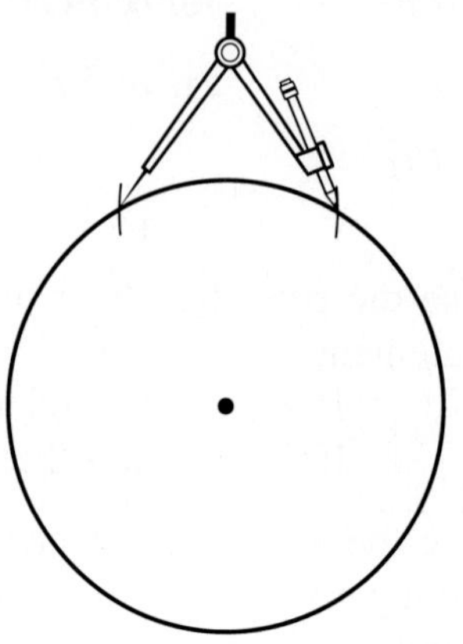

The resulting figure is a hexagon.

## SKILLS DEVELOPMENT

**CHECK UNDERSTANDING**

How are the number of sides and the number of vertices of a polygon related?

A **polygon** is a closed plane figure formed by joining three or more line segments at their endpoints. Each segment, or **side** of the polygon, intersects exactly two other segments, one at each endpoint. *No two* segments with a common endpoint are in the same line. The point at which the endpoints meet is called a **vertex of the polygon.**

| Name of polygon | Number of sides |
|---|---|
| Triangle | 3 |
| Quadrilateral | 4 |
| Pentagon | 5 |
| Hexagon | 6 |
| Heptagon | 7 |
| Octagon | 8 |
| Nonagon | 9 |
| Decagon | 10 |
| *n*-gon | *n* |

A polygon is named by the number of sides it has.

A polygon is **convex** if each line containing a side has no points in the interior of the polygon. A polygon that is not convex is called **concave.**

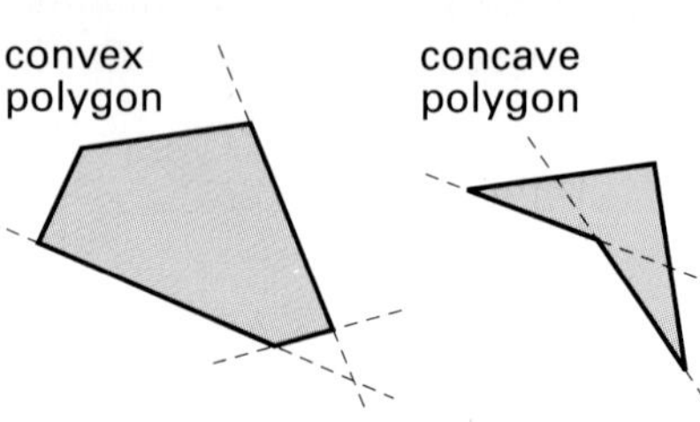

A polygon that has all sides congruent and all angles congruent is called a **regular polygon.**

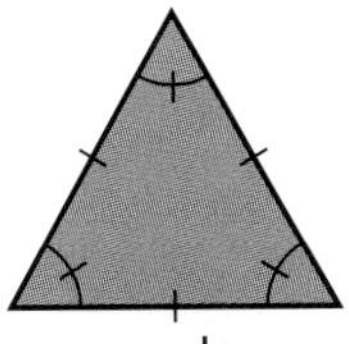
regular triangle

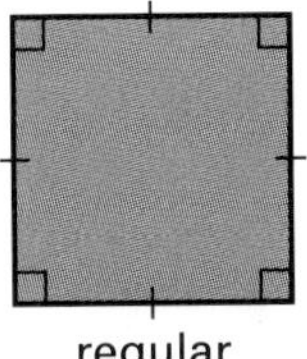
regular quadrilateral

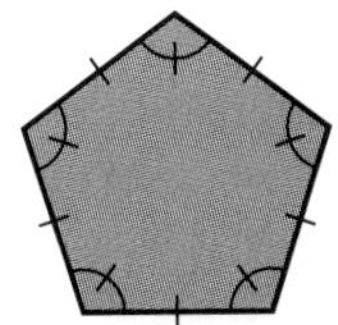
regular pentagon

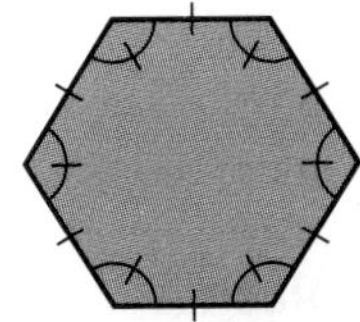
regular hexagon

JUST FOR FUN

The large hexagon shown below has 24 triangles, as drawn. These triangles, taken in groups of six, form seven smaller, overlapping hexagons. See if you can write each of the numbers 1 through 6 in the triangles so that each of the seven smaller hexagons contains all the numbers and no two consecutive numbers (for example, 5 and 6 or 6 and 1) are next to each other. Two of the numbers are already placed.

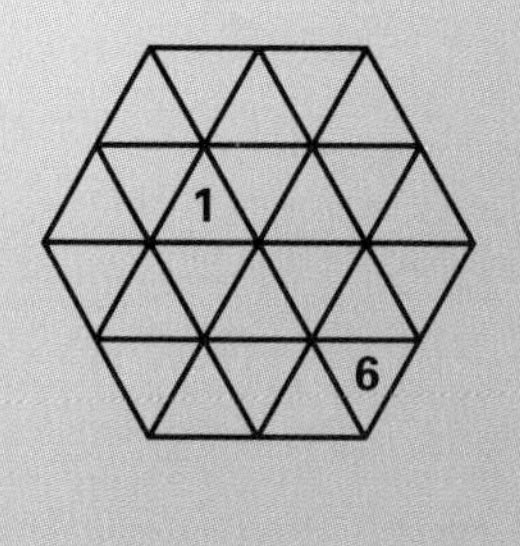


### Example 1

Identify each polygon by the number of sides. Tell whether it is *convex* or *concave, regular* or *not regular.*

**a.**
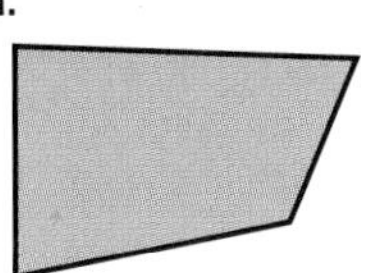
**b.**
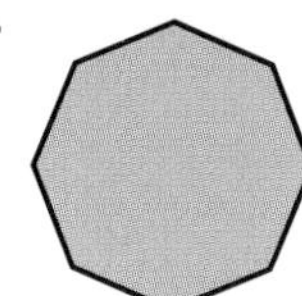
**c.**

**d.**
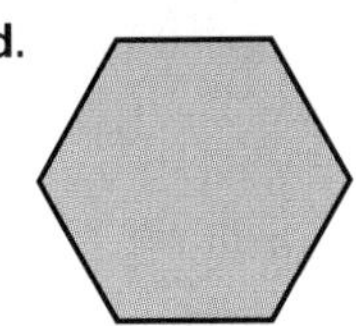

**Solution**

**a.** The figure is a quadrilateral, since it has 4 sides. No line containing a side has points on the interior, so the figure is convex. All sides are not congruent, so the figure is not regular.

**b.** The figure has 8 congruent sides and 8 congruent angles. It is a convex, regular octagon.

**c.** The figure has 5 sides of unequal length. Lines containing two sides have points in the interior, so the figure is concave and not regular.

**d.** The figure is a convex, regular hexagon. ◄

A **diagonal** of a polygon is a segment that joins two vertices but is not a side. You can find the sum of the angles of any polygon by drawing the diagonals from any vertex. The diagonals separate the interior of the polygon into nonoverlapping triangular regions. The sum of the measures of the angles of the polygon is the product of the number of triangles so formed and 180°.

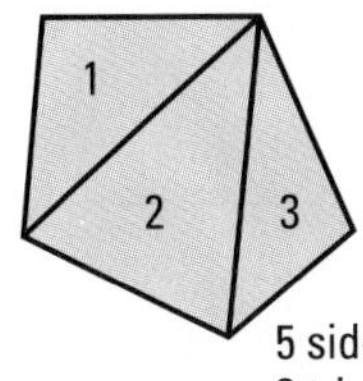

5 sides
3 triangles
sum = $3 \times 180°$

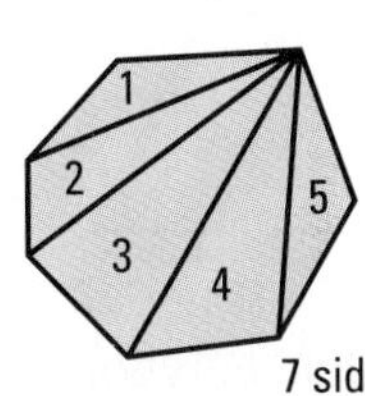

7 sides
5 triangles
sum = $5 \times 180°$

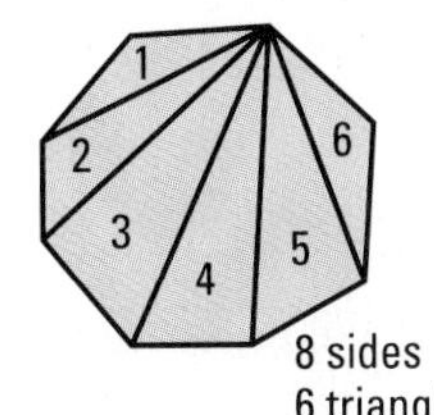

8 sides
6 triangles
sum = $6 \times 180°$

In the polygons shown above, you can see that the number of triangles is always 2 less than the number of sides of the polygon.

COMPUTER

See how the formula is used in this program, which tells you the sum of the angle measures of a polygon when you enter the number of sides.

```
10 INPUT "HOW MANY
   SIDES?";N:PRINT
20 LET D = (N - 2) * 180
30 PRINT "THE SUM OF THE
   MEASURES OF THE
   ANGLES IS "D"."
```

The sum of the angles of a polygon of $n$ sides is $(n - 2)180°$.

## Example 2

Find the sum of the angles of each polygon.

**a.** pentagon **b.** heptagon

### Solution

**a.** $n = 5$

$(n - 2)180° = (5 - 2)180° = 3 \times 180° = 540°$

The sum of the angles of the pentagon is 540°.

**b.** $n = 7$

$(n - 2)180° = (7 - 2)180° = 5 \times 180° = 900°$

The sum of the angles of the heptagon is 900°. ◄

You can find the measure of each interior angle of a regular polygon when you know the number of sides. A regular polygon of $n$ sides has $n$ angles of equal measure. The measure of each angle can be found by dividing the sum of the angles by the number of angles. According to this formula, $\frac{(n-2)180°}{n}$ equals the measure of each angle.

## Example 3

Find the measure of an interior angle of each convex polygon.

**a.** regular hexagon **b.** regular octagon

### Solution

**a.** $\frac{(n-2)180°}{n} = \frac{(6-2)180°}{6} = \frac{4 \times 180°}{6} = \frac{720°}{6} = 120°$

**b.** $\frac{(n-2)180°}{n} = \frac{(8-2)180°}{8} = \frac{6 \times 180°}{8} = \frac{1{,}080°}{8} = 135°$ ◄

## TRY THESE

Identify each polygon by its number of sides. Tell whether it is *convex* or *concave* and *regular* or *not regular*.

**1.** 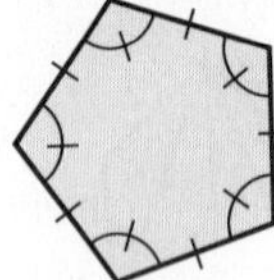 **2.** 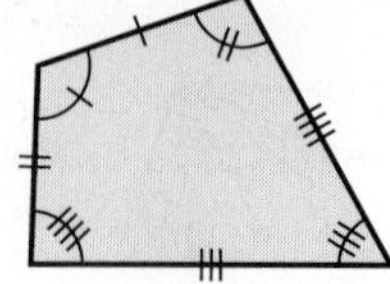 **3.** 

Find the sum of the angles of each polygon.

**4.** convex octagon **5.** convex decagon

Find the measure of an interior angle of each convex polygon.

**6.** regular pentagon **7.** regular nonagon **8.** regular 15-gon

# EXERCISES

## PRACTICE/ SOLVE PROBLEMS

For each polygon, find the sum of the interior angles.

**1.** quadrilateral **2.** pentagon **3.** hexagon **4.** octagon

**5.** decagon **6.** 13-gon **7.** 14-gon **8.** 20-gon

**9.** Find the measure of each interior angle of a regular dodecagon, or 12-sided polygon.

**10.** What is the number of sides of a regular polygon in which each interior angle measures 178°?

**11.** What is the measure of each interior angle of a convex 45-gon?

## EXTEND/ SOLVE PROBLEMS

**12.** Complete the following table for regular polygons.

| Number of sides | 10 | 15 | 30 | ■ | 60 | ■ |
|---|---|---|---|---|---|---|
| Sum of interior angle measures | ■ | 2,340° | ■ | ■ | ■ | 21,240° |
| Measure of each interior angle | 144° | ■ | 168° | 170° | ■ | ■ |

Classify each statement as *true* or *false*.

**13.** The sum of the measures of the interior angles of a 11-gon is 1,820°.

**14.** The measure of an interior angle of a regular 18-gon is 160°.

**15.** A regular polygon in which each interior angle measures 168° has 30 sides.

**16.** An interior angle of a regular polygon cannot have a measure of 148°.

**17.** The sum of the angle measures of a polygon cannot be 1,500°.

## THINK CRITICALLY/ SOLVE PROBLEMS

At each vertex of a polygon, there are a pair of supplementary angles. So, a convex polygon of $n$ vertices has $n$ , pairs of supplementary angles, one interior angle, and one exterior angle at each vertex.

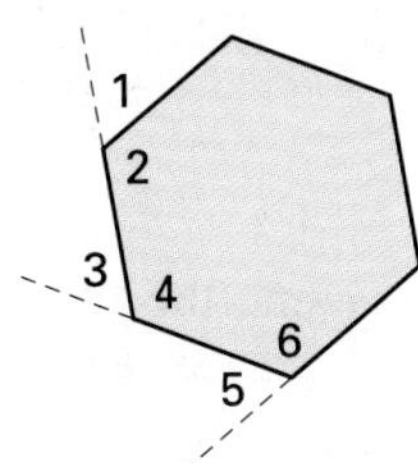


**18.** Write an expression for the sum of all the interior and exterior angles of a polygon with $n$ sides.

**19.** Solve the following equation to show that the sum of the exterior angles of any convex polygon is 360°.

(sum of interior angles) + (sum of exterior angles) = $n(180°)$

# 3-9 Circles

## EXPLORE

Look at the circle design at the right. Use a compass to copy it onto a large sheet of paper. (Hint: Recalling the construction of a hexagon will help you figure out how to do this.)

Now try constructing several more overlapping circle designs, starting from a point on the one you drew.

**CHECK UNDERSTANDING**

What is it about the circle design that is like the construction of the hexagon?

## SKILLS DEVELOPMENT

A **circle** is the set of all points in a plane that are a given distance from a fixed point in the plane. The fixed point is the **center** of the circle. The given distance is the **radius** (plural: *radii*). A radius is a segment that has one endpoint at the center and the other endpoint on the circle.

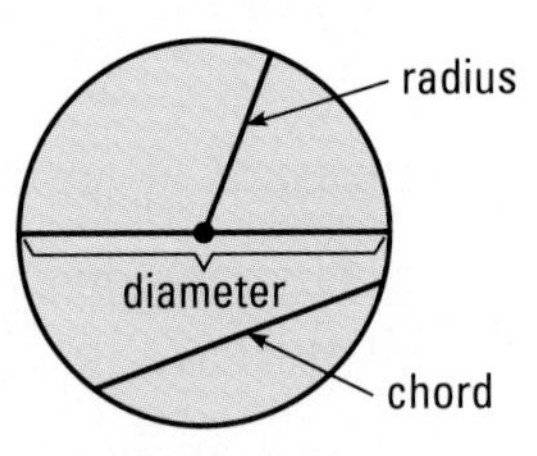


A **chord** is a segment with both endpoints on the circle.

A **diameter** is a chord that passes through the center of the circle. Any diameter is equal to two times the measure of any radius of a given circle.

### Example 1

Use letters to name the following parts of circle $P$ ($\odot P$).

**a.** three radii **b.** a diameter **c.** two chords

**Solution**

**a.** $\overline{PQ}$ or $\overline{QP}$
$\overline{RP}$ or $\overline{PR}$
$\overline{SP}$ or $\overline{PS}$

**b.** $\overline{RS}$ or $\overline{SR}$

**c.** $\overline{TU}$ or $\overline{UT}$
$\overline{RS}$ or $\overline{SR}$ ◄

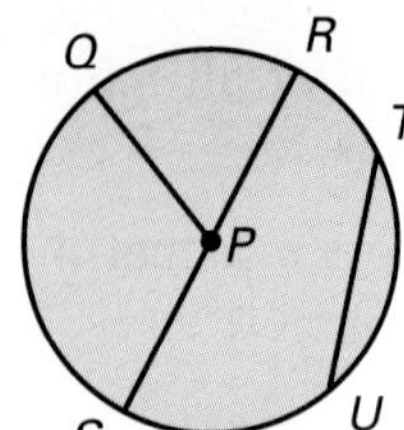

A **central angle** of a circle is an angle with its vertex at the center of a circle. In $\odot O$, $\angle AOB$ is a central angle. The measure of a central angle is always less than 180°.

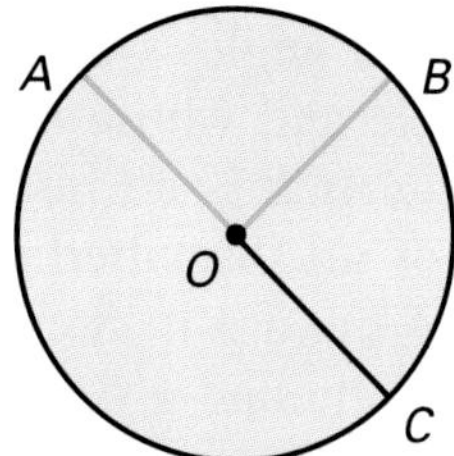


An **arc** is part of a circle.

A **semicircle** is an arc of a circle with endpoints that are the endpoints of a diameter. Three letters are used to name a semicircle. Read $\overset{\frown}{ABC}$ as "arc *ABC*" or "semicircle *ABC*."

A **minor arc** is an arc that is smaller than a semicircle. A minor arc is named by its two endpoints. Read $\overset{\frown}{AB}$ as "arc *AB*."

A **major arc** is an arc that is larger than a semicircle. A major arc is named by its three points, including its endpoints. Read $\overset{\frown}{ACB}$ as "arc *ACB*."

**CONNECTIONS**

**Concentric circles** have a common center point and radii of different lengths.

The three concentric circles shown below are bisected by a line segment. Using one continuous line—do not pick up your pencil or trace over any part of the figure—try to draw the figure.

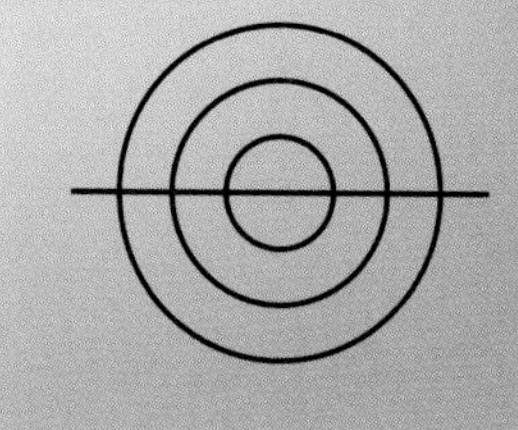

## Example 2

Identify the following for $\odot G$.

**a.** a minor arc
**b.** a major arc
**c.** a semicircle

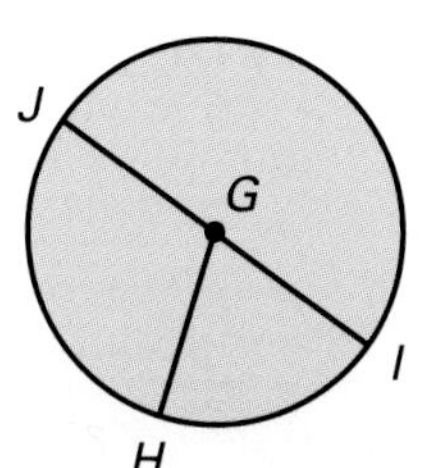


### Solution

**a.** $\overset{\frown}{JH}$ or $\overset{\frown}{HJ}$; $\overset{\frown}{HI}$ or $\overset{\frown}{IH}$ **b.** $\overset{\frown}{HJI}$ or $\overset{\frown}{IJH}$ **c.** $\overset{\frown}{IHJ}$ or $\overset{\frown}{JHI}$ ◄

Each minor arc is associated with a central angle that is said to intercept the minor arc. The measure of a minor arc is defined to be the same as the measure of its central angle.

In $\odot O$, $m\angle LOM = m\overset{\frown}{LM} = 75°$.

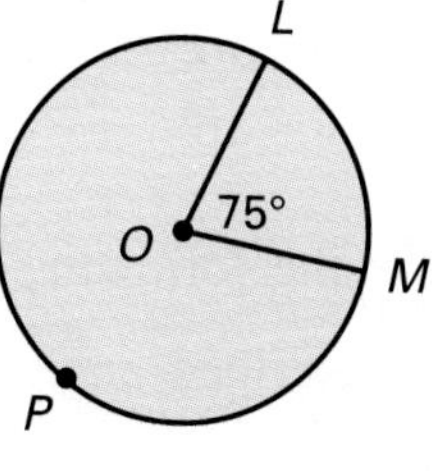


The *measure of a major arc* is found by subtracting the measure of its related minor arc from 360°.

$$\text{In } \odot O,\ m\overset{\frown}{LPM} = 360° - m\overset{\frown}{LM}$$
$$= 360° - 75° = 285°.$$

The *measure of a semicircle,* or half a circle, is 180°.

In $\odot A$, $m\overset{\frown}{BDC} = 180°$ and $m\overset{\frown}{BEC} = 180°$.

D
B
A
C
E

## Example 3

Find the measure of each arc in $\odot W$.

**a.** $m\overset{\frown}{XYZ}$ **b.** $m\overset{\frown}{YZ}$ **c.** $m\overset{\frown}{YXZ}$

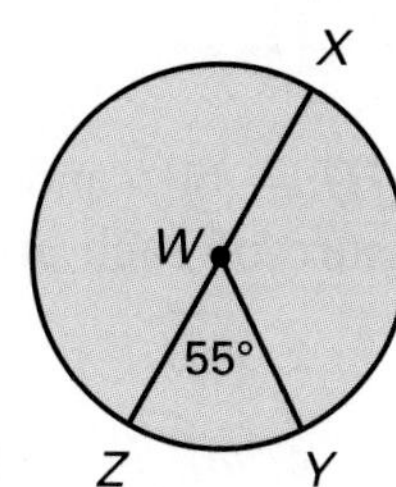


### Solution

**a.** 180° **b.** 55° **c.** 305° ◄

**JUST FOR FUN**

Try writing one of the numbers from 1 to 13 along the radii of these concentric circles so that the sum of the four numbers along each radius of the largest circle (including the number in the center) is 28 while, at the same time, the sum of the numbers around each of the three outer circles is also 28. (Three of the numbers have been positioned for you.)

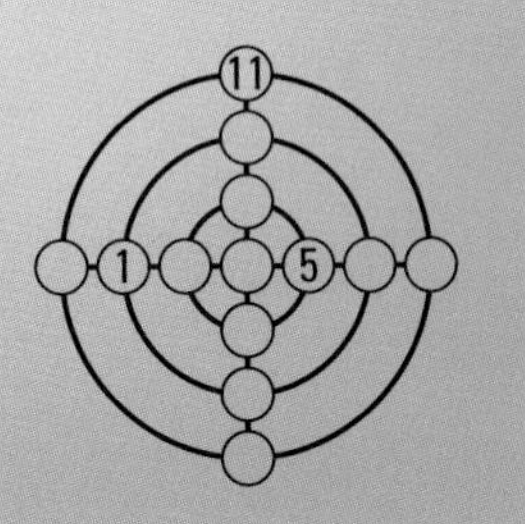


An **inscribed** angle is an angle whose vertex lies on the circle and whose sides contain chords of the circle. $\angle ACB$ is an inscribed angle. The **measure of an inscribed angle** is one-half the measure of the arc it intercepts. That is, if an inscribed angle and a central angle intercept the same arc, then the measure of the inscribed angle is half the measure of the central angle. If two inscribed angles intercept the same arc, then the angles are congruent.

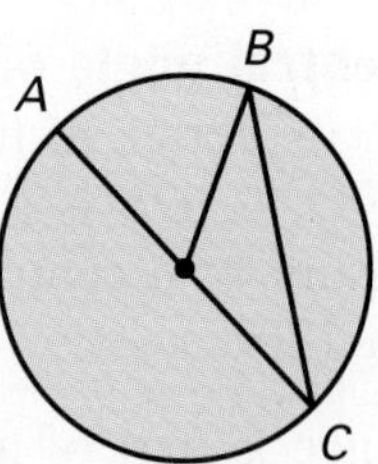


## Example 4

Identify the following for $\odot N$.

**a.** a central angle **b.** an inscribed angle

**c.** $m\angle ONQ$ **d.** $m\angle OPQ$

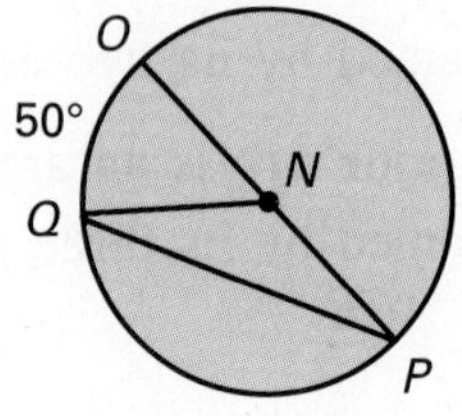


### Solution

**a.** $\angle ONQ$ **b.** $\angle OPQ$

**c.** 50° **d.** 25° ◄

## TRY THESE

Use letters to name the following for $\odot Q$.

**1.** a radius **2.** a chord

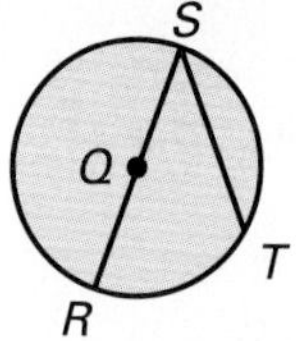


Identify the following parts of $\odot L$.

**3.** a major arc

**4.** a minor arc

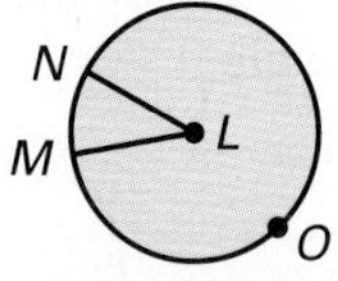


Find the measure of each arc in $\odot D$.

**5.** $m\widehat{EF}$

**6.** $m\widehat{FGE}$

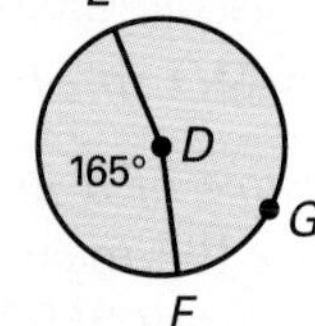


Identify the following for $\odot J$.

**7.** an inscribed angle

**8.** $m\angle OML$

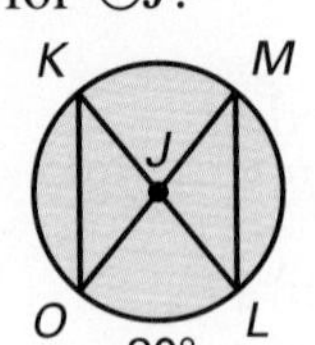


# EXERCISES

## PRACTICE/ SOLVE PROBLEMS

In $\odot H$, $m\angle KHJ = 60°$ and $m\angle JHI = 45°$. Find the measure of each arc.

**1.** $m\widehat{JK}$ **2.** $m\widehat{IJ}$

**3.** $m\widehat{IJK}$ **4.** $m\widehat{ILK}$

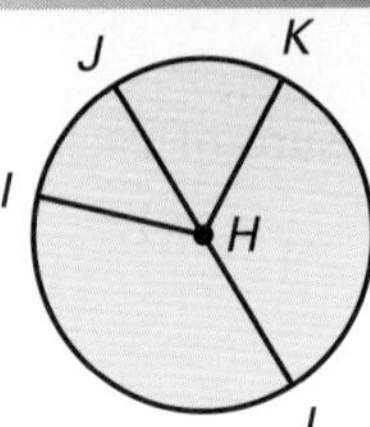

In $\odot M$, $\overline{NO}$ and $\overline{PQ}$ are diameters. Use $\odot M$ for Exercises 5–10.

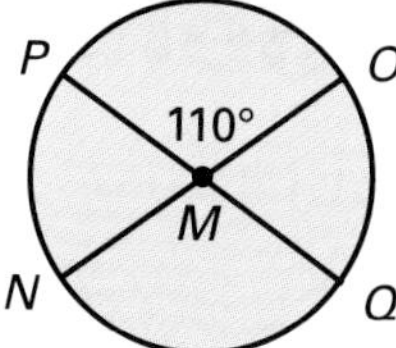


5. Name four semicircles.
6. Name four central angles.
7. Find $m\widehat{NQ}$.
8. Find $m\widehat{PNQ}$.
9. How do you know that $\overline{PM}$ is the same length as $\overline{MQ}$?
10. How do you know that $\overline{NO}$ is twice the length of $\overline{MO}$?

Classify each statement as *true* or *false*.

11. Every circle has just two radii.
12. A chord of a circle contains exactly two points on the circle.
13. The measure of a radius is always one-half the measure of a chord.
14. If a central angle and an inscribed angle are equal, their arcs will be equal.
15. If a major arc measures 200°, it is because a central angle measures 200°.

Many Native American peoples believe the circle to be a symbol of unbreakable unity. Historically, members of the Sioux, or Dakota, Nation set up their camps in huge circles to represent harmony.

Black Elk (1863–1950) of the Oglala Sioux told of the significance of the circle: "Everything the Power of the World does is done in a circle. The sky is round, and I have heard that the earth is round like a ball, and so are all the stars. . . . The sun comes forth and goes down again in a circle. The moon does the same, and both are round."

## EXTEND/ SOLVE PROBLEMS

Find the measure of each arc.

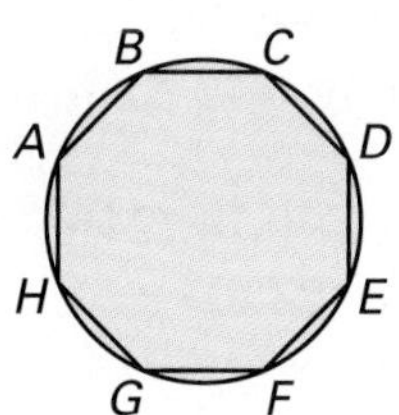


16. $m\widehat{ABD}$
17. $m\widehat{BAF}$
18. $m\widehat{DGC}$

For Exercises 19–22, draw a circle with a radius of 4 cm and a chord $\overline{MN}$. Mark points $P$, $Q$, $R$, and $S$.

19. Draw $\angle MPN$, $\angle MQN$, $\angle MRN$, and $\angle MSN$.
20. Use a protractor to measure the angles. What do you notice about their measures?
21. What is the measure of $\widehat{MN}$?
22. What conclusion can you draw from Exercise 20?

## THINK CRITICALLY/ SOLVE PROBLEMS

Make drawings to help answer Exercises 23–24.

23. If you double the measure of a minor arc, will the measure of the related major arc be doubled? Will the measure of the central angle be doubled?
24. If, in $\odot M$, $\overline{MN}$ and $\overline{MO}$ are radii that are equal in length to chord $\overline{NO}$, what is the measure of $\widehat{NO}$?

# 3-10 Problem Solving Strategies:

## MAKE A CIRCLE GRAPH

- ► READ
- ► PLAN
- ► SOLVE
- ► ANSWER
- ► CHECK

You already know that a circle graph is one means of displaying data for which comparisons between values are to be made. The whole circle represents 100% of the data. Each part, or percent of the data, is represented by a **sector.** Since there are 360° in a circle, the sum of the central angles formed by the sectors is 360°.

### PROBLEM

In a recent poll, high school students were surveyed about the person they admire most. The 950 greatest responses were as follows: *Paula Abdul,* 229; *Mom,* 209; *Dad,* 176; *Michael Jordan,* 155; *Barbara Bush,* 73; *Oprah Winfrey,* 61; *Nelson Mandela,* 47. Draw a circle graph to represent these data.

### SOLUTION

*Step 1:* Write each response as a decimal, or percent of the entire survey, by dividing by 950.

*Step 2:* Find the number of degrees in the central angle by multiplying each decimal by 360°.

*Paula Abdul:* $229 \div 950 = 0.24$ (24%) ⟶ $0.24 \times 360° = 86.4° \approx 86°$
*Mom:* $209 \div 950 = 0.22$ (22%) ⟶ $0.22 \times 360° = 79.2° \approx 79°$
*Dad:* $176 \div 950 = 0.19$ (19%) ⟶ $0.19 \times 360° = 68.4° \approx 68°$
*Michael Jordan:*
$155 \div 950 = 0.16$ (16%) ⟶ $0.16 \times 360° = 57.6° \approx 58°$
*Barbara Bush:* $73 \div 950 = 0.08$ (8%) ⟶ $0.08 \times 360° = 28.8° \approx 29°$
*Oprah Winfrey:* $61 \div 950 = 0.06$ (6%) ⟶ $0.06 \times 360° = 21.6° \approx 22°$
*Nelson Mandela:* $47 \div 950 = 0.05$ (5%) ⟶ $0.05 \times 360° = 18°$

*Step 3:* Draw a circle with a compass. To construct the central angles for the sectors, begin by drawing any radius.

**a.** Position a protractor along the radius with the "0" point of the protractor at the center of the circle. Measure and mark 86°. Draw another radius from the center to this point.

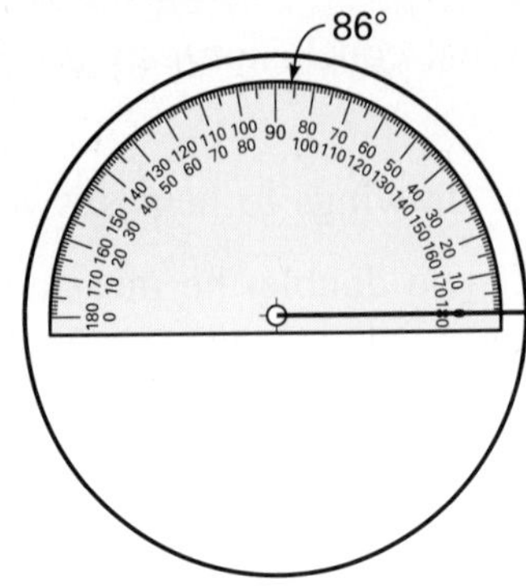

b. Position the protractor along the new radius, then measure and mark 79°, drawing another radius to this point. Draw the rest of the central angles in this way.

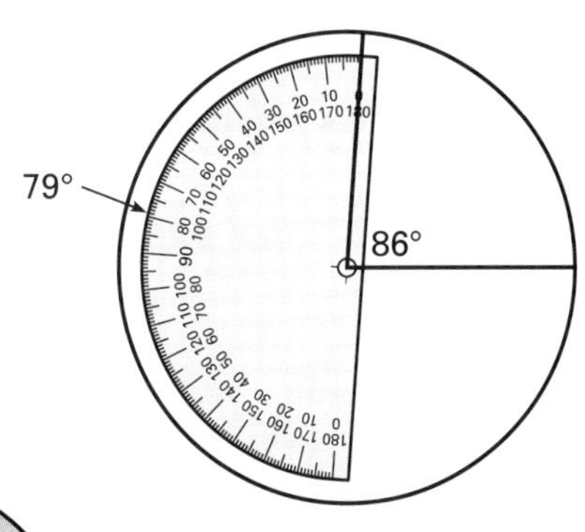


*Step 4:* Label each sector of the circle. Write a title for the graph.

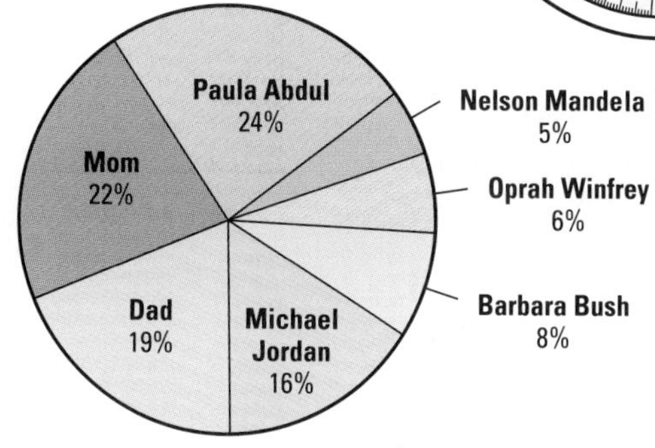


# PROBLEMS

**PROBLEM SOLVING TIP**

For each of Problems 4–5, you will have to start by finding the total number of data entries.

Make a circle graph for each set of data. Be sure to write a title for your graph. For Problems 1–3, start by finding the number of degrees in each central angle.

1.

HOW AMERICANS GET TO WORK

| Means of Travel | Percent |
|---|---|
| Drive alone | 64 |
| Carpool | 20 |
| Bus or Train | 6 |
| Walk only | 6 |
| Work at home | 2 |
| Other | 2 |

2.

ELEMENTS THAT MAKE UP THE EARTH'S SURFACE

| Element | Percent |
|---|---|
| Oxygen | 49 |
| Silicon | 26 |
| Aluminum | 7.5 |
| Iron | 4.7 |
| Calcium | 3.4 |
| Potassium | 3.4 |
| Other | 6 |

3.

MONTHLY BUDGET

| Expenses | Percent |
|---|---|
| Car | 15 |
| Travel | 19 |
| Food | 27 |
| Clothing | 23 |
| Savings | 12 |
| Miscellaneous | 4 |

**MATH IN THE WORKPLACE**

Pierre Berloquin makes his living devising math games and puzzles. Although the French writer trained to be an engineer, he became fascinated with puzzles. He soon found a new career inventing games and puzzles. He has written over twenty books, including *100 Numerical Games, 100 Geometric Games,* and *100 Logic Games.* If you enjoy games and puzzles that challenge your mind, take a look at some of Berloquin's writings.

For Problems 4–5, start by finding a decimal, or percent of the whole.

4. Here is Gameland's annual inventory for last year in their best-selling-games department.

| Game | Numbers of Sets |
|---|---|
| Monopoly® | 427 |
| Scrabble® | 85 |
| Clue® | 85 |
| Backgammon® | 225 |
| Trivial Pursuit® | 144 |

5. Here are next season's cruise destinations for a San Francisco cruise line.

| Destination | Number of Cruises |
|---|---|
| Acapulco, Mexico | 47 |
| Anchorage, Alaska | 28 |
| Honolulu, Hawaii | 53 |
| Vancouver, B.C. | 14 |
| Melbourne, Australia | 18 |

# 3-11 Properties of Three-Dimensional Figures

## EXPLORE/ WORKING TOGETHER

Trace each figure. Cut out your tracings. Roll each one to join two opposite edges. Hold the edges together as your partner tapes them. What shape does each form?

Figure 1

You could draw, then cut out, a pair of congruent shapes for the top and bottom of each figure to make the figure solid. What would the pair of shapes look like for Figure 1? for Figure 2?

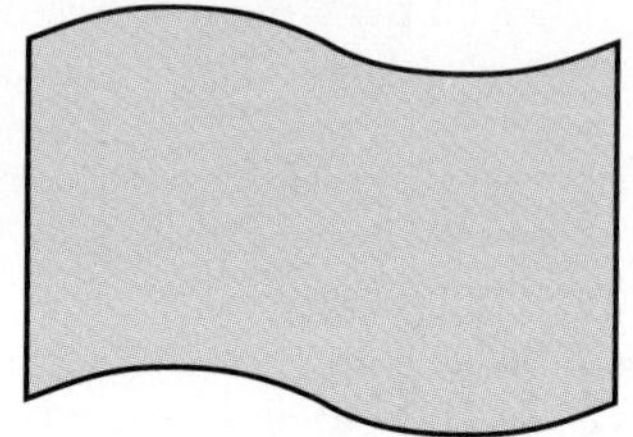
Figure 2

## SKILLS DEVELOPMENT

Some three-dimensional figures have both flat and curved surfaces. A **cylinder** is a three-dimensional shape made up of a curved region and two congruent circular bases that lie in parallel planes. The **axis** is a segment that joins the centers of the bases.

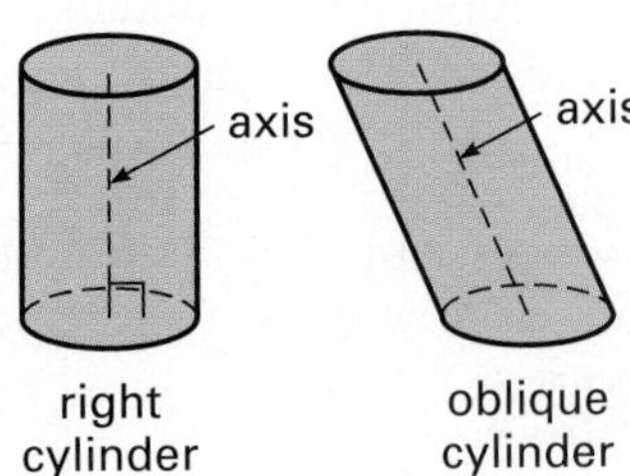


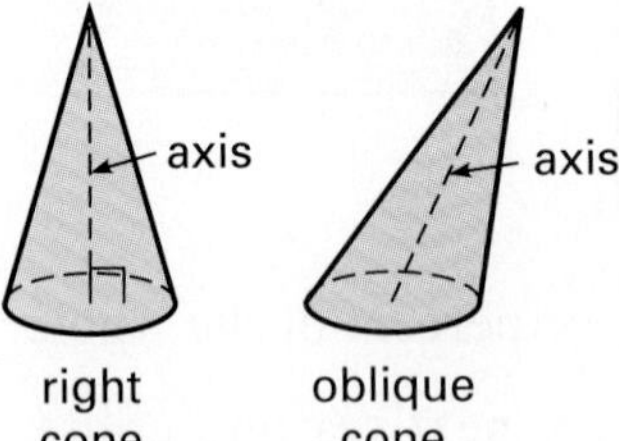


A **cone** has a curved surface and one circular base. The **axis** is a segment that joins the vertex to the center of the base.

A **sphere** is the set of all points that are a given distance from a given point, called the **center of the sphere.**

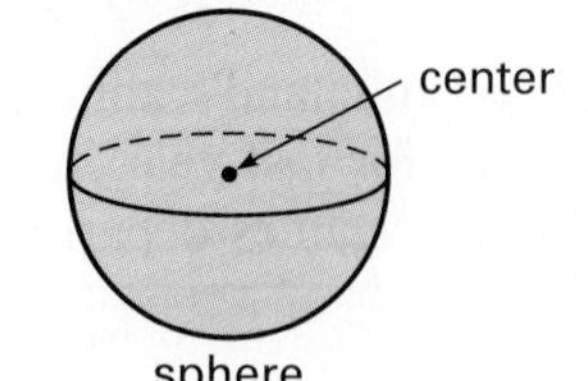


A **polyhedron** (plural: *polyhedra*) is a closed, three-dimensional figure made up of polygons alone. The polygonal surfaces are called **faces.** Two faces meet, or intersect, to form an **edge.** The point at which three or more edges intersect is called a **vertex** (plural: *vertices*).

A **prism** is a polyhedron with two identical parallel faces called **bases.** The other faces are parallelograms. A prism is named according to the shape of its bases.

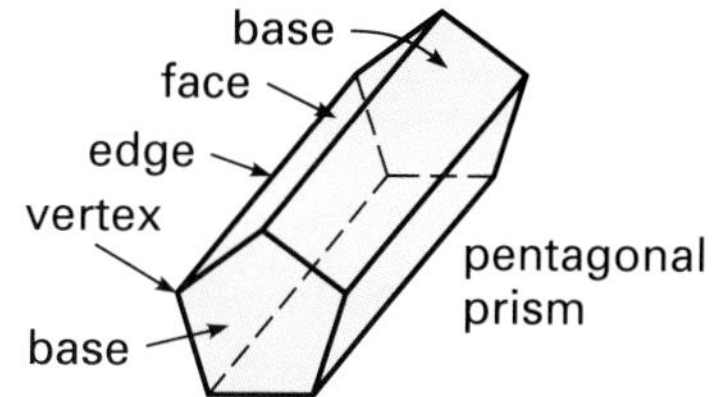


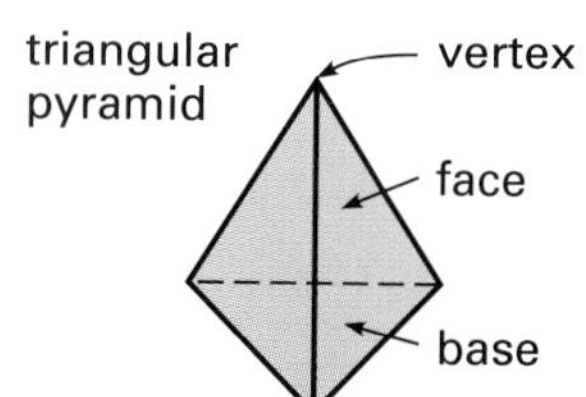


A **pyramid** is a polyhedron with only one base. The other faces are triangles that meet at a **vertex.** A pyramid is named by the shape of its base.

In prisms and pyramids, the faces that are not bases are **lateral faces.** The edges of these faces are **lateral edges.** The bases and lateral faces of these figures can be either parallel or intersecting. The lateral edges can be intersecting, parallel, or skew.

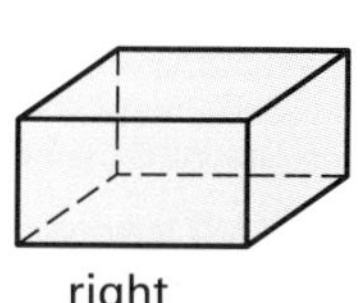
right rectangular prism

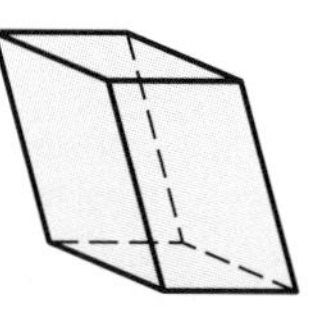
oblique rectangular prism

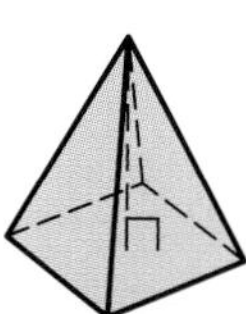
right square pyramid

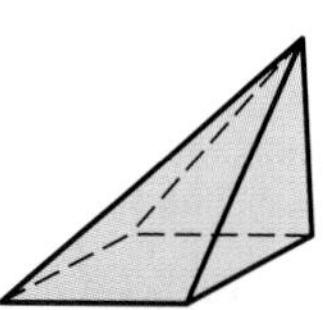
oblique square pyramid

## Example 1

Identify the figure at the right. Then identify the bases, a pair of parallel edges, a pair of intersecting edges, a pair of skew edges, and a pair of intersecting faces.

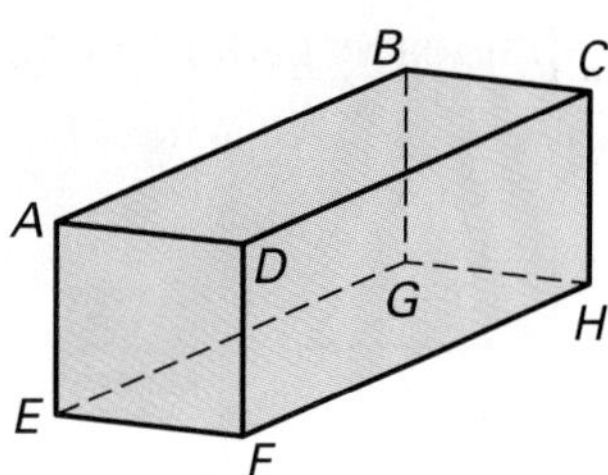


### Solution

Figure: right rectangular prism
Bases: *ADFE* and *BCHG*
Pair of parallel edges: $\overline{DC}$ and $\overline{FH}$
Pair of intersecting edges: $\overline{GH}$ and $\overline{FH}$
Pair of skew edges: $\overline{AD}$ and $\overline{CH}$
Pair of intersecting faces: *ABCD* and *CDFH* ◄

## Example 2

Draw a right triangular prism.

### Solution

*Step 1:* Draw two congruent triangles.

*Step 2:* Draw rules that connect the corresponding vertices of the triangles. Use dotted lines to show the lateral edges that cannot be seen.

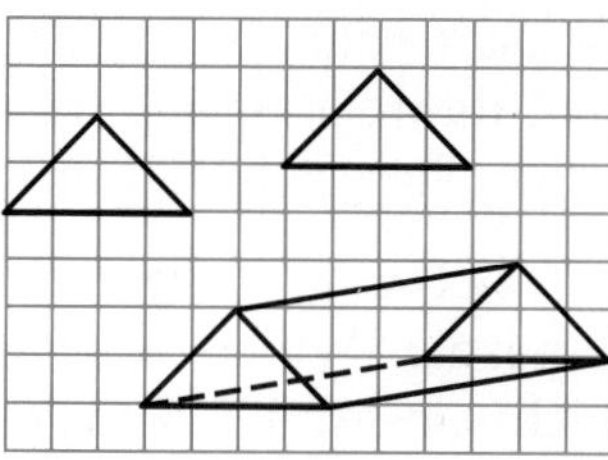

◄

CHECK UNDERSTANDING

Why is a cube also a rectangular prism?

## TRY THESE

Identify each figure. Then identify the base(s) and a pair of parallel edges, intersecting faces, and intersecting edges.

1. 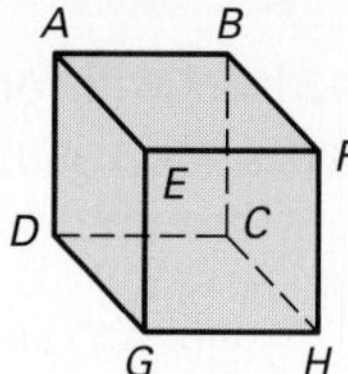


2. 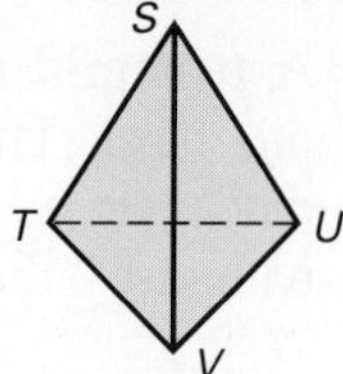


3. 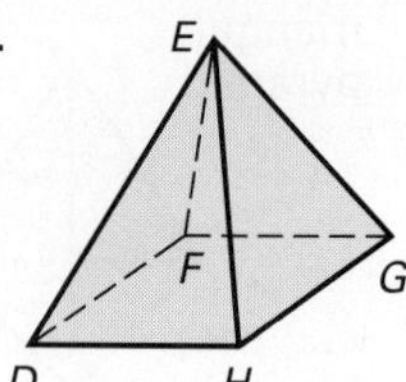


Draw each figure.

4. circular cone
5. oblique square prism
6. right hexagonal prism

# EXERCISES

### PRACTICE/ SOLVE PROBLEMS

Draw the figure.

1. square pyramid
2. rectangular prism
3. oblique cone

Identify the three-dimensional figure that would be formed by folding each pattern along the dotted lines.

4. 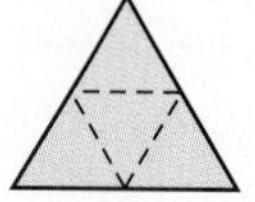

5. 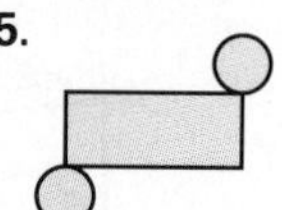

6. 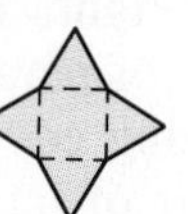

7. 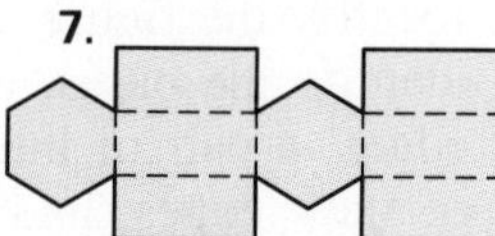

### EXTEND/ SOLVE PROBLEMS

Complete the table and extend it to include each of the polyhedra given in Exercises 9–11.

| | Polyhedron | Number of Faces ($F$) | Number of Vertices ($V$) | Number of Edges ($E$) | $F + V - E$ |
|---|---|---|---|---|---|
| 8. | rectangular prism | ■ | ■ | ■ | ■ |

9. hexagonal prism
10. pentagonal pyramid
11. square pyramid

### THINK CRITICALLY/ SOLVE PROBLEMS

12. Compare your results for Exercises 8–11. Make a generalization about the relationship between the numbers of faces, edges, and vertices in a polyhedron.

13. *USING DATA* Use the Data Index on pages 556–557 to find the puzzle "Which Way Is Up?" Study the puzzle to determine which way is "UP↑" in View 3.

- READ
- PLAN
- SOLVE
- ANSWER
- CHECK

# Problem Solving Applications:

## SPATIAL VISUALIZATION

Architects need to know what various views of a building will look like in order to draw accurate blueprints for the building.

Look at a polyhedron from different points of view, and you may see different shapes. Here are three different views of a square pyramid.

base → 

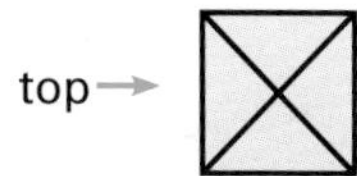

side →

Use graph paper or isometric dot paper to draw the figures for which the base, top, and side views are given in Exercises 1–3.

1. base → top → side →

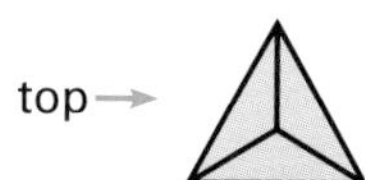

2.

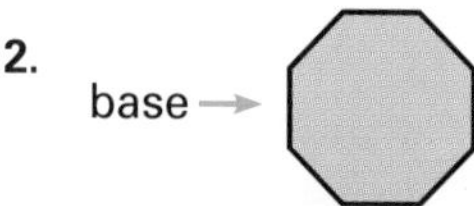

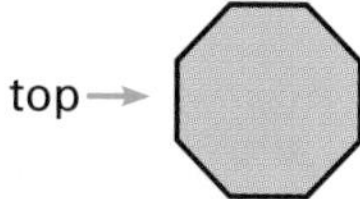

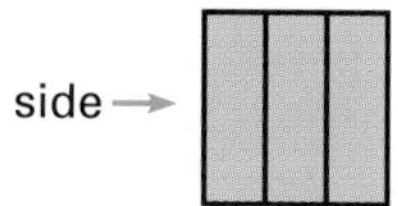

3.

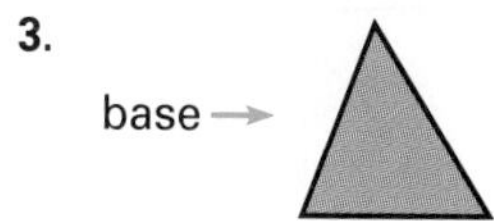

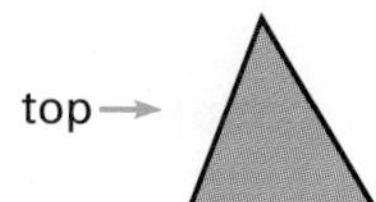

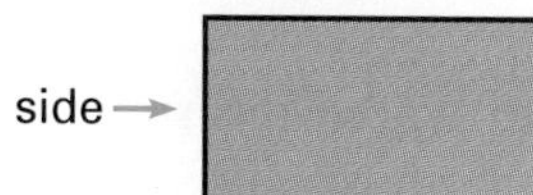

Imagine that a three-dimensional figure is cut as shown and that the two resulting pieces are separated. What shape would be formed at the cross-section?

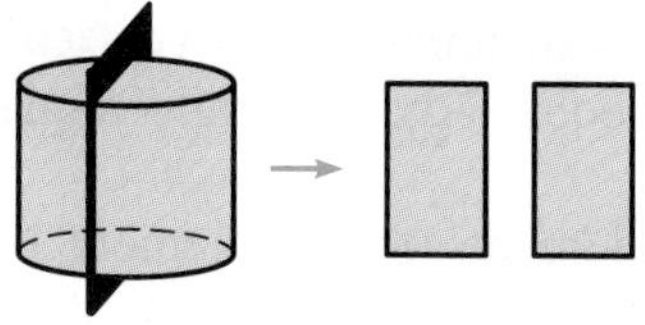

4.

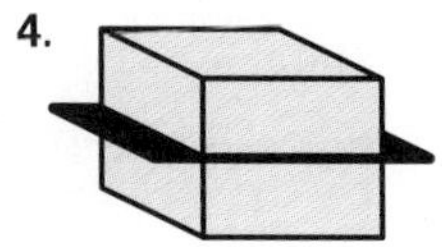

5.

6.

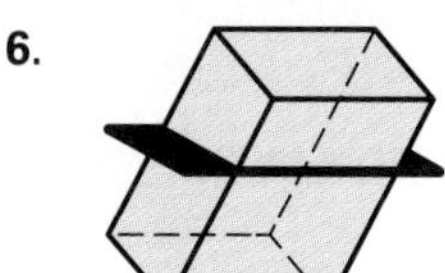

7.

8.

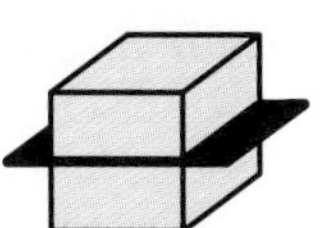

9.

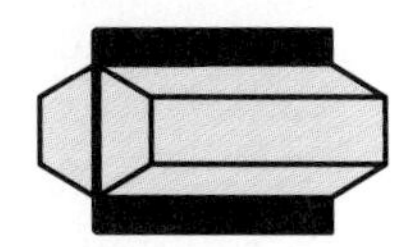

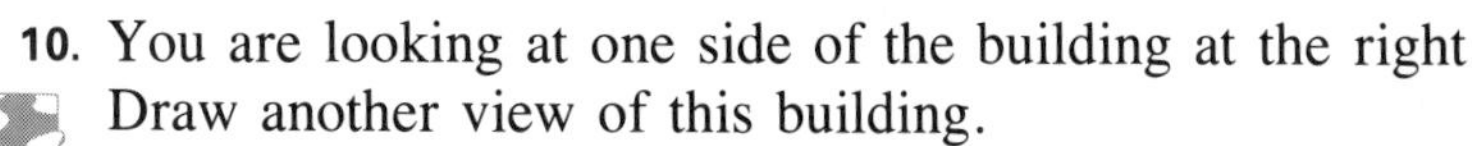

10. You are looking at one side of the building at the right. Draw another view of this building.

## CHAPTER 3 • REVIEW

Choose the term from the list that matches each definition.

1. points that lie in the same plane
2. planes that do not intersect
3. line intersecting two other lines in different points
4. parallelogram with all sides congruent
5. segment with endpoints on a circle

a. transversal
b. chord
c. rhombus
d. coplanar points
e. parallel planes

### SECTION 3–1 THE BASIC ELEMENTS OF GEOMETRY (pages 84–87)

► **Point, line,** and **plane** are basic terms in geometry.

Write the symbol for each figure.

6. P Q

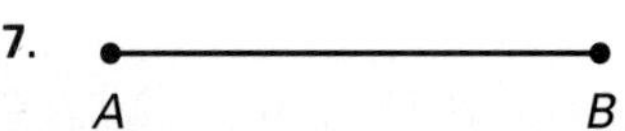

7. A B

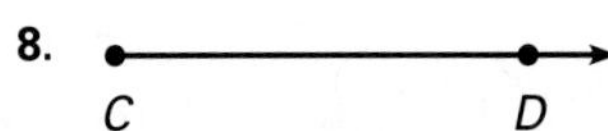

8. C D

### SECTION 3–2 ANGLES AND PERPENDICULAR LINES (pages 88–91)

► Two angles, the sum of whose measures is 90°, are **complementary.** Two angles, the sum of whose measures is 180°, are **supplementary.**
► **Adjacent angles** have a common vertex and a common side. Two intersecting lines form pairs of nonadjacent **vertical angles.**
► **Perpendicular lines** form equal adjacent right angles.

Find the measure of the complement and supplement of each angle.

9. $\angle A = 34°$
10. $\angle B = 49°$
11. $\angle C = 87°$
12. $\angle D = 12°$

### SECTION 3–3 PARALLEL LINES (pages 92–95)

► **Parallel lines** are coplanar lines that do not intersect.
► When two parallel lines are cut by a *transversal,* the *corresponding angles, alternate interior angles,* and *alternate exterior angles* formed are *congruent.*

Refer to the figure to name the following:

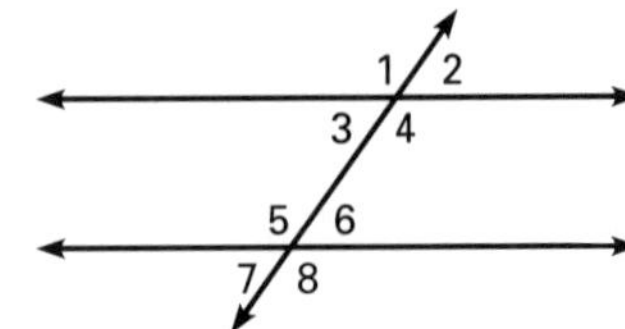


13. two pairs of corresponding angles
14. two pairs of alternate interior angles

## SECTIONS 3–4 AND 3–6 TRIANGLES, CONGRUENCE (pages 96–99, 102–105)

► The sum of the lengths of any two sides of a triangle is greater than the length of the remaining side.

Tell whether it is possible to have a triangle with sides of these lengths.

**15.** 13, 19, 8 **16.** 9, 5, 4

## SECTION 3–5 PROBLEM SOLVING (pages 100–101)

► Postulates and other geometrical statements may be put into *if–then* form.

**17.** Rephrase as an *if–then* statement: *All rectangular figures are quadrilaterals.*

## SECTIONS 3–7 AND 3–8 QUADRILATERALS AND POLYGONS (pages 106–113)

► The measure of each angle of a regular $n$-gon is $\frac{(n - 2)180°}{n}$.

Find the measure of an interior angle of the regular polygon named.

**18.** regular pentagon **19.** regular octagon

## SECTIONS 3–9 AND 3–10 CIRCLES (pages 114–119)

► A **central angle** of a circle has its vertex at the center of the circle. An **inscribed angle** has its vertex on the circle.

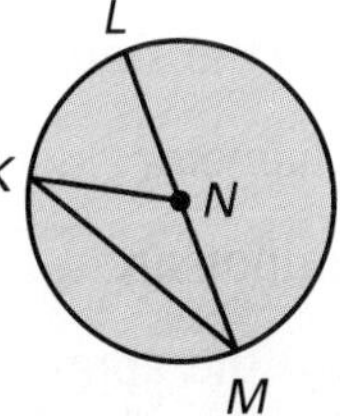


Identify these parts of $\odot N$.

**20.** a central angle **21.** an inscribed angle

## SECTION 3–11 THREE-DIMENSIONAL FIGURES (pages 120–123)

► The **prism** and the **pyramid** are **polyhedra**, figures in which each surface is a polygon. The **cylinder**, **cone**, and **sphere** are three-dimensional figures with curved surfaces.

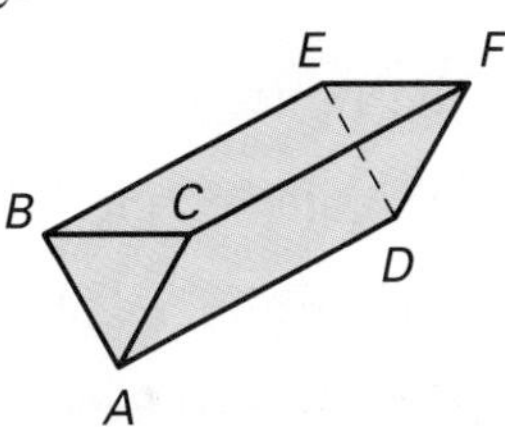


Refer to the figure to name the following.

**22.** two intersecting edges

**23.** two intersecting lateral faces

# CHAPTER 3 ● TEST

Name each figure and write the symbol for it.

**1.** 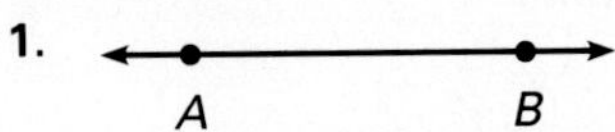


**2.** Q P

**3.** X Y

Use the figure to name the following.

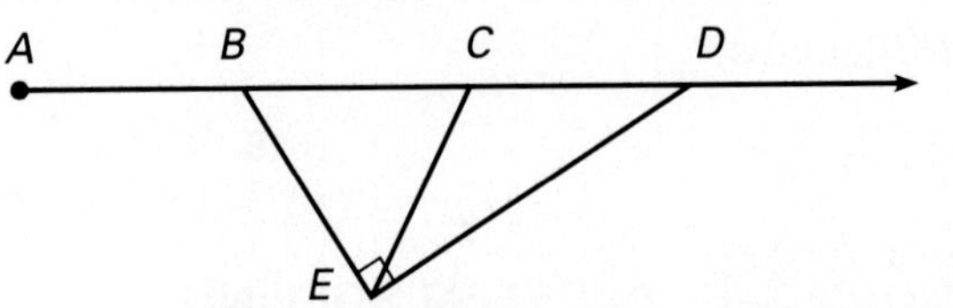


4. two obtuse angles
5. a pair of adjacent complementary angles
6. two pairs of adjacent supplementary angles
7. a pair of nonadjacent complementary angles

Use the figure to name the following. Line $k \parallel$ line $m$.

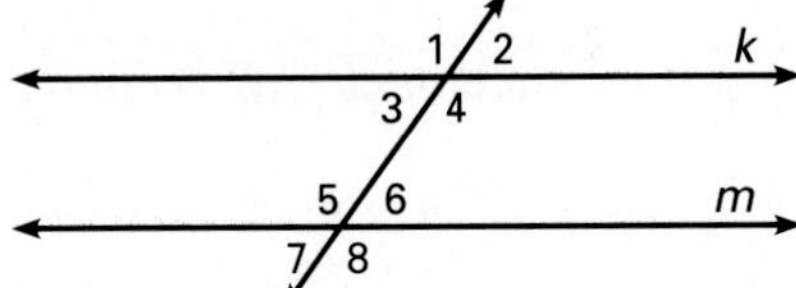


8. two pairs of alternate exterior angles
9. two pairs of same-side interior angles
10. Name the smallest and largest angles of the triangle.

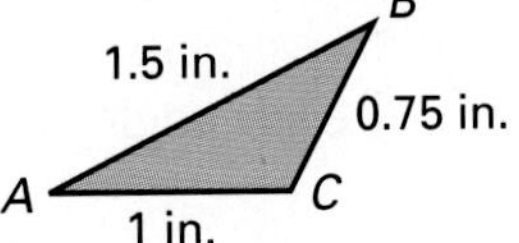


11. Write the converse of this statement: *If the sum of the measures of two angles is 90°, then the two angles are complementary.*
12. Name the corresponding angles of the two triangles shown.

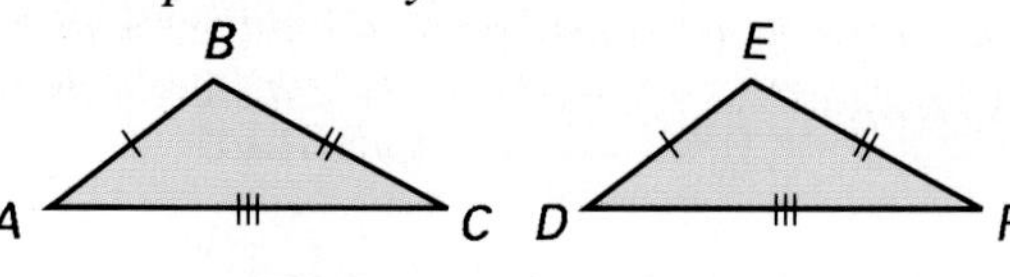


Refer to the figure to name the following.

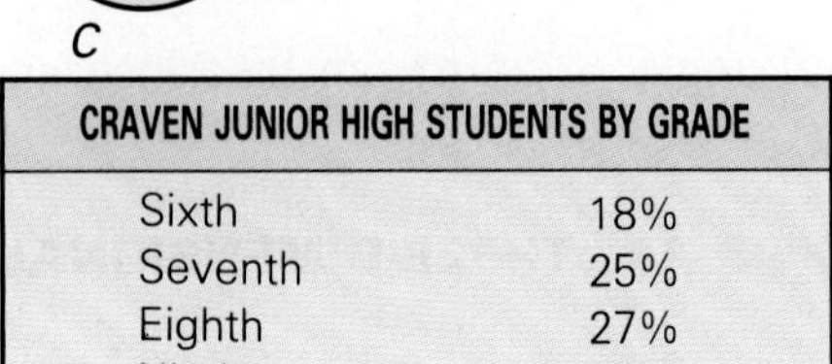


13. a diameter
14. a chord
15. a minor arc
16. an inscribed angle
17. Find the approximate number of degrees in each central angle of a circle graph that could represent each category in the table shown.

| CRAVEN JUNIOR HIGH STUDENTS BY GRADE | |
|---|---|
| Sixth | 18% |
| Seventh | 25% |
| Eighth | 27% |
| Ninth | 30% |

Refer to the figure to name the following.

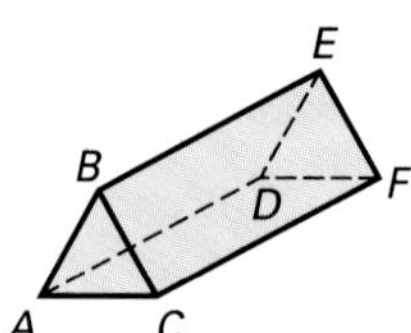


18. a pair of intersecting edges
19. a pair of parallel edges
20. a pair of parallel faces
21. a pair of edges that are skew

Simplify each expression.

1. $20 - 8 \div 2 \times 3$
2. $14 \times 3 + 5 \div 5$

Complete.

3. $8(2 + 9) = (8 \times 2) + (■ \times 9)$
4. $4.5 \times ■ = 4.5$

Find each percent to the nearest tenth.

5. 11 is what percent of 88?
6. What percent of 500 is 112?

Add or subtract.

7. $8 + (-6)$
8. $-12 - (-18)$

Multiply or divide.

9. $15 \cdot (-4)$
10. $-90 \div 6$

Simplify.

11. $4h - 13h$
12. $\frac{54a^5}{9a^2}$
13. $y^3 \cdot y \cdot y^4$

Write each number in scientific notation.

14. 462,000
15. 0.00058

Write each number in standard form.

16. $3.49 \times 10^4$
17. $2.07 \times 10^{-2}$

Name each figure, and write the symbol for it.

18.

19.

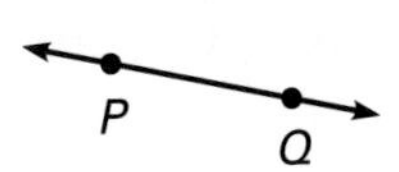


Classify each triangle as *acute, obtuse,* or *right* and as *scalene, isosceles,* or *equilateral.*

20.

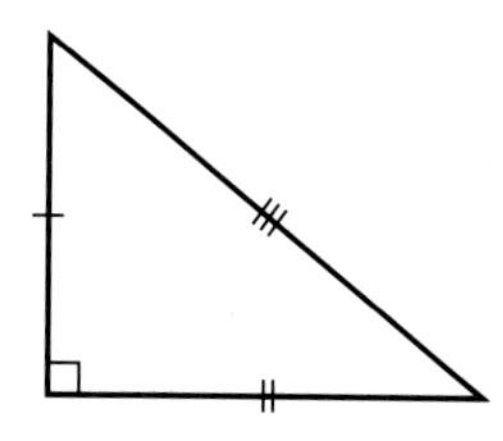

21.

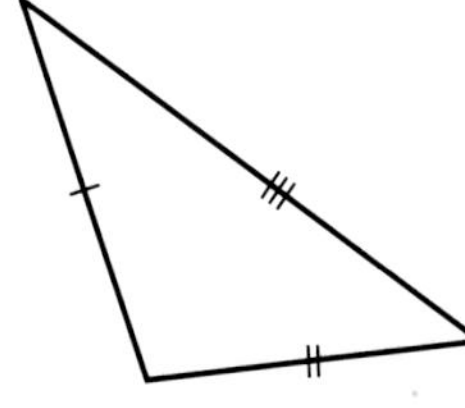

22.

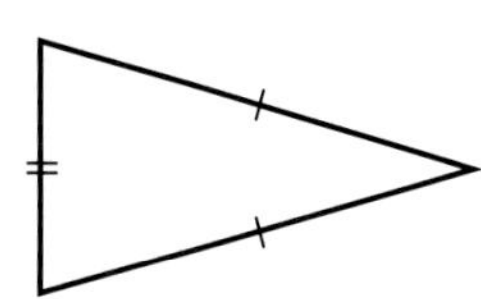

23.

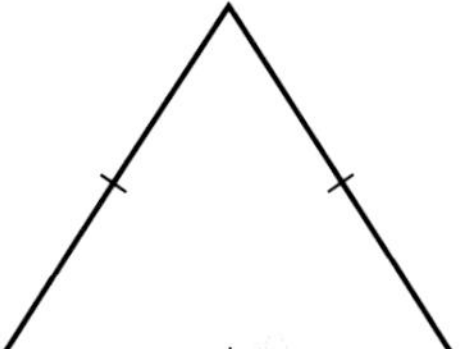

24. Write the converse of this statement:
*If a triangle is isosceles, then the triangle has two congruent sides.*

Find the sum of the interior angles of each of these polygons.

25. hexagon
26. octagon

Find the unknown angle measure.

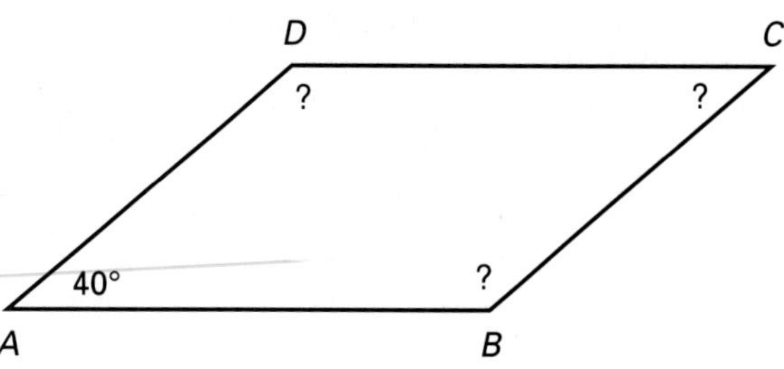


27. $\angle B$
28. $\angle C$
29. $\angle D$

# CHAPTER 3 ● CUMULATIVE TEST

Solve.

**1.** $19 + (41 + 27)$
A. 77 B. 88 C. 87 D. 78

**2.** $5 \cdot 24 \cdot \frac{1}{4}$
A. 30 B. 28 C. 20 D. 25

**3.** Which expression has the value of 24 when $x = 3$ and $y = 4$?
A. $2xy$ B. $5x + 2y$
C. $5x - 2y$ D. none of these

Find each percent to the nearest tenth.

**4.** 8 is what percent of 64?
A. 14% B. 12.5% C. 62.5% D. 8%

**5.** 45% of what number is 108?
A. 960 B. 196 C. 240 D. 153

**6.** Simplify. $(x^4)^3$
A. $x^7$ B. $x$
C. $x^{12}$ D. none of these

**7.** Write in scientific notation. 0.000943
A. $9.43 \times 10^{-4}$
B. $9.43 \times 10^{-3}$
C. $9.43 \times 10^{-6}$
D. $9.43 \times 10^{3}$

**8.** Write in standard form. $9.21 \times 10^4$
A. 921 B. 0.921
C. 92,100 D. 921,000

**9.** Evaluate $y^{-4}$ when $y = -2$.
A. $-16$ B. $\frac{1}{16}$
C. $-\frac{1}{16}$ D. 16

**10.** Identify the symbol of the figure.

R S

A. $\overline{RS}$ B. $\overleftrightarrow{RS}$
C. $\overleftrightarrow{SR}$ D. none of these

**11.** Classify $\angle HIJ$.

H, I, J, 44°

A. right
B. straight
C. acute
D. obtuse

**12.** Find the measure of $\angle 1$.

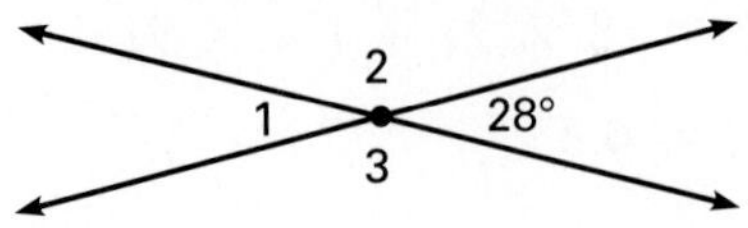


A. 28° B. 152°
C. 90° D. none of these

**13.** Identify the pair of corresponding angles.

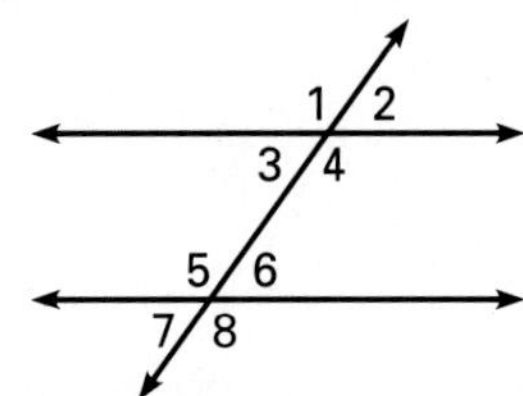


A. $\angle 3$ and $\angle 6$ B. $\angle 2$ and $\angle 4$
C. $\angle 2$ and $\angle 6$ D. $\angle 2$ and $\angle 7$

**14.** Find the value of $b$.

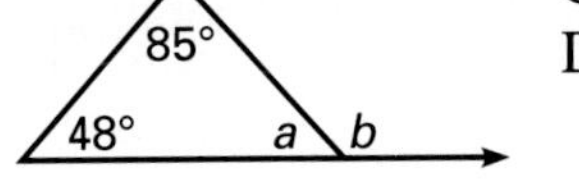


A. 47°
B. 133°
C. 85°
D. 37°

**15.** Find the measure of $\widehat{OMN}$ in $\odot P$.

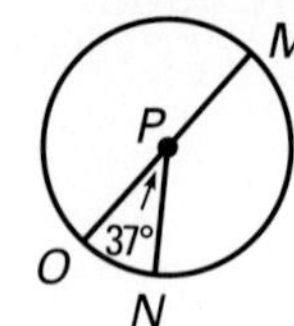


A. 323°
B. 143°
C. 37°
D. 313°

**16.** A trapezoid has
A. two pairs of parallel sides.
B. exactly one pair of parallel sides.
C. four congruent sides.
D. four right angles.

# CHAPTER 4 SKILLS PREVIEW

The names of all students in a school are placed in a box. Thirty names are drawn. These students are asked questions about their favorite sport.

1. What kind of sampling method is being used?
2. How might the results of this survey be biased?

The histogram at the right shows the amount spent by each customer in one day at a shop.

3. What was the most common amount spent?
4. How many customers spent $40 or more?
5. How many customers in all made purchases?

**CUSTOMER SPENDING**

Frequency: 0 1 2 3 4 5 6 7 8 9 10

0–9 10–19 20–29 30–39 40–49 50–59

Amount Spent (in dollars)

A car manufacturer tested a car under different driving conditions and collected the following data.

Mi/gal of gas: 22 30 15 24 16 24 23

6. Find the mean, median, mode, and range of these data.

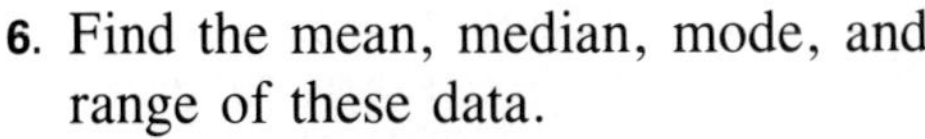

7. The manufacturer used the mode to advertise this car's average miles per gallon. Is this misleading? Explain.

A school librarian keeps track of the number of fiction books checked out each day. The data that were recorded are shown in the stem-and-leaf plot.

0|2 represents 2 books

| Stem | Leaves |
|---|---|
| 0 | 2 |
| 3 | 3 |
| 4 | 4 8 |
| 5 | 2 3 5 5 |
| 6 | 0 1 4 6 |

8. On how many days were more than 50 fiction books checked out?
9. What is the range of the number of books checked out?
10. On how many days were fewer than 45 books checked out?

Use the scatter plot at the right.

11. Is there a positive or a negative correlation between the evening time and temperature?
12. What temperature would you expect it to be at 11:00 P.M.?
13. Daniel had the tenth highest score of the 25 students who took a test. What was his percentile rank?

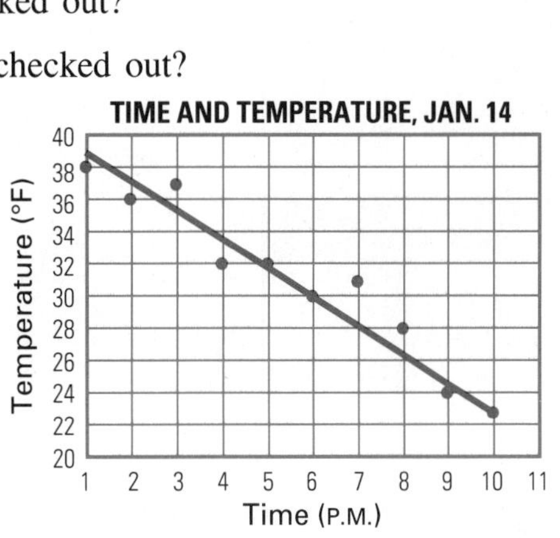

# STATISTICS

**THEME** **Entertainment**

**Statistics** is the branch of mathematics that involves collecting **data,** or pieces of information, and organizing them in such a way that they can be used to make decisions. In this chapter, you will have the opportunity to collect data, organize, and display them in frequency tables, histograms, stem-and-leaf plots, scatter plots, and box-and-whisker plots. You will use statistical measures to help describe a set of data. You will also explore how statistics can be used to estimate probabilities.

THE ELECTRONIC HOME, 1990

| Percent of U.S. homes | Facilities |
|---|---|
| 99% | Have a radio |
| 98 | Have a TV set |
| 96 | Have a color TV set |
| 92 | Have an audio system |
| 68 | Have a VCR |
| 64 | Have two or more TV sets |
| 55 | Buy basic cable |
| 52 | Own prerecorded videocassettes |
| 31 | Have a telephone answering device |
| 29 | Buy one or more pay channels |
| 25 | Have a cordless telephone |
| 23 | Have a home computer |
| 19 | Have color TV with MTS |
| 19 | Have a compact disc player |
| 19 | Have a home alarm system |
| 10 | Have a camcorder |
| 6 | Have projection TV |
| 3.5 | Have LCD TV |
| 3 | Have satellite dishes |

Source: *Universal Almanac,* 1992.

## DECISION MAKING

### Using Data

TOP 15 REGULARLY SCHEDULED NETWORK PROGRAMS, NOV. 1990

| Rank | Program name (network) | Total percent of TV households |
|---|---|---|
| 1. | Cheers (NBC) | 24.0 |
| 2. | 60 Minutes (CBS) | 21.8 |
| 3. | Murder, She Wrote (CBS) | 18.2 |
| 3. | Roseanne (ABC) | 18.2 |
| 5. | America's Funniest People (ABC) | 18.0 |
| 5. | America's Funniest Home Videos (ABC) | 18.0 |
| 7. | Murphy Brown (CBS) | 17.8 |
| 8. | Designing Women (CBS) | 17.7 |
| 9. | Empty Nest (NBC) | 17.6 |
| 10. | Bill Cosby Show (NBC) | 17.5 |
| 11. | Golden Girls (NBC) | 17.2 |
| 12. | A Different World (NBC) | 17.0 |
| 13. | NFL Monday Night Football (ABC) | 16.7 |
| 14. | Unsolved Mysteries (NBC) | 16.6 |
| 15. | Matlock (NBC) | 16.3 |

**Total U.S. TV households: 93,100,000**

Source: *Information Please Almanac,* 1991.

1. How do you think the data were collected about electronics in people's homes? Do you think each household in the United States was surveyed individually? How do you think the data from your town would compare to the data for the whole country?
2. Suppose you were the advertising director of a company that manufactured personal computers. How could you use the list of the top 15 network programs and the circle graph showing the composition of the comedy show audience to help decide on which shows to buy commercial time? If you could buy time on only one of these shows, which would you choose?

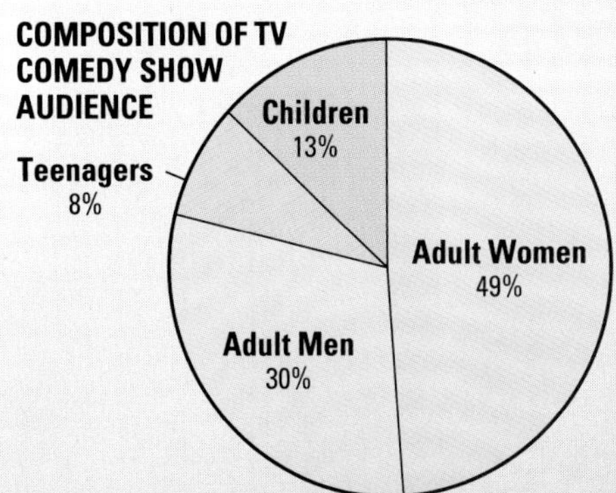

3. How does the change in the number of people going to the movies compare to the change in the price of a movie ticket? Do you think the change in one factor—the number of people going to the movies or the price of a ticket—caused the change in the other factor? What else may have been responsible for the changes?

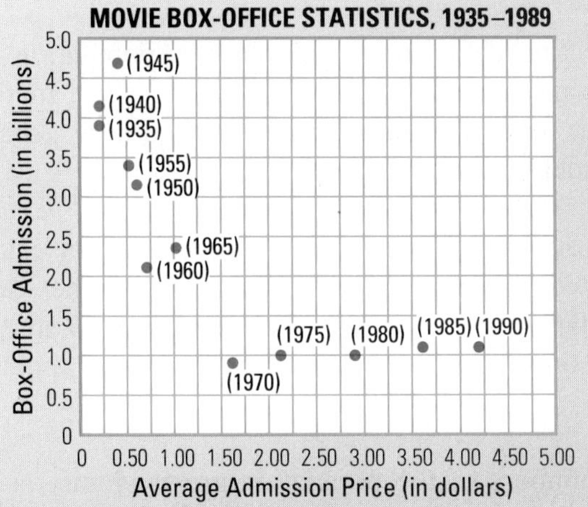


## U.S. FILM BOX-OFFICE RECEIPTS, ADMISSIONS, AND ADMISSION CHARGES, 1926 TO 1989

Despite headlines hailing 1989 as the best box-office year ever, the motion picture industry was healthier—in terms of both admissions and real-dollar earnings—in the 1930s, 1940s, and 1950s.

| | Box-Office Receipts (millions) | | | |
|---|---|---|---|---|
| Year | Current Dollars | Constant Dollars | Admissions ('000s) | Average Admission Charge |
| 1926 | $720.0 | $4,621.0 | 2,600,000 | N.A. |
| 1930 | 732.0 | 4,979.4 | 4,680,000 | N.A. |
| 1935 | 566.0 | 4,693.3 | 3,900,000 | $0.24 |
| 1940 | 735.0 | 5,964.0 | 4,160,000 | 0.24 |
| 1945 | 1,450.0 | 9,151.1 | 4,680,000 | 0.35 |
| 1950 | 1,376.0 | 6,486.0 | 3,120,000 | 0.53 |
| 1955 | 1,326.0 | 5,620.7 | 3,392,000 | 0.50 |
| 1960 | 951.0 | 3,649.8 | 2,080,000 | 0.69 |
| 1965 | 927.0 | 3,343.1 | 2,288,000 | 1.01 |
| 1970 | 1,162.0 | 3,402.1 | 920,400 | 1.55 |
| 1975 | 2,115.0 | 4,465.9 | 988,000 | 2.05 |
| 1980 | 2,748.5 | 3,789.2 | 1,021,500 | 2.69 |
| 1981 | 2,965.6 | 3,706.2 | 1,060,000 | 2.78 |
| 1982 | 3,452.7 | 4,064.5 | 1,175,400 | 2.94 |
| 1983 | 3,766.0 | 4,925.4 | 1,196,900 | 3.15 |
| 1984 | 4,030.6 | 4,406.9 | 1,199,100 | 3.36 |
| 1985 | 3,749.4 | 3,958.5 | 1,056,100 | 3.55 |
| 1986 | 3,778.0 | 3,915.9 | 1,017,200 | 3.71 |
| 1987 | 4,252.9 | 4,252.9 | 1,088,500 | 3.91 |
| 1988 | 4,458.4 | 4,282.7 | 1,084,800 | 4.11 |
| 1989 | 5,033.4 | 4,611.2 | 1,120,000 | 4.45 |

Source: *Universal Almanac*, 1992.

## Working Together

Find examples of statistical data gathered through surveys. You can find data in magazines, newspapers, and books and on television. The data may be in the form of tables, graphs, lists, or simple statements such as "Three out of four doctors recommend . . . ." For each example, discuss how a consumer or citizen is likely to use the data to make a decision. Then discuss in what occupations people would be most likely to use the data to make decisions.

# 4-1 Collecting Data

**EXPLORE**

How could you find out which recording group or artist is most popular among teenagers?

**SKILLS DEVELOPMENT**

One way to make a decision about an entire group, or **population,** is to collect data from members of that population. To do this, you need to conduct a **survey** or **poll.** Tools used are often questionnaires, interviews, or records of events. It can be very costly and can take a great amount of time to survey all members of a population. A more efficient way is to survey a representative part, or **sample,** of the population.

The first step in collecting data from a sample of a population is to choose an appropriate method of sampling. Here are some methods of sampling that might be used to determine the recording group most popular with students in your school.

**Random sampling:** Each member of the population has an equal chance of being selected. For example, the names of all the students in your school could be placed in a box. The names of 50 students would be drawn, and only those students would be asked about their favorite group.

**Cluster sampling:** Members of the population are randomly selected from particular parts of the population and surveyed in clusters. For example, a number of classrooms in your school could be selected at random. All students who have classes in those rooms during the first period of the day would be surveyed.

**Convenience sampling:** Members of a population are selected because they are readily available, and all are surveyed. For example, all 15 students who are in the school bookstore before school opens would be asked to name their favorite group.

**Systematic sampling:** Members of a population that has been ordered in some way are selected according to a pattern. For example, as all students in your school pass through the cafeteria line during one day, every tenth person would be surveyed.

## Example 1

A concert organizer wants to identify what type of music is most popular with adults in a midwestern city. What kind of sampling method is represented by each of these possibilities that the organizer has considered using?

a. Ask the first 30 adults who arrive for a concert at Symphony Hall one evening.

b. Ask all adults who shop in music stores located in an area of the city selected at random.

c. Ask all adults living in the city whose phone numbers end with the digit 3.

### Solution

a. This method represents *convenience* sampling

b. This method represents *cluster* sampling.

c. This method represents *random* sampling. ◄

Sometimes survey findings are **biased,** or not truly representative of the entire population. A biased survey can be a result of the sampling method used. For example, if the only adults surveyed about their favorite type of music are those attending a country music concert, the survey findings are likely to indicate that country music is the type of music most popular with adults. Survey findings may also be biased because some people do not respond to surveys, and some do not tell the truth on surveys. For example, suppose your teacher gives all students in your class a questionnaire to fill out about the time spent on homework each night. Some students may not fill out the questionnaire. Others may say that they spend more time on homework than they actually do.

## Example 2

A cereal manufacturer wants to identify the most popular type of breakfast food. The manufacturer decides to include a questionnaire in every tenth box of cereal that it packages.

a. What method of sampling does this represent?

b. Are the results of this survey likely to be biased? Explain.

### Solution

a. This method represents systematic sampling.

b. The survey results are likely to be biased. The only people who would be surveyed are those who eat one particular type of breakfast food, cereal. Some people who answer the questionnaire may forget to mail it, and some may not want to take the time to fill it out. The people most likely to fill out and return the questionnaire are those who want the manufacturer to know either that they have a strong liking for cereal or that they strongly disliked that particular cereal. ◄

## TRY THESE

1. The owners of a restaurant are planning a new menu and want to find out the type of sandwich meat most customers like best. What kind of sampling method is represented by each of these possibilities that the owners have considered using?
   a. Ask every fifth customer who arrives at the restaurant during one day.
   b. Ask all customers seated at the table closest to the cashier between 11:30 A.M. and 2:30 P.M. on one day.
   c. Ask all customers served during one day by a waiter chosen at random.

2. To identify the time most parents set as a curfew for their teenagers on Friday nights, all teenagers attending the 9:00 P.M. show at one movie theater were surveyed.
   a. What method of sampling does this represent?
   b. Are the results of this survey likely to be biased? Explain.

# EXERCISES

### PRACTICE/ SOLVE PROBLEMS

Suppose you want to find out the favorite sport among teenagers in your town. Name the sampling method represented by each of the following. Then give one reason why the findings that would result from using each method could be biased.

1. Ask every eighth teenager who enters a stadium for one particular game.

2. Ask eight teenagers whose names were drawn from the names of all students in your school.

3. Ask the first eight teenagers who arrive at a party.

## EXTEND/ SOLVE PROBLEMS

A town's city council is considering passing a law that would require its citizens to recycle all glass and plastic containers. The council invites the public to a meeting at which they discuss the proposed law. Before the meeting begins, the council has an interviewer ask the 50 citizens who arrive first whether they are in favor of mandatory recycling.

4. What method of sampling is being used?

5. How might the results of this survey be biased?

6. How could the sampling method be changed to produce a convenience sample that is definitely biased in favor of the proposed law?

7. How could the sampling method be changed to produce a convenience sample that is definitely biased against the law?

8. How could the sampling method be changed to produce a cluster sample that is less biased?

9. How could the sampling method be changed to produce a random sample? Would the results of this survey likely be biased?

## THINK CRITICALLY/ SOLVE PROBLEMS

Conduct your own survey! Work in groups of four or eight and follow these steps.

10. Choose a subject whose popularity you would like to investigate. Write a question that could be answered "yes" or "no" or one that will give your sample three to five choices from which to choose.

11. Test your question on members of another group to see if others understand it the way you do. If not, rewrite it and test again.

12. Explain how you will select a population from which to sample.

13. Divide your group into four individuals or pairs. Have each individual or pair plan a different sampling method—random, cluster, convenience, or systematic sampling.

14. Use your individual's or pair's sampling method to survey 20 students. Record your results.

15. Compare the results of each sampling method used by your group. Decide what conclusions can be drawn.

16. Imagine that you are writing a script for a local radio or television reporter. Describe your group's survey. Include why you selected your subject and population, how your sampling methods worked, how the data from the four samples compared, and any reasons that you believe may have caused the survey results to be biased.

# 4-2 Problem Solving Strategies:

## MAKE AN ORGANIZED LIST

- ► READ
- ► PLAN
- ► SOLVE
- ► ANSWER
- ► CHECK

Sometimes data are presented in such a way that it is difficult to find information quickly. Arranging the data in an organized way, such as a list, table, or matrix, can be helpful.

### PROBLEM

Derek, Louise, Bill, and Tina are salespeople in an electronics store. For each sale, a clerk records the salesperson's initial and records whether a VCR, TV, or stereo system (SS) was sold, and whether the customer paid with cash ($) or a check (✓) or charged the purchase (C). For example, "D-TV-✓" indicates that Derek sold a TV paid for by the customer with a check.

At the end of one weekend, the clerk's report looked like this.

| | | | |
|---|---|---|---|
| B-SS-C | L-SS-✓ | D-SS-✓ | T-TV-$ |
| T-VCR-$ | D-TV-✓ | D-VCR-C | L-VCR-✓ |
| B-TV-C | L-SS-C | B-TV-✓ | T-VCR-✓ |
| T-TV-C | B-SS-C | T-VCR-C | L-SS-C |
| L-VCR-$ | D-TV-C | B-TV-✓ | T-TV-C |
| L-SS-C | D-TV-C | B-VCR-✓ | D-VCR-$ |
| B-TV-$ | L-VCR-C | T-VCR-$ | L-TV-✓ |
| D-SS-$ | D-SS-C | | |

Based on each individual salesperson's total number of customers, who had the greatest percent of cash customers?

### SOLUTION

Organize the data in a matrix, using the salespeople's names for the row headings and other characteristics of the sales for column headings.

| | VCRs Sold | TVs Sold | Stereo Systems Sold | Customers | | |
|---|---|---|---|---|---|---|
| | | | | Cash | Check | Charge |
| Derek | 2 | 3 | 3 | 2 | 2 | 4 |
| Louise | 3 | 1 | 4 | 1 | 3 | 4 |
| Bill | 1 | 4 | 2 | 1 | 3 | 3 |
| Tina | 4 | 3 | 0 | 3 | 1 | 3 |

For each salesperson, find the total number of customers and note how many paid cash.

Derek: $2 + 2 + 4 = 8$; 2 cash  Louise: $1 + 3 + 4 = 8$; 1 cash
Bill: $1 + 3 + 3 = 7$; 1 cash  Tina: $3 + 1 + 3 = 7$; 3 cash

Then find the percent of customers who paid cash.

Derek: $\frac{2}{8} = \frac{1}{4} = 25\%$  Louise: $\frac{1}{8} = 12\frac{1}{2}\%$

Bill: $\frac{1}{7} = 14\frac{2}{7}\%$  Tina: $\frac{3}{7} = 42\frac{6}{7}\%$

So, with $42\frac{6}{7}\%$, Tina had the greatest percent of cash customers.

# PROBLEMS

Last Friday, Cinema Five sold the following numbers of tickets.
**Matinee**—Action movie: 102 adults, 38 children; Comedy: 97 adults, 88 children; Thriller: 115 adults, 61 children; Horror movie: 84 adults, 13 children; Adventure movie: 120 adults, 72 children
**Early evening show**—Thriller: 298 adults, 27 children; Adventure movie: 89 adults, 59 children; Comedy: 190 adults, 160 children; Horror movie: 225 adults, 51 children; Action movie: 177 adults, 65 children
**Late evening show**—Comedy: 152 adults, 35 children; Adventure movie: 173 adults, 2 children; Action movie: 255 adults, 3 children; Horror movie: 240 adults; Thriller: 335 adults, 11 children

1. Organize the data about Cinema Five in a matrix using the different show times as row headings. For column headings, use the five different types of movies. Then use this matrix for Exercises 2–7.
2. Which type movie was most popular?
3. Which type of film at which time had the best attendance?
4. For which show time did Cinema Five sell the most tickets?
5. Approximately what percent of Cinema Five's ticket sales were for the matinee show?
6. Write and answer one other question that can be answered using the matrix you made in Exercise 1.
7. Write one question based on data presented about Cinema Five that *cannot* be answered from the way in which you organized the data in the matrix in Exercise 1. How would you need to reorganize the data in a matrix to answer your question?

Sandy's Sandwich Shop offers small (S), medium (M), and large (L) submarine sandwiches on white or wheat rolls with 1, 2, or 3 different meats or with 0 meat (vegetarian-style). A shop employee made this chart for the submarine sandwiches sold during one hour.

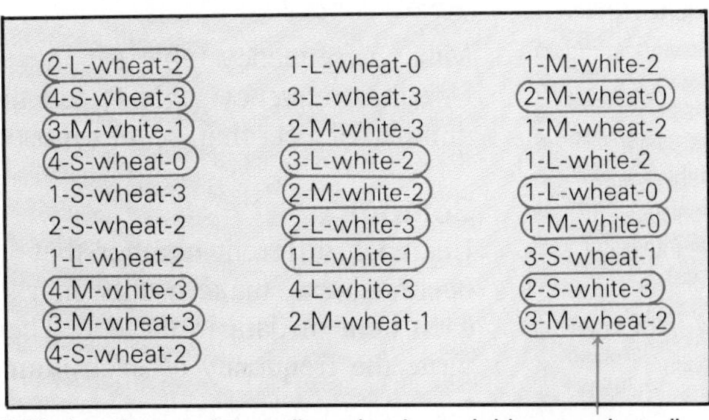


**"3-M-wheat-2" means three medium submarine sandwiches on a wheat roll with two meats. It is circled because it was sold for carry-out.**

8. Organize the data given in the employee's report in a matrix. Then write three questions about the data. Exchange matrices and questions with a classmate. Answer each other's questions.

# 4-3 Frequency Tables

## EXPLORE

A manufacturer received orders for sweatshirts in the following sizes. How many of each size were ordered? Are there any sizes that the manufacturer might want to discontinue? Why or why not?

| | | | | | | |
|---|---|---|---|---|---|---|
| small | medium | medium | large | petite | small | large |
| medium | small | petite | large | large | small | large |
| medium | large | small | medium | medium | petite | large |
| petite | X-large | large | medium | medium | small | medium |
| large | medium | medium | small | large | medium | medium |

## SKILLS DEVELOPMENT

It is very difficult to see a pattern or trend in a set of data that is not organized. For example, in the listing of sweatshirt sizes above, it is not easy to see which size is most popular. You have learned that one method of organizing data is with a matrix. Another method that is commonly used is a **frequency table.**

In a frequency table, a tally mark is used to record each time an item appears in a set of data. For every fifth tally of an item, you make a diagonal mark across four vertical marks (𝍸). The total number of tallies for an item indicates that number of times the item appears in the set of data. This number is called the **frequency** of the item.

**MATH: WHO, WHERE, WHEN**

As early as 3000 B.C., very simple frequency tables were used by the Babylonians, who pressed marks into clay tablets using a writing implement called a **stylus.** The tablets were then baked until they were hard, resulting in a permanent record.

### Example 1

Make a frequency table for these data. Use the frequency table to tell the number of pets that occurs most often.

NUMBER OF PETS PER FAMILY

| | | | | | | | |
|---|---|---|---|---|---|---|---|
| 1 | 2 | 3 | 1 | 0 | 2 | 1 | 0 |
| 1 | 0 | 1 | 4 | 1 | 2 | 0 | 0 |
| 0 | 1 | 1 | 2 | 2 | 5 | 1 | 0 |

**Solution**

List each different number that occurs. Next, make a tally mark for each item of data in the set. Then write the frequency of each number.

The number of pets that occurs most often is 1. ◄

NUMBER OF PETS PER FAMILY

| Number of Pets | Tally | Frequency |
|---|---|---|
| 0 | 𝍸 \|\| | 7 |
| 1 | 𝍸 \|\|\|\| | 9 |
| 2 | 𝍸 | 5 |
| 3 | \| | 1 |
| 4 | \| | 1 |
| 5 | \| | 1 |

Sometimes a set of numerical data contains so many different items that a frequency table with an entry for each item would not be helpful. In cases like these, you can group the data into intervals.

## Example 2

Make a frequency table for these data. Group the data into intervals of 10. Use the frequency table to answer the questions.

**a.** On how many days were there 70–79 people in attendance?

**b.** On how many days were there at least 90 people in attendance?

**c.** To the nearest percent, on what percent of the days did fewer than 80 people attend?

HEALTH CLUB ATTENDANCE IN NOVEMBER

| | | | | | |
|---|---|---|---|---|---|
| 42 | 31 | 56 | 127 | 32 | 83 |
| 94 | 26 | 72 | 84 | 121 | 67 |
| 53 | 102 | 97 | 59 | 43 | 30 |
| 74 | 86 | 92 | 127 | 108 | 90 |
| 73 | 56 | 64 | 89 | 71 | 80 |

### Solution

The least number of people is 26; the greatest is 127. Group the data into intervals.

**a.** The 70–79 row shows four days with this attendance.

**b.** Add the number of days shown for 90–99, 100–109, 110–119, and 120–129: $4 + 2 + 0 + 3 = 9$. There were at least 90 people on nine days.

**c.** Days with fewer than 80: $1 + 3 + 2 + 4 + 2 + 4 = 16$. Total number of days: 30

$\frac{16}{30} \approx 53.3\%$

Fewer than 80 people attended on about 53% of the days. ◄

HEALTH CLUB ATTENDANCE IN NOVEMBER

| Number of People | Tally | Frequency |
|---|---|---|
| 20–29 | \| | 1 |
| 30–39 | \|\|\| | 3 |
| 40–49 | \|\| | 2 |
| 50–59 | \|\|\|\| | 4 |
| 60–69 | \|\| | 2 |
| 70–79 | \|\|\|\| | 4 |
| 80–89 | ~~\|\|\|\|~~ | 5 |
| 90–99 | \|\|\|\| | 4 |
| 100–109 | \|\| | 2 |
| 110–119 | | 0 |
| 120–129 | \|\|\| | 3 |

## TRY THESE

Make a frequency table for these data. Group the data into intervals of 5. Use the frequency table to answer the questions.

NUMBERS OF CASSETTE TAPES STUDENTS OWN

| | | | | | | | | | |
|---|---|---|---|---|---|---|---|---|---|
| 2 | 3 | 5 | 6 | 17 | 3 | 23 | 2 | 12 | 19 |
| 14 | 13 | 12 | 19 | 20 | 11 | 10 | 13 | 7 | 6 |
| 8 | 13 | 16 | 4 | 6 | 7 | 1 | 9 | 3 | 7 |
| 10 | 15 | 17 | 20 | 12 | 1 | 8 | 9 | 7 | 11 |

**a.** How many students own 6–10 tapes?

**b.** How many students own more than 10 tapes?

**c.** To the nearest percent, what percent of students own more than 15 tapes?

**TALK IT OVER**

Suppose you had grouped the data into intervals of 8. Would you still have been able to answer the questions using the frequency table?

# EXERCISES

## PRACTICE/ SOLVE PROBLEMS

Make a frequency table for each set of data. Then use the table to answer the question.

1. Number of Games Bowled by League Members:
   81 78 81 81 84 81 78 84 75 84 81 81
   81 84 75 78 81 81 84 81 78 81 84 84 87
   How many league members bowled at least 81 games?

2. Attendance at Rocktown H.S. Basketball Games:
   186 162 243 487 302 189 283 491 386
   401 256 127 438 257 195 374 185 362
   To the nearest percent, at what percent of the games was the attendance less than 200?

3. *USING DATA* Use the Data Index on pages 556–557 to find entertainers' dates of birth. Group the data by decades, such as 1920–1929. In what decade were the greatest number of these entertainers born?

## EXTEND/ SOLVE PROBLEMS

Use the frequency table at the right for Exercises 4–6.

4. What is the total number of books read by this class during the grading period?

5. If students in this class were required to read from three to five books, how many students read the required number? How many read fewer books than required? How many read more books than required?

6. What percent of the class read at least six books?

**BOOK REPORTS TURNED IN BY AN ENGLISH CLASS DURING ONE GRADING PERIOD**

| Number of Book Reports | Tally | Frequency |
|---|---|---|
| 2 | \|\|\| | 3 |
| 3 | 𝍸 | 5 |
| 4 | \|\|\|\| | 4 |
| 5 | 𝍸 \| | 6 |
| 6 | \|\| | 2 |
| 7 | \|\| | 2 |
| 8 | \| | 1 |
| 9 | \| | 1 |
| 10 | \| | 1 |

## THINK CRITICALLY/ SOLVE PROBLEMS

For each math test, Mrs. Smith makes a frequency table of the class results. Marissa's test paper for one test is missing, but Mrs. Smith knows that she graded the paper and recorded it in the table.

7. If Mrs. Smith uses a frequency table such as the one in Example 1, can she tell what Marissa's score is? Explain.

8. If Mrs. Smith uses a frequency table such as the one in Example 2, can she tell what Marissa's score is? Explain.

► READ
► PLAN
► SOLVE
► ANSWER
► CHECK

# Problem Solving Applications:

## DESIGNING A SURVEY

People's television viewing habits are of great interest to television stations. The stations must plan what programs they should provide for viewers and at what times. One way stations can learn about viewing habits is to survey a sample of viewers. To get the data they need, they must ask good questions.

**Poor question:** Do you watch TV a lot?
**Good question:** How many hours of TV do you watch daily?

To write a good survey question, use these guidelines.

- Use terms most people will interpret in the same way. For example, "a lot" means different amounts to different people.
- Write a question that is easy to understand and not too long.
- A question should require only one answer. For example, you might ask, "What types of programs do you watch most often on weekdays and weekends?" A person who liked to watch comedies on weekdays and mysteries on the weekend could not answer that question with a single-word answer.

1. Work in pairs to write a good survey question that could be used to gather data about one of the following.
   a. Television sets per household
   b. How many hours of television students watch
   c. How much use VCRs receive in households
   d. Most popular type of television show

2. Use your question to survey 20 people. Organize the data you collect in a frequency table.

Another way TV stations gather data about viewing habits is to have samples of viewers keep records of the stations and shows they watch for each 15-minute interval of their viewing time.

3. Choose one night for each member of your class to gather data on his or her own viewing habits. Keep written records of the station and program you are watching every 15 minutes.

4. Make class frequency tables to organize the data you collect.

5. Use the class tables to answer these questions.
   a. Which station was watched for the most 15-minute intervals?
   b. Which show was watched by the greatest number of people?
   c. Was the most popular show also on the most popular station?

# 4-4 Measures of Central Tendency

## EXPLORE

The table at the right shows quiz scores for 10 weekly quizzes.

| Quiz | 1 | 2 | 3 | 4 | 5 | 6 | 7 | 8 | 9 | 10 |
|---|---|---|---|---|---|---|---|---|---|---|
| Ana | 9 | 7 | 10 | 9 | 8 | 8 | 9 | 10 | 2 | 9 |
| Bob | 9 | 8 | 10 | 10 | 7 | 7 | 8 | 9 | 8 | 8 |

a. What is Ana's average score for the 10 quizzes?
What is Bob's average score?

b. On how many quizzes has Ana done as well as or better than Bob?
On how many quizzes has Bob done as well as or better than Ana?

c. Which score did Ana receive most often?
Which score did Bob receive most often?

d. Which student do you think has done better? Why?

## SKILLS DEVELOPMENT

**READING MATH**

The notation $\bar{x}$ is sometimes used to denote the mean of a set of data. Scientific, statistical, and graphing calculators usually use this notation.

In statistics, there are three **measures of central tendency** you can use to describe a set of data. Each of these measures of central tendency in some way represents a central, or middle, value of the data.

- The **mean,** or arithmetic average, is the sum of the data divided by the number of data. The mean is the most appropriate measure of central tendency to use when there are not extreme values within the data.
- The **median** is the middle value of the data when the data are arranged in numerical order. (If the number of data is even, the median is the average of the *two* middle numbers.) The median is the most appropriate measure to use when there are a few extreme values within the data.
- The **mode** is the number that occurs most often in a set of data. A set of data may have one mode, more than one mode, or no mode. When you want to describe the most characteristic value of a set of data, it is most appropriate to use the mode.

**TALK IT OVER**

Suppose a set of data had a mean of 50, a median of 45, and a mode of 57. Would 45, 50, or 57 have to be numbers in this set of data?

Another number useful in describing a set of data is the **range.** The range is the difference between the greatest and least values in a set.

### Example

South Central High School's football team scored the following number of points during its games this past season.

27 32 6 24 29 30 8 26 30 32

a. Find the mean.
b. Find the median.
c. Find the mode.
d. Find the range.
e. Which measure of central tendency is the best indicator of the typical number of points scored per game?

**Solution**

a. Add the data items and divide by the number of items.

$$27 + 32 + 6 + 24 + 29 + 30 + 8 + 26 + 30 + 32 = 244$$

$$\text{number of data items} \rightarrow \frac{244}{10} = 24.4$$

The mean number of points scored per game is 24.4.

b. Rewrite the data in numerical order.

6 8 24 26 27 29 30 30 32 32

↑ middle value

Since there is an even number of data items, the median is halfway between the two middle values, 27 and 29, or 28.

c. The numbers 30 and 32 both occur twice. So, this set of data has two modes, 30 and 32.

d. The range of points is the difference between 32 and 6, or 26.

e. The best indicator of points scored per game is the median since it is not affected by the extreme values of 6 and 8, as is the mean. ◄

## TRY THESE

Round your answers to the nearest tenth.

Pat biked the following number of miles on each day of his trip.

12.3 12.8 12.4 18.4 27.1 14.9 17.5 12.7 11.4 13.5

1. Find the mean.
2. Find the median.
3. Find the mode.
4. Find the range.
5. Which measure of central tendency is the best indicator of the typical number of miles Pat biked each day?

# EXERCISES

## PRACTICE/ SOLVE PROBLEMS

The weights of the members of a girls' softball team are as follows.

120 132 156 124 109 112 120 140 138 114 115

1. Find the mean, median, mode, and range of this set of data. Round your answers to the nearest tenth.
2. Which measure of central tendency is the best indicator of the typical weight of a team member?

This table shows how much water was consumed daily by a group of people.

3. Find the mean, median, mode, and range of this set of data. Round your answers to the nearest tenth.
4. Which measure of central tendency best indicates the amount of water the average person drinks daily?

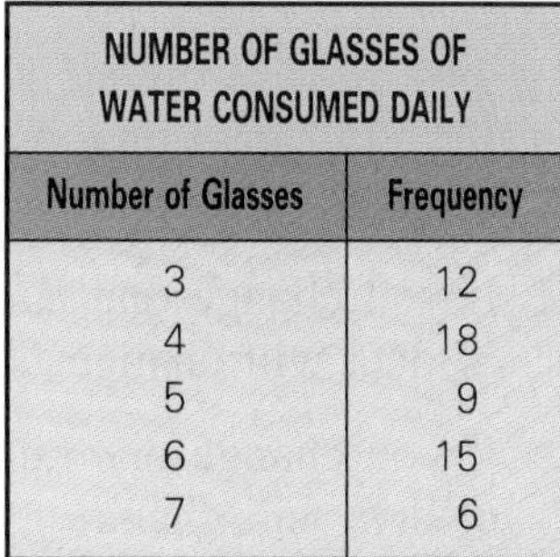

NUMBER OF GLASSES OF WATER CONSUMED DAILY

| Number of Glasses | Frequency |
|---|---|
| 3 | 12 |
| 4 | 18 |
| 5 | 9 |
| 6 | 15 |
| 7 | 6 |

**CALCULATOR**

Using the memory keys on your calculator can simplify your work when finding the mean of a set of data presented in a frequency table. To find the mean of the data in Exercise 3, you would enter:

12 [×] 3 [M+] 18 [×] 4 [M+] 9 [×] 5 [M+] 15 [×] 6 [M+] 6 [×] 7 [M+] [MR] [÷] 60 [=]

## EXTEND/ SOLVE PROBLEMS

5. Jenny spent the following on meals she ate out last month.
   $4.32 $5.16 $3.59 $2.88 $6.18
   $15.87 $8.81 $7.49 $10.00

   If the $15.87 meal were not included in the set of data, how would any of the mean, median, mode, or range of this set of data be different? By how much? Round your answers to the nearest cent.

6. Suppose you received the following scores on math tests.
   84 75 87 96 75

   If your teacher allowed you to choose the measure of central tendency that would be used to determine your grade, which measure would you choose? Explain.

7. Refer to the test scores in Exercise 6. Suppose you want to raise your average to 86. You have one more test to take.
   a. What score would you need to get if your teacher used the mean to determine your grade?
   b. Could you get an average of 86 if your teacher used the median to determine your grade. Explain.

8. Jason's math assignment was to collect a set of data and find its mean, median, mode, and range. Jason surveyed the ten families who live on his block to find out the number of children each had. The work he brought to class is shown at the right. What numbers did he forget to write in the frequency table?

| NUMBER OF CHILDREN PER FAMILY | |
|---|---|
| Number of Children | Frequency |
| 0 | ■ |
| 1 | ■ |
| 2 | ■ |
| 3 | ■ |
| 4 | ■ |

Mean: 2.2
Median: 2
Modes: 2 and 3
Range: 4

**WRITING ABOUT MATH**

In your journal, give examples of different situations in which the mean or median or mode would be the best measure of central tendency to describe the set of data.

## THINK CRITICALLY/ SOLVE PROBLEMS

Tell whether each statement is *sometimes, always,* or *never* true. If you answer *sometimes* or *never,* create sample sets of data as counterexamples to support your answers.

9. If you change one number in a set of data, the mean of the set of data will change.

10. If you change one number in a set of data, the median of that set of data will change.

11. If you change one number in a set of data, the mode of the set of data will change.

12. If you change one number in a set of data, the range of the set of data will change.

► READ
► PLAN
► SOLVE
► ANSWER
► CHECK

# Problem Solving Applications:

## USING A WEIGHTED MEAN

During the fall semester, Ricardo's math test scores were 60, 70, 80, and 90. To find the mean of his scores, Ricardo adds his scores then divides by the number of tests.

$$60 + 70 + 80 + 90 = 300; \frac{300}{4} = 75 \quad \leftarrow \text{mean}$$

Suppose Ricardo's teacher decides to give greater emphasis to the grades a student makes as the semester progresses. To do this, the teacher uses a **weighted mean** as follows. The first test is given a weight of 1, the second is given a weight of 2, the third is given a weight of 3, and the fourth is given a weight of 4. (In effect, a weight of 2 for a test score of 70 is the same as having received 70 on two tests.) Then the weighted mean is found as follows.

$$1(60) + 2(70) + 3(80) + 4(90) = 800$$

Because of the weighting there is a total of 10 scores. → $\frac{800}{10} = 80$ ← weighted mean

So, Ricardo's weighted mean score would be 80.

1. Jan is in the same math class as is Ricardo. Her test scores were 90, 80, 70, and 60. Find Jan's mean test score and her weighted mean score using the same weights 1, 2, 3, 4, as above.

2. During the spring semester, Felix's science test scores were 80, 85, 95, 92, and 86. He received 78 on the final exam.
   a. Find Felix's mean test score including the final exam.
   b. Suppose the teacher announced that the final exam would count three times as much as the regular tests. What is Felix's weighted mean score?

Many companies offer a *pension plan* to employees. Upon leaving the company, an employee may receive a regular payment determined by the plan. Suppose a company pension plan determines payments by using the employee's annual salary for his or her last five years of employment. Year 4 is given a weight of 2, and year 5 is given a weight of 4. The other three years are given a weight of 1. The employee receives an annual pension equal to 30% of his or her weighted mean salary.

Find the annual pension for each of these employees. Round answers to the nearest dollar.

| | Name | Year 1 | Year 2 | Year 3 | Year 4 | Year 5 |
|---|---|---|---|---|---|---|
| 3. | M. Royce | \$12,000 | \$14,000 | \$16,500 | \$16,500 | \$18,000 |
| 4. | T. Soong | 14,000 | 15,000 | 18,000 | 21,500 | 24,000 |
| 5. | B. Hernandez | 13,000 | 16,000 | 19,500 | 24,000 | 30,000 |

6. ***CRITICAL THINKING*** When does a weighted mean have the same value as the arithmetic mean of a set of data?

# 4-5 Problem Solving Skills:

## IDENTIFY MISLEADING STATISTICS

► READ
► PLAN
► SOLVE
► ANSWER
► CHECK

You have learned to use mean, median, and mode to interpret sets of data. Sometimes these measures of central tendency are used in misleading ways.

### PROBLEM

Furniture World has seven salespeople. The commissions they earned last week were $493, $289, $305, $299, $366, $308, and $299. The owner of the store placed this ad in the newspaper for additional salespeople.

**EXPERIENCED SALESPEOPLE**

wanted for furniture store. Earn an average weekly commission of $335. Apply in person at

**FURNITURE WORLD**
219 Plainfield Tpke.

**a.** Which measure of central tendency is used in the ad?

**b.** Does this ad give a fair picture of the salespeople's weekly commissions? Explain.

**c.** Which measure(s) of central tendency would be useful to a person considering a sales position at Furniture World? Explain.

### SOLUTION

**a.** Determine the three measures of central tendency. Arranging the data in order simplifies finding the median and the mode.

$$\$289 + \$299 + \$299 + \$305 + \$308 + \$366 + \$493 = \$2{,}359$$

mode ($299, $299)   median ($305)

$$\frac{\$2{,}359}{7} = \$337 \leftarrow \text{mean}$$

The measure of central tendency used in the ad is the mean.

**b.** No; only two of seven commissions are greater than or equal to the mean. The other five commissions are much lower than the mean.

**c.** The median and mode would be useful because five commissions are equal to or near these measures.

**CONNECTIONS**

The Latin expression *caveat emptor* means "let the buyer beware." This is a good rule to remember when making a spending decision based on survey data that has been interpreted for you.

# PROBLEMS

1. The monthly rainfall recorded, in centimeters, at Mountain Valley Resort last year was as follows.

| J | F | M | A | M | J | J | A | S | O | N | D |
|---|---|---|---|---|---|---|---|---|---|---|---|
| 0 | 2 | 7 | 12 | 18 | 24 | 25 | 25 | 22 | 6 | 3 | 0 |

Here is part of the resort's brochure:

> **BEAUTIFUL SUMMER WEATHER**
> Average monthly rainfall less than 10 centimeters!

a. Which measure of central tendency was used in the brochure?
b. Does the statement about rainfall in the brochure give a fair picture of weather conditions? Explain.
c. Which measure of central tendency would be a better indicator of rainfall during the summer?

2. A real-estate broker sold eight houses last month for $90,000, $75,000, $150,000, $80,000, $140,000, $80,000, $100,000, and $95,000.
   a. Which measure of central tendency might the broker use to tell homeowners the average selling price in order to encourage them to want to sell their houses? Explain.
   b. Which measure of central tendency should the broker use to describe the average selling price to people looking for a house?

3. Here are the salaries at a certain engineering firm that employs 40 people.
   a. Could you use any measure of central tendency to make a statement about the typical salary at this firm that would not be misleading?
   b. What other information would you need to make a statement about the typical salary that would not be misleading?

**ENGINEERING FIRM'S SALARIES**

| Position | Salary |
|---|---|
| President | $120,000 |
| Vice president | 80,000 |
| General manager | 55,000 |
| Senior engineer | 50,000 |
| Engineer | 40,000 |
| Accountant | 30,000 |
| Programmer | 30,000 |
| Administrative assistant | 18,000 |

4. A movie theater charges $7 for a ticket. The theater manager is considering raising the ticket price and asked 25 people to name the greatest amount they would be willing to pay to see a movie. The results are shown in the table.
   a. In order to justify a large increase in the ticket price, which measure of central tendency might the theater manager use? Explain.
   b. If theater patrons saw this survey, which measure of central tendency might they point out to the manager to keep the price increase low? Explain.
   c. Do you think the results of the manager's survey are biased?

**GREATEST AMOUNT PEOPLE WOULD PAY TO SEE A MOVIE**

| Amount | Frequency |
|---|---|
| $20.00 | 3 |
| 15.00 | 4 |
| 12.00 | 3 |
| 10.00 | 5 |
| 8.00 | 7 |
| 7.00 | 1 |
| 5.00 | 1 |
| 2.00 | 1 |

**CHECK UNDERSTANDING**

Suppose you decided to conduct your own survey in order to show the theater manager that the ticket price should not be increased. How would you conduct your survey differently than the manager most likely did?

# 4-6 Histograms and Stem-and-Leaf Plots

## EXPLORE

Students in one class were asked how many hours a week they exercised. All students responded to the question. The graph shows the results.

a. How many students are in the class?

b. Find the mean of the data to the nearest tenth of an hour.

c. Find the median number of hours.

d. Find the mode of the data.

e. Find the range of the data.

f. How does this graph differ from other bar graphs?

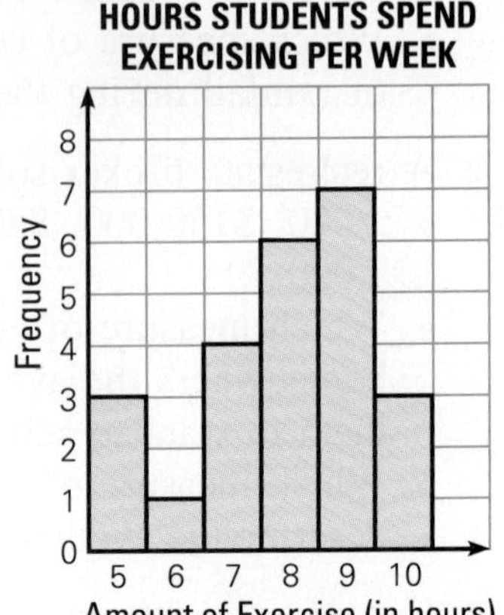


## SKILLS DEVELOPMENT

You have been working with data displayed in tables. There are many other ways to display data. For example, frequencies can be shown in a bar graph called a **histogram.** A histogram differs from other bar graphs in that no space is left between the bars and the bars usually represent grouped intervals of numbers. When you look at data displayed in a histogram, you can easily see certain characteristics of the data.

### Example 1

A gym teacher tested the number of sit-ups students in two classes could do in 1 min. The results are shown in this frequency table. Make a histogram of the data.

**SIT-UPS DONE IN 1 MINUTE**

| Number of Sit-Ups | Frequency |
|---|---|
| 0–4 | 8 |
| 5–9 | 12 |
| 10–14 | 15 |
| 15–19 | 6 |
| 20–24 | 18 |
| 25–29 | 10 |

**CHECK UNDERSTANDING**

How would the histogram have been different if intervals of 10, instead of 5, were used

**Solution**

Write the same intervals used in the frequency table along the horizontal axis and the frequency numbers along the vertical axis. ◄

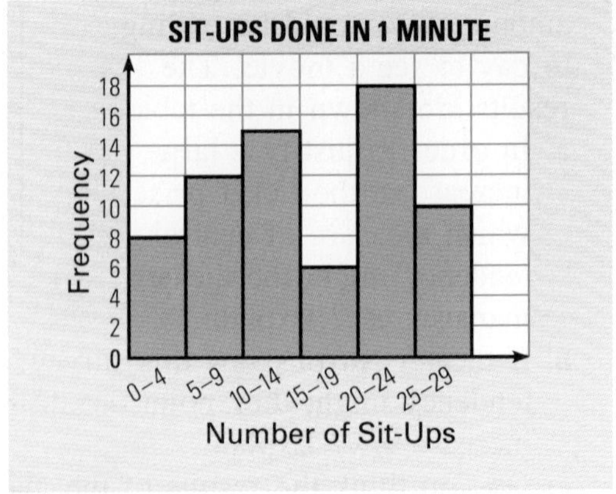

## Example 2

Use the histogram from Example 1 to answer these questions.
a. How many students were able to do 25–29 sit-ups in 1 min?
b. How many students were unable to do 10 sit-ups in 1 min?
c. Between which two consecutive intervals does the greatest increase in frequency occur? What is the increase?

### Solution

a. The bar for the interval 25–29 reaches 10 on the vertical axis. Ten students were able to do 25–29 sit-ups in 1 min.
b. Add the number of students who did 0–4 sit-ups in 1 min and the number who did 5–9 sit-ups in 1 min: $8 + 12 = 20$. Twenty students were unable to do 10 sit-ups in 1 min.
c. The greatest increase occurs between the intervals of 15–19 and 20–24. The frequencies for these intervals are 6 and 18, so the increase is $18 - 6 = 12$. ◄

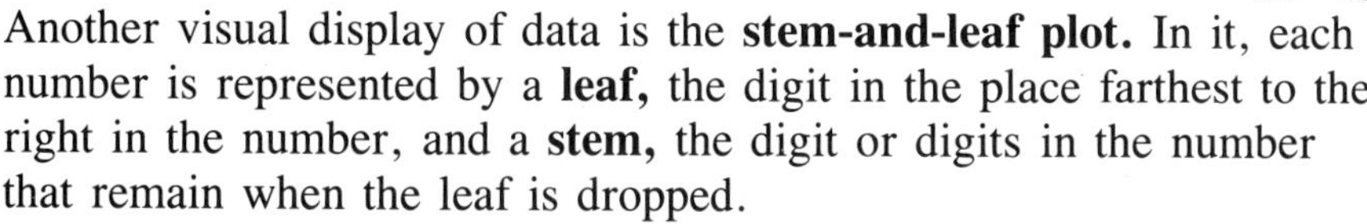

Another visual display of data is the **stem-and-leaf plot.** In it, each number is represented by a **leaf,** the digit in the place farthest to the right in the number, and a **stem,** the digit or digits in the number that remain when the leaf is dropped.

## Example 3

Organize the number of shirts sold daily for two weeks at D. Nims into a stem-and-leaf plot.

72 74 83 75 43 79 80
75 102 79 84 73 80 75

### Solution

Let the stems be the tens and hundreds of digits and the leaves be the ones digits of the data. Sort the number of shirts by the stems.

| stems | leaves |
|---|---|
| 4 | 3 |
| 7 | 2 4 5 9 5 9 3 5 |
| 8 | 3 0 4 0 |
| 10 | 2 |

Then arrange the leaves in numerical order. Write an explanation of the plot to the left of the stems. Include a title for the stem-and-leaf plot.

**NUMBER OF SHIRTS SOLD DAILY FOR TWO WEEKS AT D. NIMS**

4|3 represents 43 shirts

| | |
|---|---|
| 4 | 3 |
| 7 | 2 3 4 5 5 5 9 9 |
| 8 | 0 0 3 4 |
| 10 | 2 |

◄

**CHECK UNDERSTANDING**

What stems would you use to display each set of data in a stem-and-leaf plot?

1. 3.5, 4.6, 3.9, 4.1, 2.7, 3.4
2. 11, 9, 25, 17, 24, 19

A stem-and-leaf plot can help you interpret data. **Outliers,** or values that are much greater or less than most of the other values, as well as **clusters** of values and **gaps** between values are easy to spot.

## Example 4

Use the data in the stem-and-leaf plot from Example 3.
a. Find any outliers in the data.
b. Find any clusters or gaps.
c. Find the median.
d. Find the mode.
e. Find the range.

**TALK IT OVER**

Does displaying data in a histogram or in a stem-and-leaf plot make it any easier to find the mean? Explain.

**Solution**

a. Since 43 is much less than the other data, and 102 is much greater, these are outliers.

b. Clusters of data are in the low to middle 70s and around 79 and 80. The greatest gaps occur between the outliers and the rest of the data and between 75 and 79.

c. Since there are 14 data items, the median is halfway between the seventh and eighth pieces of data in the plot. Halfway between 7|5 and 7|9 is 7|7. So the median is 77 shirts.

d. Since 7|5 appears most often, the mode is 75 shirts.

e. The range is the difference between the greatest and least values. 10|2 and 4|3: $102 - 43 = 59$. The range is 59 shirts. ◄

## TRY THESE

An electronics company tests batteries to determine how long they last. Thirty batteries were tested until they failed. The time to failure (in hours) is recorded here.

| TIME TO FAILURE OF BATTERIES | |
|---|---|
| **Number of Hours** | **Frequency** |
| 0–9 | 8 |
| 10–19 | 9 |
| 20–29 | 7 |
| 30–39 | 4 |
| 40–49 | 2 |

1. Make a histogram of these data.
2. How many batteries lasted 19 h or less?
3. How many hours were most common for a battery to last?
4. How many batteries lasted more than 29 h?

Tickets for a rock concert were sold each day as follows.

149 253 366 169 297 421 183 256 303 427 189
229 367 520 147 168 253 146 182 305 412

5. Organize the data into a stem-and-leaf plot.
6. Name any outliers, clusters, or gaps in the data.
7. Find the median.
8. Find the mode.
9. Find the range.

**MIXED REVIEW**

1. 56 x 93
2. 427 + 829
3. 1,209 - 4,371
4. 30% of 570
5. 5,481 ÷ 87
6. $8\frac{2}{3} \div 1\frac{5}{6}$
7. A jacket originally costing \$92 is on sale for 25% off. If the sales tax rate is $7\frac{1}{4}$%, how much would someone pay for the jacket?
8. The measures of two angles of a triangle are 37° and 66°. Find the measure of the third angle.

# EXERCISES

**PRACTICE/ SOLVE PROBLEMS**

Use the histogram for Exercises 1–3.

1. How many of the council members weigh between 70 and 74 kg? How many weigh less than 70 kg?

**WEIGHTS OF CITY COUNCIL MEMBERS**

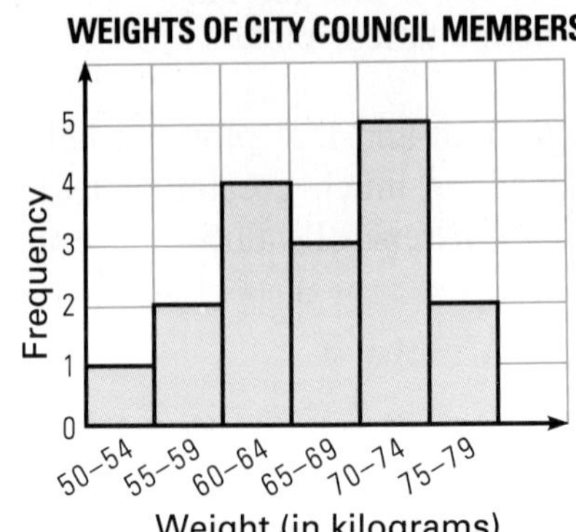

2. What is the most common weight for a council member?

3. How many council members are there?

Use these data for Exercises 4–7. Here are the percents of free throws Alan made during each basketball game he played this season.

| | | | | | | | | | |
|---|---|---|---|---|---|---|---|---|---|
| 86 | 45 | 66 | 72 | 46 | 55 | 59 | 56 | 46 | 43 |
| 24 | 64 | 47 | 43 | 41 | 79 | 53 | 60 | 39 | 57 |

4. Organize the data into a stem-and-leaf plot.

5. In how many games did Alan score on at least 60% of his free throws?

6. What is the median percent of scoring free throws Alan made per game?

7. Find the mode of this set of data.

## EXTEND/ SOLVE PROBLEMS

Refer to the histogram for Exercises 8–10.

8. Which interval contains the most evergreens?

9. Which intervals contain an equal number of trees?

10. Name an interval that contains 95% of the evergreens.

11. ***USING DATA*** Use the Data Index on pages 556–557 to find the box office gross for the 20 top-grossing feature films of 1989. Make a stem-and-leaf plot of the data. Then write a paragraph describing the data.

**HEIGHT OF EVERGREENS IN REFORESTATION PROJECT**

Frequency: 0, 2, 4, 6, 8, 10

Tree Height (in centimeters): 90–99, 100–109, 110–119, 120–129, 130–139, 140–149, 150–159

## THINK CRITICALLY/ SOLVE PROBLEMS

The employees of a certain company were asked how long it takes them to travel to work.

**Time to Travel to Work (in minutes)**

| | | | | | | | | | | | |
|---|---|---|---|---|---|---|---|---|---|---|---|
| 10 | 25 | 40 | 52 | 16 | 15 | 70 | 40 | 21 | 48 | 32 | 65 |
| 8 | 30 | 31 | 49 | 69 | 58 | 60 | 75 | 19 | 27 | 52 | 60 |

Write two questions about the data that cannot be answered by using only the histogram you made for Exercise 12.

12. Organize these data in both a histogram with intervals of 10 and a stem-and-leaf plot.

13. How do the histogram and stem-and-leaf plot you made for Exercise 12 look alike?

14. Can you find the mean, median, mode, and range for this set of data by using only the histogram you made for Exercise 12? By using only the stem-and-leaf plot?

# 4-7 Scatter Plots

## EXPLORE/ WORKING TOGETHER

How do you think the distance students live from school is related to the time it takes them to get to school?

Record the approximate number of miles each student in your class lives from school and the number of minutes it takes each student to get from home to school. Display these data in a graph using one axis for distance and the other for time. Make a point on the graph to represent each student.

a. Find the mode(s) of the distances.
b. Find the mode(s) of the times.
c. What pattern(s) do you see in the data displayed in your graph?
d. Are there any points that do not fit the pattern as do most of the points on your graph? If so, what could be the reason for this?

## SKILLS DEVELOPMENT

You have already worked with two different ways of displaying data visually. A third visual display of data is a **scatter plot.** The data in a scatter plot are represented by points, but the points are not connected. There can also be more than one point for any number on either axis.

### Example 1

Use the scatter plot at the right to answer the questions.

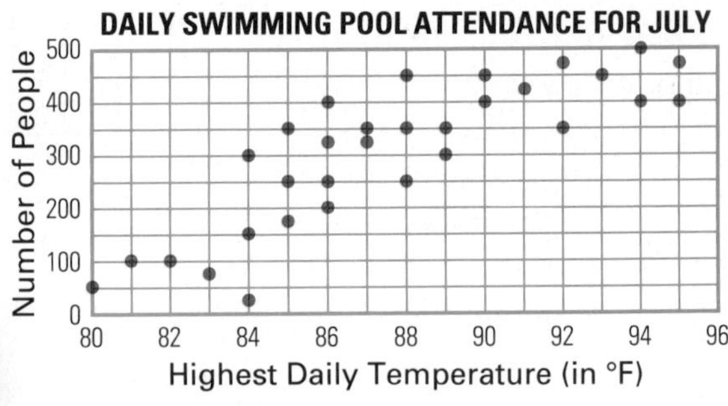

a. How many people were at the pool on the day the temperature reached 91°F?
b. Find the mode of the high temperatures.
c. Find the range of the daily attendance.
d. Find the median of the daily attendance.

**Solution**

a. Locate 91 on the horizontal axis halfway between 90 and 92. Move up to the point, then left to the vertical axis. The day the temperature reached 91°F, 425 people were at the pool.
b. Find the temperature with the greatest number of points above it. The mode of the high temperatures was 86°F.

c. Find the difference between the greatest attendance (500) and the lowest attendance (25): 500 − 25 = 475. The range of the daily attendance was 475.

d. Since there are 31 days in July, count points from the bottom of the graph to the top until the sixteenth point. The median of the daily attendance was 350. ◄

On some scatter plots, a line called the **trend line** can be drawn near most of the points. A trend line that slopes upward to the right indicates a **positive correlation** between the sets of data. A trend line that slopes downward to the right indicates a **negative correlation.**

## Example 2

Use the scatter plot at the right to answer the questions.

a. Predict the weight of an 8-year-old child.

b. Is there a positive or negative correlation between age and weight?

c. Which point lies farthest from the trend line? What could account for this piece of data?

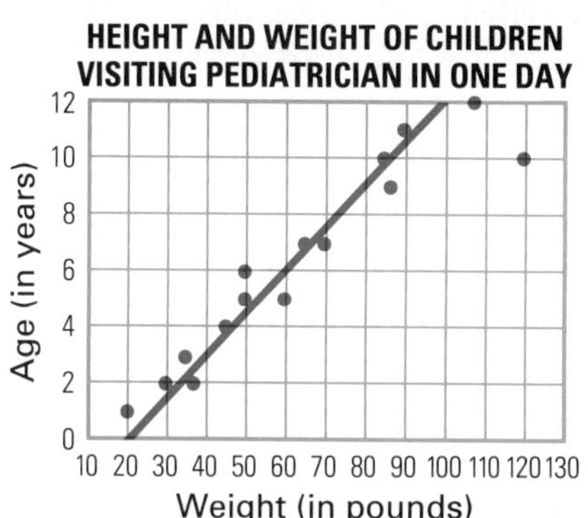

In Example 2, suppose the vertical axis was used for weight and the horizontal axis for age. Would the scatter plot indicate a positive or negative correlation between age and weight?

### Solution

a. Locate age 8 on the vertical axis. Move across to the trend line, then down to the horizontal axis. The weight should be about 73 lb.

b. The trend line slopes upward to the right, so there is a positive correlation between age and weight. As a child gets older, the child's weight increases.

c. The point for a 120-lb 10-year-old is farthest from the trend line. This could represent a child who is overweight or who is unusually tall for age 10. ◄

## TRY THESE

Use the scatter plot at the right for Exercises 1–4.

1. During the month that had 6 days of rain, how many days had above-average temperatures?

2. Find the range of days with above-average temperatures.

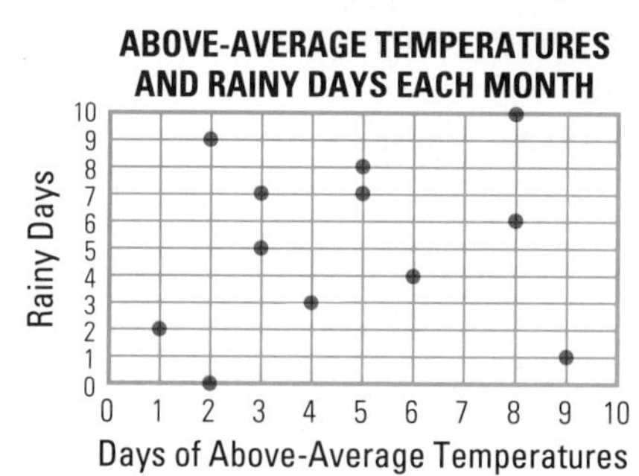


3. Find the mode of the number of rainy days each month.

4. Find the median number of days with above average temperatures.

**CALCULATOR**

Most graphing calculators have a *statistics mode* that enables data to be entered and sorted in numerical order, to calculate statistics such as the mean, or to automatically display data graphically. For example, to draw a histogram, you would select *Hist* from a statistical draw menu. The calculator defines the numerical width of the bars, then displays the graph. Consult the operating manual for the calculator you are using for specific instructions.

Use the scatter plot at the right for Exercises 5–7.

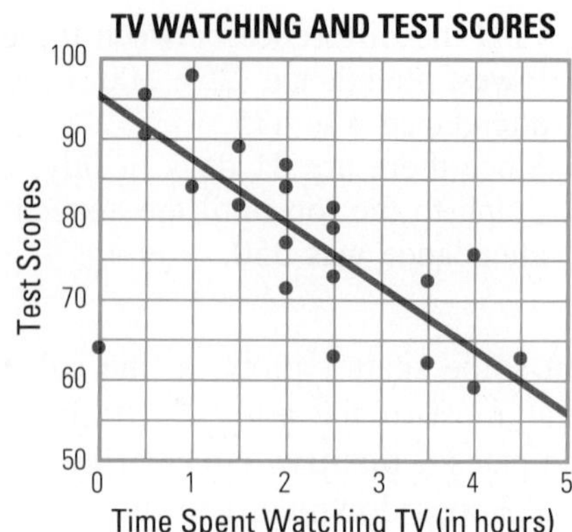


5. Predict the test score of a student who watched 3 hours of television.
6. Is there a positive or negative correlation between test scores and time spent watching television?
7. Which point lies farthest from the trend line? What could account for this piece of data?

# EXERCISES

## PRACTICE/ SOLVE PROBLEMS

Use the scatter plot at the right for Exercises 1–4.

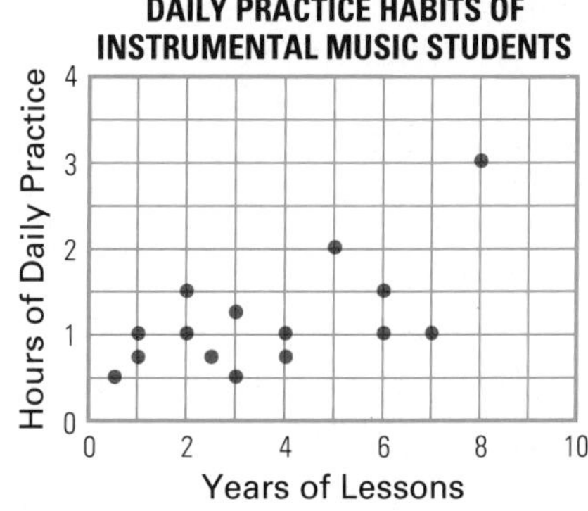


1. How many students were surveyed?
2. Find the range of daily practice times.
3. Find the mode of the number of hours practiced each day.
4. One student practices 2 h each day. How many years more of lessons has this student had than the median number of years of lessons?

Use the scatter plot at the right for Exercises 5 and 6.

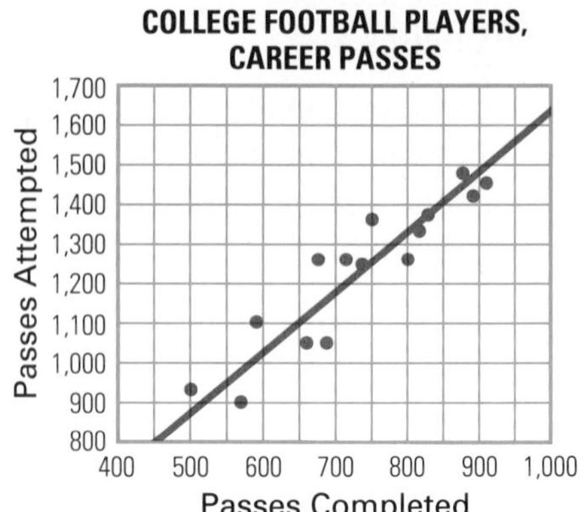


5. Suppose a player's goal is to attempt 1,600 passes. If he reaches that goal, how many completed passes would you expect him to have made?
6. Is there a positive or negative correlation between attempts and completions?

## EXTEND/ SOLVE PROBLEMS

Read through the table below to help you decide the most appropriate scales to use on a scatter plot of the data. Make the scatter plot. Then, if possible, draw a trend line on the scatter plot. Tell whether the scatter plot shows a *positive correlation, negative correlation,* or *no correlation* between the sets of data in the table. Finally, use the scatter plot to answer Exercises 7–9.

**OLYMPIC CHAMPIONS FOR WOMEN'S 100-m DASH**

| Year | Name, Country | Time |
|---|---|---|
| 1928 | Elizabeth Robinson, U.S. | 12.20 |
| 1932 | Stanislawa Walasiewicz, Poland | 11.90 |
| 1936 | Helen Stephens, U.S. | 11.50 |
| 1948 | Francina Blankers-Koen, Netherlands | 11.90 |
| 1952 | Marjorie Jackson, Australia | 11.50 |
| 1956 | Betty Cuthbert, Australia | 11.50 |
| 1960 | Wilma Rudolph, U.S. | 11.00 |
| 1964 | Wyomia Tyus, U.S. | 11.40 |
| 1968 | Wyomia Tyus, U.S. | 11.00 |
| 1972 | Renate Stecher, East Germany | 11.10 |
| 1976 | Annegret Richter, West Germany | 11.01 |
| 1980 | Lyudmila Kondratyeva, USSR | 11.06 |
| 1984 | Evelyn Ashford, U.S. | 10.97 |
| 1988 | Florence Griffith-Joyner, U.S. | 10.54 |

**7.** The Olympics were not held in 1940 and 1944 since many participating countries were engaged in World War II. If the Olympics had been held during those years, what were the winning times likely to have been?

**8.** What do you think could account for the point for 1948 being so far above the trend line?

**9.** Predict the winning time for the woman's 100-m dash in the 1992 Olympics. Then find out what the time actually was and compare it with your answer.

***USING DATA*** Use the Data Index on pages 556–557 to find numbers of TV stations and households with TV. Use these data for Exercises 10 and 11.

**10.** Make a scatter plot to relate the number of TV stations to the number of households with TV from 1950 to 1989.

**11.** Do you think the number of TV stations caused the increase in the number of households with TV or the number of households with TV caused the increase in the number of stations? Explain.

**WRITING ABOUT MATH**

Write three paragraphs explaining when it is best to use a stem-and-leaf plot, a histogram, and a scatter plot to display data.

## THINK CRITICALLY/ SOLVE PROBLEMS

Work with a partner. Research data on one or more of the following topics. For each set of data, make a scatter plot and explain if there is a *negative correlation,* a *positive correlation,* or *no correlation.* Interpret your results and draw conclusions.

**12.** prices per pint for different brands of ice cream and the number of calories in a $\frac{1}{2}$-cup serving of each brand

**13.** prices for new cars and expected highway mileage (mi/gal gasoline)

**14.** population of a country and literacy rate (expressed as a percent)

# 4-8 Quartiles and Percentiles

## EXPLORE

The test scores for all students in one class are as follows.

58 84 72 40 95 78 92 98 82 50
67 90 75 93 87 55 84 86 62 67

a. Find the median of these scores.
b. Is the median a good indicator of how well the class did as a whole on this test? Explain.
c. Find the median of all the scores below the median class score and the median of all the scores above the median class score.
d. How does using the three medians you found in Exercises **a** and **c** give a better indication of how well the class did as a whole on this test?
e. Into how many equal parts do the three medians separate the original set of data?

## SKILLS DEVELOPMENT

**READING MATH**

The abbreviations $Q_1$, $Q_2$, and $Q_3$ are sometimes used for the first, second, and third quartiles.

You have learned that the median of a set of data separates the data into two equal parts. A set of data can be further separated into quartiles. The **first quartile** is the median of the lower part of the data. The **third quartile** is the median of the upper part. The **second quartile** is another name for the median of the entire set of data. The quartiles separate the original set of data into four equal parts. Each part contains one-fourth, or 25%, of the data.

For a given set of data, the **interquartile range** is the difference between the values of the first and third quartiles.

### Example 1

Find the first quartile, median, and third quartile of the following set of scores. Then find the interquartile range.

18 29 56 42 58 31 40 28 37 46

### Solution

Write the ten scores in order from least to greatest.

median of all data, or *second quartile*
↓

18 28 29 31 37 40 42 46 56 58

↑ median of lower part, or *first quartile* (29)
↑ median of upper part, or *third quartile* (46)

The median is 38.5, halfway between the two middle scores. There are five scores below 38.5. The middle score of these is 29, so the first quartile is 29. There are five scores above 38.5. The median of these is 46, so the third quartile is 46. The interquartile range is the difference between 29 and 46, or 17. ◄

**CHECK UNDERSTANDING**

In Example 1, suppose the lowest score, 18, had not been included in this set of data. Find the first quartile, median, and third quartile.

When you take a standardized test, your score is often reported in terms of **percentiles.** A percentile ranking indicates the percent of those who took the test who achieved *below* a particular score. For example, if your score is in the 83rd percentile, it means that approximately 83% of those who took the test received a score less than or equal to yours. Only 17% of those who took the test had a score greater than yours. So, a percentile is a measure of a rank, or standing, within a group.

## Example 2

Karen took a placement test in order to be accepted into a computer course. Her score was 48th from the highest out of the 760 students who took the test. Find the percentile rank that Karen achieved.

### Solution

The total number of students who took the test was 760. Karen was 48th. So, there were 47 students who had higher scores than she did. The number of students who achieved either the same score or less than she did was 760 − 47, or 713.

Find the percent.

$$\frac{\text{number of scores less than or equal to Karen's}}{\text{total number of scores}} = \frac{713}{760} \approx 0.938 \approx 94\%$$

So, Karen's ranking was in the 94th percentile. ◄

## TRY THESE

1. Data have been obtained on the results of a test as shown at the right. Find the first quartile, the median, and the third quartile of these scores. Then find the interquartile range.

| | | | |
|---|---|---|---|
| 75 | 82 | 96 | 74 |
| 84 | 48 | 93 | 76 |
| 72 | 68 | 79 | 80 |
| 97 | 70 | 72 | 85 |
| 74 | 91 | 73 | 87 |

2. On a test, Mike had the ninth highest test score. If there were 28 students who took the test, what is Mike's percentile rank?

# EXERCISES

## PRACTICE/ SOLVE PROBLEMS

For each set of data in Exercises 1–4, find the first quartile, the median, and the third quartile.

1. 71 86 92 53 87 76 75 84 83
2. 14 20 16 15 13 12 19 18 20 18 12 10
3. 4.6 8.3 9.3 7.2 5.8 8.7 3.2 5.9 11.6
   2.7 4.8 5.9 6.3 7.2 8.6 9.7
4. 200 101 162 273 149 153 146 125 118
   129 135 142 111 156 129 171 143 172

**PROBLEM SOLVING TIP**

What information is given in Exercise 5 that you do not need in order to solve the problem?

5. Sherry took a history exam and received a score of 27 out of 50. If out of 32 students in the class she stood 19th, what is her percentile rank?

Find the percentile ranks for each of the following students who took a midterm exam.

| | Student | Standing | Number of Students Taking Exam |
|---|---|---|---|
| 6. | Felicia | 56th | 428 |
| 7. | Martha | 71st | 560 |
| 8. | Mark | 63rd | 754 |
| 9. | Bethany | 92nd | 693 |
| 10. | Robert | 44th | 167 |
| 11. | Jamal | 207th | 1,026 |

## EXTEND/ SOLVE PROBLEMS

Here are the scores received by 50 students who took an English test. Use these scores for Exercises 12–19.

| | | | | | | | | | |
|---|---|---|---|---|---|---|---|---|---|
| 28 | 37 | 27 | 12 | 40 | 31 | 18 | 25 | 34 | 15 |
| 29 | 22 | 35 | 32 | 17 | 19 | 26 | 29 | 17 | 34 |
| 38 | 28 | 21 | 24 | 16 | 30 | 25 | 22 | 31 | 23 |
| 20 | 13 | 32 | 23 | 10 | 28 | 35 | 29 | 37 | 19 |
| 31 | 26 | 36 | 24 | 19 | 29 | 24 | 33 | 28 | 35 |

12. Find the mean of the scores.
13. Find the median.
14. Find the first quartile and the third quartile.
15. Find the interquartile range for these data.

Use the English test scores on page 158 for Exercises 16–19.

16. Maria received a score of 36 on the English test. What is her percentile rank?

17. Dave received a score of 22 on the test. What is his percentile rank?

18. On the test, Angela received a score that placed her in the 84th percentile. What was her score?

19. Joel received a score that placed him in the 44th percentile on the test. What was his score?

## THINK CRITICALLY/ SOLVE PROBLEMS

20. If your score on a test is equal to the median of the scores, then in what percentile rank is your score?

21. If exactly one-quarter of your class scored higher than you did on a test, what is your percentile ranking?

22. If your score on a test is equal to the mean of all the scores, can you determine your percentile ranking? Explain.

In Kate's karate class, there are 25 students. On the first evaluation of the students' skills, 17 students scored lower than Kate and 7 scored higher. Use these data for Exercises 23 and 24.

23. Find Kate's percentile ranking on the first evaluation.

24. Suppose the instructor decided to add five points to everyone's score. How would this affect Kate's percentile rank?

25. Obtain a list of the top 20 songs for this week according to a radio station or a magazine. Select the song you like best on this list and find its percentile ranking.

26. Using a newspaper or other resource, make a list of teams for your favorite sport. Use the teams' win–loss records to rank them. Then find the percentile ranking of your favorite team.

# 4-9 Box-and-Whisker Plots

## EXPLORE

Travis is saving money to buy a cassette player. He checked with various stores in town and found the following prices for cassette players he would consider buying.

$79.99 $48.59 $89.50 $99.95
$53.50 $84.97 $87.75 $93.49

a. Write the prices in numerical order. Then draw a box around the middle half of the prices.
b. Summarize the data Travis collected by finding the range of all the prices and the range of the middle half of the prices.
c. Suppose the cassette player regularly priced at $87.75 went on sale for $59.98. How would this affect the range of all the prices and the range of the middle half of the prices?

## SKILLS DEVELOPMENT

Another way to display a set of data is with a **box-and-whisker plot.** Box-and-whisker plots can be helpful in interpreting the distribution of data.

### Example 1

Make a box-and-whisker plot for the following data.

Prices of Microwave Ovens (in dollars)
225 257 175 300 265 185 229 235 299

**Solution**
Write the data in numerical order and find the first quartile, the median, and the third quartile.

least value ↓ 175, 185, 225, 229, median ↓ 235, 257, 265, 299, greatest value ↓ 300

$$\frac{185 + 225}{2} = 205 \quad \text{(first quartile)}$$

$$\frac{265 + 299}{2} = 282 \quad \text{(third quartile)}$$

Use points to mark these values below a number line.

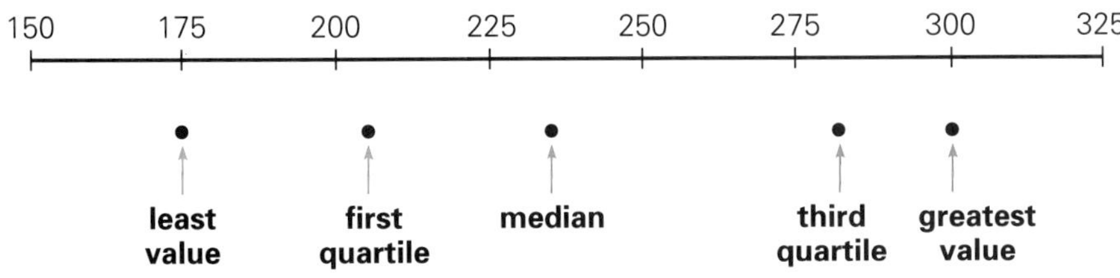


Draw a box with ends through the points for the quartiles. Then draw a vertical line through the box at the point for the median. Finally, draw *whiskers,* or lines, from each end of the box to the least and greatest values. Then give the box-and-whisker plot a title.

> **CHECK UNDERSTANDING**
>
> What percent of a set of data is represented by each of these parts of a box-and-whisker plot?
> a. the box
> b. the whisker to the left or right of the box

**PRICES OF MICROWAVE OVENS (IN DOLLARS)**

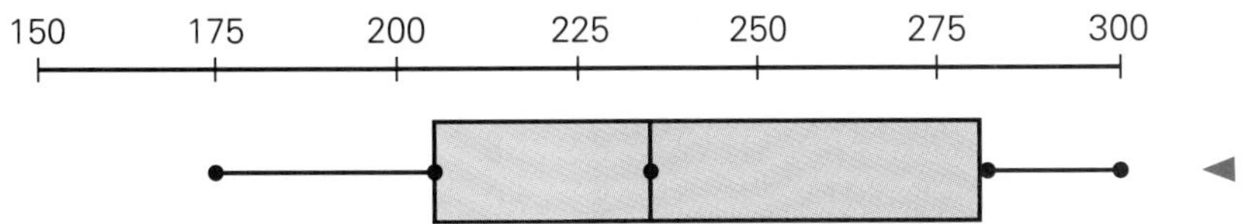


Box-and-whisker plots can also be used to compare sets of data. Outliers are marked with asterisks on this kind of plot.

## Example 2

Use these box-and-whisker plots to answer questions about the test scores of two classes.

**TEST SCORES**

0 10 20 30 40 50 60 70 80 90 100

*

Mr. Gomez's Class

Mrs. Mehaffey's Class

**a.** Which class has the highest score? Which class had the lowest score?
**b.** For which class is the range of the middle 50% of the scores greater?
**c.** For which class are the highest scores clustered more closely?

### Solution

**a.** The highest score for Mr. Gomez's class is 93. The lowest score is the outlier, 20, which is marked with an asterisk. The highest score for Mrs. Mehaffey's class is 95, and the lowest is 40. So, Mrs. Mehaffey's class has the highest score, and Mr. Gomez's class has the lowest score.

b. The range of the middle 50% of the scores for Mr. Gomez's class is 85 − 60, or 25. The range of the middle 50% of the scores for Mrs. Mehaffey's class is 90 − 60, or 30. So, the middle 50% of the scores for Mrs. Mehaffey's class has the greater range.

c. The highest 25% of the scores for Mr. Gomez's class is distributed over 8 points from 85 to 93. The highest 25% of the scores for Mrs. Mehaffey's class is distributed over 5 points from 90 to 95. This indicates that the highest scores are clustered more closely for Mrs. Mehaffey's class. ◄

**TALK IT OVER**

Which of the following can a box-and-whisker plot help you find for a set of data: mean, median, mode, range?

## TRY THESE

1. Make a box-and-whisker plot for the following data. Remember to show any outliers.

Science Test Scores

| | | | | | | | | | |
|---|---|---|---|---|---|---|---|---|---|
| 84 | 89 | 76 | 65 | 74 | 93 | 82 | 68 | 76 | 94 |
| 73 | 85 | 89 | 91 | 74 | 63 | 83 | 80 | 80 | 70 |

Use the box-and-whisker plots below for Exercises 2–3.

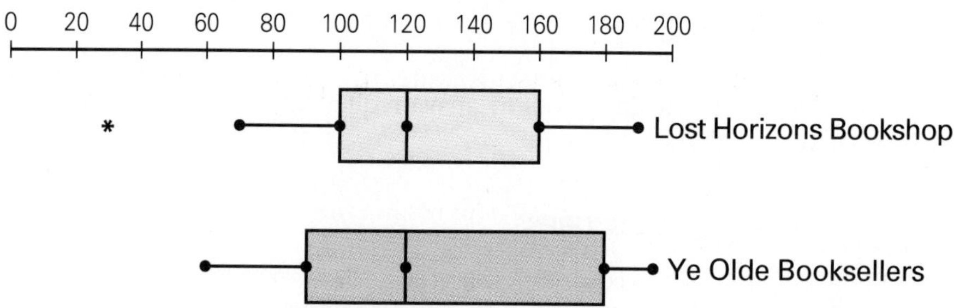


2. Which store had the most customers on any one day? Which had the fewest?

3. On the busiest 25% of the days, which store had the greater range in the number of customers?

# EXERCISES

**PRACTICE/ SOLVE PROBLEMS**

Make a box-and-whisker plot for each set of data. Be sure to show any outliers.

1. Hours Cleo worked in Her Mother's Video Store
   2 7 1 2 5 3 4 3 2 0 3 2 4 0

2. Points Scored by Basketball Team Players During Game
   11 14 0 8 0 32 3 7 4 6 5 12 4

This box-and-whisker plot shows the results of a survey of the total miles people drove while on vacation. Use this plot for Exercises 3–6.

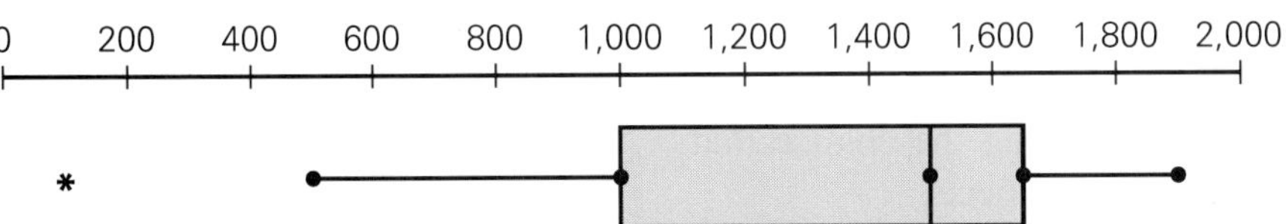


3. What are the greatest and least number of miles reported?
4. What percent of those surveyed drove 1,500 miles or more?
5. What percent of those surveyed drove less than 1,000 miles?
6. In which interval are the data most closely clustered?

## EXTEND/ SOLVE PROBLEMS

**USING DATA** Use the Data Index on pages 556–557 to find the statistics about the average daily temperatures in San Diego, California, and Milwaukee, Wisconsin.

7. Make box-and-whisker plots to compare the temperatures of these two cities.
8. Write a paragraph telling whether you would rather live in San Diego or Milwaukee and whether the temperature in each city affected your decision.

## THINK CRITICALLY/ SOLVE PROBLEMS

9. Describe a set of data that would be displayed in a box-and-whisker plot with a short box and long whiskers. Then describe a set of data that would be displayed in a box-and-whisker plot with a long box, short whiskers, and one outlier.

Obtain a list of the number of points your school's football, basketball, or other team scored in each game last season. Then use this data for Exercises 10–12.

10. Present these data in a stem-and-leaf plot and in a box-and-whisker plot.
11. Name one advantage of using each kind of plot over the other to display these data.
12. On the basis of your plots, write a paragraph that could be included in your school newspaper summarizing the team's scores.

# 4-10 Statistics and Probability

## EXPLORE/ WORKING TOGETHER

Use this scale to assign a number from 0 to 1 to each event, indicating the chance that the event will occur.

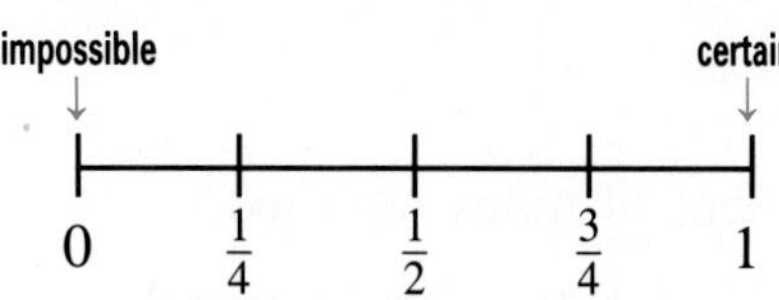


a. The sun will rise tomorrow morning.
b. People will live on the moon within ten years.
c. The President will visit your school this year.
d. You will flip a coin and get heads.
e. It will not rain at all in your town next year.
f. Your math teacher will assign homework tonight.

## SKILLS DEVELOPMENT

The chance that an event will occur is the **probability,** or likelihood, that it will happen. When you assigned a number to the chance that the sun will rise tomorrow you most likely thought about how often the sun had risen in the past. The number you assigned indicated the probability of that event.

To estimate the probability of events, you can use sets of data that are really records of observations. The probability that an event $E$ will occur is written $P(E)$ and is a number between 0 and 1. It can be estimated using this formula.

$$P(E) = \frac{\text{number of observations favorable to } E}{\text{total number of observations}}$$

### Example 1

A group of students was randomly selected for a survey. They were asked whether or not they belonged to a music organization at school. The results of the survey are in the table.

**STUDENT MEMBERSHIP IN MUSIC ORGANIZATIONS**

| | Member | Not a Member |
|---|---|---|
| Boys | 5 | 9 |
| Girls | 4 | 7 |

Your English teacher randomly assigns seats. A student *not* in the survey group is assigned a seat next to you. Estimate each probability. If not enough information is provided, write *not enough information.*
a. The student is a boy who belongs to a school music organization.
b. The student does not belong to a school music organization.
c. The student is a girl who is a member of the school band.

**Solution**

a. Here, $E$ is *a boy who belongs to a school music organization*. The data show that 5 boys are members. The total number surveyed is the sum of all the results shown in the table: $5 + 4 + 9 + 7 = 25$. Use the formula.

$$P(E) = \frac{\text{number of observations favorable to } E}{\text{total number of observations}} = \frac{5}{25} = \frac{1}{5}$$

The probability that a student is a boy who belongs to a school music organization is about $\frac{1}{5}$.

b. Here, $E$ is *a student who does not belong to a school music organization*. The data show that 9 boys and 7 girls, or 16 students, are not members. Again, 25 students were surveyed. Use the formula.

$$P(E) = \frac{\text{number of observations favorable to } E}{\text{total number of observations}} = \frac{16}{25}$$

The probability that the student does not belong to a school music organization is about $\frac{16}{25}$.

c. Here, $E$ is *a girl who is a member of the school band*. Most likely your school has other music organizations besides the band. Some of the students who said they belonged to a music organization may be band members, and some may belong to a different music organization. You cannot tell from the data how many girls are band members and how many are members of other music organizations. So, there is not enough information to estimate the probability that the girl is a member of the school band. ◄

Probabilities can also be expressed as percents. An **impossible event** has 0% chance of occurring. A **certain event** has a 100% chance of occurring.

## Example 2

The circle graph shows the various types of accidental deaths. Suppose a person died as the result of an accident. Use this graph to estimate each probability. If not enough information is provided to make an estimate, write *not enough information*.

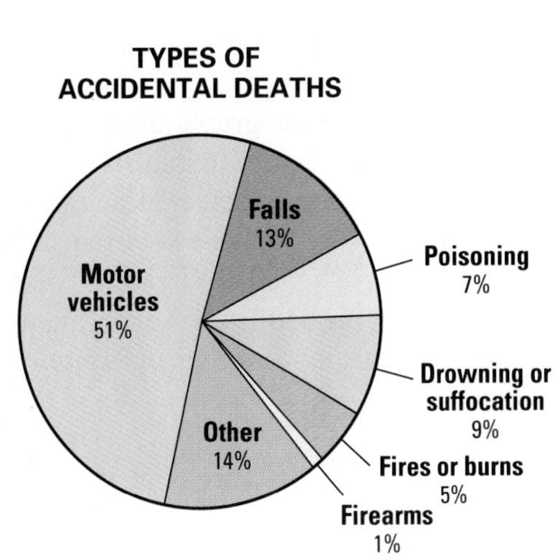


a. The accident involved a motor vehicle.
b. The accident involved an airplane.
c. The accident involved poisoning or falls.
d. The accident did not involve a fire or burns.

**CHECK UNDERSTANDING**

Suppose a meteorologist reports that there is an 80% chance of thunderstorms on Tuesday. What fraction could be used to express the probability of thunderstorms on Tuesday?

**Solution**

a. According to the graph, the probability that the accident involved a motor vehicle is about 51%.

b. The graph does not give specific data about accidents involving airplanes. Such accidents are included in the category "Other." So, there is *not enough information* to estimate the probability that the accident involved an airplane.

c. According to the graph, 7% of the accidents involved poisoning and 13% involved falls. So, the probability that the accident involved poisoning or falls is 7% + 13% = 20%.

d. According to the graph, 5% of the accidents involved fires or burns. Since all the accidents represent 100%, the percent of accidents that did not involve fires or burns is 100% − 5%, or 95%. So, the probability that the accident did not involve a fire or burns is about 95%. ◄

## TRY THESE

A group of randomly selected ninth-grade and tenth-grade students were asked whether or not they ate breakfast before coming to school. The results of the survey are shown here. Later, another student from the ninth or tenth grade was selected at random and surveyed. Estimate each probability. If not enough information is provided to make an estimate, write *not enough information.*

**SCHOOL-DAY BREAKFAST HABITS**

| | Ate Breakfast | Did Not Eat Breakfast |
|---|---|---|
| Ninth Graders | 7 | 3 |
| Tenth Graders | 4 | 6 |

1. The student is a ninth grader who did not eat breakfast before coming to school.
2. The student ate breakfast before coming to school.
3. The student had cereal for breakfast.

The circle graph shows where people listen to radios. Suppose someone is selected at random to answer, "Where was the last place you listened to a radio?" Estimate the probability of each answer or write *not enough information.*

4. In the car.
5. I'm not sure, but it was not at home.
6. At the beach.

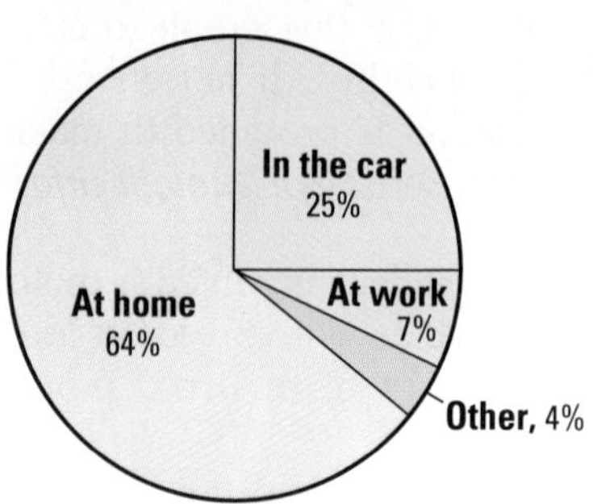

# EXERCISES

## PRACTICE/ SOLVE PROBLEMS

The circle graph shows the various types of housing units in which people in the Unied States live. A resident of the United States is selected at random. Estimate the probability that that person lives in each of the following. If not enough information is provided to make an estimate, write *not enough information.*

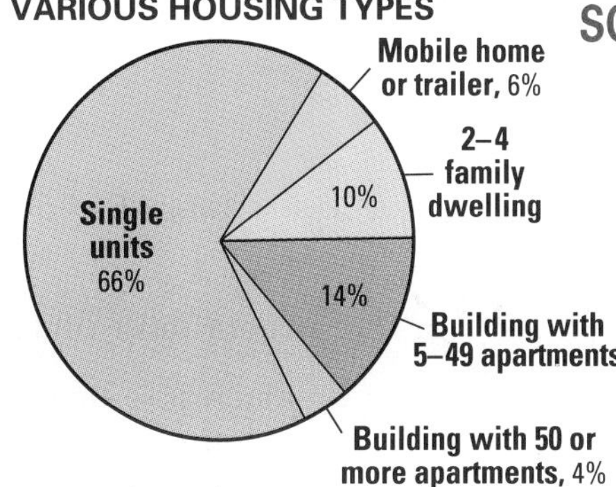


1. mobile home or trailer
2. building with more than one family or apartment
3. three-bedroom house
4. other than a single unit

## EXTEND/ SOLVE PROBLEMS

The manager of a pizza restaurant asked every fifth customer, "Do you prefer thin-crust pizza or pan pizza?" Based on this survey of 40 customers, the manager drew these conclusions.

► The probability that a teenager will order a pan pizza is about $\frac{3}{40}$.

► The probability that an adult will order a thin-crust pizza is about $\frac{3}{10}$.

► The probability that a customer will be an adult is about $\frac{7}{10}$.

5. Copy and complete the table based on the manager's conclusions.

| PIZZA PREFERENCES | | |
|---|---|---|
| | Thin Crust | Pan |
| Teenagers | ■ | ■ |
| Adults | ■ | ■ |

## THINK CRITICALLY/ SOLVE PROBLEMS

Conduct your own survey! Work in small groups.

6. Choose a subject whose frequency you would like to investigate. Write a question that will divide your school's population into two groups by age or grade. Write another question about your subject that could be answered "yes" or "no."
7. Plan a way to randomly select 20 students in your school to survey. Then use your group's questions to survey the 20 students. Record your results in a table.
8. Based on your table, make a list of four different ways the students you surveyed could be described. Select another student at random from your school. Estimate the probability that that student could be described in each of the four ways.
9. Using the same questions, survey 10 more students. How do your results compare to the probabilities you wrote in Exercise 8?

# CHAPTER 4 ● REVIEW

Choose the term from the list that completes each statement.

1. The __?__ is the difference between the greatest and least values in a set of data.
2. __?__ divide a set of data arranged in numerical order into four equal parts.
3. The middle number in a set of data arranged in numerical order is the __?__.
4. In a set of data, the number that occurs most often is the __?__.
5. A number that is much greater or much less than the other values in a set of data is known as a (an) __?__.

a. outlier
b. range
c. mode
d. quartiles
e. median

## SECTION 4–1 COLLECTING DATA (pages 132–135)

- Random, cluster, convenience, and systematic sampling are methods of collecting **data** that can be used to make decisions about a population.
- For numerous reasons, survey findings can be **biased,** or not truly representative of an entire population.

To identify the most popular current movie among teenagers, seven students seated at the same table in the school cafeteria were asked to name the most recent movie they had seen.

6. What kind of sampling method is this?
7. How might the results be biased?

## SECTIONS 4–3 AND 4–4 MEASURES OF CENTRAL TENDENCY (pages 138–145)

- The **mean** of a set of data is the sum of the data items divided by the number of items.
- The **median** is the middle value (or the mean of the two middle values) of a set of data arranged in numerical order.
- The **mode** is the number(s) that occur(s) most often in a set of data.
- The **range** is the difference between the greatest and least values in a set of data.

8. Make a frequency table for these data.
9. Find the mean, median, mode, and range of these data.
10. Would an ad stating "Most styles priced at $34" be misleading? Explain.

PRICES OF AEROBIC SHOES SOLD AT SHOE MART (IN DOLLARS)

| | | | | | | | |
|---|---|---|---|---|---|---|---|
| 49 | 34 | 37 | 28 | 39 | 44 | 34 | 49 |
| 52 | 34 | 37 | 37 | 49 | 39 | 34 | 39 |

## SECTION 4–6 HISTOGRAMS AND STEM-AND-LEAF PLOTS (pages 148–151)

- The frequency of data can be displayed in a **histogram.**
- Individual data items can be displayed in **stem-and-leaf plots.** In these, the digit farthest to the right in a number is the leaf. The other digits make up the stem.

11. Make a histogram of these data using intervals of 5 starting with 65.

12. Make a stem-and-leaf plot of these data.

DAILY HIGH TEMPERATURES (°F)

| | | | | | | | | | |
|---|---|---|---|---|---|---|---|---|---|
| 80 | 91 | 68 | 92 | 86 | 69 | 71 | 75 | 90 | 86 |
| 70 | 91 | 83 | 81 | 79 | 78 | 99 | 76 | 80 | 86 |
| 90 | 71 | 79 | 86 | 90 | 76 | 84 | 78 | 88 | 81 |

## SECTION 4–7 SCATTER PLOTS (pages 152–155)

- Data can be displayed as unconnected points on a **scatter plot.**
- A **trend line** on a scatter plot indicates a correlation between items of data and can be used to make predictions.

13. What is the correlation between points scored and minutes played?

14. Predict how many points Carrie would score if she played 24 min in one game.

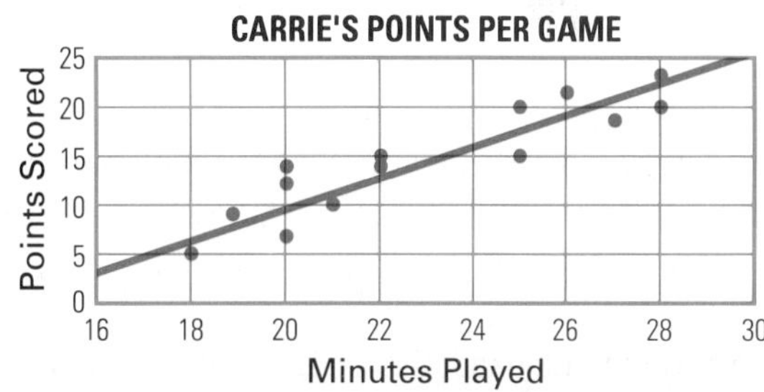


## SECTIONS 4–8 AND 4–9 QUARTILES, PERCENTILES, PLOTS (pages 156–163)

- **Box-and-whisker plots** are used to display the distribution of data.
- A **percentile** divides a group of data into two parts—those at or below a certain score and those above.

15. Make a box-and-whisker plot of the data for Exercises 8–10.

16. Out of 30 students who took a test, Darius had the fifth highest score. What is his percentile rank?

## SECTION 4–10 STATISTICS AND PROBABILITY (pages 164–167)

- **Probability** may be expressed as a number between 0 and 1 or as a percent between 0% and 100%.

17. *USING DATA* Suppose you select a person at random to ask whether he or she watched a situation comedy on TV the previous day. The person responds, "Yes." Use the data on page 131 to estimate the probability that that person is an adult male.

# CHAPTER 4 • TEST

To find out the most popular breed of dog among people today, every tenth person who entered a pet shop one Saturday was asked to name one favorite breed.

1. What kind of sampling method was being used?
2. How might the results of this survey be biased?

The histogram shows the number of hours worked each week by the part-time employees at Discount World.

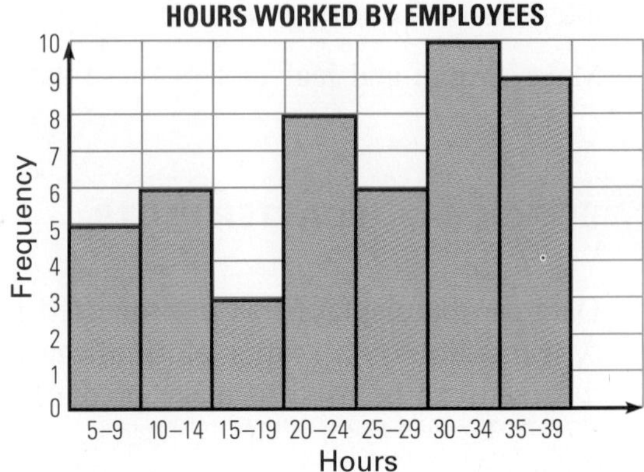


3. What is the most common number of hours worked?
4. How many employees work less than 20 h?
5. How many part-time employees work at Discount World?

A hotel charges the following daily rates for its different types of rooms.

75 120 65 84 150 79

6. Find the mean, median, mode, and range of these data.
7. The hotel used the median to advertise its average room rate. Is this misleading? Explain.

The number of games won by a certain baseball team during the past eleven seasons is shown in the stem-and-leaf plot.

**4|8 represents 48 games**

| Stem | Leaves |
|---|---|
| 4 | 7 8 9 9 |
| 5 | 0 1 8 9 |
| 6 | 1 2 |
| 8 | 3 |

8. During how many seasons did the team win 50 or fewer games?
9. What is the range of the number of games won?
10. During how many seasons did the team win more than 60 games?

Use the scatter plot at the right for Exercises 11–12.

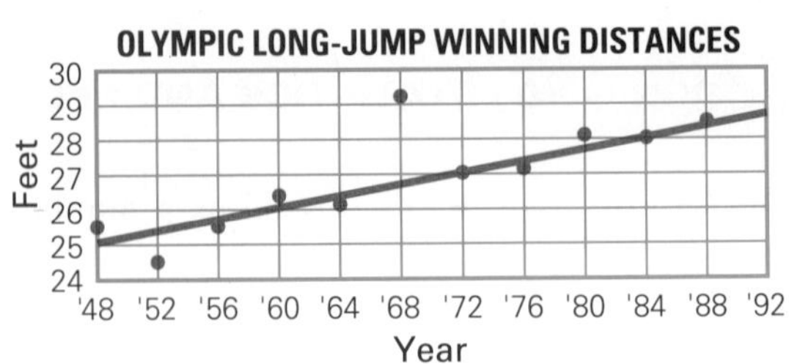


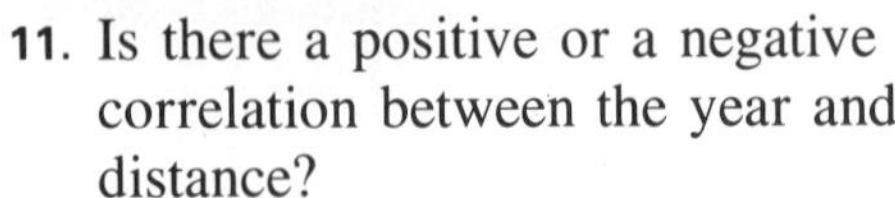

11. Is there a positive or a negative correlation between the year and distance?

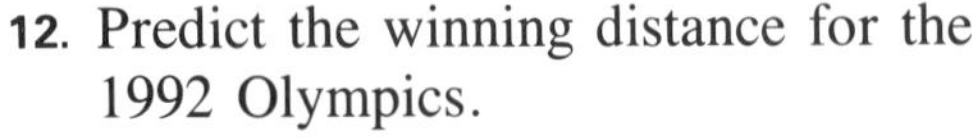

12. Predict the winning distance for the 1992 Olympics.
13. On a test, Susan had the fourth highest score. If there were 30 students who took the test, what is Susan's percentile rank?

Simplify.

1. $16 - [12 \div (2 \times 2)]$
2. $16 \cdot 2 + 6 \div 3$

Find the next three terms in each pattern.

3. 2, 6, 10, 14, ■, ■, ■
4. 4, 8, 5, 9, 6, ■, ■, ■

Write each fraction or decimal as a percent.

5. $\frac{1}{5}$
6. 0.23
7. $\frac{5}{8}$
8. 0.03

Multiply or divide.

9. $-3 \cdot 16$
10. $-75 \div (-15)$

Complete.

11. ■$(3 + 4) = (6 \times 3) + (6 \times 4)$
12. $3.7 \times$ ■ $= 0$

Simplify.

13. $n^2 \cdot n^5$
14. $\frac{28x^7}{7x^6}$
15. $(c^2)^3$

Evaluate each expression when $r = -2$ and $s = 5$.

16. $r^3 \cdot s^2$
17. $r \cdot s^3$
18. $(r^2)^{-3}$

Write the symbol for each figure.

19. Q P

20. R S

21. Write the converse of the following statement:
    If the measures of two angles are equal, then the two angles are congruent.

Name the following parts of circle $P$.

22. two radii
23. two chords

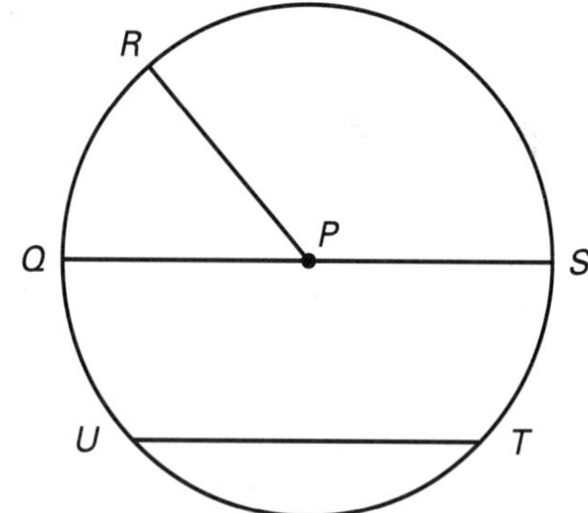


Find the sum of the interior angles of each of these polygons.

24. pentagon
25. hexagon

Use these data for Exercises 26–27.

| TEENAGERS' WEEKLY ALLOWANCE (IN DOLLARS) | | | | |
|---|---|---|---|---|
| 20 | 5 | 15 | 10 | 15 |
| 10 | 20 | 12 | 15 | 10 |
| 16 | 15 | 10 | 12 | 16 |
| 5 | 14 | 15 | 10 | 10 |

26. Make a frequency table for these data.
27. How many teenagers received an allowance of at least $10 a week?

Simplify.

1. $6 + 4 \times 6 \div 3$
   A. 15 B. 14
   C. 20 D. none of these

2. $7.2 - (0.2 \times 6)$
   A. 6 B. 3.6
   C. 9.4 D. none of these

Solve.

3. $3.4 \times 22 \times 0.2$
   A. 14.96 B. 12.96
   C. 4.4 D. none of these

4. $18 \cdot 6 \cdot \frac{1}{4}$
   A. 28 B. 27
   C. 26.5 D. none of these

Perform the indicated operations.

5. $-3.1 \cdot 2.7 \cdot 4$
   A. 33.48 B. $-33.48$
   C. 334.8 D. none of these

6. $-2 + [8.3 + (-8.7)]$
   A. $-2.4$ B. $-0.24$
   C. 0.8 D. none of these

Simplify.

7. $-(x - 2y)$
   A. $-x + 2y$ B. $x + 2y$
   C. $2xy$ D. none of these

8. $y^4 \cdot y^3$
   A. $y$ B. $y^{12}$
   C. $y^7$ D. none of these

9. Write the following in scientific notation.
   76,000,000
   A. $0.76 \times 10^8$ B. $7.6 \times 10^7$
   C. $7.6 \times 10^6$ D. none of these

10. Classify $\angle ABC$.

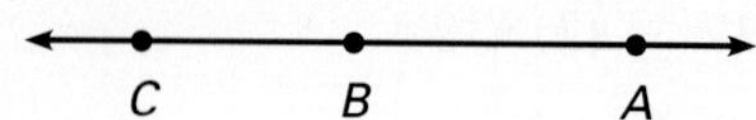


   A. right B. straight
   C. acute D. obtuse

11. Find the measure of $\angle 1$.

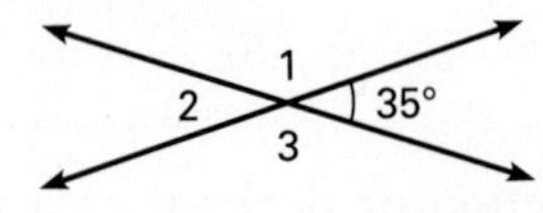


   A. 35° B. 145°
   C. 55° D. none of these

12. Identify the pair of alternate exterior angles.

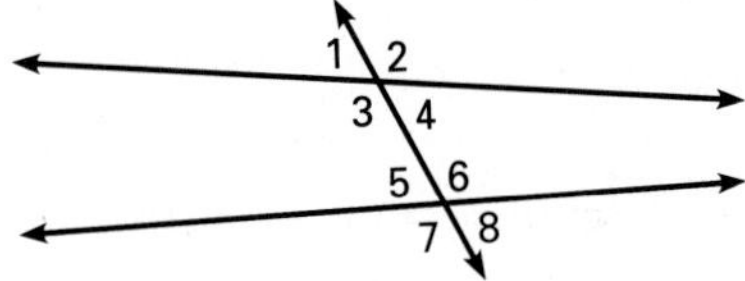


   A. 2 and 6 B. 1 and 8
   C. 1 and 4 D. 3 and 6

13. Find the value of $\angle b$.

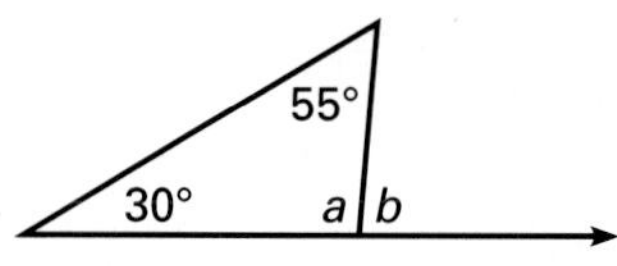


   A. 75° B. 95°
   C. 85° D. none of these

14. What are the first quartile, median, and third quartile of this set of data?

NUMBER OF CALORIES IN FROZEN ENTREES

| | | | | |
|---|---|---|---|---|
| 248 | 195 | 330 | 415 | 230 |
| 390 | 385 | 275 | 310 | |

   A. 195, 310, and 415
   B. 239, 287.5, and 357.5
   C. 230, 262.5, and 330
   D. 239, 310, and 387.5

# CHAPTER 5 SKILLS PREVIEW

Is the given number a solution of the equation? Write *yes* or *no*.

**1.** $4x = 16;\ -4$ **2.** $6 = \frac{w}{2} + 1;\ 10$ **3.** $-\frac{m}{8} = 8;\ -64$

Solve each equation.

**4.** $a + 8 = 20$ **5.** $-4z = 96$ **6.** $p + 5.7 = 3.1$

**7.** $\frac{s}{3} = -5$ **8.** $2x + 7 = 19$ **9.** $\frac{3}{4}t - 1 = 44$

**10.** $1 = 0.2y - 3$ **11.** $5w - 6 + 3w = 58$ **12.** $20 = 2x + 4 - 6x$

Solve and graph each inequality.

**13.** $8c \leq 24$ **14.** $5e + 3 > 13$

**15.** $-\frac{1}{2}g < 2$ **16.** $-7 \geq 4t + 1$

Solve.

**17.** Use the formula $d = rt$. Find the distance when $r = 45$ mi/h and $t = 3$ h.

**18.** Solve the formula $C = p - d$ for $p$.

Find each square root.

**19.** $\sqrt{49}$ **20.** $-\sqrt{81}$ **21.** $\sqrt{0.04}$ **22.** $\sqrt{\frac{9}{64}}$

Solve each equation.

**23.** $x^2 = 16$ **24.** $y^2 = 2.25$ **25.** $z^2 = \frac{1}{4}$

Write an equation and solve.

**26.** On Saturday, the zoo sold 100 more than twice the number of tickets sold on Thursday. There were 750 tickets sold on Saturday. How many tickets were sold on Thursday?

**27.** Mrs. Sanchez bought a refrigerator for \$689. She made a down payment of \$140 and paid the remaining amount in equal monthly payments of \$91.50. For how many months did she make payments?

**28.** Three less than five times a number equals 82. Find the number.

**29.** Phil has $3c + 5$ baseball cards, and Roy has $2c + 10$ cards. Write and simplify an expression for the total number of cards the boys have.

CHAPTER 5

# SOLVING EQUATIONS AND INEQUALITIES

**THEME Animals**

In this chapter, you will learn how to work with equations and formulas, both of which are ways to use data and relationships you know to determine other information. Many types of formulas are used in all fields of science. For example, the dental formulas discussed below are used by animal biologists to identify members of different groups of mammals.

Mammals generally have four main types of teeth:

- incisors – teeth used for cutting, located in the front of the jaw
- canines – sharp pointed teeth
- premolars – double-pointed teeth
- molars – grinding teeth, located in back of jaw

The four types of teeth are found in mammals in certain patterns. As a result of studying these patterns in many mammals, biologists have developed dental formulas. In the dental formula below, I stands for "incisor," C for "canine," P for "premolar," and M for "molar." The top numbers refer to the upper jaw (U) and the bottom numbers refer to the lower jaw (L). A dash (-) is used to separate the numbers that refer to the left and right sides of the jaw. Add all the top numbers to get the upper jaw total. Add all the bottom numbers to get the lower jaw total. Then add the results to get the total number of teeth.

$$I\frac{3-3}{3-3},\ C\frac{1-1}{1-1},\ P\frac{4-4}{4-4},\ M\frac{2-2}{3-3} = \frac{20}{22} = 42$$

total upper jaw (20); total lower jaw (22); 42 ← total number of teeth (upper and lower jaw)

As you can see from the table on page 175, this particular dental formula corresponds to the pattern for a wolf.

# DECISION MAKING

## Using Data

Use the table to answer the following questions.

1. Compare the total number of incisors of a wolf with that of a chipmunk.
2. Suppose an animal has 16 upper teeth but no canines. What animal could this be?
3. Look at the drawing below: (Only half the jaw is shown. Assume that the other half is identical.) Count the number of each type of teeth. Write the dental formula for the animal. Refer to the table and suggest the possible animal(s) it could be.

**DENTAL PATTERNS FOR MAMMALS**

| | Incisors | Canines | Premolars | Molars | U and L | Total | Mammal |
|---|---|---|---|---|---|---|---|
| U | 3–3 | 1–1 | 4–4 | 2–2 | 20 | 42 | wolf |
| L | 3–3 | 1–1 | 4–4 | 3–3 | 22 | | |
| U | 0–0 | 0–0 | 3–3 | 3–3 | 12 | 32 | deer, goat, bison |
| L | 3–3 | 1–1 | 3–3 | 3–3 | 20 | | |
| U | 2–2 | 0–0 | 3–3 | 3–3 | 16 | 28 | rabbit |
| L | 1–1 | 0–0 | 2–2 | 3–3 | 12 | | |
| U | 1–1 | 0–0 | 1–1 | 3–3 | 10 | 20 | chipmunk, porcupine |
| L | 1–1 | 0–0 | 1–1 | 3–3 | 10 | | |
| U | 1–1 | 0–0 | 0–0 | 3–3 | 8 | 16 | mouse, rat |
| L | 1–1 | 0–0 | 0–0 | 3–3 | 8 | | |

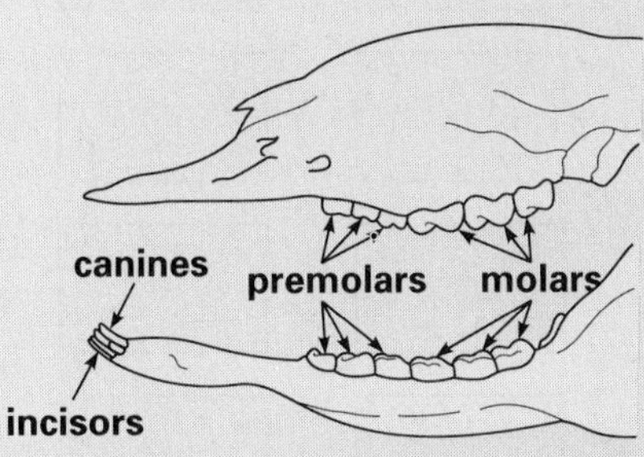

## Working Together

Work with your group to find out how many teeth dogs have and what dental formula characterizes them. You may do this through library research or by gently examining someone's pet. As a result of your findings, decide with what other animal(s) dogs could be grouped.

Remember that humans are mammals and, therefore, also have the same types of teeth discussed above. Examine your teeth and try to write the dental formula for humans. Compare results with other members of your group. If you need help, talk to a dentist, a dental hygienist, or a dental technician.

# 5-1 Working with Equations

## EXPLORE/ WORKING TOGETHER

Play this game with a partner.

a. Make a set of ten number cards by writing one of these numbers on each card: 4, −10, −15, 21, −30, 36, 48, 60, 75, 100.

b. Make a set of ten expression cards by writing one of these expressions on each card: 16 − 12, 40 ÷ (−4), −20 + 5, 16 + 5, −3 × 10, 40 − 4, 12 × 4, 30 + 30, 100 − 25, −60 + 160.

c. Shuffle the two sets of cards together, then arrange the cards face down in rows of five cards each.

d. The first player picks two cards, turns them face up, and decides if the two cards make an equality. If they do, the player writes the equality on a score sheet and keeps the cards. If the cards do not make an equality, they are replaced face down in their original positions. The next player takes a turn. Play continues until all cards have been taken. The player with the greater number of cards wins.

e. Work with your partner to make up different sets of number cards and expression cards, then play the new game.

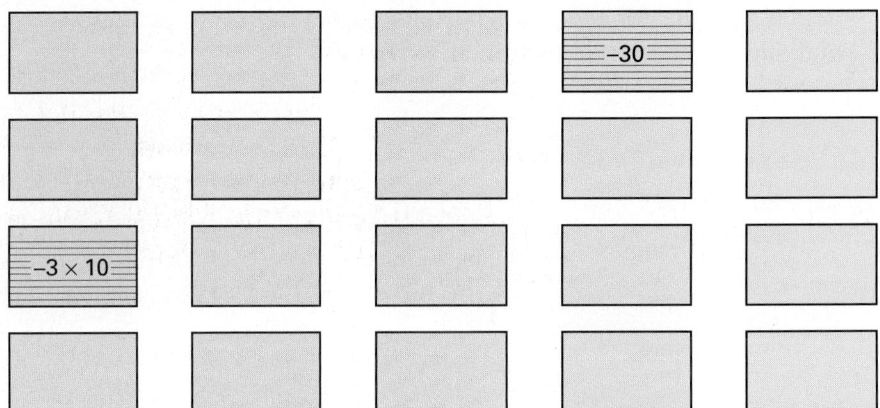


## SKILLS DEVELOPMENT

An **equation** is a statement that two numbers or expressions are equal. If the two *sides* of an equation are numerical expressions, the equation is either *true* or *false*.

**true:** $-8 + 2 = -6$ **false:** $-8 + 2 = -10$

An equation that contains one or more variables is a type of *open sentence*. An open sentence can be true or false, depending on what values are substituted for the variables. A value of the variable that makes such an equation true is called a **solution of the equation.** Equations like these can have none, one, or many solutions.

equation: $x - 3 = 11$
solution: $x = 14$

## Example 1

Is 2 a solution of the given equation?

a. $7w + 4 = 18$ b. $8x - 1 = 4x + 3$

c. $y^2 + 5 = 9$ d. $-6a^2 = 12a$

### Solution

Substitute 2 for the variable in each equation.

a. $7w + 4 = 18$
$7(2) + 4 \stackrel{?}{=} 18$
$18 = 18$ ✓
So, 2 is a solution.

b. $8x - 1 = 4x + 3$
$8(2) - 1 \stackrel{?}{=} 4(2) + 3$
$15 \neq 11$
So, 2 is not a solution.

c. $y^2 + 5 = 9$
$2^2 + 5 \stackrel{?}{=} 9$
$9 = 9$ ✓
So, 2 is a solution.

d. $-6a^2 = 12a$
$-6(2^2) \stackrel{?}{=} 12(2)$
$-24 \neq 24$
So, 2 is not a solution ◄

**CHECK UNDERSTANDING**

Tell whether each equation is *true, false,* or an *open sentence.*

1. $x + 9 = 5$
2. $16 + 23 = 39$
3. $7y = 40$
4. $8 \times (-6) = -54$

When you **solve an equation,** you find all the values of the variable that make the equation true.

**MATH: WHO, WHERE, WHEN**

The English mathematician and physician Robert Recorde (1510–1558) is credited with introducing algebra into England and with being the first to use the modern equals sign (=). Recorde decided that a symbol could avoid repetition of the words "is equal to." His equals sign was inspired by the notion of line segments of the same length.

## Example 2

Use mental math to solve each equation.

a. $p + \frac{3}{4} = 1$ b. $3w = 2.4$

### Solution

a. $p + \frac{3}{4} = 1$ What number added to $\frac{3}{4}$ equals 1?

You know that $\frac{1}{4} + \frac{3}{4} = 1$, so $p = \frac{1}{4}$.

b. $3w = 2.4$ Three times what number equals 2.4?

Since $3(8) = 24$, $3(0.8) = 2.4$, so $w = 0.8$ ◄

Sometimes, you must use opposites of absolute value to solve equations.

## Example 3

Solve each equation.

a. $-x = -7$ b. $-w = 0$ c. $-z = 5$

### Solution

Remember, $-x$ is the opposite of a number $x$.

a. $-x = -7$ The opposite of what number is −7?

The opposite of 7 is −7, so $x = 7$.

b. $-w = 0$ The opposite of what number is 0?

The opposite of 0 is 0, so $w = 0$.

c. $-z = 5$ The opposite of what number is 5?

The opposite of −5 is 5, so $z = -5$. ◄

**Example 4**

Solve each equation.

**a.** $|x| = 8$ **b.** $|c| = 0$ **c.** $|m| = -4$

**Solution**

Remember, the absolute value of a number is the distance the number is from 0 on a number line.

**a.** Think: What number is 8 units from 0? Since both 8 and $-8$ are 8 units from 0, $x = 8$ or $x = -8$.

**b.** The absolute value of 0 is 0, so $c = 0$.

**c.** The absolute value of a number is never negative. So, the equation has no solution. ◄

## TRY THESE

Which of these values, $-1$, 1, or 4, is a solution of the equation?

**1.** $y + 7 = 6$ **2.** $2x - 5 = 3$ **3.** $8m - 4 = 4$

Use mental math to solve each equation.

**4.** $t - \frac{1}{4} = \frac{1}{2}$ **5.** $6w = 3.6$ **6.** $\frac{1}{3}b = 7$

Solve each equation.

**7.** $|x| = 9$ **8.** $-t = -13$ **9.** $-b = 20$ **10.** $|a| = -12$

# EXERCISES

**PRACTICE/ SOLVE PROBLEMS**

Which of the given values is a solution of the equation?

**1.** $c + 4 = 2;\ -4, -2, 2$ **2.** $8x - 3 = 21;\ -3, 2, 3$

**3.** $m^2 - 8 = 41;\ -7, -6, 7$ **4.** $3x = 2\frac{1}{4};\ \frac{1}{2}, \frac{2}{3}, \frac{3}{4}$

**5.** $2w + 2 = 3;\ 0.1, 0.5, 1.5$ **6.** $\frac{2}{5}k = 14;\ 28, 35, 70$

Use mental math to solve each equation.

**7.** $\frac{1}{4}b = 12$ **8.** $a - 2.5 = 0$ **9.** $8s = 5.6$

**10.** $x - 24 = 25$ **11.** $-10 + w = -6$ **12.** $17 = 9 + z$

Solve each equation. If no solution exists, explain why.

**13.** $-x = 13$ **14.** $2|q| = 0$ **15.** $|k| = -10$

**16.** $-g = -0.75$ **17.** $|t| = 3.2$ **18.** $-z = -\frac{1}{3}$

**19.** $-d = 7.4$ **20.** $-4|w| = 0$ **21.** $|b| = 1\frac{4}{5}$

Which of the given values is a solution of the equation?

**EXTEND/ SOLVE PROBLEMS**

**22.** $3x^2 - 1 = 26$; $-4, -3, 3$

**23.** $-\frac{2}{3} + m = -\frac{1}{2}$; $\frac{1}{3}, \frac{1}{6}, -\frac{1}{6}$

**24.** $10a + 10 = 10$; $-2, 0, 1$

**25.** $0.5c^2 + 1 = 9$; $-4, 8, 16$

Solve each equation. If no solution exists, explain why.

**26.** $|3c| = 12$

**27.** $-|z| = -3$

**28.** $5|w| = 105$

**29.** $\left|\frac{x}{4}\right| = 20$

**30.** $|x + 1| = 7$

Solve.

**31.** A bird watcher counted 36 orioles, which was $\frac{3}{4}$ the number of robins counted. Use the equation $\frac{3}{4}r = 36$ and these values for $r$: 24, 48, 80. Find $r$, the number of robins.

**32.** The sum of Phil's and Jill's money equals the sum of Stan's and Fran's money. Jill has \$28, Stan has \$33, and Fran has \$42. Use the equation $P + 28 = 33 + 42$ and these values for $P$: 38, 47. Find $P$, the amount of money Phil has.

**THINK CRITICALLY/ SOLVE PROBLEMS**

**33.** Verify that 6 and $-6$ are solutions of the equation $x^2 = 36$. Can you find solutions of the equation $x^2 = -36$? Explain.

**34.** For what numbers $n$ does the equation $|x| = n$ have (a) exactly one solution; (b) more than one solution; (c) no solutions?

# 5-2 Solving One-Step Equations

## EXPLORE

A double-pan balance has a 10-g weight on the left side and two 5-g weights on the right side.

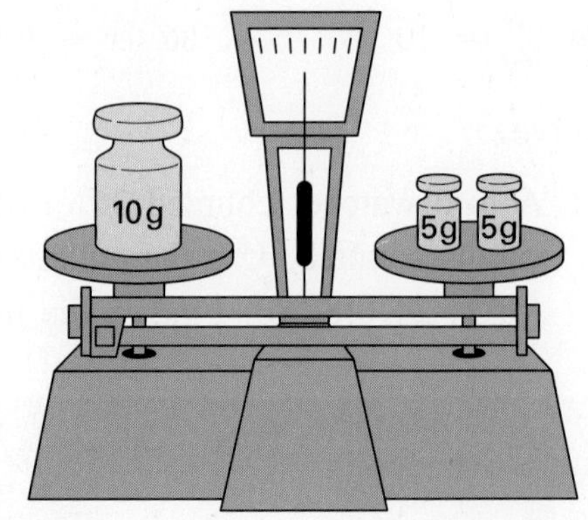


a. Is the scale in balance? How do you know?

b. Suppose you add a 5-g weight to the left side. What can you do to the right side to bring the scale into balance? What will be the total weight on each side?

c. Start again with a 10-g weight on the left and two 5-g weights on the right. Suppose you put two more 10-g weights on the left side. What is the total weight on the left? How many times the beginning weight is this?

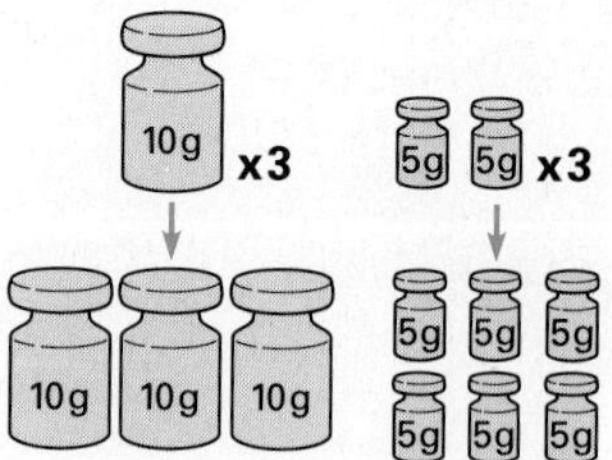


d. How many 5-g weights must you put on the right side to balance the scale now? How many 5-g weights do you have in all? How many times as many 5-g weights do you have now as when you began?

## SKILLS DEVELOPMENT

When two expressions are equal, you can add the same number to each expression and the resulting sums will be equal. This is called the **addition property of equality.**

For all real numbers $a$, $b$, and $c$, if $a = b$, then
$a + c = b + c$ and $c + a = c + b$.

For example, if $x + 1 = 4$, then $x + 1 + 5 = 4 + 5$.

You can also multiply two equal expressions by the same number. When you do, the resulting products will be equal. This is called the **multiplication property of equality.**

For all real numbers $a$, $b$, and $c$, if $a = b$, then
$ac = bc$ and $ca = cb$.

For example, if $2x = 10$, then $3(2x) = 3(10)$.

The most important use of these properties is in solving equations. If you cannot solve an equation mentally, use these properties to perform the same operations on both sides of the equation until the variable is alone on one side.

**CHECK UNDERSTANDING**

Write the reciprocal of each number.

**1.** $\frac{2}{3}$ **2.** 5 **3.** $1\frac{3}{8}$ **4.** -7

Does every real number have a reciprocal? Explain.

## Example 1

Solve each equation. Check the solution.

**a.** $x - 3 = 5$ **b.** $y + 2.7 = 6.1$ **c.** $6 - z = 14$

### Solution

Use the addition property of equality. Remember that the sum of a number and its opposite is 0, and $n + 0 = n$.

**a.** $x - 3 = 5$

Think: The opposite of −3 is 3.

Add 3 to each side.

$x - 3 + 3 = 5 + 3$
$x + 0 = 8$
$x = 8$

Check by substituting the solution into the original equation.

$x - 3 = 5$
$8 - 3 \stackrel{?}{=} 5$
$5 = 5$ ✓

The solution is 8.

**b.** $y + 2.7 = 6.1$

Think: The opposite of 2.7 is −2.7.

Add −2.7 to each side.

$y + 2.7 + (-2.7) = 6.1 + (-2.7)$
$y + 0 = 3.4$
$y = 3.4$

Check.

$y + 2.7 = 6.1$
$3.4 + 2.7 \stackrel{?}{=} 6.1$
$6.1 = 6.1$ ✓

The solution is 3.4.

**c.** $6 - z = 14$

Think: The opposite of 6 is −6.

Add −6 to each side.

$(-6) + 6 - z = (-6) + 14$
$0 + (-z) = 8$
$-z = 8$

Think: The opposite of what number is 8?

The opposite of −8 is 8, so $z = -8$.

Check.

$6 - z = 14$
$6 - (-8) \stackrel{?}{=} 14$
$6 + 8 \stackrel{?}{=} 14$
$14 = 14$ ✓

The solution is −8. ◄

**CONNECTIONS**

In Chapter 2, you learned that subtracting an integer is the same as adding its opposite. So, here is a slightly different way to solve the equation in Example **1b**.

$y + 2.7 = 6.1$
$y + 2.7 - 2.7 = 6.1 - 2.7$
$y = 3.4$

You also learned that dividing by a number is the same as multiplying by its reciprocal. Can you describe a different way to solve the equation in Example **2b**?

## Example 2

Solve each equation.

**a.** $\frac{x}{4} = 2.5$ **b.** $-7s = 35$

### Solution

Use the multiplication property of equality. Remember that the product of a number and its reciprocal is 1, and that $1 \cdot n = n$.

**a.** $\frac{x}{4} = 2.5$

Think: $\frac{x}{4}$ means $\frac{1}{4}x$.

The reciprocal of $\frac{1}{4}$ is 4.

Multiply each side by 4.

$(4)\left(\frac{x}{4}\right) = (4)2.5$
$(1)x = 10$

Check.

$\frac{x}{4} = 2.5$
$\frac{10}{4} \stackrel{?}{=} 2.5$
$2.5 = 2.5$ ✓

The solution is 10.

**b.** $-7s = 35$

Think: The reciprocal of −7 is $-\frac{1}{7}$.

Multiply each side by $-\frac{1}{7}$.

$\left(-\frac{1}{7}\right)(-7s) = \left(-\frac{1}{7}\right)(35)$
$(1)s = -5$
$s = -5$

Check.

$-7s = 35$
$-7(-5) \stackrel{?}{=} 35$
$35 = 35$ ✓

The solution is −5. ◄

## TRY THESE

Tell if you would first add (A) or multiply (M) to solve each equation, and tell what number you would use. Do not solve.

1. $3x = 12$
2. $m - 17 = 20$
3. $\frac{m}{8} = 1.6$
4. $y + \frac{2}{3} = 1$
5. $n + 3 = 0$
6. $7 - w = 12$

Solve each equation. Check the solution.

7. $4x = 48$
8. $x + 5 = 14$
9. $\frac{1}{2}c = 2.3$
10. $t - 2.5 = 3.4$
11. $8 - z = 2$
12. $5n = -45$
13. $\frac{3}{4}b = -8$
14. $\frac{2}{5} + g = \frac{1}{10}$
15. $1.5d = 30$

## EXERCISES

**PRACTICE/ SOLVE PROBLEMS**

Solve each equation. Check the solution.

1. $6w = 72$
2. $b + 8 = 29$
3. $z - 16 = 5$
4. $k + \frac{2}{3} = 2$
5. $\frac{1}{4}y = 9$
6. $1.5c = 3$
7. $12k = -14.4$
8. $-4d = 1$
9. $x + 19 = 6$
10. $p - 3.5 = 10$
11. $\frac{5}{8}n = 15$
12. $25b = -10$
13. $-52 = 13e$
14. $\frac{1}{8} = \frac{3}{4} + w$
15. $-\frac{x}{7} = -4$
16. $\frac{2}{3}m = -6$
17. $9 - a = 12$
18. $-8c = 100$

Solve.

19. A golden retriever puppy gained 35 lb during a 5-month period. To model the situation, the owner let $x$ equal "the average weight gain per month." Then the owner wrote the equation $5x = 35$. Solve the equation. On average, how much weight did the puppy gain each month?

20. Eight more than a certain number equals $-6$. To model the situation, Ana let $n$ equal "a certain number." Then she wrote the equation $n + 8 = -6$. Solve the equation. What was the original number?

## EXTEND/ SOLVE PROBLEMS

Solve and check each equation.

**21.** $3.5 - x = 6$ **22.** $-8 = 4y$ **23.** $11 + c = 4\frac{1}{2}$

**24.** $-1 = \frac{3}{4}x$ **25.** $-\frac{x}{2} = 1\frac{1}{4}$ **26.** $12 + 22 + n = 10$

**27.** $\frac{-r}{5} = 3 - 16$ **28.** $(d - 9) + 25 = 37$ **29.** $53 = -h + (4 \times 7)$

**30.** $e - 7 = |8 - 12|$ **31.** $k + |-5| = 17$ **32.** $t + 3 = |-4| - |-9|$

Find all solutions of each equation.

**33.** $|x| - 1 = 4$ **34.** $|w| + 8 = 20$

**35.** $-2 + |m| = 0$ **36.** $16 + |y| = 16$

Solve.

**37.** After Mrs. Rodriguez spent \$35.79 on groceries, she had \$16.42 left. To model this situation, she let $d$ equal "the amount of money she had originally." Then she wrote the equation $d - 35.79 = 16.42$. Solve the equation. How much money did Mrs. Rodriguez have originally?

**38.** A case of cat food weighs 18 lb and contains 24 cans. A store manager wrote the equation $24w = 18$. What does $w$ represent? Solve the equation.

## THINK CRITICALLY/ SOLVE PROBLEMS

Complete.

**39.** If $3a + 4 = 10$, then $6a + 8 = $ ■.

**40.** If $8w - 6 = 9$, then $4w - 3 = $ ■.

Solve.

**41.** If $x + 8 = 17$, find the value of $5x$.

**42.** If $w - 3 = 21$, find the value of $3w - 100$.

# 5-3 Solving Two-Step Equations

**EXPLORE** You can use algebra tiles to model equations.

▭ represents $x$. ■ represents 1.

**a.** Write the equation modeled by this diagram.

**b.** Show how to subtract 2 from each side. Subtracting 2 is the same as adding what number?

**c.** Next, show how to divide both sides into three equal groups.

**d.** What is the solution of the equation? How do you know?

**e.** Write the equation modeled by this diagram.

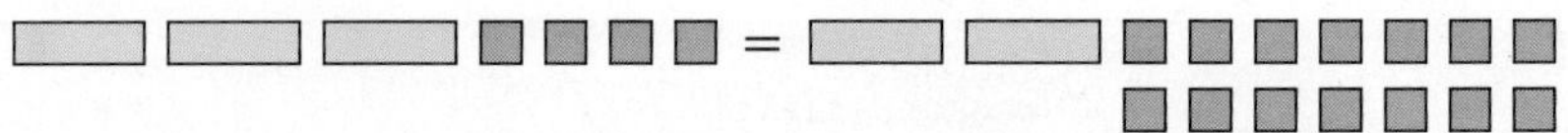

**f.** Show how to subtract 4 from each side. Subtracting 4 is the same as adding what number?

**g.** Next, show how to get one $x$ alone on the left side. What number do you subtract?

**h.** What is the solution of the equation? How do you know?

Use tiles to solve each equation.

**1.** $4x + 2 = 10$ **2.** $5x + 1 = 4x + 4$ **3.** $2x + 12 = 5x$

**SKILLS DEVELOPMENT** Some equations involve two operations. To solve these **two-step equations,** use the addition property of equality first. Then use the multiplication property.

### Example 1

Solve $\frac{x}{3} + 4 = 9$. Check the solution.

**Solution**

$$\frac{x}{3} + 4 = 9$$

$$\frac{x}{3} + 4 + (-4) = 9 + (-4) \quad \text{Add } -4 \text{ to each side.}$$

$$\frac{x}{3} = 5$$

$$3\left(\frac{x}{3}\right) = 3(5) \quad \text{Multiply each side by 3.}$$

$$x = 15$$

Check.

$$\frac{x}{3} + 4 = 9$$
$$\frac{15}{3} + 4 \stackrel{?}{=} 9$$
$$5 + 4 \stackrel{?}{=} 9$$
$$9 = 9 \checkmark$$

The solution is 15. ◄

Sometimes you simplify the equation before you can solve it.

## Example 2

Solve $7x + 6 - 2x + 3 = 34$. Check the solution.

### Solution

$$7x + 6 - 2x + 3 = 34$$ ← **$7x$ and $2x$ are like terms; 6 and 3 are like terms.**

$$5x + 9 = 34$$ ← **$7x - 2x = (7 - 2)x = 5x$ and $6 + 3 = 9$**

$$5x + 9 + (-9) = 34 + (-9)$$ ← **Add $-9$ to each side.**

$$5x = 25$$

$$\frac{1}{5}(5x) = \frac{1}{5}(25)$$ ← **Multiply each side by $\frac{1}{5}$.**

$$x = 5$$

Check.

$$7x + 6 - 2x + 3 = 34$$
$$7(5) + 6 - 2(5) + 3 \stackrel{?}{=} 34$$
$$35 + 6 - 10 + 3 \stackrel{?}{=} 34$$
$$44 - 10 \stackrel{?}{=} 34$$
$$34 = 34 \checkmark$$

The solution is 5. ◄

You can use equations to solve geometric problems.

## Example 3

Find the measures of angles $MON$ and $POQ$.

M P O $(3x + 15)°$ $(4x - 5)°$ N Q

### Solution

Angles $MON$ and $POQ$ are vertical angles, so $m\angle MON = m\angle POQ$. Use this fact to write an equation.

$$3x + 15 = 4x - 5$$

Note that a variable appears on both sides of the equation. Since variables represent numbers, you may use the addition property of equality and add a variable expression to each side of the equation.

**Add $-3x$ to each side.** → $$-3x + 3x + 15 = -3x + 4x - 5$$ ← **Combine like terms. $-3x - 3x = 0$; $-3x + 4x = (-3 + 4)x = x$**

$$15 = x - 5$$

**Add 5 to each side.** → $$15 + 5 = x - 5 + 5$$

$$20 = x$$

Substitute to find the angle measures.

$$3x + 15 = 3(20) + 15 = 60 + 15 = 75$$

$$4x - 5 = 4(20) - 5 = 80 - 5 = 75$$

The measure of $\angle MON$ is 75° and the measure of $\angle POQ$ is 75°. ◄

**PROBLEM SOLVING TIP**

In Example 3, simply finding the value of $x$ does not solve the problem. The problem asks you to find the measure of each angle. So, to complete the problem, you substitute the value of $x$ into the expression for each angle. Always read a problem carefully to make sure that you have answered the question.

**MIXED REVIEW**

Find each percent.

1. 36 is __?__% of 240?
2. 60 is __?__% of 48?

Evaluate each expression when $c = -14$.

3. $|-c|$ 4. $-(-c)$

Simplify.

5. $x^5 \cdot x^8$ 6. $y^6 \div y^{-4}$

Classify the triangle by its sides and by its angles.

7.

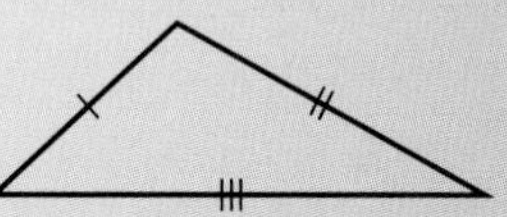

Write the converse of the statement. Tell whether the statement and its converse are *true* or *false*.

8. If two figures are squares, then they are quadrilaterals.

You will have to use the distributive property to solve some equations.

### Example 4

Solve $2(x + 7) = 4(x - 1)$. Check the solution.

**Solution**

$$\begin{aligned} 2(x + 7) &= 4(x - 1) \\ 2x + 14 &= 4x - 4 && \leftarrow \text{Use the distributive property.} \\ -2x + 2x + 14 &= -2x + 4x - 4 && \leftarrow \text{Add } -2x \text{ to each side.} \\ 0 + 14 &= 2x - 4 \\ 14 &= 2x - 4 \\ 14 + 4 &= 2x - 4 + 4 && \leftarrow \text{Add 4 to each side.} \\ 18 &= 2x \\ \tfrac{1}{2}(18) &= \tfrac{1}{2}(2x) && \leftarrow \text{Multiply each side by } \tfrac{1}{2}. \\ 9 &= x \end{aligned}$$

Check.

$$\begin{aligned} 2(x + 7) &= 4(x - 1) \\ 2(9 + 7) &\stackrel{?}{=} 4(9 - 1) \\ 2(16) &\stackrel{?}{=} 4(8) \\ 32 &= 32 \checkmark \end{aligned}$$

The solution is 9. ◄

## TRY THESE

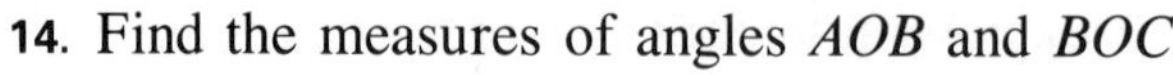

Simplify each equation. Do not solve.

1. $9b - 5 + 3b - 4 = 2$
2. $-6(y - 2) + 3(4 - y) = -8$
3. $-8 - 14x + 15x = 0$
4. $5(-2z - 7) + 2(2z + 9) = -3$

Solve each equation. Check the solution.

5. $3x + 2 = 14$
6. $\frac{m}{2} - 4 = 8$
7. $2w - 7 = 25$
8. $9z - 2z + 1 = 29$
9. $3(t - 2) = 15$
10. $-\frac{1}{4}d + 5 = 2$
11. $4(2x + 5) = 44$
12. $-6x + 8 = 4x - 42$
13. $2(y + 7) = 3(y - 1)$
14. Find the measures of angles *AOB* and *BOC*.

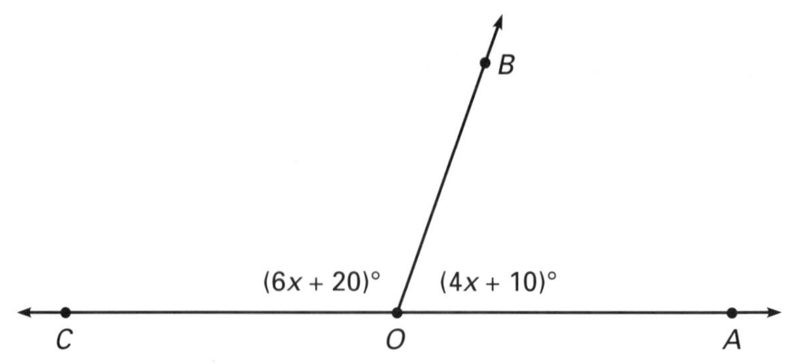

**PROBLEM SOLVING TIP**

In Exercise 14, $\angle AOB$ and $\angle BOC$ are a pair of supplementary angles, so $m\angle AOB$ plus $m\angle BOC = 180°$. Use this fact to write an equation.

# EXERCISES

## PRACTICE/ SOLVE PROBLEMS

Solve each equation. Check the solution.

1. $\frac{w}{3} + 4 = 10$
2. $3p - p + 1 = -9$
3. $5a - 10 = 0$
4. $5q + 3q - 1 = -33$
5. $\frac{3}{4}n + 2 = 8$
6. $12 = -3 - 5x$
7. $4r - 2.5 = 6.3$
8. $-2y - 2y + 3 = -13$
9. $7 = \frac{3}{8}t - 2$
10. $-12g + 8 + 3g + 3 = 29$
11. $4(b + 6) = -20$
12. $1.5z - 7 - 3.5z = 3$
13. $8(x - 1) = 0$
14. $3(2w + 5) = 45$
15. $6z - 2 = z + 13$
16. $4(c + 3) = -3(c - 2)$
17. $2(9 - d) = 4$
18. $\frac{1}{2}(14k - 8) = 10$
19. Find $m\angle ADB$ and $m\angle BDC$ in the figure at the right.
20. Find $m\angle URV$ and $m\angle TRS$ in the figure below.

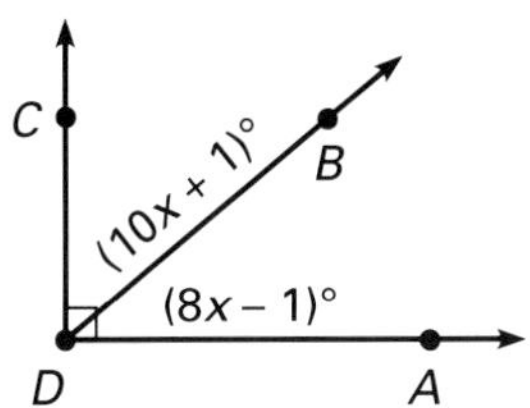


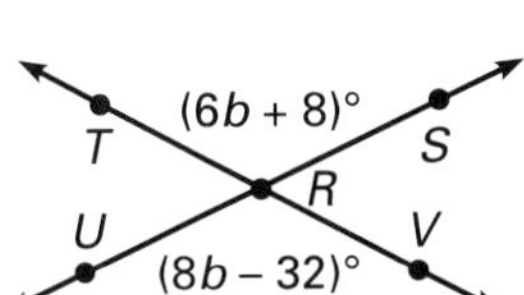


## EXTEND/ SOLVE PROBLEMS

Solve and check each equation.

21. $\frac{1}{2}(6t - 20) + 15 = -5 + 2(t + 3)$
22. $2(3 - k) = 3(2 + k)$
23. $4[1 - 3(x + 2)] + 2x = 2(x + 2)$
24. $2[3(g + 3) - (g + 2)] = 5(g + 1) + 4$
25. An architect's blueprint included triangle $JKL$. Find the measure of each angle of the triangle. What geometric relationship did you use?
26. Find the measure of angle $DAB$ and angle $BCD$ in parallelogram $ABCD$. What geometric relationship did you use?

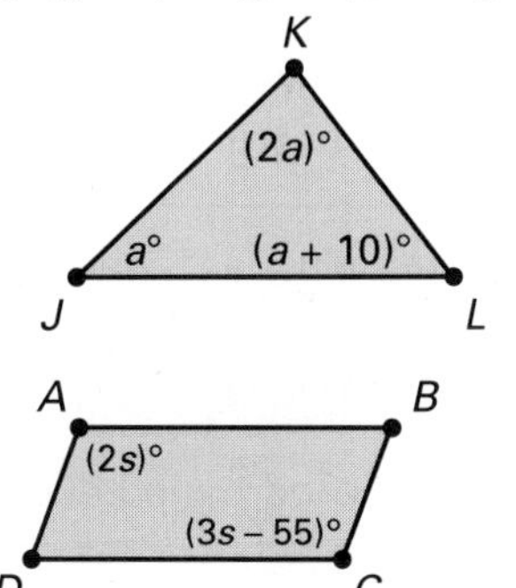

27. Twice a number, increased by 8, is the same as three times the number, decreased by 11. Find the number.

## THINK CRITICALLY/ SOLVE PROBLEMS

28. Solve the equation $2(2x - 4) + 17 = -3 + 4(x + 3)$. What did you find? What do you think this means?
29. Solve the equation $4(1 - y) + 9y = 5(y + 2)$. What did you find? What do you think this means?

# 5-4 Problem Solving Skills:
## TRANSLATING WORDS INTO SYMBOLS

- ► READ
- ► PLAN
- ► SOLVE
- ► ANSWER
- ► CHECK

A key skill in solving many problems is being able to translate word expressions into algebraic expressions. Read the problem carefully. Think about what each expression means and how the numbers are related. Choose a variable to represent an unknown quantity. Try to identify words that suggest an operation.

### PROBLEM

Write each word expression as a variable expression.

**a.** 20 mi/h more than the average running speed of a lion

**b.** the cost of a safari vacation separated into twelve equal monthly payments

### SOLUTION

**a.** Let $l$ = average running speed of a lion.
20 mi/h *more than* $l$ ← "More than" suggests addition.

Write the variable expression $l + 20$.

**b.** Let $c$ = cost of the safari vacation.
$c$ *separated* into twelve equal parts ← "Separated" suggests division.

Write the variable expression $c \div 12$ or $\frac{c}{12}$.

This table shows certain words and phrases that suggest which operation to use in solving a problem.

| | |
|---|---|
| + | add, plus, and, sum of, increased by, more than, total of |
| − | subtract, less, minus, difference of, decreased by |
| × | multiplied by, product of, times, twice, doubled, tripled |
| ÷ | divided by, quotient of, gives, is, results in, separated into |

# PROBLEMS

Which operation would you use to translate each of the following?

1. two less than eight
2. three more than ten
3. twice a number
4. half of a number

Match each word expression with an algebraic expression.

5. a number $n$ decreased by 8
6. 8 divided by a number $n$
7. the sum of 8 and a number $n$
8. 8 decreased by $n$
9. the product of 8 and $n$
10. the result of dividing $n$ by 8
11. $n$ multiplied by $-8$
12. the quotient of $n$ and $-8$

a. $\frac{n}{8}$
b. $8n$
c. $n - 8$
d. $-8n$
e. $8 \div n$
f. $8 - n$
g. $8 + n$
h. $n \div (-8)$

Write a variable expression for each of the following. Be sure to tell what the variable represents.

13. forty dollars more than last week's price
14. twice Gina's age
15. fifteen points less than Karen's score
16. the interest on a car loan separated into 36 equal payments
17. the number of hours times $7.50, the hourly wage

Write a word expression that could be represented by each variable expression. Be sure to tell what the variable represents.

18. $5 + p$
19. $4w$
20. $2a + 3$
21. $s \div 12$

Nancy works at the Big B Market where she earns $6 per hour. Write an expression to show how much she earns for working each amount of time.

22. $n$ hours
23. $(n + 3)$ hours
24. $(4n - 1)$ hours
25. $\frac{n}{2}$ hours

Write an expression in cents for the value of the following.

26. $n$ nickels
27. $3x$ quarters
28. $\frac{m}{2}$ dimes
29. $(4n + 10)$ pennies

# 5-5 Problem Solving Strategies:

## WRITE AND SOLVE AN EQUATION

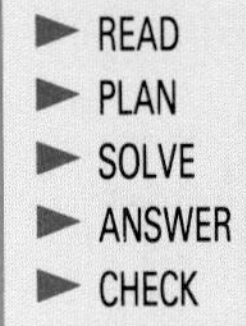


Writing and solving an equation is a very powerful problem solving strategy. To write equations, you first translate the words of the problem into algebraic expressions. You then use the situation described in the problem to state how the numbers and expressions are related.

### PROBLEM

A zookeeper checked the budget one month and found that the amount allotted for animal feed was \$3,000 more than one-third of the total monthly expenses. If \$8,400 was allotted for animal feed that month, what were the total monthly expenses?

### SOLUTION

*Given*: Cost of feed for the month was \$8,400.

Cost of feed was \$3,000 more than one-third of the total monthly expenses.

*Find*: Total monthly expenses

Let $t$ represent the total monthly expenses, in dollars. ← **The variable represents the unknown quantity.**

Then, $\frac{1}{3}t + 3{,}000$ represents the feed cost, in dollars.

You know that the actual monthly feed cost was \$8,400. So, write an equation.

$$\frac{1}{3}t + 3{,}000 = \$8{,}400$$

Now solve the equation.

$$\frac{1}{3}t + 3{,}000 + (-3{,}000) = 8{,}400 + (-3{,}000)$$
$$\frac{1}{3}t = 5{,}400$$
$$3\left(\frac{1}{3}t\right) = 3(5{,}400)$$
$$t = 16{,}200$$

Check. $\frac{1}{3}(16{,}200) + 3{,}000 \stackrel{?}{=} 8{,}400$
$5{,}400 + 3{,}000 \stackrel{?}{=} 8{,}400$
$8{,}400 = 8{,}400$ ✓

The zoo's total monthly expenses were $16,200.

# PROBLEMS

Choose the letter of the equation that represents the problem situation correctly. Do not solve the problem.

1. The length of a snake less 5 cm is 110 cm. How long is the snake? Let $l$ represent the length of the snake.
   **a.** $5 - l = 110$ **b.** $l + 5 = 110$ **c.** $l - 5 = 110$

2. The total cost of 6 zoo admissions and a $7 ride pass is $52. What is the cost of admission to the zoo? Let $c$ represent the cost of admission.
   **a.** $6c + 7 = 52$ **b.** $6c = 52 + 7$ **c.** $6c - 52 = 7$

Solve each problem by using an equation.

3. Tashi wants to buy a camera that costs $175. This is $37 more than three times what he has saved so far. How much has he saved so far?

4. Twice the number of lions at the zoo plus 1 gives a total of 23. How many lions are at the zoo?

5. After exercising, Jordana's average pulse rate was 112 beats per minute. This was 34 beats less than double her average rate when resting. What is Jordana's average at-rest pulse rate?

6. The sum of five times Lee's age and 18 is 93. How old is Lee?

7. At the zoo gift shop, a stuffed panda bear costs $5 more than a stuffed monkey. Together, the two items cost $52. How much does each item cost?

**USING DATA** Use the Data Index on pages 556–557 to find information on how long animals live.

8. Dana's cockatoo is 2 years older than one-fifth the longest recorded lifespan for the bird's species. How old is Dana's bird?

# 5-6 Solving Inequalities

## EXPLORE

**TALK IT OVER**

What does each symbol represent?

$<$ $\leq$ $>$ $\geq$

Make up a strategy for remembering the meaning of each.

**a.** The graph of the inequality $x < 2$ is shown by Graph A. Describe the graph. Explain why there is an open circle at the 2 point.

A. −6 −5 −4 −3 −2 −1 0 1 2 3 4 5 6

**b.** Graph each of the following inequalities on a number line. Label each graph with the given letter.

Graph B $x > 2$

Graph C $x = 2$

**c.** Which graphs could you combine to get the graph of $x \geq 2$? Draw the graph.

**d.** Which graphs could you combine to get the graph of $x \leq 2$? Draw the graph.

**e.** Is 3 a solution of $x \leq 2$? How do you know?

**f.** How many solutions are there for $x \leq 2$?

**g.** Write an inequality for which $x = 3$ is a solution.

## SKILLS DEVELOPMENT

**READING MATH**

Words such as "at least," "at most," "greater than," "less than," "maximum," and "minimum" often indicate that an inequality will be used in solving a problem.

A mathematical sentence that contains one of the symbols $<$, $>$, $\leq$, or $\geq$ is called an **inequality.** Like an equation, an inequality may be *true, false,* or *an open sentence.*

| **true:** | **false:** | **open sentence:** |
|---|---|---|
| $-4 \leq -4$ | $-4 < -4$ | $m < -4$ |

An inequality that is an open sentence may be true or false, depending on what values are substituted for the variable. A value of the variable that makes such an inequality true is called a **solution of the inequality.** Inequalities like these often have an infinite number of solutions.

inequality: $x + 2 \leq -3$

solutions: all real numbers less than or equal to $-5$

When you **solve an inequality,** you find all values of the variable that make the inequality true by using techniques similar to those used to solve an equation. For example, if you add the same number to each side of an inequality, the *order* of the inequality remains the same. This is called the **addition property of inequality.**

For all real numbers $a$, $b$, and $c$:

If $a < b$, then $a + c < b + c$, and $c + a < c + b$.

If $a > b$, then $a + c > b + c$, and $c + a > c + b$.

## Example 1

Solve and graph $n + 5 \geq 8$.

### Solution

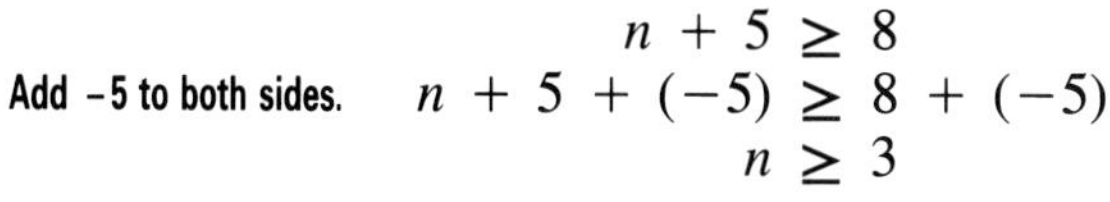

$$n + 5 \geq 8$$

Add $-5$ to both sides. $$n + 5 + (-5) \geq 8 + (-5)$$

$$n \geq 3$$

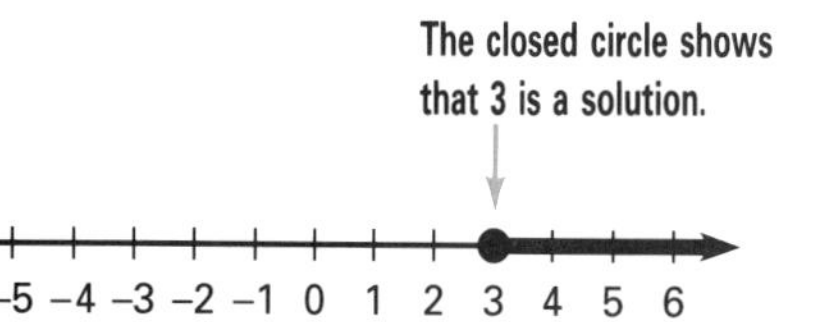


You can also multiply each side of an inequality by the same number, but it is very important to be aware of the sign of the number. If you multiply each side of an inequality by the same *positive* number, the order of the inequality remains the same. However, if you multiply each side of an inequality by the same *negative* number, the order of the inequality is reversed. This is called the **multiplication property of inequality.**

For all real numbers $a$ and $b$ and for $c > 0$:
If $a < b$, then $ac < bc$, and $ca < cb$.
If $a > b$, then $ac > bc$, and $ca > cb$.
For all real numbers $a$ and $b$ and for $c < 0$:
If $a < b$, then $ac > bc$, and $ca > cb$.
If $a > b$, then $ac < bc$, and $ca < cb$.

**TALK IT OVER**

In some books, the figure formed by the graph of an inequality of the form $x > b$ or $x < b$ is called a *half-line.* The graph of $x \geq b$ or $x \leq b$ is called a *ray*. Can you think of an inequality whose graph is a line segment that includes the endpoints?

## Example 2

Solve and graph each inequality.

**a.** $3x \leq 9$ **b.** $-\frac{1}{4}y \leq 2$

### Solution

**a.**

$$3x \leq 9$$

Multiply each side by $\frac{1}{3}$. $$\frac{1}{3}(x) \leq \frac{1}{3}(9)$$

$$x \leq 3$$

**b.**

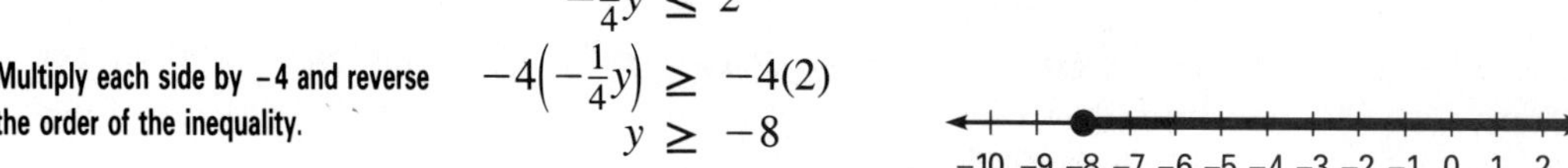

$$-\frac{1}{4}y \leq 2$$

Multiply each side by $-4$ and reverse the order of the inequality. $$-4\left(-\frac{1}{4}y\right) \geq -4(2)$$

$$y \geq -8$$



## Example 3

Solve and graph $-2x - 3 < -1$.

### Solution

$$-2x - 3 < -1$$

Add 3 to each side. $$-2x - 3 + 3 < -1 + 3$$

$$-2x < 2$$

Multiply each side by $-\frac{1}{2}$ and reverse the order of the inequality. $$\frac{-2x}{-2} > \frac{2}{-2}$$

$$x > -1$$

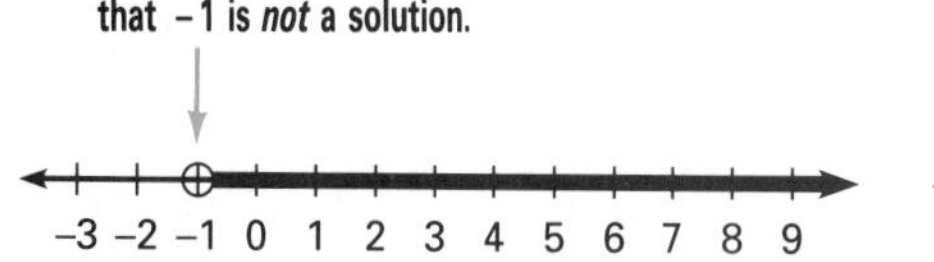

### Example 4

Nita is training for a bicycle race. She plans to ride at least 100 mi each week. Because a wheel broke, she could ride only 7 mi on Monday. What is the least number of miles she must average daily for the next six days to achieve her goal?

**Solution**

Write and solve an inequality that represents the situation.

Let $m$ = number of miles Nita rides each day.
Let $6m$ = number of miles for the next six days.

$$100 \le 6m + 7$$
$$100 + (-7) \le 6m + 7 + (-7)$$
$$93 \le 6m$$
$$\frac{1}{6}(93) \le \frac{1}{6}(6m)$$
$$15.5 \le m \text{ or } m \ge 15.5$$

Nita must average at least 15.5 mi each day for the next 6 days. ◄

**COMPUTER**

One use of the computer is for the solution of equations. The program below solves equations of the form: $Ax + B = Cx + D$. The program includes the possibilities that all real numbers are solutions or that there is no solution.

```
10 PRINT "FOR THE
   EQUATION AX + B =
   CX + D, WHAT ARE"
20 PRINT "A, B, C, AND D";
30 INPUT A, B, C, D
40 IF A = C AND B = D
   THEN PRINT "ALL REAL
   NUMBERS ARE
   SOLUTIONS."
50 IF A = C AND B < > D
   THEN PRINT "NO
   SOLUTION."
60 IF A < > C THEN PRINT
   "THE VALUE OF X IS";
   (D - B)/(A - C);"."
```

## TRY THESE

Start with the first inequality. Tell what steps you would take to arrive at the second inequality.

1. $3x - 14 < 7$
   $x < 7$
2. $-\frac{w}{3} + 5 \ge -1$
   $w \le 18$

Solve and graph each inequality.

3. $z - 4 < -2$
4. $3m \ge 12$
5. $8 - 2p \le 10$
6. $-9 > 4n + 7$

Solve.

7. Manuel wants to read a 150-page book over a five-day period, so he planned to read 30 pages each day. However, on the first day he read 38 pages. What is the least number of pages he must average for the next four days to reach his goal?

## EXERCISES

**PRACTICE/ SOLVE PROBLEMS**

Solve and graph each inequality.

1. $6n + 5 < -25$
2. $2 - c > -2$
3. $2 \ge -\frac{1}{2}x$
4. $7k + 4 \ge -10$
5. $4z + 8 > -2$
6. $5 + 4y < 37$
7. $14 < 5e + 4$
8. $4 - 3q > 13$
9. $16 \ge 3c + 7$

10. $\frac{2}{3}b - 3 \leq -1$ 11. $10 \leq -2x + 5$ 12. $0.8x - 7 \leq 0.2$

13. $3 - 9t > 30$ 14. $2 - \frac{1}{4}a \geq 4$ 15. $17 \geq 5 - 6x$

Solve.

16. Roy is on an 1,800-calorie-a-day diet. He consumed 450 calories at breakfast. What must Roy's average caloric intake be for lunch and dinner in order for him to maintain his diet?

17. The maximum load for an elevator is 2,400 lb. A crate weighing 300 lb is put on the elevator. Suppose that the average weight of a passenger is 150 lb. How many passengers can get on the elevator along with the crate?

## EXTEND/ SOLVE PROBLEMS

Solve and graph each inequality.

18. $3x - 6x \geq 12$ 19. $4(c + 1) > -3$

20. $2p + 3.7 < p - 1.5$ 21. $1 + m \leq 3 - 2m$

22. $7t - (9t + 1) \geq -5$ 23. $6 - 4b > 4 - 3b$

24. $5x + 4 > 3x + 2$ 25. $2(x + 1) < 4(x + 3)$

26. Three-fourths of a number $n$, decreased by 11, is greater than or equal to 4. What values are possible for $n$?

27. Jenna input 18 pages of a report in the first hour and 23 pages in the second hour. How many pages must she input in the next hour to maintain an average number of at least 20 pages per hour for the three-hour period?

## THINK CRITICALLY/ SOLVE PROBLEMS

28. Write a problem that could be solved using the inequality $3c + 25 \leq 100$.

29. a. Solve and graph each of the following inequalities.
    A. $x > 3$ B. $x < -3$

    b. Study Graphs A and B. How do you think the graphs are related to the graph of the absolute-value inequality $|x| > 3$?

30. Solve and graph $|y + 1| \geq 4$.

31. If $a > b$ and $c > d$, then is $a - c > b - d$ *sometimes, always,* or *never* true? Use examples to justify your answer.

32. If $m \geq n$ and $-m \geq -n$, what do you think is true about $m$ and $n$?

# 5-7 Working with Formulas

## EXPLORE

The Thrifty Taxi Company charges passengers a fixed amount plus an amount based on the number of miles traveled. The charge for each mile is the same. Use the information in the table.

| Number of Miles | 1 | 2 | 3 | 4 |
|---|---|---|---|---|
| Cost | $2.45 | $3.40 | $4.35 | $5.30 |

a. How much does each additional mile add to the cost of a ride?
b. What is the fixed amount? How did you find your answer?
c. Write a formula to calculate the cost $C$ of traveling $m$ miles.
d. Use your formula. How much would a 7-mi ride cost?
e. If you knew that a ride cost $11.00, how could you find the number of miles? Find it.

## SKILLS DEVELOPMENT

A **formula** is an equation stating a relationship between two or more quantities. For example, the distance that you travel is equal to your average speed, or rate, multiplied by the time you spend traveling at that rate.

$$\text{distance} = \text{rate} \times \text{time}$$
$$d = rt$$

Often you can evaluate a variable in a formula directly, using given information. For example, suppose that a car travels at 55 mi/h for 3 h.

$$d = rt$$
$$d = (55)3 = 165$$ The distance traveled is 165 mi.

At other times, however, you must solve for a variable in a formula.

### Example 1

The density ($D$) of a substance is the mass ($M$) per unit volume ($V$) of the substance. The formula for density is $D = \frac{M}{V}$. The density of cork is 0.24 g/cm$^3$. Find the mass of a cork board that has a volume of 800 cm$^3$.

Change 192 g to kilograms.

**Solution**

$$D = \frac{M}{V}$$
$$0.24 = \frac{M}{800}$$ Substitute the known values of the variables into the formula.
$$800(0.24) = 800\left(\frac{M}{800}\right)$$ Solve the equation.
$$192 = M$$

The mass of the cork board is 192 g. ◄

### Example 2

The formula $T = \frac{n}{4} + 40$ relates the temperature ($T$) in degrees Fahrenheit to the number of times ($n$) a cricket chirps in 1 minute. Use the formula to find how often a cricket chirps at 55°F.

**Solution**

$$T = \frac{n}{4} + 40$$

$$55 = \frac{n}{4} + 40 \quad \leftarrow \text{Substitute 55 for } T.$$

$$55 + (-40) = \frac{n}{4} + 40 + (-40)$$

$$15 = \frac{n}{4}$$

$$4(15) = 4\left(\frac{n}{4}\right)$$

$$60 = n$$

At a temperature of 55°F, a cricket will chirp 60 times per minute. ◄

Sometimes you must solve a formula for a variable that is not alone on one side of the equals sign.

### Example 3

Solve the formula $d = rt$ for $t$.

**Solution**

To solve a formula for a given variable, rewrite the formula so that the variable stands alone on one side of the equals sign.

$$d = rt$$

$$\frac{1}{r}(d) = \frac{1}{r}(rt) \quad \leftarrow \text{Multiply both sides of the equation by } \frac{1}{r} \text{ to get } t \text{ alone on one side.}$$

$$\frac{d}{r} = t$$

The formula shows that time can be found by dividing distance by rate. ◄

## TRY THESE

Solve.

1. An iron bar has a mass of 474 g. The volume of the bar is 60 cm³. Determine the density of iron.
2. Jesse rode his bicycle for $2\frac{1}{2}$ h and traveled a distance of $26\frac{1}{4}$ mi. What speed did Jesse average on the trip?
3. Solve the formula $y = kx$ for $x$.

# EXERCISES

## PRACTICE/ SOLVE PROBLEMS

1. Sound travels at the rate of 330 m/s. If a firecracker explodes 3,960 m away from you, how long does it take you to hear it?

2. Renny and Dave drove a distance of 424 mi. Their travel time was 8 h. What was their average speed for that part of the trip?

3. The relationship between the length $l$ of the human male humerus (upper arm bone) and the height $H$ of the male is given by the formula $H = 2.89l + 70.64$, where $l$ and $H$ are measured in centimeters. How tall is a man with a 38-cm humerus?

Solve each formula for the indicated variable.

4. $I = prt$, for $r$

5. $w = 5h - 190$, for $h$

6. ***USING DATA*** Use the Data Index on pages 556–557 to find how fast insects fly. What is the greatest distance a hawkmoth can travel in 0.5 h?

## EXTEND/ SOLVE PROBLEMS

In a psychology course, the following formula is used to convert the number of points scored on a test, $R$ (raw score), to a percent.

$$P = \frac{7}{8}R + 20$$

7. What percent score would be equivalent to a raw score of 84?

8. Solve the formula for $R$.

9. What raw score is equivalent to a score of 62%?

The formula $d = \frac{s+s^2}{20}$ is used to determine the approximate stopping distance ($d$) in feet for a car traveling $s$ mi/h on a dry road. Use the formula to complete the table.

| **Speed in mi/h ($s$)** | 20 | 25 | 30 | 40 | 55 |
|---|---|---|---|---|---|
| **Stopping distance in feet ($d$)** | **10.** ■ | **11.** ■ | **12.** ■ | **13.** ■ | **14.** ■ |

## THINK CRITICALLY/ SOLVE PROBLEMS

Use the following facts to complete the table. Lois jogged to the end of the park at a rate of 12 km/h. She walked back at a rate of 4 km/h. Her total time traveling was 1 h. She followed the same route both ways.

| | **Rate** | **Time** | **Distance** |
|---|---|---|---|
| Jogging | 12 | $t$ | **15.** ■ |
| Walking | 4 | **16.** ■ | **17.** ■ |

18. How long did Lois spend walking?

► READ
► PLAN
► SOLVE
► ANSWER
► CHECK

# Problem Solving Applications:

## ENERGY COSTS

Many appliances, such as refrigerators, TV sets, and hair driers run on electricity. The amount of electricity used is usually measured by a device called an electrical meter. Utility bills reflect the cost for the amount of electrical energy used. The amount used depends on the **power** required by the appliances and how long they are used. You can find the amount of energy an appliance uses with this formula.

$$E = P \times T$$
$$\text{energy} = \text{power} \times \text{time}$$

The power needed to operate the appliance usually appears on the appliance itself. Electrical power ($P$) is expressed in watts (W) or kilowatts (kW). The unit of electrical energy ($E$) is the kilowatt-hour (kWh). One kilowatt-hour is 1,000 watts of power used for one hour. Utility companies bill for the number of kilowatt hours used. You can calculate your electric bill by multiplying the energy used by the cost per kilowatt-hour.

| Appliance | Power Usage (Watts per Hour) |
|---|---|
| Hair drier | 1,000 |
| Microwave oven | 700 |
| TV set (color) | 200 |
| Refrigerator (15 $ft^3$) | 620 |
| Stereo | 110 |
| 100-watt bulb | 100 |

1. The Changs watch their color TV an average of 4 hours per day. How many kilowatt-hours of energy do they use daily?
2. Suppose the Changs pay \$0.13/kWh for electricity. What is the weekly cost of running their TV to the nearest cent?
3. If the rate charged for electricity is \$0.11/kWh, what is the cost of operating a refrigerator for the month of April? What assumption did you make?
4. Sharla figured that the weekly cost of using her hair drier is \$0.38. If the rate charged is \$0.10/kWh, how many hours does Sharla use her drier each week?
5. Suppose the rate charged for electricity is \$0.12 from 8 A.M. to 5 P.M., but only \$0.09 from 5 P.M. to 8 A.M. If you listen to your stereo 2 h daily, how much can you save yearly by listening after 5 P.M.?
6. Find out the power rating for some appliances you use and the rate charged for electricity in your area. Estimate the amount of time you use each appliance monthly and determine the cost. Which appliances cost the most to operate?

# 5-8 Squares and Square Roots

## EXPLORE

If you wanted to make a square quilt with an area of 196 square units, what dimensions would the quilt have? Here's one way to find out.

a. Write a formula for the area of a square with a side of length $s$.
b. Use graph paper to draw a square with an area of 196 square units. Tell how you found the correct square.
c. What is the length of each side of your square?
d. What number squared is 196?
e. Can you use graph paper to find a square with an area of exactly 210 square units? Explain.

## SKILLS DEVELOPMENT

A number, $a$, is a **square root** of another number, $b$, if $a^2 = b$. Every positive real number has two square roots.

$256 = (16)^2$ — **16 is the positive or principal square root of 256. You write $\sqrt{256} = 16$**

$256 = (-16)^2$ — **-16 is the negative square root of 256. You write $-\sqrt{256} = -16$**

The number 0 has only one square root: $\sqrt{0} = 0$.
Numbers such as $\frac{9}{16}$ and 0.0121 also have two square roots, one positive and one negative.

$$\sqrt{\frac{9}{16}} = \frac{3}{4} \text{ and } -\sqrt{\frac{9}{16}} = -\frac{3}{4}$$
$$\sqrt{0.0121} = 0.11 \text{ and } -\sqrt{0.0121} = -0.11$$

READING MATH

The symbol $\sqrt{\ }$ is called a *radical sign*. An expression such as $\sqrt{a}$ is called a *radical*. An expression written beneath the radical sign is called the *radicand*.

### Example 1

Find each square root.

a. $\sqrt{81}$ b. $-\sqrt{\frac{4}{25}}$ c. $\sqrt{0.01}$

### Solution

a. $9 \times 9 = 81$, so $\sqrt{81} = 9$ b. $-\frac{2}{5} \times \left(-\frac{2}{5}\right) = \frac{4}{25}$, so $-\sqrt{\frac{4}{25}} = -\frac{2}{5}$
c. $0.1 \times 0.1 = 0.01$, so $\sqrt{0.01} = 0.1$ ◄

## Example 2

Simplify each expression.

a. $\sqrt{49} + \sqrt{16}$ b. $\sqrt{5^2 + 12^2}$

### Solution

a. $\sqrt{49} = 7$ and $\sqrt{16} = 4$, so $\sqrt{49} + \sqrt{16} = 7 + 4 = 11$

b. $5^2 = 25$ and $(12)^2 = 144$, so $\sqrt{5^2 + 12^2} = \sqrt{25 + 144} = \sqrt{169}$
Since $13 \times 13 = 169$, $\sqrt{169} = 13$ ◄

**MATH: WHO, WHERE, WHEN**

Johann Zacharias Dase (1824–1861), a German famous for his ability to calculate rapidly, was considered by many as the most extraordinary calculative prodigy who ever lived. He was able to find the square root of a 100-digit number mentally in 52 s.

A **perfect square** is an integer whose square roots are integers. The square root of any positive integer that is not a perfect square is an **irrational number.** So, numbers such as $\sqrt{2}$, $-\sqrt{17}$, and $\sqrt{38}$ are irrational.

## Example 3

Between which two consecutive integers does $\sqrt{29}$ lie?

### Solution

Find the perfect squares.

$$\begin{array}{ccccc} \sqrt{25} & < & \sqrt{29} & < & \sqrt{36} \\ \downarrow & & \downarrow & & \downarrow \\ 5 & < & \sqrt{29} & < & 6 \end{array}$$

So, $\sqrt{29}$ is between 5 and 6. ◄

You can estimate the square root of an irrational number by using a table of square roots or a calculator to find an *approximate* value.

## Example 4

Approximate $\sqrt{29}$ to the nearest thousandth.

### Solution

Use a calculator. Enter this key sequence: [2] [9] [√]
The calculator display shows 5.3851648.
To the nearest thousandth, $\sqrt{29} = 5.385$. ◄

**TALK IT OVER**

Do you think $\sqrt{29}$ is closer to 5 or to 6? Why?

## Example 5

The area of a square is 54 $m^2$. What is the approximate length of each side? Round your answer to the nearest tenth of a meter.

### Solution

Since the area is 54 $m^2$ and the figure is square, the measure of the length of each side is equal to $\sqrt{54}$.

Use a table of square roots or a calculator to find $\sqrt{54}$. To the nearest tenth, $s \approx 7.3$
The length of each side of the square is approximately 7.3 m. ◄

$s^2 = 54$

$s$

$s$

TALK IT OVER

Suppose your calculator does not have a square-root key $\boxed{\sqrt{\ }}$. How could you go about finding the square root of a number?

## TRY THESE

Find each square root.

1. $\sqrt{\frac{64}{121}}$ 2. $-\sqrt{\frac{169}{225}}$ 3. $\sqrt{\frac{9}{400}}$ 4. $\sqrt{\frac{144}{441}}$

5. $\sqrt{3.61}$ 6. $\sqrt{0.04}$ 7. $\sqrt{0.25}$ 8. $-\sqrt{0.0484}$

Simplify each expression.

9. $\sqrt{36} + \sqrt{16}$ 10. $\sqrt{81} + \sqrt{121}$ 11. $\sqrt{144} - \sqrt{49}$ 12. $\sqrt{324} - \sqrt{196}$

13. $\sqrt{24^2 + 18^2}$ 14. $\sqrt{15^2 + 8^2}$ 15. $\sqrt{35^2 - 21^2}$ 16. $\sqrt{15^2 - 12^2}$

Use a calculator or a table of square roots. Approximate each square root to the nearest thousandth.

17. $\sqrt{28}$ 18. $\sqrt{41}$ 19. $\sqrt{73}$ 20. $\sqrt{99}$

Find the consecutive integers between which each square root lies.

21. $\sqrt{7}$ 22. $\sqrt{18}$ 23. $\sqrt{27}$ 24. $\sqrt{53}$

25. The area of a square greeting card is 41 in.$^2$ What is the length of each side to the nearest hundredth of an inch?

## EXERCISES

### PRACTICE/ SOLVE PROBLEMS

Find each square root.

1. $\sqrt{\frac{9}{25}}$ 2. $\sqrt{\frac{49}{64}}$ 3. $\sqrt{\frac{121}{196}}$ 4. $-\sqrt{\frac{1}{625}}$

5. $\sqrt{2.89}$ 6. $-\sqrt{529}$ 7. $\sqrt{0.0001}$ 8. $\sqrt{0.81}$

Use a calculator or a table of square roots. Approximate each square root to the nearest thousandth.

9. $\sqrt{20}$ 10. $\sqrt{139}$ 11. $-\sqrt{292}$ 12. $\sqrt{404}$

Find the consecutive integers between which each square root lies.

13. $\sqrt{11}$ 14. $\sqrt{45}$ 15. $\sqrt{70}$ 16. $\sqrt{158}$

17. $\sqrt{13}$ 18. $\sqrt{30}$ 19. $\sqrt{88}$ 20. $\sqrt{112}$

Simplify.

21. $\sqrt{36} + \sqrt{64}$ 22. $\sqrt{42^2 + 56^2}$ 23. $\sqrt{169} - \sqrt{81}$

24. $\sqrt{10^2 + 24^2}$ 25. $\sqrt{17^2 - 8^2}$ 26. $\sqrt{144} + \sqrt{400}$

27. The square root of a number is one more than twice 6.5. What is the number?

28. The area of a square rug is 58 ft². Find the measure of a side. Round your answer to the nearest tenth of a foot.

## EXTEND/ SOLVE PROBLEMS

Estimate each square root. Then use a calculator to approximate each square root to the nearest thousandth.

29. $\sqrt{8.4}$
30. $\sqrt{46}$
31. $\sqrt{69.9}$
32. $\sqrt{235}$

Simplify each expression when $x = 4$. Do not approximate irrational roots.

33. $\sqrt{x}$
34. $\sqrt{2x + 1}$
35. $\sqrt{7x - 3}$
36. $\sqrt{9x} + \sqrt{4x}$
37. $\sqrt{25x} - \sqrt{x}$
38. $\sqrt{x + 3^2}$
39. $\sqrt{2x - 1^2}$
40. $\sqrt{10x}$
41. $\sqrt{x^2 + 2x + 1}$

42. A square poster will have an area of 2.56 m². What will be the length of each side?

43. The area of a square quilt is 230 in.² Find the perimeter of the quilt to the nearest tenth of an inch.

44. A rectangular table is made up of two square tables pushed together. Each of the square tables has an area of 156.25 ft². What are the dimensions of the rectangular table?

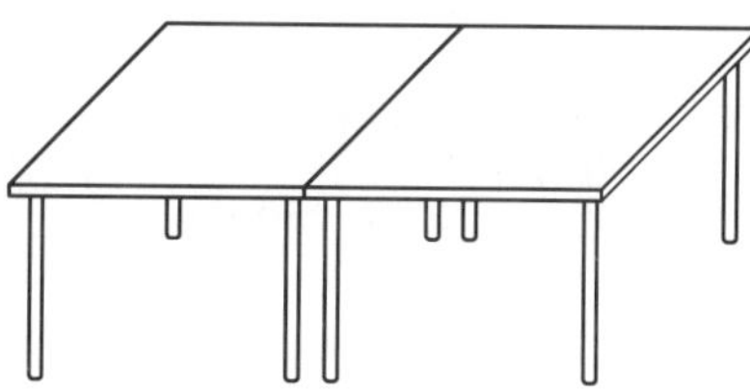

## THINK CRITICALLY/ SOLVE PROBLEMS

Complete.

45. $\sqrt{9 \times 4} = \sqrt{36} = ■$ and $\sqrt{9} \times \sqrt{4} = ■ \times ■ = ■$.

46. $\sqrt{25 \times 4} = \sqrt{100} = ■$ and $\sqrt{25} \times \sqrt{4} = ■ \times ■ = ■$.

47. $\sqrt{25 \times 16} = \sqrt{400} = ■$ and $\sqrt{25} \times \sqrt{16} = ■ \times ■ = ■$.

48. Study your results for Exercises 45–47. What property of square roots is suggested? Use variables in your statement of the property.

Complete.

49. $\sqrt{\frac{36}{9}} = \sqrt{4} = ■$ and $\frac{\sqrt{36}}{\sqrt{9}} = \frac{■}{■} = ■$.

50. $\sqrt{\frac{400}{4}} = \sqrt{100} = ■$ and $\frac{\sqrt{400}}{\sqrt{4}} = \frac{■}{■} = ■$.

51. What property of square roots is suggested by Exercises 49–50? Use variables in your statement of the property.

# 5-9 Solving Equations Involving Squares

## EXPLORE

An artist has 64 square mosaic tiles with which to create a square design. Each tile is a square measuring 1 in. on a side.

a. What equation can the artist use to find the number of tiles for one side of the square?

b. What is $\sqrt{64}$?

c. Does the equation $x^2 = 64$ have more than one solution? Explain. Why isn't the negative square root a solution for this problem?

d. If $x^2 = 0$, how many solutions can you find? Explain.

e. If $x^2 = -9$, how many solutions can you find? Explain.

## SKILLS DEVELOPMENT

To solve an equation that contains an expression such as $x^2$ or $5y^2$ on one side and a non-negative real number on the other, recall that all positive real numbers have two square roots.

$$x^2 = 25$$
$$x = \sqrt{25} \text{ or } x = -\sqrt{25}$$
$$x = 5 \text{ or } x = -5$$

So, the solutions are 5 and $-5$.

**PROBLEM SOLVING TIP**

Sometimes the wording of the problem will tell you which square root you should choose. Other times, only one square root makes sense as a solution for the problem.

### Example 1

Solve $x^2 = 1.21$. Check the solutions.

### Solution

$$x^2 = 1.21$$ **Think: The square of what number equals 1.21?**

$$x = 1.1 \text{ or } x = -1.1$$ ← **You can write $x = \pm 1.1$.**

Check each value.

$$x^2 = 1.21 \qquad\qquad x^2 = 1.21$$
$$(1.1)^2 \stackrel{?}{=} 1.21 \qquad\qquad (-1.1)^2 \stackrel{?}{=} 1.21$$
$$(1.1)(1.1) \stackrel{?}{=} 1.21 \qquad\qquad (-1.1)(-1.1) \stackrel{?}{=} 1.21$$
$$1.21 = 1.21 \checkmark \qquad\qquad 1.21 = 1.21 \checkmark$$

The solutions are 1.1 and $-1.1$. ◄

## Example 2

Solve $y^2 = \frac{16}{81}$. Check the solutions.

### Solution

$$y^2 = \frac{16}{81}$$
$$y = \pm\frac{4}{9}$$

Check each value.

$$y^2 = \frac{16}{81} \qquad\qquad y^2 = \frac{16}{81}$$
$$\left(\frac{4}{9}\right)^2 \stackrel{?}{=} \frac{16}{81} \qquad\qquad \left(-\frac{4}{9}\right)^2 \stackrel{?}{=} \frac{16}{81}$$
$$\left(\frac{4}{9}\right)\left(\frac{4}{9}\right) \stackrel{?}{=} \frac{16}{81} \qquad\qquad \left(-\frac{4}{9}\right)\left(-\frac{4}{9}\right) \stackrel{?}{=} \frac{16}{81}$$
$$\frac{16}{81} = \frac{16}{81} \checkmark \qquad\qquad \frac{16}{81} = \frac{16}{81} \checkmark$$

The solutions are $\frac{4}{9}$ and $-\frac{4}{9}$. ◄

## Example 3

Solve $z^2 = 20$. Check the solutions.

### Solution

$$z^2 = 20$$
$$z = \pm\sqrt{20}$$

Check each value.

$$z^2 = 20 \qquad\qquad z^2 = 20$$
$$(\sqrt{20})^2 \stackrel{?}{=} 20 \qquad\qquad (-\sqrt{20})^2 \stackrel{?}{=} 20$$
$$20 = 20 \checkmark \qquad\qquad 20 = 20 \checkmark$$

The solutions are $\sqrt{20}$ and $-\sqrt{20}$. ◄

**CHECK UNDERSTANDING**

Why does $(\sqrt{20})^2 = 20$?

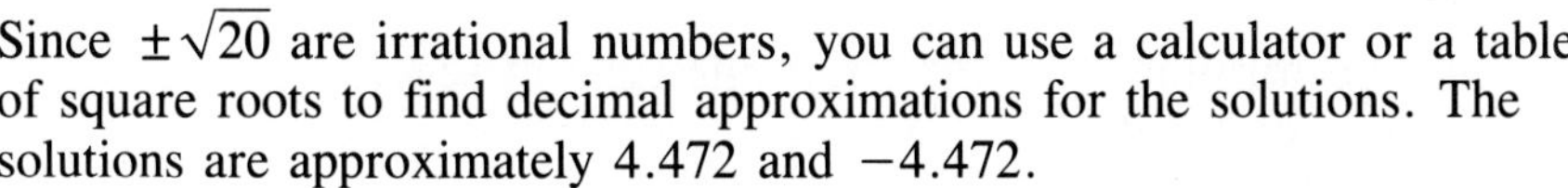

Since $\pm\sqrt{20}$ are irrational numbers, you can use a calculator or a table of square roots to find decimal approximations for the solutions. The solutions are approximately 4.472 and −4.472.

Sometimes you must use the addition or multiplication properties of equality to get $x^2$ alone on one side of the equation.

## Example 4

Solve $x^2 - 6 = 19$. Check the solutions.

### Solution

$$x^2 - 6 = 19$$
$$x^2 - 6 + 6 = 19 + 6 \quad \leftarrow \text{Add 6 to each side.}$$
$$x^2 = 25$$
$$x = \pm 5$$

Check each value.

$$x^2 - 6 = 19 \qquad\qquad x^2 - 6 = 19$$
$$(5)^2 - 6 \stackrel{?}{=} 19 \qquad\qquad (-5)^2 - 6 \stackrel{?}{=} 19$$
$$25 - 6 \stackrel{?}{=} 19 \qquad\qquad 25 - 6 \stackrel{?}{=} 19$$
$$19 = 19 \checkmark \qquad\qquad 19 = 19 \checkmark$$

The solutions are 5 and −5. ◄

### Example 5

The formula $d = 16t^2$ gives the distance ($d$) in feet traveled by a freely falling object dropped from a resting position. The time of the fall in seconds is $t$. Find the time in seconds for an object to fall 484 ft.

**Solution**

$$d = 16t^2$$
$$484 = 16t^2 \quad \leftarrow \text{Substitute 484 for } d.$$
$$\frac{1}{16}(484) = \frac{1}{16}(16t^2) \quad \leftarrow \text{Multiply each side by } \frac{1}{16}.$$
$$30.25 = t^2$$
$$\pm 5.5 = t$$

Since $t$ represents time, only the positive square root makes sense. So, it takes 5.5 s for an object to fall 484 ft. ◄

## TRY THESE

Solve each equation. Check the solutions.

1. $x^2 = 36$
2. $y^2 = 2.25$
3. $z^2 = \frac{49}{100}$
4. $c^2 = -7$
5. $x^2 + 8 = 89$
6. $2x^2 = 288$
7. $2.4b^2 = 0$
8. $m^2 = -36$
9. $3x^2 - 5 = 25$
10. Using the formula $d = 16t^2$, find the time in seconds for an object to fall 1,600 ft.

## EXERCISES

**PRACTICE/ SOLVE PROBLEMS**

Solve and check each of the following equations. Do not approximate irrational square roots.

1. $x^2 = 169$
2. $y^2 = \frac{36}{121}$
3. $z^2 = 25$
4. $b^2 = 17$
5. $c^2 = 59$
6. $d^2 = \frac{1}{4}$
7. $m^2 = -1.8$
8. $a^2 - 1 = 15$
9. $x^2 + 23 = 279$
10. $3n^2 = 300$
11. $2w^2 = 1.62$
12. $4y^2 = 200$

Solve each equation. Then, using a calculator or a table of square roots, approximate the solutions to the nearest thousandth.

**13.** $x^2 = 8$ **14.** $d^2 = 47$ **15.** $m^2 = 2.6$

**16.** $2b^2 = 70$ **17.** $x^2 - 3 = 102$ **18.** $5w^2 = 90$

**19.** The formula $b = 0.06v^2$ gives the distance $b$ (in feet) needed to stop a car after the brakes are applied. The speed at which a car is traveling when the brakes are applied, in miles per hour, is $v$. If a car traveled 96 ft after the brakes were applied, how fast was it going when the driver applied the brakes?

**20.** Tina has three identical square pieces of fabric with a total area of 588 in.[2] Write and solve an equation to find the dimensions of each piece of fabric.

## EXTEND/ SOLVE PROBLEMS

Solve each equation. Then, using a calculator or a table of square roots, approximate the solutions to the nearest thousandth.

**21.** $a^2 = \frac{49}{10{,}000}$ **22.** $k^2 = 2{,}500$ **23.** $y^2 = 0.0144$

**24.** $c^2 = 67$ **25.** $7f^2 = 1.75$ **26.** $x^2 - 8 = 30$

**27.** $4p^2 = 328$ **28.** $3m^2 + 2 = 50$ **29.** $6y^2 - 9 = 117$

**TALK IT OVER**

Try to compute $\sqrt{-1}$ on your calculator. What does the display show? Explain.

**30.** The length of a rectangle is equal to twice its width. The area of the rectangle is 72 m². Write and solve an equation to find the dimensions of the rectangle.

## THINK CRITICALLY/ SOLVE PROBLEMS

**31.** Solve and check $2(y - 7)^2 = 128$. Explain each step.

**32.** (a) Find the time in seconds for an object to fall each distance: 200 ft, 400 ft, 600 ft. Round each answer to the nearest hundredth.

(b) Compare your results for Exercise 32a. Does it take twice as long for an object to fall twice as far? Is the difference in time between 600 ft and 400 ft the same as between 400 ft and 200 ft? What conclusions can you draw?

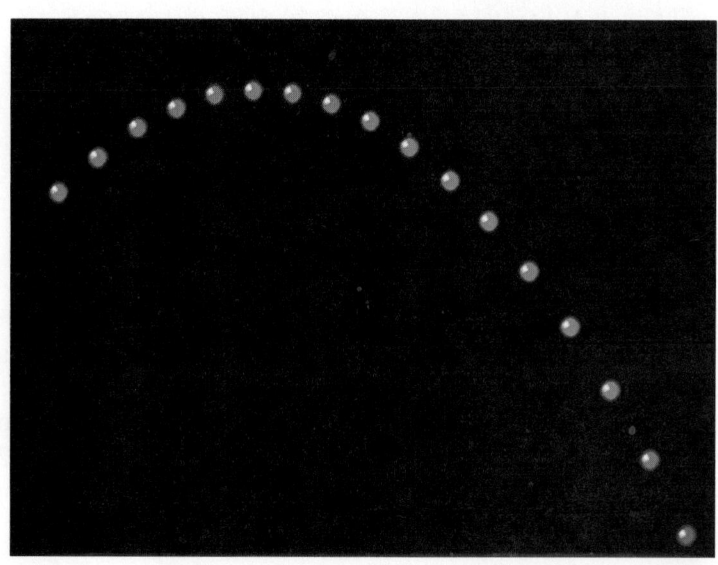

# CHAPTER 5 ● REVIEW

Choose a phrase from the list that completes each statement.

1. A value of the variable that makes an equation true is called a(n) __?__ of the equation.
2. When you add the same number to each side of an inequality, the result is a(n) __?__ inequality.
3. An equation stating a relationship between two or more quantities is called a(n) __?__.
4. Squaring a number and finding that number's square root are __?__ operations.

a. formula
b. inverse
c. solution
d. equivalent

## SECTION 5–1 WORKING WITH EQUATIONS (pages 176–179)

► An equation is a statement that two numbers or expressions are equal.
► An equation may have none, one, or many solutions.

Is the given number a solution of the equation? Write *yes* or *no*.

5. $2p - 3 = 8$; 7
6. $-4x = 6$; $-1.5$
7. $9 = \frac{z}{3} + 5$; 12

## SECTIONS 5–2 AND 5–3 SOLVING EQUATIONS (pages 180–187)

► You can add (subtract) the same number to (from) each side of an equation and/or multiply (divide) each side by the same number.
► Perform the same operations on each side until the variable is alone on one side of the equation.

Solve.

8. $x + 7 = 3$
9. $-4m = 3.2$
10. $3w - 5 = 16$
11. $\frac{c}{2} + 8 = -1$
12. $25 = 40n + 5$
13. $2z - 11 + 9z = 11$

## SECTIONS 5–4 AND 5–5 PROBLEM SOLVING (pages 188–191)

► To solve a problem by writing an equation, look for key words to help you decide how the numbers in the problem are related.

14. Ned's full-time annual salary was double his previous salary as a part-time employee. After a year full time, his salary rose \$2,000 to \$29,000. How much did Ned earn as a part-time worker?
15. Ten more than three times a number is equal to 88. Find the number.

## SECTION 5–6 SOLVING INEQUALITIES (pages 192–195)

► Use the addition and multiplication properties to solve inequalities.

Solve and graph each inequality.

**16.** $3m \leq -3$ **17.** $4z - 4 \geq 4$ **18.** $-5 \leq \frac{x}{2} + 1$

## SECTION 5–7 WORKING WITH FORMULAS (pages 196–199)

► When working with a formula, substitute for known values of the variables. Then solve the resulting equation.

**19.** In the 1896 Olympics, Thomas Burke of the United States won the men's 400-meter run with a time of 54.2 s. Find his average rate.

**20.** Solve the formula $P = \frac{F}{A}$ for $F$.

**21.** *USING DATA* Use the dental formula and the table on page 175.

An animal with 16 teeth has 4 incisors, 0 canines, and 12 molars. How many premolars does it have? What animal could it be?

## SECTION 5–8 SQUARES AND SQUARE ROOTS (pages 200–203)

► The inverse of squaring a number is finding the square root.
► Every positive real number has a positive and a negative square root. The symbol $\sqrt{\ }$ denotes the positive or principal square root.
► Use a calculator or a table of square roots to approximate irrational square roots.

Find each square root. Approximate irrational roots to the nearest thousandth.

**22.** $\sqrt{625}$ **23.** $\sqrt{73}$ **24.** $-\sqrt{\frac{1}{900}}$ **25.** $\sqrt{0.0289}$

Find the consecutive integers between which each square root lies.

**26.** $\sqrt{3}$ **27.** $\sqrt{12}$ **28.** $-\sqrt{55}$ **29.** $\sqrt{180}$

## SECTION 5–9 SOLVING EQUATIONS INVOLVING SQUARES (pages 204–207)

► When an equation contains an expression such as $x^2$ or $3y^2$ on one side and a non-negative real number on the other, solve by finding the square root of each side. Remember to use both square roots.

Solve each equation.

**30.** $x^2 = 121$ **31.** $3q^2 = 243$ **32.** $2a^2 - 1 = -0.02$

# CHAPTER 5 • TEST

Is the given number a solution of the equation? Write *yes* or *no*.

1. $-3x = 12; -4$
2. $11 = 3 + \frac{c}{2}; 16$
3. $-\frac{b}{6} = 6; 36$

Solve each equation.

4. $d - 9 = 9$
5. $48 = -8n$
6. $w - 7.3 = -6.2$
7. $\frac{x}{6} = -2$
8. $5g - 13 = 32$
9. $-\frac{2}{3}n + 10 = 24$
10. $0 = 0.5t - 1$
11. $8g - 10 + g = 71$
12. $35 = 3x + 1 - 5x$

Solve and graph each inequality.

13. $7d \geq 21$
14. $3f - 2 < 4$
15. $-\frac{x}{2} + 1 < 3$
16. $-10 \leq 3k + 8$

Solve.

17. Express the temperature 95°F in degrees Celsius. Use the formula $F = \frac{9}{5}C + 32$.
18. Solve the formula $p = \frac{a}{2} + 110$ for $a$.

Find each square root.

19. $\sqrt{121}$
20. $-\sqrt{400}$
21. $\sqrt{1.69}$
22. $\sqrt{\frac{4}{81}}$

Solve each equation.

23. $x^2 = 100$
24. $m^2 = 0.36$
25. $w^2 = \frac{9}{196}$

Write an equation and solve.

26. José bought a shirt for \$12.99 and three pairs of socks that each cost the same amount. The total cost of José's purchases, before sales tax, was \$19.74. How much did each pair of socks cost?
27. The number of students in the computer club is 4 fewer than twice the number of students in the photography club. Together, the clubs have 101 members. How many members does each club have?
28. Last month, Tanya received \$495, which included a \$75 employee-of-the-month award. Her hourly wage is \$8.40. How many hours did Tanya work?
29. Tapes cost $(2a + b)$ dollars and compact discs cost $(6a - b)$ dollars. Write and simplify an expression to show the total cost of 3 tapes and 2 compact discs.

Simplify.

1. $7 - 2 \times 3$
2. $8 + 3 \times 4 - 2$
3. $5 \div 5 \times 5 - 5 + 5$

Use the bar graph for Exercises 4–6.

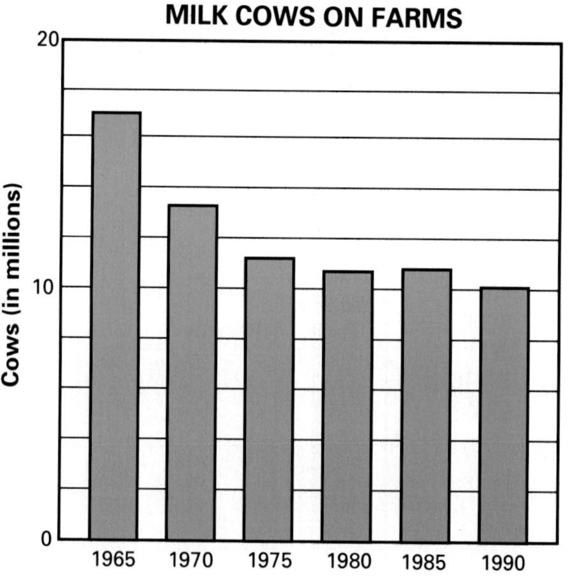


4. About how many milk cows were on farms in 1990?
5. During which 5-year interval did the greatest decrease in the number of cows occur?
6. During which 5-year interval did the number of cows decrease by about 2 million?

Find each answer.

7. $-72 \div 8$
8. $-8 + (-9)$
9. $2 - (-18)$
10. $-3 \times -12$

Solve.

11. The measure of an angle is 58°. Find the measures of its complement and supplement.

Use the figure below for Exercises 12–13.

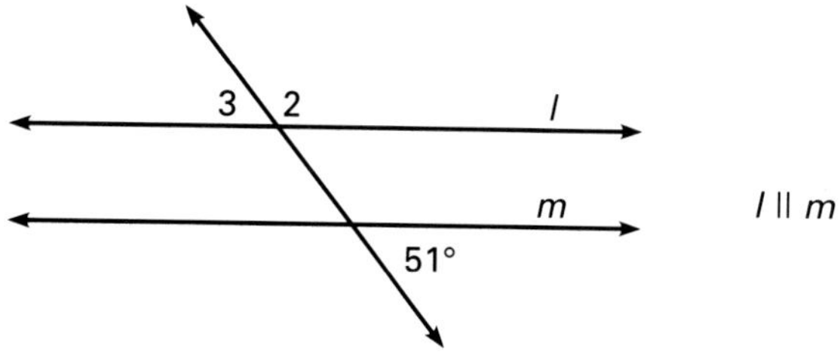


12. Find the measure of $\angle 2$.
13. Find the measure of $\angle 3$.

Identify each polygon by its number of sides. Tell whether each is *convex* or *concave* and *regular* or *not regular*.

14. 

15. 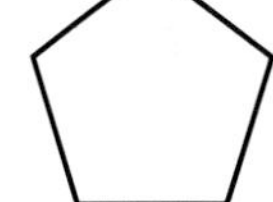

Use this frequency table for Exercises 16–18.

**FISH CAUGHT**

| Number of Pounds | Frequency |
|---|---|
| 7 | 6 |
| 8 | 7 |
| 10 | 3 |
| 13 | 1 |

16. Find the median.
17. Find the mode.
18. Find the range.

Solve each equation.

19. $b + 15 = 12$
20. $5(c + 3) - 4 = -29$
21. $3x^2 + 2 = 110$
22. Dan jogged for 1.2 h and covered a distance of 6.6 mi. What was Dan's jogging rate? (Use the distance formula $d = rt$.)

# CHAPTER 5 • CUMULATIVE TEST

1. 15 is what percent of 300?
A. 50% B. 5% C. 2% D. 200%

2. 40% of what number is 36?
A. 14.4 B. 72 C. 9 D. 90

3. What number is 110% of 110?
A. 233 B. 12 C. 121 D. 1,210

Use the line graph for Exercises 4–6.

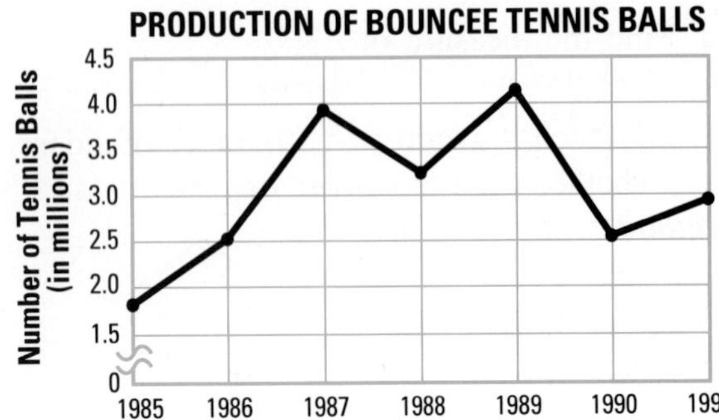


4. In which year was production about 3.1 million tennis balls?
A. 1986 B. 1987 C. 1988 D. 1991

5. Between which two years did the number of tennis balls produced show the greatest increase?
A. 1985–1986 B. 1986–1987
C. 1988–1989 D. 1989–1990

6. To the nearest half million, about how many Bouncee tennis balls were produced during the years 1989–1991?
A. 6.5 million B. 8 million
C. 9.5 million D. 12.5 million

7. Add. $-9 + (-16)$
A. $-25$ B. 7 C. $-7$ D. 25

8. Subtract. $-8 - (-13)$
A. $-5$ B. 21 C. 5 D. $-21$

9. Divide. $120 \div (-15)$
A. 8 B. $-7$ C. $-8$ D. 105

10. Multiply. $-4 \times (-3.5)$
A. $-7.5$ B. 14 C. $-1.5$ D. $-14$

11. Angle *XYZ* measures 77°. What is the measure of the complement of angle *XYZ*?
A. 13° B. 23°
C. 103° D. 113°

12. What is the name given to a parallelogram with all four sides congruent?
A. rectangle B. rhombus
C. square D. trapezoid

13. Find the sum of the angles of a convex octagon.
A. 1,800° B. 1,440°
C. 1,080° D. 720°

Use the table below for Exercises 14–15.

**POINTS SCORED PER SOFTBALL GAME**

| Number of Points | Frequency |
|---|---|
| 0 | 2 |
| 1 | 5 |
| 2 | 4 |
| 4 | 5 |
| 5 | 1 |
| 6 | 2 |
| 8 | 1 |

14. In how many games were fewer than 3 points scored?
A. 11 B. 9 C. 0 D. 3

15. In what percent of the games did the team score more than 4 points?
A. 57% B. 25%
C. 20% D. 75%

16. Solve. $4(x - 1) + 7 = -21$
A. 6 B. $-6\frac{3}{4}$
C. $-4\frac{1}{2}$ D. $-6$

17. Solve. $x^2 + 9 = 90$
A. $\pm 3$ B. 81
C. $\pm 9$ D. $\pm\sqrt{99}$

18. Use the formula $C = \frac{5}{9}(F - 32)$ to change 50°F to a Celsius temperature.
A. 10°C B. 15°C
C. 18°C D. 90°C

# CHAPTER 6 SKILLS PREVIEW

Find the unknown measure for each figure.

1. 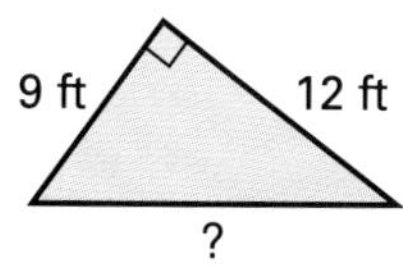


2. 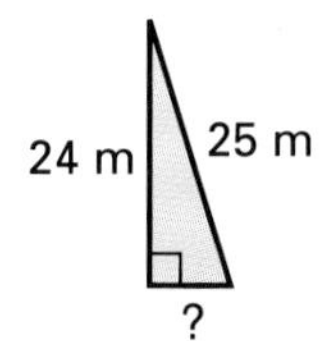


3. 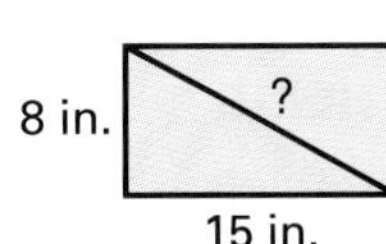


4. 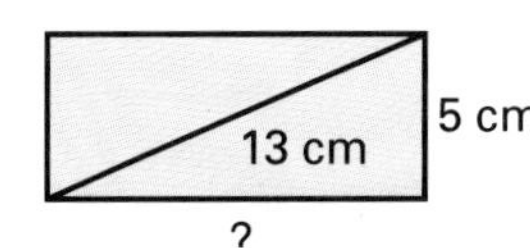


5. A rectangular garden is 12 yd long and 25 ft wide. What is the length of the fencing that would be needed to enclose the garden?

6. A randomly thrown dart hits the target shown at the right. What is the probability that it lands in the shaded area of the target?

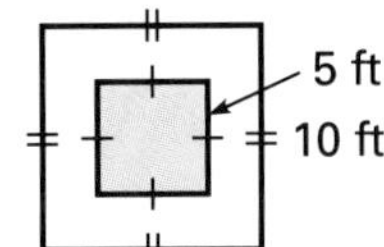


Find the area of each figure. Use $\pi \approx \frac{22}{7}$.

7. 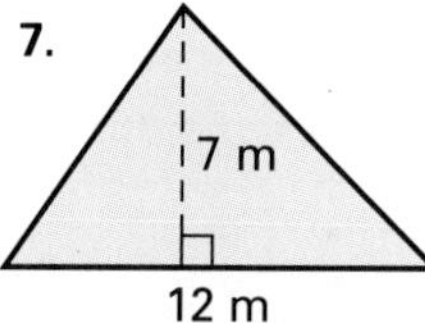


8. 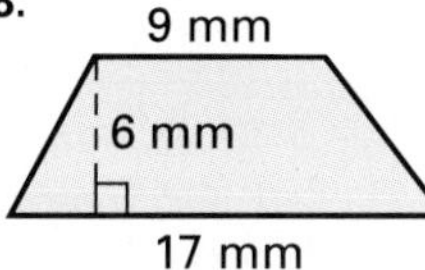


9. 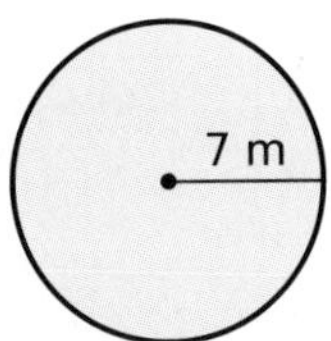


10. 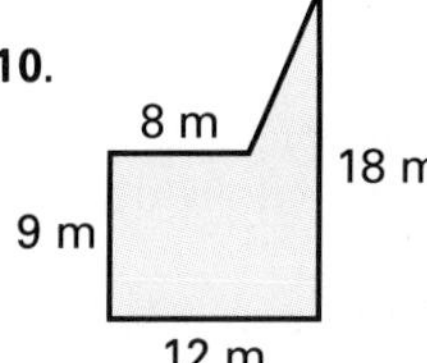


Find the surface area of each figure. Use $\pi \approx 3.14$.

11. 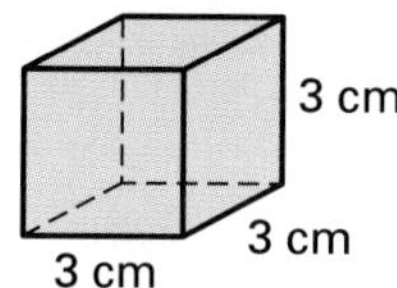


12. 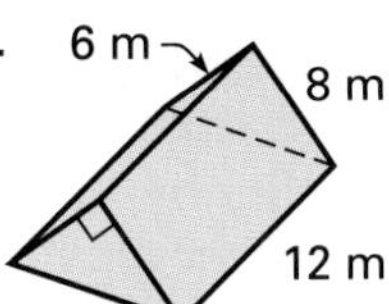


13. 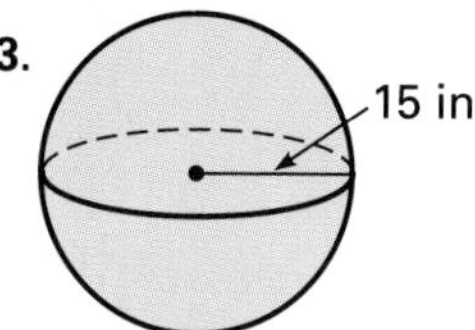


14. 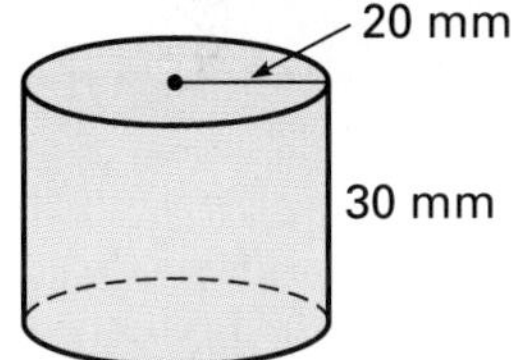


Find the volume of each figure. Use $\pi \approx 3.14$. Round your answers to the nearest tenth.

15. 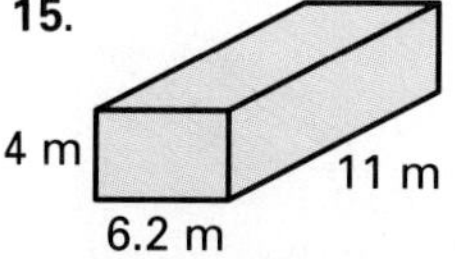


16. 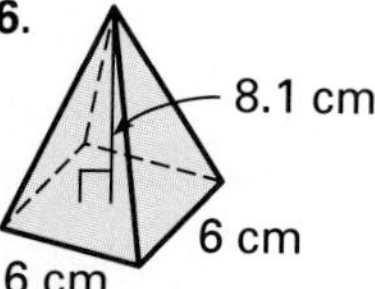


17. 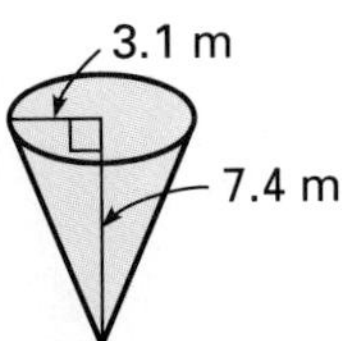


18. 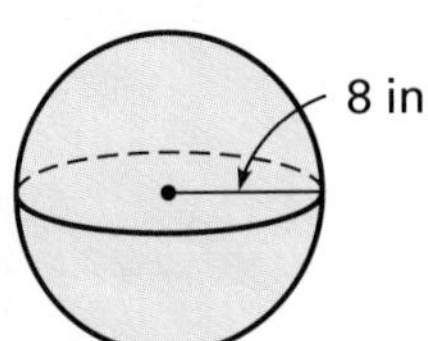


19. If you had to find the length of fencing needed to enclose a circular swimming pool with a radius of 5 m, which formula would you use?
   **a.** $A = \pi r^2$ **b.** $C = 2\pi r$ **c.** $V = \frac{4}{3}\pi r^3$

20. Which of these measurements could be made using a ruler whose smallest division is 1 mm?
   **a.** 7.5 mm **b.** 21.3 cm **c.** 17 mm **d.** 4.21 cm

CHAPTER 6

# USING FORMULAS IN GEOMETRY

**THEME** **Plant Life**

Knowing the ways to find length, area, and volume are skills that can be applied to many activities. These skills are basic requirements for specialized careers such as engineering, but they can also save time and effort in everyday undertakings such as home improvement and gardening. In this chapter, you will solve problems that involve finding dimensions, areas, and volumes of geometrical shapes.

**GARDEN PLANTING GUIDE**

| Plant | Height | Distance Needed Between Plants | Garden Hints for Best Results |
|---|---|---|---|
| Strawflower | 3 ft | 12 in. | Sunny area, moist soil |
| Moss Rose | 6 in. | 9 in. | Sunny area, tolerates dry soil |
| Stock | 18 in. | 15 in. | Sunny area, average soil |
| Poppy | 16 in. | 12 in. | Partial shade, average soil |
| Nasturtium | 12 in. | 12 in. | Sunny area, prefers dryish soil |
| Petunia | 15 in. | 8 in. | Warm, sunny area, average soil |
| Cosmos | 4 ft | 18 in. | Sun, partial shade |
| Snapdragon | 20 in. | 12 in. | Sunny area, average soil |
| Dill | 24 in. | 18 in. | Partial shade, average soil |
| Basil | 24 in. | 12 in. | Sunny area, moist soil |
| Radish | * | 1 in. | Cool weather, ready in 29 days |
| Tomato | 6 ft | $2\frac{1}{2}$ ft | Needs stakes, ready in 54 days |
| Cucumber | vine | 4 ft | Sunny area, pick when 8 in. long |
| Corn | 8 ft | 2 ft | Warm and wet, ready in 120 days |

*root vegetable

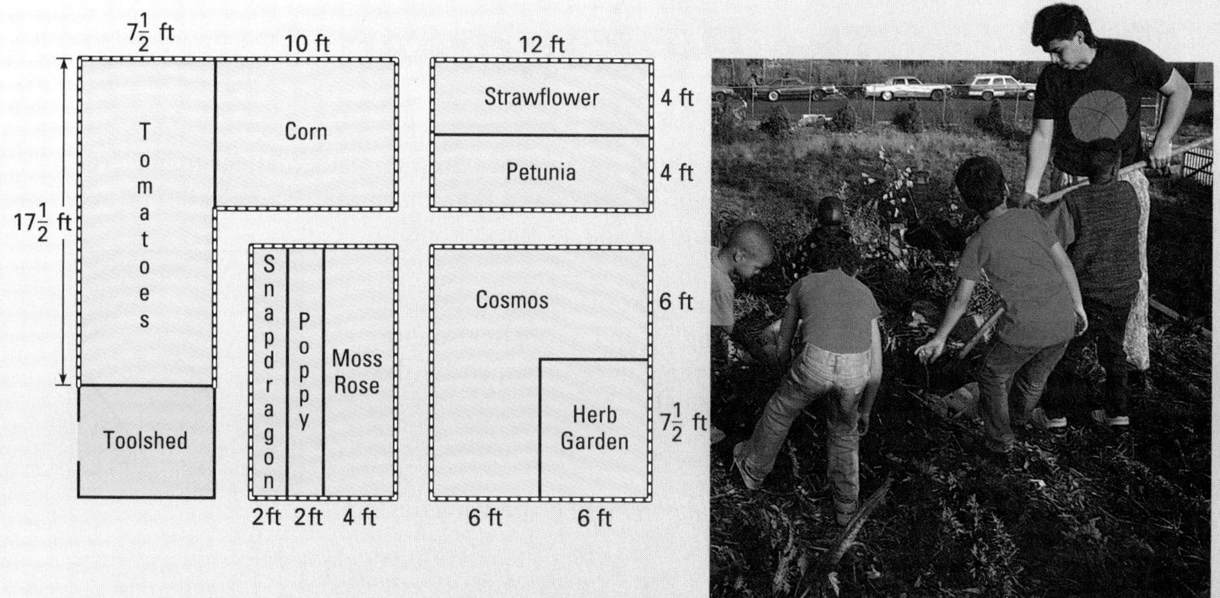


## DECISION MAKING

### Using Data

Use the information in the table and the garden plan to answer the following questions.

1. What length of fencing would you need to enclose the vegetable patch? You do not need any fencing along the toolshed walls.
2. What is the total area allotted for the strawflower and petunia beds together?
3. How many plants would you put in each bed? (Assume you will plant only dill in the herb garden.) Use dots to show each arrangement. What decisions did you make?

### Working Together

Work in a group to plan your own garden. Use additional reference sources and gardening catalogs, or visit nurseries to choose the plants you would like to grow. Make sure that the plants are suited to your climate, and find out the conditions that help them flourish.

Draw a plan of your garden, labeling areas for different plants. Give the dimensions of your garden and the number of each type of plant you plan to grow. (*Hints:* Allow access room for weeding and picking the plants. In general, a garden looks better when the taller plants are growing behind the shorter ones.)

# 6-1 The Pythagorean Theorem

## EXPLORE

Scholars believe that an ancient Chinese mathematical manuscript called the *Zhōubì* is between 2,000 and 3,000 years old. The work includes a figure similar to the one below.

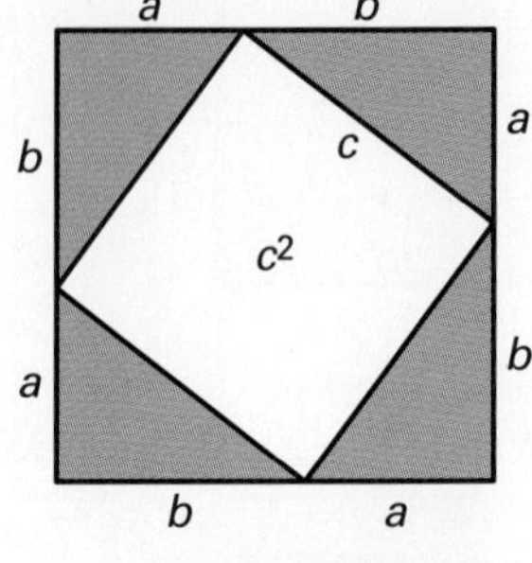


The figure is a square that encloses four congruent triangles and a smaller square. The region of the figure occupied by the triangles is shaded.

Notice what happens when the triangles are rearranged.

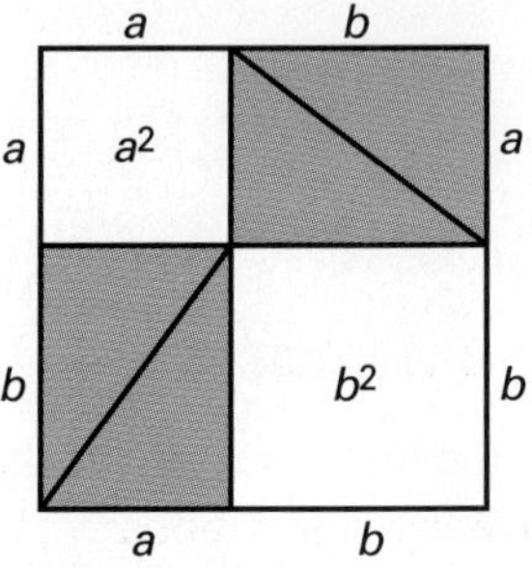


The entire figure stays the same size. The region of the figure occupied by the triangles remains shaded.

You know that the two figures have the same area and that the size of the triangles has not changed. What can you say about the area of the unshaded regions in each figure? How do the unshaded regions relate to the length of the sides of the triangles?

## SKILLS DEVELOPMENT

In a right triangle, the side opposite the right angle is called the **hypotenuse.** The other two sides are called the **legs.** About 2,500 years ago, a Greek mathematician named Pythagoras developed a proof that the relationship between the hypotenuse and the legs is true for *all* right triangles.

> In any right triangle, the square of the length of the hypotenuse is equal to the sum of the squares of the lengths of the legs.

In right triangle *ABC*, you can state the relationship of the lengths of the sides as follows:

$$c^2 = a^2 + b^2$$

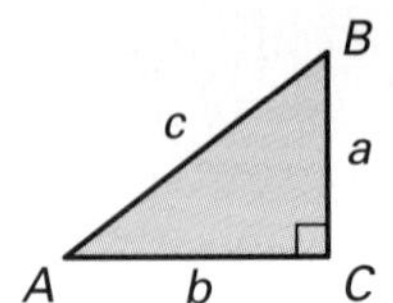


This property is named after Pythagoras and is called the **Pythagorean Theorem.**

## Example 1

Find the length of the hypotenuse of triangle *ABC*.

### Solution

Use the Pythagorean Theorem.

$$c^2 = a^2 + b^2$$
$$c^2 = 8^2 + 15^2$$
$$c^2 = 64 + 225$$
$$c^2 = 289$$
$$c = \sqrt{289} = 17$$

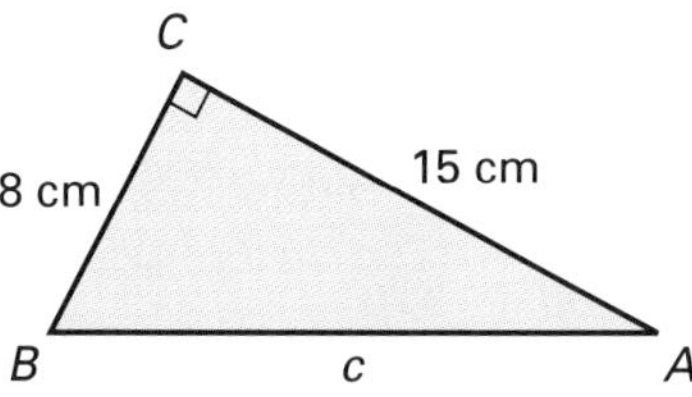

The length of the hypotenuse is 17 cm. ◄

**CHECK UNDERSTANDING**

Use the Pythagorean Theorem. Write expressions for $a^2$ and $b^2$.

## Example 2

In triangle *MNO*, find the length *n* to the nearest tenth of a meter.

### Solution

For triangle *MNO*, $9^2 = 4^2 + n^2$.
Subtract $4^2$ from each side of the equation.

$$n^2 = 9^2 - 4^2$$
$$n^2 = 81 - 16$$
$$n^2 = 65$$
$$n = \sqrt{65} \approx 8.062$$
$$n \approx 8.1$$

N
4 m
9 m
M n O

← **Use a table of square roots or a calculator to find $\sqrt{65}$.**

The length *n* is about 8.1 m. ◄

**CONNECTIONS**

In Chapter 3 you learned how to form the converse of a statement. In the case of the Pythagorean Theorem, the converse is also a true statement.

*If the square of one side of a triangle is equal to the sum of the squares of the other two sides, then the triangle is a right triangle.*

The converse provides a method of deciding if a given triangle is, in fact, a right triangle.

## Example 3

The longest side of a triangular sail measures 6 m, and the base of the sail measures 3 m. What is the minimum height, to the nearest tenth, of the mast of the sailboat?

### Solution

Draw a diagram of the problem. The triangle will be a right triangle, and the longest side is the hypotenuse.

$$c^2 = a^2 + b^2$$
$$a^2 = c^2 - b^2$$
$$a^2 = 6^2 - 3^2$$
$$a^2 = 36 - 9 = 27$$
$$a = \sqrt{27} \approx 5.196$$
$$\approx 5.2 \text{ (nearest tenth)}$$

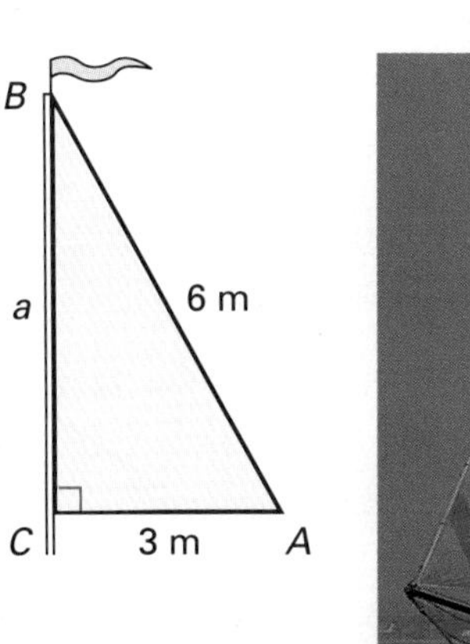


The height of the mast must be at least 5.2 m. ◄

**MATH: WHO, WHERE, WHEN**

Ancient Egyptian engineers needed to be able to construct right angles for their buildings. They employed men called *rope stretchers*. These men worked in pairs, using a loop of rope that had been knotted to mark off twelve equal sections.

The men would drive a post into the ground through one of the knots. Then one man would hold the rope at a knot three sections to one side of the post, and the other would hold a knot four sections to the other side of the post.

When the men stretched the loop of rope into a triangle, the sides measured 3, 4, and 5 sections long. Because $3^2 + 4^2 = 5^2$, the triangle was a right triangle. The angle formed at the post was opposite the hypotenuse and was the right angle.

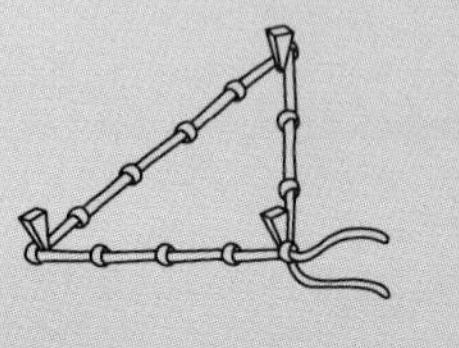

## Example 4

What is the length of the diagonal of rectangle *ABCD*?

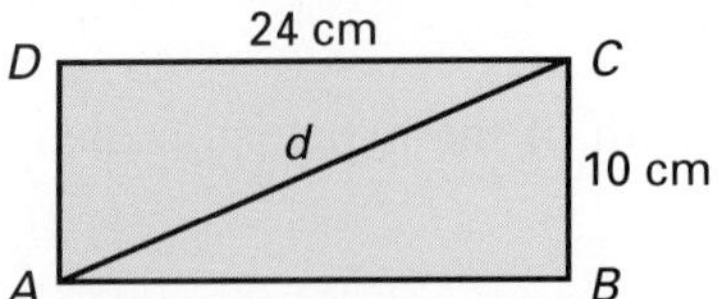


**Solution**

Since the diagonal forms the hypotenuse of a right triangle, the Pythagorean Theorem can be used to solve the problem.

$$d^2 = 10^2 + 24^2$$
$$d^2 = 100 + 576$$
$$d^2 = 676$$
$$d = \sqrt{676} = 26$$

The length of the diagonal of rectangle *ABCD* is 26 cm. ◄

## TRY THESE

Use the Pythagorean Theorem to find the unknown length. Round your answer to the nearest tenth.

1. 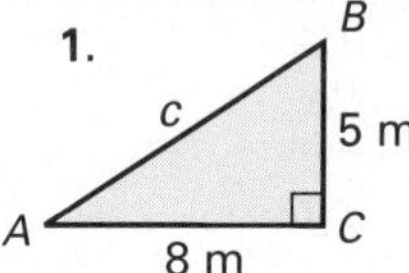


2. 3. 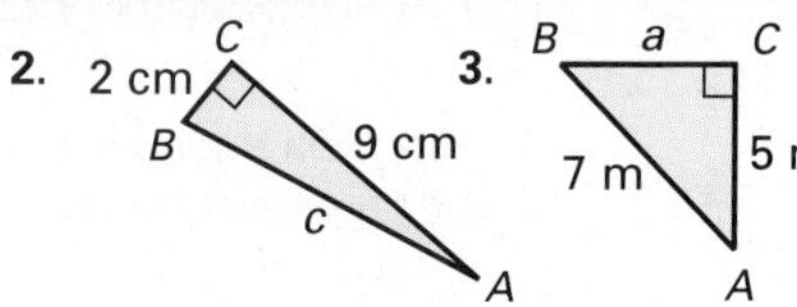


4. 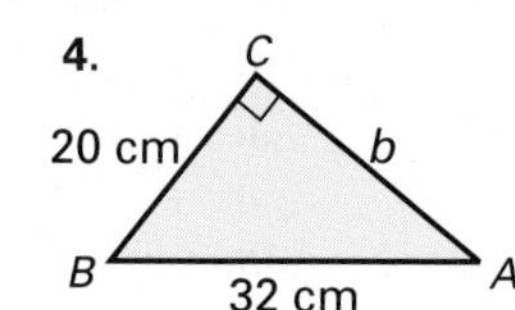


5. A flagpole is supported by a wire cable connected to its highest point. The cable is 40 ft long and is attached to the ground 15 ft from the base of the pole. How tall is the flagpole?

6. How long is the diagonal of a square with sides of 4 cm?

# EXERCISES

**PRACTICE/ SOLVE PROBLEMS**

Find the length of the hypotenuse in each right triangle. Round your answer to the nearest tenth.

1. 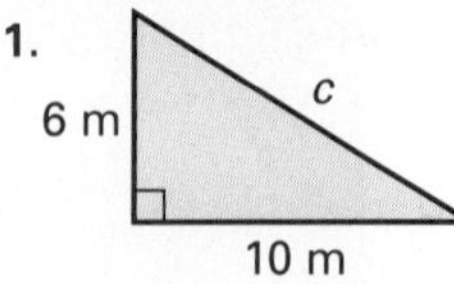


2. 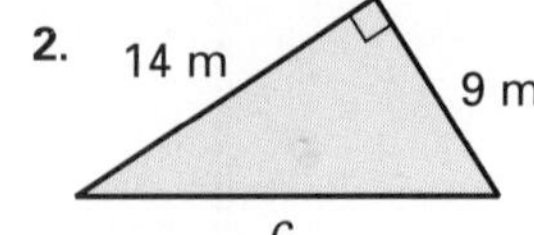


3. 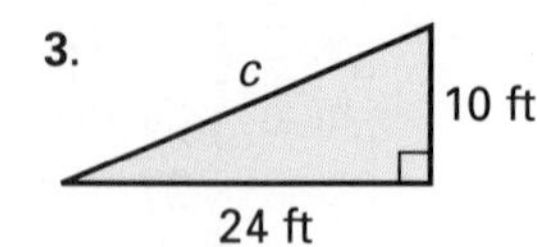


Find the unknown length. Round your answer to the nearest tenth.

4. 8 cm, x, 12 cm

5. 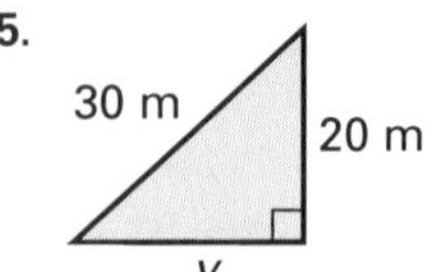


6. 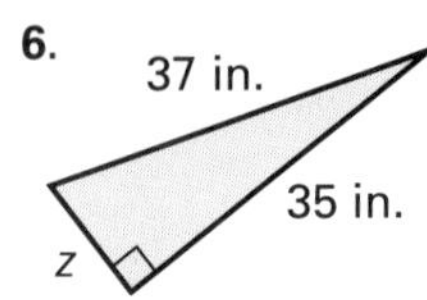

7. 8. 9.

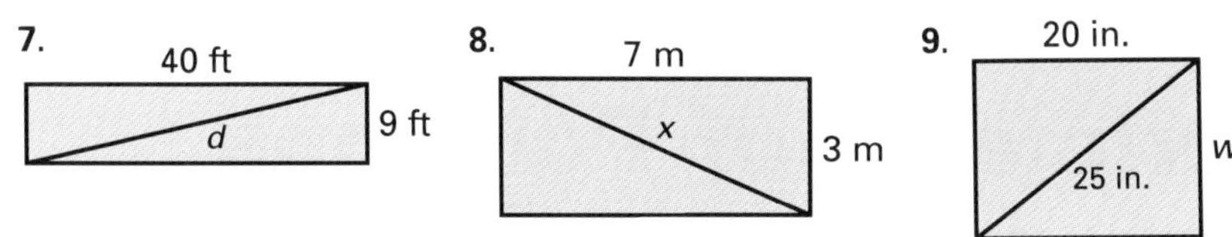


Find each unknown measurement. Round your answer to the nearest tenth.

10. height of the ramp

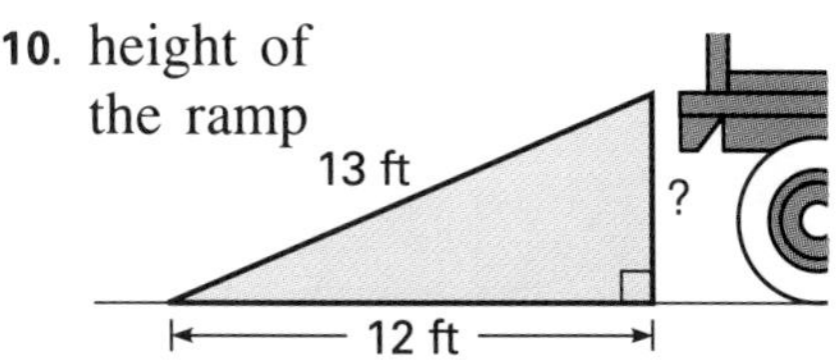


11. length of the pond

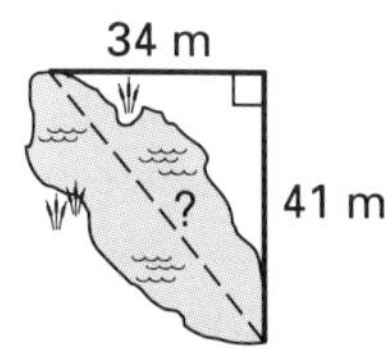


12. A string of pennants is 65 ft long. It is stretched from the top of a tree to a point on the ground 39 ft from the base of the tree. How tall is the tree?

**ESTIMATION TIP**

You can use square root tables that go only to 100 to estimate square roots of greater numbers. Mentally, divide the number by 100. Then look up the square root of the number closest to your quotient. Multiply the square root by 10, and you'll have a good estimate.

For example, you can estimate $\sqrt{795}$.

$795 \div 100 = 7.95 \approx 8$

$\sqrt{8} \approx 2.828$

$2.828 \times 10 = 28.28 \approx 28$

$\sqrt{795} \approx 28$

## EXTEND/ SOLVE PROBLEMS

Find the unknown length in each figure. Round your answer to the nearest tenth.

13. 14. 15.

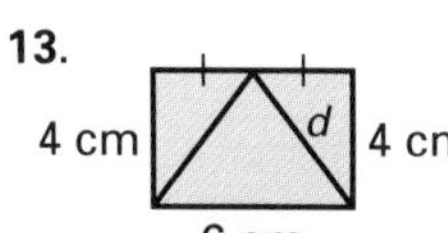


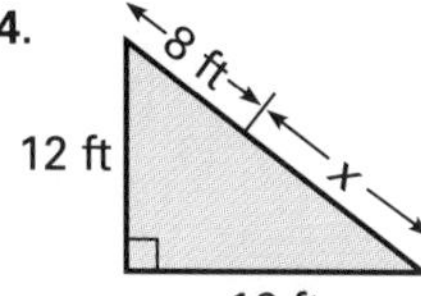


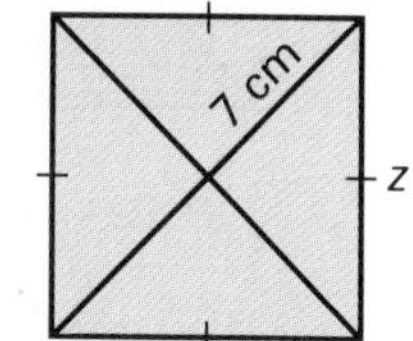


Solve. Round your answer to the nearest tenth.

16. The sides of a tent measure 7 ft from the peak to the ground. If the base of the tent is 6 ft wide, how tall is the tent?

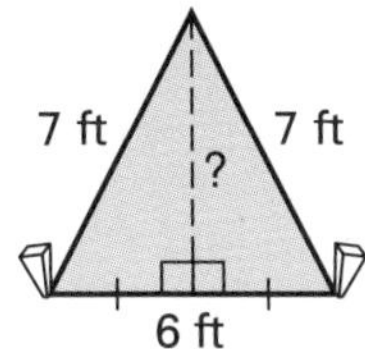


17. A baseball diamond is a square with sides of 90 ft. What is the shortest distance between first and third bases?

18. If you traveled 24 miles north and then 7 miles west, what is the shortest distance you could travel to return to your starting point?

## THINK CRITICALLY/ SOLVE PROBLEMS

19. A lumber mill is cutting lengths of wood with square cross-sections that have 12-in. sides. What is the minimum diameter of the tree trunks that the lumber mill can use?

20. Evaluate the expressions $2n$, $n^2 - 1$, and $n^2 + 1$ for $n = 2, 3, 4$, and 8. Study the three numbers you obtained for each value of $n$. What do you notice? How can each set of three numbers be related to the Pythagorean Theorem?

# 6-2 Perimeter of Polygons

## EXPLORE

At a picnic, there are nine identical square tables. Each table can seat four people, one per side. To seat larger groups, the tables can be pushed together with one or more entire sides touching. If all the tables are used, what is the greatest number of people that can sit together at once? What is the least number of people that can sit together at once?

## SKILLS DEVELOPMENT

The distance around a plane figure is called its **perimeter** and is the sum of the lengths of the figure's sides. In each formula below, $P$ represents the perimeter of the figure.

triangle: sides $a$, $b$, $c$ — $P = a + b + c$

rectangle: length $l$, width $w$ — $P = 2l + 2w$

square: side $s$ — $P = 4s$

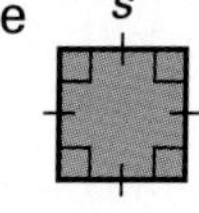

**READING MATH**

The word perimeter is derived from two ancient Greek words—*peri,* meaning "all around," and *metron,* meaning "measure." So, perimeter means "the measure all around" a figure or place.

Other words derived from these and other Greek words are *periscope* (a device that can "look all around") and *thermometer* (an instrument to "measure heat"). Can you think of any other words derived from *peri* and *metron*?

### Example 1

How much fencing would you need to enclose a square field with sides that measure 84 m?

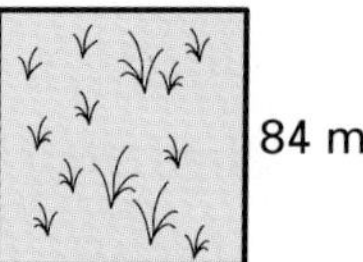


**Solution**

Use the formula $P = 4s$, where $s$ is the length of each side.

$$P = 4s$$
$$P = 4(84) = 336$$

You would need 336 m of fencing. ◄

### Example 2

Find the width of a rectangular vegetable garden if its perimeter is 16.4 m and its length is 5.05 m.

**Solution**

Use the formula $P = 2l + 2w$, where $l$ is the length and $w$ is the width.

$$P = 2l + 2w$$
$$16.4 = 2(5.05) + 2w \quad \text{Substitute.}$$
$$16.4 = 10.1 + 2w \quad \text{Add } -10.1 \text{ to each side.}$$
$$6.3 = 2w \quad \text{Multiply each side by } \tfrac{1}{2}.$$
$$3.15 = w$$

◄

## Example 3

Find the perimeter of the right triangle.

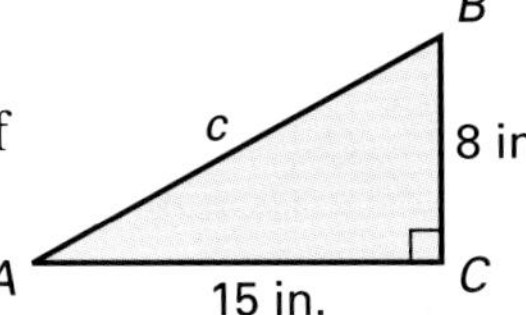


**Solution**

Since you are given the lengths of only two of the sides, you can use the Pythagorean Theorem to find the missing length.

$$c^2 = a^2 + b^2$$
$$c^2 = 8^2 + 15^2$$
$$c^2 = 64 + 225$$
$$c^2 = 289$$
$$c = \sqrt{289} = 17$$

Now add the lengths of the sides to find the perimeter.

$$P = a + b + c$$
$$P = 8 + 15 + 17 = 40$$

The perimeter of the right triangle is 40 in. ◄

The sides of a regular polygon each have the same measure. The perimeter $P$ of a regular polygon with $n$ sides of length $s$ can be found using the formula $P = ns$.

## Example 4

Find the perimeter of each regular polygon.

**a.**

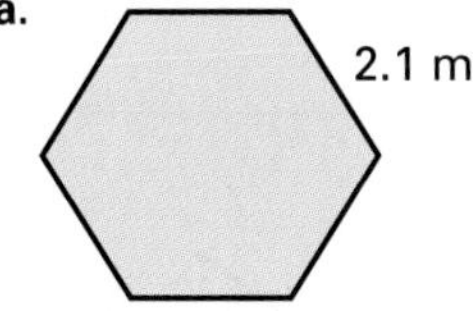


**b.**

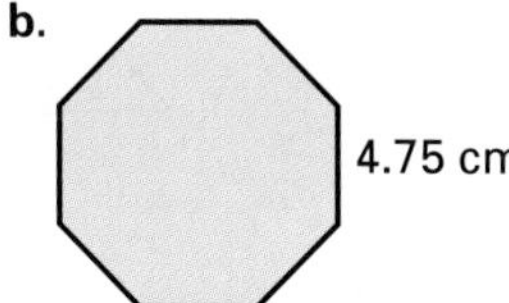


**Solution**

**a.** $P = ns$
$P = 6(2.1) = 12.6$

The perimeter is 12.6 m.

**b.** $P = ns$
$P = 8(4.75) = 38$

The perimeter is 38 cm. ◄

### MIXED REVIEW

Estimate each answer.

1. $\frac{4}{9} \times 687$
2. $7{,}193 \div 91$
3. 63% of 318
4. $\frac{4}{13} \times 449$

Simplify.

5. $x^5 \cdot x^8$
6. $(b^3)^4$
7. $w^{-2} \cdot w^{-5}$
8. $\frac{c^4}{c^{-6}}$

Solve and graph each.

9. $x + 6 \leq 4$
10. $-2m - 3 > -5$

## TRY THESE

Find the perimeter of each square or rectangle.

**1.**

$2\frac{1}{2}$ in.

**2.**

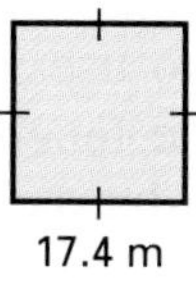

17.4 m

**3.**

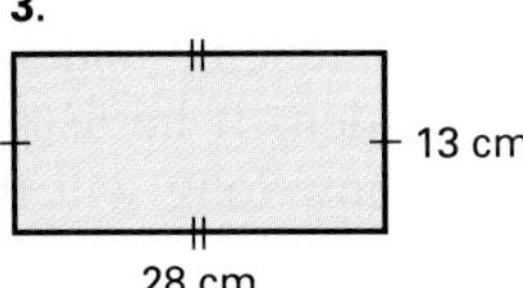


**4.**

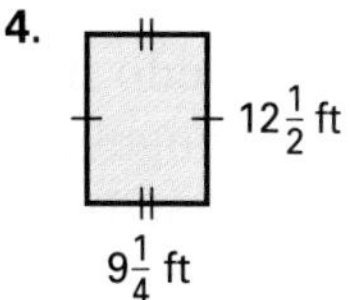


### CHECK UNDERSTANDING

What is true about the perimeter of all of these figures?

—a square with sides of 4 cm
—a rhombus with sides of 4 cm
—a rectangle with length of 6 cm and width of 2 cm

Find the perimeter of each triangle.

5. 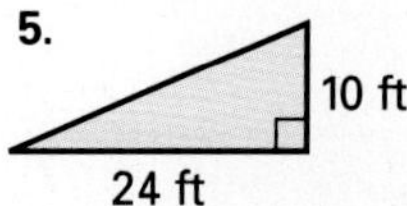


6. 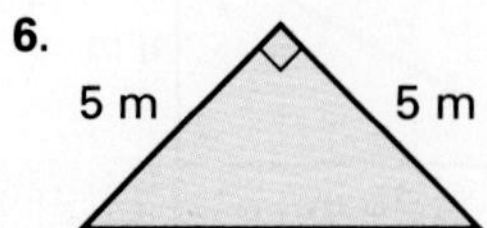


7. 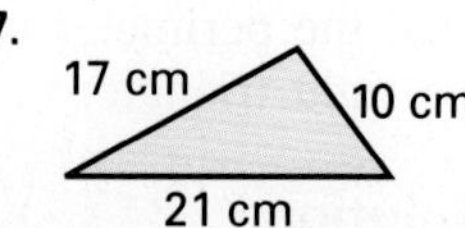


Find the perimeter of each regular polygon.

8. 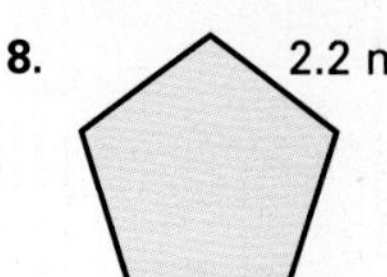


9. 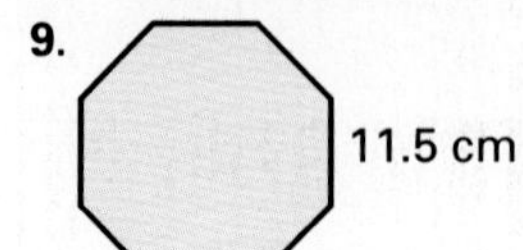


10. 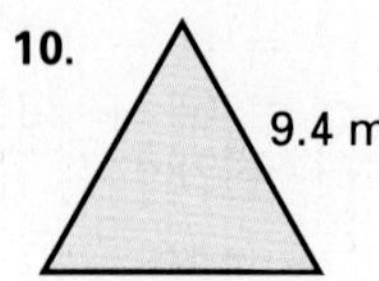


11. A rectangular greenhouse is $12\frac{1}{2}$ ft long and has a perimeter of 42 ft. What is the width of the greenhouse?

# EXERCISES

## PRACTICE/ SOLVE PROBLEMS

Find the perimeter of each figure.

1. 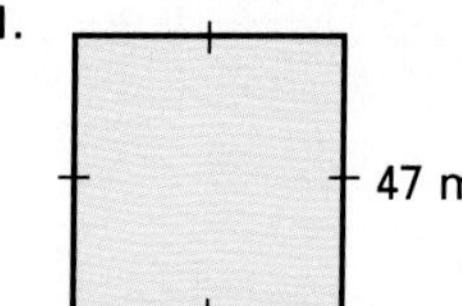


2. 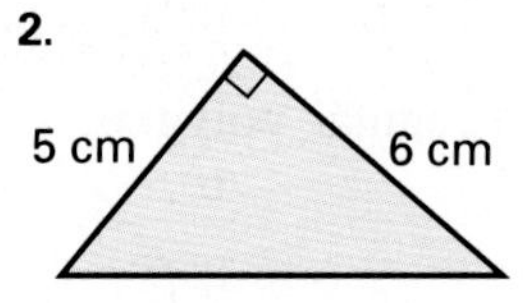


3. 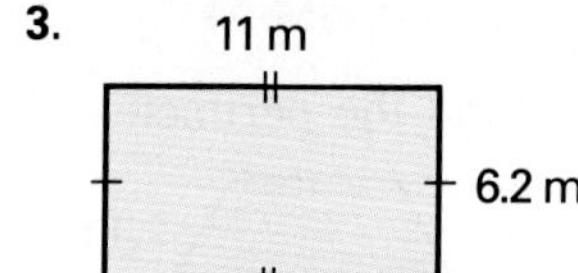


4. 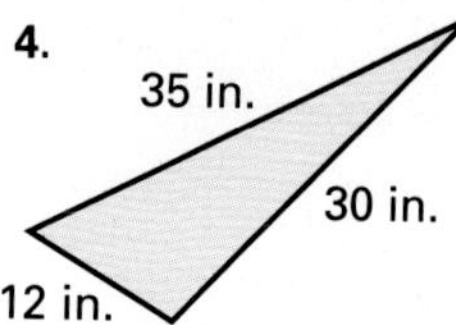


5. 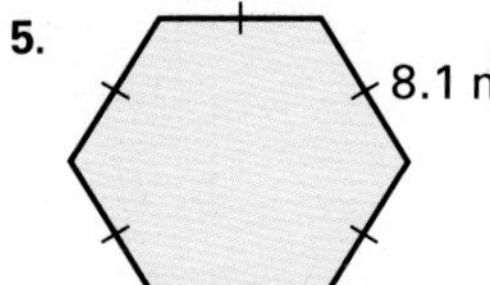


6. 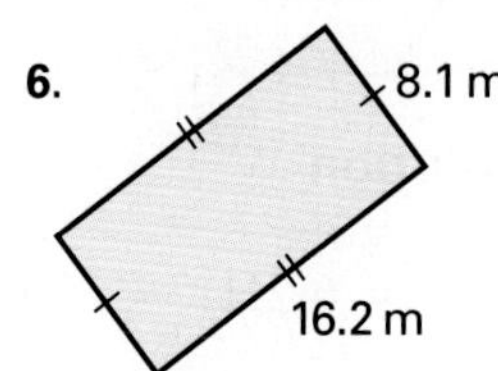


7. 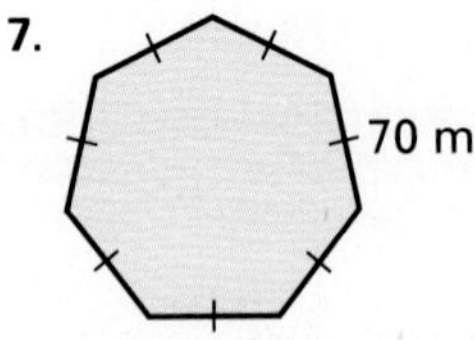


8. 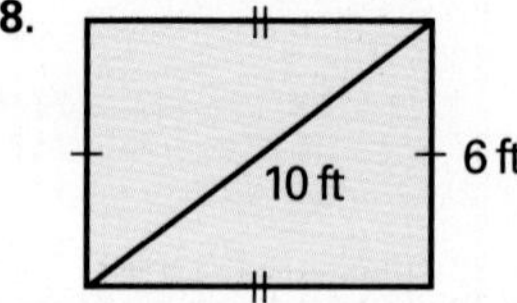


9. 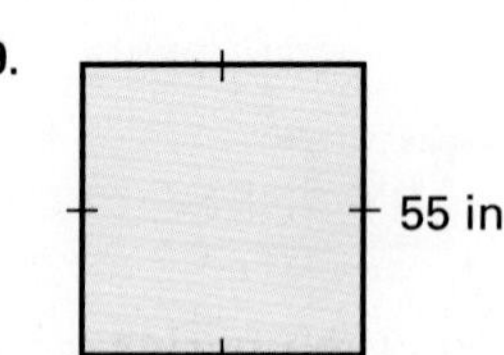


10. Renaldo has 84 ft of fencing and wants to use it to enclose a rectangular herb garden. If he uses all the fencing and makes the garden 16 ft wide, how long will the garden be?

11. The fountain at a park is surrounded by a pond in the shape of a regular octagon with sides of 2.8 m. A border of flowers has been planted around the pond. How long is the border?

## EXTEND/ SOLVE PROBLEMS

Find the perimeter of each figure.

12.

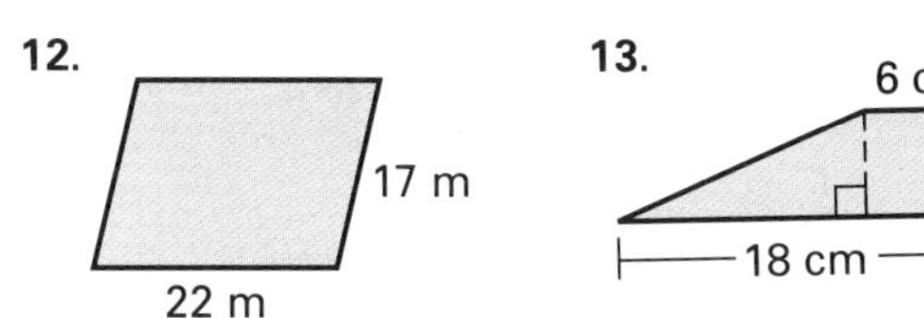


13.

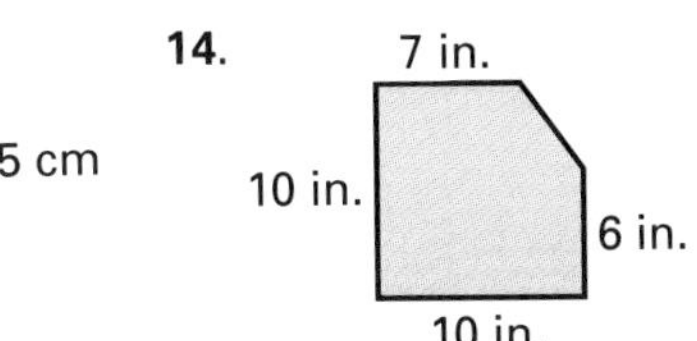


14.

7 in.
10 in.
6 in.
10 in.

15. A rectangular garden runs 14 ft along a wall. Corey uses a total of 36 ft of fencing to enclose the rest of the garden. What are the dimensions of the garden?

16. The perimeter of a regular polygon measures $31\frac{1}{2}$ ft. Each side is $4\frac{1}{2}$ ft long. How many sides does the polygon have?

17. Leigh uses a total of 120 m of fencing to enclose two separate rectangular plots of land. Each plot is 10 m wide. One of the plots is 25 m long. How long is the other plot?

## THINK CRITICALLY/ SOLVE PROBLEMS

18. Each side of a rectangular park is a whole number of miles long. The perimeter of the park is 20 miles. What are the possible dimensions of the park?

19. A landscaper plants 20 shrubs along the perimeter of a garden that has the shape of a rectangle. The shrubs are planted 10 ft apart. How many different lengths are possible for the perimeter of the rectangle?

20. A rectangular forest plot that is twice as long as it is wide will be divided into three equal rectangular sections. The two outer sections are to be fenced. Which way should the plot be divided to use the least fencing—along its length or along its width?

# 6-3 Problem Solving Skills:

## USE LIKE UNITS OF MEASURE

- READ
- PLAN
- SOLVE
- ANSWER
- CHECK

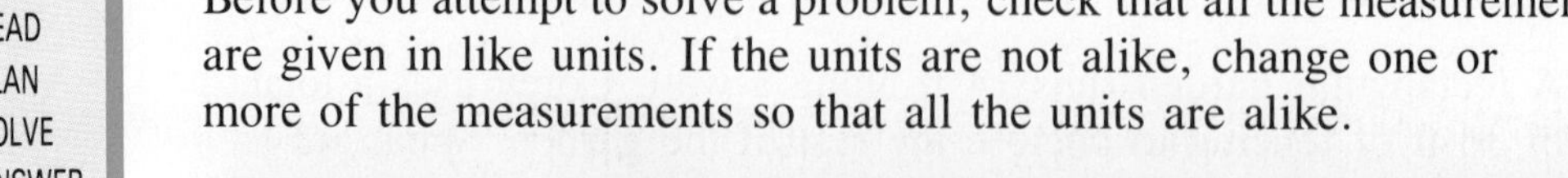

Before you attempt to solve a problem, check that all the measurements are given in like units. If the units are not alike, change one or more of the measurements so that all the units are alike.

**PROBLEM SOLVING TIP**

Sometimes a problem that involves unlike units will indicate which units to use in your solution. If there are no directions given, choose the units that seem most appropriate for the solution.

In the problem about the football field, a solution given in either feet or yards is acceptable.

### PROBLEM

A rectangular football field has side lines 120 yd long. Each of the end lines is 160 ft long. What is the perimeter of the football field?

### SOLUTION

Before you try to find the perimeter of the football field, look at the units of measure given. The side lines are measured in yards, while the end lines are measured in feet. So, before using the formula for the perimeter of a rectangle, you must change units so they are alike.

**CHECK UNDERSTANDING**

How many yards are equal to 1,040 ft? How many feet are equal to $346\frac{2}{3}$ yd?

**Change to feet.**

length = $120 \times 3 = 360$
width = $160$

$$P = 2l + 2w$$
$$P = 2(360) + 2(160)$$
$$P = 720 + 320$$
$$P = 1{,}040$$

**Change to yards.**

length = $120$
width = $160 \div 3 = 53\frac{1}{3}$

$$P = 2l + 2w$$
$$P = 2(120) + 2\left(53\frac{1}{3}\right)$$
$$P = 240 + 106\frac{2}{3}$$
$$P = 346\frac{2}{3}$$

The perimeter of the football field is 1,040 ft, or $346\frac{2}{3}$ yd.

# PROBLEMS

1. Jean-François is going to wrap waterproof tape around the base of a rectangular aquarium that is 2 ft long and 11 in. wide. What is the least length of tape he needs?
2. The top of a ladder is placed so that it reaches a windowsill that is 4 m above the ground. The foot of the ladder is secured at a point 90 cm from the base of the wall. How long is the ladder?
3. Francine used 176 ft of fencing to enclose a rectangular paddock for her horse. If the paddock is 10 yd wide, how long is it?

4. Every morning, Melva jogs once around the park near her house. The park is a rectangle measuring 1.2 km in length and 750 m in width. How far does Melva jog each week?
5. Elsa is edging each of her two flower beds with a single row of bricks. One flower bed is a square with sides measuring 4 yd, and the other is a regular octagon with sides measuring 5 ft 6 in. Which of the flower beds will need more bricks for its border?

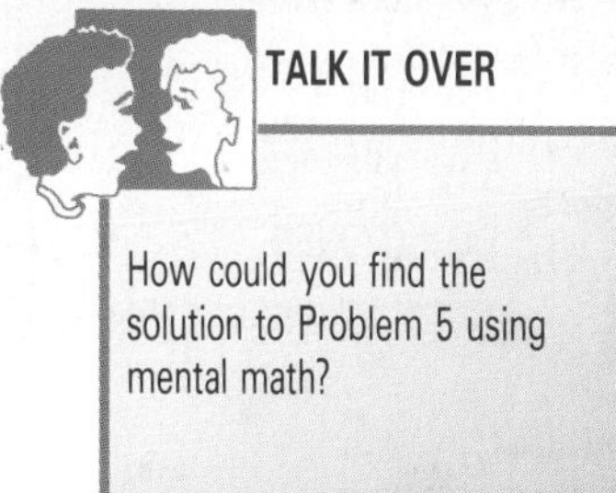


How could you find the solution to Problem 5 using mental math?

6. A square rug with a perimeter of 56 ft is placed exactly in the center of a square room that has a perimeter of 32 yd. How wide is the border between the edges of the rug and the walls?
7. The distance between the opposite corners of a rectangular room is 20 ft. The room is 4 yd wide. What is the perimeter?

Use this diagram of a rectangular tablecloth for Exercises 8–10.

8. What length of red ribbon is used on the tablecloth?
9. What length of blue ribbon is used on the tablecloth?
10. To the nearest tenth of an inch, what length of green ribbon is used on the tablecloth?

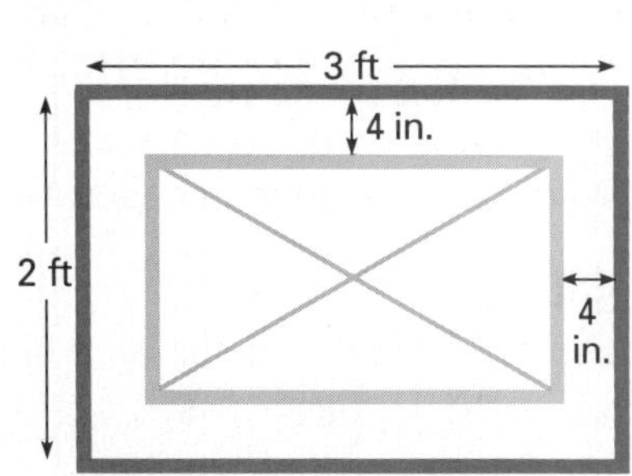


11. ***CRITICAL THINKING*** The center of a rectangular table is 260 cm from each corner of the table. The table is exactly 2 m wide. How long is the table?

# 6-4 Area of Polygons

## EXPLORE

On a blank sheet of paper, accurately draw a trapezoid. Cut around the perimeter of the figure, and then carefully fold it in half by matching the parallel sides. Cut along the fold line, and then arrange the two cut-out figures to form a parallelogram.

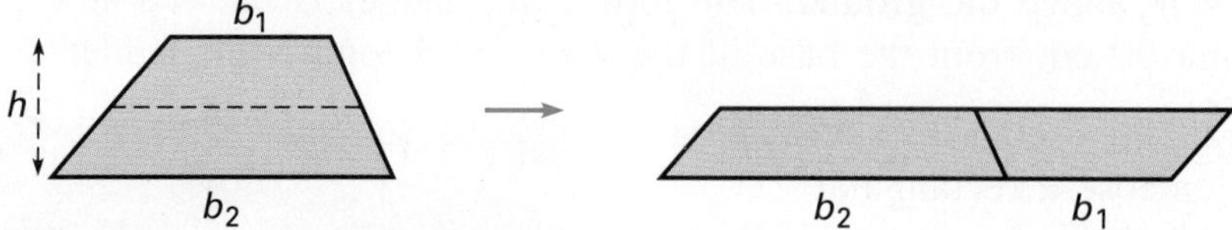


What is the height of the parallelogram? What is the length of its base? How do the areas of the parallelogram and the trapezoid compare?

## SKILLS DEVELOPMENT

Here are some special polygons and the formula used to find the area, $A$, of each.

square

rectangle

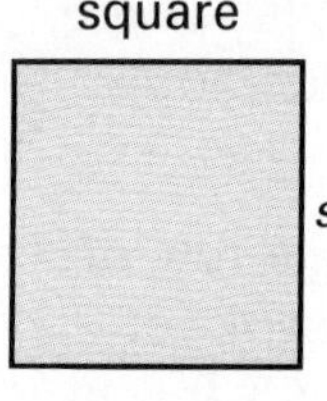


parallelogram

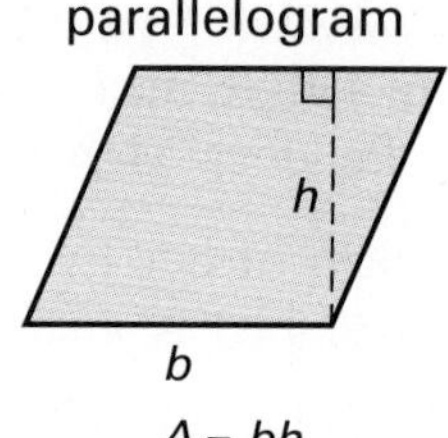


triangle

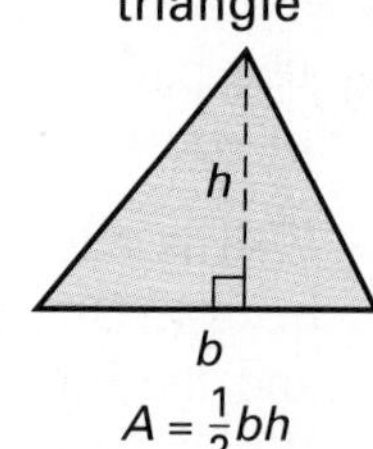


$s$

$A = s^2$ $\quad A = lw$ $\quad A = bh$ $\quad A = \frac{1}{2}bh$

**TALK IT OVER**

The area formula for a trapezoid is sometimes stated as the product of the height and the average of the bases. Explain.

In the Explore activity, you discovered how a trapezoid can be "rearranged" to form a parallelogram. You can use the area formula for a parallelogram to find the area formula for a trapezoid.

$$A = bh$$
$$A = (b_1 + b_2) \times \tfrac{1}{2}h, \text{ or}$$
$$A = \tfrac{1}{2}h(b_1 + b_2)$$

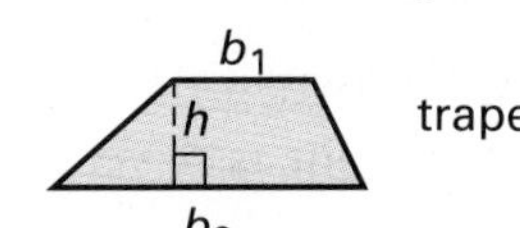


trapezoid

### Example 1

One section of the roof of a greenhouse needs replacement. What is the area of the glass that will be used?

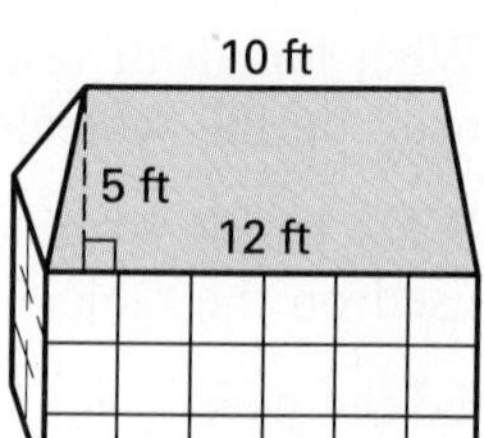


**Solution**

Use the formula for the area of a trapezoid.

$$A = \tfrac{1}{2}h(b_1 + b_2)$$
$$A = \tfrac{1}{2} \times 5 \times (12 + 10)$$
$$A = \tfrac{1}{2} \times 5 \times 22 = 55$$

The area of the needed glass will be 55 $ft^2$. ◄

## Example 2

Find the unknown measurements in each figure.

**a.**

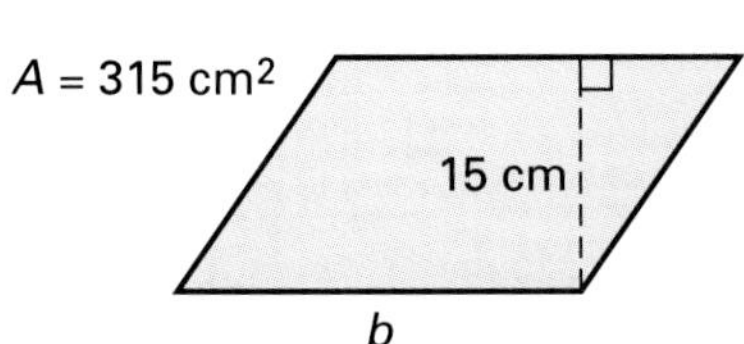


**b.**

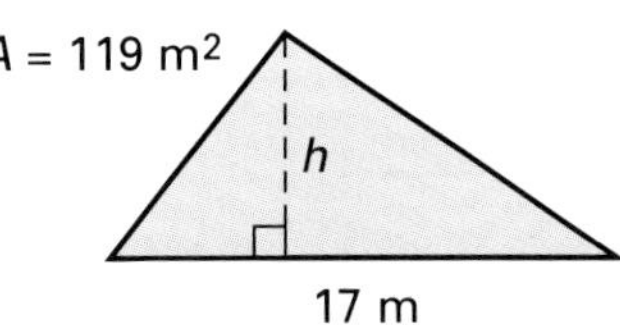


**Solution**

**a.** $A = bh$

$315 = b \times 15$

$\frac{315}{15} = b$

$21 = b$

The length of the base $b$ is 21 cm.

**b.** $A = \frac{1}{2}bh$

$119 = \frac{1}{2} \times 17 \times h$

$119 = 8.5h$

$\frac{119}{8.5} = h$

$14 = h$

The height is 14 m. ◄

## Example 3

Find the area of the rectangle.

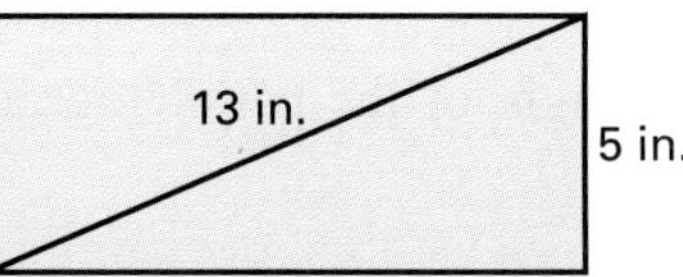


**Solution**

$c^2 = a^2 + b^2$ ← First, use the Pythagorean Theorem to find the length of the rectangle.

$13^2 = a^2 + 5^2$

$169 = a^2 + 25$

$a^2 = 169 - 25 = 144$

$a = \sqrt{144} = 12$

The length of the rectangle is 12 in.

$A = lw$ ← Next, use the formula for the area of a rectangle.

$A = 12 \times 5 = 60$

The area of the rectangle is 60 in.$^2$ ◄

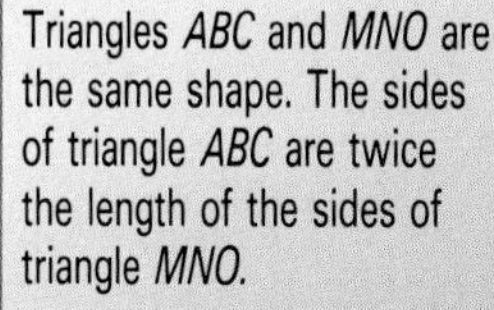

**TALK IT OVER**

Triangles *ABC* and *MNO* are the same shape. The sides of triangle *ABC* are twice the length of the sides of triangle *MNO*.

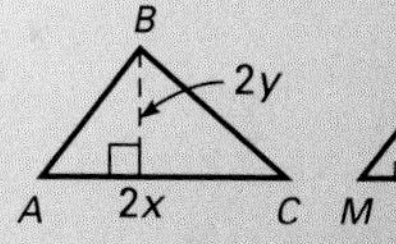


Would you expect the area of triangle *ABC* to be twice the area of triangle *MNO*?

Check your answer by finding the area of each triangle.

## Example 4

Find the area of the figure.

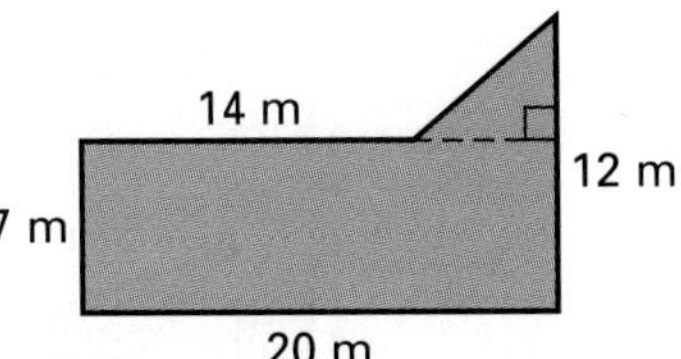


**Solution**

The figure can be divided into a rectangle and a triangle.

| rectangle | triangle |
|---|---|
| $A = lw$ | $A = \frac{1}{2}bh$ |
| $A = 20 \times 7$ | $A = \frac{1}{2} \times 6 \times 5 = 15$ |
| $A = 140$ | |

The length of the base is $20 - 14 = 6$ m. The height is $12 - 7 = 5$ m.

The total area of the figure is the sum of 140 m$^2$ and 15 m$^2$. The area of the figure is 155 m$^2$. ◄

## TRY THESE

Find the area of each figure.

1\.

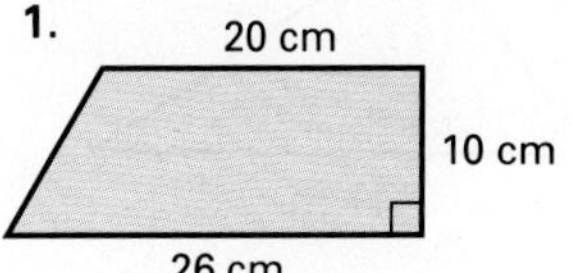


2\.

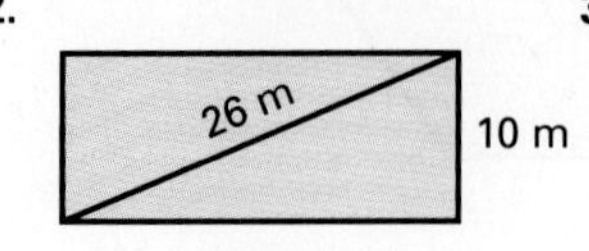


3\.

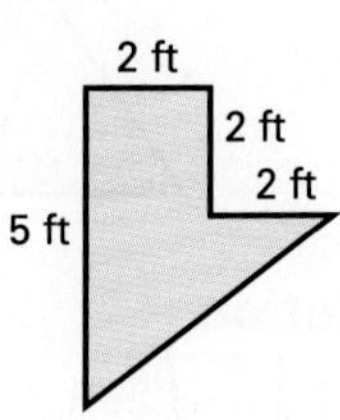


Find each unknown measure.

4\.

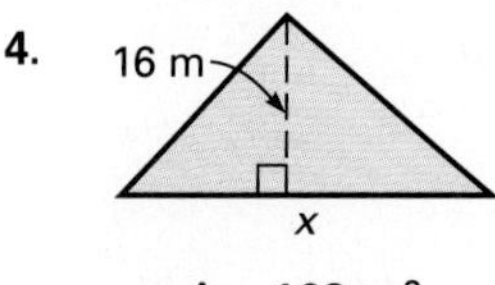


A = 168 m$^2$

5\. 24.4 cm, y

A = 280.6 cm$^2$

6\.

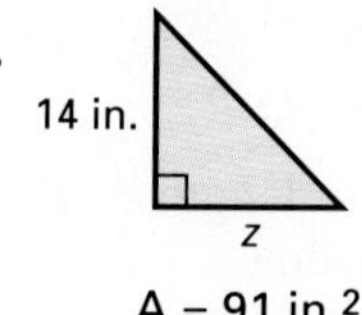


A = 91 in.$^2$

7\. A parallelogram has an area of 414 cm$^2$. Its base measures 23 cm. What is the height of the parallelogram?

# EXERCISES

**PRACTICE/ SOLVE PROBLEMS**

Find the area of each figure.

1\.

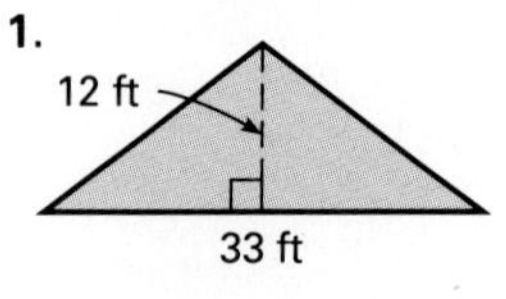


2\.

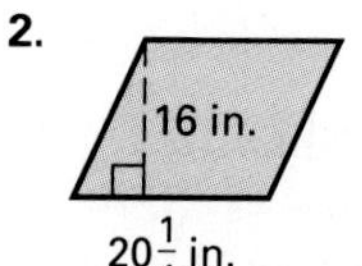


3\.

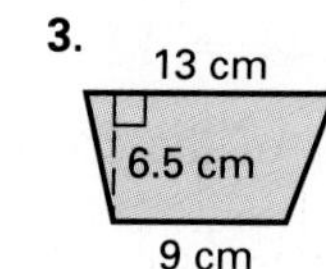


4\.

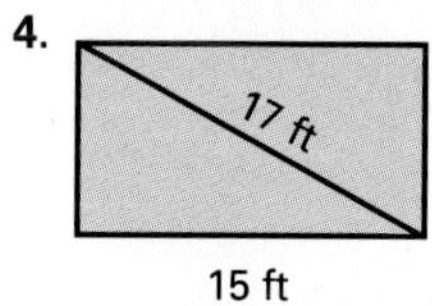


5\.

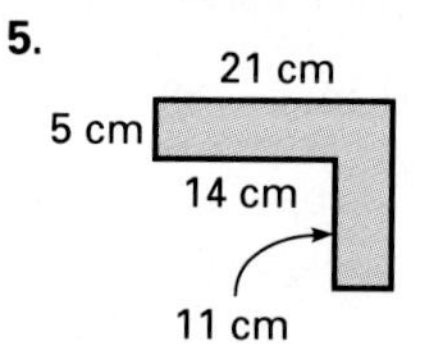


6\.

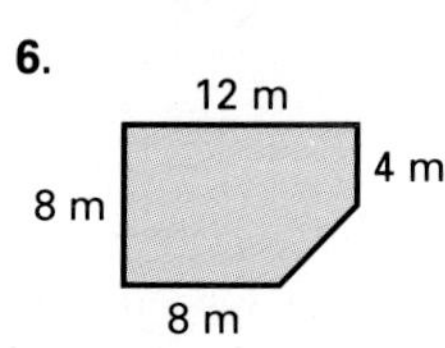


Find each unknown measure.

7\. 17 m, 8 m, m

A = 120 m$^2$

8\.

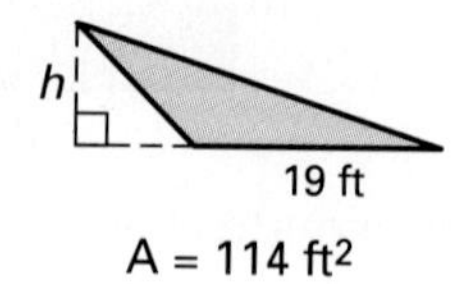


A = 114 ft$^2$

9\.

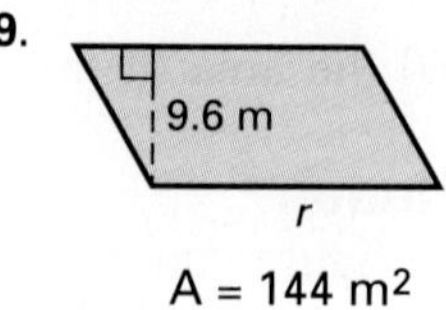


A = 144 m$^2$

10\. What is the height of a triangle that has a base of 6.2 cm and an area of 4.65 cm$^2$?

11\. The shortest distance between opposite corners of a rectangular garden is 30 ft. If the garden is 18 ft wide, what is its area?

## EXTEND/ SOLVE PROBLEMS

Find the area of each figure.

12.

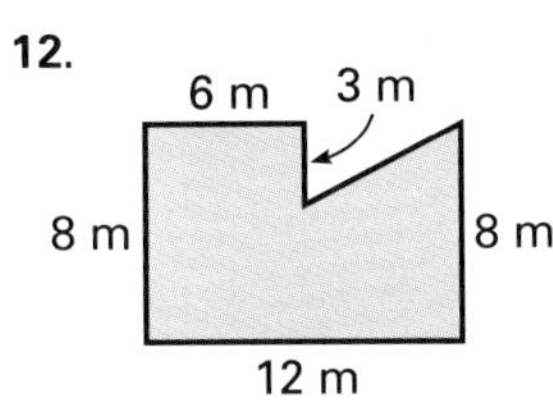


13.

14.

Find the area of the shaded region of each figure.

15.

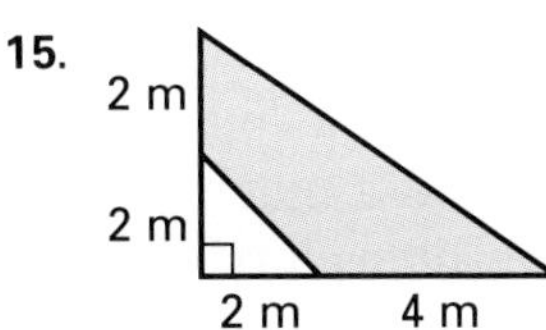


16.

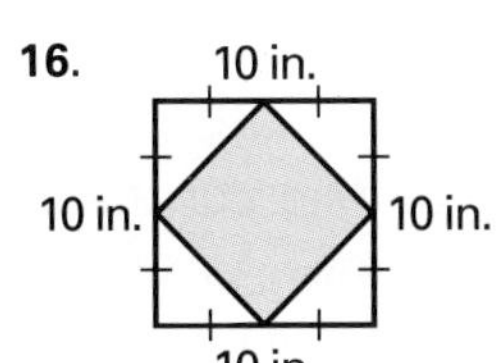


17.

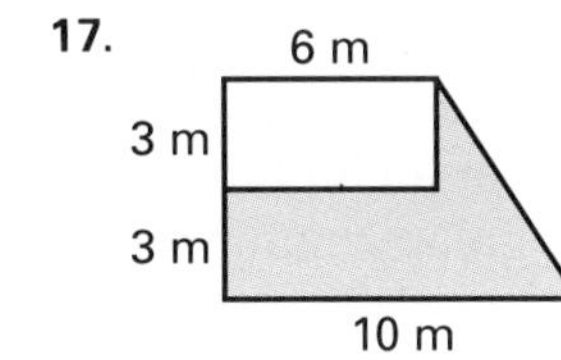


18. In the figure at the right, the vertex of the triangle is exactly at the center of the flag. What fraction of the area of the flag is blue?

19. What fraction of the area of the flag is red?

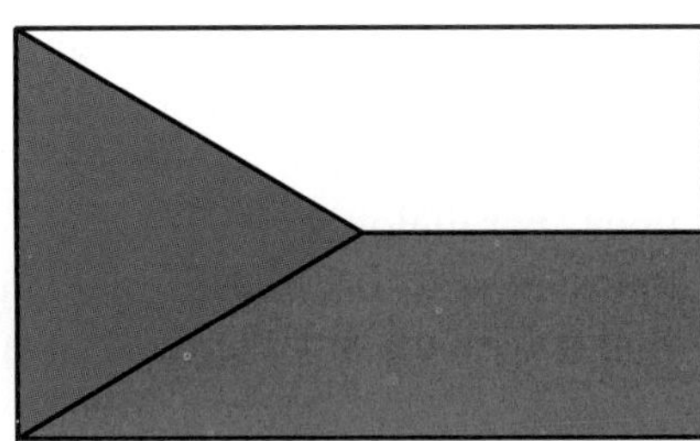

## THINK CRITICALLY/ SOLVE PROBLEMS

20. A square and a rectangle that is not a square each have the same area. Which figure has the longer diagonal?

21. In the figure at the right, the area of the rectangle *XBYZ* is $\frac{1}{3}$ the area of rectangle *ABCD* and *X* is the midpoint of $\overline{AB}$. How does the length of $\overline{BY}$ compare to the length of $\overline{BC}$?

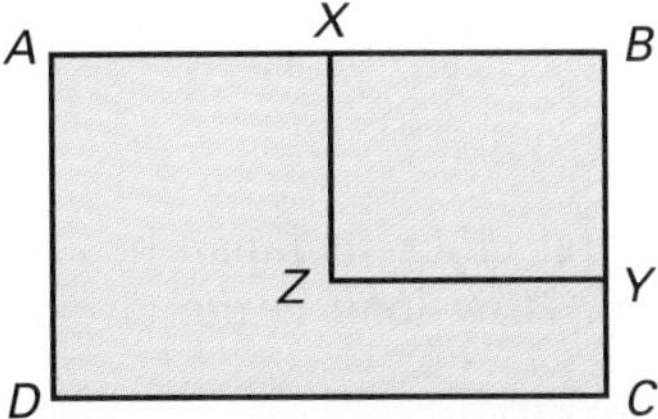

# 6-5 Circumference and Area of Circles

## EXPLORE

a. Use a compass to draw a large circle on a sheet of paper. Then use a protractor to divide the circle into six equal sectors. Carefully cut out the sectors.

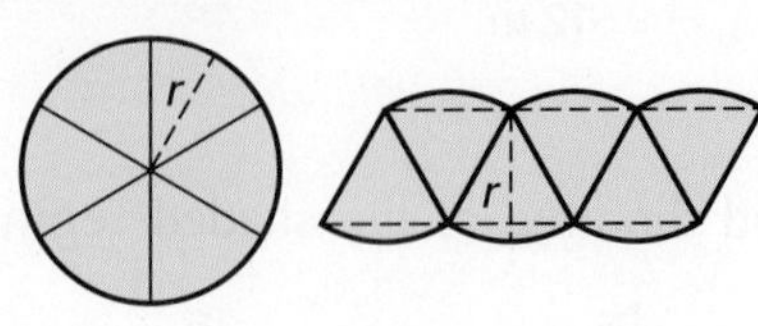


b. Rearrange the sectors to form a shape that resembles a "bumpy parallelogram." Suppose you had cut the circle into a much greater number of equal sectors. As the number of sectors increases, what happens to the shape of the figure you can form from the sectors?

c. The distance around the circle is the circumference $C$. How does the "base" of your parallelogram compare with the circumference of the circle you drew?

d. How does the area of the circle you drew compare to the area of your "parallelogram"? What is the "height" of your "parallelogram"? Show how to use the formula for the area of a parallelogram to find a formula for the area of a circle.

## SKILLS DEVELOPMENT

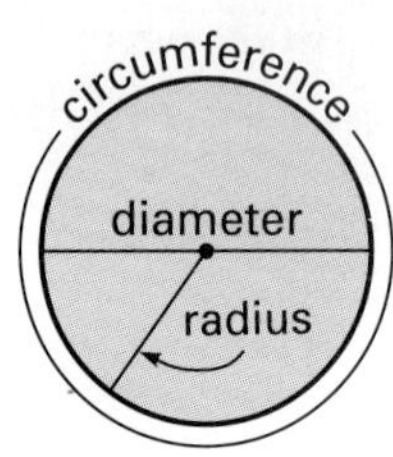


The distance around a circle is its **circumference.** To find the circumference $C$ of a circle, use one of these formulas.

$C = \pi d$ or $C = 2\pi r$

To find the area of a circle, use this formula. $A = \pi r^2$

The formulas for the circumference and area of circles can be used to solve problems about semicircles and concentric circles.

### Example 1

A floral clock in Japan has the world's largest clock face. Its diameter is 21 m. How long is the circumference of the clock face?

**Solution**

Use the formula.

$C = \pi d$

$C \approx 3.14 \times 21$

$C \approx 65.94$

**Since $\pi$ is irrational, use an approximate value for $\pi$, 3.14 or $\frac{22}{7}$.**

The length of the circumference is approximately 65.94 m. ◄

The diameter of a circle divides the circle into two semicircles. The circumference and area of a semicircle measure half the circumference and half the area of a circle that has the same radius.

### Example 2

The straight edge of a semicircular garden measures 14 ft. What is the area of the garden?

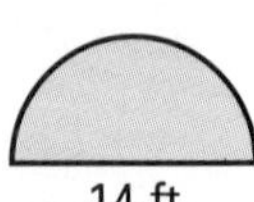

## Solution

The straight edge is the diameter of the semicircle, so its radius is half of 14 ft, or 7 ft. The area of the semicircle will be half the area of a circle with a radius of 7 ft.

$$A = \pi r^2$$
$$A \approx \frac{22}{7}(7)(7)$$
$$A \approx 154$$
$$\frac{1}{2}A \approx \frac{1}{2}(154) \approx 77$$

The area of the garden is approximately 77 ft$^2$. ◄

**PROBLEM SOLVING TIP**

In Example 2, why is it easier to use the value of $\frac{22}{7}$ for $\pi$ rather than the value of 3.14?

In Example 3, why is it easier to use the value of 3.14?

## Example 3

What is the radius of a circular garden with an area of 28.26 m$^2$?

## Solution

Use the formula $A = \pi r^2$. Substitute 28.26 for $A$.

$$28.26 \approx 3.14 \times r^2$$
$$\frac{28.26}{3.14} \approx r^2$$
$$r \approx \sqrt{9} = 3$$

The radius of the garden is approximately 3 m. ◄

## Example 4

What is the area of the shaded region of each figure?

**a.**

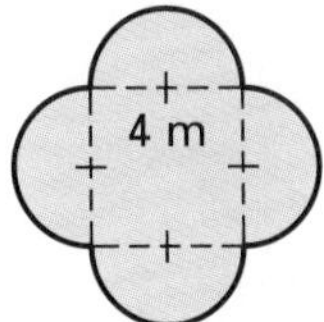


**b.**

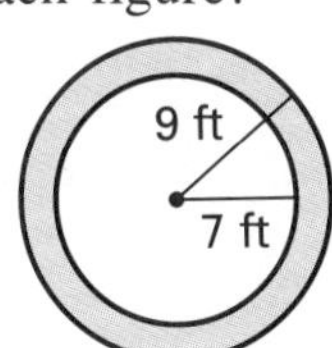

**TALK IT OVER**

Can you find a shorter way to solve Example **4b**?

## Solution

**a.** Add the areas of the square and the four semicircles. The radius of each semicircle is half the side of the square or 2 m.

Find the area of the square.

$$A = s^2$$
$$A = 4 \times 4 = 16$$

Find the area of each semicircle.

$$A = \pi r^2$$
$$A \approx 3.14(2^2)$$
$$A \approx 12.56$$
$$\frac{1}{2}A \approx 6.28$$

Find the total area.

$$A \approx 16 + 4(6.28)$$
$$A \approx 41.12$$

The area is about 41.12 m$^2$.

**b.** The shaded area is the area of the larger circle minus the area of the smaller circle.

The area of the larger circle is

$$A = \pi r^2$$
$$A \approx 3.14(9^2)$$
$$A \approx 254.34$$

The area of the smaller circle is

$$A = \pi r^2$$
$$A \approx 3.14(7^2)$$
$$A \approx 153.86$$

Subtract.

$$A \approx 254.34 - 153.86$$
$$A \approx 100.48$$

The area is about 100.48 ft$^2$. ◄

**JUST FOR FUN**

Suppose that the distance from the center of a watch face to the tip of the second hand is one half inch. How far does the tip of the second hand travel in one year? Complete this chart to help you find the answer.

| Time | Distance |
|---|---|
| 1 min | ■ in. |
| 1 h | ■ ft |
| 1 day | ■ yd |
| 1 year | ■ mi |

## TRY THESE

Find the circumference, area, or radius as indicated.

**1.**

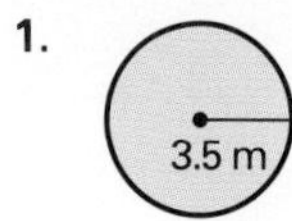


C ≈ ■

**2.**

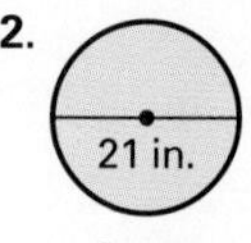


C ≈ ■

**3.**

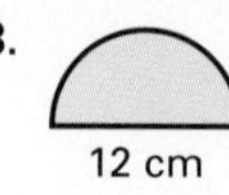

A ≈ ■

**4.**

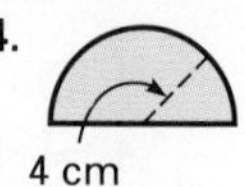


A ≈ ■

**5.**

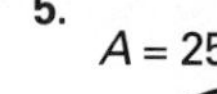


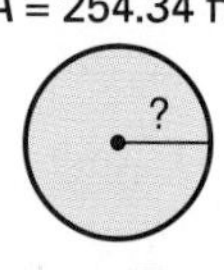


$r$ ≈ ■

**6.** 6 m

4 m

shaded $A$ ≈ ■

**7.**

$A$ ≈ ■

**8.** 8 m

$A$ ≈ ■

# EXERCISES

## PRACTICE/ SOLVE PROBLEMS

1. What is the circumference of a circle with radius 7.7 cm?
2. What is the area of a semicircle with diameter 20.8 m?
3. What is the radius of a circle with an area of 15.7 cm²?

Find the area of the shaded region of each figure.

**4.**

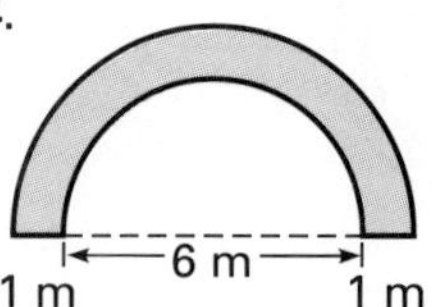


**5.**

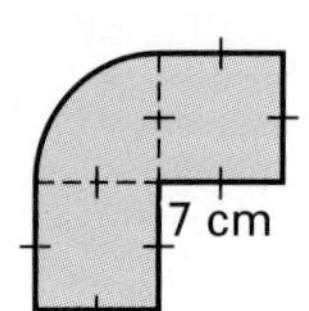


**6.**

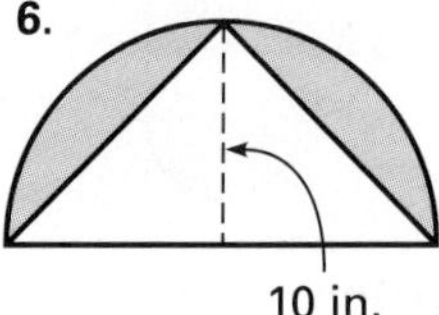


## EXTEND/ SOLVE PROBLEMS

7. The radius of a flowerpot at its rim is 3 in. Flowerpots of this size will be placed, without stacking, into a square box whose sides measure 2 ft. How many will fit?
8. A semicircular garden stretches for 5 m along a straight wall. A border 1 m wide is paved around the edge of the garden. What is the area of the paved border?

## THINK CRITICALLY/ SOLVE PROBLEMS

Find the area of the shaded region of each figure.

**9.**

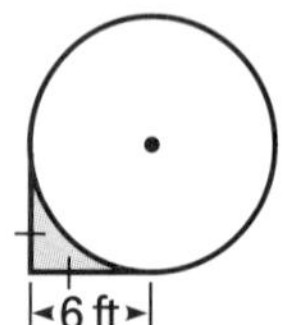


**10.** 10 m

10 m

**11.**

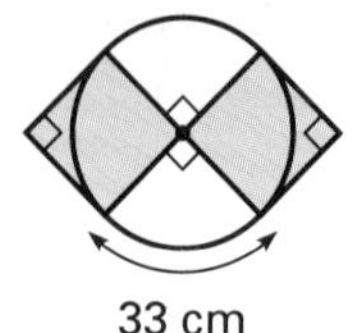

- ► READ
- ► PLAN
- ► SOLVE
- ► ANSWER
- ► CHECK

# Problem Solving Applications:

## HOW MUCH PAINT?

If you are preparing to paint the outside of a house, the first thing you should pick up is a pencil!

Whichever type of paint you choose, you'll need to know how much of it you'll use. You can estimate the amount of paint by finding the total area to be painted.

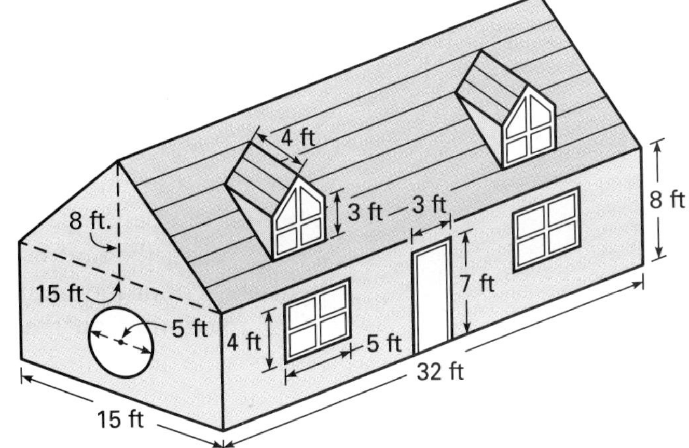


First, find the total wall area of the house. Assume that the roof and dormer tops are shingled.

1. **Front wall:** a rectangle $A = lw$ Area: ■ ft²
2. **Side wall:** a rectangle and a triangle $A = lw + \frac{1}{2}bh$ Area: ■ ft²
3. **Dormers:** 4 congruent triangles $A = 4 \times \frac{1}{2}bh$ Area: ■ ft²
4. Assume that the front and back walls of the house are identical and that the side walls are identical. There are no dormers in the back. What is the total wall area?

Next, you will need to subtract the areas of the windows and doors. Find their total area.

5. **Door:** a rectangle $A = lw$ Area: ■ ft²
6. **Front windows:** 2 congruent rectangles $A = 2 \times lw$ Area: ■ ft²
7. **Side window:** a circle $A = \pi r^2$ Area: ■ ft²
8. The back door and back windows are the same as those in the front. Both sides have identical circular windows. What is the total area to be subtracted?
9. What is the total area that will be painted?

A gallon of high-quality paint covers about 400 ft². A gallon of less expensive paint covers about 250 ft² and may require two coats.

10. How many gallons of paint will the house require if you use one coat of high-quality paint? How many gallons if you use two coats of less expensive paint?

# 6-6 Area and Probability

## EXPLORE/ WORKING TOGETHER

On a blank rectangular sheet of paper that measures $8\frac{1}{2}$ in. by 11 in., very lightly draw a square with sides of 3 in. The square can be anywhere on the paper.

Turn the paper over so that the location of the square is not visible. Have each of your classmates, with eyes closed, and try to mark a point on the paper. If the mark is not on the paper, the person tries again. When the entire class has marked the paper, hold it up to the light and count the number of points that fall within the square. What fraction of all the points fell within the square? Compare your results with those of the rest of the class. What fraction of the total area of the paper does the area of the square cover? Is there any relationship to your results?

## SKILLS DEVELOPMENT

**TALK IT OVER**

Does probability tell you what *will* happen or what is *likely* to happen? How does this relate to why results varied in the Explore activity?

If point $A$ is picked randomly in a given region $M$, the probability of its being within region $N$ can be found by comparing the areas of the two regions.

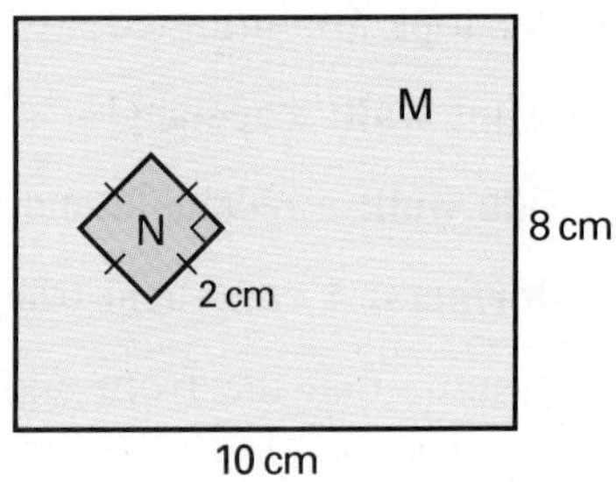


$$P(A \text{ lies within region } N) = \frac{\text{area of } N}{\text{area of } M} = \frac{2^2}{10(8)} = \frac{4}{80} = 0.05$$

So, the probability that a point $A$ lies within region $N$ is 0.05.

### Example 1

A rectangular garden measures 25 ft by 40 ft. In the garden there is a square fish pond with sides of 5 ft. If the wind carries a seed into the garden, what is the probability that it lands in the pond?

#### Solution

$P(\text{seed landing in pond}) = \dfrac{\text{area of pond}}{\text{area of garden}}$

| area of pond | area of garden |
|---|---|
| $A = s^2$ | $A = lw$ |
| $A = 5^2 = 25$ | $A = 25(40) = 1{,}000$ |

$P(\text{seed landing in pond}) = \dfrac{25}{1{,}000} = 0.025$

The probability that the seed lands in the pond is 0.025. ◄

## Example 2

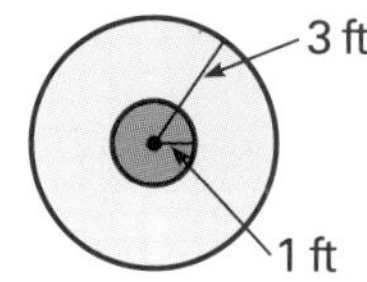


A penny is dropped into an empty wishing well. The bottom of the well is painted as shown.

**a.** What is the probability that the penny lands in the red region?
**b.** What is the probability that the penny lands in the blue region?

### Solution

**a.** $P(\text{red}) = \frac{\text{area of red region}}{\text{area of entire region}}$

$= \frac{\pi \times 1^2}{\pi \times 3^2} = \frac{1}{9}$

Notice that you do not have to calculate each area because $\pi$ is a common factor.

The probability that the penny lands in the red region is $\frac{1}{9}$.

**b.** Since $P(\text{red or blue}) = 1$, $P(\text{blue}) = 1 - P(\text{red})$.

$P(\text{blue}) = 1 - \frac{1}{9}$

$= \frac{8}{9}$

The probability that the penny lands in the blue region is $\frac{8}{9}$. ◄

## Example 3

A farmer scatters 200 seeds in a rectangular field that measures 80 m by 50 m. If 46 of the seeds land in a pool that lies in the field, what is the probable area of the pond?

### Solution

$P(\text{seed landing in the pond}) = \frac{\text{area of pond}}{\text{area of field}}$

Since 46 of the 200 seeds landed in the pond, the probability of a single seed landing in the pond is $\frac{46}{200}$ or $\frac{23}{100}$.

So, $\frac{23}{100} = \frac{\text{area of pond}}{\text{area of field}}$

$\frac{23}{100} = \frac{\text{area of pond}}{4{,}000}$

The area of the rectangular field is $80 \times 50 = 4{,}000\ m^2$.

Let $x$ = area of pond. Solve for $x$.

$\frac{23}{100} = \frac{x}{4{,}000}$

$4{,}000\left(\frac{23}{100}\right) = 4{,}000\left(\frac{x}{4{,}000}\right)$

$920 = x$

The probable area of the pond is 920 $m^2$. ◄

**CONNECTIONS**

A probability can be expressed in several different ways. For example, each of these statements gives the same information. *There is a 25% chance of rain today. The probability that it will rain today is 0.25 or $\frac{25}{100}$. The chances that it will rain today are 25 out of 100.*

So, the probability can be expressed as a decimal, a fraction, a percent, or a ratio.

Complete.

The value of a probability cannot be greater than each of the following.

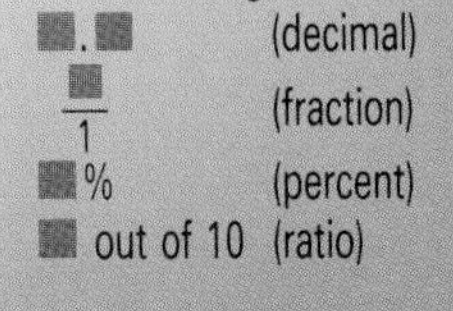

## TRY THESE

Find the probability that a point selected at random in each figure is in the shaded region.

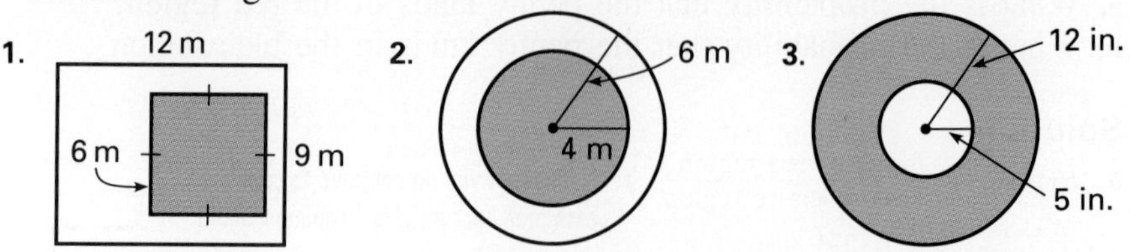


4. A potter makes a 12-in. round plate. Around the rim, there is a 1-in.-wide border. If a bubble forms in the clay when the plate is fired, what is the probability that the bubble is located on the border? Round your answer to the nearest hundredth.
5. A rectangular dart board measures 6 m by 4 m. There is a red circle painted on the board. Out of 150 darts thrown randomly at the board, 40 landed in the red circle. What is the probable area of the red circle?

## EXERCISES

**PRACTICE/ SOLVE PROBLEMS**

Find the probability that a point selected at random in each figure is in the shaded region.

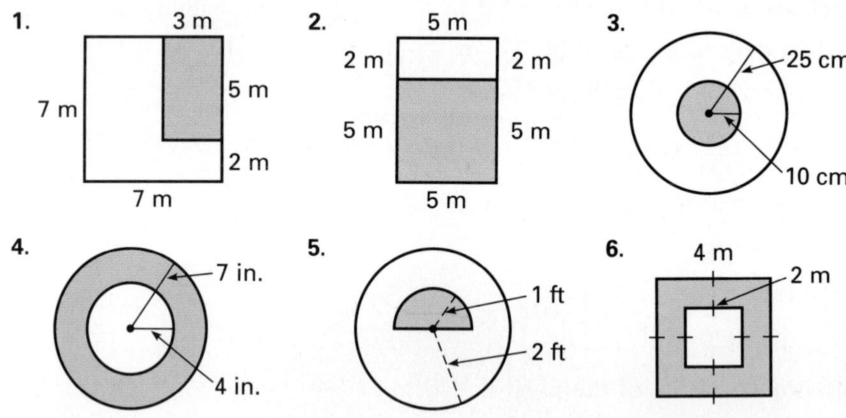


7. One wall of a building faces a baseball field. The wall has a length of 35 ft and a height of 24 ft. There are 9 square windows in the wall, each with sides of 4 ft. If a baseball hit out of the field strikes the wall, what is the probability that it breaks a window?
8. Of 40 randomly shot arrows that hit a circular target, 3 hit the bull's-eye. If the target has a diameter of 140 cm, what is the probable area of the bull's-eye?

## EXTEND/ SOLVE PROBLEMS

Find the probability that a point selected at random in each figure is in the shaded region.

**9.**

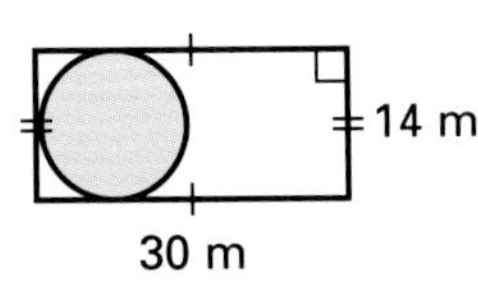


**10.**

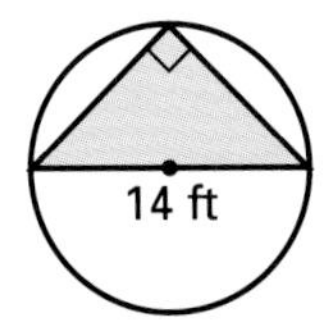


**11.**

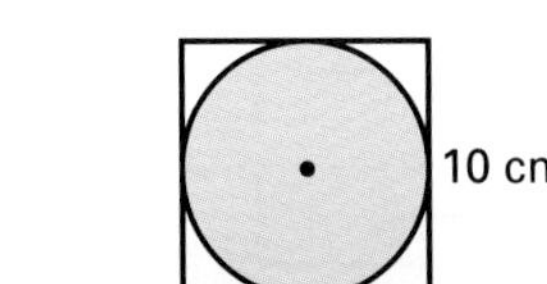


**12.**

14 m

30 m

**13.**

14 ft

**14.**

10 cm

**15.** At a large picnic, a man loses his wallet in a square field that is 50 m on a side. Each of the people at the picnic searches an area 10 m square. Assuming that someone finds the wallet, what is the probability that the man who lost the wallet will be the one who finds it?

*USING DATA* Refer to the Data Index on pp. 556–557 to find information about the areas of certain states.

**16.** If a meteor lands in Colorado, what is the probability of its landing in a forested area? Round to the nearest hundredth.

**17.** Alan picks two points at random on a map. One is in Utah, and the other is in Arizona. Which point has the greater probability of being in forested land?

## THINK CRITICALLY/ SOLVE PROBLEMS

**18.** The rectangular bed of a truck measures 6 ft wide by 10 ft long. A single layer of square flower trays, each with sides of 2 ft, is loaded into the bed and covers the bottom completely. One of the trays contains marigolds. What is the probability that there are marigolds in the row closest to the cab of the truck?

**19.** A dime has a radius of 9 mm. The squares on a checkerboard have sides of 3.6 cm. If you throw a dime onto the checkerboard, what is the probability that it will land completely within a square, without touching any boundary of a square? (*Hint:* Draw a diagram of a single square. To be completely within the square, the center of the dime must be more than 9 mm from the edges.)

# 6-7 Surface Area of Three-Dimensional Figures

**EXPLORE**

Each of the two-dimensional patterns can be folded to form a polyhedron. Match each pattern to the polyhedron it would form.

**1.** 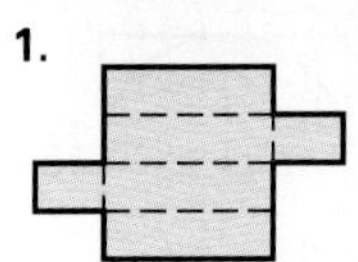 **2.** 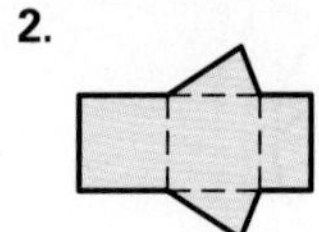 **3.** 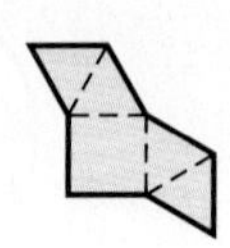 **4.** 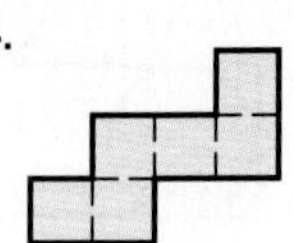

**a.**  **b.** **c.** 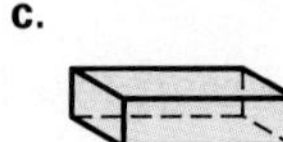 **d.** 

**SKILLS DEVELOPMENT**

The **surface area** of a polyhedron is the sum of the areas of all its bases and faces.

### Example 1

Find the surface area of the rectangular prism.

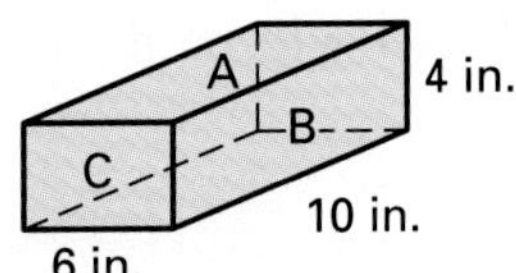


**Solution**

A rectangular prism has three pairs of congruent faces. The surface area is the sum of the areas of all the faces.

$SA = 2(\text{area } A) + 2(\text{area } B) + 2(\text{area } C)$

area of $A$ $\quad A = lw$
$A = 10 \times 6$
$A = 60$

A 6 in. 10 in.

area of $B$ $\quad A = lw$
$A = 10 \times 4$
$A = 40$

B 4 in. 10 in.

area of $C$ $\quad A = lw$
$A = 6 \times 4$
$A = 24$

C 4 in. 6 in.

$$SA = 2(60) + 2(40) + 2(24)$$
$$= 120 + 80 + 48 = 248$$

The surface area is 248 in.$^2$ ◄

Example 1 illustrates the formula for the surface area of a rectangular prism, where $l$ is the length, $w$ is the width, and $h$ is the height.

$$SA = 2lw + 2lh + 2wh \text{ or}$$
$$SA = 2(lw + lh + wh)$$

## Example 2

Find the surface area of the triangular prism shown at the right.

4 m
3 m
E
F
6 m
G
5 m

### Solution

The triangular prism shown has one pair of congruent triangular faces and three rectangular faces, where area $D$ is the base.

$$SA = \text{area } D + \text{area } E + \text{area } F + 2(\text{area } G)$$

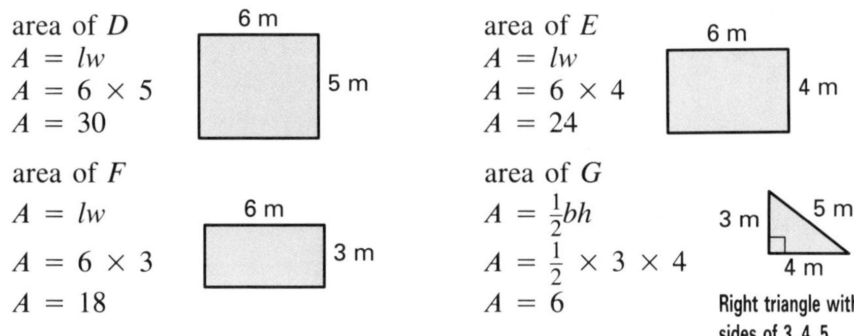


area of $D$
$A = lw$
$A = 6 \times 5$
$A = 30$

area of $E$
$A = lw$
$A = 6 \times 4$
$A = 24$

area of $F$
$A = lw$
$A = 6 \times 3$
$A = 18$

area of $G$
$A = \frac{1}{2}bh$
$A = \frac{1}{2} \times 3 \times 4$
$A = 6$

Right triangle with sides of 3, 4, 5

$SA = 30 + 24 + 18 + 2(6) = 84$

The surface area of the triangular prism measures 84 $m^2$. ◄

**TALK IT OVER**

When does a triangular prism have more than one pair of congruent surfaces?

A square pyramid with four congruent triangular faces is called a **regular square pyramid.** If you know the dimensions of just one triangular face, you can calculate the surface area of the pyramid.

## Example 3

Find the surface area of the regular square pyramid shown at the right.

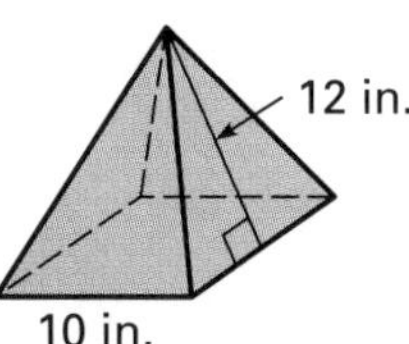


### Solution

$SA = 4(\text{area of triangular face}) + \text{area of square base}$

area of triangular face
$A = \frac{1}{2}bh$
$A = \frac{1}{2} \times 10 \times 12$
$A = 60$

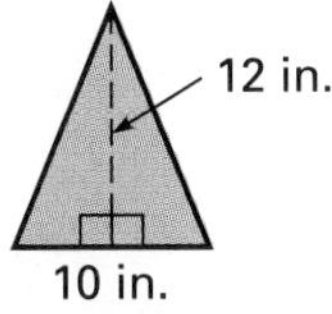


area of square base
$A = s^2$
$A = 10^2$
$A = 100$

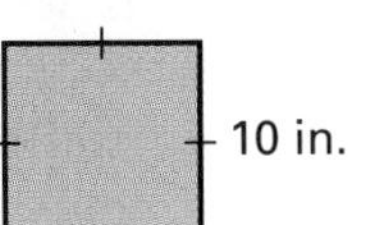


$SA = 4(60) + 100 = 340$

The surface area is 340 $in.^2$ ◄

**PROBLEM SOLVING TIP**

When you are asked to find the surface area of a figure, think of how the figure would look if it were cut apart. Decide if there are any surfaces that are congruent. Draw and label a diagram of each surface for which you need to find the area.

## TRY THESE

Find the surface area of each polyhedron. Assume that all pyramids are regular pyramids.

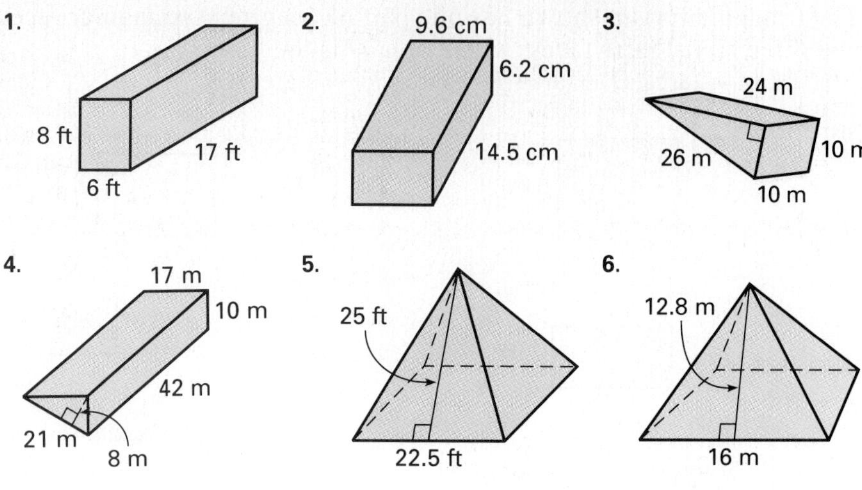


## EXERCISES

**PRACTICE/ SOLVE PROBLEMS**

Find the surface area of each polyhedron. Assume that all pyramids are regular pyramids.

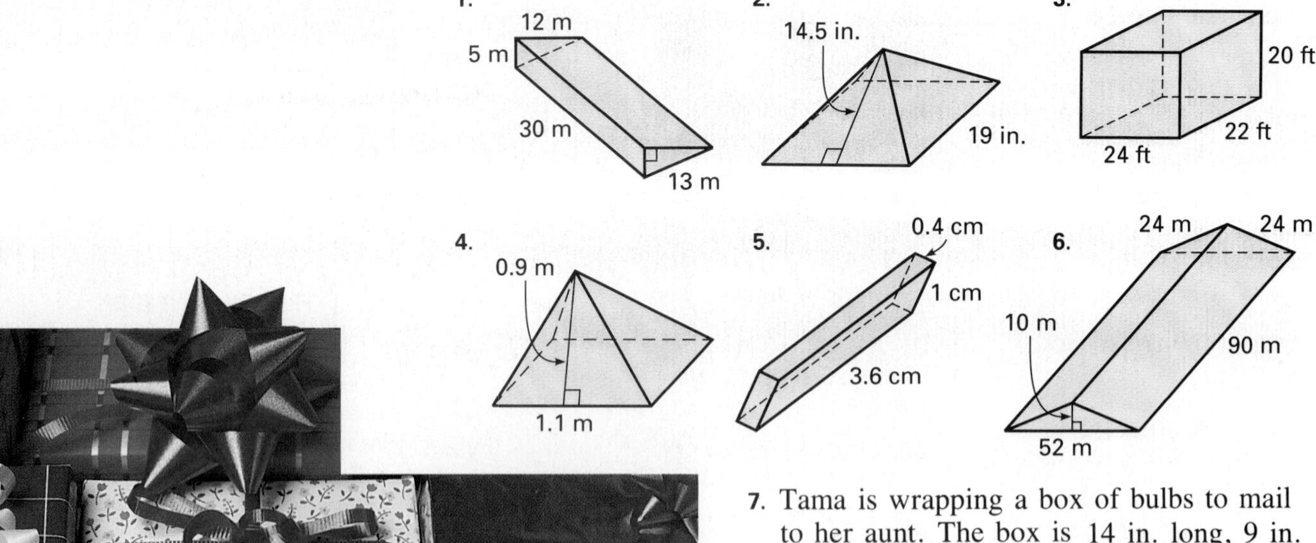


7. Tama is wrapping a box of bulbs to mail to her aunt. The box is 14 in. long, 9 in. wide, and 8 in. high. What is the minimum area of wrapping paper that she will need to cover the box?

## EXTEND/ SOLVE PROBLEMS

Find the surface area of each polyhedron to the nearest whole number. Assume all pyramids are regular pyramids.

8. 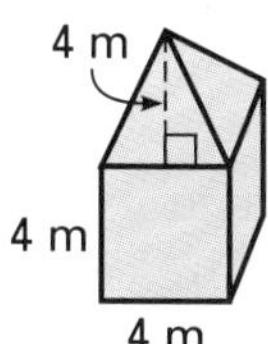


9. 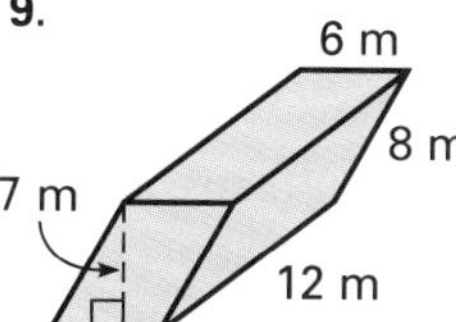


10. 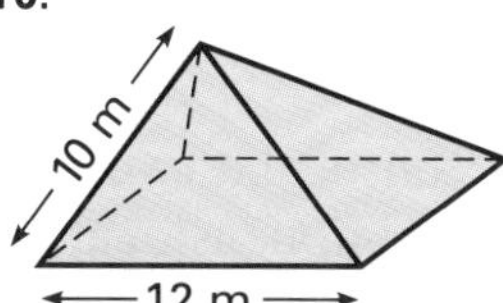


11. The Great Pyramid at Giza in Egypt has a square base about 756 ft long. The height of each triangular face is about 612 ft. What is the surface area of the faces of the Great Pyramid? (Note that the base is not one of the faces of the pyramid.)

12. A triangular prism made of glass is 10.5 cm long. Each end is an equilateral triangle with sides of 4 cm. To the nearest whole number, what is the surface area of the prism? (*Hint:* Use the Pythagorean Theorem to find the height of the triangles.)

13. What is the total area of glass (including the roof) that would be used to build the greenhouse at the right?

3 ft
6 ft
8 ft
10 ft

14. The diagram at the right shows the Washington Monument. What is its total surface area excluding its base? Round your answer to the nearest tenth.

16.8 m
10.5 m
152.5 m
16.8 m

## THINK CRITICALLY/ SOLVE PROBLEMS

15. Write a formula to show the surface area of a regular square pyramid with a base length $b$ and height of the triangular face $l$.

16. How would the surface area of a cube change if the length of the side were (a) doubled? (b) tripled?

17. Suppose the length of a side of a regular square pyramid were doubled and the height of each triangular face were doubled, how would the surface area change?

# 6-8 Formulas for Surface Area

**EXPLORE**

Each of the two-dimensional patterns below can be folded to form a three-dimensional figure. Match each pattern with a figure.

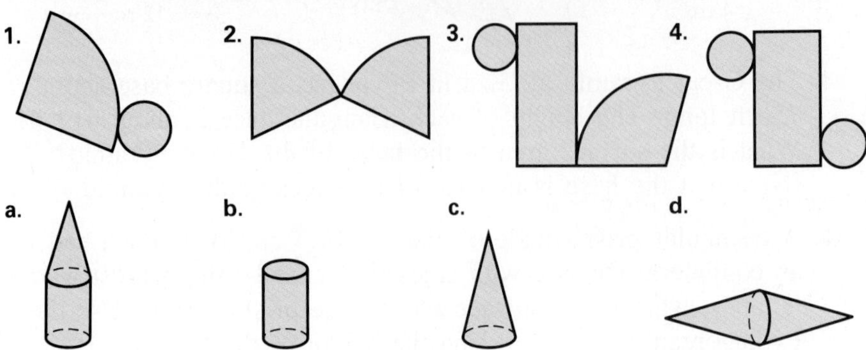


**SKILLS DEVELOPMENT**

A cylinder is a three-dimensional figure with two congruent circular bases and a curved surface. To find the **surface area** of a cylinder, add the area of the curved surface to the sum of the areas of the two bases.

### Example 1

Find the surface area of the cylinder shown at the right.

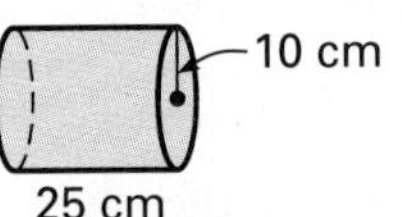


**Solution**

The curved surface can be laid flat to form a rectangle. The length of the rectangle is equal to the circumference of the cylinder. The width of the rectangle is equal to the height of the cylinder.

Think: $A = lw$

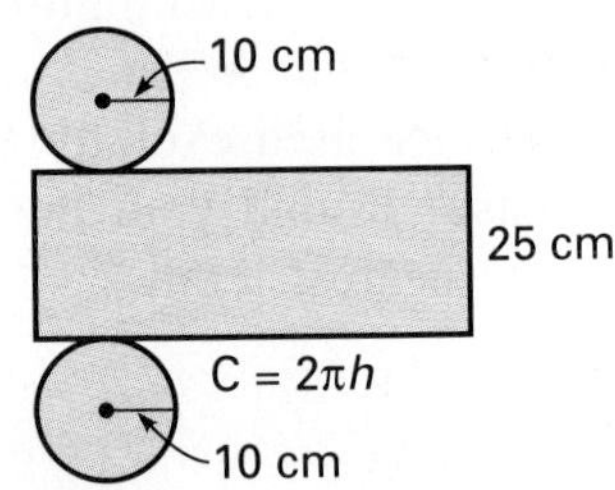

**PROBLEM SOLVING TIP**

A useful way of writing the formula for the surface area of a cylinder is as follows.

$2\pi r(h + r)$

Can you find a different way to write the formula for the surface area of a cylinder?

**area of the curved surface**

$A = 2\pi rh$

$A \approx 2 \times 3.14 \times 10 \times 25$

$A \approx 1{,}570$

**area of each circular base**

$A = \pi r^2$

$A \approx 3.14 \times 100$

$A \approx 314$

There are two circular bases, so: $2 \times 314 = 628$.

The surface area of the cylinder ($SA$) $\approx 1{,}570 + 628 \approx 2{,}198$.

The surface area of the cylinder is approximately equal to 2,198 cm$^2$. ◄

Example 1 illustrates the formula for the surface area of a cylinder.

$$SA = 2\pi rh + 2\pi r^2$$

A cone is a three-dimensional figure with a curved surface and one circular base. The height of a cone is the length of a perpendicular line drawn from its tip to its base. To find the surface area of a cone, you need to know its **slant height** ($s$), the length of a line drawn from its tip to its base along the side of the figure. Then the formula for the surface area of a cone with a radius $r$ is $SA = \pi rs + \pi r^2$.

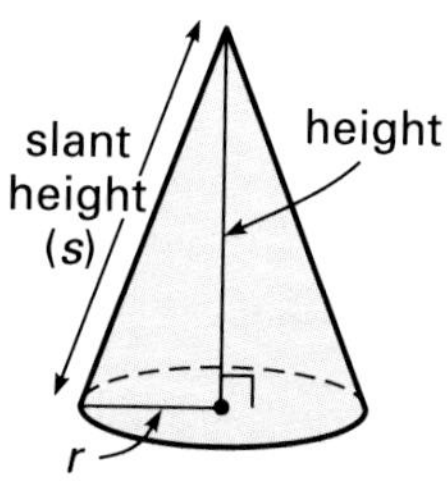


### Example 2

Find the surface area of the cone shown at the right.

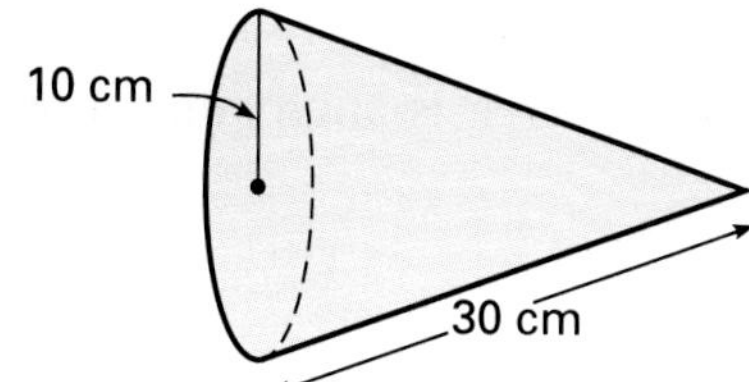


**Solution**

$SA = \pi rs + \pi r^2$
$SA = \pi \times 10 \times 30 + \pi \times 10^2$
$SA \approx 3.14 \times 300 + 3.14 \times 100$
$SA \approx 942 + 314$
$SA \approx 1{,}256$

The surface area of the cone is approximately equal to 1,256 cm$^2$. ◄

The formula for the surface area of a sphere with a radius $r$ is
$SA = 4\pi r^2$.

### Example 3

A jeweler is going to gold plate the surface of a copper sphere. The sphere has a radius of 3.5 cm. What area will the jeweler gold plate? Round your answer to the nearest whole number.

**Solution**

Find the surface area of the sphere.

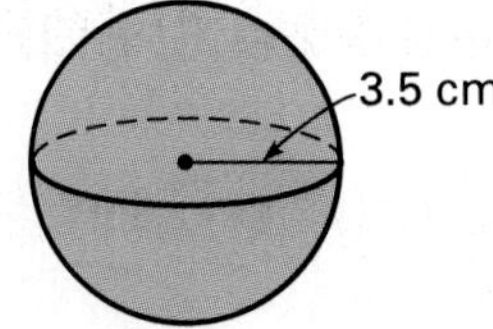


$SA = 4\pi r^2$
$SA \approx 4 \times 3.14 \times (3.5)^2$
$SA \approx 153.9$
$SA \approx 154$

The jeweler will gold plate about 154 cm$^2$. ◄

### Example 4

Find the surface area of the figure at the right. Round your answer to the nearest whole number.

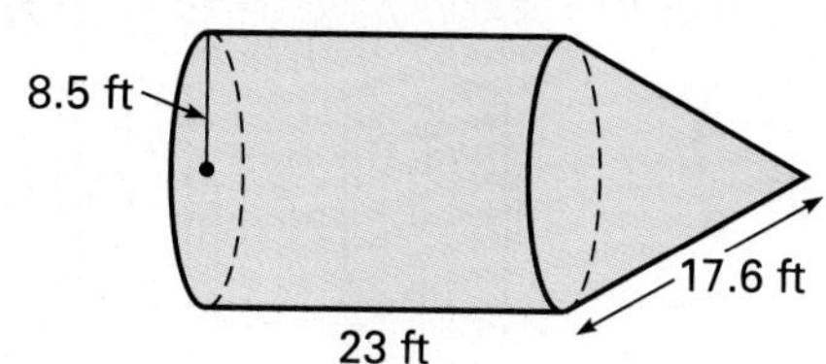

**Solution**

The figure is made up of a cylinder with one base and the curved surface of a cone.

$SA$ = area of circular base + area of curved surface of cylinder + area of curved surface of cone

| **area of base** | **area of curved surface of cylinder** | **area of curved surface of cone** |
|---|---|---|
| $A = \pi r^2$ | $A = 2\pi rh$ | $A = \pi rs$ |
| $A \approx 3.14 \times (8.5)^2$ | $A \approx 2 \times 3.14 \times 8.5 \times 23$ | $A \approx 3.14 \times 8.5 \times 17.6$ |
| $A \approx 226.9$ | $A \approx 1{,}227.7$ | $A \approx 469.7$ |

$SA \approx 226.9 + 1{,}227.7 + 469.7$

$SA \approx 1{,}924$

The surface area is approximately equal to 1,924 ft$^2$. ◄

## TRY THESE

Find the surface area of each figure. Use $\pi \approx 3.14$. Round your answer to the nearest whole number.

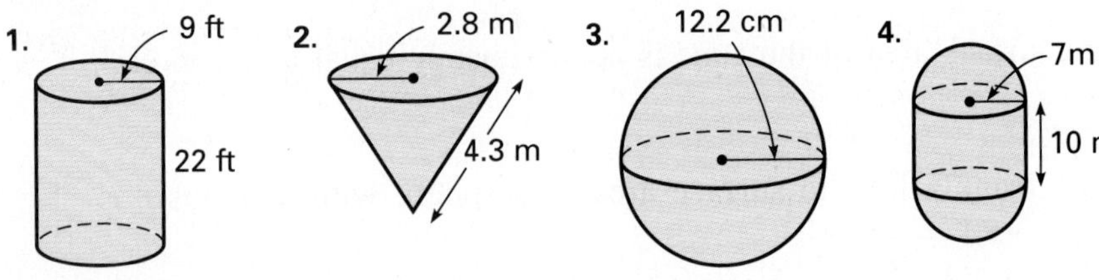


**CALCULATOR TIP**

When you are doing several calculations that involve $\pi$, you can store the value 3.14 in memory. Then, to compute a value such as $49\pi$, enter this key sequence.

49 [x] [MR] [=] 153.86

Some calculators have a special key for $\pi$.

## EXERCISES

**PRACTICE/ SOLVE PROBLEMS**

Find the surface area of each figure. Use $\pi \approx 3.14$. Round your answer to the nearest whole number.

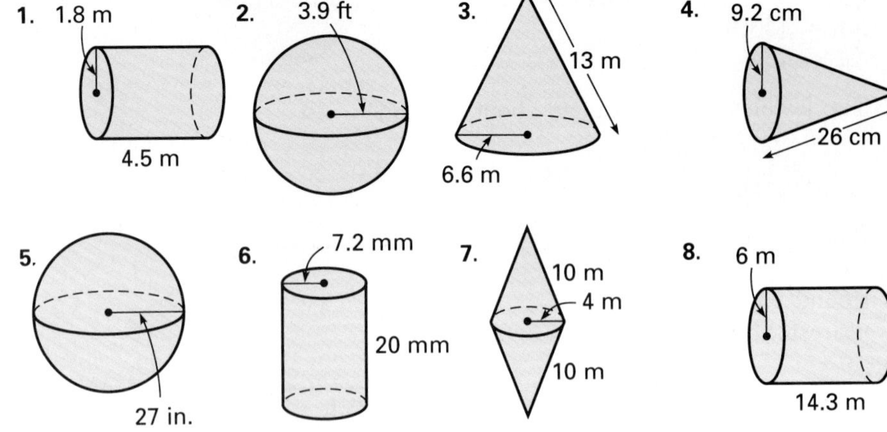

9.

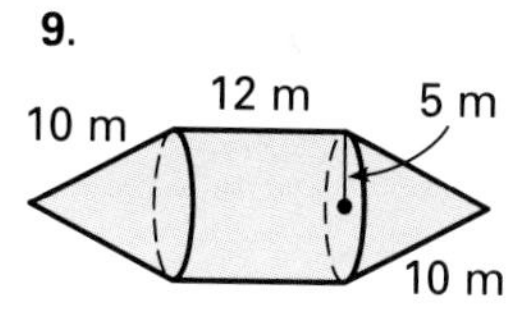


10.

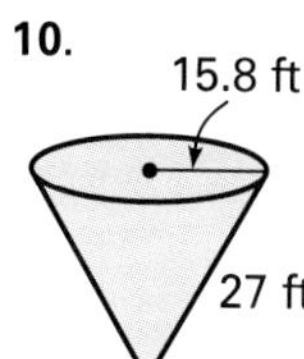


11.

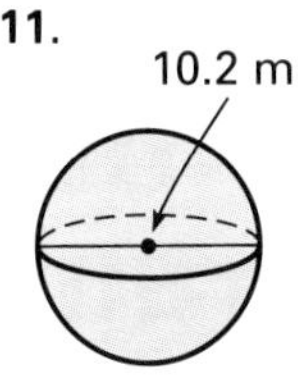


12.

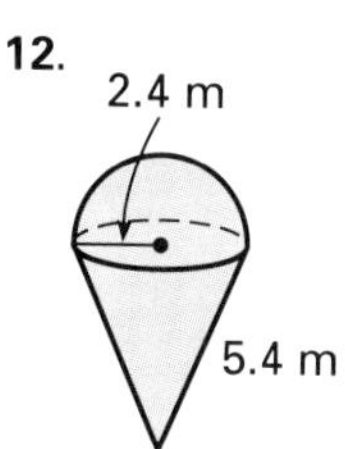

13. The curved surface of a cylindrical can is covered by a paper label. The can has a diameter of 75 mm and a height of 11 cm. What is the area of the paper label to the nearest whole number?

14. A log is 4 m long and has a diameter of 1 m. What is the surface area of the log to the nearest whole number?

## EXTEND/ SOLVE PROBLEMS

Find the surface area of each figure. Use $\pi \approx 3.14$. Round your answer to the nearest whole number.

15.

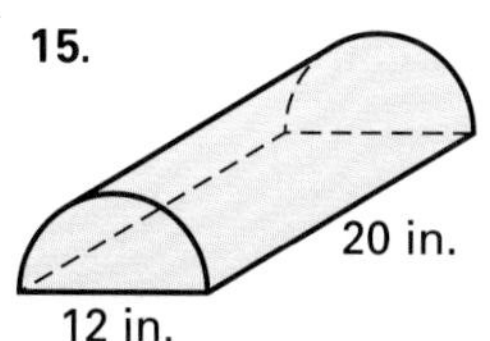


16.

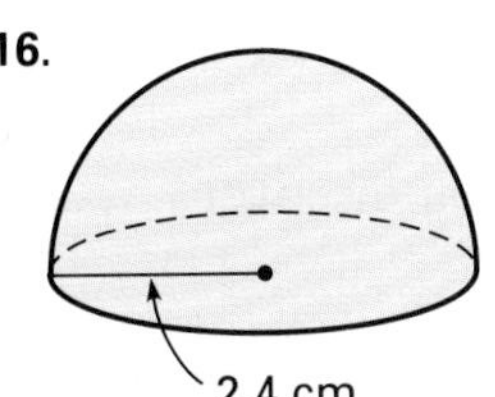


17.

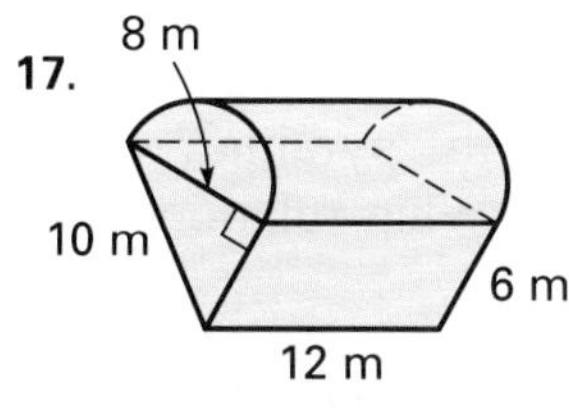


18. The length of the equator is about 25,000 miles. To the nearest million miles, what is the surface area of the earth? Assume that the earth is a sphere. (*Hint:* Use scientific notation.)

19. The surface area of a cone is 440 in.$^2$ The radius of the cone measures 7 in. What is its slant height? (Use $\pi \approx \frac{22}{7}$.)

20. A hole is drilled through a solid cube that has sides of 4 cm. The diameter of the hole is 2 cm. To the nearest whole number, what is the total surface area of the resulting solid figure?

### JUST FOR FUN

Suppose three people each mark a point on the same orange. What do you think the probability is that all the marks will be in the same hemisphere?

Try it out with two friends. You may be surprised to find that the probability is 1! Any three points on a sphere will always be located within the bounds of a hemisphere.

## THINK CRITICALLY/ SOLVE PROBLEMS

21. Write a formula that could be used to find the surface area of a sphere with a circumference of $C$ units.

22. A cone has a height $h$ and a radius $r$. Find an expression for the slant height. Then write a formula for the surface area of the cone using this expression.

23. Which would cause a greater increase in the surface area of a cylinder—doubling the height or doubling the radius of the base? Explain.

# 6-9 Volume of Prisms and Pyramids

**EXPLORE/ WORKING TOGETHER**

You can use models to discover the relationship between the volume of a cube and the volume of a pyramid with the same base and height. Use sheets of heavy paper, a ruler, tape, and centimeter cubes.

a. Use the patterns below to make a cube open at the top and a square pyramid open at the base.

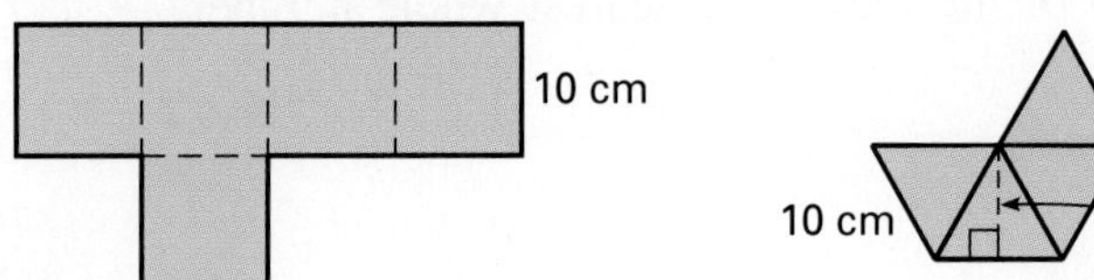


b. Completely fill the cube with centimeter cubes. Record the number of cubes you used.

c. Using the same number of cubes, how many times can you fill the pyramid?

d. What relationship did you discover?

**SKILLS DEVELOPMENT**

Recall that **volume** is a measure of the number of cubic units needed to fill a region of space. To find the volume ($V$) of any prism, multiply the area of the base ($B$) by the height ($h$) of the prism.

volume of a prism: $V = Bh$

### Example 1

Find the volume of each prism shown.

a. 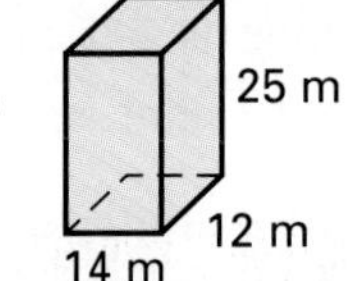


b. 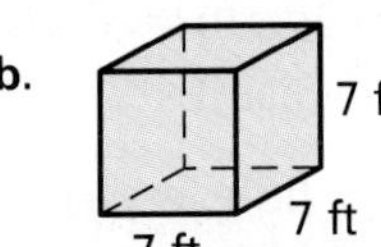


**Solution**

a. The base is a rectangle.

$B = lw$

$B = 14 \times 12 = 168$

Use the volume formula.

$V = Bh$

$V = 168 \times 25$

$V = 4{,}200$

The volume is 4,200 $m^3$.

b. The base is a square.

$B = s^2$

$B = 7^2 = 49$

Use the volume formula.

$V = Bh$

$V = 49 \times 7$

$V = 343$

The volume is 343 $ft^3$. ◄

You can use these formulas to find the volume of a rectangular prism and the volume of a cube with side $s$.

volume of a rectangular prism: $V = lwh$

volume of a cube: $V = s^3$

### Example 2

Find the volume of the triangular prism shown.

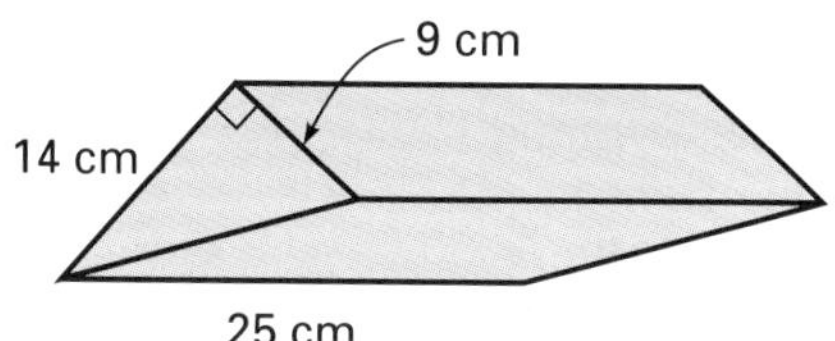


**Solution**

The base is a right triangle.

$B = \frac{1}{2}bh$

$B = \frac{1}{2} \times 9 \times 14$

$B = 63$

Be careful not to confuse the height of the base triangle with the height of the triangular prism.

Use the volume formula.

$V = Bh$

$V = 63 \times 25$

$V = 1{,}575$

The volume is 1,575 cm³. ◄

To find the volume of a pyramid, multiply $\frac{1}{3}$ of the area of its base ($B$) by its height ($h$).

volume of a pyramid: $V = \frac{1}{3}Bh$

### Example 3

Find the volume of the rectangular pyramid shown.

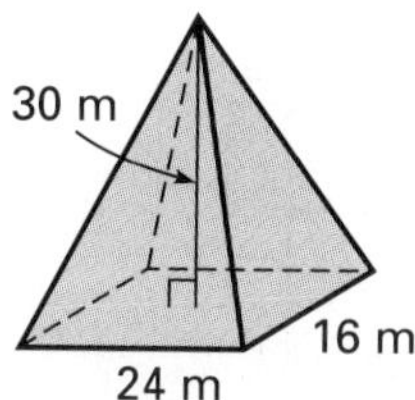


**Solution**

The base is a rectangle.

$B = lw$

$B = 24 \times 16$

$B = 384$

Use the volume formula.

$V = \frac{1}{3}Bh$

$V = \frac{1}{3} \times 384 \times 30$

$V = 3{,}840$

The volume is 3,840 m³. ◄

**CHECK UNDERSTANDING**

Suppose that a prism and a pyramid have the same base and height. The volume of the prism is 240 cm³. What is the volume of the pyramid?

### Example 4

Find the volume of the figure at the right.

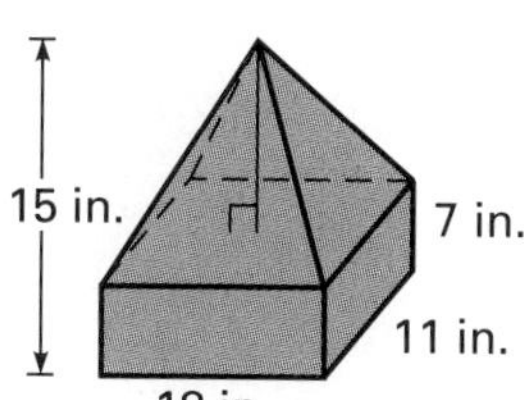


**Solution**

$V$ = volume of the rectangular prism + volume of the pyramid.

**volume of prism**

$V = Bh$

$V = (18 \times 11) \times 7$

$V = 1{,}386$

**volume of pyramid**

$V = \frac{1}{3}Bh$

$V = \frac{1}{3}(18 \times 11) \times 8$

$V = 528$

Think: The height of the pyramid is the height of the figure minus the height of the prism. $15 - 7 = 8$

Total volume = 1,386 + 528 = 1,914

The volume is 1,914 in.³ ◄

## Example 5

The triangular pyramid shown at the right has a volume of 7,038 m³. Find the height of the pyramid.

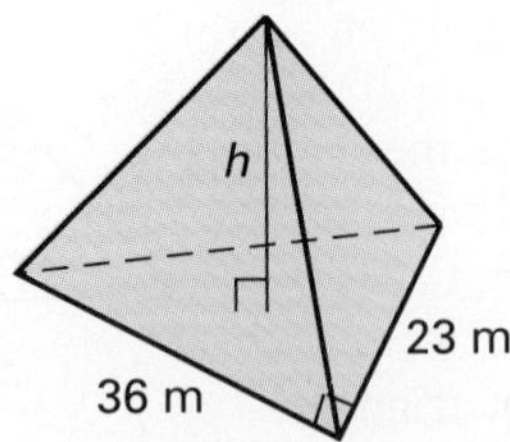

**Solution**

The base is a triangle.

$$B = \frac{1}{2}bh$$
$$B = \frac{1}{2}(36)(23)$$
$$B = 414$$

Substitute in the volume formula.

$$V = \frac{1}{3}Bh$$
$$7{,}038 = \frac{1}{3}(414)h$$
$$7{,}038 = 138h$$
$$51 = h$$

The height of the pyramid is 51 m. ◄

**TALK IT OVER**

A cubical box has edges that each measure 1 m, or 100 cm. Find the volume of the box in cubic meters and in cubic centimeters. How many cubic centimeters are in 1 m³? Express the last answer as a power of 10.

How can you use exponents to express each of the following?

number of cubic inches in 1 ft³
number of cubic feet in 1 yd³
number of cubic millimeters in 1 cm³

## TRY THESE

Find the volume of each figure.

1. 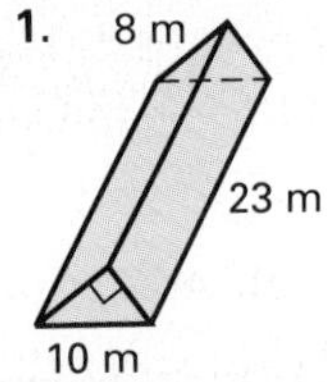


2. 9 mm, 10.3 mm, 18 mm

3. 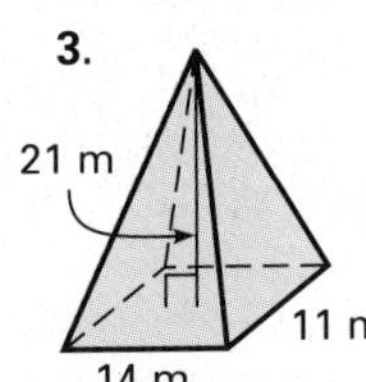


4. 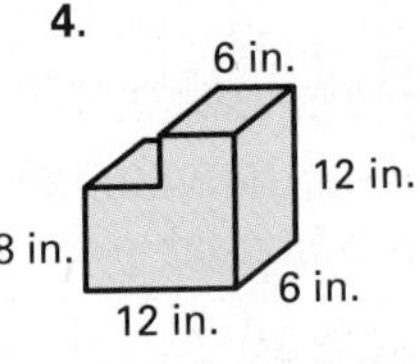


5. A box that is 15 in. high is a prism with a square base. It has a volume of 540 in.³ What is the length of the sides of its base?

# EXERCISES

**PRACTICE/ SOLVE PROBLEMS**

Find the volume of each figure.

1. 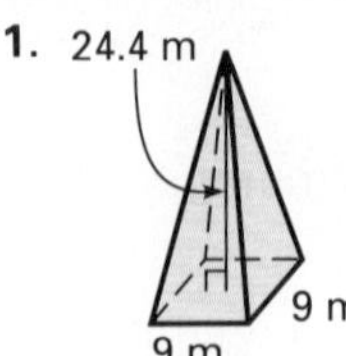


2. 7 m, 6 m, 9.5 m

3. 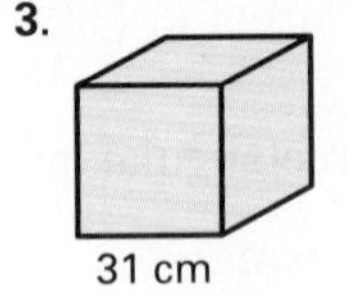


4. 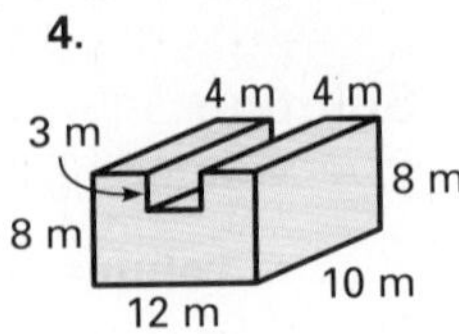


5. 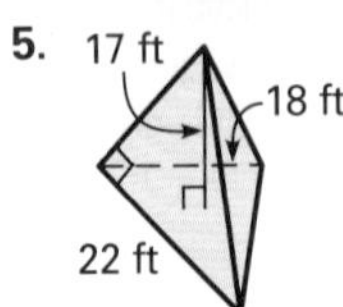


6. 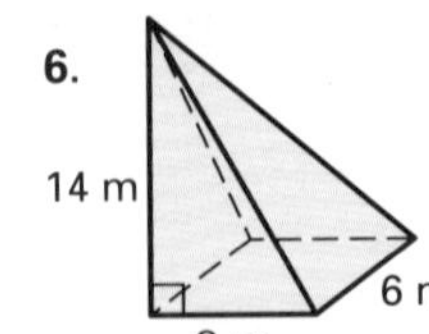


7. 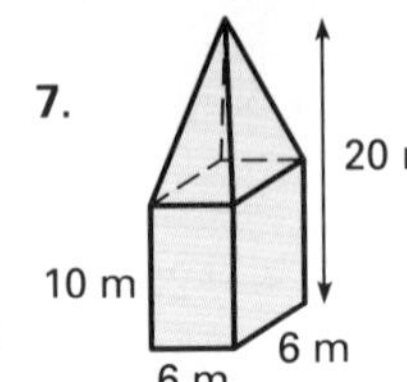


8. 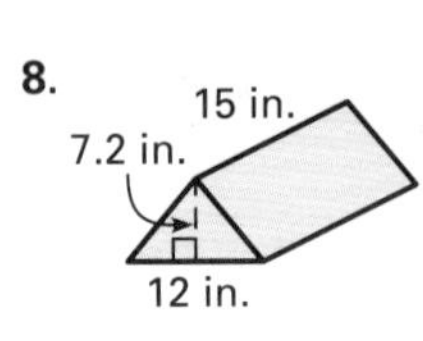

9. The tent shown at the right has a volume of 105 ft³. What is the height of the tent?

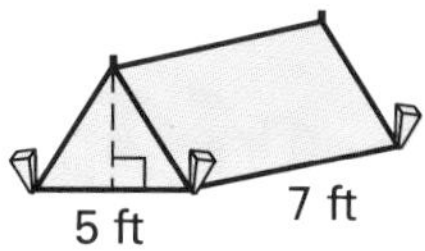


10. A cereal box is 8 in. long and 4 in. wide. Its volume is 368 in.³ What is the height of the box?

## EXTEND/ SOLVE PROBLEMS

Find the volume of each figure.

11. 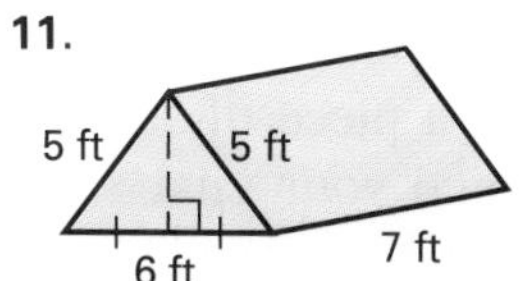


12. 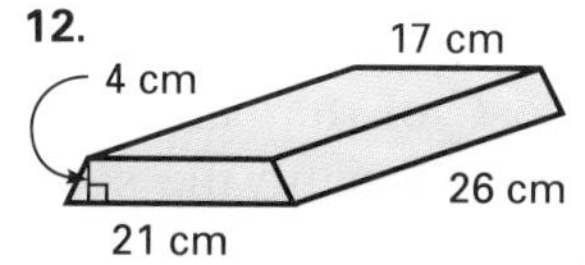


13. 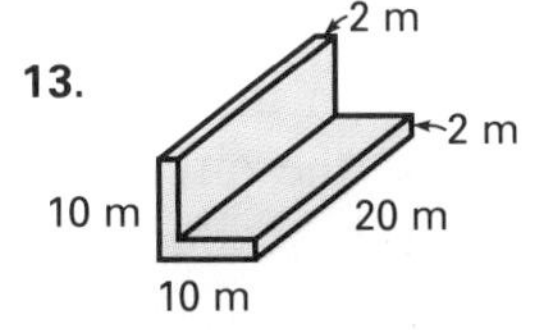

14. A dump truck has a bed that is 12 ft long and 8 ft wide. The walls of the bed are $4\frac{1}{2}$ ft high. When the truck is loaded, no material can be higher than the walls of the bed. If topsoil costs $18 per cubic yard, what is the cost of a truckload of topsoil?

15. When water freezes, its volume increases by about 10%. A tin container that measures 1 ft by 10 in. by 9 in. is exactly half full of water. It is left outside on a winter day, and all of the water freezes. What is the approximate volume of ice in the tin?

16. The sides of a cube each measure 1 ft. If each side were increased by 1 in., by how many cubic inches would the volume of the cube increase?

17. An aquarium has a length of 15 in. and a width of 11 in. A rock put into the aquarium causes the water level to rise by 2 in. The rock is completely submerged. What is the volume of the rock?

## THINK CRITICALLY/ SOLVE PROBLEMS

Suppose you have 8 separate centimeter cubes. The total volume is 8 cm³ and the total surface area is 48 cm². For each exercise, assume that the cubes are arranged to form a single shape (no loose cubes).

18. Arrange 8 cubes to get a surface area of 34 cm². Arrange them to get a surface area of 28 cm². Draw sketches.

19. How can you arrange 8 cubes to get the least possible surface area? What is this surface area? What is the volume?

20. What arrangement do you think will have the least possible surface area for 27 cubes? What is the surface area? What is the volume?

21. Suppose you wanted to make a cardboard box with a volume of 1,000 cm³. What dimensions would you give to the box in order to use the least amount of cardboard?

# 6-10 Volume of Cylinders, Cones, and Spheres

## EXPLORE/ WORKING TOGETHER

A person's **vital capacity** is the measure of the volume of air held in his or her lungs. You can find your vital capacity by taking a deep breath and then exhaling as much air as possible into a balloon.

Push in on the end of the balloon so that it forms a sphere. Then measure the circumference $C$ of the balloon. (Wrap a piece of string around the balloon and then measure the string.) Find your vital capacity by using the formula $VC = \frac{C^3}{6\pi^2}$.

## SKILLS DEVELOPMENT

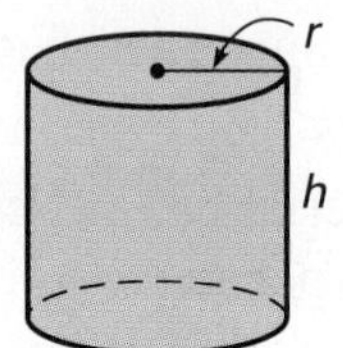


Recall that the general formula for finding the volume of a prism is $V = Bh$, where $B$ is the area of the base. Since the base of a cylinder is circular, replace $B$ in this formula with $\pi r^2$.

volume of a cylinder: $V = B \times h$

$V = \pi r^2 \times h$, or $V = \pi r^2 h$

### Example 1

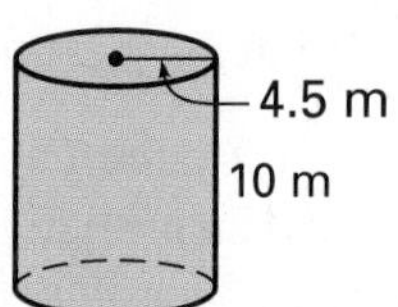


Find the volume of the cylinder at the right.

**Solution**

$V = \pi r^2 h$

$V \approx 3.14 \times (4.5)^2 \times 10$ ← **Substitute 4.5 for *r* and 10 for *h*.**

$V \approx 635.85$

The volume of the cylinder is approximately 635.85 $m^3$. ◀

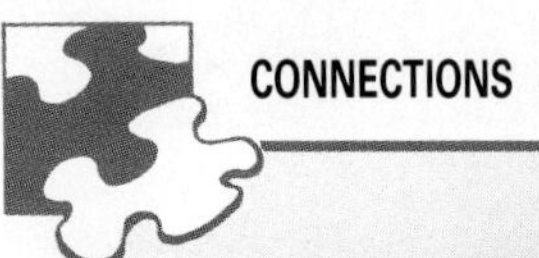

**CONNECTIONS**

How is the formula for the volume of a cone similar to the formula for the volume of a pyramid?

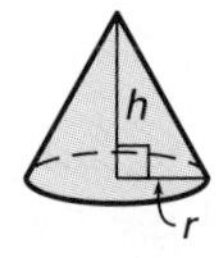


Note how the formula for the volume of a cone is related to the formula for the volume of a cylinder.

volume of a cone: $V = \frac{1}{3}\pi r^2 h$

The volume of a cone with a given radius and height is $\frac{1}{3}$ the volume of a cylinder with the same radius and height.

### Example 2

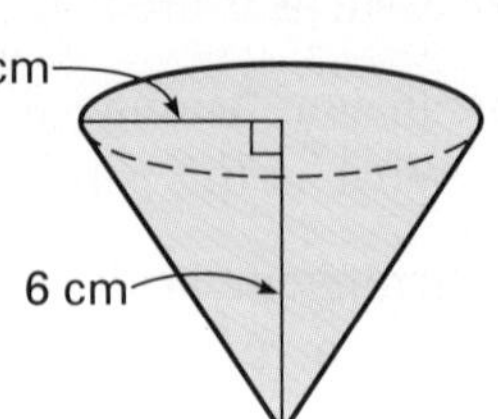


Find the volume of the cone at the right.

**Solution**

$V = \frac{1}{3}\pi r^2 h$

$V \approx \frac{1}{3} \times 3.14 \times 4^2 \times 6$ ← **Substitute 4 for *r* and 6 for *h*.**

$V \approx 100.48$

The volume of the cone is approximately 100.48 $cm^3$. ◀

Use this formula for the volume of a sphere: $V = \frac{4}{3}\pi r^3$

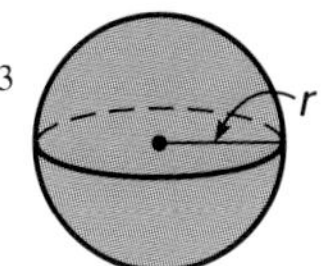


### Example 3

The diameter of the planet Pluto is about 1,500 mi. What is its volume?

**Solution**

Assume that Pluto is a sphere. Its radius is half of 1,500 mi, or 750 mi.

$$V = \frac{4}{3}\pi r^3$$

$$V \approx \frac{4}{3} \times 3.14 \times (750)^3 \approx 1{,}766{,}250{,}000$$

The volume of Pluto is approximately 1,766 million mi³. ◄

### Example 4

Find the volume of the figure at the right.

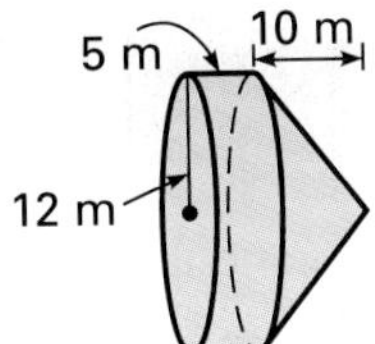


**Solution**

The volume is the sum of the volumes of a cylinder and a cone.

| **volume of cylinder** | **volume of cone** |
|---|---|
| $V = \pi r^2 h$ | $V = \frac{1}{3}\pi r^2 h$ |
| $V \approx 3.14 \times 12^2 \times 5$ | $V \approx \frac{1}{3} \times 3.14 \times 12^2 \times 10$ |
| $V \approx 2{,}260.8$ | $V \approx 1{,}507.2$ |

Total volume $\approx 2{,}260.8 + 1{,}507.2 \approx 3{,}768$

The volume of the figure is approximately 3,768 m³. ◄

**CALCULATOR TIP**

In Example 3, you may find that the product has more digits than a calculator can display.

One way to use a calculator to find $\frac{4}{3} \times 3.14 \times (750)^3$ is as follows.

First, find $\frac{4}{3} \times 3.14 \times 750^2$.

$\frac{4}{3} \times 3.14 \times 750^2 = 2{,}355{,}000$

Divide the number by 1,000.

$2{,}355{,}000 \div 1{,}000 = 2{,}355$

Multiply the result by 750.

$2{,}355 \times 750 = 1{,}766{,}250$

Since you divided by 1,000, you now must multiply by 1,000 to get the actual answer.

$1{,}766{,}250 \times 1{,}000 = 1{,}766{,}250{,}000$

You can use this method in many cases where you have to find large products.

## TRY THESE

Find the volume of each. Round answers to the nearest whole number.

1. 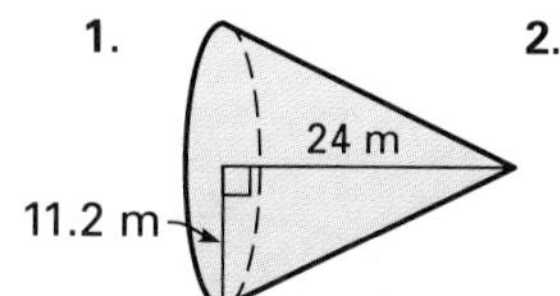


2. 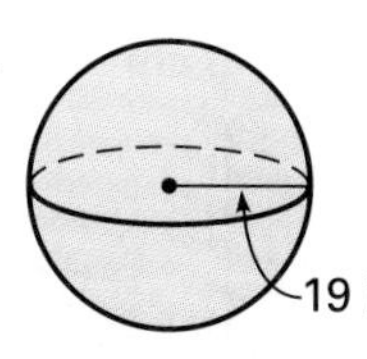


3. 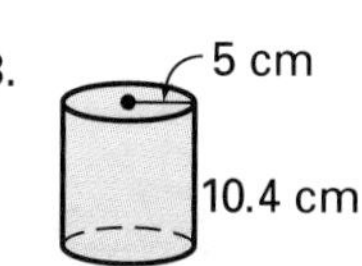


4. 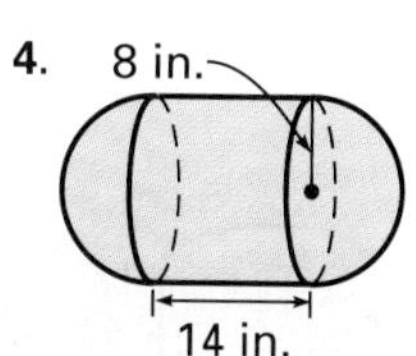


# EXERCISES

**PRACTICE/ SOLVE PROBLEMS**

Find the volume of each. Round answers to the nearest whole number.

1. 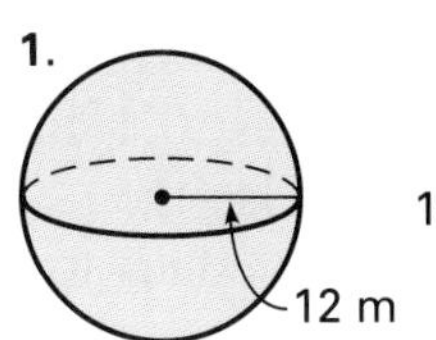


2. 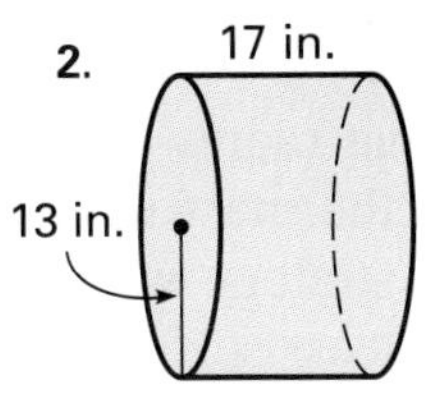


3. 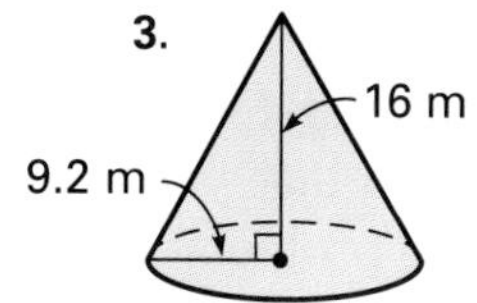


4. 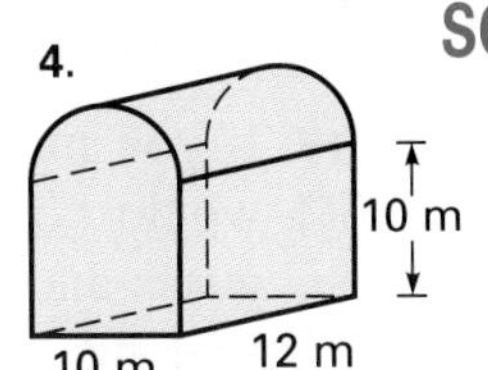

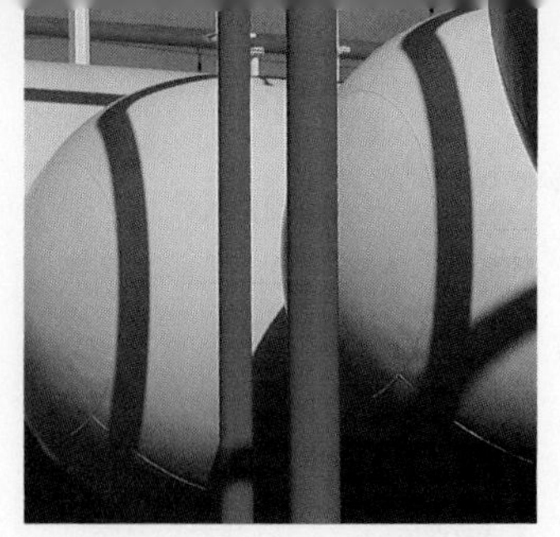

5. A gasoline storage tank is a cylinder with a radius of 10 ft and a height of 6 ft. How many cubic feet of gasoline will it hold?

6. A sphere has a volume of $523\frac{1}{3}$ in.$^3$ What is the radius of the sphere?

## EXTEND/ SOLVE PROBLEMS

Find the volume of each. Round answers to the nearest tenth.

7.

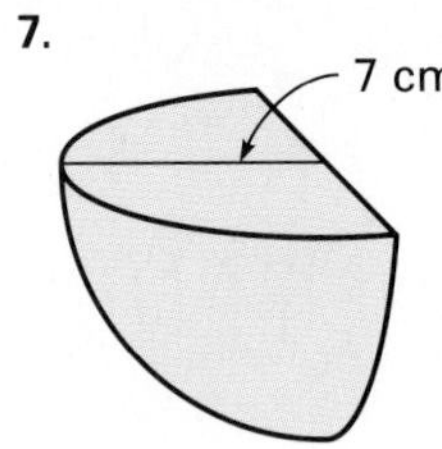


8.

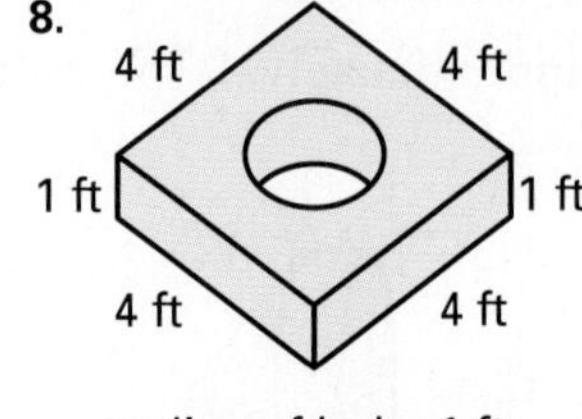


radius of hole: 1 ft

9.

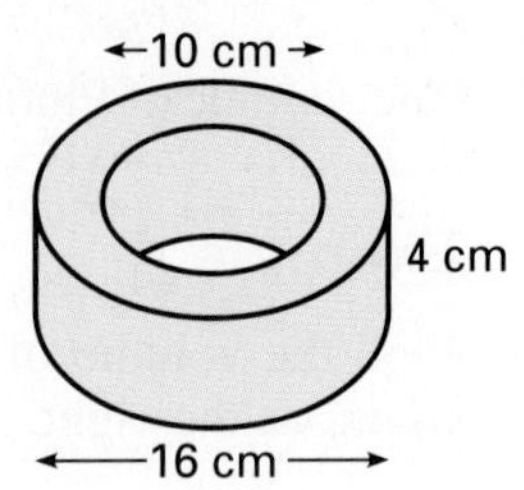


10. Cork weighs 15 lb per cubic foot. How much does a cork sphere with a radius of 6 in. weigh?

11. A sphere with a radius of 10 in. fits exactly into a cubical box. What percent of the volume of the box is not taken up by the sphere?

12. A chemist pours 188.4 cm$^3$ of a liquid into a glass cylinder that has a radius of 2 cm. How many centimeters deep is the liquid?

## THINK CRITICALLY/ SOLVE PROBLEMS

The diagrams of the cans shown below each has a volume of 1,000 cm$^3$. (1,000 cm$^3$ = 1 L) Use these diagrams for Exercises 13–15.

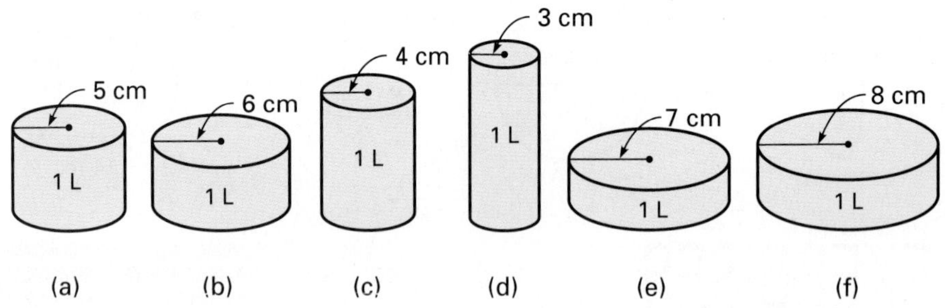


(a) (b) (c) (d) (e) (f)

13. Find the height of each can.

14. Which two cans have the least surface area?

15. If you manufactured 1-L cans, what radius (to the nearest 0.1 cm) would you use so that you saved the most material? (Use the guess-and-check problem solving strategy with several values to find the radius that results in the least surface area.)

- ► READ
- ► PLAN
- ► SOLVE
- ► ANSWER
- ► CHECK

# Problem Solving Applications:

## RAINFALL MEASUREMENTS

Meteorologists measure rainfall using an instrument called a **rain gauge**—an open-topped cylinder placed in an unsheltered location so that rain collects in it. The gauge is calibrated to show the amount of rain, in inches, that has fallen. The term *one inch of rain* means that the average depth of rainfall in an area was 1 in.

**SOME RECORD RAINFALLS FOR 24-HOUR PERIODS**

| Location | Date | Rainfall (inches) |
|---|---|---|
| La Reunion, Indian Ocean | 3/15/52 | 73.62 |
| Kilauea Plantation, Hawaii | 1/24/56 | 38.00 |
| Alvin, Texas | 7/25/79 | 43.00 |
| Angoon, Alaska | 10/12/82 | 15.20 |
| Yankeetown, Florida | 9/5/50 | 38.70 |

If 1 in. of rain fell over an area of 1 $mi^2$, how much water would that be in cubic feet? How much water would that be in gallons? What would be the weight (in tons) of that amount of water?

1. To find how much water would fall over an area of 1 $mi^2$, you could start by thinking about the problem in this way.
   $$1 \text{ mi}^2 = 5{,}280 \text{ ft} \times 5{,}280 \text{ ft or } 27{,}878{,}400 \text{ ft}^2$$
   To find the amount of water that would fall on an area this size, multiply the area by $\frac{1}{12}$ (1 in. = $\frac{1}{12}$ ft). This will give the volume of rain in cubic feet ($ft^3$). Explain your answer.
2. To find the amount of rain in gallons, you need to know that 1 $ft^3$ = 7.48 gal. So, how many gallons of water would fall on an area of 1 $mi^2$ if that area received 1 in. of rain?
3. One gallon of water weighs about 8.3 lb. How much would all the water that falls over an area of 1 $mi^2$ during a 1-in. rainfall weigh? How can you express that weight in tons? Round answers to the nearest whole number.

Use the data in the table at the top of the page that lists some record rainfalls for 24-hour periods.

4. Find the volume of water, in cubic feet, that fell over an area of 1 $mi^2$ of Kilauea Plantation, Hawaii, and over an area of 1 $mi^2$ of Alvin, Texas.
5. Find the volume of water, in gallons, that fell over an area of 1 $mi^2$ of Angoon, Alaska. Round your answer to the nearest whole number.
6. Find the weight of water, in tons, that fell over an area of 1 $mi^2$ of Yankeetown, Florida. Round your answer to the nearest whole number.

### MIXED REVIEW

1. Find 24% of 100.
2. What percent of 48 is 42?
3. 150% of what number is 51?
4. Express $2.63 \times 10^{-5}$ in standard form.
5. Express 370,800,000 in scientific notation.

Find the measure of each.

6. ∠BOC
7. ∠COD
8. ∠DOA

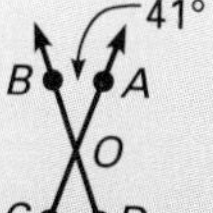


Find the first quartile, median, and third quartile.

9. 4.3, 4.9, 5.1, 4.9, 5.2, 5.2, 5.3, 5.4, 5.3, 5.3, 5.4, 5.8, 5.6, 5.7, 5.6, 5.8, 6.9, 5.9, 6.1, 6.2
10. 66, 85, 67, 85, 68, 87, 75, 87, 77, 77, 90, 90, 82, 83, 92, 92, 95, 84, 92, 79

# 6-11 Problem Solving Strategies:

## CHOOSE A LENGTH, AREA, OR VOLUME FORMULA

- READ
- PLAN
- SOLVE
- ANSWER
- CHECK

You can often use a formula to solve problems that involve geometric figures. But first you must decide which formula is the right one for the problem. You may need to use a formula to find a **length,** such as the perimeter or circumference of a garden, an **area,** such as the size of a field, or a **volume,** such as the amount a container will hold.

### PROBLEM

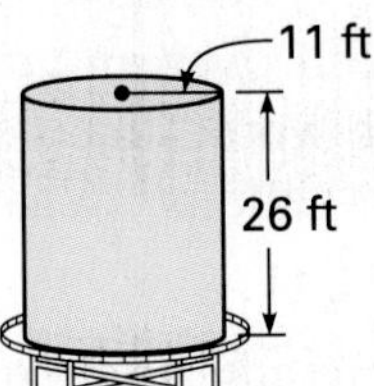


The tank of a cylindrical water tower is 26 ft high and has a radius of 11 ft. Around the base of the tank is a walkway 2 ft wide.

**a.** About how many gallons of paint would it take to paint the exterior of the tank, if 1 gal of paint will cover 300 $ft^2$?

**b.** There is a railing around the walkway. How long is it?

**c.** How many gallons of water does the tank hold? (The volume of 1 gal is approximately 0.1337 $ft^3$.)

### SOLUTION

**a.** Each gallon of paint covers an **area** of 300 $ft^2$. Use the formula for the surface area of a cylinder.

$$SA = 2\pi r^2 + 2\pi rh$$
$$SA \approx (2 \times 3.14 \times 11^2) + (2 \times 3.14 \times 11 \times 26)$$
$$SA \approx 2{,}556$$

The surface area is approximately 2,556 $ft^2$.

number of gallons of paint $= 2{,}556 \div 300 \approx 8.5$

It would take 8.5 gal of paint to paint the exterior of the tank.

**b.** The **length** of the railing is the circumference of a circle with a radius 2 ft greater than the radius of the tank. Use the formula for the circumference of a circle.

$$C = 2\pi r$$
$$C \approx 2 \times 3.14 \times (11 + 2) \approx 81.6$$

The railing is approximately 81.6 ft long.

**c.** The number of gallons the tank will hold is the **volume** of the tank. Use the formula for the volume of a cylinder.

$$V = \pi r^2 h$$
$$V \approx 3.14 \times 11^2 \times 26 \approx 9{,}878.4$$

volume in gallons $\approx 9{,}878.4 \div 0.1337 \approx 73{,}884.8$

The tank holds approximately 73,884.8 gal.

**CHECK UNDERSTANDING**

In question **a** of the problem you are asked to find the number of gallons of paint it would take to paint the exterior of the tank. A gallon is a unit of volume, so why wouldn't you use a formula for volume to answer the question?

# PROBLEMS

Use $\pi \approx 3.14$. Round answers to the nearest tenth.

1. A Montezuma cypress tree has a radius of $18\frac{3}{4}$ ft. If it were fenced in so that there was a border 8 ft wide around the trunk of the tree, how many feet of fencing would be needed?
2. South African ironwood is the world's heaviest wood, weighing about 90 lb per cubic foot. How much would an 8-in. cube of ironwood weigh?
3. How many square inches of paper are needed for the label of a cylindrical soup can that is 7 in. high and has a radius of $2\frac{1}{4}$ in.?
4. The windows of a building have a total area of 32.4 $m^2$. Each window is a rectangle measuring 1.2 m by 1.8 m. How many windows are there in the building?
5. A pound of grass seed covers an area of 250 $ft^2$. How many pounds of seed would you need for a rectangular lawn measuring 35 ft by 50 ft?
6. The roof of a shed is a square pyramid with sides of 2.4 m. The height of each triangular face is 2 m. How many square feet of tar paper would it take to cover the roof?
7. A circular swimming pool has a diameter of 17 ft and a depth of 6 ft. If the water level is 1 ft below the rim, how many cubic feet of water are in the pool?
8. A cardboard Halloween hat is made up of a cone with radius 5 in. and slant height 14 in. and a 4-in.-wide circular brim. How many square inches of cardboard were used to make the hat?

You can use the program below to find the surface area and volume of a cylinder and a cone of the same height *h*, and each having a base with radius *r*.

```
10 PRINT "WHAT IS THE
   RADIUS OF THE BASE
   OF THE CYLINDER?"
   :INPUT R
20 PRINT "WHAT IS THE
   HEIGHT OF THE
   CYLINDER?":INPUT H
30 PRINT "THE SURFACE
   AREA OF THE CYLINDER
   IS ABOUT ";2 * 3.14 *
   R ^ 2 + 2 * 3.14 * R * H
   ;"."
40 PRINT "THE VOLUME OF
   THE CYLINDER IS ABOUT
   ";3.14 * R ^ 2 * H ;"."
50 PRINT "THE SURFACE
   AREA OF THE CONE IS
   ABOUT ";3.14 * R ^ 2 +
   SQR(H ^ 2 + R ^ 2) *
   3.14 ;"."
60 PRINT "THE VOLUME OF
   THE CONE IS ABOUT "
   ;(3.14 * R ^ 2 * H)/3 ;"."
```

***USING DATA*** Lumber is cut to certain standard sizes, but when the lumber is dried and planed, its dimensions decrease. The names of the sizes of boards specify the dimensions *before* they are dried and planed. Refer to the Data Index on pp. 556–557 to find the table listing the actual dimensions of finished lumber.

9. A patio is built using 15 one-by-ten boards that are each 12 ft long. How many square feet is the patio?
10. A steel band binds a stack of two-by-fours that is 8 boards wide and 12 boards high. There are two possible lengths for the steel band, depending on how the boards are arranged. What are they?
11. A lumber yard has 124 four-by-six boards that are each 10 ft long. What is the minimum amount of space that the lumber yard needs to store the boards? Round your answer to the nearest whole number.

# 6-12 Precision in Measurement

## EXPLORE/ WORKING TOGETHER

What is the smallest division on your ruler? In metric units, it is probably 1 mm. In customary units, it is probably $\frac{1}{16}$ in. or $\frac{1}{32}$ in.

Use your ruler to draw a line segment, and then mark two points on the segment with a pencil. Measure the distance between the two points as accurately as you can, using the smallest units shown on your ruler. Record the measurement on a separate piece of paper.

Now exchange papers with a classmate and measure each other's distance between points. Compare your measurements.

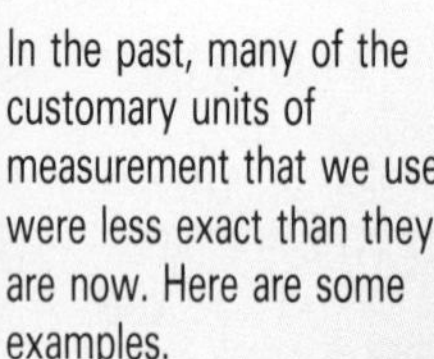

**MATH: WHO, WHERE, WHEN**

In the past, many of the customary units of measurement that we use were less exact than they are now. Here are some examples.

**foot:** the length of a human foot
**inch:** the length of three grains of barley placed end to end
**mile:** the length of one thousand paces
**acre:** the area of land that could be plowed by a team of oxen in one day

Nowadays, units of measurement are defined quite precisely. The length of a meter is defined as being equal to 1,650,763.73 wavelengths of the orange-red light of excited krypton of mass number 86!

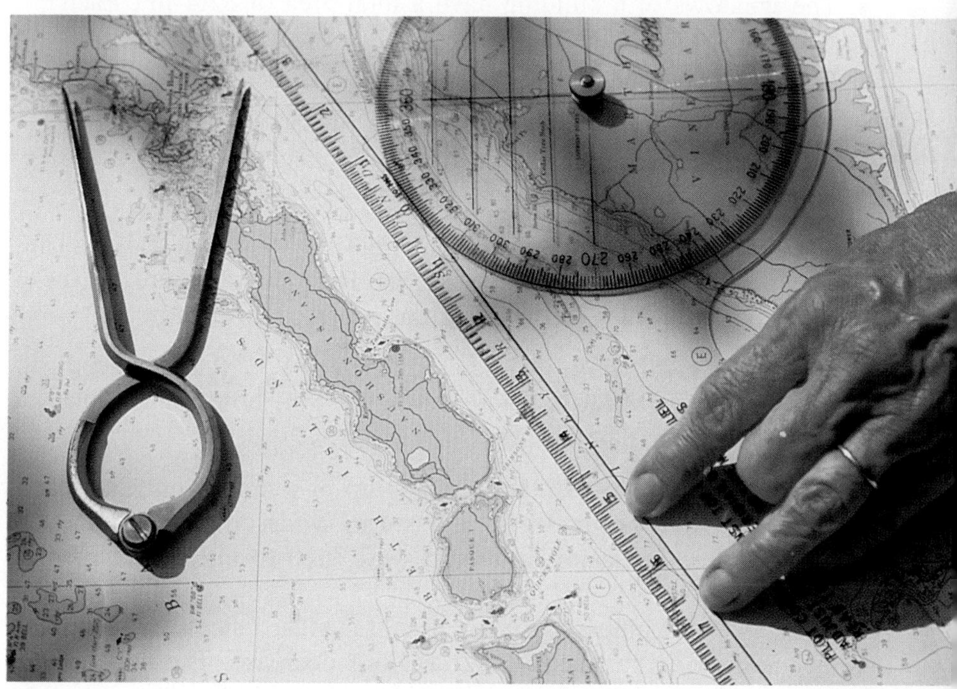

## SKILLS DEVELOPMENT

The precision of a measurement is determined by the smallest unit on the measuring instrument. The smallest unit is called the **unit of precision.** When you record a measurement, the unit of precision is indicated by the place value of the last **significant digit** of the measurement. You can recognize which digits of a measurement are significant by following these guidelines.

Digits other than zero are always significant.
A zero is significant only under the following conditions.
    If it is between two significant digits, as in 2.09.
    If it is the last digit of a decimal, as in 4.0.
    If it is underlined, as in 1,$\underline{0}$00.

### Example 1

Give the number of significant digits in each measurement. What is the last significant digit?

a. 24.9 cm  b. 306.0 m

c. 830 kg  d. 0.06 m

**Solution**

a. None of the digits is a zero, so the digits are all significant. There are three significant digits. The last one is 9.

b. The final zero is significant because it is the last digit of a decimal. The first zero is also significant because it is between significant digits. There are four significant digits. The last significant digit is the final 0.

c. The final zero is not significant. There are two significant digits. The last significant digit is 3.

d. The zeros are not significant. There is one significant digit. The last significant digit is 6. ◄

### Example 2

Give the unit of precision for each measurement.

a. 50.0 cm  b. 15,0̲00 m  c. 0.704 L

**Solution**

a. The last significant digit is the final 0. The unit of precision is 0.1 cm.

b. The last significant digit is the underlined 0. The unit of precision is 100 m.

c. The last significant digit is the 4. The unit of precision is 0.001 L. ◄

**CHECK UNDERSTANDING**

A measurement of 240 cm on a ruler marked in 1-cm units has 2, 4, and 0 as significant digits because there are 240 1-cm units in the measurement. If the ruler were marked in 10-cm units, what would be the significant digits of a 240-cm measurement?

When you add or subtract with measurements, your answer should be rounded to the place of the *least* precise measurement.

### Example 3

Solve. Round your answer to the correct unit of precision.

a. 41.89 m + 6.3 m  b. 12,00̲0 ft − 1,234 ft

**Solution**

a.
```
   41.89   ← precise to 0.01 m
 +  6.3    ← precise to 0.1 m
 -------
   48.19
```

The unit of least precision is 0.1 m. Round the answer to the nearest tenth of a meter: 48.2 m.

b.
```
   12,000   ← precise to 10 ft
 −  1,234   ← precise to 1 ft
 --------
   10,766
```

The unit of least precision is 10 ft. Round the answer to the nearest ten feet: 10,770 ft. ◄

## TRY THESE

Give the number of significant digits in each measurement.

1. 17.00 cm
2. 600 gal
3. 5.01 m
4. 0.002 kg

Give the unit of precision for each measurement.

5. $54,\underline{0}00$ m
6. 0.030 km
7. 31,508 mi
8. 7.009 L

Solve. Round your answer to the correct unit of precision.

9. 4,540 lb + 700 lb
10. 113.56 km − 69.019 km

# EXERCISES

### PRACTICE/ SOLVE PROBLEMS

Give the number of significant digits in each measurement.

1. 0.032 m
2. 0.030 m
3. 17.001 km
4. 380 lb
5. 10,000 mi
6. $6,8\underline{0}0$ gal
7. 0.008 m
8. 500.1 L

Give the unit of precision for each measurement.

9. 7.0 m
10. 7,200 mi
11. 1.010 L
12. 85 mL
13. 320 g
14. 0.0035 km
15. $15,40\underline{0}$ mi
16. 14.05 cm

Solve. Round your answer to the correct unit of precision.

17. 2,502 km + 494 km
18. 371 h − 200 h

19. A redwood tree from the Pacific Coast reaches about 250 ft in height. A young redwood is 56 ft tall. About how much taller will it grow? Round your answer to the correct unit of precision.
20. North Carolina has a total area of about 52,700 $mi^2$. South Carolina has a total area of about 31,000 $mi^2$. What is the combined area of the two states? Round your answer to the correct unit of precision.

## EXTEND/ SOLVE PROBLEMS

Find the less precise measurement in each pair.

21. 6,280 mi and 11 mi
22. 5.0 g and 158 g
23. 3,000 g and 2,600 g
24. 48.44 m and 0.002 km
25. 20.0 L and 0.10 L
26. 18.0 cm and 180 mm

Use the data in the table for Exercises 27–29. Round your answers to the correct unit of precision.

**UTAH'S NATIONAL PARKS**

| Park | Area |
|---|---|
| Arches National Park | 73,100 acres |
| Bryce Canyon National Park | 35,835 acres |
| Canyonlands National Park | 337,570 acres |
| Capitol Reef National Park | 240,000 acres |
| Zion National Park | 146,600 acres |

27. What is the combined area of Capitol Reef and Zion national parks?
28. How much greater is the area of Arches than Bryce Canyon?
29. What is the total area of the national parks in Utah?

## THINK CRITICALLY/ SOLVE PROBLEMS

30. Many digital clocks display only hours and minutes. The minute display changes every 60 seconds. For example, if the actual time is 10 h 41 min 01 s, the clock displays 10:41. If the actual time is 10 h 41 min 59 s, the clock still displays 10:41. In the last case, the clock does not display the time to the nearest minute, which is 10:42.

    Assume that there is a telephone time service that tells you the exact time every 10 seconds. How could you set a digital clock so that its display was precise to the nearest minute?

31. In a report about Saturn, Eric wrote that the diameter of Saturn's large moon called Titan is 5,150 km, while the diameter of the tiny moon, Mimas, is only 329 km. Eric compared the moons by stating that the diameter of Titan is 4,821 km greater than the diameter of Mimas. Was Eric's statement scientifically correct? Explain.

# CHAPTER 6 • REVIEW

Complete each formula at the left, and match it with the letter of the correct figure in the box. A letter may be chosen more than once.

| | | |
|---|---|---|
| 1. ■ $= 2\pi r$ | 4. ■ $= \pi rs + \pi r^2$ | a. pyramid   e. cone |
| 2. ■ $= \frac{4}{3}\pi r^3$ | 5. ■ $= \frac{1}{2}bh$ | b. sphere   f. prism<br>c. circle   g. trapezoid |
| 3. ■ $= 2\pi r^2 + 2\pi rh$ | 6. ■ $= \frac{1}{3}\pi r^2 h$ | d. triangle   h. cylinder |

## SECTION 6–1 THE PYTHAGOREAN THEOREM (pages 216–219)

► For all right triangles, $c^2 = a^2 + b^2$, where $c$ is the length of the hypotenuse and $a$ and $b$ are the lengths of the legs.

7. What is the length of the hypotenuse of a right triangle whose legs measure 13 cm and 6 cm? Round your answer to the nearest tenth.

## SECTIONS 6–2 AND 6–4 PERIMETER AND AREA (pages 220–223, 226–229)

► The perimeter of a polygon is the sum of the lengths of its sides.

► Use these formulas to find the areas of these polygons.

rectangle: $A = lw$    parallelogram: $A = bh$

triangle: $A = \frac{1}{2}bh$    trapezoid: $A = \frac{1}{2}h(b_1 + b_2)$

Find the perimeter and area of each figure.

8. 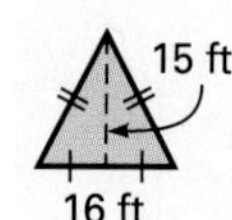


9. 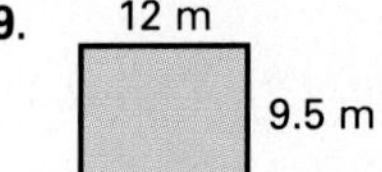


10. 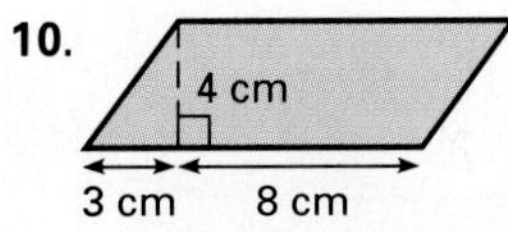


## SECTION 6–5 CIRCUMFERENCE AND AREA OF CIRCLES (pages 230–233)

► To find the circumference of a circle, use the formula $C = 2\pi r$.

► To find the area of a circle, use the formula $A = \pi r^2$.

11. Find the circumference of the figure and the area of the shaded region. Use $\pi \approx \frac{22}{7}$.

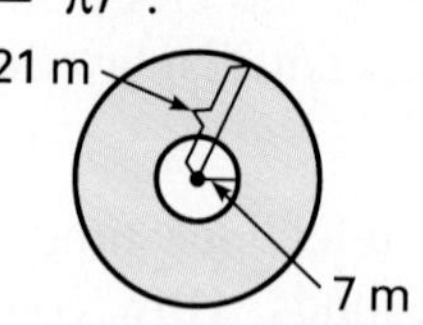


## SECTION 6–6 AREA AND PROBABILITY (pages 234–237)

► To find the probability of a random point in a figure being within a given region, divide the area of the given region by the total area of the figure.

12. A paper airplane glides into a 100-ft-square park. What is the probability that it lands in a circular pond that has a radius of 20 ft?

## SECTIONS 6–7 AND 6–8 SURFACE AREA (pages 238–245)

► Surface areas can be found using these formulas.

rectangular prism: $SA = 2(lw + lh + wh)$ sphere: $SA = 4\pi r^2$

cylinder: $SA = 2\pi r^2 + 2\pi rh$ cone: $SA = \pi rs + \pi r^2$

Find the surface area of each figure to the nearest tenth. Use $\pi \approx 3.14$.

**13.** 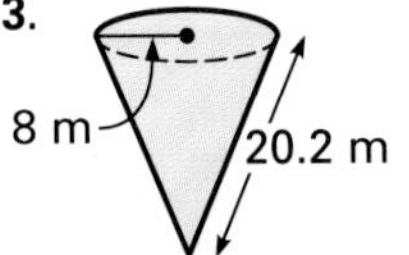


**14.** 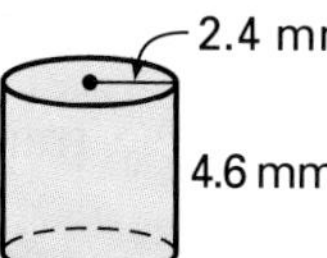


**15.** 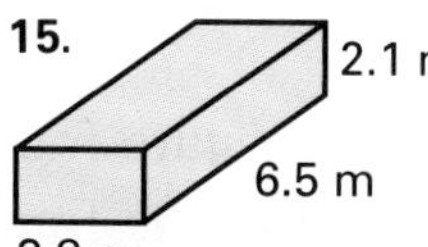


**16.** 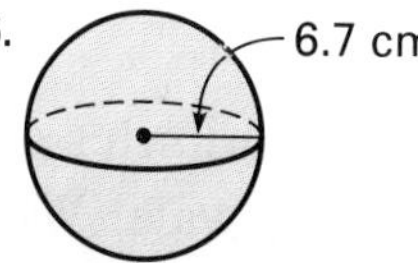


## SECTIONS 6–9 AND 6–10 VOLUME (pages 246–253)

► Volumes can be found using these formulas.

prism: $V = Bh$ pyramid: $V = \frac{1}{3}Bh$ cylinder: $V = \pi r^2 h$

cone: $V = \frac{1}{3}\pi r^2 h$ sphere: $V = \frac{4}{3}\pi r^3$

Find the volume of each figure to the nearest tenth. Use $\pi \approx 3.14$.

**17.** 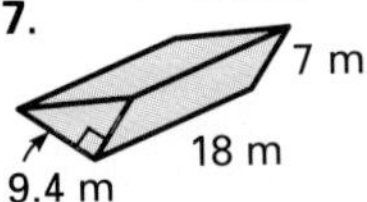


**18.** 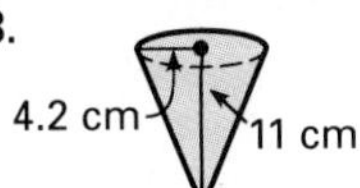


**19.** 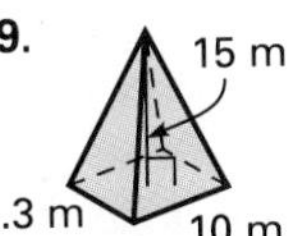


**20.** 6.1 mm 3.8 mm

## SECTIONS 6–3 AND 6–11 PROBLEM SOLVING (pages 224–225, 254–255)

► You may need to use a length, area, or volume formula to solve a problem.

**21.** A rectangular room measures 12 yd by 22 ft. How many square yards of carpet will it take to carpet the entire room?

## SECTION 6–12 PRECISION IN MEASUREMENT (pages 256–259)

► The place value of the last significant digit of a measurement indicates the precision of the measurement.

**22.** Find the number of significant digits and the unit of precision of 0.205 m.

***USING DATA*** Use the table on page 214 to solve Exercise 23.

**23.** Nasturtiums are planted in a border around a semicircular garden with a radius of 4 ft. About how many plants are used? Draw a diagram to illustrate.

# CHAPTER 6 ● TEST

Find each unknown measurement. Round your answer to the nearest tenth.

1. 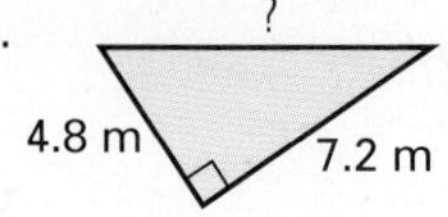


2. 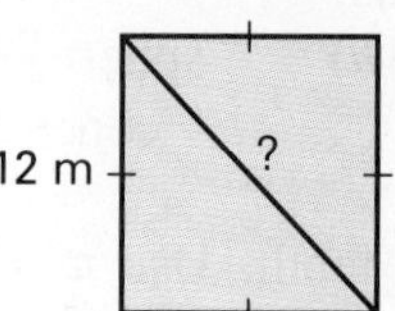


3. 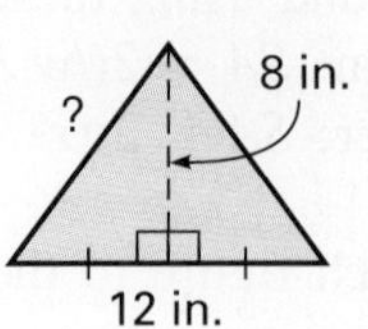


4. 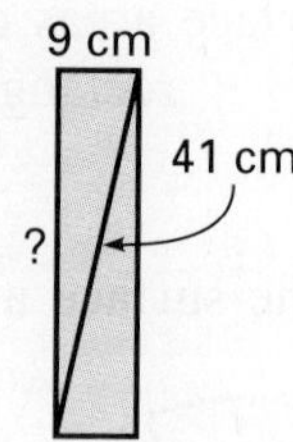


5. The perimeter of a rectangular picture frame is $4\frac{1}{2}$ ft. The frame is 10 in. wide. What is the height of the frame?

6. A circular pond with a radius of 14 ft lies in a rectangular field that measures 115 ft by 60 ft. If the wind carries a seed into the field, what is the probability that the seed lands in the pond? Round your answer to the nearest hundredth.

Find the area of each figure. Use $\pi \approx 3.14$. Round your answer to the nearest tenth.

7. 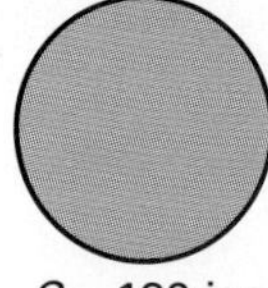


8. 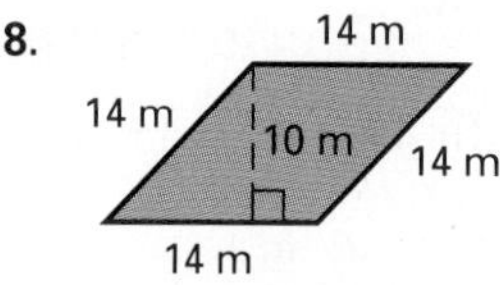


9. 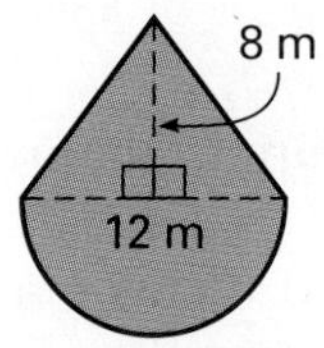


10. 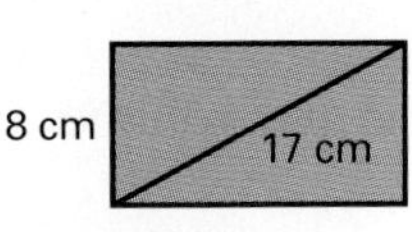


Find the surface area of each figure. Use $\pi \approx 3.14$. Round your answer to the nearest tenth.

11. 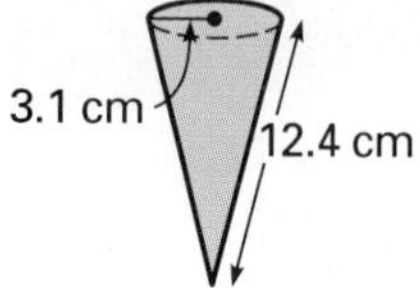


12. 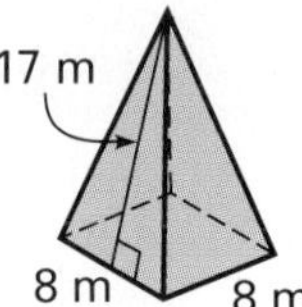


13. 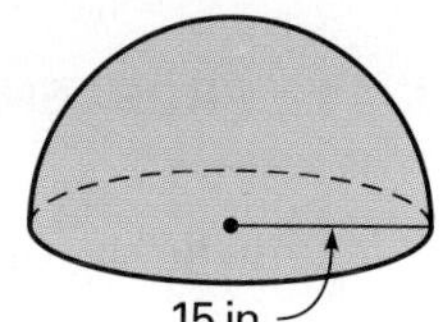


14. 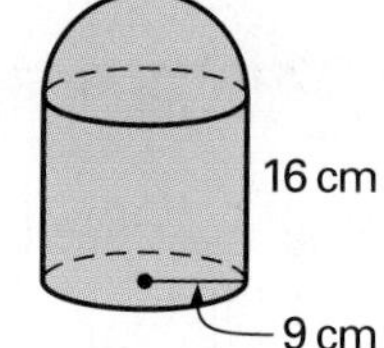


Find the volume of each figure. Use $\pi \approx 3.14$. Round your answer to the nearest tenth.

15. 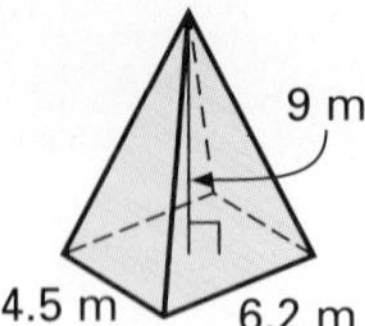


16. 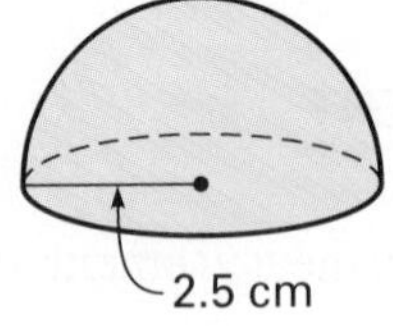


17. 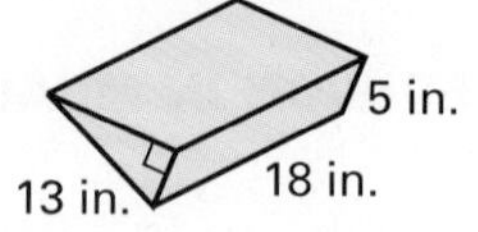


18. 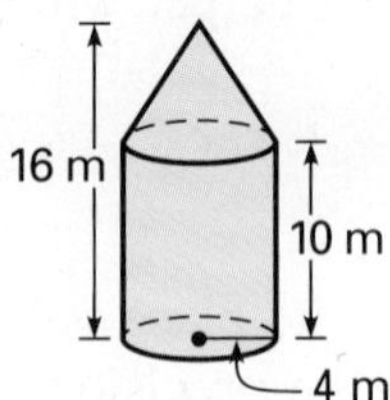


19. To find the number of square feet of wrapping paper needed to cover a box shaped like a rectangular prism, which formula would you use?
 a. $P = 2l + 2w$  b. $V = lwh$  c. $SA = 2(lw + lh + wh)$

20. Which of the following measurements is the most precise?
 a. 7 cm  b. 0.07 m  c. 70 mm  d. 7.0 cm

# CHAPTER 6 ● CUMULATIVE REVIEW

Find the next three terms in each pattern.

1. 4, 5, 7, 10, 14, ■, ■, ■
2. 168, 84, 48, 24, ■, ■, ■
3. 1, $1\frac{1}{2}$, $2\frac{1}{2}$, 3, 4, $4\frac{1}{2}$, ■, ■, ■

4. Complete the function table.

| $w$ | $3w - 5$ |
|---|---|
| −6 | ■ |
| −2 | ■ |
| 0 | ■ |
| 7 | ■ |
| 19 | ■ |

Graph each on a number line.

5. all real numbers less than −2
6. the whole numbers from 3 to 8

Write each number in scientific notation.

7. 418,000
8. 0.00053

9. At 7 A.M. the outside temperature was −2°F. By noon the temperature had reached 11°F. By how many degrees did the temperature change?

10. Name the longest side and shortest side of this triangle.

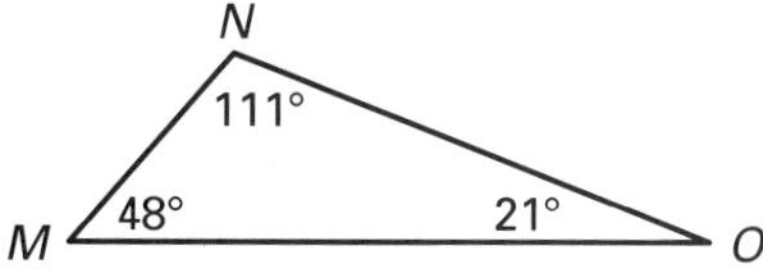


11. Find the appropriate number of degrees in each central angle of a circle graph that could represent the data in the table at the right.

| SALES BY CATEGORY | |
|---|---|
| Rock | 35% |
| Rap | 24% |
| Country | 20% |
| Classical | 13% |
| Jazz | 8% |

Use the scatter plot for Exercises 12 and 13.

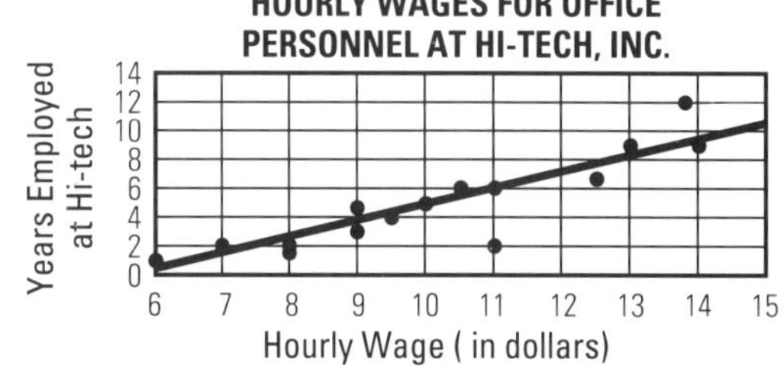


12. Is there a positive or a negative correlation between the years an employee has worked at Hi-tech, Inc. and hourly wages?

13. Suppose an employee earns \$12 an hour. How many years has that person probably worked at Hi-tech, Inc.?

14. On a test, Laurie had the seventh highest score. If 32 students took the test, what is Laurie's percentile rank?

Solve each equation.

15. $29 = 4(x + 1) - 9x$
16. $12 = 0.6y - 18$
17. $m^2 = 0.36$

Find each square root.

18. $-\sqrt{\frac{16}{49}}$
19. $\sqrt{0.09}$

20. For dinner, Rhea consumed 150 calories more than twice the number of calories she consumed for lunch. Rhea consumed 800 calories for dinner. How many calories had she consumed for lunch?

21. Find the perimeter of a rectangle that is 2.4 m long and 68 cm wide.

Find the area of each figure. Use 3.14 for $\pi$.

22.
23.

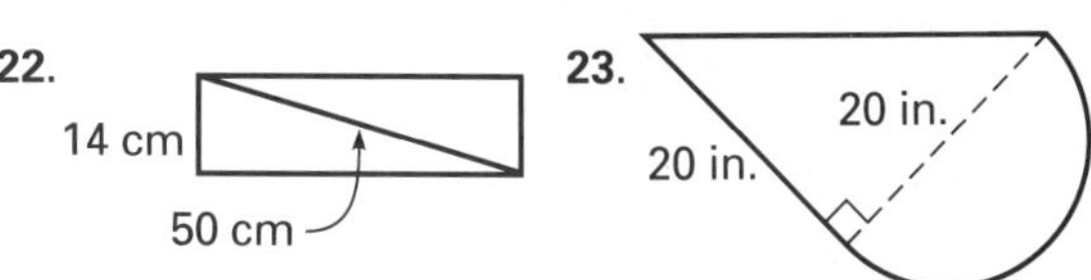

1. Simplify.
$18 \div 3 \times 4 + 4 - 9$
A. 6 B. 19 C. 3.8 D. 1

2. Choose the best estimate.
$38.7 \times 4.9$
A. 120 B. 160 C. 200 D. 250

3. Write $\frac{13}{50}$ as a percent.
A. 26% B. 2.6% C. 13% D. 39%

4. Evaluate the expression $x^3y^{-4}$ when $x = 3$ and $y = -2$.
A. $-432$ B. $\frac{9}{16}$ C. $\frac{27}{16}$ D. $\frac{-8}{81}$

5. Write $3.4 \times 10^{-6}$ in standard form.
A. 3,400,000 B. 0.000034
C. 340,000 D. 0.0000034

6. During a 5-hour period, the temperature dropped from a high of 87°F to a low of 62°F. If the change in temperature took place at a steady rate, how would you describe the average hourly change?
A. −25°F B. −3°F
C. −5°F D. 5°F

Use the figure below for Exercises 7 and 8.

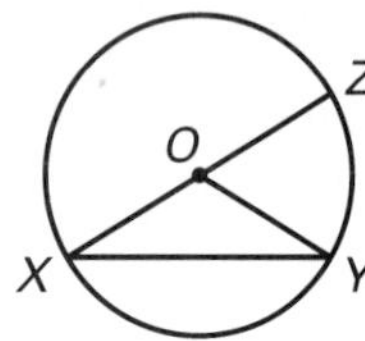


7. Identify a central angle in ⊙$O$.
A. $\angle XYO$ B. $\angle XOY$
C. $\angle ZYX$ D. $\angle OXY$

8. Identify an inscribed angle in ⊙$O$.
A. $\angle XOY$ B. $\angle OYX$
C. $\angle YOZ$ D. $\angle ZXY$

9. Identify the figure that is the set of all points that are a given distance from a given point in space.
A. sphere B. cylinder
C. circle D. none of these

Use the bar graph for Exercises 10 and 11.

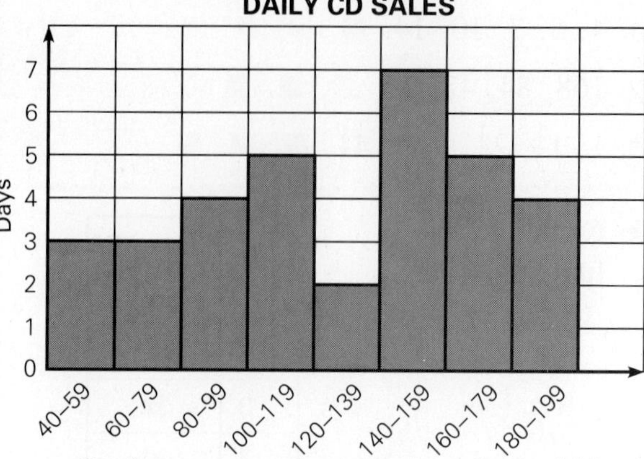


10. On how many days were at least 140 CDs sold?
A. 16 B. 9 C. 7 D. 3

11. On approximately what percent of the days were fewer than 100 CDs sold?
A. 17% B. 29%
C. 30% D. 40%

12. Solve. $3(c + 4) = 6(c - 1)$
A. $\frac{5}{3}$ B. 6
C. 2 D. −6

13. Which expression can be used to represent the product of twice a number $n$ and 5?
A. $5n$ B. $2n + 5$
C. $2n \div 5$ D. $10n$

14. Lenny hiked $2\frac{1}{4}$ h at an average speed of $3\frac{1}{2}$ mi/h. How far did he hike?
A. $5\frac{3}{4}$ mi B. $7\frac{7}{8}$ mi
C. $1\frac{1}{4}$ mi D. none of these

15. Find the area of a circle with a diameter that measures 60 cm. Use 3.14 for $\pi$.
A. 94.2 cm$^2$ B. 188.4 cm$^2$
C. 2,826 cm$^2$ D. 11,304 cm$^2$

16. The surface area of a cube is 337.5 in.$^2$ Find the length of a side of the cube.
A. 7.5 in. B. 18.38 in.
C. 56.25 in. D. 2,025 in.

# CHAPTER 7 SKILLS PREVIEW

Simplify.

1. $10a^2b^2 + 8a^2b + 3a^2b - 6a^2b^2$
2. $(2x^2y - 4x + 2y) + (3x^2y - 5x - 3y)$
3. $(9p^2q - 6pq + 2) - (2p^2q + 4pq)$
4. $(-2c^2)(13c^3)$
5. $(-2x^3y^2)^3$
6. $-3h(-h^2 + 4h + 8)$
7. $3(a^2 + ab) - 5(a^2 - ab)$
8. $4x(3x - 2) + 2x(3x - 2)$
9. $(x + 5)(x + 3)$
10. $(a - 4)(a + 7)$
11. $(y + 3)(y - 3)$
12. $(c - 4)^2$
13. $4(y + 2)(y - 5) - 3(y - 3)(y + 2)$

Use the figure at the right for Exercises 14 and 15.

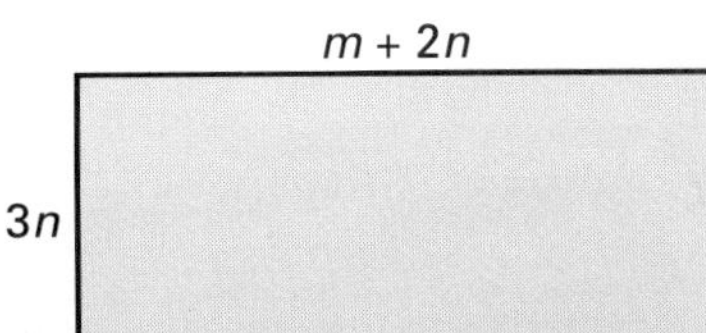


14. Write an expression for the perimeter of the rectangle.
15. Write an expression for the area of the rectangle.

16. Bernard spent \$1.44 for a pen, then spent half of his remaining money on stamps. He then bought a used book for \$2.75. If he had \$4.03 left, how much did he have to begin with?

Find the GCF.

17. 21 and 28
18. $15a^3b^2$ and $10ab^2$

Find the LCM.

19. 15 and 30
20. $9m^4n^3$ and $12m^3n^2$

Factor each polynomial.

21. $12x + 9y$
22. $3hk - 9h^2k^3$
23. $18x^2y^2 + 9xy^2 - 27x^3y$

Simplify.

24. $\dfrac{14xy^3z^4}{21xyz^2}$
25. $\dfrac{3w - 6w^2}{9w}$
26. $\dfrac{e - e^2 + e^5}{e}$
27. $\dfrac{2x^3y + 6x^2y}{2xy}$

Factor each polynomial.

28. $p^2 - 10p + 25$
29. $c^2 + 18c + 81$
30. $k^2 - 81$

CHAPTER

# POLYNOMIALS

**THEME** **Transportation**

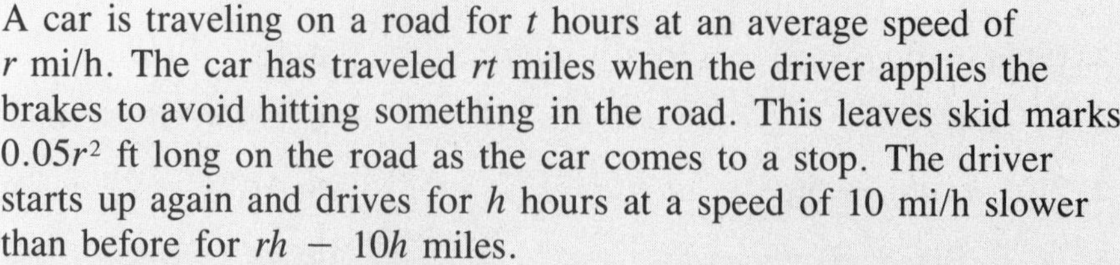

A car is traveling on a road for $t$ hours at an average speed of $r$ mi/h. The car has traveled $rt$ miles when the driver applies the brakes to avoid hitting something in the road. This leaves skid marks $0.05r^2$ ft long on the road as the car comes to a stop. The driver starts up again and drives for $h$ hours at a speed of 10 mi/h slower than before for $rh - 10h$ miles.

Algebraic expressions, such as $rt$, $0.05r^2$, and $rh - 10h$, contain combinations of numbers, variables, and exponents. These expressions are called *polynomials.* In this chapter, you will learn to perform operations on polynomials.

The chart lists fuel-economy mileage for six classes of late-model passenger cars.

**BEST MILEAGE BY SIZE CLASS, PASSENGER CARS (1992)**

| Size Class | City Driving (mi/gal) | Highway Driving (mi/gal) |
|---|---|---|
| Two-seater | 41 | 46 |
| Minicompact | 28 | 38 |
| Subcompact | 53 | 58 |
| Compact | 32 | 40 |
| Mid-size | 24 | 31 |
| Large | 20 | 26 |

## DECISION MAKING

### Using Data

Use the information in the chart on page 266 to write an algebraic expression for each of Exercises 1–4.

1. Write an expression for the number of miles the compact car can travel in the city on $g$ gal of gas. Write an expression for the number of miles this car can travel in the city on $g + 5$ gal.
2. How many miles can the minicompact car travel in the city on $x$ gal of gas?
3. How many more highway miles can the two-seater travel than the large car on $y$ gal of gas?
4. Write an expression for the number of gallons the subcompact car would use in traveling $m$ highway miles.

### Working Together

Survey at least ten automobile drivers. Ask them to estimate how many city miles and highway miles they drive in a year. Record the responses. Find the mean for each of the two kinds of estimates in order to make a conjecture about the number of miles that the average car is driven in a year, both in the city and on the highway.

Check local gas stations for the price of one gallon of regular unleaded gas. Record the prices, then find the mean. Use the mean to make another conjecture about the average annual cost of gas for city driving and for highway driving for each class of car listed in the table. Record your results.

Why might someone be interested in knowing your results?

# 7-1 Adding and Subtracting Polynomials

## EXPLORE

You can use algebra tiles to model variable expressions.

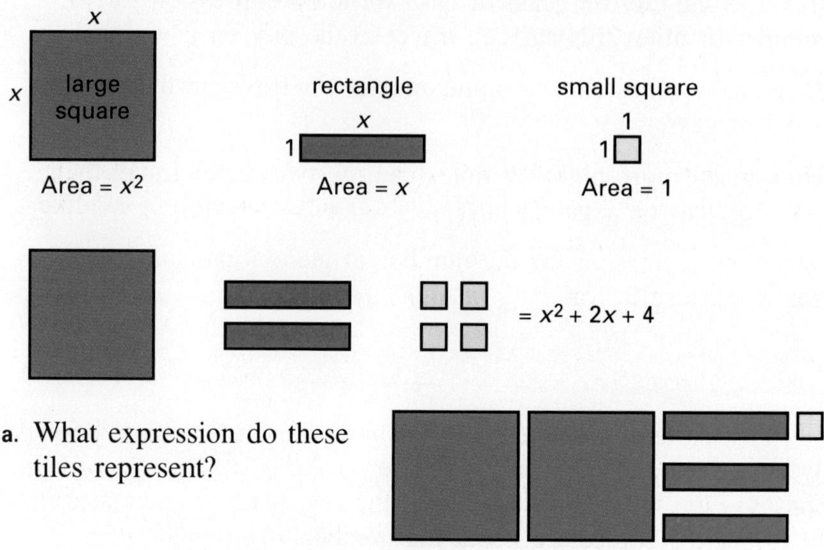


a. What expression do these tiles represent?

b. Use tiles to model the expression $3x^2 + x + 2$.

c. Add three more rectangular tiles to your model. What expression do the tiles now represent?

d. Remove two large square tiles from your second model. What expression do the remaining tiles represent?

## SKILLS DEVELOPMENT

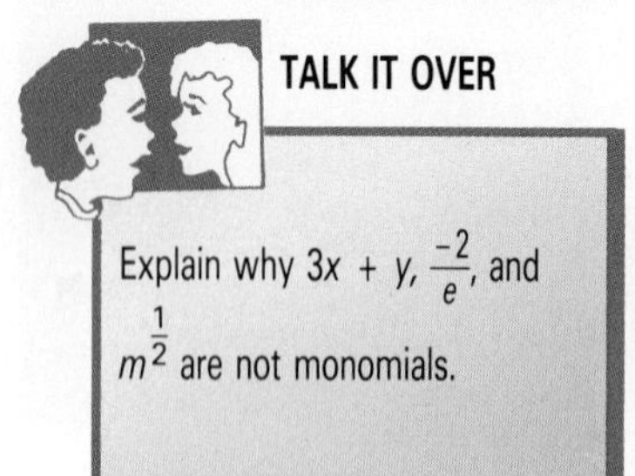

**TALK IT OVER**

Explain why $3x + y$, $\frac{-2}{e}$, and $m^{\frac{1}{2}}$ are not monomials.

A **monomial** is an expression that is a number, a variable, or a product of a number and one or more variables with whole number exponents. Each of these expressions is a monomial.

$-11 \qquad k \qquad mx^4 \qquad 7p^2f$

The number in a monomial is called the **coefficient.** A monomial that contains no variables, such as $-11$, is called a **constant.**

A sum of monomials is called a **polynomial.** Each monomial is called a **term** of the polynomial. A polynomial with two terms is called a **binomial.** A polynomial with three terms is called a **trinomial.** A monomial can be thought of as a polynomial with one term.

binomials: $8n + (-n^2)$ $\qquad xy + 3x^3y$
(↑ term, ↑ term)

trinomials: $a^2 + (-4ab) + 4b^2$ $\qquad 6m + (-9n) + 1$
(↑ term, ↑ term, ↑ term)

A polynomial is written in *standard form* when its terms are ordered from the greatest to the least powers of one of the variables.

**CHECK UNDERSTANDING**

Sometimes polynomials are written with subtraction signs. How could you write the polynomials below to show that each is really a *sum* of monomials?

1. $6b - b^2$
2. $c^2 - 2cd + 3e^2$
3. $4p - 2q + 5$

## Example 1

Write each polynomial in standard form for the variable $x$.

**a.** $3x + 5 + 7x^3 - 2x^2$ **b.** $-8x^2y^3 + 4xy^4 + x^3$

**Solution**

Order the terms from greatest to least powers of $x$.

**a.** $3x + 5 + 7x^3 - 2x^2 = 7x^3 - 2x^2 + 3x + 5$

**b.** $-8x^2y^3 + 4xy^4 + x^3 = x^3 - 8x^2y^3 + 4xy^4$ ◀

Terms like $3xy^2$ and $-7xy^2$ that differ only in their coefficients are called *like terms*. You simplify a polynomial by combining all like terms so that only unlike terms remain.

## Example 2

Simplify.

**a.** $4k^2 - 3k + 5k^2$ **b.** $2x^2y + 4x^2 + 3xy^2 - 7x^2y$

**Solution**

**a.** $4k^2 - 3k + 5k^2 = 4k^2 + 5k^2 - 3k$ ← Apply the commutative property to rearrange, or collect, like terms.

$= (4 + 5)k^2 - 3k$ ← Combine like terms by applying the distributive property.

$= 9k^2 - 3k$

**b.** $2x^2y + 4x^2 + 3xy^2 - 7x^2y = 2x^2y - 7x^2y + 4x^2 + 3xy^2$

$= (2 - 7)x^2y + 4x^2 + 3xy^2$

$= -5x^2y + 4x^2 + 3xy^2$ ◀

To add polynomials, write the sum and simplify by combining like terms.

## Example 3

Simplify.

**a.** $7n + (5 - 3n)$ **b.** $(3x^2y - 7x + 2y) + (5x^2y + 2x - 3y)$

**Solution**

To simplify, use the associative and commutative properties as necessary to collect like terms and then use the distributive property to combine like terms.

**a.** $7n + (5 - 3n) = (7n - 3n) + 5$

$= (7 - 3)n + 5$

$= 4n + 5$

**b.** $(3x^2y - 7x + 2y) + (5x^2y + 2x - 3y)$

$= 3x^2y + (-7x) + 2y + 5x^2y + 2x + (-3y)$

$= (3x^2y + 5x^2y) + (-7x + 2x) + [2y + (-3y)]$

$= (3 + 5)x^2y + (-7 + 2)x + (2 - 3)y$

$= 8x^2y + (-5x) + (-y)$

$= 8x^2y - 5x - y$ ◀

**PROBLEM SOLVING TIP**

You may find it easier to add or subtract polynomials by lining up like terms vertically.

$$\begin{array}{r} 3x^2y - 7x + 2y \\ +5x^2y + 2x - 3y \\ \hline 8x^2y - 5x - y \end{array}$$

To subtract one polynomial from another, add the opposite of the polynomial being subtracted.

### Example 4

Simplify.

a. $8y - (5y + 3)$    b. $(2xy^2 - 5xy + 6) - (xy^2 + 3xy)$

**Solution**

Change each term of the polynomial being subtracted to its opposite. Then follow the same procedure as for adding polynomials.

a. $8y - (5y + 3) = 8y + [-5y + (-3)]$
$= [8y + (-5y)] + (-3)$
$= [8 + (-5)]y + (-3)$
$= 3y - 3$

b. $(2xy^2 - 5xy + 6) - (xy^2 + 3xy)$
$= (2xy^2 - 5xy + 6) + [-xy^2 + (-3xy)]$
$= [2xy^2 + (-xy^2)] + [(-5xy) + (-3xy)] + 6$
$= (2 - 1)xy^2 + (-5 - 3)xy + 6$
$= 1xy^2 + (-8xy) + 6$
$= xy^2 - 8xy + 6$ ◄

## TRY THESE

Write each polynomial in standard form for the variable $x$.

1. $9x^2 - 2x + 4 + 3x^3$
2. $-3 - 3x - 3x^2$
3. $x^2y^2 - y^3x$
4. $y^3 + 2xy^2 - 3x^2y - 5x^3$

Simplify.

5. $9p - 4p$
6. $k^2 + 2k - 3k - k^3$
7. $8y^3 - 4y^2 - 2y^3$
8. $5x - 2xy + 3x + 4xy$
9. $(8h^2 - 2h) + (3h^2 + 5h)$
10. $(3x^2y + 2xy - 5) + (9x^2y + 8xy)$
11. $(4a^2b + 8a) - (7a^2b - 3a)$
12. $(-2jk^2 + 6jk) - (5jk^2 + 3jk)$
13. From $14mn^2 + 6mn - 2$, subtract $-3mn^2 + 2mn$.

## EXERCISES

**PRACTICE/ SOLVE PROBLEMS**

Write each polynomial in standard form for the variable $x$.

1. $2 + x^2 + 3x^3$
2. $3x^3 + 2x^4 + x + x^2$
3. $-x - x^5 + x^3$
4. $7x - 1 + 3x^2 - 4x^3$
5. $xy^4 + x^2y^2 + x^3y$
6. $6 - 5m^3x$

Simplify.

7. $3n - 5n$

8. $4x^2 + 9x^2$

9. $7w + 5w - 4w - 8w$

10. $-5c + 7d - 3d + 6$

11. $2x^3 + 4x + 7x$

12. $e^2 - 4e - 2e^2$

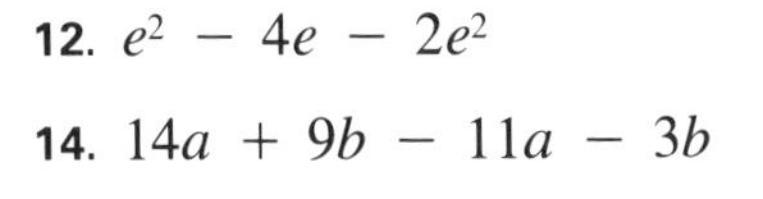

13. $-4n^2 + 2n + 6n + 8n^2$

14. $14a + 9b - 11a - 3b$

15. $6x - 3xy + 4xy - 5x$

16. $-8hk + 10k + 3hk + 11hk$

17. $(5x + 2) + (3x - 4)$

18. $(2m + 3m^2) + (5m^2 + 6)$

19. $(-k^2 + 6k) + (k^2 - 5k)$

20. $(8m + 7n - 6) + (7m - 6n + 10)$

21. $(12p + 7) - (8p + 2)$

22. $(-3a + 5b) - (2a - 6b)$

23. $(x^2 - 5x + 7) + (3x^2 + 7x - 9)$

24. $(6k^2 + 7k - 4) - (9k^2 - 11k + 7)$

25. Jason said that yesterday he earned $6r + 28t$ dollars and that today he earned $9r + 18t$ dollars. Write an expression for the total amount Jason earned in the two days.

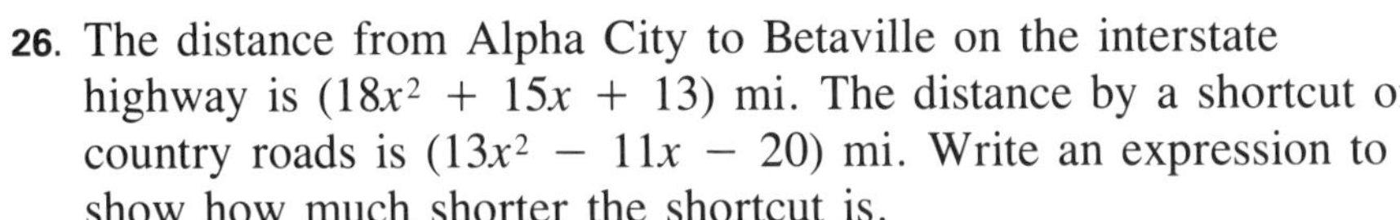

26. The distance from Alpha City to Betaville on the interstate highway is $(18x^2 + 15x + 13)$ mi. The distance by a shortcut on country roads is $(13x^2 - 11x - 20)$ mi. Write an expression to show how much shorter the shortcut is.

## EXTEND/ SOLVE PROBLEMS

Simplify.

27. $2.7m + 3.9n - 4.6m + 7.7n + 6.2 - 8.5$

28. $\frac{3}{4}x^2 - \frac{1}{2}x + \frac{3}{8} - \frac{5}{8}x^2 + \frac{1}{4}x - \frac{1}{2}$

29. $7a^2b^2 + b^3 - b^2 - 4 - 3a^2b^2 + b^2$

30. $(10a^3 + 8a^2b - 4ab^2 - 7b^3) - (12a^3 - 3a^2b + 5ab^2 + 13b^3)$
 $+ (-15a^3 - 9a^2b + ab^2 + 7b^3)$

31. Write and simplify an expression for the perimeter of each. By how much does the perimeter of **b** exceed the perimeter of **a**?

**a.**

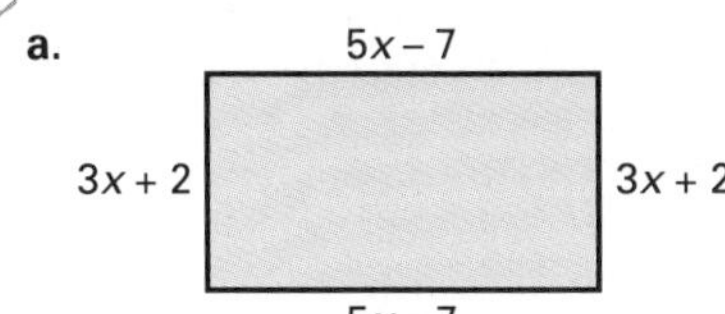


**b.**

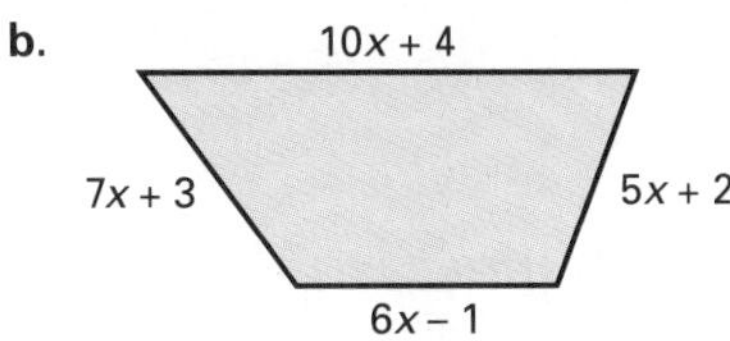


## THINK CRITICALLY/ SOLVE PROBLEMS

32. Write two polynomials whose sum is $8p^2 - 6p + 4$.

33. Write two polynomials whose difference is $3x^2 - 5x + 7$.

34. The perimeter of the end of any box mailed parcel post must not exceed 102 in. Find the maximum possible value of $x$ for the perimeter of the shaded face of this box.

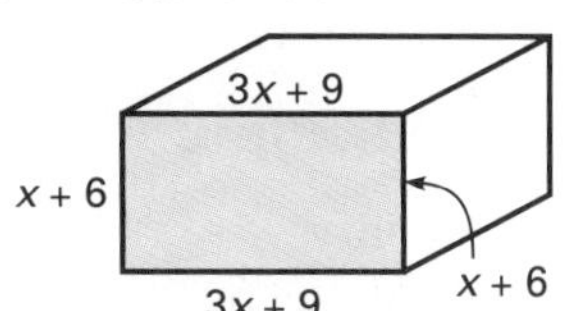

# 7-2 Multiplying Monomials

## EXPLORE

Use rectangular algebra tiles to model monomial multiplication. Assign the value $a$ to the length of a tile and the value $b$ to the width.

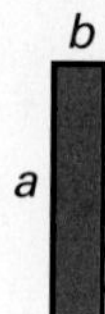


a. What is the area of each rectangular tile?
b. Model a rectangle with length $3a$ and width $2b$. How many tiles did you use? What is their total area?
c. Write the area of the rectangle you modeled as a product of its length and width.
d. Use your model to find the product of $3a$ and $2b$ as a monomial.
e. Model each product. Write the product as a monomial.
   $a \times 2b$ $\quad$ $4a \times 3b$ $\quad$ $2a \times 5b$ $\quad$ $4a \times 2b$
f. Use your results to write a rule for multiplying two monomials.

## SKILLS DEVELOPMENT

You can use the commutative and associative properties of multiplication to find a product of two monomials.

**CHECK UNDERSTANDING**

Show how to use the commutative and associative properties to rewrite $(5x)(7y)$ as $35xy$ in Example **1a**.

### Example 1

Simplify.

a. $(5x)(7y)$ $\quad$ b. $(-2h)(4k)$ $\quad$ c. $\left(-\frac{1}{3}mn\right)(-12x)$

### Solution

a. $(5x)(7y) = (5)(7)(x)(y)$
$= 35xy$

b. $(-2h)(4k) = (-2)(4)(h)(k)$
$= -8hk$

c. $\left(-\frac{1}{3}mn\right)(-12x) = \left(-\frac{1}{3}\right)(-12)(mn)(x)$
$= 4mnx$ ◄

Recall the *product rule for exponents:*

$$a^m \times a^n = a^{m+n}$$

You can use the product rule for exponents, together with the commutative and associative properties of multiplication, to find a product of two monomials.

### Example 2

Simplify.

a. $(5e^2)(-6e^3)$ $\quad$ b. $(4m^2n^3)(-3mn^4p)$

**Solution**

a. $(5e^2)(-6e^3) = (5)(-6)(e^2 \times e^3)$
$= -30e^{2+3}$
$= -30e^5$

b. $(4m^2n^3)(-3mn^4p) = (4)(-3)(m^2 \times m)(n^3 \times n^4)(p)$
(*m* means $m^1$)
$= -12(m^{2+1})(n^{3+4})p$
$= -12m^3n^7p$ ◄

**CHECK UNDERSTANDING**

In Example **2b,** why is it that you *cannot* add the exponents in the expression $-12m^3n^7p$?

Recall also the following rules for exponents.

| **power rule** | **power of a product rule** |
|---|---|
| $(a^m)^n = a^{mn}$ | $(ab)^m = a^mb^m$ |

Use these rules, along with the commutative and associative properties of multiplication, to simplify monomials involving powers.

## Example 3

Simplify.

a. $(5k^3)^2$

b. $(-3w^5y^6)^4$

**Solution**

a. $(5k^3)^2 = (5)^2(k^3)^2$
$= 25k^{3\times 2}$
$= 25k^6$

b. $(-3w^5y^6)^4 = (-3)^4(w^5)^4(y^6)^4$
$= 81w^{5\times 4} \cdot y^{6\times 4}$
$= 81w^{20}y^{24}$ ◄

**WRITING ABOUT MATH**

Write a few sentences explaining the difference in meaning between the expressions $a^m \times a^n$ and $(a^m)^n$.

## TRY THESE

Simplify.

1. $(9e)(3f)$
2. $(-5x)(8y)$
3. $\left(\frac{1}{2}a\right)(-16bc)$
4. $(12n^4)(4n^{12})$
5. $(-2c^5de^2)(-3c^2d^4)$
6. $(-2p^5)^4$
7. $(4a^3b^2c^5)^3$
8. What is the area of a rectangle that has a length of $4ab^2c$ in. and a width of $3a^3b$ in.?

# EXERCISES

**PRACTICE/ SOLVE PROBLEMS**

Simplify.

1. $(3p)(7q)$
2. $(-2a)(4b)$
3. $(6m)(-2n)$
4. $(-xy)(-5hk)$

5. $9(-2a)$

6. $(-16w)\left(-\frac{3}{4}t\right)$

7. $(3.6k)(-2.5e)$

8. $(2a^5)(5b^6)$

9. $(-5m^2)(4m)$

10. $(8xy)(-3xy)$

11. $\left(\frac{2}{3}h^3\right)(-9h^2)$

12. $(4.3xy^2)(0.6y^4)$

13. $(a^5b^2)(a^3b^4)$

14. $(-20x^5y)\left(-\frac{3}{4}x^2y^2\right)$

15. $8e(-2e^3f)$

16. $(-abc)(abc)$

17. $(14hkn)(3.5hm)$

18. $\left(-\frac{2}{5}pq^2\right)\left(\frac{15}{16}p^2q\right)$

19. $(-3x)^2$

20. $(2h^3)^3$

21. $(2ab)^5$

22. $(-5x^2y)^2$

23. $(m^2n^5)^4$

24. $(-abc^2)^6$

Write and simplify an expression for the area of each figure.

25.

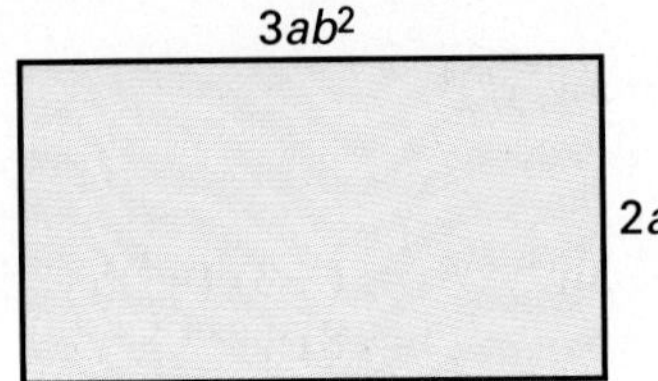


26.

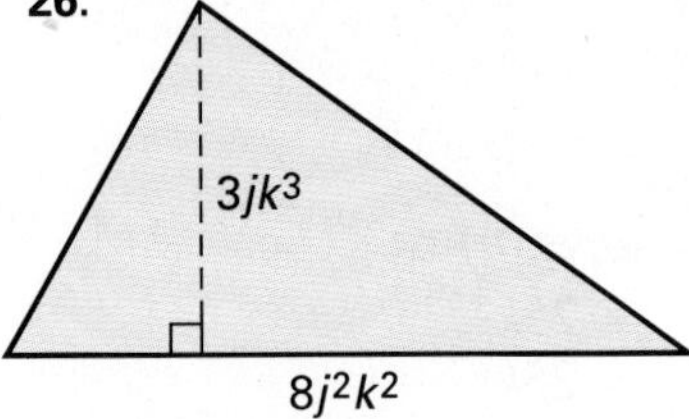


27. Find the area of a rectangular parking lot that is $4.6r$ yards long and $3.5r^3$ yards wide.

## EXTEND/ SOLVE PROBLEMS

Simplify.

28. $(m^3)(m^2)(m^6)$

29. $(-2a^5)(2a^4)(-2a^7)$

30. $(2y)(3y)^2$

31. $(-5x^2)(2x^3)^3$

32. $(4hk)^3(-h^3k^2)$

33. $(-a^2b)^5(a^2b)^4$

34. $(-x^3y^2)(3x^4yz)(-5xy^4z^3)$

35. $(4h^2)^2(2h^3)^3$

36. $(-3a^2b)^3(2ab^3)^2(ab^3)$

37. $(-4e^3f)(-e^2f^2)^3\left(\frac{1}{2}ef^4\right)^2$

Write and simplify an expression for the area of each figure.

38.

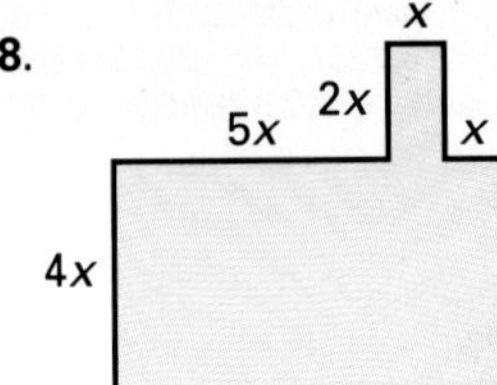


39.

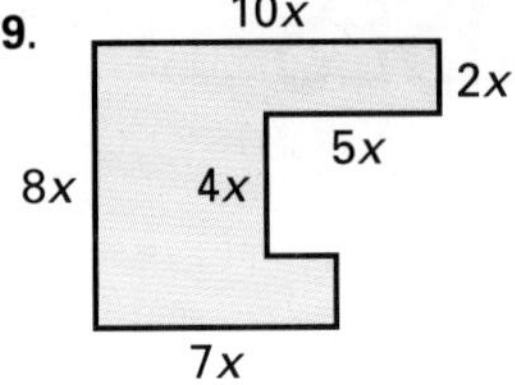

40. Write and simplify an expression for the area of the shaded region.

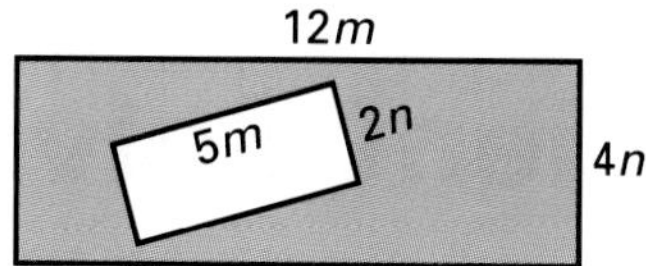


Write and simplify an expression for the volume of each prism.

41.

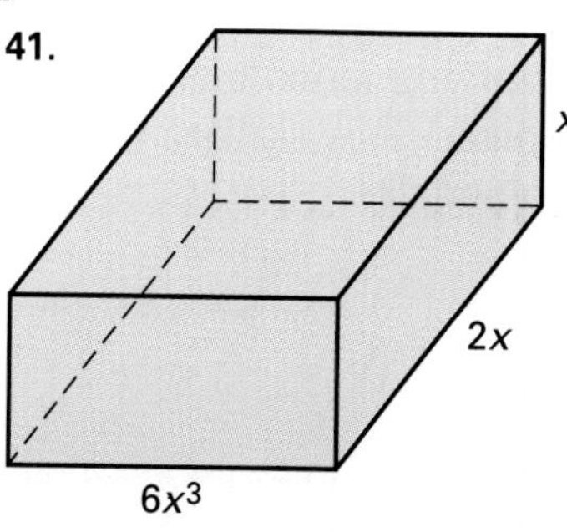


42.

$10m^2n^3$

6m

$4m^2n$

43.

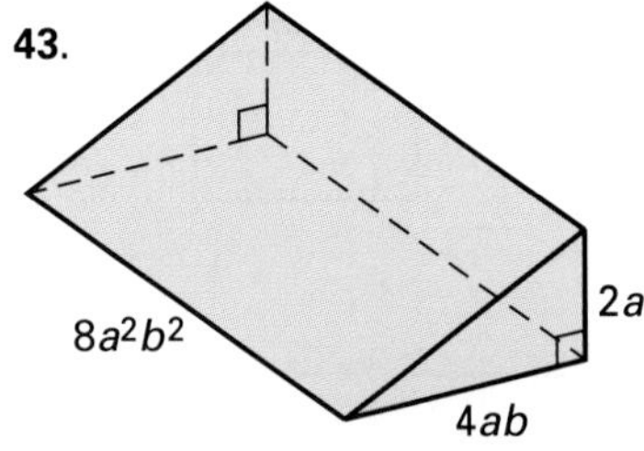


44.

$9h^2k$

$(2k^2)^2$

$(4hk^3)^2$

## THINK CRITICALLY/ SOLVE PROBLEMS

Solve each equation for $n$.

45. $p^8 \times p^n = p^{11}$ 46. $(k^2)^n = k^6$ 47. $n^8 = e^{16}$

48. A rectangle has an area that can be expressed as $64a^6b^9$. If its width can be expressed as $4a^2b$, how can its length be expressed?

49. For the product $12x^2$, find nine pairs of monomial factors that have positive integral coefficients.

# 7-3 Problem Solving Skills:

## USE AN AREA MODEL

- ► READ
- ► PLAN
- ► SOLVE
- ► ANSWER
- ► CHECK

In the last lesson you learned rules for multiplying monomials. You can also use area models to multiply monomials. This skill can then be extended to help you in multiplying polynomials.

### PROBLEM

Multiply $(2m)(3n)$.

### SOLUTION

Draw a rectangle whose dimensions are $m$ and $n$. Using the area formula for rectangles, you know that its area is $mn$ square units.

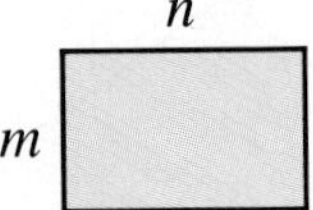


Extend the rectangle so that its dimensions are $2m$ and $3n$. You can divide the new rectangle into six smaller rectangles, each of which has an area of $mn$ square units. So, the area of the new rectangle is $6mn$ square units.

$3n$

| $2m$ | | $n$ | $n$ | $n$ |
|---|---|---|---|---|
| | $m$ | $mn$ | $mn$ | $mn$ |
| | $m$ | $mn$ | $mn$ | $mn$ |

Use care when the dimensions of the rectangle include exponents. To multiply $(2c^3)(4e^2)$ you would use this model.

$4e^2$

| $2c^3$ | | $e^2$ | $e^2$ | $e^2$ | $e^2$ |
|---|---|---|---|---|---|
| | $c^3$ | $c^3e^2$ | $c^3e^2$ | $c^3e^2$ | $c^3e^2$ |
| | $c^3$ | $c^3e^2$ | $c^3e^2$ | $c^3e^2$ | $c^3e^2$ |

$$\text{area} = c^3e^2 + c^3e^2 + c^3e^2 + c^3e^2 + c^3e^2 + c^3e^2 + c^3e^2 + c^3e^2$$
$$= 8c^3e^2$$

# PROBLEMS

Write the monomial multiplication that each model illustrates. Then write the product.

**1.** 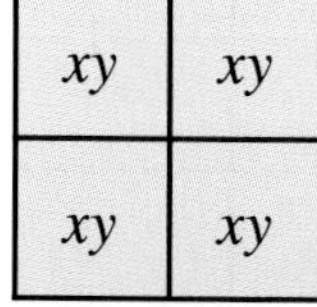


**2.**

| $a^2$ | $a^2$ | $a^2$ | $a^2$ |
|---|---|---|---|

**3.**

| $m^2n^3$ | $m^2n^3$ | $m^2n^3$ |
|---|---|---|
| $m^2n^3$ | $m^2n^3$ | $m^2n^3$ |

Show how to use an area model to find the product.

**4.** $(3p)(4p)$ **5.** $(6x)(2y)$

**6.** $(3a^2)(3a)$ **7.** $(m^2)(4n^4)$

**8.** $(2x^3)(5y^3)$ **9.** $(4a^5)(2bc)$

You can use this area model to multiply a binomial by a monomial. Use the model for Exercises 10–14.

$3x$ $2$
$x$ $x$ $x$ $1$ $1$
$2y$ $y$ $y$

**10.** What is the length of the large rectangle? What is the width?

**11.** Express the area of the large rectangle by writing a variable expression for its length times its width.

**12.** Express the area of each of the two types of small rectangles by writing a variable expression for the length times the width.

**13.** Express the area of the large rectangle by writing it as a sum of the areas of the small rectangles.

**14.** Write the polynomial that is the product of the factors you wrote in Exercise 11.

Use an area model to find the product.

**15.** $2x(x + 1)$ **16.** $3a(2b + 2)$

**17.** $m(n + 3)$ **18.** $3x(3y + 2)$

# 7-4 Multiplying a Polynomial by a Monomial

## EXPLORE

a. Express the area of the large rectangle by writing a variable expression for its length times its width.

b. Add the areas of the small squares and rectangles to find the area of the large rectangle.

c. Compare your answers to **a** and **b** above. What property is illustrated?

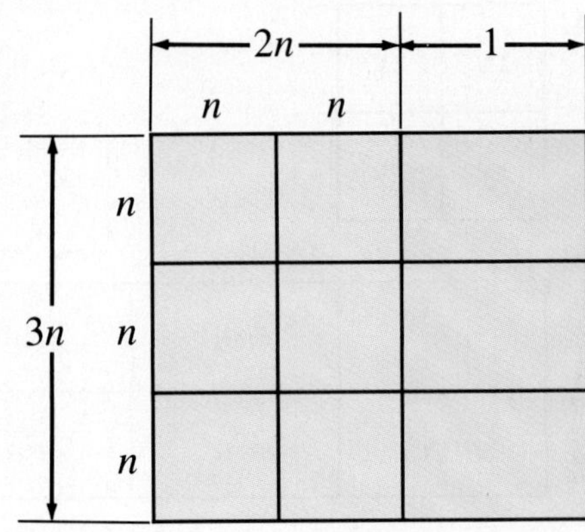


## SKILLS DEVELOPMENT

The distributive property gives a method you can use to multiply a polynomial by a monomial.

distributive property: $a(b + c) = ab + ac$

monomial polynomial

$a(b - c) = ab - ac$

**READING MATH**

The distributive property states that when a sum or difference of terms is multiplied by a factor, the factor must be *distributed,* through multiplication, over each term.

$a(b + c) = ab + ac$
Factor $a$ is *distributed* over both terms, $b$ and $c$.

### Example 1

Simplify.

a. $5n(2n - 3)$

b. $-6k(-k^2 + 2k + 5)$

**Solution**

a. $5n(2n - 3) = 5n(2n) - 5n(3)$ ← **Use the distributive property.**
$= 10n^2 - 15n$ ← **Multiply each pair of monomials.**

b. $-6k(-k^2 + 2k + 5) = -6k(-k^2) + (-6k)(2k) + (-6k)(5)$
$= 6k^3 - 12k^2 - 30k$ ◄

### Example 2

Simplify.

a. $4(x^2 + xy) - 3(x^2 - xy)$

b. $7m(2m - 4) + 2m(2m - 4)$

**Solution**

a. $4(x^2 + xy) - 3(x^2 - xy) = (4x^2 + 4xy) - (3x^2 - 3xy)$
$= 4x^2 + 4xy - 3x^2 + 3xy$
$= (4x^2 - 3x^2) + (4xy + 3xy)$
$= x^2 + 7xy$

**b.** $7m(2m - 4) + 2m(2m - 4) = [7m(2m) - 7m(4)] + [2m(2m) - 2m(4)]$
$= (14m^2 - 28m) + (4m^2 - 8m)$
$= 14m^2 - 28m + 4m^2 - 8m$
$= (14m^2 + 4m^2) + (-28m - 8m)$
$= 18m^2 - 36m$ ◄

**PROBLEM SOLVING TIP**

When you are doing simplifications involving two or more multiplications, look for binomials that are alike. For instance, in Example **2b** the term $(2m - 4)$ appears twice. Think of the binomial $(2m - 4)$ as a "chunk," and use the distributive property to simplify.

$7m(2m - 4) + 2m(2m - 4)$
$= (7m + 2m)(2m - 4)$
$= 9m(2m - 4)$
$= 18m^2 - 36m$

## Example 3

The cost of renting a four-wheel-drive vehicle at Vacation Rentals is \$43 per day plus \$0.08 per mile. The cost $C$ can be expressed by the formula $C = 43 + 0.08m$, where $m$ represents the number of miles the vehicle is driven. Wilderness Camp rented a fleet of $v$ vehicles for one day. Write a formula for the total cost of renting all the vehicles. Assume all vehicles travel the same number of miles.

### Solution

total cost = number of vehicles × cost per vehicle

$C = v(43 + 0.08m)$

$C = 43v + 0.08mv$ ◄

## TRY THESE

Simplify.

1. $6x(x + 4)$
2. $2n(8 - n^2)$
3. $12k^3(k^2 + k^4 + 3)$
4. $-4p(p^2 - 2p + 5)$
5. $5(2x^2 + xy) + 4(x^2 - 3xy)$
6. $14a(7a - 6) - 11a(7a - 6)$
7. The fare for a ride in a Maroon Cab is \$2 for the first mile and \$0.75 for each additional mile. The fare $F$ can be expressed by the formula $F = 2 + 0.75m$, where $m$ represents the number of miles driven after the first mile. The Smith for President Committee engaged $c$ cabs for a motorcade through the city. Write and simplify a formula for the total fare.

# EXERCISES

## PRACTICE/ SOLVE PROBLEMS

Simplify.

1. $3m(2m + n)$
2. $11x(2x - y)$
3. $2a(3a + 4)$
4. $4k(5k - 7)$
5. $9c(3c + c^2 + 6c^3)$
6. $5x(3x + 2x^2)$
7. $-4e(3e^2 + 5e - 3)$
8. $7x(1 + 2x + x^2)$
9. $-3y(y^3 - y^2 - y)$
10. $4m(6m^2 + 3mn + n^2)$
11. $2(h^2 + 5) + 3(h^2 - 2)$
12. $8(2x - 3) - 3(2x - 3)$
13. $14(w^2 - 4) - 9(w^2 - 4)$
14. $6(a - 2b) - 3(a - 2b)$
15. The formula $C = 8 + 2h$ gives the cost $C$ of renting a rototiller for $h$ hours. Write and simplify the cost of renting $r$ rototillers for $h$ hours.
16. Rose works 36 h per week and earns $8.50 per hour. Write a variable expression in simplest form to represent her weekly earnings if her hourly wage is increased by $d$ dollars per hour.
17. Marcus drove 5 h at an average rate of 56 mi/h. Write a variable expression in simplest form that represents how far he would have driven in the same amount of time if he had driven $m$ fewer miles per hour.
18. A farmer sold 8,000 bushels of wheat for $3.25 per bushel. A year later the farmer sold the same quantity of wheat at $c$ cents more per bushel. Write a variable expression in simplest form that represents the amount the farmer was paid for wheat in the second year.

Simplify.

19. $\frac{1}{2}e(24e^4 - 20e^3 + 8e)$
20. $(x^3 - 9.2x^2 - 4.6x)2.5x^2$
21. $2y(y^3 + 3y^2 - 5y) - 5y(-3y^3 - 2y^2 + 3y)$

## EXTEND/ SOLVE PROBLEMS

Write and simplify an expression for the area of each figure.

22.

4m

6m + 3

23.

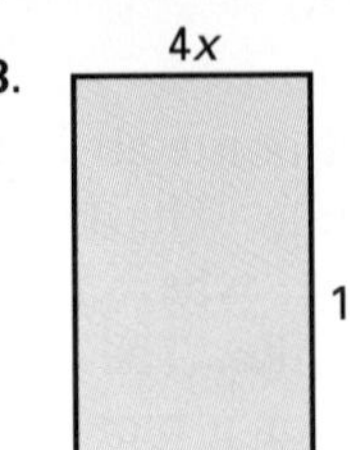


24.

8x² + 3y

4x

14x² – 9y

25. Steve usually types at a rate of 60 words per minute. One day he typed for 20 min at a rate of $w$ more words per minute than his normal rate. Then he typed for 20 min at a rate of $w$ fewer words per minute than his usual rate.
   a. Write a variable expression in simplest form that represents how many more words he typed in the first 20 min than in the second 20 min.
   b. How many words did he type in all?

26. On Monday a share of ABC stock sold for $x$ dollars. On Tuesday the price rose 3 dollars per share. On Wednesday the price fell 5 dollars per share.
   a. Jill bought 20 shares of ABC on Monday, 30 shares on Tuesday, and 40 shares on Wednesday. Write a variable expression in simplest form that represents the total cost of Jill's purchases.
   b. It cost Maria $225 to buy 9 shares of ABC on Tuesday and 6 shares on Wednesday. Find the cost of a share of ABC on Monday.

27. A picture frame measures 16 in. by 12 in. Write and simplify an expression for the area of the frame if the longer side is increased by $n$ inches.

**THINK CRITICALLY/ SOLVE PROBLEMS**

28. Write and simplify an expression that represents the area of the shaded region.

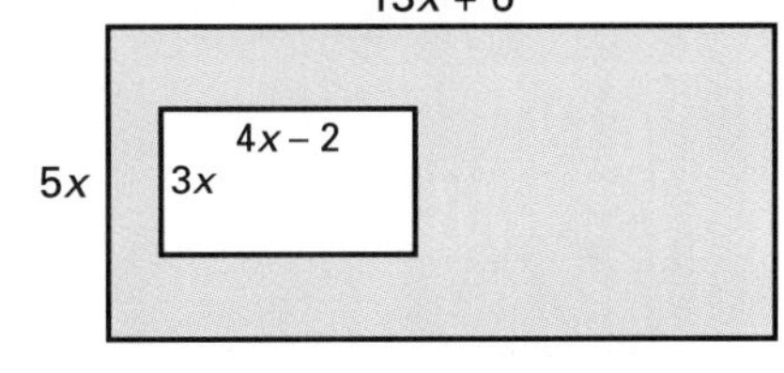


29. Write an expression in simplest form that represents the surface area of the rectangular prism.

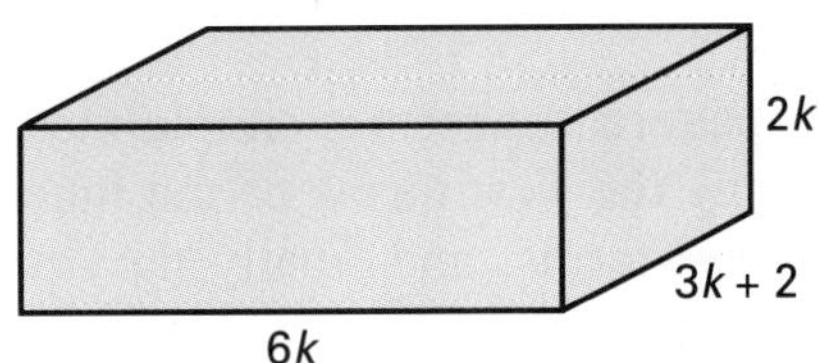

# 7-5 Multiplying Binomials

**EXPLORE**

a. Express the area of the figure by writing a variable expression for its length times its width.

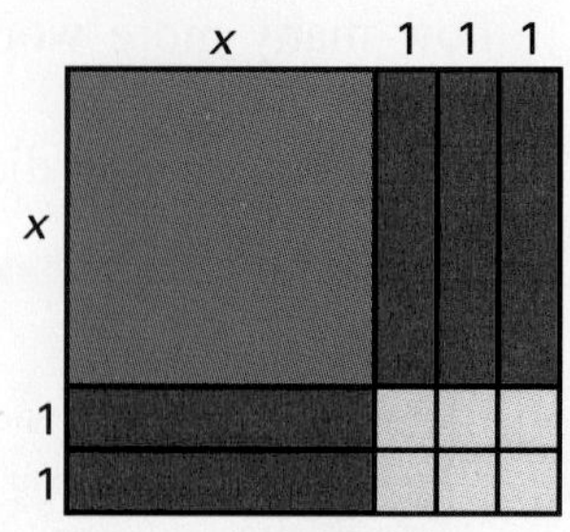


b. Write the area of the large rectangle as a sum of the areas of the squares and rectangles it contains. Simplify the expression.

c. Use algebra tiles to model the product $(x + 1)(x + 4)$. What is the product?

**SKILLS DEVELOPMENT**

You can find the product of two binomials by repeatedly applying the distributive property.

### Example 1

Find the product $(x + 3)(x + 7)$.

**Solution**

$(x + 3)(x + 7) = (x + 3)(x) + (x + 3)(7)$ ← Use the distributive property to distribute the binomial $(x + 3)$ over both $x$ and 7.

$= x^2 + 3x + 7x + 21$ ← Use the distributive property twice. Simplify.

$= x^2 + 10x + 21$ ◄

Notice that each term in the second step is the product of a different pair of monomials from the original expression.

$x^2$ 21

$(x + 3)(x + 7) = x^2 + 3x + 7x + 21$

$3x$ $7x$

This pattern suggests a simple method for multiplying two binomials: Multiply the *first* terms of the binomials, then the *outer* terms, then the *inner* terms, and finally the *last* terms. Write these four products as a sum and simplify.

first last

$$(x + 6)(x + 4) = x^2 + 4x + 6x + 24$$
$$= x^2 + 10x + 24$$

inner outer

Many people use the memory device *FOIL* to remember this shortcut.

| F | O | I | L |
|---|---|---|---|
| first | outer | inner | last |

Remember: You can rewrite a subtraction as addition of the opposite.

**TALK IT OVER**

In Example **2b,** the sum of the "outer" and "inner" terms is zero, resulting in a product with only two terms ($m^2$ and $-49$). Describe a pattern in the original binomials that will always produce this result.

## Example 2

Simplify.

**a.** $(x + 5)(x - 8)$ **b.** $(m + 7)(m - 7)$ **c.** $(k - 9)^2$

### Solution

Use the FOIL method and the distributive property.

F L I O

**a.** $(x + 5)(x - 8) = (x + 5)[x + (-8)]$
$= x(x) + (-8x) + 5x + 5(-8)$
$= x^2 - 3x - 40$

**b.** $(m + 7)(m - 7) = (m + 7)[m + (-7)]$
$= m(m) + (-7m) + 7m + 7(-7)$
$= m^2 - 49$

**c.** $(k - 9)^2 = (k - 9)(k - 9)$ ← **Write the terms indicated by the exponent 2.**
$= [k + (-9][k + (-9)]$
$= k(k) + (-9k) + (-9k) + (-9)(-9)$
$= k^2 - 18k + 81$ ◄

After working with factors involving subtraction for a while, you may not need to rewrite subtraction as addition of the opposite.

**COMPUTER**

This program will square binomials of the form $X + A$.

```
10  PRINT "TO SQUARE
    THE BINOMIAL X + A,
    WHAT IS A";
20  INPUT A
30  LET B = 2 * A
40  IF B < 0 THEN 70
50  PRINT "X^2 +";B;
    "X +";A * A
60  GOTO 80
70  PRINT "X^2 -";ABS(B);
    "X +";A * A
80  PRINT
90  PRINT "ANY MORE
    BINOMIALS TO SQUARE
    (1 = YES, 2 = NO)";
100 INPUT W
110 IF W = 1 THEN 20
```

## Example 3

Simplify.

$5(n + 3)(n - 4) - 2(n - 6)(n + 5)$

### Solution

Use the FOIL method and the distributive property as often as needed.

$$5(n + 3)(n - 4) - 2(n - 6)(n + 5)$$
$$= 5[n^2 - 4n + 3n + 3(-4)] - 2[n^2 + 5n - 6n - 6(5)]$$
$$= 5(n^2 - n - 12) - 2(n^2 - n - 30)$$
$$= 5n^2 - 5n - 60 - 2n^2 + 2n + 60$$
$$= 3n^2 - 3n$$ ◄

## TRY THESE

Find the product.

1. $(x + 2)(x + 5)$
2. $(n - 7)(n + 3)$
3. $(p - 6)(p + 6)$
4. $(h - 4)^2$
5. Simplify. $4(y + 3)(y + 2) + 3(y + 4)(y + 1)$

# EXERCISES

### PRACTICE/ SOLVE PROBLEMS

**COMPUTER TIP**

You can adapt the program on page 283 to square binomials of the form $Ax + B$ in Exercises 17–22. One change would be to INPUT values $A$ and $B$ in line 20. Then, if $C$ is used as the variable for the middle term of the trinomial, what other changes would have to be made in the program?

Find the product.

1. $(a + 3)(a + 2)$
2. $(c + 4)(c - 5)$
3. $(p - 3)(p + 1)$
4. $(x - 5)(x - 3)$
5. $(e + 2)^2$
6. $(c + 5)(c - 5)$
7. $(6 + x)(6 - x)$
8. $(m - 10)(m + 8)$
9. $(7 + t)(12 + t)$
10. $(b + 12)^2$
11. $(w - 9)(w - 8)$
12. $(20 + x)(-4 + x)$

Simplify.

13. $(y - 4)(y + 6) + (y - 9)(y + 3)$
14. $x(x + 1)(x - 3)$
15. $3(x + 4)(x + 2) - 2(x + 1)(x + 3)$
16. Write an expression for the area of a square if the measure of each side is $5 - v$.

### EXTEND/ SOLVE PROBLEMS

Find the product. Use the FOIL method.

17. $(2x + 3)(3x + 2)$
18. $(4k - 4)(3k - 5)$
19. $(5c + 1)(5c - 1)$
20. $(7y + 3)(7y - 4)$
21. $(6x + 5)(5x + 6)$
22. $(3v + 5)(5v - 4)$

Simplify.

23. $(3k - 1)(2k + 1) - (k - 1)^2$
24. $(5h - 7)^2 - (7 - 5h)^2$

### THINK CRITICALLY/ SOLVE PROBLEMS

25. Find the product. $(x + 1)^3$
26. Find the product. $(2c - 3)^3$

Write an expression in simplest form for the volume of each.

27.

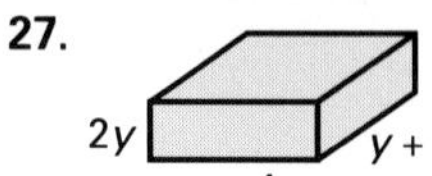


28.

3n
n − 2
n + 6

29.

2n + 1
2n
3n − 1

► READ
► PLAN
► SOLVE
► ANSWER
► CHECK

# Problem Solving Applications:

## USING POLYNOMIALS TO SOLVE AREA PROBLEMS

### PROBLEM

A rectangular swimming pool is twice as long as it is wide. A walkway surrounding the pool is 2 yd wide and has an area of 196 $yd^2$. Find the dimensions of the pool.

### SOLUTION

Make a drawing. Let $x$ represent the width of the pool. Your sketch should show the other dimensions in terms of $x$.

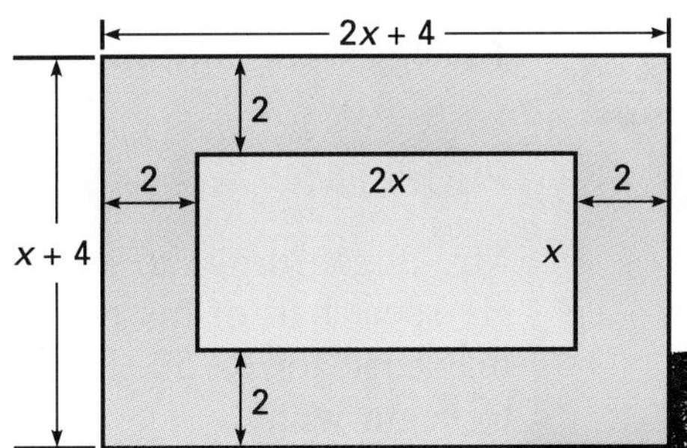


area of walkway = area of pool and walkway − area of pool

$$
\begin{aligned}
196 &= (2x + 4)(x + 4) - (2x)(x) \\
196 &= 2x^2 + 8x + 4x + 16 - 2x^2 \\
196 &= 12x + 16 \\
180 &= 12x \\
15 &= x
\end{aligned}
$$

The width of the pool is 15 yd. The length is twice the width, or 30 yd.

Solve.

1. A rectangular playground is four times as long as it is wide. The area of a 3-ft-wide sidewalk surrounding the playground is 1,236 $ft^2$. Find the dimensions of the playground.
2. A photo is 6 in. longer than it is wide. A $1\frac{1}{2}$-inch frame surrounds the photo. If the area of the frame is 99 $in.^2$, what are the dimensions of the photo?
3. A rectangle is three times as long as it is wide. If the length is increased by 1 cm and the width decreased by 1 cm, the new rectangle has an area 15 $cm^2$ less than that of the original rectangle. Find the original dimensions.
4. The most common size for a color transparency used by photographers is 12 mm longer than it is wide. A cardboard border 8 mm wide on two sides and 14 mm wide on two sides surrounds the transparency as shown. The area of the border is 1,796 $mm^2$. Find the dimensions of the transparency.

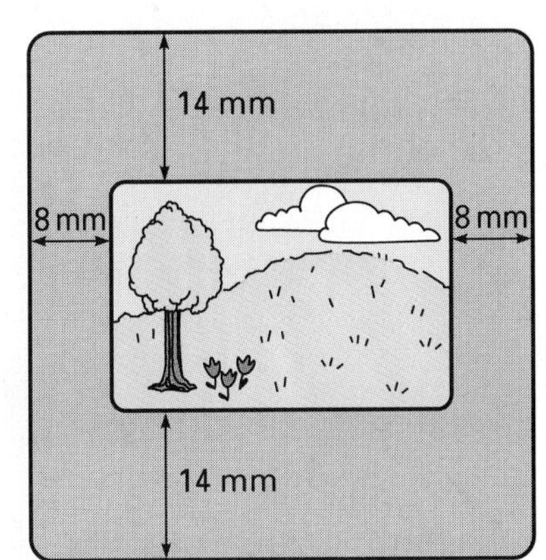

# 7-6 Problem Solving Strategies:
## WORK BACKWARDS

- READ
- PLAN
- SOLVE
- ANSWER
- CHECK

When a problem takes place over a period of time, you often solve it by working forward from the beginning of the problem to the end. Sometimes, however, you are told what happened at the end and asked to find out what happened at the beginning. You can use the strategy *work backwards* to solve this type of problem.

### PROBLEM

Mr. Bogen drove to a filling station, where he spent \$16 on gas. He spent half of his remaining cash for lunch, and then he bought a magazine for \$2.50. He had \$9.50 left. How much cash did he have to begin with?

### MIXED REVIEW

Write each fraction or decimal as a percent.

**1.** 0.2 **2.** $\frac{1}{8}$ **3.** 0.02

Find the product or quotient.

**4.** $-6(7)$ **5.** $-88 \div -11$

**6.** Find the value of *x*.

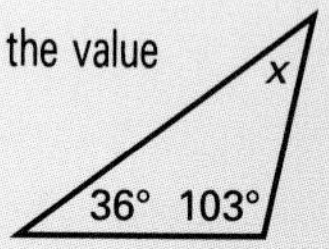


**7.** Solve. $\frac{m}{3} - 5 = -3$

**8.** Find the circumference of a circle with a radius of 6 in. Use 3.14 for $\pi$.

### SOLUTION

The diagram illustrates the steps described in the problem.

Beginning amount: ? → Subtract \$16.00. → Divide by 2. → Subtract \$2.50. → Ending amount: \$9.50

To find the beginning amount, work backwards from the end by reversing each step.

| | |
|---|---|
| Ending amount: | \$9.50 |
| *Add* \$2.50: | \$9.50 + \$2.50 = \$12.00 |
| *Multiply* by 2: | \$12.00 × 2 = \$24.00 |
| *Add* \$16.00: | \$24.00 + \$16.00 = \$40.00 |

So Mr. Bogen had \$40 to begin with.

Check: \$40 − \$16.00 = \$24.00
\$24.00 ÷ 2 = \$12.00
\$12.00 − \$2.50 = \$9.50 ✓

# PROBLEMS

1. After leaving her home in the morning, Gwen drove 25 mi at an average speed of 50 mi/h, then stopped 40 min for breakfast. Resuming her trip, she drove 90 mi at an average speed of 60 mi/h, arriving at 11 A.M. What time did she leave home?

2. If you multiply Wilbur's age by 8, add 8, divide by 8, and then subtract 8, the result is 8. How old is Wilbur?

3. Between 1980 and 1985 the price of a Peerless digital watch decreased \$40. From 1985 to 1987 the price halved. Inflation increased the price by \$5 in 1988, but by 1993 the price had fallen to \$7, which is one third of the 1988 price. Find the price of a Peerless watch in 1980.

**USING DATA** Refer to the Data Index on pages 556–557 to find the table listing the longest highway tunnels in the United States.

4. Paula drove at an average rate of 60 mi/h through a tunnel and then 18.85 mi farther to a rest stop. After a 10-min rest she noted that it had been exactly 30 min since she had started through the tunnel. Which tunnel was it?

5. Pete is reading a book with 480 pages. When he has read three times as many pages as he already has read, he will be 144 pages from the end. How many pages has he read?

6. Each time a dropped ball bounces it returns to a height $\frac{2}{3}$ the height of the previous bounce. After the third bounce the ball returns to a height of 4 ft. From what height was it dropped?

Work backwards to find the factors that were multiplied to produce the given product.

7. $x^2 + 5x = x(\blacksquare + \blacksquare)$

8. $m^2n - m^3 = m^2(\blacksquare - \blacksquare)$

9. $7c^2 + 21c - 42 = 7(\blacksquare + \blacksquare - \blacksquare)$

10. $9a^3 + 15a^2 - 6a = 3a(\blacksquare + \blacksquare - \blacksquare)$

11. $8w^4y - 10w^3k^2 + 2w^2 = 2w^2(\blacksquare - \blacksquare + \blacksquare)$

12. $24x^3 + 32x^2 - 40x + 16 = \blacksquare(\blacksquare + \blacksquare - \blacksquare + \blacksquare)$

# 7-7 Factors and Multiples

**EXPLORE**

Use 24 square tiles or centimeter cubes.

a. Find as many ways as you can to arrange the 24 tiles to form a rectangle with no tiles left over. Write the length and width of each rectangle.

b. Repeat question **a** using 23 tiles.

c. What important difference did you discover about the numbers 24 and 23?

**SKILLS DEVELOPMENT**

When two or more numbers or expressions are multiplied, each is a **factor** of the product.

Since $7 \times 6 = 42$, 7 and 6 are factors of 42.

Since $(a^3)(5a) = 5a^4$, $a^3$ and $5a$ are factors of $5a^4$.

A whole number greater than 1 that has exactly two factors, 1 and the number itself, is a **prime number.** A whole number greater than 1 that is not a prime number is a **composite number.**

When a whole number is written as a product of prime factors, the product is the **prime factorization** of the number. There is only one prime factorization of a number, although the order in which the factors are given may vary.

### Example 1

Write the prime factorization of 40.

**Solution**

Express 40 as a product of its smallest prime factor and another number. Continue until you have a product in which all factors are prime.

$$\begin{aligned} 40 &= 2 \cdot 20 \\ &= 2 \cdot 2 \cdot 10 \\ &= 2 \cdot 2 \cdot 2 \cdot 5 \\ &= 2^3 \cdot 5 \end{aligned}$$

40
2 20
2 10
2 5 ◄

Two or more integers may have some factors that are the same. The greatest integer that is a factor of each of the integers is called their **greatest common factor (GCF).** You can find a GCF by using the prime factorization of each integer.

### Example 2

Find the GCF of 36 and 90.

**Solution**

Write the prime factorization of each. Underline the common factors.

$36 = \underline{2}^2 \cdot \underline{3}^2 \qquad 90 = \underline{2} \cdot \underline{3}^2 \cdot 5$

Choose the *least* power of each underlined factor, then multiply.

$2 \cdot 3^2 = 18$

So, the GCF of 36 and 90 is 18. ◄

You can also find the GCF of two or more monomials. The GCF is the product when the GCF of the coefficients is multiplied by the least power of the *common* variable factors.

## Example 3

Find the GCF of $27a^4b^3$ and $45ab^5c$.

**Solution**

$27a^4b^3 = \underline{3}^3 \cdot \underline{a}^4 \cdot \underline{b}^3$ ← Underline common prime factors and common variable factors.

$45ab^5c = \underline{3}^2 \cdot 5 \cdot \underline{a} \cdot \underline{b}^5 \cdot c$

$\text{GCF} = 3^2 \cdot a \cdot b^3 = 9ab^3$ ← Choose the *least* power of each underlined factor.

So, the GCF of $27a^4b^3$ and $45ab^5c$ is $9ab^3$. ◄

When you multiply an integer by 1, 2, 3, and so on, you obtain **multiples** of the integers.

multiples of 18: 18, 36, 54, 72, 90, 108, 126, . . .

multiples of 24: 24, 48, 72, 96, 120, 144, 168, . . .

Two or more integers will always have some multiples that are the same. The least integer that is a multiple of each of the integers is called their **least common multiple (LCM).** By inspecting the list of multiples above, for example, you can see that the LCM of 18 and 24 is 72. You can find the LCM of two or more integers by using prime factorization. You can also find the LCM of two or more monomials. The LCM is the product when the LCM of the coefficients is multiplied by the greatest power of *all* variable factors.

## Example 4

Find the LCM of each set.

**a.** 18 and 24 **b.** $4m^5n^2p$ and $30m^2n^4$

**Solution**

**a.** Write the prime factorization of each integer. $18 = 2 \cdot 3^2 \quad 24 = 2^3 \cdot 3$
Choose the greatest power of each factor, then multiply. $2^3 \cdot 3^2 = 72$
So, the LCM of 18 and 24 is 72.

**b.** $4m^5n^2p = \underline{2}^2 \cdot \underline{m}^5 \cdot n^2 \cdot \underline{p}$ ← Underline the *greatest* power of all prime factors and all variable factors.

$30m^2n^4 = 2 \cdot \underline{3} \cdot \underline{5} \cdot m^2 \cdot \underline{n}^4$

$\text{LCM} = 2^2 \cdot 3 \cdot 5 \cdot m^5 \cdot n^4 \cdot p = 60m^5n^4p$ ← Multiply the underlined factors.

So, the LCM of $4m^5n^2p$ and $30m^2n^4$ is $60m^5n^4p$. ◄

You can solve certain problems by finding the GCF or the LCM.

### Example 5

Two groups of folk dancers are to be arranged on stage side by side in the same number of rows. There are 56 dancers in one group and 48 dancers in the other group. What is the greatest number of rows into which each group of folk dancers can be arranged?

**Solution**

Find the greatest common factor of 56 and 48.

$56 = 2 \cdot 2 \cdot 2 \cdot 7 \qquad 48 = 2 \cdot 2 \cdot 2 \cdot 2 \cdot 3$

The GCF of 56 and 48 is $2^3$ or 8.

The greatest number of rows is 8. ◄

## TRY THESE

1. Write the prime factorization of 63.
2. Find the GCF: 84, 120
3. Find the GCF: $24m^2$, $32m^3n$
4. Find the LCM: 12, 21
5. Find the LCM: $8k^3$, $12k^2y$

## EXERCISES

**PRACTICE/ SOLVE PROBLEMS**

Write the prime factorization.

1. 32
2. 21
3. 54
4. 65
5. 152

Find the GCF of each set.

6. 56, 84
7. 20, 12
8. 52, 65
9. 48, 84
10. $4a^2$, $6ab$
11. $24p^2$, $60p$
12. $12x^3y$, $15x^2y^3$

Find the LCM of each set.

13. 10, 45
14. 15, 18
15. 9, 7
16. 36, 54
17. $8xy^2$, $6x^3y$
18. $20hk$, $50k$
19. $12pt^2w$, $35p^2w^3$
20. Some locusts hatch every 17 years. Century plants bloom every 10 years. Both occurred in 1992. When will they occur again in the same year?
21. One neon sign flashes every 4 seconds. Another flashes every 10 seconds. How long after they flash together will they flash together again?

22. If 36 spruces and 24 pines are to be planted so that there are the same number of rows of each kind of tree, what is the greatest number of rows into which each group can be arranged?

## EXTEND/ SOLVE PROBLEMS

Find the GCF of each set.

23. 12, 15, 18
24. 16, 40, 56
25. 42, 56, 35
26. $8a^2bc^3$, $12ab^2c^2$, $24ab^4c^3$
27. $35mn^2p^3$, $14m^3n^2p^3$, $28m^2n^2p^2$
28. $36x^3y^3z^2$, $24x^2y^2z^3$, $48x^3y^3z^3$

Find the LCM of each set.

29. 6, 8, 12
30. 14, 18, 21
31. 18, 32, 72
32. $9r^2s$, $75rs^2t^2$, $50r^2s^2t^2$
33. $16k^2l^3$, $52j^3k^3l^4$, $32j^2k$
34. A wood carver has three pieces of wood measuring 63 in., 91 in., and 105 in. He wants to cut the wood into equal pieces without wasting any. Find the length of the longest pieces of equal length that the carver can cut from the three pieces.
35. A golf course runs 336 yd along a lake and 528 yd beside a wildlife preserve. The course designer wants to place a row of equally spaced trees all along the side of the course that borders on the preserve and the lake. The designer wants to place one tree between the lake and the preserve and have the maximum possible distance between trees. Find the distance between trees.

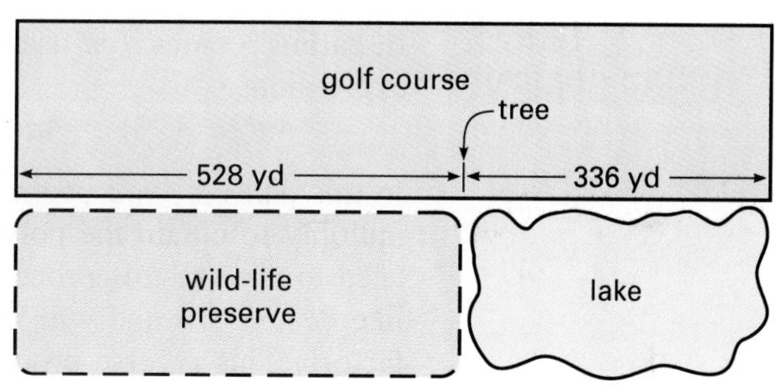


## THINK CRITICALLY/ SOLVE PROBLEMS

36. *Twin primes* are two prime numbers, such as 17 and 19, that differ by 2. Find all twin primes less than 50.

If two numbers have 1 as their GCF, they are **relatively prime.** Tell whether the numbers are relatively prime.

37. 12, 25
38. 14, 35
39. 21, 39
40. 19, 30
41. 64, 66
42. 60, 141
43. What is the GCF of two prime numbers? Explain.
44. Find all pairs of numbers less than 100 that have a GCF of 24.
45. What is the LCM of 1 and any other number?
46. What is the LCM of two prime numbers?

# 7-8 Common Monomial Factoring

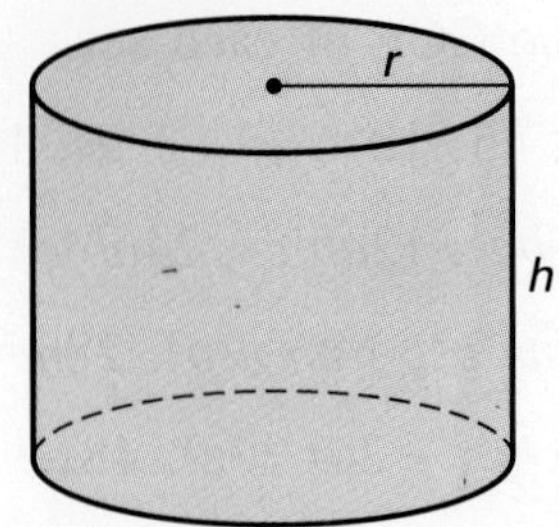


## EXPLORE

The formula for the surface area ($SA$) of a cylinder is usually given as $SA = 2\pi r^2 + 2\pi rh$, where $r$ is the radius of the cylinder and $h$ is the height.

The formula can also be written $SA = 2\pi r(r + h)$.

**a.** Find the surface area of a cylinder with a radius of 3.6 cm and a height of 8.4 cm using both formulas. Use 3.14 for $\pi$.

**b.** Which formula was easier to use? Explain.

**c.** What advantage might there be in writing a polynomial in a different form?

## SKILLS DEVELOPMENT

In earlier lessons you used the distributive property to simplify expressions.

$$4a(2a + 3) = 4a(2a) + 4a(3) = 8a^2 + 12a$$

In this process, you begin with the factors $4a$ and $2a + 3$, then multiply to obtain the polynomial $8a^2 + 12a$. Sometimes you will need to reverse this process. That is, you will be given a polynomial like $8a^2 + 12a$ and will be asked to work backwards to find the factors. This reverse process is called **factoring** a polynomial. To factor a polynomial like $8a^2 + 12a$, you find the GCF of its monomial terms.

### Example 1

Factor each polynomial.

**a.** $10n + 6$ **b.** $4c^2 - 12c^5$ **c.** $9h^2k^2 - 12hk^2 + 24h^3k$

### Solution

**a.** Find the GCF of $10n$ and 6.

$10n = 2 \cdot 5 \cdot n \qquad 6 = 2 \cdot 3$

The GCF is 2.

Use the GCF and the distributive property to rewrite the polynomial.

$$10n + 6 = 2 \cdot 5n + 2 \cdot 3$$
$$= 2(5n + 3)$$

Check by multiplying.

$2(5n + 3) = 2(5n) + 2(3) = 10n + 6$ ✓

So, $10n + 6 = 2(5n + 3)$.

b. Find the GCF of $4c^2$ and $12c^5$.

$4c^2 = 2^2c^2$ $\qquad$ $12c^5 = 2^2 \cdot 3 \cdot c^5$

The GCF is $2^2c^2$, or $4c^2$.

$4c^2 - 12c^5 = 4c^2(1 - 3c^3)$

Check: $4c^2(1 - 3c^3) = 4c^2 - 12c^5$ ✓

So, $4c^2 - 12c^5 = 4c^2(1 - 3c^3)$.

c. Find the GCF for each term.

$9h^2k^2 = 3^2h^2k^2$

$12hk^2 = 2^23hk^2$

$24h^3k = 2^33h^3k$

The GCF is $3hk$.

$9h^2k^2 - 12hk^2 + 24h^3k = 3hk(3hk - 4k + 8h^2)$

Check: $3hk(3hk - 4k + 8h^2) = 9h^2k^2 - 12hk^2 + 24h^3k$ ✓

So, $9h^2k^2 - 12hk^2 + 24h^3k = 3hk(3hk - 4k + 8h^2)$. ◄

## Example 2

The formula for the surface area ($SA$) of a rectangular prism with length $l$, width $w$, and height $h$ is $SA = 2lw + 2wh + 2lh$. Rewrite the formula by factoring.

**Solution**

The GCF of $2lw + 2wh + 2lh$ is 2.

$2lw + 2wh + 2lh = 2(lw + wh + lh)$

So, $SA = 2(lw + wh + lh)$. ◄

## TRY THESE

Factor each polynomial.

1. $7w - 21$
2. $4c^2 + 6c^3$
3. $9a^2 - 6a$
4. $16xy + 12x$
5. $25mn^3 - 15mn^2 + 5m$
6. The amount ($A$) in dollars returned on a principal of $P$ dollars invested at an annual interest rate of $i$ percent is given by the formula $A = P + Pi$. Rewrite the formula by factoring.

# EXERCISES

## PRACTICE/ SOLVE PROBLEMS

Factor each polynomial.

1. $8mn - 8mp$
2. $12k + 15$
3. $4x - 20$
4. $7e^2 + 21e$
5. $9p^3 + 27p^2$
6. $x^2 - xy$
7. $12k^2 - 42k$
8. $5y^4 - 20y^3$
9. $w^5 - w^4$
10. $100c - 200$
11. $3xy^2 + 6x$
12. $-7xy - 56xz$
13. $v^5 - 6v^4 + 3v^3$
14. $5n^3 - 30m^2 - 15$
15. $3a^6 - 5a^3 + 2a^2$
16. $xy + xz + 2x$
17. $12x^4y^4 + 3x^3y^2 - 6x^2y^2$
18. $2xya - 4xyb + 6xyc$
19. $x^2 + 6xy - x$
20. $8h^2 - 16h + 24$
21. $50m^2 + 125mn + 25n^2$
22. $15a^3b + 20a^2b - 10ab$
23. $64x^6 - 48x^4 + 24x^2$
24. $6m^3n^3 + 3m^3n^2 + m^2n^2$

Rewrite each formula.

25. The formula for the perimeter ($P$) of a rectangle with length $l$ and width $w$ is $P = 2l + 2w$. Rewrite the formula by factoring.
26. The formula for the surface area ($SA$) of a cone with radius $r$ and slant height $h$ is $SA = \pi r^2 + \pi rh$. Rewrite the formula by factoring.
27. The formula for the number of diagonals ($D$) that can be drawn in a polygon with $n$ sides is $D = \frac{1}{2}n^2 - \frac{3}{2}n$. Rewrite the formula by factoring.
28. The formula for the sum ($S$) of an arithmetic progression with first term $a$, last term $l$, and $n$ terms is $S = \frac{1}{2}an + \frac{1}{2}ln$. Rewrite the formula by factoring.

## EXTEND/ SOLVE PROBLEMS

Factor each polynomial.

29. $18m^3n^2 + 45m^2n^3 + 27m^4n - 54m^2n^2$
30. $35x^5y - 40x^3y^2 + 10x^2 + 45x^4y^3$
31. $48ab - 40a^3b^2 + 24a^2b^3 + 28a^2b^2$
32. $x^2y^3z - x^2y^2z^2 + xy^4z - xy^3z^2$

Write an expression for the perimeter of each figure. Then factor the expression.

**33.** 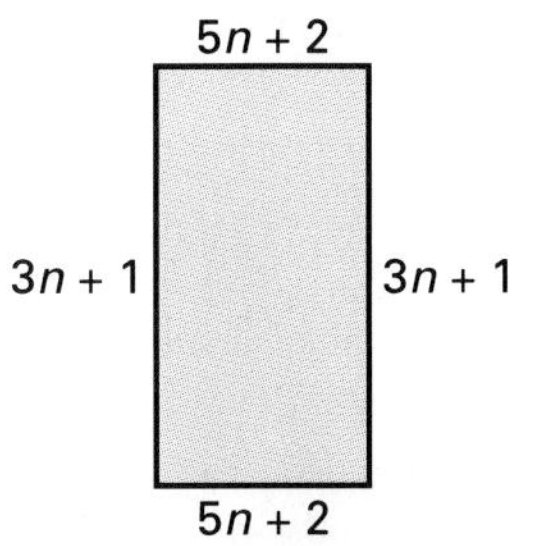


**34.** 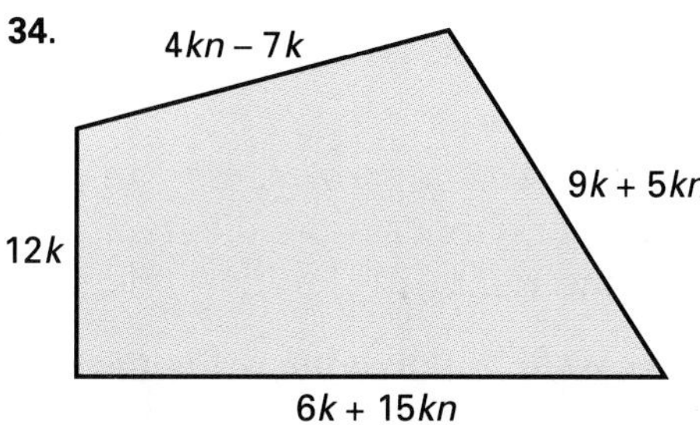


**35.** 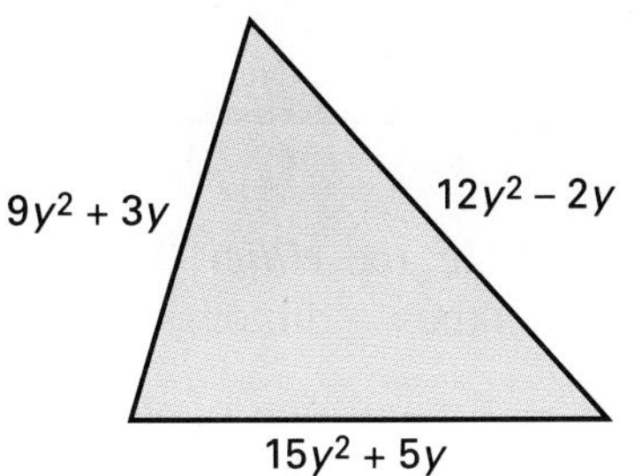


**36.** 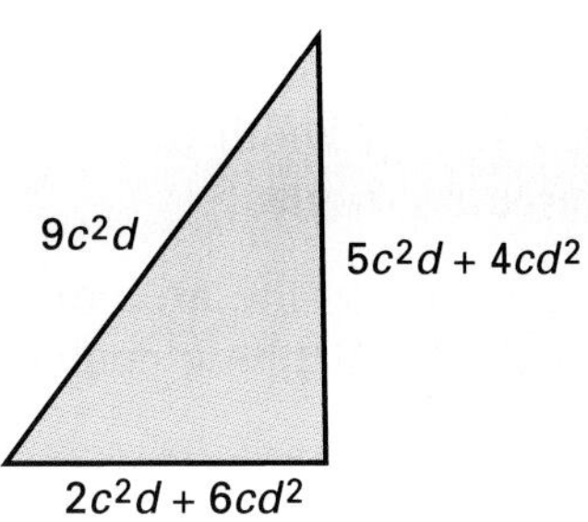


**37.** The figure gives the measurements of two paintings hanging on a wall. Write an expression for the total area of the paintings and then write it in factored form.

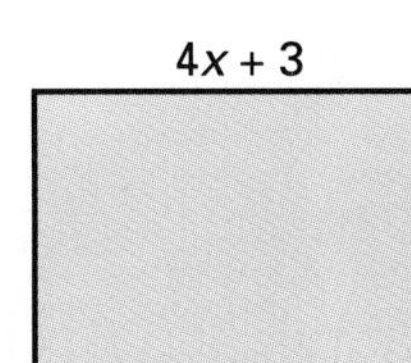


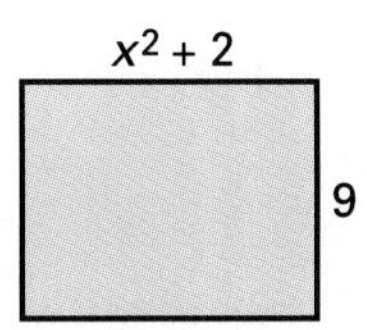


## THINK CRITICALLY/ SOLVE PROBLEMS

If two polynomials have 1 as their only common factor, they are *relatively prime.* Tell whether the polynomials are relatively prime.

**38.** $2,\ 3x$ **39.** $mn,\ n^2$ **40.** $4ab,\ 3ab$

**41.** $c + c^2,\ 5c^4$ **42.** $k,\ k + 5$ **43.** $2m + 2,\ km^2 + km$

Find each product.

**44.** $(x + 5)(x - 5)$ **45.** $(n + 7)(n - 7)$

**46.** $(p + 4)(p - 4)$ **47.** $(k + 9)(k - 9)$

Study the pattern in the products for Exercises 44–47. Then factor each polynomial.

**48.** $v^2 - 9$ **49.** $w^2 - 64$

**50.** $m^2 - 36$ **51.** $c^2 - 100$

**52.** $4x^2 - 25$ **53.** $36y^2 - 121z^6$

# 7-9 Dividing by a Monomial

## EXPLORE/ WORKING TOGETHER

Work with a partner. Use algebra tiles to duplicate the diagram at the right.

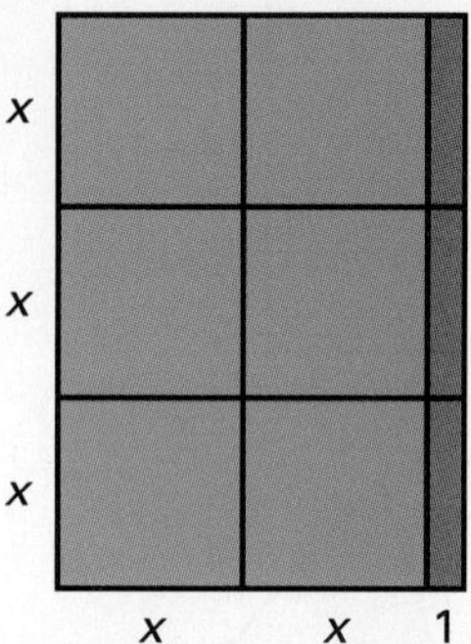


**a.** Write a polynomial for the area of the large rectangle.

**b.** Write monomials for the length and width of the large rectangle.

**c.** If you knew the area of the large rectangle and *either* the length *or* the width, how could you find the unknown dimension?

**d.** Suppose the area of a rectangle is $x^2 + 4x$ and one of its dimensions is $x$. Use algebra tiles and try to "build" the rectangle. Now find the unknown dimension. Finally, write an equation that describes the work you did.

## SKILLS DEVELOPMENT

Recall how the quotient rule for exponents works.

$$\frac{a^m}{a^n} = a^{m-n}$$

You can use the quotient rule to divide one monomial by another. Throughout this section, it is assumed that no denominator has a value of zero.

**CHECK UNDERSTANDING**

In Example **1a**, what is the GCF of $12mn$ and $9m$? How is dividing a monomial by a monomial like simplifying a fraction?

### Example 1

Simplify.

**a.** $\dfrac{12mn}{9m}$ **b.** $\dfrac{-14h^7k^5}{18hk^3}$

### Solution

**a.** $\dfrac{12mn}{9m} = \dfrac{2^2 3mn}{3^2 m}$ ← **Write the prime factorization of each coefficient.**

$= \dfrac{4n}{3}$ ← **To simplify, divide the numerator and denominator by any common factors.**

**b.** $\dfrac{-14h^7k^5}{18hk^3} = \dfrac{(-1)(2)(7)h^7k^5}{(2)(3^2)hk^3}$

$= \dfrac{(-1)(7)h^6k^2}{3^2}$ ← **Simplify.**

**$2^{1-1} = 2^0 = 1$**
**$h^{7-1} = 7^6$**
**$k^{5-3} = k^2$**

$= \dfrac{-7h^6k^2}{9}$ ◄

You know how to use the distributive property to multiply each term of a polynomial by a monomial.

$$3n(4n + 7) = 3n(4n) + 3n(7)$$
$$= 12n^2 + 21n$$

Since division is the inverse of multiplication, you can reverse this process to divide a polynomial by a monomial.

$$3n(4n + 7) = 12n^2 + 21n, \text{ so } (12n^2 + 21n) \div 3n = 4n + 7.$$

To divide a polynomial by a monomial, divide each term of the polynomial by the monomial. You can use the GCF and the quotient rule for exponents to simplify the quotient $(12n^2 + 21n) \div 3n$.

$$\frac{12n^2 + 21n}{3n} = \frac{12n^2}{3n} + \frac{21n}{3n}$$
$$= \frac{12}{3} \cdot \frac{n^2}{n} + \frac{21}{3} \cdot \frac{n}{n}$$
$$= 4 \cdot n^{2-1} + 7 \cdot 1$$
$$= 4n + 7$$

Divide each pair of coefficients by their GCF, 3.
Divide each pair of variable parts by their GCF, $n$.

## Example 2

Simplify.

**a.** $\frac{8n + 12}{4}$ **b.** $\frac{9x^4 - 12x^3 + 15x^2}{3x^2}$ **c.** $\frac{3a^2b + 9ab^2}{3ab}$

### Solution

**a.** $\frac{8n + 12}{4} = \frac{8n}{4} + \frac{12}{4} = 2n + 3$

**b.** $\frac{9x^4 - 12x^3 + 15x^2}{3x^2} = \frac{9x^4}{3x^2} - \frac{12x^3}{3x^2} + \frac{15x^2}{3x^2} = 3x^2 - 4x + 5$

**c.** $\frac{3a^2b + 9ab^2}{3ab} = \frac{3a^2b}{3ab} + \frac{9ab^2}{3ab} = a + 3b$ ◄

**TALK IT OVER**

For any real numbers $a$, $b$, and $c$, where $c$ is not equal to zero, $\frac{a + b}{c} = \frac{a}{c} + \frac{b}{c}$.

Discuss how this rule is applied in the solution of Example **2**.

## TRY THESE

Simplify.

1. $\frac{-36ab}{12b}$
2. $\frac{24m^5n^6}{30m^2n^6}$
3. $\frac{-15x^3y^2z}{20x^3y}$
4. $\frac{64x^2yz^3}{8x^2z^2}$
5. $\frac{-42a^4b^5c}{7a^4bc}$
6. $\frac{-81m^3n^3p^3}{-9mn^3p^3}$
7. $\frac{6p - 12r}{6}$
8. $\frac{16a^2 + 4ab}{4a}$
9. $\frac{4xyz^2 - 4yz^2}{yz^2}$
10. $\frac{14k^3m^2 - 7km}{7km}$
11. $\frac{6h^5 - 2h^4 + 10h^3}{2h^3}$
12. $\frac{15c^3d^2 - 5cd + 10c}{-5c}$

# EXERCISES

**PRACTICE/ SOLVE PROBLEMS**

Simplify.

1. $\frac{8x^2}{2x}$
2. $\frac{12a^6}{12a^4}$
3. $\frac{-20y^3}{14y^3}$
4. $\frac{66uv^2}{-11v}$
5. $\frac{6ab}{12ac}$
6. $\frac{-16p^2q}{4p}$
7. $\frac{-150x^3y^2}{-10x^2y}$
8. $\frac{60x^4yz^2}{-12z}$
9. $\frac{110a^2bc^5}{10ac^2}$
10. $\frac{-30m^3n}{-6mn}$
11. $\frac{42a^2bc^3}{14a^2bc^3}$
12. $\frac{-24m^7n^9}{16m^3n^6}$
13. $\frac{4a + 20}{4}$
14. $\frac{9h - 21}{3}$
15. $\frac{15a^2 - 10a}{5a}$
16. $\frac{21n^2 + 35n^3}{-7n}$
17. $\frac{14x^2 - 21x^3}{7x^2}$
18. $\frac{a - 3a^3 + 5a^5}{a}$
19. $\frac{36c^6 + 30c^5 - 54c^4}{6c^3}$
20. $\frac{25w^3 - 15w^2 - 30w}{5w}$
21. $\frac{-24x^5 + 8x^4 - 64x^3}{-8x^3}$
22. $\frac{12m^3n^4 - 20m^4n^3 + 32mn^2}{4mn^2}$
23. $\frac{3x^2yz - 4xy^2z + 2xyz^2 - xyz}{xyz}$
24. $\frac{12ab - 21a^2b + 27ab^2 - 6a^2b^2}{-3ab}$

**EXTEND/ SOLVE PROBLEMS**

Write an expression for the unknown dimension of each rectangle.

25. area $= 36ab$

26. area $= 64mnp$

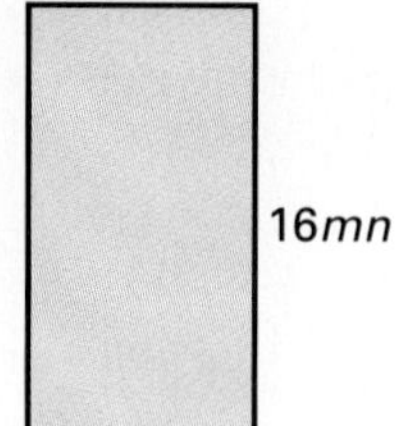


27. A rectangular garden has an area of $48pq$ square units. The width is $8q$ units. Write an expression for the length.

28. The area of a rectangular door is $72mn$ square units. If the length is $9n$ units, what is the width?

29. The total cost of a fleet of compact cars is $\$56mn$. Each car retails for $\$8n$. How many cars are there in the fleet?

30. How many tape decks can you buy for $\$48xy^2$ if each deck sells for $\$3y$?

**MENTAL MATH TIP**

If you learn to think in "chunks," you can factor many polynomials mentally. Think of the *largest* monomial that is a factor of each monomial in the polynomial.

To factor $20x^2 + 15x$, think:

What is the largest monomial factor of $20x^2$ and $15x$? Answer: $5x$.

Think: $\frac{20x^2}{5x} = 4x$ and $\frac{15x}{5x} = 3$

So, $20x^2 + 15x = 5x(4x + 3)$.

31. The volume of a rectangular prism is $135mxy + 45my^2$ cubic units. The length is $5y$ units and the height is $9m$ units. Write an expression for the width.

## THINK CRITICALLY/ SOLVE PROBLEMS

32. The area of a rectangle is $8x^2y^2$. The width of the rectangle is $xy$. Write an expression for the perimeter of the rectangle.

33. A square has a perimeter of $28x + 40y$. Express the area as a polynomial.

34. A rectangle has a perimeter of $16x + 8 + 4y$ units and a width of $2y$ units. Write an expression for its area.

To divide a polynomial by a polynomial, factor the numerator and denominator. Then divide by common factors.

Simplify.

35. $\frac{4n + 12}{n + 3}$ 36. $\frac{2x - 12}{x - 6}$ 37. $\frac{24 - 8m}{3 - m}$

38. $\frac{5a + xa}{5 + x}$ 39. $\frac{yp^2 - 7p^2}{y - 7}$ 40. $\frac{3k^2 - 12k}{k - 4}$

41. Simplify. $\frac{9x + 9y + mx + my}{x + y}$

42. Solve. $\frac{3x^2 + 6xy}{x + 2y} = 21$

**PROBLEM SOLVING TIP**

In Exercise 41, group the first two terms and the last two terms of the numerator together. Factor each pair of terms.

# 7-10 Factoring Perfect Squares and Differences of Squares

## EXPLORE

**a.** Describe a characteristic shared by the following numbers: 1, 25, 100, 49

**b.** Simplify each of the following. Describe the pattern in the products.

$(x + 1)^2$ $(n - 5)^2$ $(k + 10)^2$ $(p - 7)^2$

## SKILLS DEVELOPMENT

**TALK IT OVER**

Why is there a factor of 2 in the middle term of a perfect square trinomial?

You may recall from Chapter 5 that integers such as 1, 4, and 9 are called *perfect squares* because they have square roots that are integers. A trinomial that results from squaring a binomial is called a **perfect square trinomial.**

$$(w + 6)^2 = (w + 6)(w + 6) = w^2 + 6w + 6w + 36$$
$$= w^2 + 12w + 36$$

So, $w^2 + 12w + 36$ is a perfect square trinomial. It fits the patterns of all perfect square trinomials.

- The first term is a perfect square. $w^2 = w \cdot w$
- The last term is a perfect square. $36 = 6 \cdot 6$
- The middle term is *twice* the product of the square roots of the first and last terms. $12w = (2)(6)w$

### Example 1

Tell whether the trinomial is a perfect square trinomial.

**a.** $n^2 + 10n + 25$ **b.** $k^2 + 7k + 49$

### Solution

**a.** The first term is a perfect square. $n^2 = n \cdot n$

The last term is a perfect square $25 = 5 \cdot 5$

The middle term is twice the product of the square roots of the first and last terms. $10n = (2)(5)n$

The trinomial is a perfect square trinomial.

**b.** First term: $k^2 = k \cdot k$
Last term: $49 = 7 \cdot 7$
Middle term: $7k \neq (2)(7)k$
The trinomial is not a perfect square trinomial. ◄

You can use the patterns in a perfect square trinomial to factor it.

## Example 2

Factor each polynomial.

**a.** $n^2 + 10n + 25$ **b.** $y^2 - 20y + 100$

### Solution

**a.** First establish that the trinomial is a perfect square trinomial.
Find the square root of the first term: $n^2 = n \cdot n$
Find the square root of the last term: $25 = 5 \cdot 5$
Use the square roots to write the factors:

$$n^2 + 10n + 25 = (n + 5)(n + 5) = (n + 5)^2$$

**b.** The trinomial is a perfect square trinomial.
$y^2 = y \cdot y$ $100 = 10 \cdot 10$
$y^2 - 20y + 100 = (y - 10)(y - 10) = (y - 10)^2$
(Since the sign of the middle term in the trinomial is negative, use a negative sign in each factor.) ◄

A different pattern develops in the product of the sum and difference of some monomials, such as $(y + 9)(y - 9)$.

sum of y and 9 ↓ difference of y and 9 ↓

$$(y + 9)(y - 9) = y^2 - 9y + 9y - 81 = y^2 - 81$$

A polynomial such as $y^2 - 81$ is called the **difference of two squares.** If you recognize that a polynomial is a difference of two squares, you can easily work backwards to find the factors.

## Example 3

Factor. $k^2 - 9$

### Solution

$k^2 - 9$ — First establish that the polynomial is the difference of two squares.

$k \cdot k$ $3 \cdot 3$

To factor a difference of two squares, write the sum and the difference of the square roots of the terms.

$$k^2 - 9 = (k + 3)(k - 3)$$ ◄

### MIXED REVIEW

1. Find 35% of 72.
2. What percent of 48 is 18?
3. 63 is 75% of what number?
4. Subtract: $-12 - (-15)$
5. Add: $(-8) + (-3) + 2$
6. Solve. $8x + 9 = -7$
7. Find $n$.

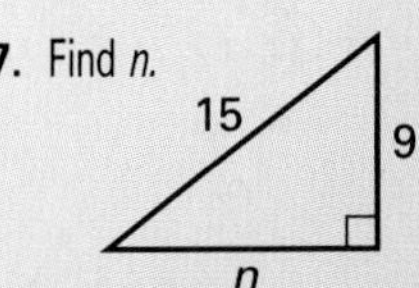


8. Find the area of a triangle with base 14 cm and height 8 cm.

## TRY THESE

Tell whether the trinomial is a perfect square trinomial.

1. $c^2 - 5c + 25$
2. $h^2 + 8h + 16$

Factor each polynomial if possible.

3. $m^2 + 18m + 81$
4. $f^2 - 2f + 1$
5. $p^2 + 49$

# EXERCISES

## PRACTICE/ SOLVE PROBLEMS

Tell whether the trinomial is a perfect square trinomial.

1. $x^2 + 6x + 9$
2. $y^2 + 14y + 49$
3. $n^2 + 5n + 25$
4. $p^2 - 20p + 40$
5. $x^2 + 24x + 144$
6. $m^2 - 40m + 100$

Factor each polynomial if possible.

7. $c^2 - 4c + 4$
8. $k^2 - 18k + 81$
9. $y^2 + 22y + 121$
10. $v^2 - 20v + 40$
11. $e^2 - 6e + 9$
12. $f^2 + 12f + 36$
13. $n^2 - 100$
14. $x^2 - 64$
15. $h^2 - 25$
16. $y^2 - 1$
17. $c^2 - 50$
18. $w^2 + 49$
19. $p^2 - 144$
20. $k^2 - 4$
21. $m^2 + 9$
22. $n^2 - 16$
23. $t^2 - 400$
24. $z^2 - 81$

## EXTEND/ SOLVE PROBLEMS

Factor each polynomial.

25. $3m^2 - 27$
26. $n^2p - 25p$
27. $2m^2 + 32m + 128$
28. $x^2m^2 - 14x^2m + 49x^2$
29. $3ak^2 - 300a$
30. $4y^2 + 72y + 324$

**USING DATA** Refer to the Data Index on pages 556–557 to find the table listing the longest highway tunnels in the United States.

31. The length of the largest tunnel in Belgium, the Schelde Tunnel, is the square root of the length of the Eisenhower Tunnel in the United States. How long is the Schelde Tunnel?

## THINK CRITICALLY/ SOLVE PROBLEMS

Factor each polynomial.

32. $9k^2 - 49$
33. $4n^2 - 25$
34. $4x^2 + 12x + 9$
35. $25x^2 - 40x + 16$
36. $49p^2 - 100$
37. Suppose the area measures of these squares are both perfect squares. If the difference in the areas is 32 square units, find the values of $x$ and $y$.

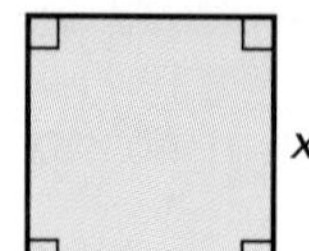


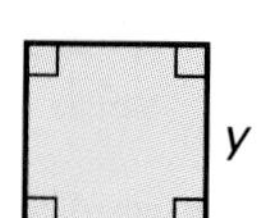

- READ
- PLAN
- SOLVE
- ANSWER
- CHECK

# Problem Solving Applications:

## USING POLYNOMIALS TO ANALYZE NUMBER TRICKS

Think of a number. Add 9. Multiply by 3. Subtract 15. Divide by 3. Subtract your original number. The result? It will be 4 every time!

You can use polynomials to analyze number tricks like this one and discover why they work.

Let $n$ = your number.

| **Steps** | **Polynomial** |
|---|---|
| Think of a number. | $n$ |
| Add 9. | $n + 9$ |
| Multiply by 3. | $3(n + 9) = 3n + 27$ |
| Subtract 15. | $3n + 27 - 15 = 3n + 12$ |
| Divide by 3. | $\frac{3n + 12}{3} = \frac{3n}{3} + \frac{12}{3} = n + 4$ |
| Subtract your original number. | $n + 4 - n = 4$ |

As you can see, the result will be 4 no matter what the original value of $n$ is.

 Use polynomials to show why each trick works.

1. Think of a number. Subtract 7. Multiply by 3. Add 30. Divide by 3. Subtract the original number. The result is always 3.
2. Think of any non-zero number. Subtract 6. Multiply by 5. Add 30. Divide by the original number. The result is always 5.
3. Think of a number. Add 5. Multiply by 6. Divide by 2. Subtract 15. Divide by 3. Describe the result. Use polynomials to explain why the result is always the same.
4. Make up your own number trick. Use polynomials to show why the trick works.

**JUST FOR FUN**

A magic square is a square in which the numbers in each row, column, and diagonal have the same sum. In the square below, what number will produce a magic square when substituted for $a$? What is the "magic" sum?

| $(2a - 1)^2$ | $a$ | $3a + 1$ |
|---|---|---|
| $a^2$ | 6 | $a^3$ |
| $a + 3$ | 10 | 3 |

## CHAPTER 7 • REVIEW

Choose the word from the list that completes each statement.

1. The number 19 is an example of a(n) ___?___.
2. In $3xy^2$, 2 is a(n) ___?___.
3. The number 8 is a(n) ___?___ of 32.
4. Use the ___?___ to rewrite $2(x + y)$ as $2x + 2y$.
5. $2m^3 + 4$ is an example of a(n) ___?___.

a. binomial
b. distributive property
c. prime number
d. factor
e. exponent

### SECTION 7–1 ADDING AND SUBTRACTING POLYNOMIALS (pages 268–271)

► A polynomial is written in *standard form* when its terms are ordered from the greatest to the least powers of one of the variables.
► Simplify a polynomial by combining all like terms so that only unlike terms remain.

6. Write $8xy^2 - 7x^2y + 5$ in standard form for the variable $x$.
7. Simplify. $8k^2 - 5k + 8k - 4k^2$

### SECTION 7–2 MULTIPLYING MONOMIALS (pages 272–275)

► To find the product of two monomials, use the commutative and associative properties of multiplication and simplify.
► Use the product rule for exponents to find a product of two monomials that have the same base.
► Use the power rule for exponents and the power of a product rule for exponents to simplify monomials that involve powers.

Simplify.

8. $(-3m)(2m^2n)$
9. $4x(2x^2)^3$

### SECTIONS 7–3 AND 7–6 PROBLEM SOLVING (pages 276–277, 286–287)

► You can use area models to multiply monomials.

10. Use an area model to find the product $(3h)(4h)$.

### SECTION 7–4 MULTIPLYING A POLYNOMIAL BY A MONOMIAL (pages 278–281)

► Use the distributive property and the rules for exponents to multiply a polynomial by a monomial.

Simplify.

11. $8m(2m^2 + 3m - 4)$
12. $3x(2x^2y + xy + 5y^2)$

## SECTION 7–5 MULTIPLYING BINOMIALS (pages 282–285)

► To multiply two binomials, write the products of the first terms, the outer terms, the inner terms, and the last terms (FOIL), then simplify.

Find the product.

**13.** $(e + 3)(e + 7)$

**14.** $3(w - 6)(w + 6)$

## SECTION 7–7 FACTORS AND MULTIPLES (pages 288–291)

► The greatest common factor (GCF) of two or more monomials is the product of the GCF of the coefficients and the least power of the *common* variable factors.

► The least common multiple (LCM) of two or more monomials is the product of the LCM of the coefficients and the greatest power of *all* variable factors.

**15.** Find the GCF of $24a^2b^4$ and $40ab^3$.

**16.** Find the LCM of $6x^3y$ and $8xy^2$.

## SECTION 7–8 COMMON MONOMIAL FACTORING (pages 292–295)

► To *factor* an expression, use the GCF and the distributive property.

Factor each polynomial.

**17.** $12x - 16$

**18.** $xy^2 - x^2y + xy$

## SECTION 7–9 DIVIDING BY A MONOMIAL (pages 296–299)

► To simplify the quotient of two monomials, divide both monomials by their GCF.

Simplify.

**19.** $\frac{6xy^3}{8xy}$

**20.** $\frac{-9c + 21c^3}{3c}$

**21.** $\frac{3p^4q + 12p^3q^2 + 15p^2q}{3pq}$

## SECTION 7–10 FACTORING PERFECT SQUARES (pages 300–303)

► Use these patterns to factor perfect square trinomials and polynomials that are differences of squares.

$a^2 + 2ab + b^2 = (a + b)^2$ $\quad a^2 - 2ab + b^2 = (a - b)^2$ $\quad a^2 - b^2 = (a + b)(a - b)$

Factor each polynomial.

**22.** $k^2 + 12k + 36$

**23.** $c^2 - 20c + 100$

**24.** $x^2 - 64$

**USING DATA** Use the chart on page 266 for Exercises 25 and 26. Suppose a car traveled $2rt + 16t^2$ highway miles.

**25.** Rewrite the expression $2rt + 16t^2$ by factoring.

**26.** With $r = 52$ and $t = 3$, the car used 12 gallons of gas. Classify the car by size.

# CHAPTER 7 ● TEST

Simplify.

1. $12x^2y^3 - 4xy^2 + 8xy^2 - 24x^2y^3$
2. $(3a^2b - 2a + 3b) + (4a^2b - 3a + 2b)$
3. $(6r^2s + 2rs - 3) - (4r^2s - 2rs)$
4. $(3a^3)(-4a^4)$
5. $(-4a^3b^4)^2$
6. $-2b(-4b^2 - b + 3)$
7. $4(x^2 + xy) - 3(x^2 - xy)$
8. $2n(4n - 1) + 3n(4n - 1)$
9. $(b + 4)(b + 2)$
10. $(p - 4)(p + 6)$
11. $(a + 8)(a - 8)$
12. $(k - 5)^2$
13. $3(x + 3)(x - 2) - 2(x - 1)(x + 4)$

Use the figure at the right for Exercises 14 and 15.

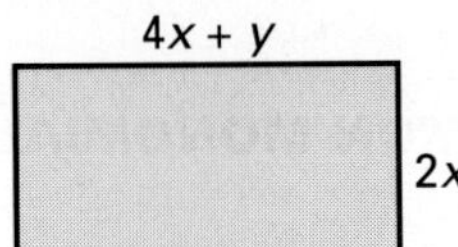


14. Write an expression for the perimeter of the rectangle.
15. Write an expression for the area of the rectangle.
16. If you divide Ana's age by 6, add 5, multiply by 3, and subtract 7, the result is 29. How old is Ana?

Find the GCF.

17. 28 and 42
18. $8x^2y$ and $10xy^2$

Find the LCM.

19. 15 and 20
20. $6k^3p^2$ and $9kp^3$

Factor each polynomial.

21. $6m + 9$
22. $4x^2y - 8x^3y^2$
23. $12h^2k^3 + 6hk^2 - 30h^4k$

Simplify.

24. $\dfrac{10a^2b^3c}{12a^2b}$
25. $\dfrac{-15p + 20p^2}{25p}$
26. $\dfrac{x + x^2 + x^3}{x}$
27. $\dfrac{4a^2b + 8a^3b^2}{4a^2}$

Factor each polynomial.

28. $x^2 - 4x + 4$
29. $f^2 + 16f + 64$
30. $y^2 - 100$

1. Use the function rule $f(x) = x - \frac{2}{3}$. What is the value of $f(x)$ when $x = -\frac{1}{3}$?

2. What is the function rule for the function table shown at the right?

| $x$ | $y$ |
|---|---|
| 1 | $-\frac{1}{3}$ |
| 2 | $-\frac{2}{3}$ |
| $-2$ | $\frac{2}{3}$ |
| 10 | $-3\frac{1}{3}$ |

3. A compact disc is on sale for \$19.50. If the sale price is 65% of the regular price, what is the regular price?

Evaluate each expression when $x = -4$ and $y = 2$.

4. $|x| + |y|$

5. $x - y$

6. $x - (-y)$

Perform the indicated operations.

7. $15.9 - (-7.5)$

8. $\left(-3\frac{1}{2}\right)\left(\frac{1}{5}\right)$

9. $8\frac{1}{2} \div (-3)$

Use the figure below for Exercises 10–12.

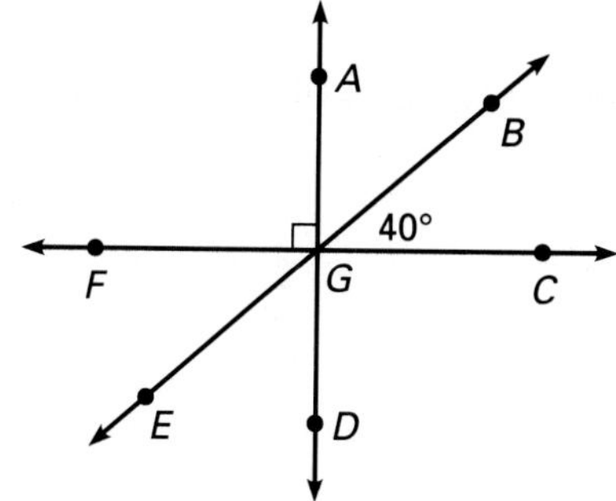


10. What is the measure of $\angle AGE$?

11. Name two angles that each measure 50°.

12. Name an angle congruent to $\angle AGE$.

Use the information and stem-and-leaf plot below to answer Exercises 13–16.

For 20 school days before the senior play was to be performed, the cast kept track of the number of tickets sold daily. The data are shown in the stem-and-leaf plot.

| Stem | Leaf |
|---|---|
| 1 | 7 |
| 2 | 0 5 6 8 |
| 3 | 1 2 7 7 7 9 |
| 4 | 0 3 3 5 8 |
| 5 | 6 7 9 |
| 6 | 2 |

1|7 represents 17 tickets

13. How many of the days during the 20-day period had ticket sales of fewer than 20 tickets?

14. How many of the days had ticket sales of more than 50 tickets?

15. What is the range of the number of tickets sold daily during the 20-day period?

16. What is the mode for the number of tickets sold daily?

Solve and graph each inequality.

17. $5n + 2 < -8$

18. $2x + 3 \geq 5$

Find the surface area of each figure.

19.

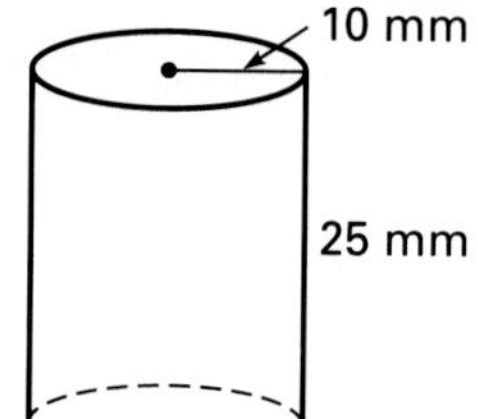


20.

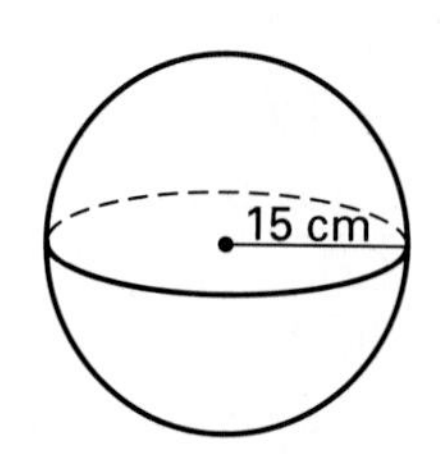


Simplify.

21. $7(a^2 + ab) - 4(a^2 - ab)$

22. $\dfrac{-35x^2y^2 + 20xy^2}{5xy}$

1. What property is demonstrated by the equation $\frac{1}{2}\left(\frac{2}{3} + 4\right) = \left(\frac{1}{2} \cdot \frac{2}{3}\right) + \left(\frac{1}{2} \cdot 4\right)$?
   A. associative B. distributive
   C. muliplicative identity D. additive identity

2. Given the sequence 1, 2, 4, 8, 16, . . . and using $n$ to represent the last of the given terms, which of the following expressions *cannot* be used to represent the next term in the sequence?
   A. $n + 2$ B. $2n$
   C. $n + n$ D. $3n - n$

3. Which is the value of the expression $-|x| + y$ when $x = -3$ and $y = -4$?
   A. $-7$ B. $-1$
   C. 1 D. 7

4. Which is the graph of the inequality $x \leq 1$?
   A. 0 1 2 B. 0 1 2
   C. 0 1 2 D. 0 1 2

5. Which is the result of simplifying the expression $(3a - 1) - (2a + 5)$?
   A. $a + 4$ B. $-a + 4$
   C. $a - 6$ D. $-a - 6$

Use the figure at the right for Exercises 6 and 7.

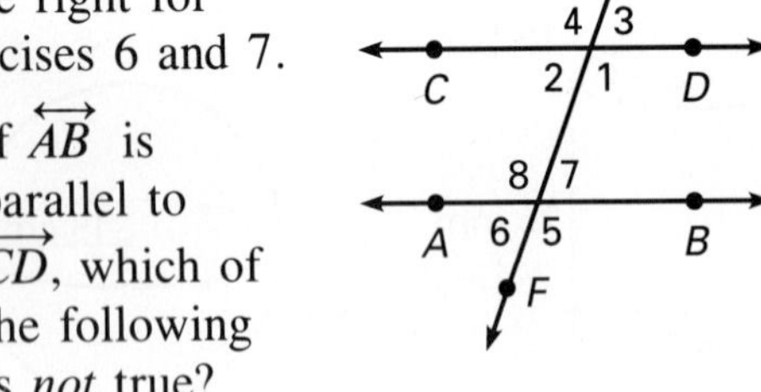


6. If $\overleftrightarrow{AB}$ is parallel to $\overleftrightarrow{CD}$, which of the following is *not* true?
   A. $\angle 2 = \angle 3$ B. $\angle 2 = \angle 6$
   C. $\angle 3 = \angle 7$ D. $\angle 4 = \angle 6$

7. Which is a pair of alternate exterior angles?
   A. $\angle 1$ and $\angle 4$ B. $\angle 4$ and $\angle 3$
   C. $\angle 4$ and $\angle 5$ D. $\angle 3$ and $\angle 5$

8. Which of the following *cannot* be the lengths of the sides of a triangle?
   A. 6 cm, 3 cm, 4 cm
   B. 11 cm, 5 cm, 7 cm
   C. 10 cm, 8 cm, 2 cm
   D. 12 cm, 6 cm, 8 cm

Use this set of data for Exercises 9–12.

10, 12, 15, 15, 16, 16, 18, 20, 23, 25

9. What is the mean for the given data?
   A. 15 B. 16
   C. 17 D. none of these

10. What is the median for the given data?
    A. 15 B. 16
    C. 17 D. none of these

11. What is the mode for the given data?
    A. 15 B. 16
    C. 15 and 16 D. none of these

12. What is the range for the given data?
    A. 15 B. 16
    C. 17 D. none of these

13. Which is a graph of $3x + 2 \geq -4$?
    A. −3 −2 −1 B. −3 −2 −1
    C. −3 −2 −1 D. −3 −2 −1

14. What is the volume of the rectangular pyramid?
    A. 1,000 ft³ B. 1,500 ft³
    C. 2,000 ft³ D. 3,000 ft³

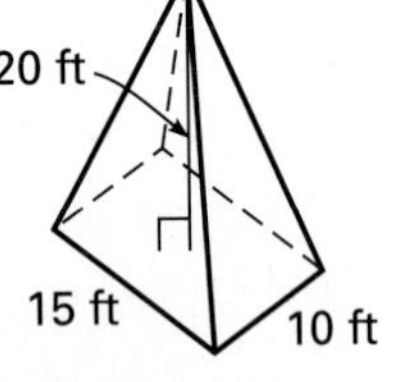


15. Which is the result of simplifying $3(y + 2)(y - 1) - 2(y - 2)(y + 5)$?
    A. $y^2 - 3y - 26$ B. $y^2 - 3y + 14$
    C. $y^2 - 3y - 16$ D. none of these

16. Which is the result of factoring $36a^2b^3 + 12ab^2 - 18a^3b^2$?
    A. $6ab(6ab^2 + 2b - 3a^2b)$
    B. $6a^2b(6b^2 - ab)$
    C. $6ab^2(6ab + 2 - 3a)$
    D. none of these

# CHAPTER 8 SKILLS PREVIEW

For Exercises 1–3, give the coordinates of the point or points shown on the graph at the right.

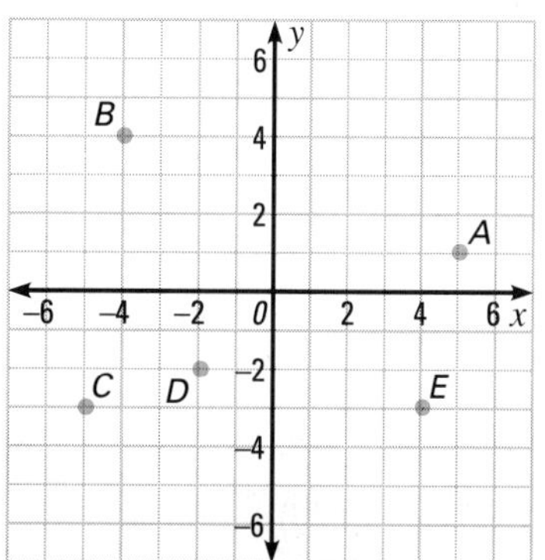


1. point $A$

2. point $B$

3. two points in the third quadrant

4. Find the midpoint $P$ of a line segment whose endpoints are $A(5, 8)$ and $B(3, 4)$.

Use the given values to find five solutions of each equation. Write the solutions as ordered pairs.

5. $y = x + 4$ for $x = -4, -2, 0, 2, 4$

6. $y = 2x - 2$ for $x = -2, -1, 0, 1, 2$

Graph the function for the domain of real numbers.

7. $y = 2$

8. $y = x - 2$

Maria begins to walk along a road at a point 3 miles from her high school. She walks at a constant rate of 2 miles an hour. Her distance from school, $d$, is a function of the amount of time she walks, $t$. An equation representing this function is $d = 2t + 3$, where $t$ is the number of hours Maria walks.

9. Graph the function.

10. How far is Maria from school after 2 h?

Refer to the graph at the right. Complete each ordered pair so that it corresponds to a point on the graph.

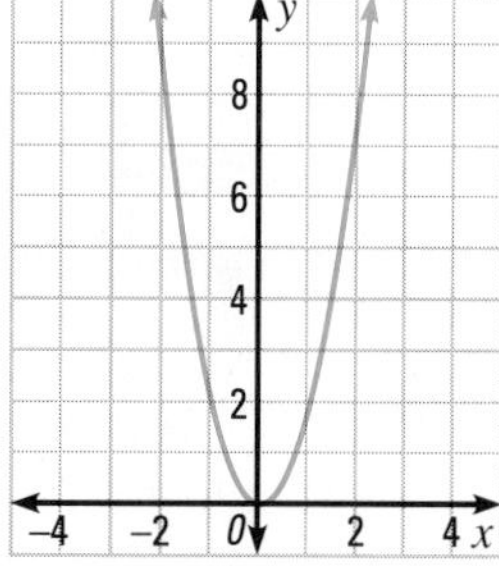


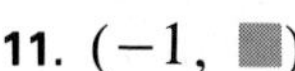

11. (−1, ■)

12. (0, ■)

13. (■, 8)

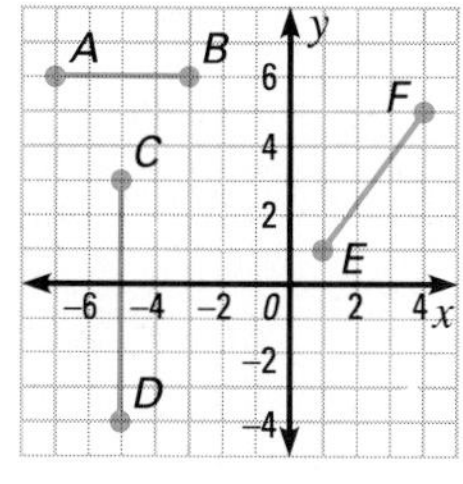


Refer to the graph at the left to calculate the distance between each set of points.

14. $A$ and $B$

15. $C$ and $D$

16. $E$ and $F$

Tell whether the ordered pair is a solution of the inequality.

17. (1, 4); $y > x + 2$

18. (2, 1); $y \geq 2x - 1$

19. Graph the inequality $y \leq x - 3$.

# CHAPTER 8 GRAPHING FUNCTIONS

**THEME Uses of Science**

If you have ever tried to find a place located on a map you may have been working with a coordinate grid. Scientists use coordinate grids to graph certain kinds of data—those that can be represented as functions.

In this chapter, you will solve equations that represent functions. You will also see how a relationship expressed as an equation may be visualized on a coordinate grid. You will draw the graphs of functions represented by equations and functions that are represented by inequalities. You will also interpret the graphs of functions in order to solve word problems.

A chemical solution is a mixture in which one substance is dissolved in another. The substance that is dissolved is the **solute.** The substance in which the solute is dissolved is the **solvent.** The maximum amount of a solute that can be dissolved in a given amount of solvent at a given temperature is the **solubility** of that solute.

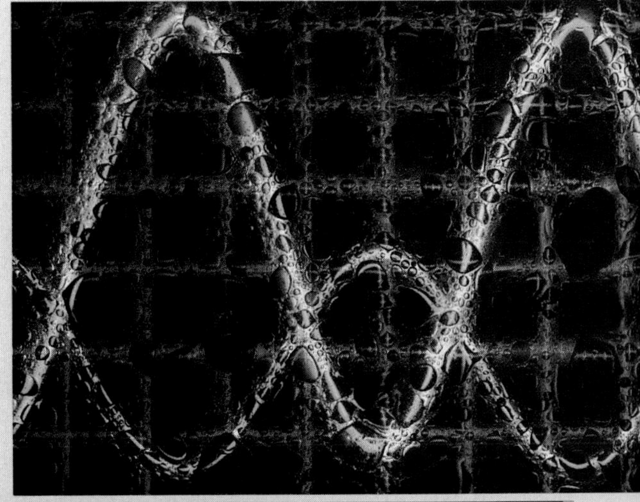

The chart shows the solubility, in grams, of three different chemical solutes in 100 g of water at four different temperatures.

**SOLUBILITY OF SOME CHEMICAL SUBSTANCES IN WATER**

| Compound | Formula | Solubility (in grams per 100 g water) 0°C | 20°C | 60°C | 100°C |
|---|---|---|---|---|---|
| Potassium nitrate | $KNO_3$ | 13.9 | 31.6 | 109 | 245 |
| Potassium bromide | KBr | 53.6 | 65.3 | 85.5 | 104 |
| Sodium chloride | NaCl | 35.7 | 35.9 | 37.1 | 39.2 |

The data in the chart can be displayed by graphs called **solubility curves.**

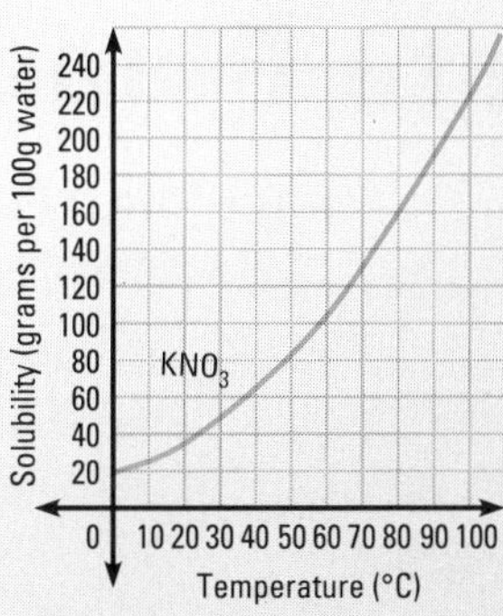


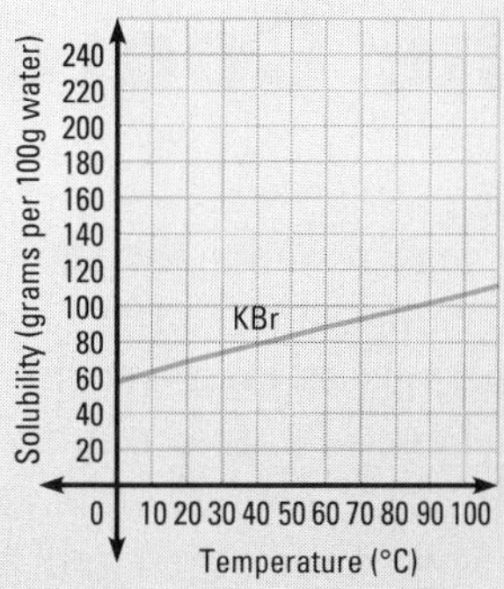


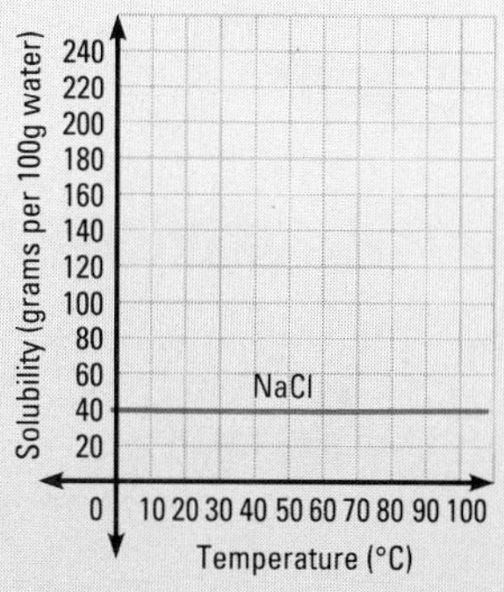


## DECISION MAKING

### Using Data

Use the chart on page 310 for Exercises 1 and 2.

1. How much sodium chloride (NaCl) can dissolve in 100 g of water at 60°C?
2. How much potassium bromide (KBr) can dissolve in 50 g of water at 0°C?

Use the solubility curves for Exercises 3–5.

3. At 0°C, is NaCl more soluble or less soluble than $KNO_3$?
4. About how much more soluble is $KNO_3$ than NaCl at 100°C?
5. What is the approximate temperature at which $KNO_3$ and KBr are equally soluble?

### Working Together

Do these experiments to determine how to increase the rate at which a solute dissolves. You will need a source of hot and cold water, two cups, two spoons, and eight sugar cubes. Have one member of your group time each experiment.

The timing should begin as you simultaneously drop one sugar cube into a cup of hot water and another into a cup of cold water. Record the time needed for the sugar in each cup to dissolve. Empty the cups. Repeat the experiment, this time stirring the contents of the cups at the same rate until the sugar dissolves. Record the time for each. Crush each of the remaining four sugar cubes separately. Repeat the procedure twice more, first without stirring, then with stirring.

Discuss your findings. How can you increase the solubility of a solute? Why do you think this is so?

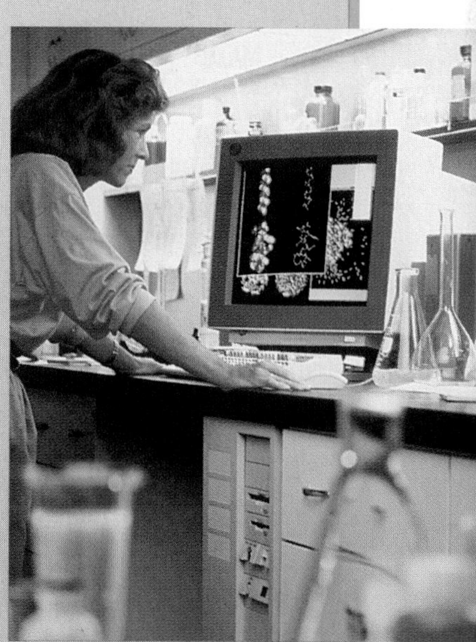

# 8-1 The Coordinate Plane

## EXPLORE

### READING MATH

The prefix *quad-* means four. How can this help you remember the meaning of the word *quadrant*? What do you think *quadrisect* means? Make a list of other *quad-* words you know.

The map at the right shows the regions and some attractions at Four-World Fun Park. To give directions to different places, start at the Main Gate. First tell how many units to move to the east or west; then tell how many units to move to the north or south. For example, to go from the Main Gate to the Souvenir Shop, go east 2 units, then north 1 unit.

FOUR-WORLD FUN PARK

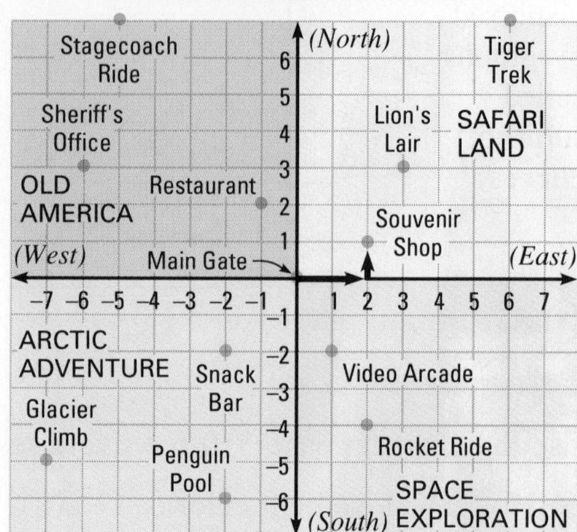


a. Give directions to go from the Main Gate to Tiger Trek.
b. Give directions to go from the Main Gate to Glacier Climb.
c. If you went east 2 units from the Main Gate and then south 4 units, where would you be?
d. Between which two regions would you travel if you went directly from the Restaurant to Lion's Lair?

## SKILLS DEVELOPMENT

The system used to identify locations on a map is much like the mathematical system called a coordinate plane. On a **coordinate plane,** two number lines are drawn perpendicular to each other. The horizontal number line is called the ***x*-axis.** The vertical number line is called the ***y*-axis.** The axes (plural of *axis*) separate the plane into four regions called **quadrants.** The point where the axes cross is called the **origin.**

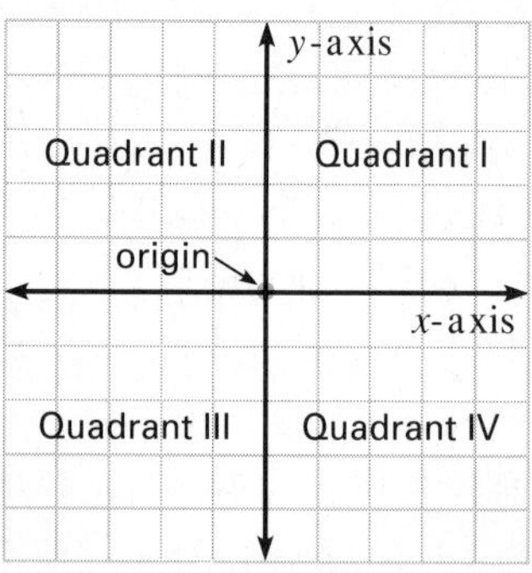

### MATH: WHO, WHERE, WHEN

Benjamin Banneker (1731–1806) was an African-American farmer, surveyor, mathematician, and astronomer. He helped survey the site on which the District of Columbia was to be developed. Surveying is the technique of measuring to determine the position of points or of marking points and boundaries. Surveying is believed to have originated in ancient Egypt.

In the figure at the right, point $M$ is 4 units to the left of the origin and 2 units up from it. So the point is identified as $(-4, 2)$, where $-4$ is the ***x*-coordinate** and 2 is the ***y*-coordinate.** The order of the coordinates is important because, for example, $(2, -4)$ identifies point $N$, and $(-4, 2)$ identifies point $M$. For this reason, the coordinates for a point, such as $(2, -4)$ and $(-4, 2)$ are called *ordered pairs.*

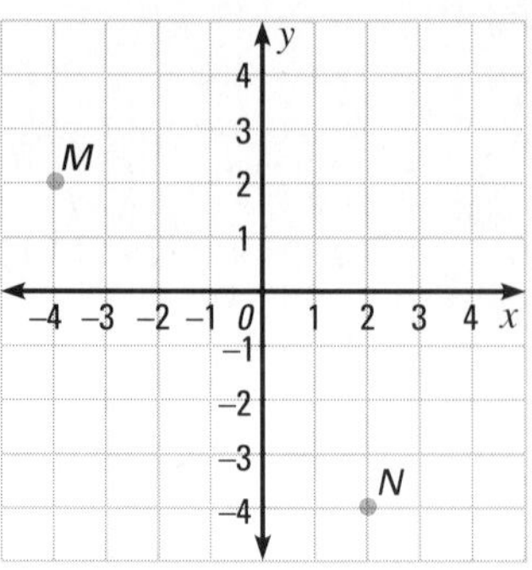

When you read the coordinates and graph points in the plane, you need to be aware of directions. The table shows the direction in which to move in order to locate a point based on the sign of each coordinate in the ordered pair.

| | ***x*-coordinate** | ***y*-coordinate** |
|---|---|---|
| **positive** | move *right* → | move *up* ↑ |
| **negative** | move *left* ← | move *down* ↓ |

## Example 1

Give the coordinates of the point or points.

**a.** point $C$

**b.** two points in the fourth quadrant

### Solution

**a.** Point $C$ is 2 units to the left of the origin and 3 units up from the origin. So, the coordinates of $C$ are $(-2, 3)$.

**b.** Points $D$ and $H$ are in the fourth quadrant. Their coordinates are $D(2, -2)$ and $H(4, -3)$. ◄

## Example 2

Graph points $X(1, 4)$, $Y(-3, 0)$, and $Z(-2, -4)$.

### Solution

Point $X$ is 1 unit to the right of the origin and 4 units up. Point $Y$ is 3 units to the left of the origin and 0 units up or down. Point $Z$ is 2 units to the left of the origin and 4 units down.

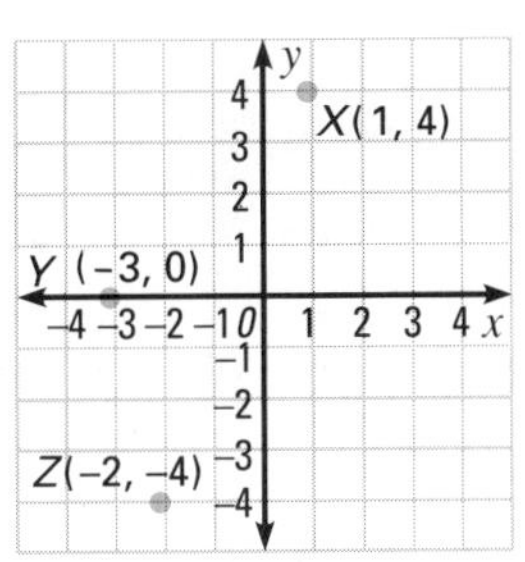


◄

## Example 3

Find the midpoint of the line segment whose endpoints are $M(2, 5)$ and $N(-4, 1)$.

### Solution

To find the midpoint $P$ of line segment $MN$, find the average values of the $x$- and $y$-coordinates of the endpoints.

$$\text{coordinates of } P = \left(\frac{2 + (-4)}{2}, \frac{5 + 1}{2}\right) = (-1, 3)$$

So, $P(-1, 3)$ is the midpoint of line segment $MN$. ◄

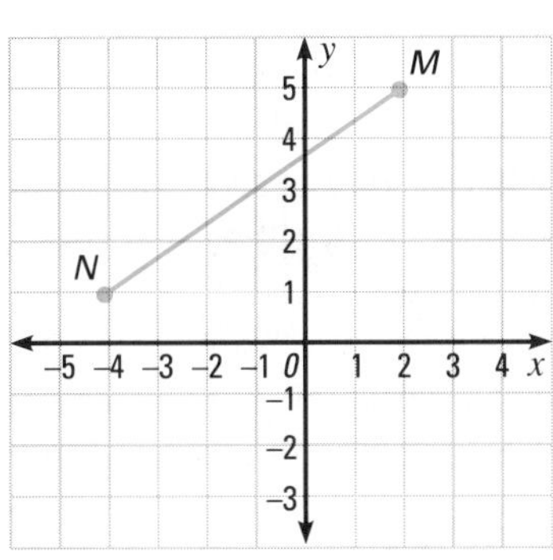


**CONNECTIONS**

The ideas of vertical and horizontal axes play an important role in the culture of Native American groups, such as the Sioux. The Sioux believed that there were four horizontal dimensions of space (north, south, east, west) and two vertical dimensions (sky and earth). Together, these made six dimensions, with the seventh, symbolizing the human spirit, at the center where all the dimensions meet. This explains why seven was considered a sacred number in these cultures.

CHECK UNDERSTANDING

Why is it that, in Example 4, you can count units to find the length of the base and the height of triangle *ABC*? Explain why you cannot count units to find the length of $\overline{AC}$.

### Example 4

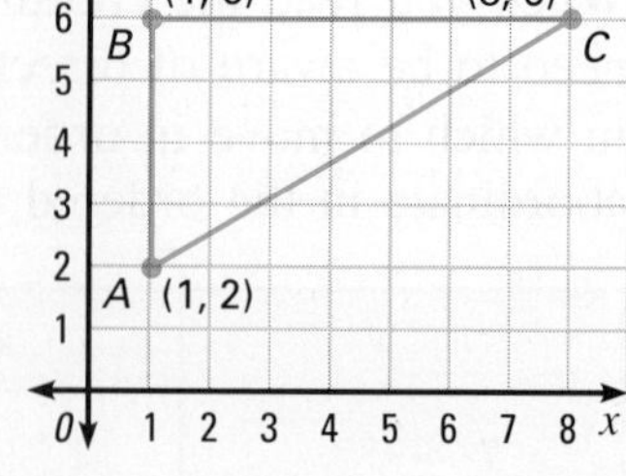


Find the area of the triangle whose vertices are $A(1, 2)$, $B(1, 6)$, $C(8, 6)$.

**Solution**

Triangle *ABC* is a right triangle. Count units to find the length of the base $\overline{AB}$ and the height $\overline{BC}$. The length of $\overline{AB}$ is 4 units. The length of $\overline{BC}$ is 7 units. Then use the formula for the area of a triangle.

$$A = \frac{1}{2}bh$$

$$A = \frac{1}{2}(4)(7)$$

$$A = 14$$

The area of triangle *ABC* is 14 square units. ◄

## TRY THESE

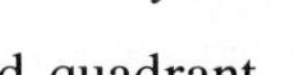

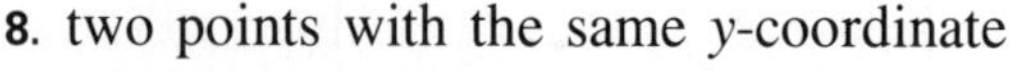

Refer to the figure for Exercises 1–8. Give the coordinates of the point or points.

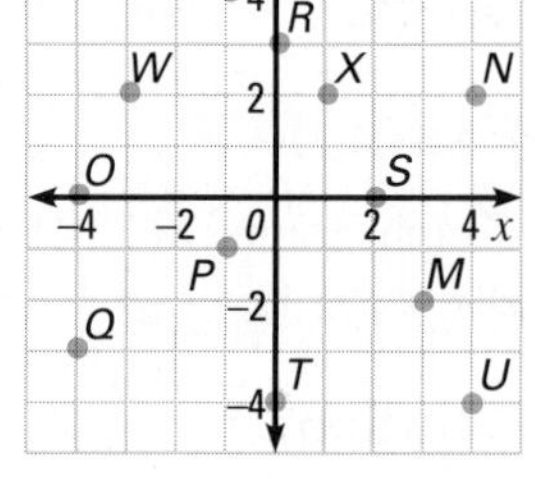


1. point *X*
2. point *W*
3. point *M*
4. point *S*
5. origin
6. a point on the *y*-axis
7. a point in the third quadrant
8. two points with the same *y*-coordinate

Graph each point on a coordinate plane.

9. $A(3, 5)$
10. $B(-2, 4)$
11. $C(4, -1)$
12. $D(0, 2)$
13. $E(-3, 0)$
14. $F(-5, -5)$
15. Find the midpoint of the line segment whose endpoints are $K(4, 7)$ and $L(10, 11)$.
16. Find the area of the triangle whose vertices are $X(0, -5)$, $Y(8, -5)$, and $Z(8, -1)$.

# EXERCISES

**PRACTICE/ SOLVE PROBLEMS**

For Exercises 1–8, refer to the figure at the top of page 315. Give the coordinates of the point or points.

1. point *J*
2. point *K*
3. two points on the *x*-axis

4. two points with the same $y$-coordinate

5. a point in the second quadrant

6. a point in the fourth quadrant

7. a point whose coordinates are equal

8. a point whose coordinates are opposites

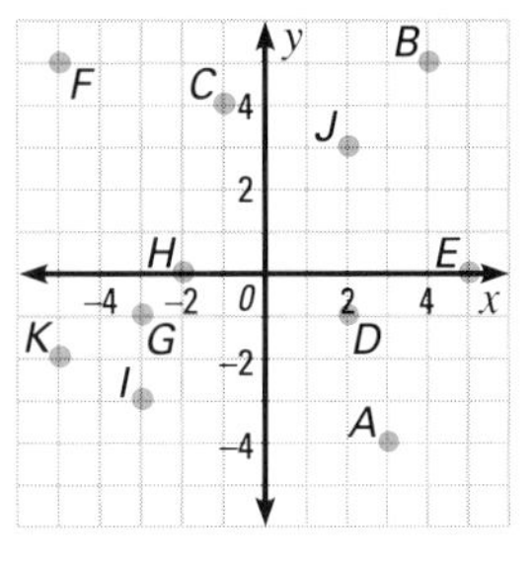

Graph each point on a coordinate plane.

9. $S(5,\ 1)$
10. $T(3,\ 0)$
11. $U(0,\ -4)$
12. $V(-2,\ 3)$
13. $W(4,\ -7)$
14. $X(-6,\ -1)$
15. $Y(0,\ 0)$
16. $Z(8,\ -8)$

17. a. On a coordinate plane, sketch the rectangle with $Q(3,\ 2)$ and $R(-4,\ -3)$ as endpoints of a diagonal.
    b. Find the perimeter of the rectangle you drew.
    c. Find the area of the rectangle you drew.

18. In the third quadrant of a coordinate plane, sketch a right triangle with an area equal to 25 square units.

## EXTEND/ SOLVE PROBLEMS

Graph each set of points on a coordinate plane. Then join the points in order, identify the geometric figure, and find its area.

19. $A(-2,\ 3)$, $B(4,\ 3)$, $C(4,\ -3)$, $D(-2,\ -3)$

20. $E(2,\ 0)$, $F(6,\ 0)$, $G(5,\ 3)$, $H(3,\ 3)$

21. Write the values in the table at the right as ordered pairs. Then graph the points that the ordered pairs represent. What pattern do you see? Write two other ordered pairs that fit the pattern.

| x | y |
|---|---|
| −1 | −2 |
| 0 | 0 |
| 1 | 2 |
| 2 | 4 |

22. The vertices of a triangular island are located at points $P(3,\ 5)$, $Q(5,\ 7)$, and $R(7,\ 3)$.
    a. Sketch the island on a coordinate plane.
    b. Will an aircraft on a course passing through points (1, 2), (4, 4), and (7, 6) fly over the island?

## THINK CRITICALLY/ SOLVE PROBLEMS

23. Point $X$ has coordinates $(-4,\ 5)$.
    a. Give the coordinates of three other points on a horizontal line containing point $X$.
    b. Give the coordinates of three other points on a vertical line containing point $X$.

24. Plot the following points: $A(-7, 14)$, $B(2, 14)$, $C(2, 5)$, $D(15, -6)$, $E(16, -8)$, $F(15, -11)$, $G(16, -13)$, $H(8, -14)$, $I(7, -10)$, $J(3, -8)$, $K(2, -8)$, $L(0, -5)$, $M(-2, -3)$, $N(-2, -1)$, $O(-3, 0)$, $P(-4, 2)$, $Q(-8, 9)$. Join the points in order with line segments. The polygon formed is a simplified map of what state? Try to create a coordinate map of another state.

# 8-2 Equations and Functions

## EXPLORE

a. Look at this set of ordered pairs:

(1, 5) (2, 10) (5, 25) (−1, −5) (−3, −15)

Describe the relationship you see between the $x$-coordinate and $y$-coordinate of each pair.

b. Graph the ordered pairs given in **a.** What do you notice?

c. Give two other ordered pairs that seem to fit this pattern, then graph them.

d. Think about all the possible ordered pairs that fit this pattern. Could there be two ordered pairs with the same $x$-coordinates but different $y$-coordinates?

## SKILLS DEVELOPMENT

You have already solved certain equations with one variable. There are other equations, however, that have two variables. A **solution of an equation with two variables** is an ordered pair of numbers that makes the equation true. Equations with two variables usually have an infinite number of solutions. You will see that graphs provide a convenient way of representing all the solutions.

To find some of the solutions of an equation with two variables, begin by choosing a value for $x$. Substitute that value into the equation and solve to find the corresponding value of $y$. Make a table to keep track of the ordered pairs that are solutions of the equation.

### Example 1

Find five solutions of the equation $y = -2x + 1$. Use −2, −1, 0, 1, and 2 as values for $x$.

#### Solution

For each of the given values of $x$, find the value of $y$. Then write the ordered pairs.

| $x$ | $y$ | |
|---|---|---|
| −2 | 5 | ← When $x = -2$, $y = -2(-2) + 1$, or 5. |
| −1 | 3 | |
| 0 | 1 | |
| 1 | −1 | |
| 2 | −3 | |

The ordered pairs (−2, 5), (−1, 3), (0, 1), (1, −1), and (2, −3) are five solutions of $y = -2x + 1$. ◄

In Section 1–4 you learned that a function matches each number in a given set of numbers with exactly one number in a second set of numbers. You represented functions using function tables and function rules, and you used the function notation $f(x)$ to write function rules. Another way to write a function rule is by using an equation with two variables. For example, if the function rule is $f(x) = x + 3$, you can also write it as $y = x + 3$. The set of all possible values of $x$ is called the **domain** of the function, and the set of all possible values of $y$ is called the **range** of the function. The values of $y$ are also known as the **values of the function.**

When you find solutions of an equation with two variables, you find a set of ordered pairs. Since the equation represents a function, you can draw the **graph of the function** by graphing the points that correspond to the ordered pairs.

**CHECK UNDERSTANDING**

Give the range for each of the functions in Examples 2 and 3.

## Example 2

Graph $y = 3x + 2$ when the domain is $\{-2, -1, 0, 1, 2\}$.

### Solution

Make a table to show the ordered pairs. Then graph the points that correspond to the ordered pairs.

| x | y |
|---|---|
| −2 | −4 |
| −1 | −1 |
| 0 | 2 |
| 1 | 5 |
| 2 | 8 |

← When $x = -2$, $y = 3(-2) + 2$, or $-4$.

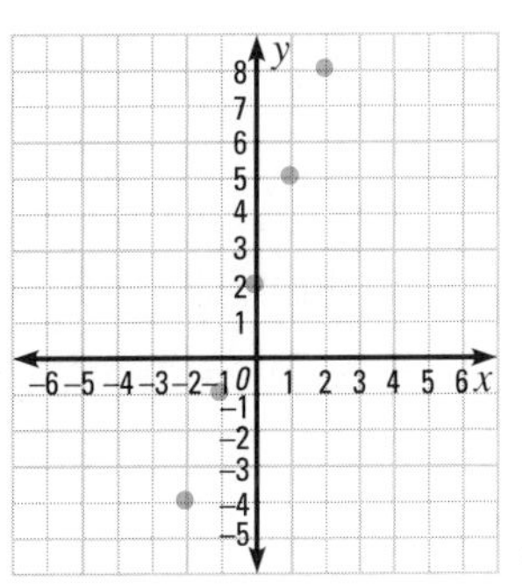


## Example 3

Graph $y = -4$ when the domain is $\{-2, -1, 0, 1, 2\}$.

### Solution

For any value of $x$, the value of $y$ is always equal to $-4$.

| x | y |
|---|---|
| −2 | −4 |
| −1 | −4 |
| 0 | −4 |
| 1 | −4 |
| 2 | −4 |

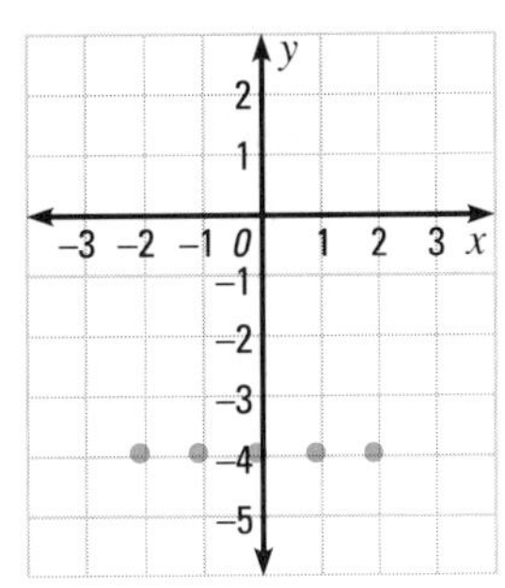

**WRITING ABOUT MATH**

Write a short paragraph comparing the word *range* as it was used in this section to the meaning of the word as it was used in statistics (Chapter 4). How are the meanings alike? How are they different?

For a function given by an equation with two variables, the value of $y$ depends on the value of $x$ chosen. Many relationships in everyday life can be represented by mathematical functions.

### Example 4

The area of a square depends on the length of a side. Let $y$ = the area and let $x$ = the length of a side. Then the area function is given by $y = x^2$. Graph $y = x^2$ when the domain is $\{1, 1.5, 2, 2.5, 3\}$.

**TALK IT OVER**

In Example 4, would it make sense to include negative values in the domain? Why or why not? Could there be negative values in the range of this function?

**Solution**

Make a table. Then use the ordered pairs to graph the function.

| $x$ | $y$ |
|---|---|
| 1 | 1 |
| 1.5 | 2.25 |
| 2 | 4 |
| 2.5 | 6.25 |
| 3 | 9 |

← When $x = 1.5$, $y = (1.5)^2$, or 2.25.

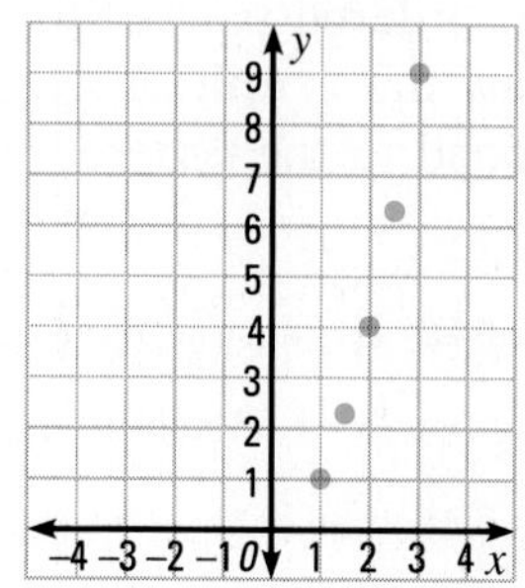


## TRY THESE

Use the given values to find five solutions of each equation. Write the solutions as ordered pairs.

1. $y = 2x - 1$ for $x = -4, -2, 0, 1, 2$
2. $y = -3x + 5$ for $x = -5, -3, 0, 2, 4$

Graph each function when the domain is $\{-2, -1, 0, 1, 2\}$.

3. $y = 4x + 2$ 4. $y = 7$ 5. $y = -2x - 2$ 6. $y = 2x^2$

7. The area of a circle is a function of the length of the radius. Let $y$ be the area and let $r$ be the length of the radius. Then the area function is given by $y = \pi r^2$. Graph $y = \pi r^2$ when the domain is $\{1, 2, 3, 4, 5\}$. Round values of $y$ to the nearest whole number.

# EXERCISES

**PRACTICE/ SOLVE PROBLEMS**

Use the given values to find five solutions of each equation. Write the solutions as ordered pairs.

1. $y = 2x + 3$ for $x = -3, -1, 0, 2, 4$
2. $y = -4x - 1$ for $x = -2, -1, 0, 1, 2$

3. $y = \frac{1}{2}x + 3$ for $x = -6, -4, 0, 2, 8$

4. $y = 0.1x$ for $x = -10, -5, 3, 7, 20$

Graph each function when the domain is $\{-2, -1, 0, 1, 2\}$.

5. $y = 3x$ 6. $y = -2x + 7$ 7. $y = -3$

8. $y = 5x + 1$ 9. $y = 4x - 6$ 10. $y = x^2 - 2$

11. If a car is moving at a constant speed of 50 mi/h, then the distance it covers is a function of amount of time it is moving. Let $y$ be the distance in miles and let $x$ be the time in hours. Then the distance function is $y = 50x$. Graph the function when the domain is $\left\{\frac{1}{4}, \frac{1}{2}, 1, 1\frac{1}{2}, 2\right\}$.

## EXTEND/ SOLVE PROBLEMS

| $x$ | $y$ |
|---|---|
| 0 | 0 |
| 3 | 6 |
| 6 | 24 |
| 9 | 54 |
| 12 | 96 |
| 15 | 150 |

12. Refer to the function table at the right. Which rule for this function is correct?
    a. $y = \frac{2}{3}x$ b. $y = \frac{2}{3}x^2$

13. Write two more ordered pairs for the function you selected in Exercise 12. Use $x = 18$ and $x = 21$.

14. Graph the function you selected in Exercise 12 when the domain is $\{0, 3, 6, 9, 12, 15, 18, 21\}$. (*Hint:* Let one interval along the $y$-axis represent 20 units.)

15. The energy produced by vibrating objects is carried through the air as sound waves. The frequency of a sound is the number of vibrations per second, and the wavelength of a sound is a function of its frequency. Since the speed of sound is about 330 m/s, the wavelength function is $y = \frac{330}{x}$ where $y$ is the wavelength in meters and $x$ is the frequency in vibrations per second. Graph this function for the domain $\{10, 20, 30, 40, 50, 60, 70, 80, 90, 100\}$. Round values of $y$ to the nearest whole number.

**CONNECTIONS**

Low sounds are created by low-frequency waves with very long wavelengths. High sounds are created by high-frequency waves with very short wavelengths.

Most people cannot hear sound frequencies above 20,000 vibrations per second *(ultrasonic waves)* or below 20 vibrations per second *(infrasonic waves)*. Bats, however, can detect frequencies as high as 100,000 vibrations per second.

## THINK CRITICALLY/ SOLVE PROBLEMS

Assume that the domain of each function in Exercises 16–20 is the set of real numbers. Give the range of the function.

16. $y = |x|$ 17. $y = -|x|$ 18. $y = -x^2$ 19. $y = x^3$ 20. $y = -x^3$

21. Is the relationship that matches a number with its square root a function? Explain.

22. Graph the function $y = \frac{10}{x}$ when the domain is $\{-10, -8, -6, -4, -2, 2, 4, 6, 8, 10\}$. Explain why 0 cannot be part of the domain. As $|x|$ gets larger, what happens to the values of $y$? Is there any value for $x$ for which $y$ is equal to 0? Explain.

# 8-3 Linear Functions

## EXPLORE

Look closely at points $P$ and $Q$.

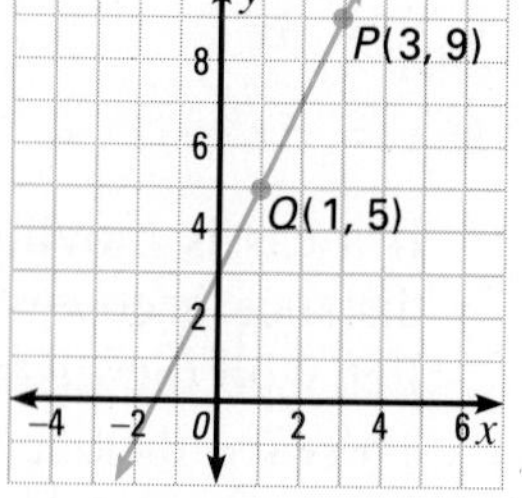


a. Determine whether the coordinates of these points are solutions of the equation $y = 2x + 3$.

b. Give the coordinate pairs for three other points that are solutions of $y = 2x + 3$.

c. Do you think there are points on the line that are *not* solutions of $y = 2x + 3$. Explain.

d. Are there any breaks in the line? Tell why or why not.

## SKILLS DEVELOPMENT

A block-party association has decided that it needs 12 bottles of cola for a party. It plans to buy some regular cola and some diet cola. This can be stated by $d = 12 - r$, where $d$ represents the number of bottles of diet cola and $r$ represents the number of bottles of regular cola.

Show the solutions of the equation in a function table.

| $r$ | $d$ |
|---|---|
| 0 | 12 |
| 1 | 11 |
| 2 | 10 |
| . | . |
| 10 | 2 |
| 11 | 1 |
| 12 | 0 |

The equation expresses $d$ as a function of $r$. Let the horizontal axis of a graph represent bottles of regular cola, $r$, and let the vertical axis represent bottles of diet cola, $d$.

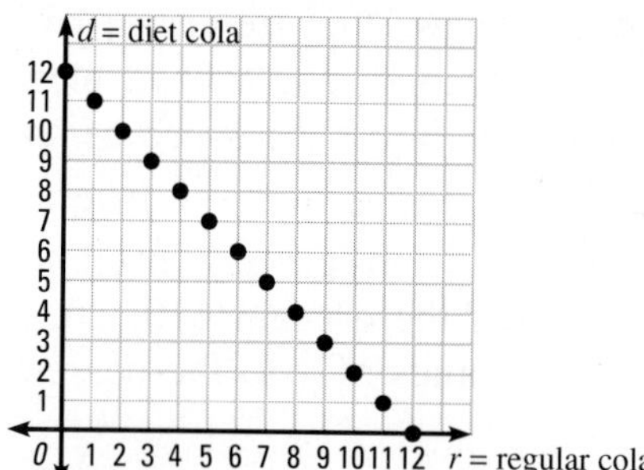


You can see from the graph that there are 13 possible combinations of regular and diet cola and that the number of diet colas decreases as the number of regular colas increases.

In this situation, the domain of the function is the set of positive whole numbers. No negative values or fractional values are included since a negative number of bottles would make no sense and a fractional part of a bottle cannot be bought.

Other mathematical situations may require a graph that includes negative numbers and/or fractions. In such cases, the domain of the function is the domain of all real numbers.

Suppose that, instead of bottles of cola, you wanted to find all real numbers $d$ and $r$ whose sum is 12. The function $d = 12 - r$ can now be extended to include the domain of all real numbers. The function table can include other coordinate pairs that are solutions of the equation $d = 12 - r$, such as $(-1, 13)$ and $\left(5\frac{1}{2}, 6\frac{1}{2}\right)$.

The new coordinate pairs are graphed and the points are connected to show that this graph is *continuous*. That is because, for all real numbers, every point along the line satisfies the equation $d = 12 - r$.

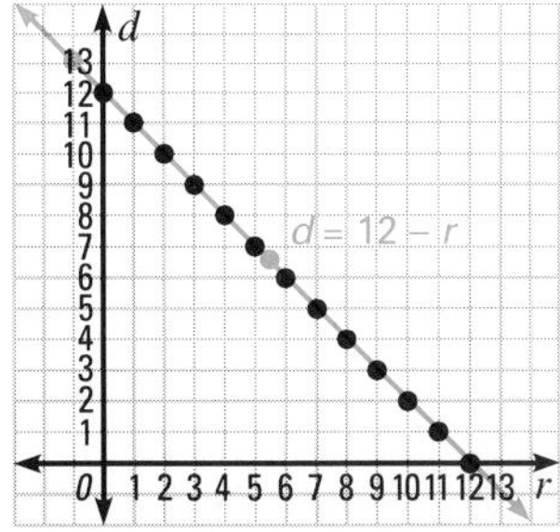


## Example 1

Graph $y = 4x - 3$ when the domain is the set of real numbers.

### Solution

First make a table to show the ordered pairs. Choose at least three values of $x$ from the domain.

| x | y |
|---|---|
| −1 | −7 |
| 0 | −3 |
| 1 | 1 |

Then graph the points that correspond to the ordered pairs. Draw a line to connect the points.

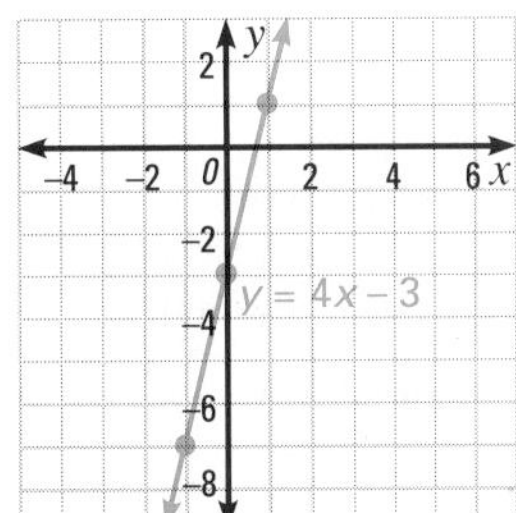


Since all the points on the graph of the equation $y = 4x - 3$ lie on one line, the equation is called a **linear equation.** A function that can be represented by a linear equation is called a **linear function.** If the domain of a linear function is not indicated, assume the domain to be the set of real numbers.

## Example 2

Graph $2x + y = 4$.

### Solution

Start by solving the equation for $y$.

$$2x + y = 4$$
$$y = -2x + 4 \quad \leftarrow \textbf{Add } -2x \textbf{ to each side.}$$

Make a table for the ordered pairs. Then graph the points and draw a line connecting them.

| x | y |
|---|---|
| −1 | 6 |
| 0 | 4 |
| 2 | 0 |

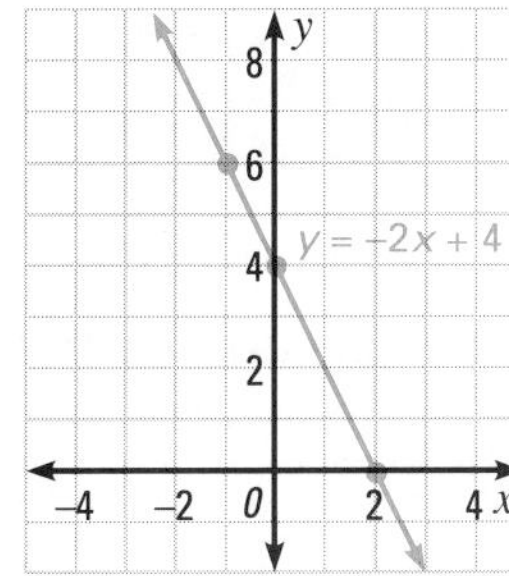


... the graph of one or more functions. On most calculators of this type, the function to be graphed must be given in the form $y = f(x)$. So, in order to enter the function $x + y = 8$, you must first solve for $y$.

$x + y = 8$

$y = -x + 8$ **(Add $-x$ to each side.)**

The calculator has a screen, called a **window,** which displays a portion of the coordinate grid. For example, the screen may show a window in which the $x$-values range from −10 to +10 and the $y$-values range from 0 to 20. (Note that intervals are usually not numbered on the calculator grid.) You can enter your choice of values to define the window. If you do not do this, the calculator automatically uses its *default window.* The operating manual will describe the default window used by your calculator. Often you must adjust the window carefully to see your graph.

Use a graphing calculator to graph each function. Tell whether you could use the default window. If not, note the window you used.

1. $y = x + 4$
2. $2x + y = 25$

## Example 3

Deena begins walking along a straight road at a point 2 mi from school. She walks at a constant rate of 3 mi/h. Her distance from school, $d$, is then a function of the amount of time she walks, $t$. An equation that represents this function is $d = 3t + 2$, where $t$ is the number of hours Deena walks.

**a.** Graph the function $d = f(t)$.

**b.** How far is Deena from school after $1\frac{1}{2}$ h?

### Solution

**a.**

| $t$ | $d$ |
|---|---|
| 0 | 2 |
| 1 | 5 |
| 2 | 8 |

Deena's starting point, 2 mi from school, is at (0, 2), which is a solution of the equation.

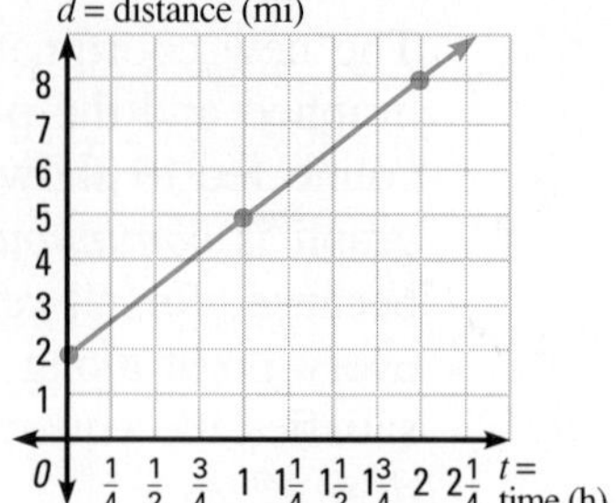

**TALK IT OVER**

Why are three ordered pairs all you need to graph an equation?

**b.** When $t = 1\frac{1}{2}$, $d = 6\frac{1}{2}$. After $1\frac{1}{2}$ h, Deena is $6\frac{1}{2}$ mi from school. ◄

## TRY THESE

Graph the function when the domain is the set of real numbers. Start by making a table of at least three ordered pairs.

**1.** $y = 3x - 6$ **2.** $y = 2x - 6$ **3.** $y = 4$

Graph the function. Start by solving the equation for $y$.

**4.** $y - 2x = -3$ **5.** $3 = y + \frac{1}{2}x$

**6.** When the snow is 6 in. deep on a ski slope, a snow-making machine is turned on. The machine makes snow at a rate of 2 in./h. The depth of the snow, $d$, is then a function of the amount of time the machine is running, $t$. An equation that represents this function is $d = 2t + 6$.

**a.** Graph the function.

**b.** How deep is the snow after the machine has been on for $2\frac{1}{4}$ h?

**CHECK UNDERSTANDING**

In Exercise 6, how many inches of snow did the snow-making machine make in $2\frac{1}{4}$ h?

# EXERCISES

**PRACTICE/ SOLVE PROBLEMS**

Graph the function when the domain is the set of real numbers.

**1.** $y = -2$ **2.** $y = \frac{3}{2}x - 2$ **3.** $y = 2x + 5$

**4.** $y = \frac{1}{2}x - 1$ **5.** $y + x = 1$ **6.** $y + \frac{3}{2}x = 2$

When a power boat is 10 m from shore, the boat begins moving at a constant rate of 15 m/s. The distance from shore, $d$, is a function of the amount of time the boat travels, $t$. An equation that represents this function is $d = 15t + 10$, where $t$ is the number of seconds the boat travels.

7. Graph the function.
8. How far is the boat from shore after 2.5 s?
9. How long would it take the boat to travel 62.5 m from shore?

## EXTEND/ SOLVE PROBLEMS

Graph the function.

10. $-x + 3y = 0$  11. $x - 5y = 0$  12. $y = 0.5x$

According to a scientific supply catalog, the cost of a chemical in cents, $C$, is represented by the equation $C = 3V$, where $V$ represents the amount of liquid in milliliters.

13. Graph the function.
14. What is the cost of 15 mL of the chemical? What is the cost of 60 mL?
15. How much of the chemical can be purchased for $0.39?

**USING DATA** Refer to the Data Index on pages 556–557 to find the number of calories burned by people of different weights.

16. Make a graph in which the number of calories burned during scuba diving is a function of the weight of individuals who weigh 110 lb, 154 lb, and 198 lb.

## THINK CRITICALLY/ SOLVE PROBLEMS

The perimeter of a rectangle is 16 in.

17. Write an equation that expresses the length of the rectangle, $l$, as a function of its width, $w$.
18. Graph the function.
19. Assume that the value of $l$ must be greater than or equal to the value of $w$. Describe the domain and the range of the function.

The freezing point of water is expressed in degrees Celsius as 0°C and in degrees Fahrenheit as 32°F. The formula $F = \frac{9}{5}C + 32$ is used to find the equivalent Fahrenheit temperature for a given Celsius temperature. The relationship between equivalent temperatures can be expressed as a linear function.

20. Graph the function.
21. What Fahrenheit temperature is equivalent to 20°C?

# 8-4 Nonlinear Functions

**EXPLORE**

For each graph below, make a table of ordered pairs. Graph the points that correspond to the ordered pairs. Join the points with a smooth line or curve.

**a.** Graph $y = x$ when the domain is $\{-2, -1, 0, 1, 2\}$.

**b.** Graph $y = x^2$ when the domain is $\{-2, -1, 0, 1, 2\}$.

**c.** How would you describe the difference between your graph for question **a** and your graph for question **b**?

**d.** Suppose that you were asked to graph each of these equations.

**A.** $y = x + 3$ **B.** $y = x^2 + 1$ **C.** $y = 3x^2$ **D.** $y = 2x$

Which graphs do you think would be like your graph for question **a**? Which would be like your graph for question **b**? Give reasons for your answers.

**SKILLS DEVELOPMENT**

Not all equations with two variables represent linear functions. In this section, you will look at some **nonlinear functions.**

### Example 1

The graph at the right shows a nonlinear function. Complete each ordered pair so that it corresponds to a point on the graph.

**a.** (3, ■) **b.** (−2, ■)
**c.** (■, 2) **d.** (■, −3)

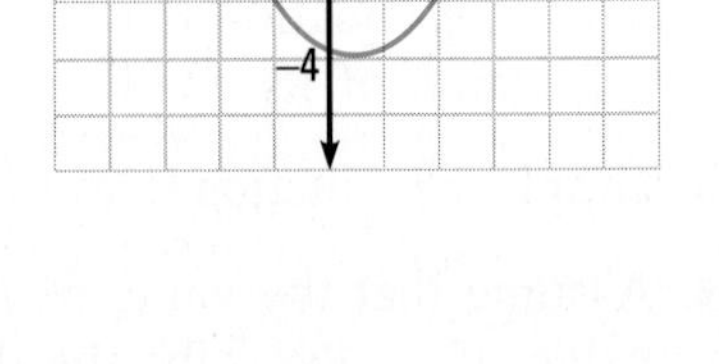


**Solution**

For coordinates **a** and **b**, find the $y$-value that corresponds to the given $x$-value.

**a.** (3, −1) **b.** (−2, −1)

For coordinates **c** and **d**, note that for each given $y$-value there are two possible $x$-values.

**c.** Both $x = -3$ and $x = 4$ correspond to the $y$-value of 2. So, (−3, 2) and (4, 2) are both solutions.

**d.** Both $x = -1$ and $x = 2$ correspond to the $y$-value of −3. So, (−1, −3) and (2, −3) are both solutions. ◄

When you graph a nonlinear function, construct a table of ordered pairs. Use enough points to know how to draw the graph.

## Example 2

Graph $y = x^2 - 2$.

**Solution**

Find at least five ordered pairs.

| $x$ | $y$ |
|---|---|
| −2 | 2 |
| −1 | −1 |
| 0 | −2 |
| 1 | −1 |
| 2 | 2 |

Graph the points that correspond to the ordered pairs. Draw a smooth curve through the points.

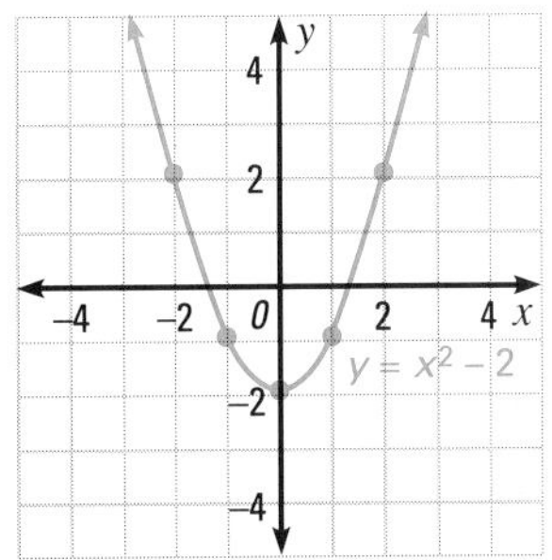


◄

Graphs of nonlinear functions can be used to solve many types of problems.

## Example 3

A ball is thrown into the air. The height of the ball, $h$, is a function of the amount of time, $t$, that the ball is in the air. An equation that represents this function is $h = -t^2 + 25t$, where $h$ is the height in feet and $t$ is the number of seconds that the ball is in the air.

**a.** At what times does the ball reach a height of 100 ft?
**b.** At what time does the ball reach its maximum height?
**c.** At what time does the ball hit the ground?

**Solution**

Draw a graph to answer the questions. Construct a table of ordered pairs.

| $t$ | $h$ |
|---|---|
| 0 | 0 |
| 5 | 100 |
| 10 | 150 |
| 15 | 150 |
| 20 | 100 |
| 25 | 0 |

Draw the graph. Let the horizontal axis represent time and let the vertical axis represent height.

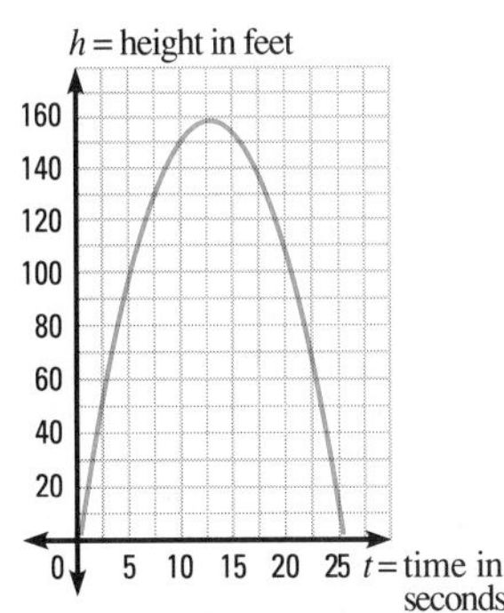


**a.** Look at the graph to find when the height of the ball is 100 ft. When $t = 5$, $h = 100$, and when $t = 20$, $h = 100$. So, the ball has a height of 100 ft after 5 s and then again after 20 s.
**b.** The maximum height of the ball is the $h$-value at the top of the curve. The maximum height occurs when $t = 12.5$ s.
**c.** The ball hits the ground when the curve again shows a height of 0. So, the ball hits the ground when $t = 25$ s. ◄

**CALCULATOR**

To graph a nonlinear function, use the same steps as for a linear function. First be sure the function is in the form $y = f(x)$, then key it in. On many graphs, the [∧] key is used to indicate powers. So, to graph $y = x^2$, you would enter

[X] [∧] 2.

Decide if you need a special window and enter it. Then use the graphing key.

Once the graph appears on the screen, a special feature found on most graphing calculators can help you explore the graph. The *trace* feature displays a blinking cursor at the middle $x$-value on the screen. Using the direction keys [►] or [◄], you can move the cursor from one graphed point to another, and the $x$- and $y$-coordinates for each point are displayed at the bottom of the screen.

Graph $y = x^2 - 1$ on a graphing calculator. Use the trace feature to find two pairs of integer coordinates.

**CHECK UNDERSTANDING**

Which of the following are quadratic equations?

1. $y = 3x^2 + 5$
2. $y = 2x^3 + x^2 + 1$
3. $y = 4x^2 + 2x + 3$
4. $y = 8x + 25$

**WRITING ABOUT MATH**

Write a few sentences describing the similarities among graphs of quadratic equations.

The equations you worked with in Examples 1 through 3 are called *quadratic equations*. A **quadratic equation in $x$** is an equation that involves an $x^2$ term but involves no term with a higher power of $x$.

A function that can be represented by a quadratic equation is called a **quadratic function.** As with linear functions, if the domain of a quadratic function is not specified, it is understood that the domain is the set of real numbers.

There are many other types of nonlinear functions besides quadratic functions.

### Example 4

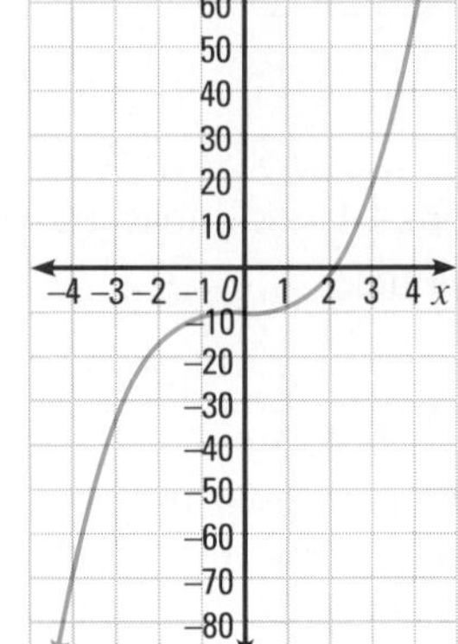


Refer to the graph at the right.

a. When $x = 0$, what does $y$ equal?
b. For what value of $x$ does $y = 54$?
c. When $y$ is negative, what is true about $x$? When $x$ is positive, what is true about $y$?

**Solution**

a. When $x = 0$, $y = -10$.
b. When $x = 4$, $y = 54$.
c. When $y$ is negative, $x$ is also negative. When $x$ is positive, $y > -10$. ◄

## TRY THESE

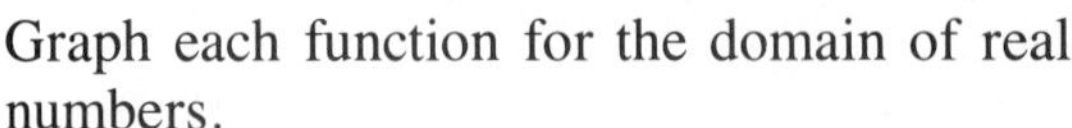

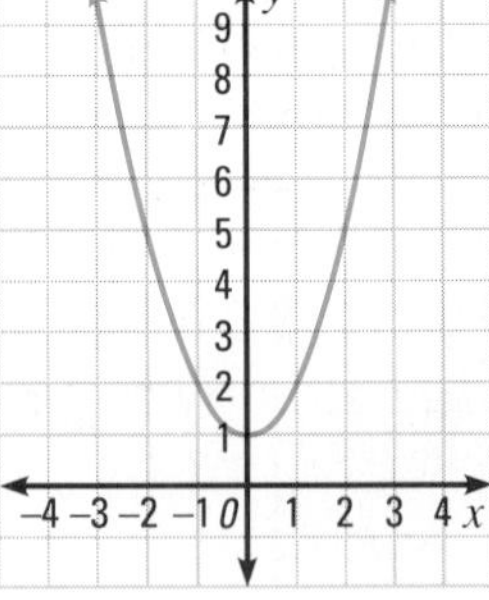


Refer to the graph at the right. Complete each ordered pair so that it corresponds to a point on the graph.

1. $(-1, ■)$
2. $(3, ■)$
3. $(■, 1)$
4. $(■, 5)$

Graph each function for the domain of real numbers.

5. $y = x^2 + 2$
6. $y = 2x^2 - 5$
7. An arrow is shot into the air. The height, $h$, of the arrow is a function of the amount of time, $t$, that the arrow is in the air. The equation that represents this function is $h = -5t^2 + 20t$ where $h$ is the height in meters and $t$ is the time in seconds.
   a. At what times does the arrow reach a height of 15 m?
   b. At what time does the arrow reach its maximum height?
   c. At what time does the arrow hit the ground?

# EXERCISES

## PRACTICE/ SOLVE PROBLEMS

Graph each function for the domain of real numbers.

1. $y = x^2 + 4$
2. $y = 3x^2 - 7$
3. $y = -2x^2$

4. The height of a ball thrown into the air is given by the equation $h = -t^2 + 10t$, where $h$ is the height in feet and $t$ is the time in *quarter seconds*.
   a. Graph this function for the domain $\{0, 1, 2, 3, 4, 5, 6, 7, 8, 9, 10\}$.
   b. At what time does the ball reach its maximum height? (Remember that time is measured in quarter seconds.)
   c. What is the maximum height?
   d. How many seconds is the ball in the air?

5. The unit cost, $C$, in dollars of producing different numbers, $n$, of certain items is shown by the graph at the right.
   a. What is the unit cost when $n = 3$?
   b. For what number of items is the unit cost equal to $40?

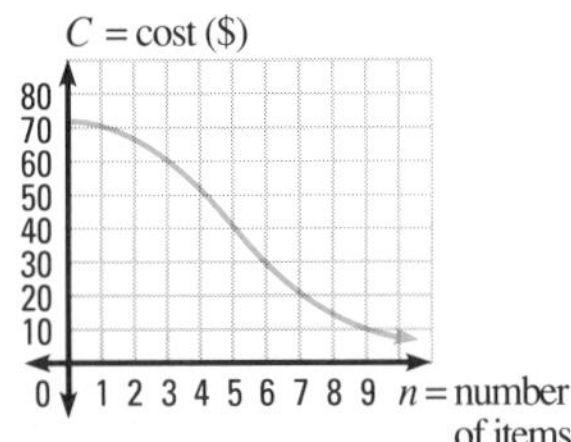


## EXTEND/ SOLVE PROBLEMS

6. A taxicab company charges $1.00 for the first $\frac{1}{2}$ mi of a ride and $0.25 for each additional $\frac{1}{4}$ mi or part thereof.
   a. What is the cost of a 1-mile ride?
   b. Draw a graph that shows the cost of rides from 0 mi to 2 mi in length.
   c. Describe your graph. Can you think of an appropriate name for this type of function?

## THINK CRITICALLY/ SOLVE PROBLEMS

7. Draw and label the graphs of the equations $y = x^2$, $y = x^2 + 2$, and $y = x^2 - 3$ on the same set of axes.
   a. What do you notice?
   b. What function would a graph exactly the same shape as $y = x^2$ but 10 units above it represent?

8. How could you use what you know about the graph of $y = x^2$ to graph $y = (x + 2)^2$?

# 8-5 Problem Solving Strategies:

## USE A GRAPH

- READ
- PLAN
- SOLVE
- ANSWER
- CHECK

Nonlinear functions can be used to model many real-world situations. The graphs of these functions can help you both to understand the relationship between the variables and to make predictions.

Using *function graphing software,* you can instruct a computer to draw a graph using a function rule $y = f(x)$ for a given domain of $x$. You can also input a set of ordered pairs, have the computer select the best intervals along the axes, then graph and connect the points. By studying the computer-generated graph, you can decide if there is a functional relationship in the ordered pairs. The computer can provide a *hard copy,* or printout, of the graph.

### PROBLEM

Suppose that you invest $1,000 in a bank account that yields 5% interest compounded quarterly. The total amount of money in the account, $A$, is a function of the time the money remains in the account, $t$. The table at the left shows this function for the first 20 years that the money is in the account.

a. About how much money was in the account halfway through the fifth year?

b. In which year will the amount in the account reach $4,000?

| $t$ (years) | $A$ (dollars) |
|---|---|
| 1 | 1,050 |
| 2 | 1,104 |
| 3 | 1,160 |
| 4 | 1,219 |
| 5 | 1,282 |
| 6 | 1,347 |
| 7 | 1,415 |
| 8 | 1,488 |
| 9 | 1,563 |
| 10 | 1,643 |
| 11 | 1,727 |
| 12 | 1,815 |
| 13 | 1,907 |
| 14 | 2,005 |
| 15 | 2,107 |
| 16 | 2,214 |
| 17 | 2,327 |
| 18 | 2,445 |
| 19 | 2,570 |
| 20 | 2,701 |

### SOLUTION

**$1,000 INVESTED AT 5% COMPOUNDED QUARTERLY**



Use the ordered pairs $(t, A)$ from the table to draw a graph.

a. To find how much money was in the account halfway through the fifth year, examine the part of the graph *between* the graphed point for the fifth and the sixth years, then estimate a value. A good estimate is about $1,315.

b. To find the year in which the amount in the account will reach $4,000, predict a value of the variable $A$ that is *beyond* what is shown on the graph. Assuming that the function continues to behave in the same way, you can extend the graph to find when $A$ will equal $4,000. Some computer graphing programs can do this automatically. You can also copy the graph and extend it accurately and smoothly to see that, after approximately 28 years, there will be about $4,000 in the account.

# PROBLEMS

If a computer with function graphing software is available, use it to draw the graphs. If not, draw the graphs manually.

1. The table below shows data for the earth's population as a function of time.

| Time, *t* (the year) | 1650 | 1700 | 1750 | 1800 | 1850 | 1900 | 1950 | 1975 |
|---|---|---|---|---|---|---|---|---|
| Population, *P* (in billions) | 0.5 | 0.6 | 0.7 | 0.9 | 1.1 | 1.6 | 2.5 | 4.1 |

   a. Draw a graph of this function. Show years along the horizontal axis and the population along the vertical axis.
   b. Use your graph to estimate the population in 1875 and in 1925.
   c. Estimate the earth's population for the year 2000.
   d. When will the earth's population be about 8.5 billion?

2. Bacteria often increase in number under suitable temperature conditions. The table below shows data that were collected hourly for a certain bacteria culture.

| Time, *t* (hours) | 1 | 2 | 3 | 4 | 5 | 6 | 7 | 8 | 9 |
|---|---|---|---|---|---|---|---|---|---|
| Number of bacteria, *n* | 25 | 40 | 65 | 100 | 150 | 200 | 300 | 500 | 900 |

   a. Draw a graph of the number of bacteria as a function of time. Use $t$ for the horizontal axis and $n$ for the vertical axis.
   b. Estimate the number of bacteria after 10 hours.
   c. When will the number of bacteria exceed 3,000?

3. USING DATA The *half-life* of a substance is the amount of time it takes for half the atoms in a sample to decay. So, the amount of the substance that remains is a function of time. For example, iodine-131 has a half-life of about 8 days. If you start with a 20-g sample of iodine-131, after 8 days there would be 10 g remaining. Use the Data Index on pages 556–557 to find the half-life of bismuth-210.
   a. Suppose you begin with a 256-g sample of bismuth-210. Copy and complete the table below to show data for nine half-lives.

| Days elapsed | 0 | ■ | ■ | ■ | ■ | ■ | ■ | ■ | ■ | ■ |
|---|---|---|---|---|---|---|---|---|---|---|
| Amount of substance remaining (grams) | 256 | 128 | ■ | ■ | ■ | ■ | ■ | ■ | ■ | ■ |

   b. When will there be less than 0.01 g remaining?

**CONNECTIONS**

All living things contain a relatively constant amount of carbon-14, a radioactive isotope with a half-life of 5,730 years. Decaying carbon-14 is constantly replaced by new carbon-14 in a living organism. However, after an organism dies, its carbon-14 decays without replacement. By measuring the amount of carbon-14 remaining in a fossil or skeleton, scientists can determine the approximate age of the specimen. This method is called *carbon-14 dating.*

# 8-6 Problem Solving Skills:

## INTERPRET GRAPHS

- ► READ
- ► PLAN
- ► SOLVE
- ► ANSWER
- ► CHECK

The graph of a function represents values of the domain plotted against values of the range. The graphs of some functions are straight lines. The graphs of other functions are curved lines. Graphs of yet other functions combine straight and curved lines.

### PROBLEM

This bottle is being filled with liquid at a constant rate of flow. The chart below shows the time it takes, in seconds, to fill the bottle to a given height, in centimeters. In the accompanying graph, the data from the chart are used to plot the time taken against the height of the water in the bottle.

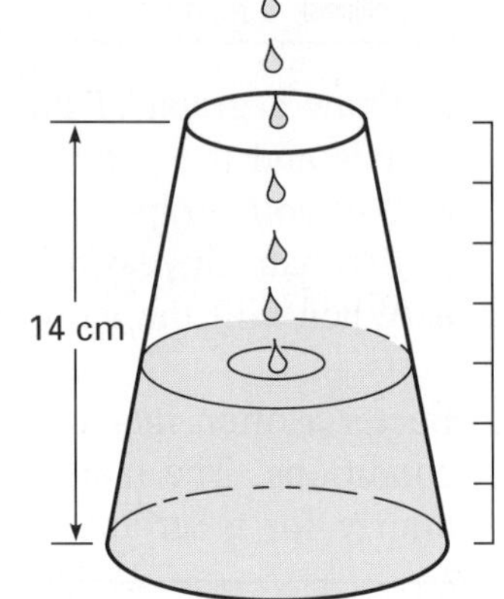


| Time, $t$ (seconds) | Height, $h$ (centimeters) |
|---|---|
| 5 | 0.6 |
| 10 | 1.2 |
| 15 | 1.9 |
| 20 | 2.8 |
| 25 | 4 |
| 30 | 6 |
| 35 | 8.5 |

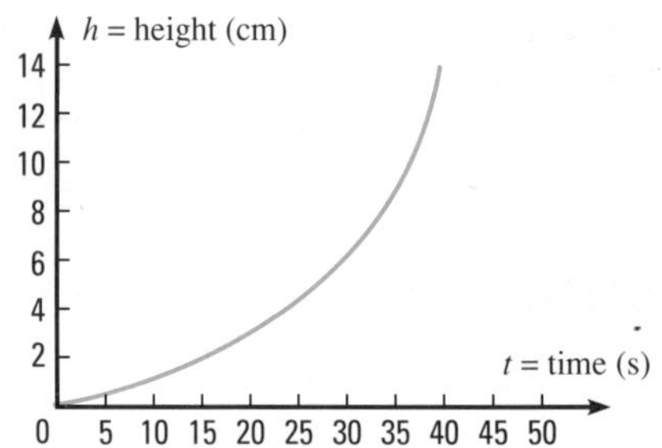


a. Between which values of $t$ does the bottle fill most slowly?
b. Between which values of $t$ does the bottle fill most rapidly?
c. How does the shape of the curve reveal the relationship between the shape of the bottle and the time it takes to fill it?

### SOLUTION

a. In relation to the time, the height of the liquid in the bottle rises most slowly between 0 s and 20 s.
b. The height of the liquid rises most rapidly between 20 s and 40 s.
c. At 40 s the bottle is filled. The graph shows that, because the bottle is wider at the bottom and narrows toward the top, it takes more time to fill the lower part of the bottle and less time to fill the upper part of the bottle.

# PROBLEMS

1. The cylindrical container shown at the right is to be filled with liquid. The time, $t$, in seconds, that it takes to fill the container is plotted along the $x$-axis. The height of the liquid, $h$, in centimeters, is plotted along the $y$-axis. For a steady flow of liquid, which of the following graphs do you think shows the relationship between height and time?

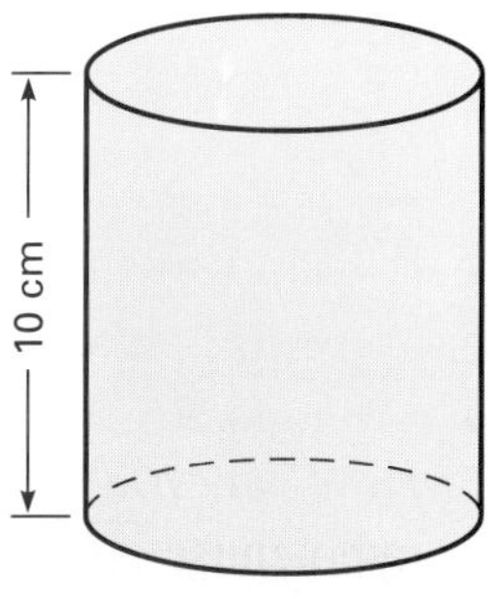


**a.**

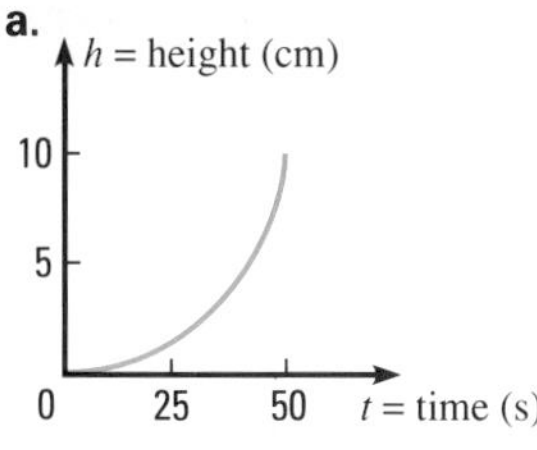


**b.**

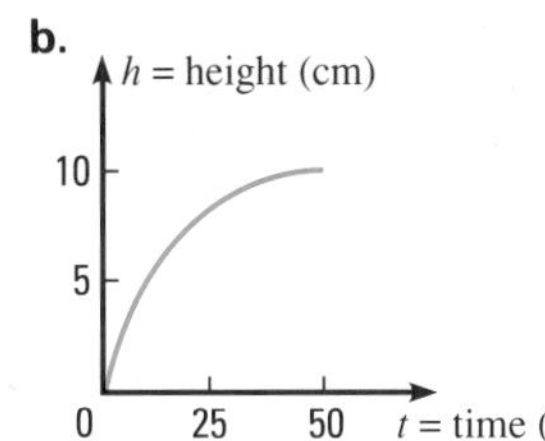


**c.**

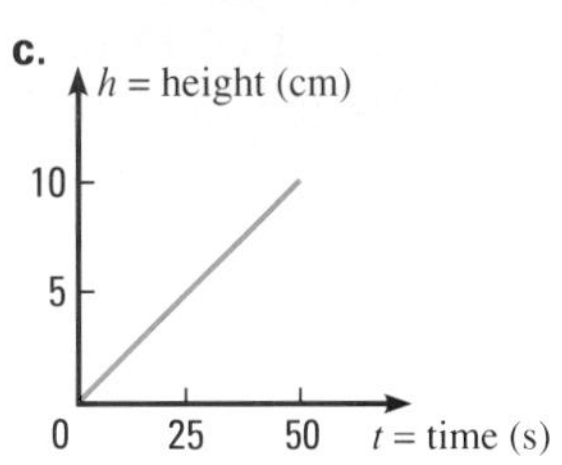


Each container shown in Problems 2–4 is to be filled with water flowing at a constant rate. Which of the three graphs shown for each container best represents the relationship between time ($t$) and corresponding height ($h$) of the liquid?

**2.**

**a.**

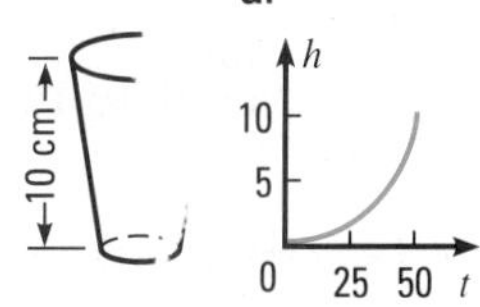


**b.**

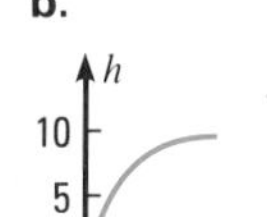


**c.**

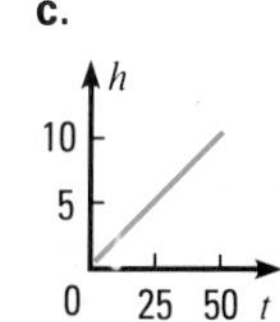


**3.**

**a.**

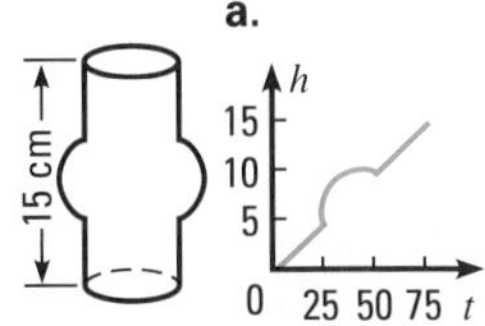


**b.**

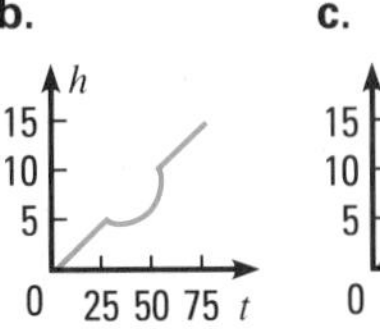


**c.**

**4.**

**a.**

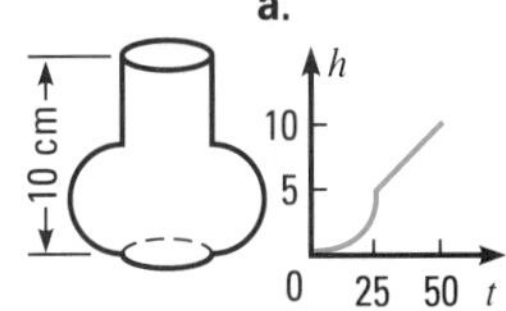


**b.**

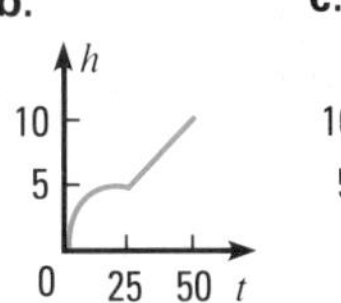


**c.**

Three bottles hidden from view are being filled. For each bottle, the time in seconds, $t$, is plotted against the corresponding height, $h$, of the liquid in the bottle. Using the graphs as a source of information, sketch what you think is the shape of each bottle.

**5.**

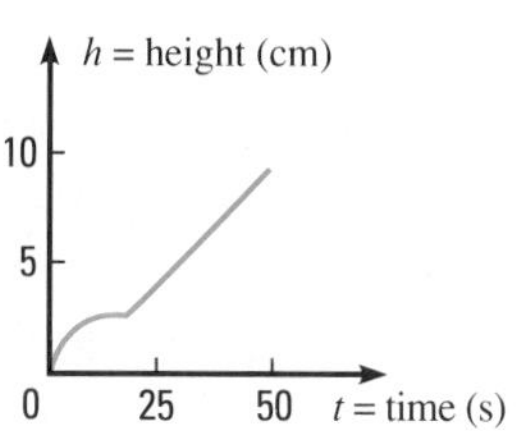


**6.**

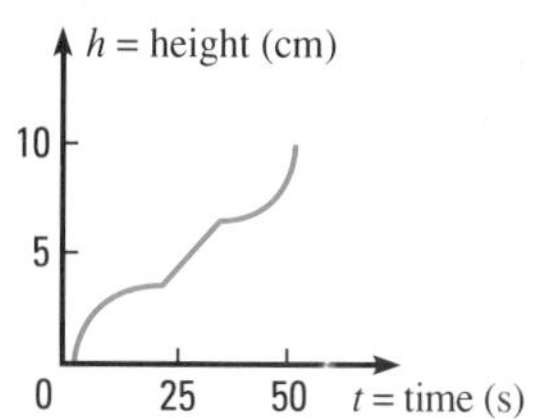


**7.**

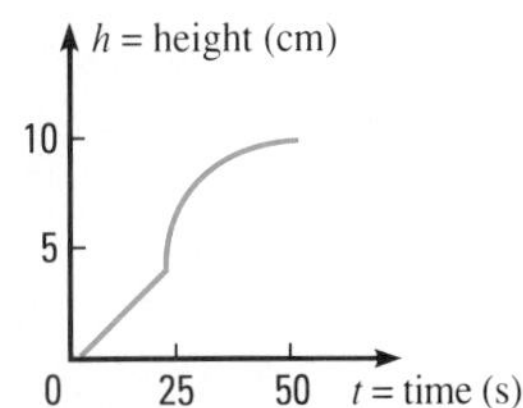

# 8-7 Distance on the Coordinate Plane

## EXPLORE/ WORKING TOGETHER

Study the graph at the right.

1. Count units to find the length of $\overline{AB}$.
2. Count units to find the length of $\overline{BC}$.
3. Triangle $ABC$ is a right triangle. How can you find the length of the hypotenuse $\overline{AC}$? What is the length?
4. On a coordinate plane, draw a different right triangle, $XYZ$. Exchange right triangles with a partner. Find the length of the hypotenuse of your partner's triangle.

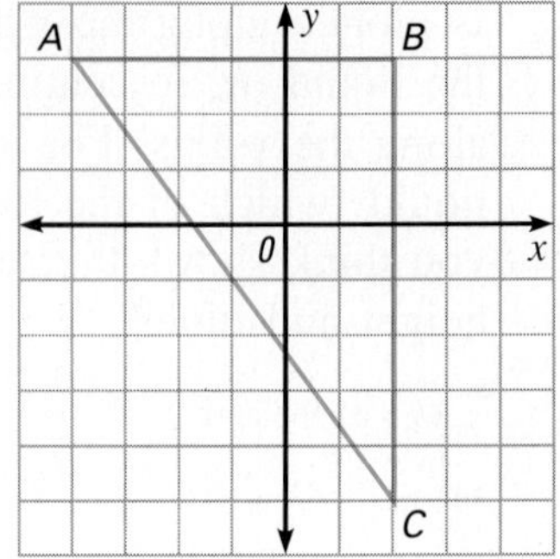


## SKILLS DEVELOPMENT

In the Explore activity, you counted units to find the length of a segment, or the distance between the two endpoints of a segment that is parallel to the $x$-axis or $y$-axis. Now you will see that you can also calculate the distance between those points algebraically.

### Example 1

Refer to the graph at the right to calculate the distance between each set of points.

**a.** $B$ and $C$ **b.** $P$ and $Q$

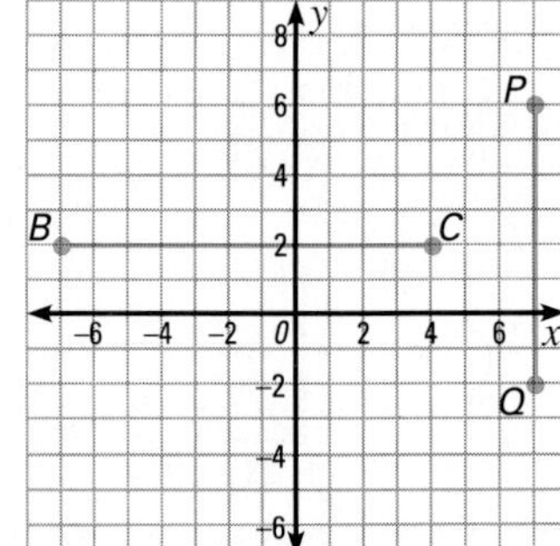


### Solution

**a.** Points $B(-7, 2)$ and $C(4, 2)$ are on a line parallel to the $x$-axis. Calculate the absolute value of the difference between the $x$-coordinates of $B$ and $C$.

$$|4 - (-7)| = |4 + 7| = 11 \quad \text{or}$$
$$|-7 - 4| = |-11| = 11$$

The distance between $B$ and $C$ is 11 units.

**b.** Points $P(7, 6)$ and $Q(7, -2)$ are on a line parallel to the $y$-axis. Calculate the absolute value of the difference between the $y$-coordinates of $P$ and $Q$.

$$|-2 - 6| = |-8| = 8 \quad \text{or}$$
$$|6 - (-2)| = |8| = 8$$

The distance between $P$ and $Q$ is 8 units. ◄

You can use the Pythagorean Theorem (page 216) to find the distance between two points not on an axis or on a line parallel to an axis.

**MIXED REVIEW**

Simplify.

1. $4 - (-9)$
2. $\frac{4}{5} + \frac{2}{3}$
3. $46 \div 2.3$
4. $2\frac{1}{6} - \frac{3}{4}$
5. $4.007 - 0.03$
6. $\frac{2}{5} \times 2\frac{1}{2}$

Solve.

7. Find the average of the lengths 13 cm, 10 cm, and 18 cm.
8. Simplify. $3x + x + 4x$

## Example 2

Calculate the distance between points $D$ and $E$ on the graph shown.

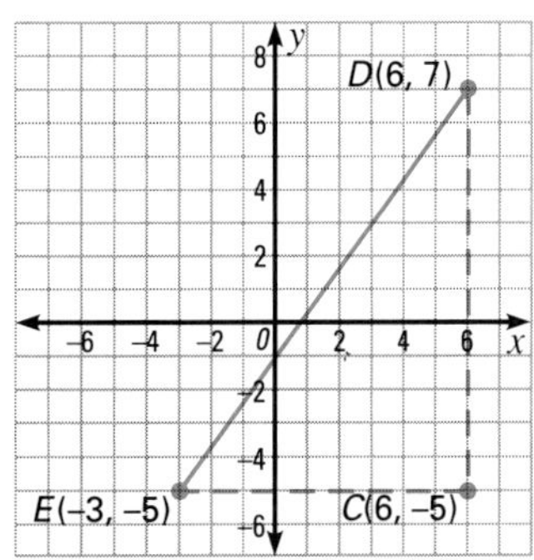


### Solution

Draw a dashed horizontal segment from $(-3, -5)$ to $(6, -5)$. Draw a dashed vertical segment from $(6, 7)$ to $(6, -5)$. Label the intersection point $C$. Triangle $EDC$ is a right triangle with hypotenuse $\overline{DE}$.

Use the Pythagorean Theorem to find the length of $\overline{DE}$.

$$DE = \sqrt{(CD)^2 + (CE)^2} = \sqrt{[7 - (-5)]^2 + [6 - (-3)]^2}$$
$$= \sqrt{(12)^2 + (9)^2} = \sqrt{144 + 81} = \sqrt{225} = 15$$

The distance from $D$ to $E$ is 15 units. ◀

Use the **distance formula** to find the distance between any two points.
For any points $P_1(x_1, y_1)$ and $P_2(x_2, y_2)$,
$P_1P_2 = \sqrt{(x_2 - x_1)^2 + (y_2 - y_1)^2}$.

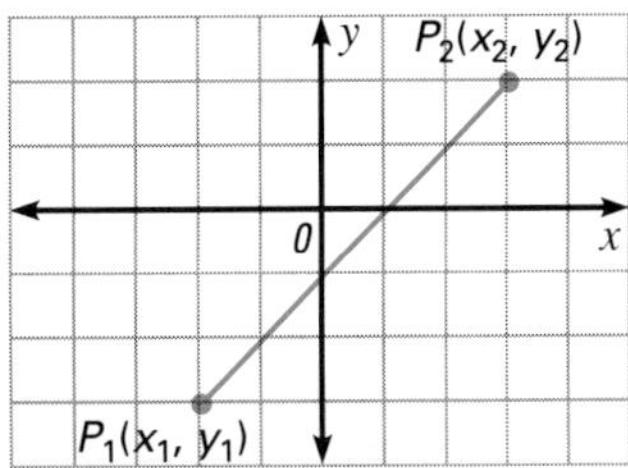


## Example 3

Calculate the distance between points $M(-2, -3)$ and $N(2, 3)$. Round to the nearest tenth.

### Solution

Substitute the values for the $x$- and $y$-coordinates of $M$ and $N$ in the distance formula.

$$MN = \sqrt{(x_2 - x_1)^2 + (y_2 - y_1)^2}$$
$$MN = \sqrt{(-2 - 2)^2 + (-3 - 3)^2} = \sqrt{(-4)^2 + (-6)^2}$$
$$= \sqrt{16 + 36} = \sqrt{52} = 7.211$$

The distance between $M$ and $N$ is 7.2 units. ◀

You can use the distance formula to find the equation of a circle, if you know the coordinates of any point on the circle.

For a circle with its center at the origin of a coordinate plane and a radius of 4, you can use the distance formula to find the distance from the center $(0, 0)$ to any point $P(x, y)$ on the circle.

$$\sqrt{(x - 0)^2 + (y - 0)^2} = 4$$
$$\sqrt{x^2 + y^2} = 4$$
$$x^2 + y^2 = 16$$

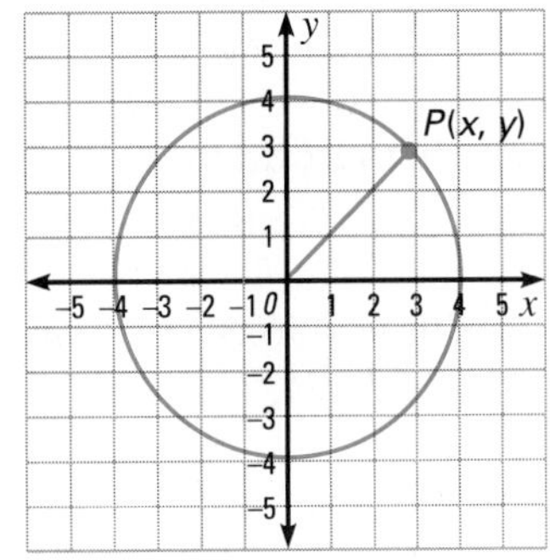


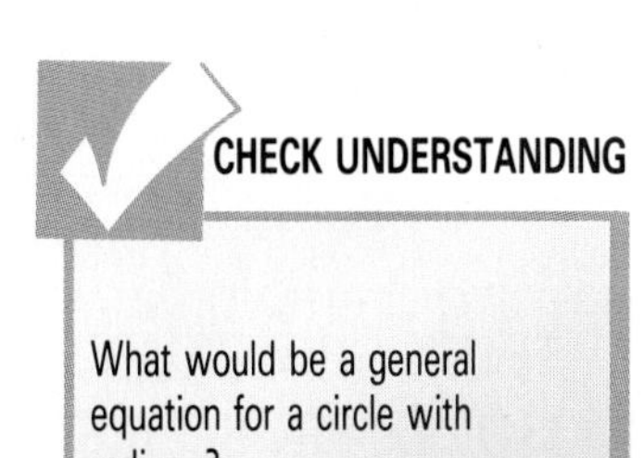


What would be a general equation for a circle with radius $r$?

## TRY THESE

For Exercises 1 and 2, refer to the graph at the right.

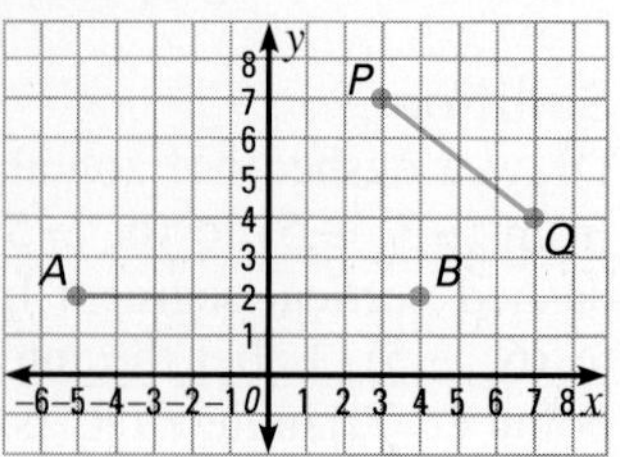

1. Calculate the length of segment $AB$.
2. Calculate the length of segment $PQ$.
3. Use the distance formula to calculate the distance between points $D(-8, -3)$ and $E(4, 2)$.

# EXERCISES

**PRACTICE/ SOLVE PROBLEMS**

Calculate the length of each segment.

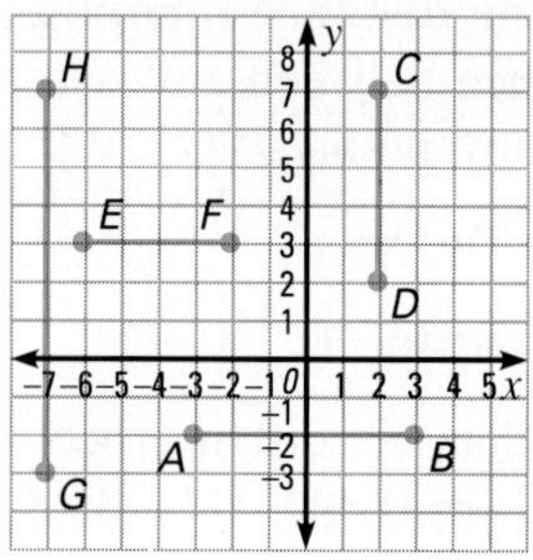

1. $\overline{AB}$
2. $\overline{CD}$
3. $\overline{EF}$
4. $\overline{GH}$

Use the distance formula to calculate the distance between each pair of points. Round answers to the nearest tenth.

5. $A(6, 2)$ $B(5, -1)$
6. $C(-2, -1)$ $D(-2, 9)$
7. $E(-8, 3)$ $F(4, -2)$
8. $G(2, -5)$ $H(0, 6)$

Find an equation for a circle with each radius, centered at the origin.

9. 3
10. 6
11. 8
12. 14

**EXTEND/ SOLVE PROBLEMS**

13. The vertices of a triangle are $A(2, 3)$, $B(5, 7)$, and $C(8, 3)$. Find the length of each side. What type of triangle is this?

Find an equation for each circle described.

14. center: (2, 0), radius: 7
15. center: (0, 3), radius: 10

**THINK CRITICALLY/ SOLVE PROBLEMS**

16. Three points are marked on the coordinate plane: $P(-3, -5)$, $Q(-2, 2)$, and $R(6, 4)$. Which point is closer to $Q$, $P$ or $R$?
17. Is the point (3, 3) *inside, on,* or *outside* the circle with equation $x^2 + y^2 = 9$?

▶ READ
▶ PLAN
▶ SOLVE
▶ ANSWER
▶ CHECK

# Problem Solving Applications:

## COORDINATES AND MAPS

Placing a coordinate plane over a map allows you to calculate distances between places on the map. To find distance, you need to know what distance each unit on the grid represents.

A coordinate plane is superimposed over a map. On this grid, 1 unit represents 2 km.

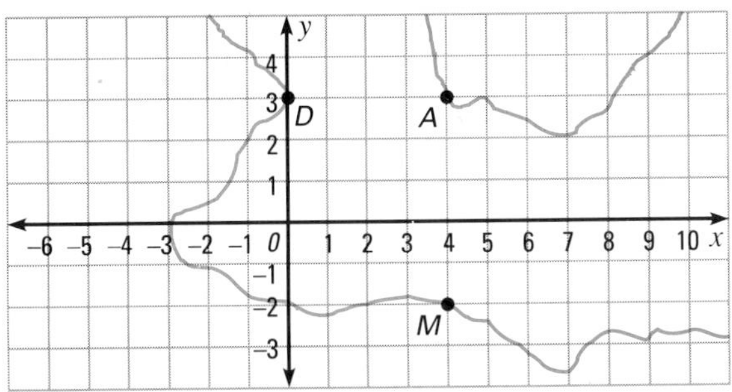


1. Find the distance from the airport, $A$, to the dock, $D$.
2. How far is it from the airport, $A$, to the marina, $M$?
3. What is the distance by boat from the marina to the dock?

On the coordinate plane superimposed over the map below, 1 unit represents 27 km.

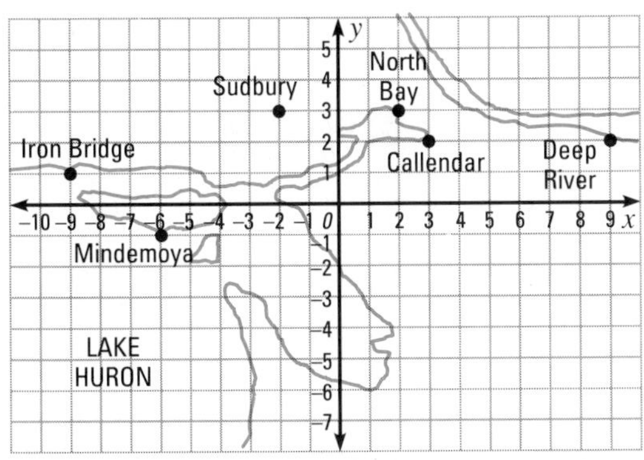


4. What is the flying distance from North Bay to Iron Bridge?
5. How far is Sudbury from Deep River?

A coordinate plane is superimposed over the map of a harbor. On this grid, 1 unit represents 200 m.

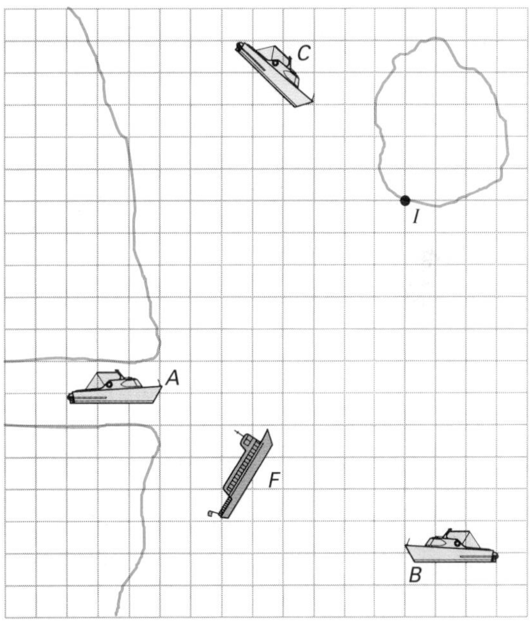


6. Calculate the distance between Boat $A$ and Boat $B$. (*Hint:* Decide where to place the origin.)
7. How far is Boat $C$ from the mouth of the canal?
8. How far is the ferry, $F$, from point $I$ on the island?

# 8-8 Graphing Inequalities on the Coordinate Plane

## EXPLORE

The line shown on the coordinate plane at the right is the graph of the linear equation $y = x$. Note the shaded and unshaded parts of the plane.

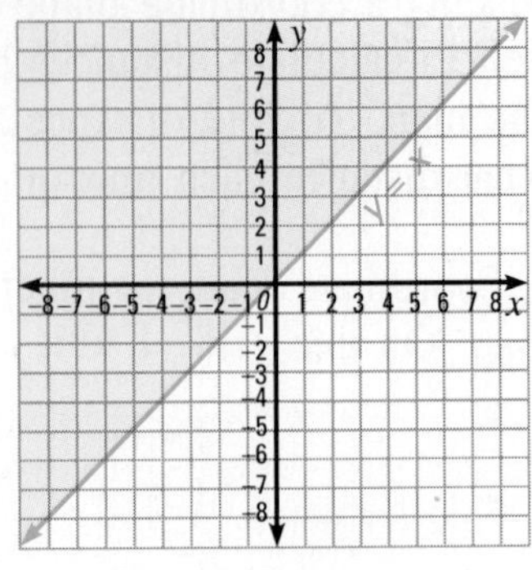


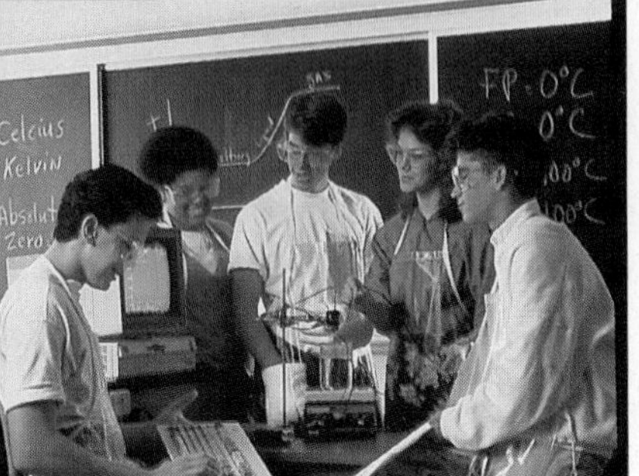

a. What is the relationship between the $x$-coordinate and the $y$-coordinate for each point on the line?

b. Which part of the plane contains the points $(-2, 1)$ and $(-4, -2)$?

c. What is the relationship between the $x$-coordinate and the $y$-coordinate for each point in the blue region?

d. Which part of the plane contains the points $(2, -2)$ and $(-3, -5)$?

e. What is the relationship between the $x$-coordinate and the $y$-coordinate for each point in the unshaded region of the plane?

## SKILLS DEVELOPMENT

The plane at the right shows the graph of the linear equation

$$y = 2x - 1.$$

The graphed line separates the coordinate plane into two regions. Each of the regions on both sides of a line in the coordinate plane is called an **open half-plane.** The line separating the half-planes forms the **boundary,** or edge, of each half-plane.

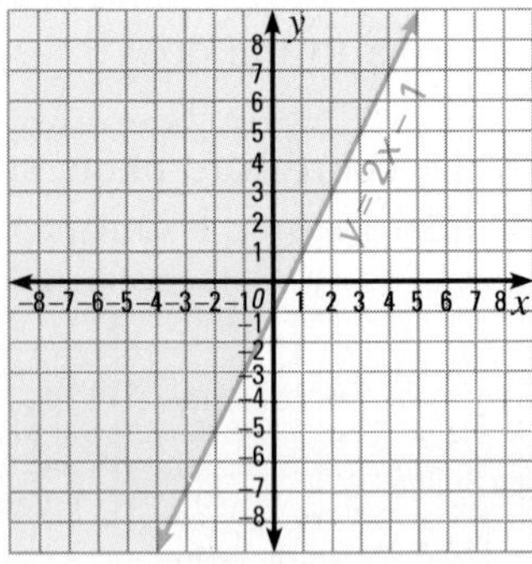


The shaded half-plane *above* the line is the graph of the solution set of the inequality

$$y > 2x - 1$$

The unshaded half-plane *below* the line is the graph of the solution set of the inequality

$$y < 2x - 1$$

Any ordered pair of real numbers that makes an inequality true is **a solution of the inequality.**

## Example 1

Tell whether the ordered pair is a solution of the inequality.

**a.** (2, 6); $y > 2x - 1$ **b.** (−2, 4); $y < 2x - 1$

### Solution

**a.** Substitute the value 2 for $x$ and the value 6 for $y$ in the inequality.

$$y > 2x - 1$$
$$6 \overset{?}{>} 2(2) - 1$$
$$6 \overset{?}{>} 4 - 1$$
$$6 > 3 \checkmark$$

The ordered pair (2, 6) is a solution of $y > 2x - 1$.

**b.** Substitute the value −2 for $x$ and the value 4 for $y$ in the inequality.

$$y < 2x - 1$$
$$4 \overset{?}{<} 2(-2) - 1$$
$$4 \overset{?}{<} -4 - 1$$
$$4 \not< -5$$

The symbol $\not<$ means "is not less than."

The ordered pair (−2, 4) is not a solution of $y < 2x - 1$. ◄

The **graph of an inequality** is the set of all ordered pairs that make the inequality true. The graph of the inequality $y \geq 2x - 1$ is the upper half-plane *and* the line. The graph of the inequality $y \leq 2x - 1$ is the lower half-plane *and* the line. Each of these graphs is called a **closed half-plane.**

To graph an inequality with two variables, first graph the equation related to the inequality. If the boundary is not a part of the graph of the inequality, then draw a dashed line. If the boundary is a part of the graph, then draw a solid line. Then graph the inequality as a shaded region above or below the line.

MATH IN THE WORKPLACE

Graphing inequalities is a way of solving difficult problems in business and industry. Suppose, for example, that a food company produces two kinds of fruit cooler by mixing apple and cranberry juice in different amounts. What is the greatest profit it can make on each cooler for the least cost from a given amount of each type of juice and the costs of labor and equipment?

The solutions to such problems, known as *linear programming problems,* require graphing inequalities with many variables.

Narenda Karmarkar, a researcher at AT&T's Bell Laboratories, developed a method that allowed a computer to solve such a problem involving 800,000 variables in only 10 hours.

## Example 2

Graph the inequality. $y < -2x + 3$

### Solution

Make a table of ordered pairs that can be used to graph the line.

| x | y |
|---|---|
| −1 | 5 |
| 0 | 3 |
| 1 | 1 |

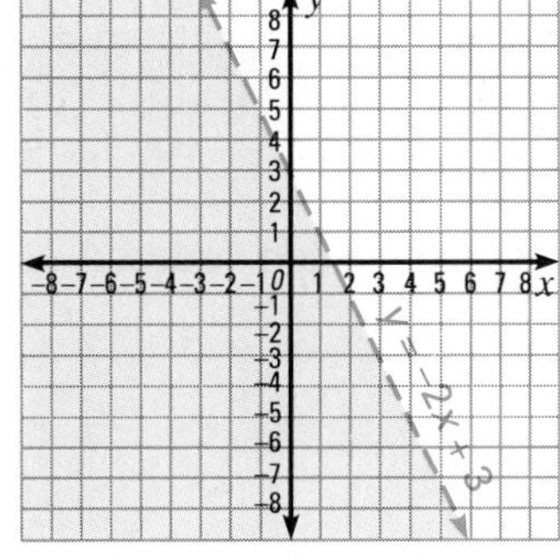


Use the ordered pairs to graph the equation $y = -2x + 3$. Draw the graph as a dashed line. Then shade the lower half-plane. The resulting open half-plane is the graph of $y < -2x + 3$. ◄

Use the following table to help you graph inequalities in two variables.

| Form of Inequality | Boundary | Shaded Region |
|---|---|---|
| $y > mx + b$ | dashed | above the line |
| $y \geq mx + b$ | solid | above the line |
| $y < mx + b$ | dashed | below the line |
| $y \leq mx + b$ | solid | below the line |

When an inequality has the form $Ax + By < C$, for example, you cannot use the inequality sign as a clue as to which half-plane to shade. You can transform the inequality into an inequality with $y$ alone on one side of the inequality sign and then graph it. However, it is often easier just to test a point to see which inequality is satisfied by the point. For example, if (0, 0) is not on the boundary, use it as a test point.

### Example 3

Graph $2x + y \leq -4$.

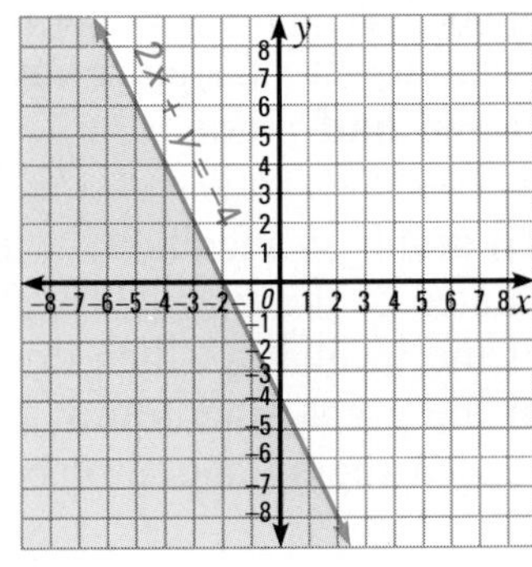


**Solution**

Graph the equation of the boundary $2x + y = -4$. Use a solid line. Then test the point (0, 0).

$$2(0) + 0 \overset{?}{\leq} -4$$
$$0 + 0 \overset{?}{\leq} -4$$
$$0 \not\leq -4$$

The point (0, 0) does not satisfy the inequality and is above the boundary. Shade the lower half-plane and draw a solid line. ◄

## TRY THESE

Tell whether each ordered pair is a solution of the inequality.

1. $(4, 1)$; $y > x - 5$
2. $(3, 2)$; $y \leq 2x - 1$
3. $(-3, -1)$; $y \geq x + 4$
4. $(1, 1)$; $y < 3x$
5. $(3, 3)$; $y \leq 2x - 3$
6. $(-1, 5)$; $y \leq 4x - 3$

Graph each inequality.

7. $y < x + 4$
8. $y \geq 3x - 1$
9. $y \leq 2x + 1$
10. $x + y \geq 2$
11. $2x - y < 2$
12. $x + y \leq 6$

# EXERCISES

## PRACTICE/ SOLVE PROBLEMS

State whether the graph of the inequality has a *dashed* or a *solid* line.

**1.** $y \leq x - 3$ **2.** $y > 3x - 2$ **3.** $2x - y \geq 4$ **4.** $x + 2y < 1$

Use (0, 0) or some other coordinate pair to determine whether the solution sets of the inequality are *above* or *below* the boundary.

**5.** $y - x < 0$ **6.** $2x + y \geq 4$ **7.** $3y - x > 4$ **8.** $-2y > x + 1$

Graph each inequality.

**9.** $y < 4$ **10.** $x \geq -1$ **11.** $y < -x$

**12.** $y \geq -x + 2$ **13.** $y \geq 3 - x$ **14.** $y \leq 1 - 2x$

## EXTEND/ SOLVE PROBLEMS

Write each statement as an inequality. Then graph the inequality.

**15.** The $y$-coordinate of a point is equal to or less than 3 more than the $x$-coordinate.

**16.** The sum of the $x$-coordinate and the $y$-coordinate is less than 8.

**17.** The $y$-coordinate of a point decreased by three times the $x$-coordinate is greater than or equal to 2.

Graph each inequality.

**18.** $y \leq \frac{1}{2}x + 1$ **19.** $y < -\frac{1}{3}x - 1$ **20.** $y < -\frac{4}{3}x$

Write an inequality for each graph shown.

**21.**

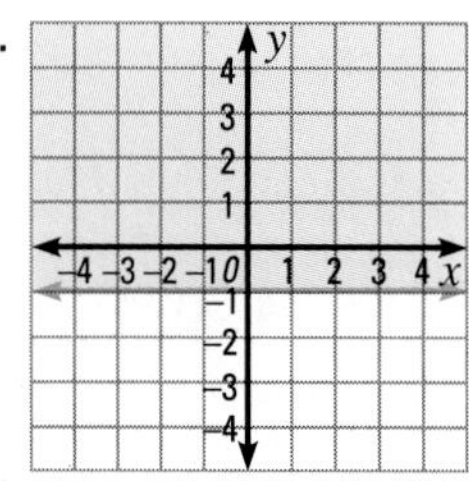

**22.**

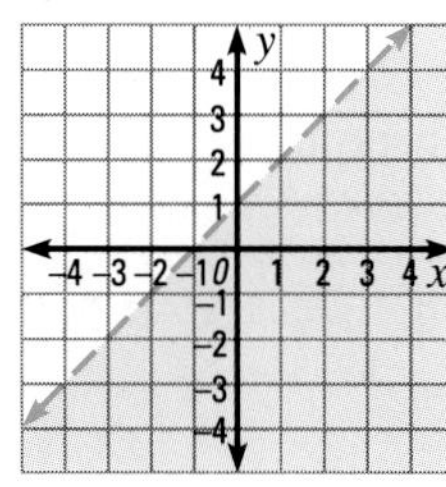

**23.**

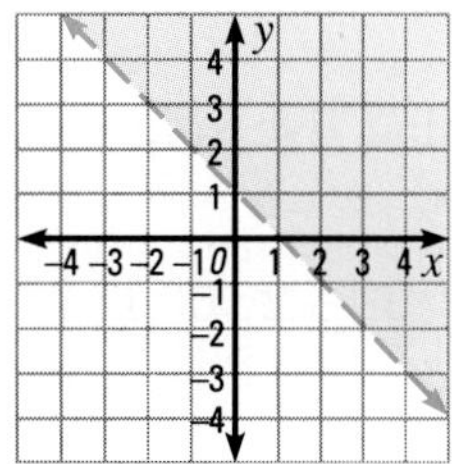

## THINK CRITICALLY/ SOLVE PROBLEMS

Graph each equality or inequality.

**24.** $y = |x|$ **25.** $y > |x|$ **26.** $y \leq |x|$ **27.** $|y| < 3$ **28.** $|y| \geq 4$

Graph each inequality.

**29.** $x^2 + y^2 < 9$ **30.** $x^2 + y^2 > 9$

## CHAPTER 8 ● REVIEW

Choose the word from the list that completes each statement.

1. On a coordinate plane, the horizontal number line is called the __?__.
2. The set of all possible values of $y$ is called the __?__ of the function.
3. Equations with two variables usually have a(n) __?__ number of solutions.
4. A(n) __?__ equation in $x$ is an equation involving no term with a higher power of $x$ than $x^2$.
5. The region on either side of a line in the coordinate plane is called a(n) __?__ half-plane.

a. range
b. infinite
c. open
d. quadratic
e. $x$-axis

### SECTION 8–1 THE COORDINATE PLANE (pages 312–315)

► On a coordinate plane, the horizontal number line is called the **$x$-axis;** the vertical number line is called the **$y$-axis.** The axes separate the plane into four **quadrants.** The point where the axes cross is called the **origin.**

6. Give the coordinates of point $C$, 3 units to the right of the origin and 2 units down from the origin.
7. Find the midpoint, $P$, of the line segment whose endpoints are $K(5, 2)$ and $L(1, 2)$.

### SECTION 8–2 EQUATIONS AND FUNCTIONS (pages 316–319)

► A function such as $f(x) = x + 2$ can be written as $y = x + 2$. The set of all possible values of $x$ is the **domain** of the function; the set of all possible values of $y$ is the **range.**

► The solutions of an equation with two variables are sets of ordered pairs. To graph a function, graph the points that correspond to the ordered pairs.

Use the given values to find five solutions of each equation. Write the solutions as ordered pairs.

8. $y = x - 4$ for $x = -3, -2, 0, 1, 2$
9. $y = 2x + 2$ for $x = -2, -1, 0, 1, 2$

### SECTIONS 8–3 AND 8–4 LINEAR AND NONLINEAR FUNCTIONS (pages 320–327)

► An equation such as $y = 2x + 3$ is called a **linear equation.** A linear equation represents a **linear function.**

► Some equations with two variables are **nonlinear functions,** an example of which is a quadratic function, represented by a quadratic equation.

► To graph a nonlinear function, construct a table of values and draw a smooth curve through the points.

Graph the function when the domain is the set of real numbers.

**10.** $y = x - 1$

**11.** $y = 2x$

Use the graph at the right to complete each ordered pair so that it corresponds to a point on the graph.

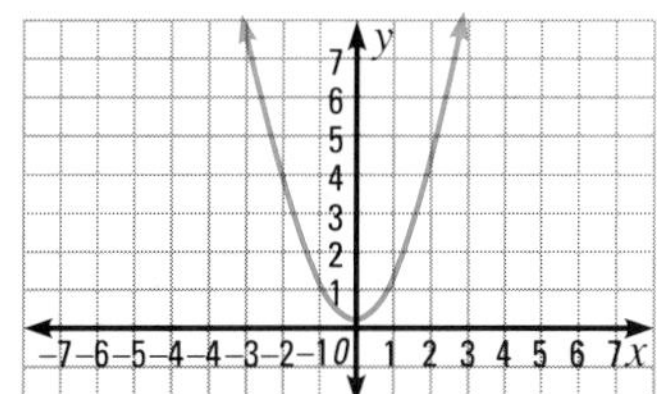

**12.** (1, ■)

**13.** (3, ■)

## SECTIONS 8–5 AND 8–6 PROBLEM SOLVING (pages 328–331)

► The graphs of linear functions are straight lines; those of nonlinear functions are curved lines. The graphs of some other functions combine both kinds of lines.

**14.** The bottle below is to be filled with water. Which graph best represents the relationship between time and the corresponding height of the liquid, for a constant flow of water?

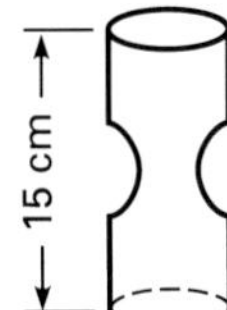


a.

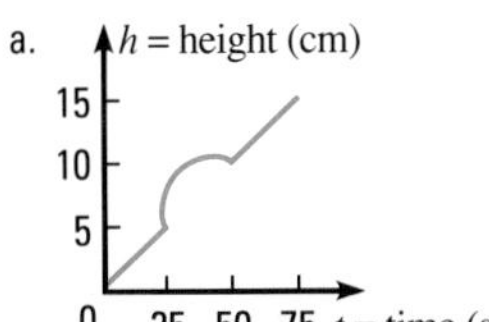


b.

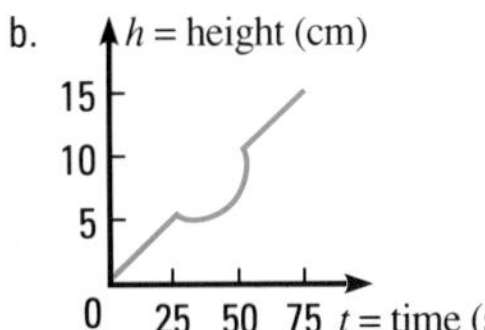


c.

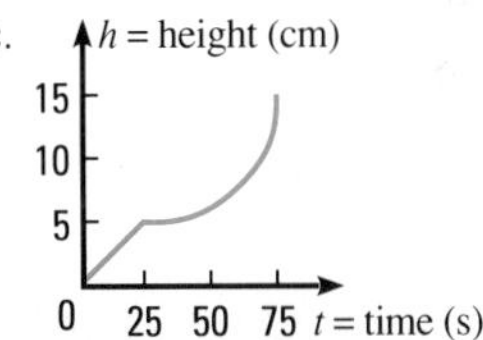


*USING DATA* Refer to the solubility chart on page 310 for Exercise 15.

**15.** At 0°C, sugar has a solubility of 179 g. At 100°C, its solubility is 487 g. About how many times more soluble is sugar than $KNO_3$ at 0°C? About how many times more soluble is sugar than $KNO_3$ at 100°C?

## SECTION 8–7 DISTANCE ON THE COORDINATE PLANE (pages 332–335)

► To find the distance between any two points, use the distance formula.

$$P_1P_2 = \sqrt{(x_2 - x_1)^2 + (y_2 - y_1)^2}$$

**16.** Use the distance formula to calculate the distance between point $A(2, 3)$ and point $B(6, 6)$.

**17.** Use the distance formula to find an equation for a circle of radius 6, centered at the origin.

## SECTION 8–8 INEQUALITIES ON THE COORDINATE PLANE (pages 336–339)

► The **graph of an inequality** is the set of all ordered pairs that make the inequality true. Any one of these pairs is a solution of the inequality.

Tell whether the ordered pair is a solution of the inequality.

**18.** (2, 5); $y < 2x + 2$

**19.** (1, 4); $y \leq x - 4$

# CHAPTER 8 ● TEST

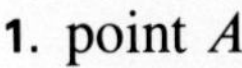

For Exercises 1–3, give the coordinates of the point or points shown on the graph at the right.

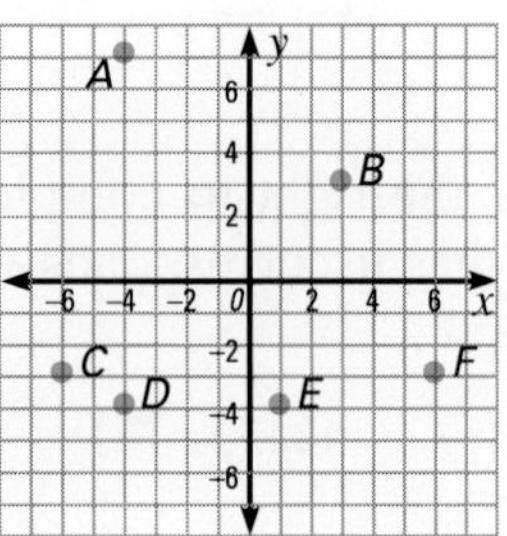


1. point $A$
2. point $B$
3. two points in the fourth quadrant

4. Find the midpoint $P$ of a line segment whose endpoints are $A(6, 3)$ and $B(2, 5)$.

Use the given values to find five solutions of each equation. Write the solutions as ordered pairs.

5. $y = x - 3$ for $x = -4, -2, 0, 2, 4$
6. $y = 2x + 3$ for $x = -2, -1, 0, 1, 2$

Graph the function for the domain of real numbers.

7. $y = -3$
8. $y = x + 2$

Carlo begins walking along a road at a point 1 km from the public library. He walks at a constant rate of 2.5 km an hour. His distance from the library, $d$, is a function of the amount of time he walks, $t$. An equation representing this function is $d = 2.5t + 1$, where $t$ is the number of hours Carlo walks.

9. Graph the function.
10. How far is Carlo from the library after 2 h?

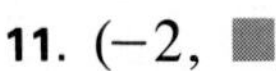

Refer to the graph at the right. Complete each ordered pair so that it corresponds to a point on the graph.

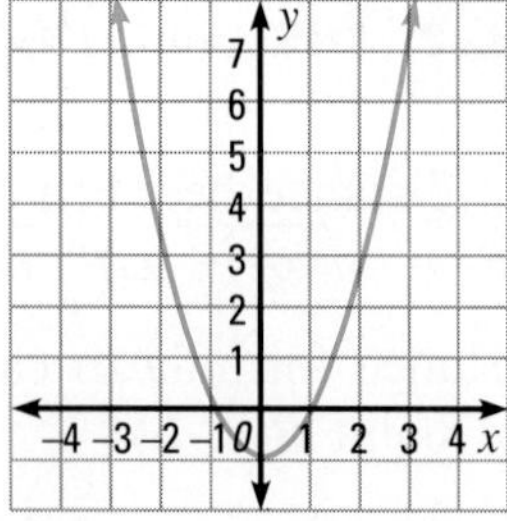


11. (−2, ■)
12. (0, ■)
13. (■, 8)

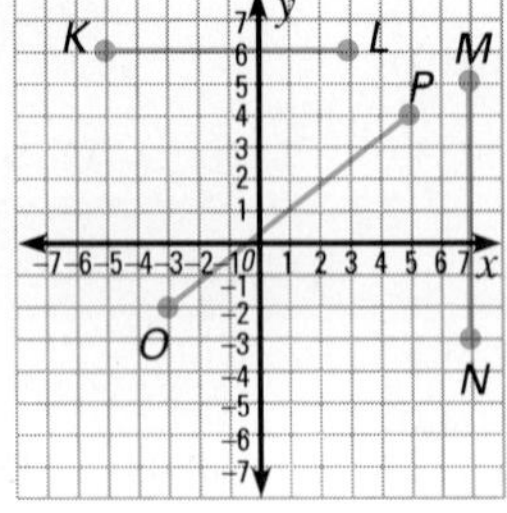


Refer to the graph at the left to calculate the distance between each set of points.

14. $K$ and $L$
15. $M$ and $N$
16. $O$ and $P$

Tell whether the ordered pair is a solution of the inequality.

17. (1, 4); $y > 2x - 2$
18. (3, 3); $y \leq x - 3$
19. Graph the inequality $y < 2x - 4$.

# CHAPTER 8 ● CUMULATIVE REVIEW

1. Use the table to find the function rule.

| x | y |
|---|---|
| 2 | 7 |
| 4 | 11 |
| 6 | 15 |
| 8 | 19 |
| 10 | 23 |

2. Write the phrase "8 more than 3 times a number" as a variable expression.

Evaluate each expression when $a = -6$ and $b = 2$.

3. $b + |a|$

4. $a - b^3$

5. Write $2.06 \times 10^4$ in standard form.

6. The measures of two angles of a triangle are 61° and 29°. Tell whether the triangle is *acute, right,* or *obtuse*.

In the figure below, *ABCD* is a parallelogram.

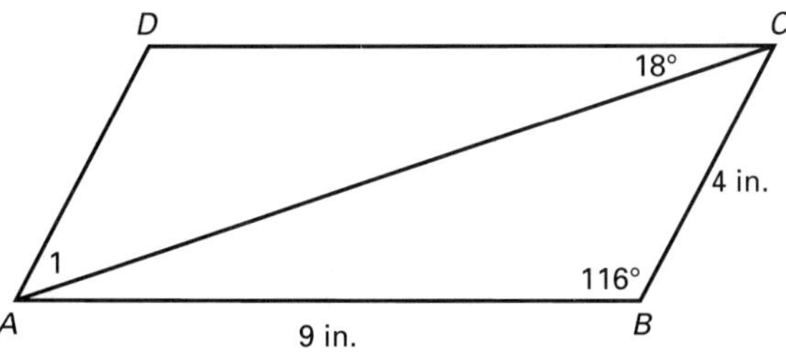


7. Find the length of $\overline{DC}$.

8. Find $m\angle 1$.

Use these data for Exercises 9 and 10.

**Bowler's Scores**

156 145 148 145 184 176

9. Find the mean.

10. Find the mode.

The circle graph shows the greatest problems U.S. residents believe are facing the country. Suppose a resident of the United States is selected at random.

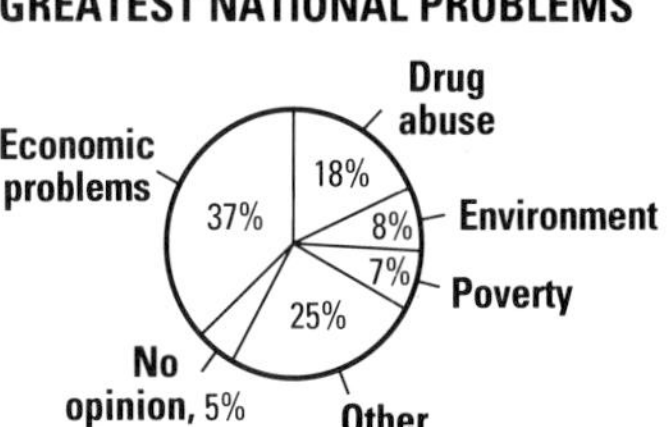


11. Estimate the probability that the person would name drug abuse or poverty as the greatest problem facing the country.

12. Estimate the probability that the same person who named drug abuse or poverty would *not* name the environment as the greatest problem.

Solve each equation.

13. $5(2w + 7) = 45$

14. $39 = 7(c - 3) + 8c$

15. The length of one side of a regular hexagon is 5.9 cm. Find the perimeter.

16. Find the surface area of the triangular prism shown here.

5 in.
3 in.
5 in.
4 in.

17. The diameter of a cylinder is 26 cm and the height is 30 cm. Find the volume of the cylinder. Use $\pi = 3.14$.

18. Find the product. $(8xy^2)(-3x^2z)$

Simplify. No denominator equals zero.

19. $\dfrac{(2a^2b^4)^3}{ab}$

20. $\dfrac{-36x^2y^2z^4}{4xyz^3}$

Find the missing coordinate so that the resulting ordered pair is a solution of $y - 2x = 4$.

21. (3, ■)

22. (−5, ■)

1. Which equation would you use to find 8% of 140?
A. $n = 8 \cdot 140$  B. $n = 0.8 \cdot 140$
C. $n = 0.08 \cdot 140$  D. $n = 140 \div 0.08$

2. On a number line, which number is to the left of $-1.5$?
A. $-2$  B. $-1.2$  C. 0  D. 2

3. Solve. $3^4 \cdot 5 - 2^5 \div 8$
A. 403  B. 131
C. 46.625  D. 401

4. Which set of line segments of the given lengths *cannot* be the sides of a triangle?
A. 14 cm, 11.3 cm, 5.6 cm
B. 10m, 3m, 5m
C. 8 ft, 7 ft, 9 ft
D. 5 in., 12 in., 13 in.

5. Find the unknown angle measure in this polygon.
A. 141°  B. 83°
C. 97°  D. 457°

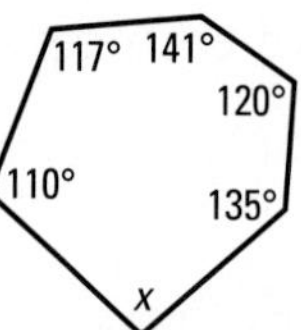


6. Imagine a circle graph showing this data.

**Favorite After-School Snack**

| Snack | pizza | yogurt | fruit | chips |
|---|---|---|---|---|
| Percent | 45% | 27% | 10% | 18% |

What is the approximate measure of the central angle for the sector that represents yogurt?
A. 97°  B. 36°  C. 27°  D. 162°

Use the scatter plot for Exercises 7 and 8.

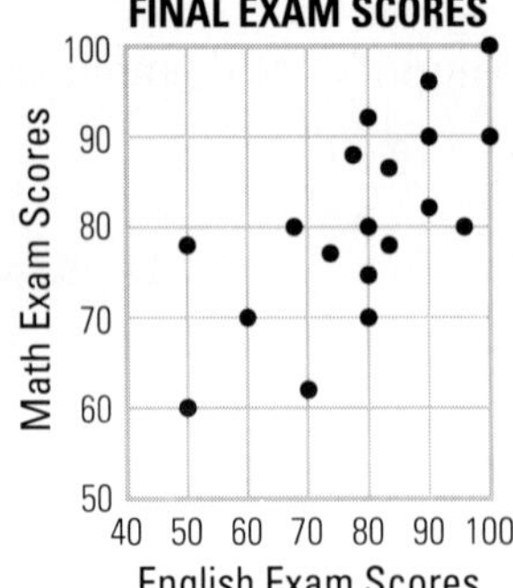


7. How many students received an 80 on their math final exam?
A. 0
B. 2
C. 4
D. none of these

8. What is the range of English final exam scores?
A. 40  B. 50  C. 60  D. 80

9. Solve. $3(4b + 2) - 2b = 46$
A. 4  B. 4.3  C. $3\frac{1}{3}$  D. $-4$

10. Find $-\sqrt{92}$ to the nearest tenth.
A. 8,464  B. $-9.2$
C. 9.6  D. $-9.6$

11. Find the area of the shaded region.

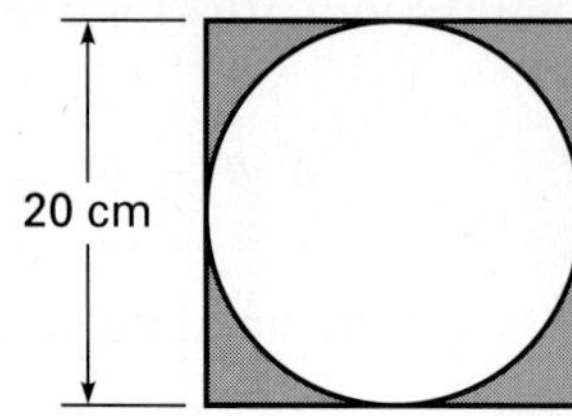


A. 86 cm²  B. 368.6 cm²
C. 314 cm²  D. 856 cm²

12. Find the volume of a regular square pyramid with base edges 12 ft and height 8 ft.
A. 1,152 ft³  B. 336 ft³
C. 384 ft³  D. 768 ft³

13. Find the product.
$$(-c^2d)(5cd^2)(7c^3d^2)$$
A. $-5c^6d^6$  B. $-35c^6d^5$
C. $-35c^5d^6$  D. $35c^7d^4$

14. Simplify. The denominator does not equal zero.
$$\frac{8xy(-4x^2y^3)}{-16x^2y^2}$$
A. $\frac{1}{2}xy^2$  B. $2x^2y$
C. $2xy^2$  D. $-2xy$

15. Which of the ordered pairs is *not* a solution of $3x = y + 4$?
A. (5, 11)  B. (−2, 2)
C. (0, −4)  D. $\left(\frac{1}{3}, -3\right)$

# CHAPTER 9 SKILLS PREVIEW

Find the slope of each line from the given information.

1. 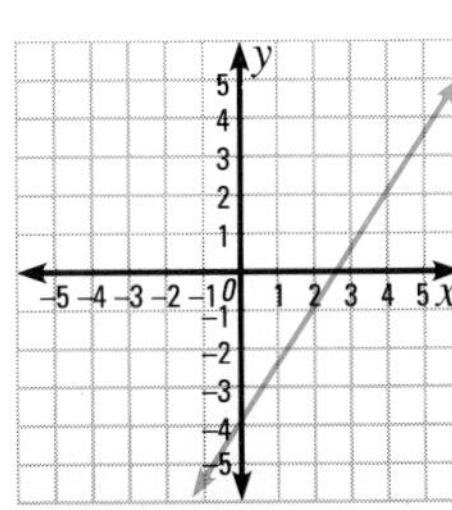

2. $A(-1, 5)$ and $B(2, 1)$

3. $-2x + 5y = -15$

4. $y = -4x + 3$

Identify the slope and $y$-intercept for the line with the given equation. Then identify the slope of a line parallel to it and the slope of a line perpendicular to it.

5. $y = -3x + 2$

6. $y = -\frac{2}{5}x - 3$

7. $4x - 3y = 12$

Graph each equation.

8. $y = 2x - 1$

9. $4x + 2y = -6$

10. $-3x + 5y = -5$

Write an equation of each line from the given information.

11. 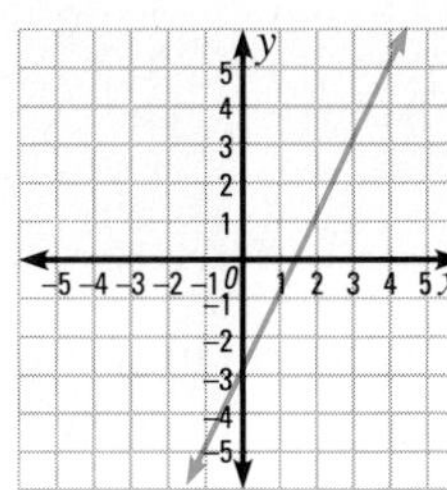

12. $m = -4 \quad b = 7$

13. $m = -1 \quad P(3, -1)$

14. $K(-5, 1) \quad L(7, -2)$

Use a graph to solve each system.

15. $-5x + y = -1$
$15x - 3y = -9$

16. $-2x + y = 5$
$x - 4y = -13$

17. $4x - 2y > -6$
$x + 3y \leq -6$

Solve.

18. $x + y = 9$
$-3x + 2y = -2$

19. $5x - 2y = 16$
$3x + 2y = 0$

20. $5x - 4y = 9$
$-2x + 3y = -5$

21. Abdul and Yasir's ages together total 21. Abdul is 3 years younger than Yasir. How old is each boy?

22. Trisha has 23 nickels and dimes. The number of nickels is 2 more than twice the number of dimes. How many of each coin does she have?

# SLOPE AND SYSTEMS

**THEME** **Building and Construction**

In this chapter, you will find the slope of lines, graph lines using slope, and use slope to predict the number of solutions a system of equations has.

The verb **construct** means to make or form by combining parts. One of the parts that is combined in building a house is the stairs. Builders rely on the concept of slope to design the stairs and plan the quantity of materials they will need to build them.

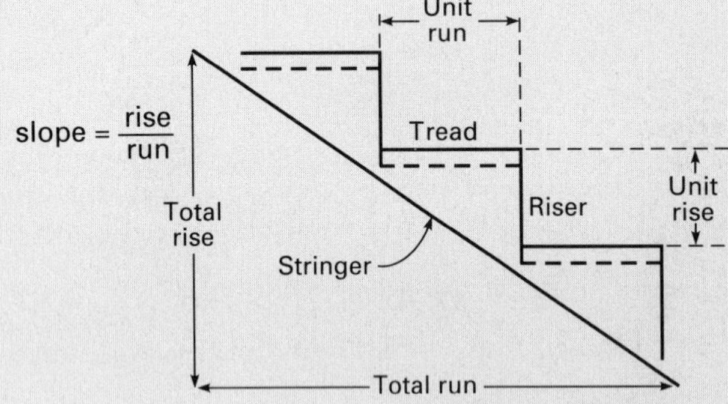


The slope of the stairs—the relationship between the riser and the tread—is critical for both ease of climbing and safety. Builders use guidelines that are based on what is considered to be an average climbing stride. The average rise in typical house stairs is 7 in. or $7\frac{1}{2}$ in.

## GUIDELINES FOR SIZE OF STAIRS

1. Rise + tread + rise should be about 25 in.
2. Rise + tread should be about 17 to 18 in.
3. Rise × tread should be about 75 in.

To build stairs, you need four basic calculations to find measurements for the stairway layout.

**CALCULATIONS**

| TO FIND: | YOU SHOULD: |
|---|---|
| number of risers | divide total rise by 7 (round answer) |
| rise of a step | divide total rise by the number of steps |
| tread of a step | use Guideline 2 |
| stairway span (total run) | multiply tread measure by number of treads |

## DECISION MAKING

### Using Data

1. Use the guidelines to determine the "ideal" measurements for the riser and the tread of stairs. Would a riser of 4 in. and a tread of 14 in. be acceptable? Why or why not?

2. Why is the total rise divided by 7 to calculate the number of treads? What is the relationship between the number of treads and the number of risers?

3. If the total rise for a stairway is 8 ft 4 in., calculate each of the following.
   a. number of risers
   b. rise of a step
   c. number of treads
   d. tread width

### Working Together

Your group's task is to measure a variety of stairways. Record your information in chart form. Be sure to include the total rise, the number of risers, the rise of a step, the run of a tread, the number of treads, the tread width, and the location and purpose of the stairway. Write a paragraph to summarize your findings and your reasons to explain the similarities and/or discrepancies.

# 9-1 Slope of a Line

## EXPLORE/ WORKING TOGETHER

Work with a partner. Match each equation below with its graph at the right.

You can check your own work. If you write the letters for the correct graphs in the order of the given equations, they will spell the main topic of this lesson.

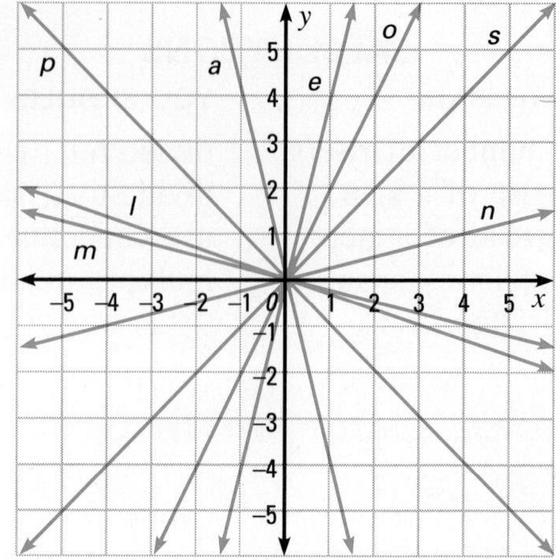


$x = y \qquad -x = 3y \qquad 2x = y \qquad x = -y \qquad 4x = y$

## SKILLS DEVELOPMENT

Often the meaning of a mathematical term is related to its meaning in everyday life. For instance, you may already know about the slope of a roof or the slope of a hill.

**CONNECTIONS**

Skiing is a sport that uses the term "slope." What does it mean when someone says, "I'm going to the slopes"? What are some of the names that ski slopes have that indicate how difficult the hill is and/or how good the skier should be?

The **slope** of a segment is the ratio of its *rise* (vertical distance) compared with its *run* (horizontal distance).

Rise

Run

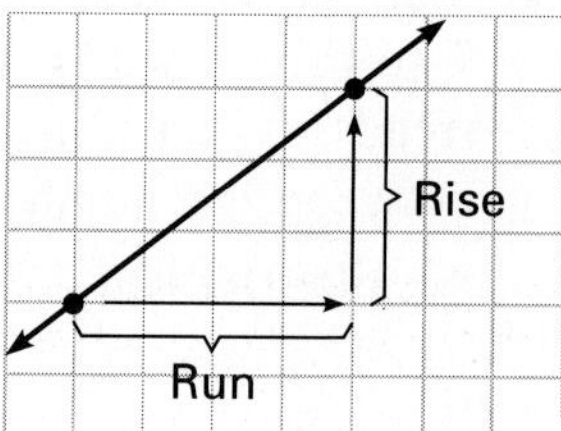


$$\text{slope} = \frac{\text{rise}}{\text{run}} = \frac{\text{change in } y\text{-coordinates}}{\text{change in } x\text{-coordinates}}$$

Using a grid, you can count the units between any two points on a line to find its slope. Conversely, given a point on a line and the slope of that line, you can locate other points on the line. You can also find the slope of a line algebraically.

### Example 1

Find the slope of $\overleftrightarrow{AB}$.

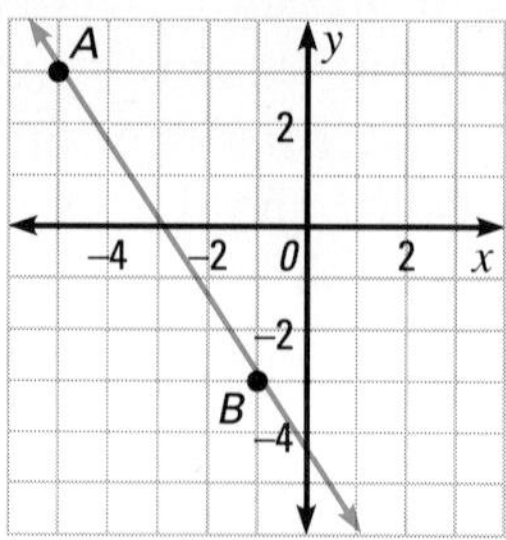

### Solution

$$\text{Slope of } \overleftrightarrow{AB} = \frac{\text{rise}}{\text{run}} = \frac{\text{down 6}}{\text{right 4}} = \frac{-6}{+4} = -\frac{3}{2}$$

The slope of $\overleftrightarrow{AB}$ is $-\frac{3}{2}$. ◄

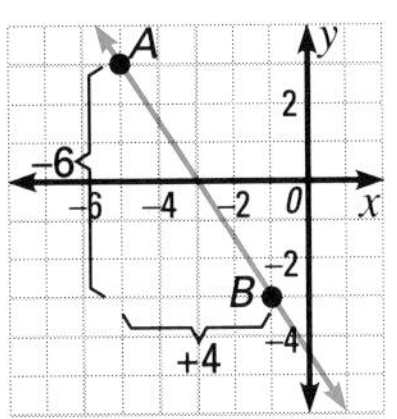


## Example 2

Graph the line that passes through the point (2, 1) and has a slope of $\frac{5}{4}$.

### Solution

To graph the line, first plot the point (2, 1). The slope is $\frac{5}{4}$, so for each 5 units of rise, the run must be 4 units in the direction that creates a positive slope. So, if you go 5 units up, you must go 4 units right; if you go 5 units down, you must go 4 units left. Recall that if you divide a number by a number that has the same sign, the result is positive. ◄

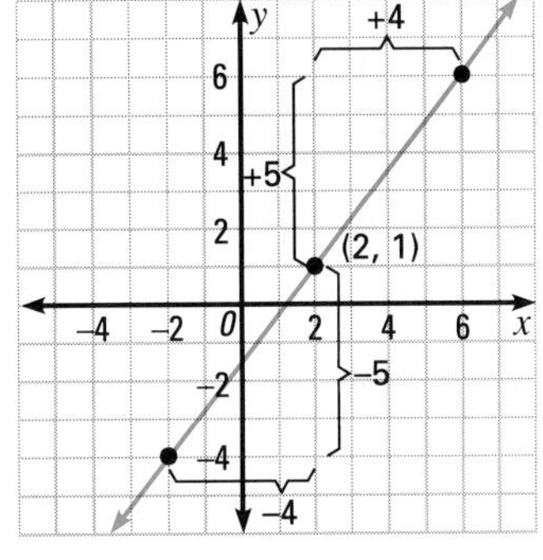


**CHECK UNDERSTANDING**

Look back at the graphs in Example 1. How would the slope have changed if you had moved left and up from *B* to *A* instead of moving down and right from *A* to *B*? Explain.

## Example 3

Find the slope of $\overleftrightarrow{PQ}$ containing these points.

**a.** $P(-2, 6)$ and $Q(1, -3)$ **b.** $P(4, -3)$ and $Q(4, 3)$

### Solution

**a.** $\text{slope} = \frac{\text{change in } y\text{-coordinates}}{\text{change in } x\text{-coordinates}} = \frac{y_2 - y_1}{x_2 - x_1} = \frac{-3 - 6}{1 - (-2)} = \frac{-9}{3} = -3$

The slope of $\overleftrightarrow{PQ}$ is $-3$.

**b.** $\frac{y_2 - y_1}{x_2 - x_1} = \frac{3 - (-3)}{4 - 4} = \frac{6}{0}$ ← **Remember, you cannot divide by zero. We say that the slope of $\overleftrightarrow{PQ}$ is *undefined*.** ◄

**TALK IT OVER**

Suppose you chose (1, −3) to be $(x_1, y_1)$ and (−2, 6) to be $(x_2, y_2)$ in Example 3. What would happen to the slope? Can you explain why? Can you predict what the slope would be if you chose two other points on the line?

## Example 4

Find the slope of the line with the given equation.

**a.** $2x - 3y = 6$ **b.** $y = -3x + 5$

### Solution

**a.** $2x - 3y = 6$

| | |
|---|---|
| $x = 0$ | $y = 0$ |
| $y = -2$ | $x = 3$ |
| $(0, -2)$ | $(3, 0)$ |

← **Find two points on the line.** →

**b.** $y = -3x + 5$

| $x$ | $y$ | $(x, y)$ |
|---|---|---|
| 0 | 5 | (0, 5) |
| 1 | 2 | (1, 2) |

← **Find the slope.** →

**a.** $\frac{0 - (-2)}{3 - 0} = \frac{2}{3}$

The slope is $\frac{2}{3}$.

**b.** $\frac{5 - 2}{0 - 1} = \frac{3}{-1} = -3$

The slope is $-3$. ◄

## TRY THESE

Find the slope of each of the lines shown.

1. 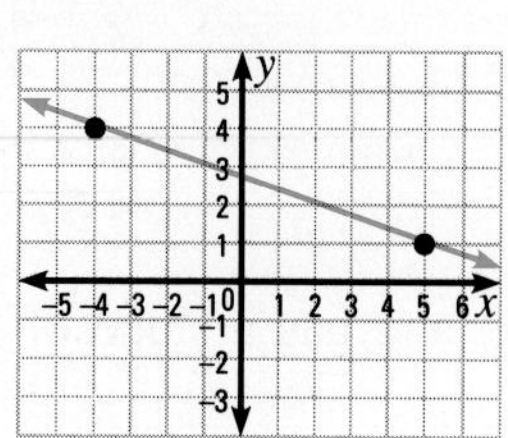

2. 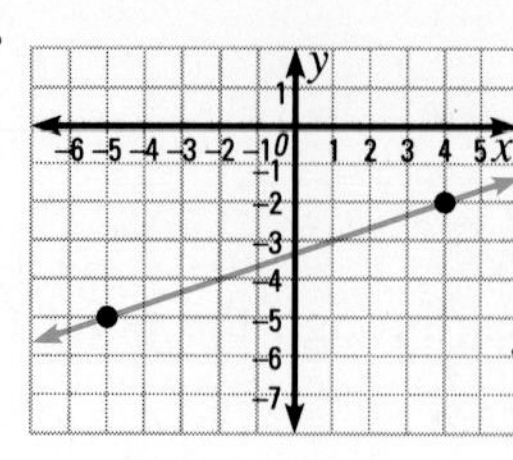

3. 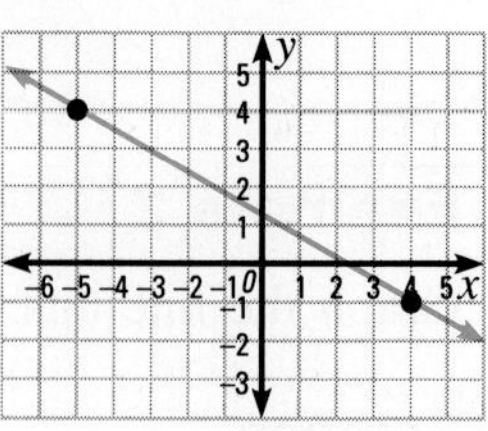

Graph the line that passes through the given point $P$ and has the given slope.

4. $P(3, 5)$ slope $= -\frac{4}{5}$
5. $P(-2, 1)$ slope $= \frac{3}{4}$
6. $P(4, -1)$ slope $= 2$
7. $P(1, -3)$ slope $= -3$

Find the slope of the line containing the given points.

8. $A(2, 3)$ and $B(8, 6)$
9. $X(-4, 3)$ and $Y(-8, 6)$
10. $C(4, 7)$ and $D(5, 2)$
11. $M(6, 8)$ and $N(3, 6)$

Find the slope of the line with the given equation.

12. $3x - 2y = 12$
13. $y = -4x + 5$
14. $-6x - 9y = 18$
15. $y = 3x - 7$

# EXERCISES

### PRACTICE/ SOLVE PROBLEMS

Find the slope of each line segment shown.

1. $\overline{AB}$
2. $\overline{CD}$
3. $\overline{EF}$
4. $\overline{GH}$
5. $\overline{IJ}$
6. $\overline{KL}$
7. $\overline{MN}$
8. $\overline{OP}$

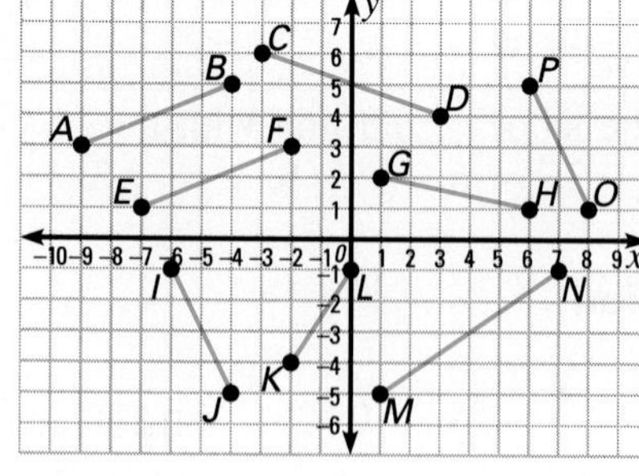


Graph the line that passes through the given point $P$ and has the given slope.

9. $P(3, 4)$ slope $= \frac{1}{5}$
10. $P(-3, 2)$ slope $= -\frac{2}{3}$
11. $P(1, -5)$ slope $= -\frac{3}{4}$
12. $P(-2, -1)$ slope $= \frac{5}{2}$
13. $P(-2, 0)$ slope $= \frac{2}{5}$
14. $P(1, -3)$ slope $= -\frac{1}{3}$

Find the slope of the line.

15. $5x - y = 10$

16. $-4x + 2y = -8$

17. $y = -3x + 2$

18. $y = 7x - 4$

## EXTEND/ SOLVE PROBLEMS

Find the slope of the line that contains the given points. Then plot the points and draw the lines.

19. $A(-2, 5)$ $B(6, 5)$

20. $C(3, -1)$ $D(3, 4)$

21. $E(-4, 1)$ $F(-4, 3)$

22. $G(-5, 2)$ $H(6, 2)$

***USING DATA*** For Exercises 23–28, use the Data Index on pages 556–557 to find the table showing the building codes standards for slope and safety.

Determine which of these stairs meet the safety specifications.

23.

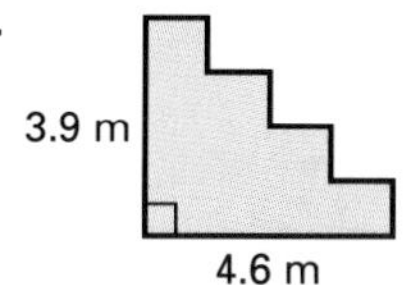


24.

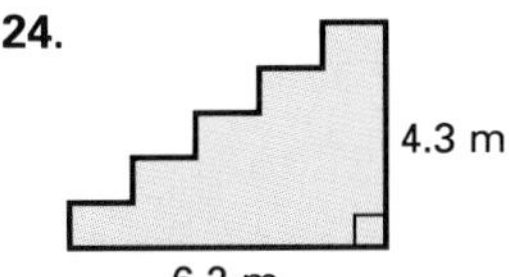


25.

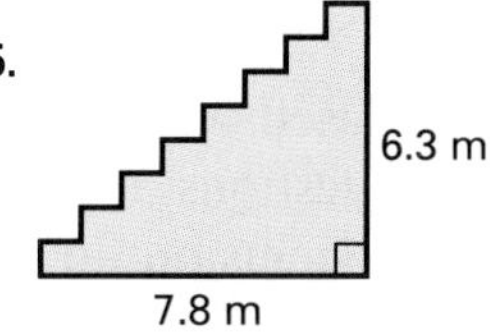


26. Some streets in San Francisco are on hills with a run of 9 m and a rise of 4.2 m. Would it be safe to park your car on one of those streets?

27. The Kells' driveway has a run of 1.2 m and a rise of 0.4 m. Does it meet the safety specifications?

28. A ramp is to be built at the library for wheelchair accessibility. When a grid is placed over the architect's plans, the top of the ramp has coordinates (72, 4). The bottom of the ramp has coordinates (22, 1). Will the ramp meet safety specifications?

## THINK CRITICALLY/ SOLVE PROBLEMS

Find the slope of the line that contains the given points. Then plot the points and draw the line.

29. $A(-2, 1)$ $B(-5, -3)$

30. $C(7, -6)$ $D(4, 3)$

31. $E(-1, 2)$ $F(3, -2)$

32. $G(4, 1)$ $H(-2, -3)$

Use Exercises 19–22 and 29–32 to help you with Exercises 33–34.

33. How can you tell just from the coordinates whether a line is horizontal or vertical?

34. Suppose you are examining a line from left to right. Complete.
   a. If the line goes *up*, the slope is always __?__. (sign)
   b. If the line goes *down*, the slope is always __?__. (sign)

# 9-2 Slopes of Parallel and Perpendicular Lines

**EXPLORE/ WORKING TOGETHER**

1. a. Plot the following points.
   $A(-7, 4)$ $B(3, 12)$ $C(-5, -2)$ $D(5, 6)$
   b. Draw the line segments $AB$ and $CD$. What do you notice?
   c. Find the slopes of the line segments. What is true?

2. Investigate the diagram.
   a. Which pairs of lines appear to be parallel?
   b. Find the slopes of the lines that appear to be parallel. What is true?

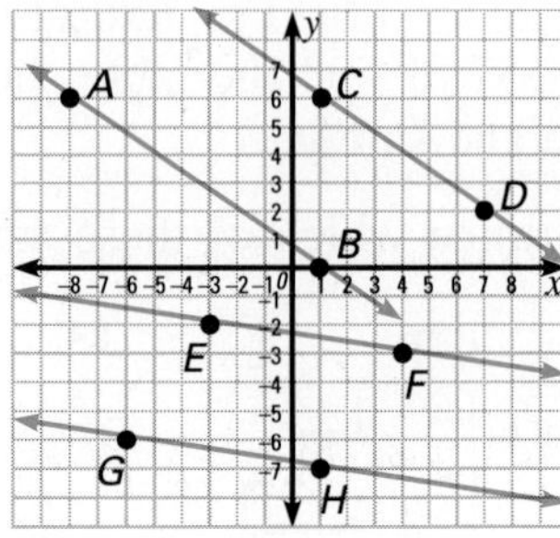


3. Based on your results in questions 1 and 2, what conclusion do you draw about the slopes of parallel lines?

4. a. Plot the following points.
   $P(-1, 3)$ $Q(5, -6)$ $R(6, 1)$ $S(-3, -5)$
   b. Draw the line segments $PQ$ and $RS$. What do you notice?
   c. Find the slope of the line segments. What is the relationship of the slopes to each other?
   d. Multiply the slopes. What is the product?

5. Investigate the diagram.
   a. Which pairs of lines appear to be perpendicular?
   b. Find their slopes. What is the relationship of the slopes to each other?
   c. Multiply the slopes. What is the product?

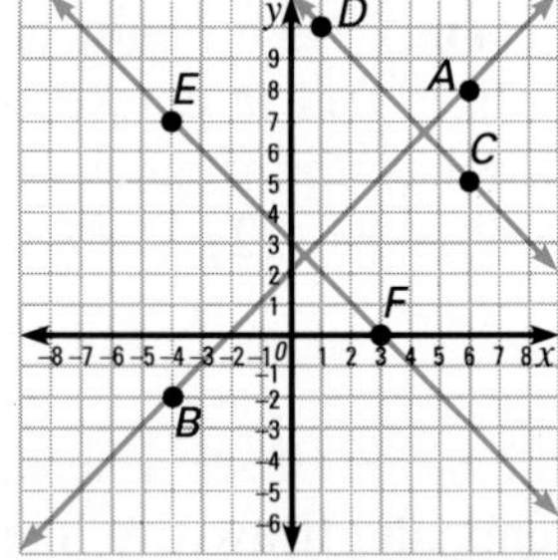


6. Based on your results in questions 4 and 5, what conclusion do you draw about the slopes of perpendicular lines?

**SKILLS DEVELOPMENT**

As you can see from the Explore activities, you can sometimes tell these relationships between two lines if you know their slopes.

- ► If two lines are parallel, they have the same slope or both slopes are undefined.
- ► If two lines have the same slope, they are parallel.

From your work with fractions, you should remember that two rational numbers that have a product of 1 are called reciprocals. If two rational numbers have a product of $-1$, they are called **negative reciprocals.**

Again using what you discovered in the Explore activities, you should realize that these relationships are true.

- ► If two lines are perpendicular, the product of their slopes is $-1$. (You can also say that the slopes are negative reciprocals of each other.)
- ► If the slopes of two lines are negative reciprocals of each other, then the lines are perpendicular.

Further, you can use what you have learned to find the slope of a line that is either parallel to or perpendicular to a given line.

## Example 1

For each pair of points on a line, identify the slope of a line parallel to the given line and a line perpendicular to it.

**a.** $A(6, 3)$ and $B(4, 6)$ **b.** $C(-1, -5)$ and $D(3, -2)$
**c.** $E(0, 5)$ and $F(-1, 6)$ **d.** $G(-4, -2)$ and $H(-5, -2)$

**CHECK UNDERSTANDING**

Can you think of a case in which two lines are perpendicular but the product of their slopes is *not* $-1$? (*Hint:* Think of a special case.) Explain your answer.

### Solution

Use $m$ to represent slope. Then $m = \frac{y_2 - y_1}{x_2 - x_1}$.

The slope of the $\parallel$ line is $m$ and the slope of the $\perp$ line is $\frac{-1}{m}$.

| | Line | $\parallel$ Line | $\perp$ Line |
|---|---|---|---|
| **a.** | $\frac{3-6}{6-4} = \frac{-3}{2} = -\frac{3}{2}$ | $-\frac{3}{2}$ | $\frac{2}{3}$ |
| **b.** | $\frac{-5-(-2)}{-1-3} = \frac{-3}{-4} = \frac{3}{4}$ | $\frac{3}{4}$ | $-\frac{4}{3}$ |
| **c.** | $\frac{6-5}{-1-0} = \frac{1}{-1} = -1$ | $-1$ | $1$ |
| **d.** | $\frac{-2-(-2)}{-5-(-4)} = \frac{0}{-1} = 0$ | $0$ | undefined ◄ |

## Example 2

Points $A(9, 11)$ and $B(-4, -2)$ are on $\overleftrightarrow{AB}$. Points $C(6, 5)$ and $D(1, 10)$ are on $\overleftrightarrow{CD}$. Are $\overleftrightarrow{AB}$ and $\overleftrightarrow{CD}$ *parallel, perpendicular,* or *neither*?

### Solution

slope of $\overleftrightarrow{AB} = \frac{-2 - 11}{-4 - 9} = \frac{-13}{-13} = 1$ **Find the slope of each line.**

slope of $\overleftrightarrow{CD} = \frac{10 - 5}{1 - 6} = \frac{5}{-5} = -1$

$(1)(-1) = -1$ **Find the product of the slopes.**

$\overleftrightarrow{AB}$ and $\overleftrightarrow{CD}$ are perpendicular. ◄

### Example 3

The vertices of a triangle are $A(3, 6)$, $B(-2, -4)$, and $C(-5, 2)$. Determine if $\triangle ABC$ is a right triangle. Explain your answer.

**Solution**

Plot the points and draw the triangle. Sides $AC$ and $BC$ appear to be perpendicular.

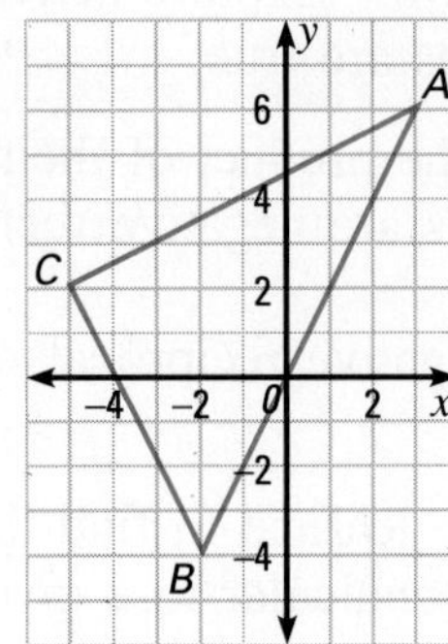


Find the slope of $\overline{AC}$ and $\overline{BC}$. Then find the product of the slopes.

| Slope | $\overline{AC}$ | $\overline{BC}$ |
|---|---|---|
| $\frac{\text{rise}}{\text{run}}$ | $\frac{1}{2}$ | $\frac{-2}{1} = -2$ |

Product $= \left(\frac{1}{2}\right)(-2) = -1$

Since the product of the slopes $= -1$, $\overline{AC} \perp \overline{BC}$.
Since $\overline{AC} \perp \overline{BC}$, $\angle C$ is a right angle.
Since $\angle C$ is a right angle, $\triangle ABC$ is a right triangle. ◄

## TRY THESE

For each pair of points on a line, identify the slope of a line parallel to the given line and a line perpendicular to it.

1. $A(7, 2)$ and $B(5, 7)$
2. $C(1, 5)$ and $D(-2, 3)$
3. $E(0, -2)$ and $F(-3, -5)$
4. $G(1, 3)$ and $H(-2, 3)$

Points $A$ and $B$ are on line $AB$. Points $C$ and $D$ are on line $CD$. Are lines $AB$ and $CD$ *parallel, perpendicular,* or *neither*?

5. $A(7, 4)$ $B(3, -1)$
   $C(2, 8)$ $D(-2, 3)$
6. $A(5, 12)$ $B(9, 9)$
   $C(6, 5)$ $D(1, 7)$

The vertices of $\triangle ABC$ are given. Determine if $\triangle ABC$ is a right triangle. Explain your answer.

7. $A(1, 6)$, $B(7, 4)$, $C(-1, -1)$
8. $A(0, 9)$, $B(4, 5)$, $C(-2, -1)$

## EXERCISES

**PRACTICE/ SOLVE PROBLEMS**

For each pair of points on a line, identify the slope of a line parallel to the given line and a line perpendicular to it.

1. $M(-2, 3)$ and $N(2, 1)$
2. $P(-7, -3)$ and $Q(0, -1)$
3. $F(-3, 3)$ and $G(2, -4)$
4. $T(5, -1)$ and $U(-3, -1)$

Points $R$ and $S$ are on $\overleftrightarrow{RS}$. Points $D$ and $E$ are on $\overleftrightarrow{DE}$. Are $\overleftrightarrow{RS}$ and $\overleftrightarrow{DE}$ *parallel, perpendicular,* or *neither*?

**5.** $R(-3, 5)$ $S(1, 2)$
$D(1, 12)$ $E(-2, 8)$

**6.** $R(-4, -2)$ $S(-1, -5)$
$D(7, 3)$ $E(9, 1)$

**7.** $R(7, -2)$ $S(-1, -2)$
$D(-3, 4)$ $E(5, 4)$

**8.** $R(-1, 6)$ $S(2, -4)$
$D(5, -3)$ $E(-1, -7)$

**9.** Patrice plotted $W(-3, -3)$, $X(-4, 3)$, $Y(5, 4)$, and $Z(4, 0)$ on a coordinate grid. Then she connected the four points. Was the resulting figure a parallelogram? Why or why not?

## EXTEND/ SOLVE PROBLEMS

To answer Exercises 18–19, review the properties of parallel and perpendicular lines as well as the properties of other geometric figures you worked with in Chapter 3.

Determine the value of $x$ so that the line containing the given points is parallel to another line whose slope is also given.

**10.** $A(-5, -1)$ and $B(x, -4)$
slope of other line $= \frac{1}{2}$

**11.** $C(3, -5)$ and $D(x, 9)$
slope of other line $= -2$

**12.** $E(0, 4)$ and $F(x, 4)$
slope of other line $= 0$

**13.** $G(-2, -3)$ and $H(x, 6)$
slope of other line $= \frac{9}{4}$

Determine the value of $y$ so that the line containing the given points is perpendicular to another line whose slope is also given.

**14.** $A(4, 8)$ $B(6, y)$
slope of other line $= -\frac{1}{3}$

**15.** $C(5, -2)$ $D(-7, y)$
slope of other line $= 2$

**16.** $E(2, 3)$ $F(6, y)$
slope of other line $= 4$

**17.** $G(1, 3)$ $H(4, y)$
slope of other line $= \frac{3}{2}$

**18.** The vertices of quadrilateral $JKLM$ are $J(-4, 1)$, $K(0, 2)$, $L(3, -1)$, and $M(2, -5)$. Is $JKLM$ a trapezoid? Explain.

**19.** Jon drew plans to construct a square walk. On the grid paper used, the coordinates were $A(0, -6)$, $B(-10, -8)$, $C(-4, 4$ and $D(-14, 2)$. Were the coordinates correct? Explain.

Determine if each statement is *always, sometimes,* or *n* Explain your answer. Wherever possible, support you with an illustration.

**20.** Parallel lines are horizontal.

**21.** Vertical lines are parallel.

**22.** Two sides of a triangle can have the same slope.

**23.** Slopes of the diagonals of a square are the same.

**24.** If the slopes of consecutive sides of a parallelog reciprocals, the parallelogram is a square.

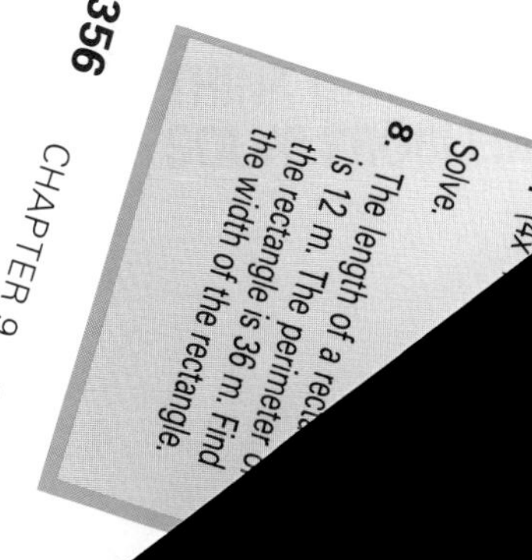

# 9-3 Slope-Intercept Form of an Equation

## EXPLORE/ WORKING TOGETHER

With a partner, study the graphs at the right and their corresponding equations.

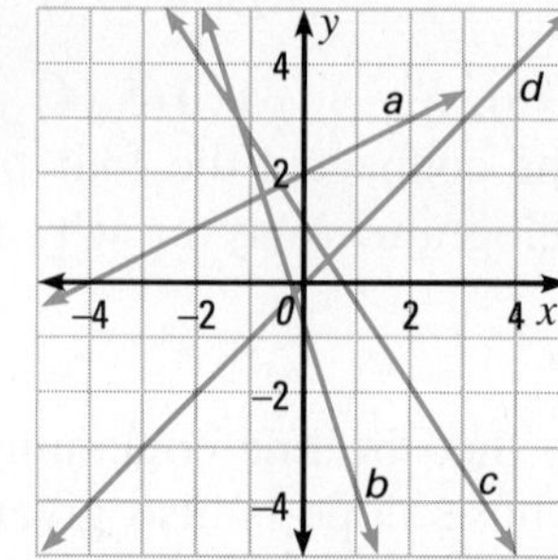

a. $y = \frac{1}{2}x + 2$

b. $y = -3x - 1$

c. $y = -\frac{3}{2}x + 1$

d. $y = x$

Find the slope of each line. Find the point where each line crosses the $y$-axis. Compare these with the equations for the lines. What do you notice?

## SKILLS DEVELOPMENT

When an equation is written in the form $y = mx + b$, $m$ represents the slope of the line and $b$ represents the $y$-intercept. The $y$-intercept is the point where the graph of the line crosses the $y$-axis. Since the $x$-coordinate for *every* point on the $y$-axis is 0, the coordinates of the $y$-intercept are always $(0, b)$. By using the $y = mx + b$ form of an equation, which is called the **slope-intercept form,** you can easily determine the slope and $y$-intercept of a line, then use this information to graph the equation.

**MIXED REVIEW**

Solve each equation or inequality.

1. $3x + 2 = 38$
2. $-4x - 5 = 27$
3. $2x - 3 \geq 7$
4. $-5x + 1 < -14$

Simplify.

5. $3a - 2b - 5a + 9b$
6. $-7r + 6t - 8r - 6t$

… $+ 2y) - (7x - 3y)$

…angle

### Example 1

Identify the slope and $y$-intercept for the line with the given equation.

a. $y = 3x - 2$ b. $3x + 2y = 6$

### Solution

a. $y = 3x - 2$ ← **Use the form $y = mx + b$ to read the slope, $m$, and the $y$-intercept, $b$.**

$m = 3;\ b = -2$

For $y = 3x - 2$, the slope is 3 and the $y$-intercept is $-2$.

b. Sometimes you need to use algebra to rewrite the equation in slope-intercept form.

$$\begin{aligned} 3x + 2y &= 6 \\ -3x + 3x + 2y &= -3x + 6 \\ \frac{2y}{2} &= \frac{-3x + 6}{2} \\ y &= -\frac{3}{2}x + 3 \end{aligned}$$

**To write the equation in slope-intercept form, you must solve the equation for $y$.**

$m = -\frac{3}{2};\ b = 3$

…r $3x + 2y = 6$, the slope is $-\frac{3}{2}$ and the $y$-intercept is 3. ◄

Using the slope-intercept form of an equation makes graphing the line easier than the method you used in Chapter 8.

### Example 2

Graph each equation.

**a.** $y = 3x - 1$

**b.** $2x + 3y = 6$

**Solution**

First, using the $y = mx + b$ form of the equation, plot the point $(0, b)$. Then plot two other points on that line.

**a.** $y = 3x - 1$ is already in slope-intercept form.

Since $b = -1$, $(0, b)$ is $(0, -1)$.

Since the slope is positive 3, locate one point that is 3 units up and 1 unit to the right from $(0, -1)$. Then locate another point 3 units down and 1 unit to the left from the original point $(0, -1)$.

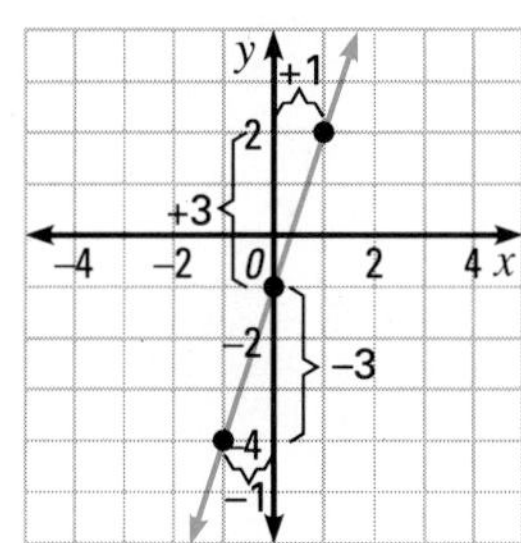


**b.** Put $2x + 3y = 6$ in slope-intercept form: $y = -\frac{2}{3}x + 2$.

Since $b = 2$, $(0, b)$ is $(0, 2)$.

Since the slope is negative $\frac{2}{3}$, locate one point that is 2 units up and 3 units to the left from $(0, 2)$. Then locate another point that is 2 units down and 3 units to the right from the original point $(0, 2)$.

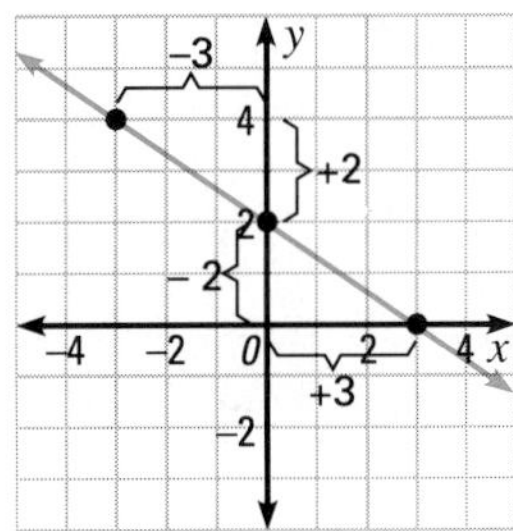


### Example 3

Determine whether each pair of lines is *parallel, perpendicular,* or *neither.*

**a.** $y = 5x - 2$
$y = -\frac{1}{5}x + 3$

**b.** $2x - y = 3$
$-4x + 3y = -6$

**PROBLEM SOLVING TIP**

When working with two different equations, you can identify the slopes by using $m_1$ for the first equation and $m_2$ for the second equation.

**Solution**

Use the $y = mx + b$ form to compare the slopes.

**a.** $y = 5x - 2 \qquad m_1 = 5$

$y = -\frac{1}{5}x + 3 \qquad m_2 = -\frac{1}{5}$

$m_1 \neq m_2$
$m_1 \cdot m_2 = -1$

Since $m_1 \cdot m_2 = -1$, the lines are perpendicular.

**b.** $2x - y = 3$
$y = 2x - 3 \qquad m_1 = 2$
$-4x + 3y = -6$
$y = \frac{4}{3}x - 2 \qquad m_2 = \frac{4}{3}$

$m_1 \neq m_2$
$m_1 \cdot m_2 \neq -1$

Since $m_1 \neq m_2$ and $m_1 \cdot m_2 \neq -1$, the lines are neither parallel nor perpendicular.

## Try These

Identify the slope and the $y$-intercept for each line.

1. $y = -2x + 5$
2. $y = \frac{3}{4}x - 1$
3. $5x - 2y = 8$
4. $3x + 4y = 12$

Graph each equation.

5. $y = 2x + 3$
6. $y = \frac{3}{2}x - 1$
7. $3x + 2y = 4$
8. $-x - 5y = 10$

Determine if the two lines are *parallel, perpendicular,* or *neither.*

9. $y = 3x + 7$
   $y = 3x - 2$
10. $y = \frac{2}{3}x + 1$
    $y = -\frac{3}{2}x - 4$
11. $4x - 5y = 15$
    $2x - 3y = 12$
12. $5x + 3y = 9$
    $-10x - 6y = 12$

# EXERCISES

**PRACTICE/ SOLVE PROBLEMS**

Identify the slope and the $y$-intercept for each line.

1. $y = -4x + 3$
2. $y = -\frac{2}{5}x + 1$
3. $y = \frac{7}{3}x - 2$
4. $3x + 2y = -4$
5. $-2x - 5y = 5$
6. $-5x + 4y = -8$

Graph each equation.

7. $y = -3x + 2$
8. $y = \frac{2}{3}x - 4$
9. $y = \frac{1}{2}x - 3$
10. $y = -\frac{5}{4}x + 1$
11. $3x + 4y = -8$
12. $-x + 3y = 12$
13. $-4x - 2y = 6$
14. $2x - 5y = -15$

Determine if the two lines are *parallel, perpendicular,* or *neither.*

15. $y = -2x - 7$
    $y = -\frac{1}{2}x + 4$
16. $y = 3x + 5$
    $y = 3x - 2$
17. $y = \frac{4}{5}x + 3$
    $y = -\frac{4}{5}x - 1$
18. $-3x + 2y = 6$
    $2x + 3y = 9$
19. $x + y = 10$
    $3x + 3y = -12$
20. $3x + 7y = 21$
    $2x - 8y = 16$

21. An aircraft takes off following the path shown on the map. Another plane has $y = \frac{1}{3}x + 5$ as the equation of its path of take-off. On the same map, graph the path of the second plane. Will the planes crash? Explain.

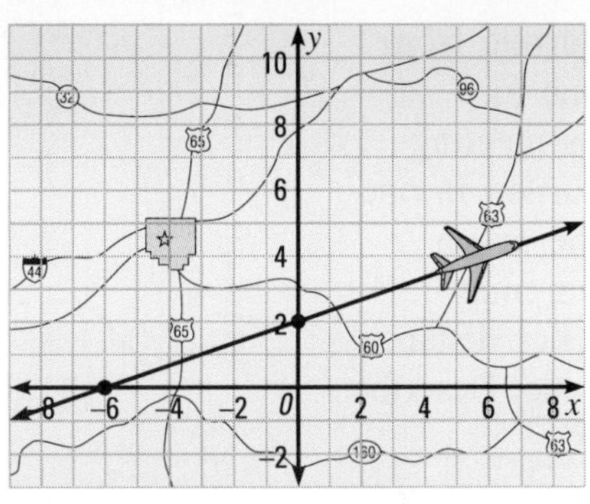

22. Could segments on the lines $2x - 3y = -18$ and $2x - 3y = 9$ be sides of a parallelogram? Explain.

## EXTEND/ SOLVE PROBLEMS

Graph each equation.

23. $y = 2$

24. $y = -3$

25. $x = 4$

26. $x = -1$

27. Will the graphs of equations that have the form $y =$ a constant be *parallel, perpendicular,* or *neither*?

28. Will the graphs of equations that have the form $x =$ a constant be *parallel, perpendicular,* or *neither*?

29. Will the graph of $y = 4$ be *parallel, perpendicular,* or *neither* to the graph of $x = -3$?

## THINK CRITICALLY/ SOLVE PROBLEMS

Graph each equation.

30. $y = \frac{1}{3}x - 2$

31. $-2x = -6y - 12$

32. $-3x + 9y = -18$

33. What are the slope and $y$-intercept for each of Exercises 30–32?

34. Make a general statement that summarizes your answer to Exercise 33.

# 9-4 Problem Solving Skills:

## INTERPRETING SLOPE

► READ
► PLAN
► SOLVE
► ANSWER
► CHECK

A graph illustrates how two quantities change relative to one another. By analyzing the shape of the graph you can interpret its meaning. Here slope is used to show a rate.

### PROBLEM

Kerrie drove from her home to a business meeting and returned in one day. The graph plots her time against her distance from home. Describe the trip.

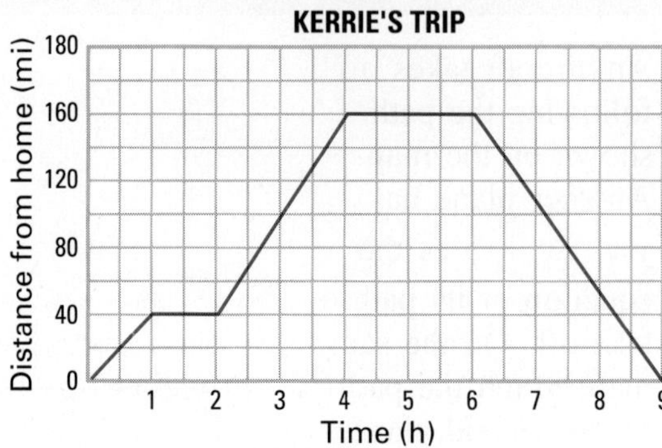


### SOLUTION

The trip can be divided into five parts as shown in the graph at the right.

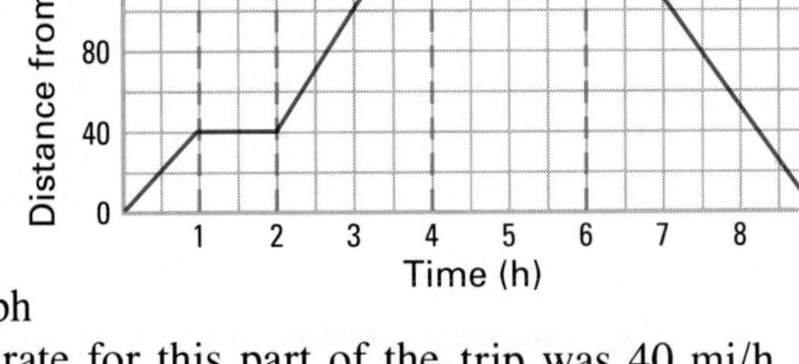


I. Kerrie drove 40 mi in 1 h at a steady rate. The slope of this part of the graph is 40. Her average rate for this part of the trip was 40 mi/h.

II. The horizontal line indicates that distance did not change for 1 h because the slope is zero. (Kerrie must have stopped driving.)

III. The slope of the next segment is $\frac{160 - 40}{4 - 2}$, or 60. Kerrie is traveling faster than she was in Section I. Her average speed or rate for this section is 60 mi/h.

IV. Kerrie spent 2 h at the business meeting.

V. This line segment has a negative slope. Since speed cannot be negative, you need a different interpretation. Kerrie is returning home (distance is approaching zero). The slope of this segment is $\frac{160 - 0}{9 - 6}$ or $53\frac{1}{3}$. Her average speed for the last part of the trip was $53\frac{1}{3}$ mi/h.

**MATH: WHO, WHERE, WHEN**

The first person to plot points on a coordinate grid and to analyze how quantities change relative to one another was the French mathematician René Descartes (1596–1650). It is said that Descartes was observing a fly on a tiled ceiling when it occurred to him to locate the fly by measuring its distance from the horizontal and vertical lines separating the tiles.

# PROBLEMS

Find the average speed (mi/h).

1. 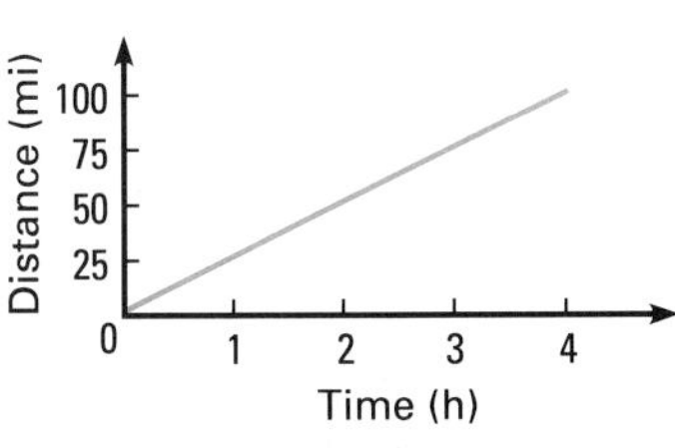


2. 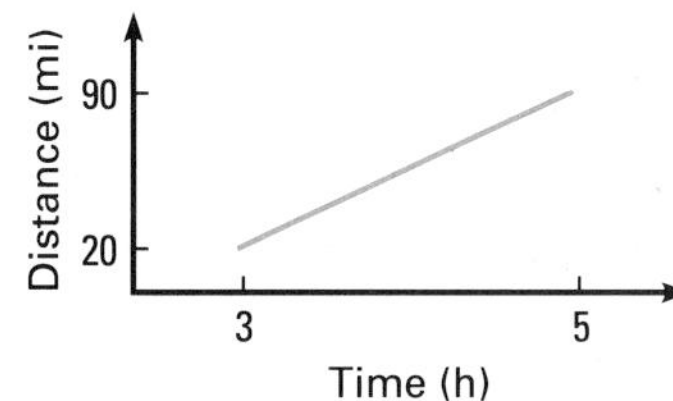


3. 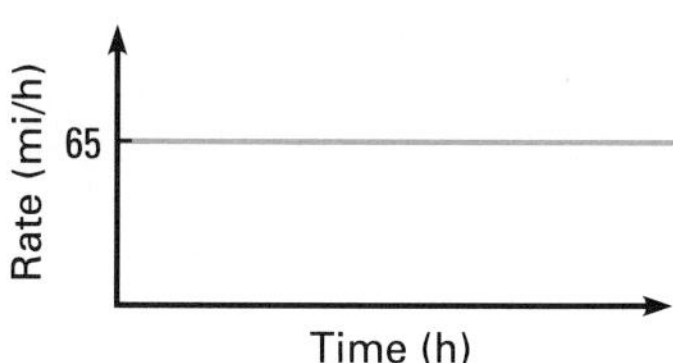


4. 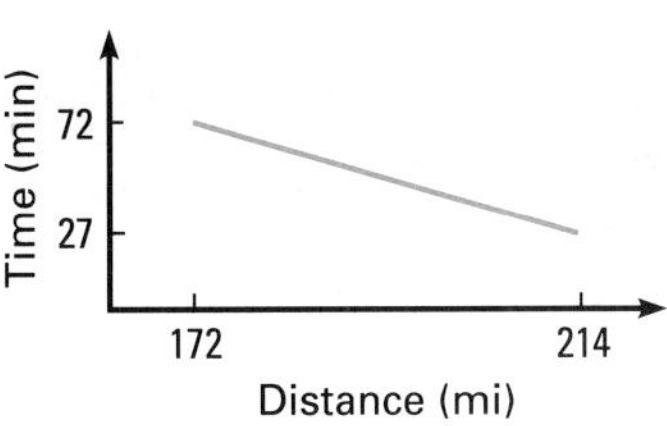


A bathtub fills with water. Use the graph to solve.

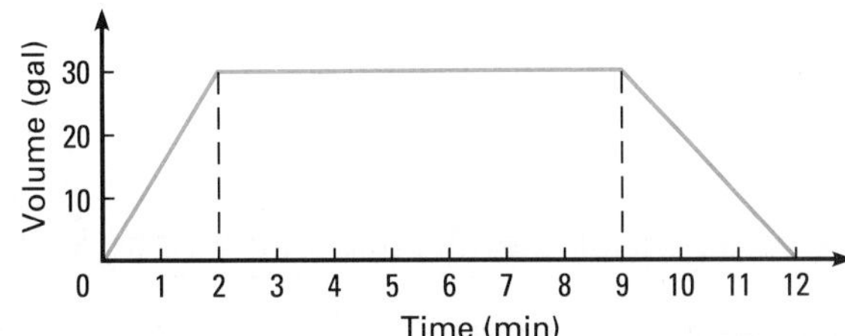


5. How long did it take to fill the tub?
6. How much water was in the full tub?
7. At what rate did the tub fill?
8. How long was the tub full?
9. At what rate did the tub drain?

10. The furnace in Brent's house is controlled by a thermostat. Brent turned the furnace on one cold winter morning. Describe what happened over the next three hours.

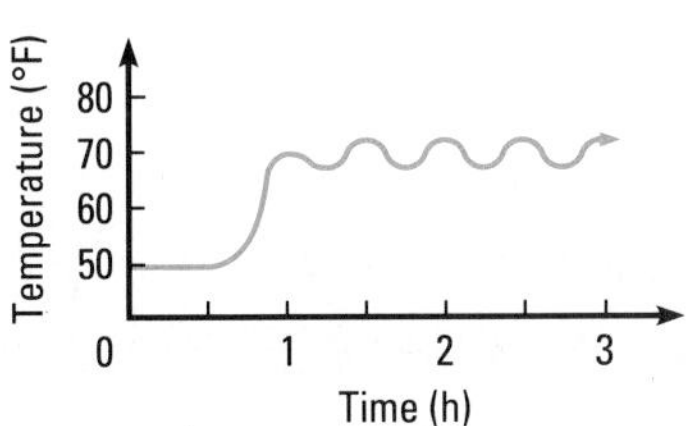


Annette, Kai, and Marco hiked separately from their campsite to the lake. Tell whose hike is described.

11. a steady rate all the way
12. slow at first, then fast
13. fast at first, then slow

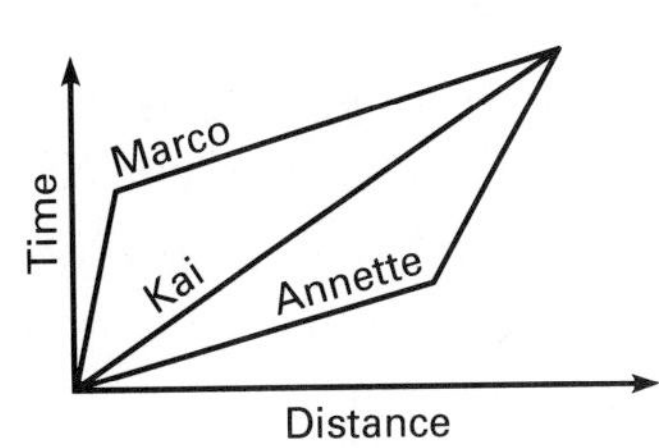

# 9-5 Writing Equations for Lines

## EXPLORE

Each graph at the right represents the solution for a linear equation. Match each graph with the correct slope and $y$-intercept.

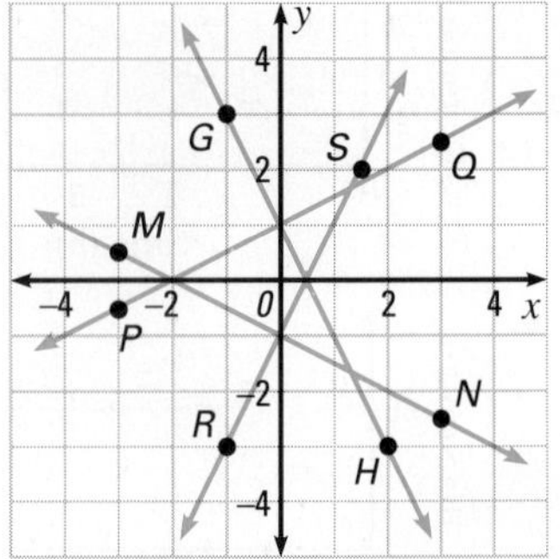


**a.** slope $= 2$, $y$-intercept $= -1$ ___?___

**b.** slope $= -\frac{1}{2}$, $y$-intercept $= -1$ ___?___

**c.** slope $= -2$, $y$-intercept $= 1$ ___?___

**d.** slope $= \frac{1}{2}$, $y$-intercept $= 1$ ___?___

## SKILLS DEVELOPMENT

To write an equation in the form $y = mx + b$, you need to know the slope $m$ and the $y$-intercept $b$. Suppose you know $m$ and $b$.

### Example 1

Write an equation of the line whose slope is $-\frac{2}{3}$ and $y$-intercept is $-1$.

**Solution**

Given: $m = -\frac{2}{3}$, $b = -1$

$y = mx + b$ Substitute for $m$ and $b$.

$y = -\frac{2}{3}x + (-1)$

An equation of the line is $y = -\frac{2}{3}x - 1$. ◄

You may have to find the slope of the line or its $y$-intercept algebraically.

### Example 2

Write an equation of the line whose slope is 3 and which contains the point $P(4, -1)$.

**Solution**

You know that $m = 3$, $x = 4$, and $y = -1$. So, you must find $b$.

$$\begin{aligned} y &= mx + b && \text{Substitute for } m, x, \text{ and } y.\\ -1 &= 3(4) + b && \text{Solve for } b.\\ -1 &= 12 + b \\ -1 - 12 &= 12 - 12 + b \\ -13 &= b \end{aligned}$$

$$y = 3x - 13 \qquad \text{Substitute for } m \text{ and } b.$$

An equation of the line is $y = 3x - 13$. ◄

Sometimes you will need to find both the slope and the $y$-intercept.

## Example 3

Write an equation of the line that contains the points $R(-1, 5)$ and $S(1, -3)$.

**Solution**

You know that two points on the line are $R(-1, 5)$ and $S(1, -3)$. So, you must find $m$ and $b$.

$$m = \frac{y_2 - y_1}{x_2 - x_1} = \frac{5 - (-3)}{-1 - 1} = \frac{8}{-2} = -4 \quad \text{Find } m \text{ algebraically.}$$

$$\begin{aligned} y &= mx + b && \text{Substitute for } m, x, \text{ and } y. \\ 5 &= -4(-1) + b && \text{Solve for } b. \\ 5 &= 4 + b \\ -4 + 5 &= -4 + 4 + b \\ 1 &= b \\ y &= -4x + 1 && \text{Substitute for } m \text{ and } b. \end{aligned}$$

An equation of the line is $y = -4x + 1$. ◄

Often you will be able to write the equation for a line if you are given the graph of the line.

## Example 4

Write an equation of the line whose graph is shown at the right.

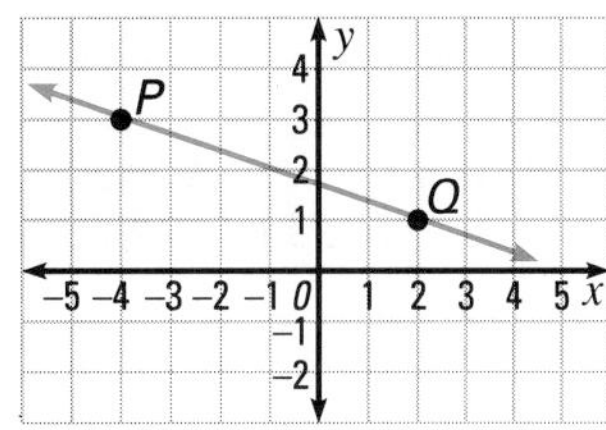


**Solution**

You know that $P(-4, 3)$ and $Q(2, 1)$ are two points on the line. So find $m$ and $b$.

$$m = \frac{y_2 - y_1}{x_2 - x_1} = \frac{3 - 1}{-4 - 2} = \frac{2}{-6} = -\frac{1}{3}$$

$$\begin{aligned} y &= mx + b \\ 1 &= -\tfrac{1}{3}(2) + b \\ 1 &= -\tfrac{2}{3} + b \\ \tfrac{2}{3} + 1 &= -\tfrac{2}{3} + \tfrac{2}{3} + b \\ 1\tfrac{2}{3} &= b \end{aligned}$$

An equation for the line is $y = -\frac{1}{3}x + 1\frac{2}{3}$, or $y = -\frac{1}{3}x + \frac{5}{3}$. ◄

The computer program below will print the equation of a line in slope-intercept form when the slope of the line and the coordinates of a point on that line are INPUT.

```
10 PRINT "WHAT IS THE SLOPE OF THE LINE AND THE"
20 PRINT "COORDINATES OF A POINT ON THE LINE";
30 INPUT M, X1, Y1
40 LET B = Y1 - M * X1
50 PRINT "EQUATION IS Y = ";
60 IF M = 0 THEN 240
70 IF B = 0 THEN 140
80 IF B < 0 THEN 190
90 IF M > 1 THEN 120
100 PRINT "X +"; B
110 GOTO 250
120 PRINT M;"X +"; B
130 GOTO 250
140 IF M > 1 THEN 170
150 PRINT "X"
160 GOTO 250
170 PRINT M;"X"
180 GOTO 250
190 IF M < -1 THEN 220
200 PRINT "X -"; ABS(B)
210 GOTO 250
220 PRINT M;"X -"; ABS(B)
230 GOTO 250
240 PRINT B
250 PRINT
260 PRINT "ANY MORE LINES (1=YES, 2=NO)";
270 INPUT W
280 IF W = 1 THEN 30
```

MIXED REVIEW

Find the sum or difference.

1. -6 - (-8)
2. 9 + (-7)
3. -23 + (-12)
4. 17 - (-8)

Find the mean, median, and mode.

5. -2, 5, 3, -2, 1
6. 6, 11, -3, -5, 11
7. 3, -10, -1, 17, 2, -8

Solve.

8. A ladder is leaning against the side of a building. The foot of the ladder is 18 ft from the building. The top of the ladder is 24 ft up the side of the building. What is the height of the ladder?

## TRY THESE

Write an equation of the line whose slope and $y$-intercept are given.

1. $m = -2,\ b = 3$
2. $m = \frac{3}{4},\ b = 2$
3. $m = \frac{5}{3},\ b = -1$
4. $m = -\frac{1}{2},\ b = -5$

Write an equation of the line whose slope and a point on the line are given.

5. $m = 4,\ P(-1, 2)$
6. $m = -1,\ P(3, -4)$
7. $m = -3,\ P(5, -2)$
8. $m = 2,\ P(-1, -3)$

Write an equation of the line that contains the given points.

9. $A(-1, 1)$ and $B(1, 5)$
10. $C(4, 2)$ and $D(-1, 7)$
11. $E(4, 3)$ and $F(1, -3)$
12. $G(1, 5)$ and $H(-2, -4)$

Write an equation of the lines whose graphs are shown below.

13.

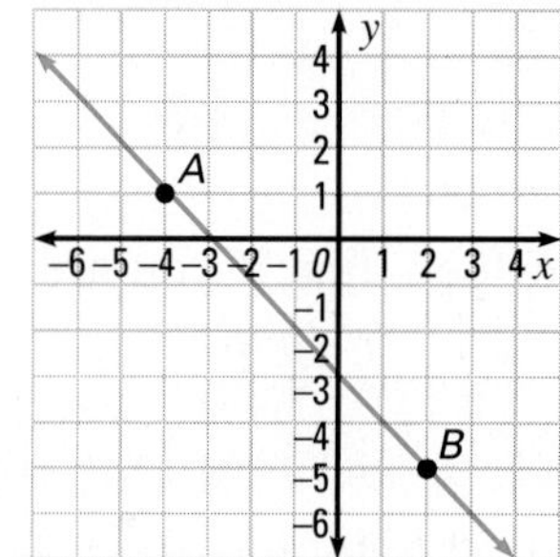


14.

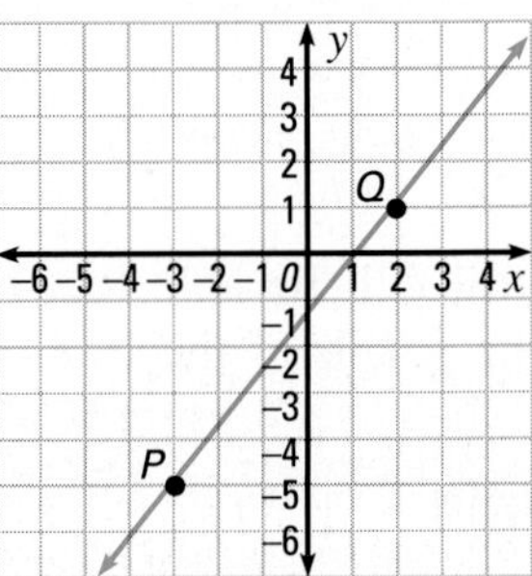


# EXERCISES

## PRACTICE/ SOLVE PROBLEMS

Write an equation of the line whose slope and $y$-intercept are given.

1. $m = -7,\ b = 3$
2. $m = 4,\ b = -1$
3. $m = -\frac{3}{5},\ b = -4$
4. $m = \frac{2}{3},\ b = 6$

Write an equation of the line whose slope and a point on the line are given.

5. $m = 3,\ P(1, 5)$
6. $m = 2,\ P(-2, -7)$
7. $m = 2,\ P(3, -2)$
8. $m = 4,\ P(-1, 4)$

Write an equation of the line that contains the given points.

9. $D(2, 3)$ and $E(1, 1)$
10. $U(3, -6)$ and $V(1, -4)$
11. $P(3, 7)$ and $B(2, 2)$
12. $M(0, -1)$ and $K(1, 2)$

Write an equation of the lines whose graphs are shown below.

13. 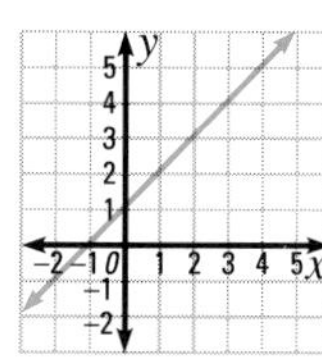

14. 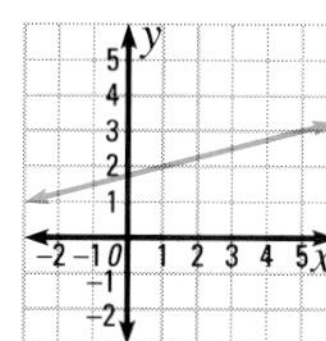

15. 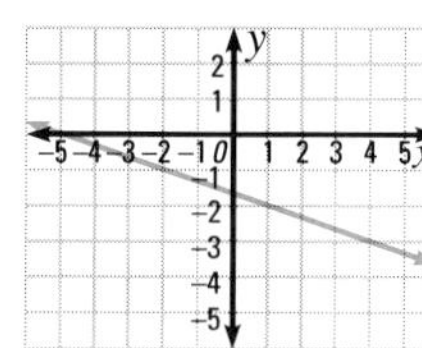

## EXTEND/ SOLVE PROBLEMS

Write an equation for the line with the given information.

16. It has slope $\frac{5}{2}$ and passes through $A(0, -3)$.

17. It has slope 0 and passes through $P(-3, 2)$.

18. It has an undefined slope and passes through $R(4, 1)$.

19. It is parallel to the graph of $y = -2x - 3$ and passes through $S(3, 1)$.

20. It is perpendicular to the graph of $y = -\frac{1}{4}x + 8$ and passes through $D(2, 3)$.

21. A boat travels on the path shown on the grid. Write an equation for the path of the boat.

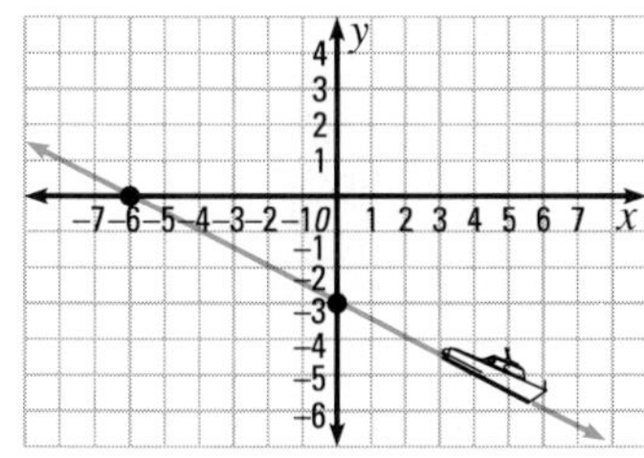

22. The path of a jet is shown on the grid. Write an equation for the path of the jet.

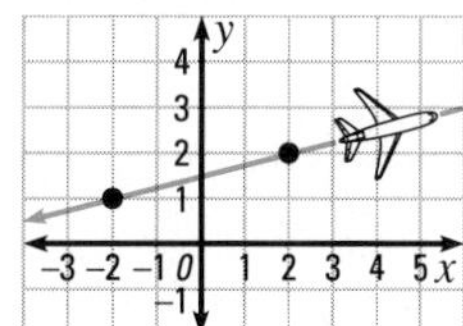

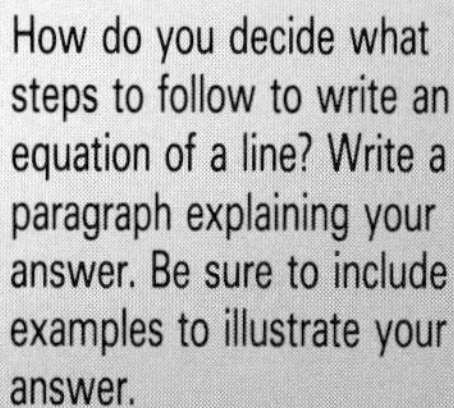

**WRITING ABOUT MATH**

How do you decide what steps to follow to write an equation of a line? Write a paragraph explaining your answer. Be sure to include examples to illustrate your answer.

## THINK CRITICALLY/ SOLVE PROBLEMS

The general form of the linear equation is $Ax + By = C$, where $A$, $B$, and $C$ are real numbers.

23. In any given equation, how many of the values $A$, $B$, and $C$ can equal 0? Explain, giving examples to support your answer.

24. If $A = 0$, describe the resulting line. Draw an example.

25. If $B = 0$, describe the resulting line. Draw an example.

26. If $C = 0$, describe the resulting line. Draw an example.

27. If $A$ and $C$ equal 0, what line is the graph of the equation?

28. If $B$ and $C$ equal 0, what line is the graph of the equation?

29. Write the general form of the linear equation in slope-intercept form.

30. Express the slope of the line $Ax + By = C$ in terms of $A$ and $B$.

# 9-6 Systems of Equations

## EXPLORE/ WORKING TOGETHER

Use the graph to explore solutions of equations.

a. An equation for line *a* is $x = 2$. Write three ordered pairs that are on line *a*. What do you notice?

b. An equation for line *c* is $y = 3$. Write three ordered pairs that are on line *c*. What do you notice?

c. Find an ordered pair that is a solution of both equations.

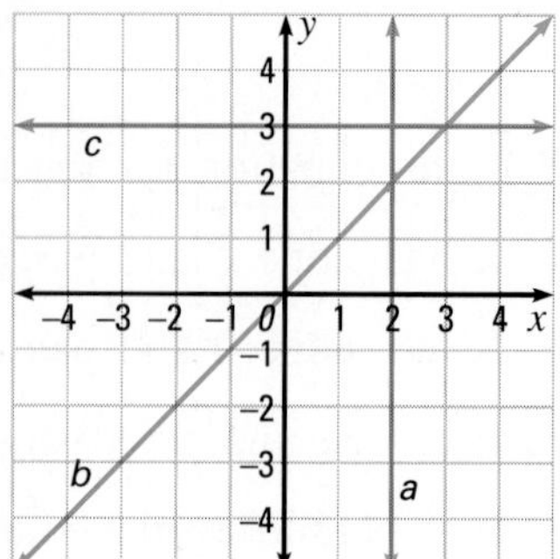


d. An equation for line *b* is $y = x$. Can you find an ordered pair that is a solution for all three equations? Why or why not?

## SKILLS DEVELOPMENT

An ordered pair is a solution of a **system of equations** if it is a solution for each of the equations. To solve a system of equations using graphs, graph each equation on the same pair of axes. All points of intersection are **solutions.** If the lines are parallel, there are no solutions.

### Example 1

State whether the given ordered pair is a solution of the system of equations.

$$(-1, 6) \qquad \begin{aligned} 2x + y &= 4 \\ x - y &= -7 \end{aligned}$$

**Solution**

$$\begin{aligned} 2x + y &= 4 \\ 2(-1) + 6 &= 4 \\ -2 + 6 &= 4 \\ 4 &= 4 \end{aligned}$$

← Substitute −1 for *x* and 6 for *y* in each equation. →

$$\begin{aligned} x - y &= -7 \\ -1 - 6 &= -7 \\ -7 &= -7 \end{aligned}$$

Since $(-1, 6)$ is a solution for each of the equations, it is a solution of the system of equations. ◄

You can solve a system of equations by graphing both equations on one set of axes.

### Example 2

Solve each system of equations.

a. $y = -2x + 4$
$y = x - 2$

b. $y = 3x - 1$
$y = 3x + 2$

## Solution

First, graph each equation using $m$ and $b$.
Then read the solution from the graph.

**a.** $y = -2x + 4$ $\quad b = (0, 4)$, $m = -2$

$y = x - 2$ $\quad b = (0, -2)$, $m = 1$

**b.** $y = 3x - 1$ $\quad b = (0, -1)$, $m = 3$

$y = 3x + 2$ $\quad b = (0, 2)$, $m = 3$

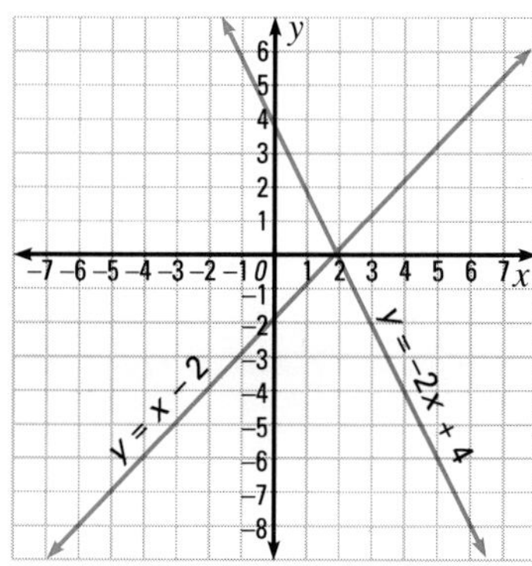


The solution is (2, 0).

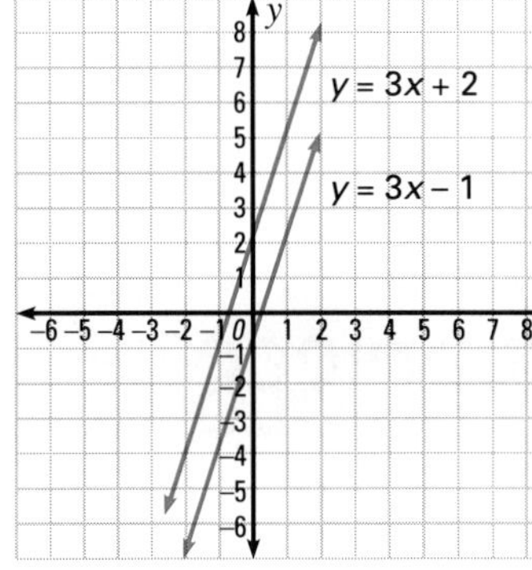


The lines are parallel, so there is no solution.

To check the solution, substitute (2, 0) in each equation.

$y = -2x + 4$ $\qquad y = x - 2$

$0 \stackrel{?}{=} -2(2) + 4$ $\qquad 0 \stackrel{?}{=} 2 - 2$

$0 = 0$ ✓ $\qquad 0 = 0$ ✓ ◄

**CHECK UNDERSTANDING**

How can you detect an error in graphing just by looking at the graph and the sign of the slope?

## Example 3

Solve the system of equations.

$$x + 3y = 6$$
$$3x + 2y = 4$$

## Solution

First, put each equation into $y = mx + b$ form.

$x + 3y = 6$, $\quad y = -\frac{1}{3}x + 2$

$3x + 2y = 4$, $\quad y = -\frac{3}{2}x + 2$

Then graph each equation using $m$ and $b$.

$y = -\frac{1}{3}x + 2$
$b = (0, 2)$
$m = -\frac{1}{3}$

$y = -\frac{3}{2}x + 2$
$b = (0, 2)$
$m = -\frac{3}{2}$

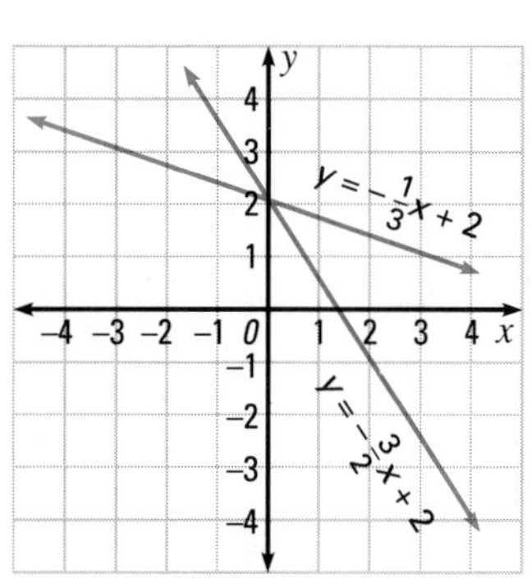


The solution is (0, 2). ◄

**CHECK UNDERSTANDING**

Explain how you would check the solution of Example 3.

## TRY THESE

State whether the given ordered pair is a solution of the system of equations.

1. (2, 1) $\quad 3x - 2y = 6$
   $\quad x + y = 2$

2. (2, 5) $\quad -4x + y = -3$
   $\quad 2x + y = 9$

Solve each system of equations.

3. $y = 2x - 6$
   $y = x - 4$

4. $2x + 3y = 3$
   $3x + 4y = 5$

# EXERCISES

### PRACTICE/ SOLVE PROBLEMS

State whether the given ordered pair is a solution of the system of equations.

1. (2, −3) $\quad 2x - y = 10$
   $\quad x + 2y = -5$

2. (1, 0) $\quad 3x + y = 3$
   $\quad x - 4y = 1$

Solve each system of equations.

3. $y = -x + 7$
   $y = 2x - 5$

4. $3x - 2y = 4$
   $2x + y = 5$

5. One of the points on Liz's graph of the line $x + 4y = -5$ has coordinates that are the same. Karen's graph of the line $3x + y = -4$ also contains a point with the same coordinates. Are the points with the same coordinates the same for both lines?

### EXTEND/ SOLVE PROBLEMS

6. The position of an accident on a highway is located at $P(4, -6)$ on a map grid. Four emergency vehicles travel on paths given by the following equations. Which, if any, of the vehicles will arrive exactly at the site of the accident?
   a. fire truck: $x - 12 = 2y$
   b. ambulance: $y = -10 - x$
   c. fire chief: $2x + 5 = 3y$
   d. police car: $x - 5y = 34$

### THINK CRITICALLY/ SOLVE PROBLEMS

Solve each system of equations. State the number of solutions.

7. $y = 4x - 2$
   $x - 4y = 2$

8. $y = 7 - 3x$
   $6x + 2y = 14$

9. $y = 2x - 8$
   $2x - y = 10$

10. Use the results in Exercises 7–9 to describe how you can determine the number of solutions for a system of equations by comparing the slopes and the $y$-intercepts.

READ
PLAN
SOLVE
ANSWER
CHECK

# Problem Solving Applications:

## THE BREAK-EVEN POINT

The Gramatan Goggles Co. plans to carry a new deluxe line of safety goggles. Expenses to introduce the new goggles will be \$1,000. Their cost for the goggles will be \$100 a pair. They use the formula $y = 100x + 1,000$ to determine expenses for $x$ pairs of goggles. Since they plan to sell the goggles for \$150 a pair, they use the formula $y = 150x$ to determine income.

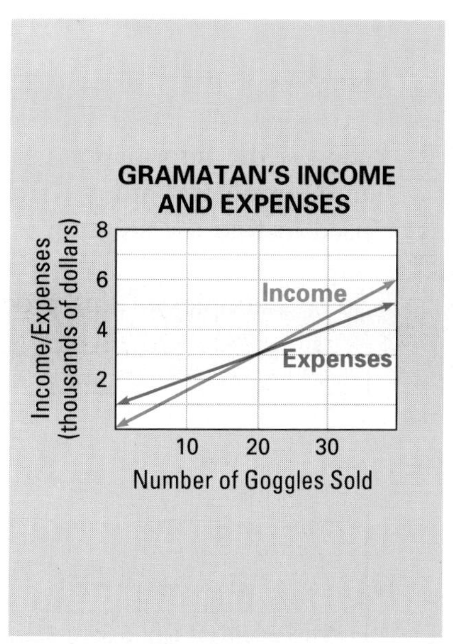


By graphing the two equations on the same set of axes, Gramatan Goggles Co. can determine their **break-even point,** the point where their income and their expenses are equal. This is the point where the graphs intersect.

1. How many pairs of goggles must Gramatan sell to break even?
2. What is their income when they break even?
3. If they sell only 10 pairs of goggles, what is their loss?
4. If they sell 40 pairs of goggles, what is their profit?

There is a moderate line of goggles that would cost Gramatan \$50 a pair to make. This line would sell for \$75 a pair. Gramatan would use the formula $y = 50x + 1000$ for their expenses and $y = 75x$ for their income.

5. Make a graph to show the break-even point for the moderate line.
6. What are Gramatan's expenses at the break-even point?
7. Which line is more profitable for Gramatan? Why?

Boring Drill Co. makes sets of drill bits. They calculate their expenses for $x$ number of sets using the formula $y = 40x + 200$. Since they sell the sets for \$48, they use the formula $y = 48x$ to calculate their income.

8. Make a graph to show the break-even point.
9. How many sets must be sold to break even?
10. Is it cost effective to sell 30 sets? Why or why not?

# 9-7 Problem Solving Strategies:
## USING A SYSTEM OF EQUATIONS

- READ
- PLAN
- SOLVE
- ANSWER
- CHECK

Some problems give you enough information so that you can write two equations. After reading the problem, choose variables that suggest the information they represent. Then translate the information in the problem into a system of equations. This system can then be used to find the solution to the problem.

Some examples of this process are shown in the chart below.

**PROBLEM SOLVING TIP**

Notice that one equation deals with the number of coins and one deals with the value of the coins. Remember to multiply each value by 100 so that all amounts are in cents.

| Sentence | Define Variables | Equation |
|---|---|---|
| The sum of two numbers is 18. | Let $s$ = smaller number<br>$l$ = larger number | $s + l = 18$ |
| The larger of the numbers is 3 less than twice the smaller number. | | $l = 2s - 3$ |
| The width of a rectangle is 18 m less than the length of the rectangle. | Let $w$ = width<br>$l$ = length | $w = l - 18$ |
| The perimeter is 84. | | $2l + 2w = 84$ |
| A coin bank has 16 nickels and dimes. | Let $n$ = number of nickels<br>$d$ = number of dimes | $n + d = 16$ |
| The value of the coins in the bank is $0.95. | | $5n + 10d = 95$ |

### PROBLEM

A work crew of 5 bricklayers and 3 carpenters earned $891 for a job. Another crew of 12 bricklayers and 4 carpenters earned $1,748 on the same job. Write a system of equations that could be used to find the wage of each type of worker.

### SOLUTION

Define variables. Let $b$ = wage of each bricklayer
$c$ = wage of each carpenter

Translate into a system of equations.

$$5b + 3c = 891$$
$$12b + 4c = 1{,}748$$

# PROBLEMS

For each of the following problems, define variables and write a system of equations that could be used to solve the problem.

1. A construction crew of men and women has 52 workers. The number of female workers is one less than twice the number of male workers. How many members of the crew are female?
2. The sum of Kelly's and Bob's ages is 29 years. Kelly's age exceeds Bob's by 5 years. How old is each person?
3. The width of a rectangle is 9 m less than the length. The perimeter of the rectangle is 26 m. What are the dimensions of the rectangle?
4. Yuki paid \$5.05 for 3 muffins and 4 coffees. Dave paid \$4.90 for 4 muffins and 2 coffees. How much did each item cost?
5. Two vans carry 30 workers from the parking lot to the job site. One van carries 6 fewer than twice the number of workers in the other van. How many workers does each van transport?
6. Todd's and Elena's ages together equal 41. Twice Todd's age is 7 years more than Elena's age. How old is each person?
7. A number increased by $\frac{1}{2}$ of another number is 26. The sum of the numbers is 46. What is the larger number?
8. Heidi has \$53 consisting of one-dollar bills and five-dollar bills. She has one more five-dollar bill than the number of one-dollar bills. How many bills of each type does she have?

9. The number of gallons of super unleaded gas sold at Mike's is five more than twice the number of gallons of regular unleaded gas. The total number of gallons of gas sold is 12,074. How many gallons of the regular unleaded gas were sold?
10. Two angles are supplementary. One angle measures 30° more than the other. Find the measure of both angles.
11. Fleetwood Construction Company is building a walkway in the shape of an isosceles triangle. The combined length of the two congruent sides of the walkway is 5 m longer than the length of the third side. The perimeter of the walkway is 35 m. How long is each side of the walkway?
12. At football games, the Student Council sells cups of coffee for \$0.60 and cups of cocoa for \$0.90. During one game they collected \$288 for a total of 330 drinks. How many cups of coffee did they sell?

# 9-8 Solving Systems by Substitution

## EXPLORE/ WORKING TOGETHER

The two ships shown on the map at the right will stop in the middle of the strait in order to meet and exchange some cargo.

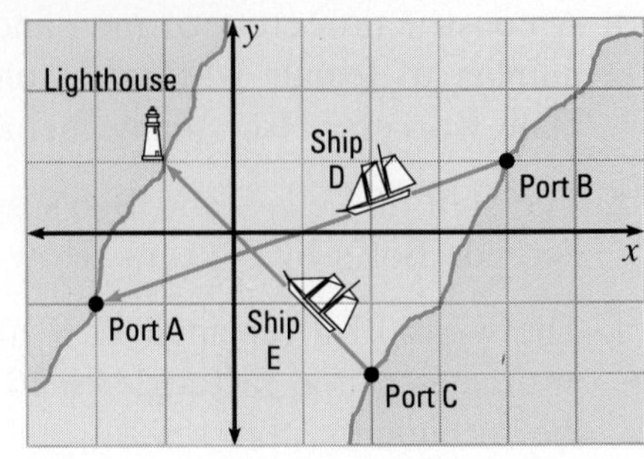


Work with a partner. Make a guess at the coordinates of the meeting place. Compare your solutions with those of other classmates. Did you all agree on the coordinates? Why or why not?

## SKILLS DEVELOPMENT

When the solution for a system of equations is difficult to read from the graph, you can use algebraic methods to solve the system of equations. One of these methods is **substitution.** In any algebraic method, you need to eliminate one of the variables so you will have one equation in one variable to solve. Here are the steps to follow using the substitution method.

1. Solve one of the equations for one variable in terms of the other.
2. Substitute that expression in the other equation and solve.
3. Substitute that value in one of the original equations and solve.
4. Check in both of the original equations.

### Example 1

Solve. $x - 2y = 3$
$x + y = 6$

**CHECK UNDERSTANDING**

In Example 1, if you solve for $x$ instead of $y$, how does the solution change?

### Solution

$$x + y = 6$$
$$y = -x + 6$$

← Solve the second equation for $y$ in terms of $x$.

$$x - 2y = 3$$
$$x - 2(-x + 6) = 3$$
$$x + 2x - 12 = 3$$
$$3x - 12 = 3$$
$$3x - 12 + 12 = 3 + 12$$
$$\frac{3x}{3} = \frac{15}{3}$$
$$x = 5$$

← Write the first equation. Substitute $(-x + 6)$ for $y$. Solve for $x$.

Choose one of the original equations.

$$\begin{aligned} x + y &= 6 \\ 5 + y &= 6 \\ -5 + 5 + y &= 6 - 5 \\ y &= 1 \end{aligned}$$

← Substitute 5 for $x$.
← Solve for $y$.

Check $x = 5$ and $y = 1$ in each original equation.

$$\begin{aligned} x - 2y &= 3 \\ 5 - 2(1) &= 3 \\ 5 - 2 &= 3 \\ 3 &= 3 \ \checkmark \end{aligned} \qquad \begin{aligned} x + y &= 6 \\ 5 + 1 &= 6 \\ 6 &= 6 \ \checkmark \end{aligned}$$

The solution is (5, 1). ◄

## Example 2

Solve.

$$\begin{aligned} x - 2y &= -9 \\ -2x + 4y &= -13 \end{aligned}$$

### Solution

$$\begin{aligned} x - 2y &= -9 && \leftarrow \text{Solve for } x. \\ x &= 2y - 9 \\ -2x + 4y &= -13 && \leftarrow \text{Substitute for } x. \\ -2(2y - 9) + 4y &= -13 \\ -4y + 18 + 4y &= -13 \\ 18 &= -13 && \leftarrow \text{False statement.} \end{aligned}$$

There is no solution. ◄

**CHECK UNDERSTANDING**

If you solved the system of equations in Example 2 using graphs, what kind of lines would you have?

## Example 3

M & K Construction Company sent 29 workers out in two crews. One crew had 5 more than twice the number of workers in the other crew. How many workers were in each crew?

### Solution

Define each of the variables. Write and solve a representative system of equations.

Let $l$ = number of workers in larger crew
$s$ = number of workers in smaller crew

There are 29 workers. $\quad l + s = 29$
The larger crew has 5 more than twice the smaller. $\quad l = 2s + 5$

$$\begin{aligned} l + s &= 29 && \leftarrow \text{Substitute } 2s + 5 \text{ for } l. \\ 2s + 5 + s &= 29 && \leftarrow \text{Solve for } s. \\ 3s + 5 &= 29 \\ 3s + 5 - 5 &= 29 - 5 \\ \frac{3s}{3} &= \frac{24}{3} \\ s &= 8 \\ l + s &= 29 && \leftarrow \text{Substitute for } s. \\ l + 8 &= 29 && \leftarrow \text{Solve for } l. \\ l + 8 - 8 &= 29 - 8 \\ l &= 21 \end{aligned}$$

The larger crew has 21 workers. The smaller crew has 8 workers. ◄

## TRY THESE

Solve each system of equations. Check the solutions.

1. $2x - y = 14$
   $x + y = 7$
2. $x - 2y = 7$
   $3x - y = 6$
3. $x - 3y = -4$
   $-2x + 6y = 7$
4. $-5x + y = 3$
   $20x - 4y = -2$
5. A yellow or a blue hard hat was worn by each of the 16 workers at a particular job site. The number of yellow hard hats was one more than twice the number of blue hard hats. How many of each color hard hat was worn?

# EXERCISES

### PRACTICE/ SOLVE PROBLEMS

Solve each system of equations. Check the solutions.

1. $x + y = 4$
   $3x - 2y = 2$
2. $-2x + y = 1$
   $x + y = 1$
3. $-2x + y = 2$
   $4x - 2y = 3$
4. $3x + 2y = 11$
   $-2x + y = -5$
5. $2x + y = 7$
   $x - y = 2$
6. $3x - y = 1$
   $-12x + 4y = -3$
7. $x + 2y = 4$
   $3x + y = 7$
8. $4x + y = 17$
   $x + 2y = 6$
9. $x + y = 5$
   $2x - y = 4$
10. $x + 5y = 9$
    $3x - 2y = -7$
11. $5x + 2y = 10$
    $x + y = 5$
12. $9x - 6y = 12$
    $-3x + 2y = 5$
13. $2x - 7y = -2$
    $-4x + 14y = 3$
14. $-2x + y = -3$
    $3x - 3y = -18$
15. On a construction site there are 14 cranes and trucks. Three times the number of cranes is 2 more than twice the number of trucks. How many cranes are there on the site?

**16.** At the food truck, Joe bought 3 hot dogs and 4 drinks for \$10. Cindy paid \$5 for 1 hot dog and 3 drinks. Find the cost of each hot dog and each drink.

**17.** Jim bought 4 jars of jelly and 6 jars of peanut butter. Adam bought 3 jars of the same jelly and 5 jars of the same peanut butter. Jim paid \$19.32 and Adam paid \$15.67. What is the cost of a jar of peanut butter?

## EXTEND/ SOLVE PROBLEMS

Solve each system of equations by substitution.

**18.** $2x - y = 4$
$-4x + 2y = -8$

**19.** $2x + y = 3$
$-12x - 6y = -18$

**20.** $3x - y = -4$
$-12x + 4y = 8$

**21.** $-5x + 4y = -12$
$10x - 8y = -16$

**22–25.** Now solve each of Exercises 18–21 by graphing.

Complete to make each statement true.

**26.** When the solution of a system of equations results in a true statement, the lines are __?__ and the number of solutions is __?__.

**27.** When the solution of a system of equations results in a false statement, the lines are __?__ and the number of solutions is __?__.

## THINK CRITICALLY/ SOLVE PROBLEMS

Solve each system of equations.

**28.** $x + y - z = -4$
$x = -2y$
$y - z = -6$

**29.** $a + b + c = 1$
$b = 2c$
$-a + c = -5$

**30.** $r + s + t = 0$
$r = -3t$
$s + t = 3$

**31.** $j + k + l = 7$
$l = k + 2$
$j - k = -4$

**32.** The sum of the ages of Jan, Kyle, and Luke is 38. Jan is 6 years older than Luke. Luke is 2 years younger than Kyle. Find the age of each.

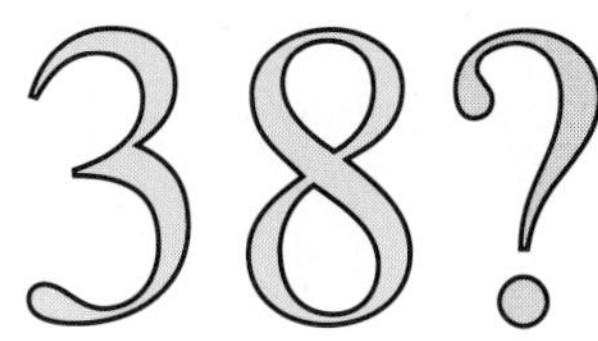

**PROBLEM SOLVING TIP**

To solve a system of three equations in three variables, try to eliminate one variable and get two equations in two variables. For example, in Exercise 28, you can see that $x = -2y$. So, you can eliminate $x$ by substituting $-2y$ for $x$ in the first equation. Then solve the first and third equations for $y$ and $z$. Give your answer by listing all three variables and their values.

# 9-9 Solving Systems by Adding, Subtracting, or Multiplying

## EXPLORE

You can use the addition property of opposites to explore other algebraic methods of solving a system of equations. Recall that when opposites are added, the sum is always 0.

$$-x + x = 0 \qquad 5a + (-5a) = 0$$

For each of the following, explain what you would do to the first term to make the two terms opposites.

**a.** $m, \ -m$ **b.** $r, \ r$ **c.** $-2z, \ -4z$

**d.** $5g, \ -15g$ **e.** $3x, \ \frac{1}{3}x$ **f.** $-\frac{4}{7}q, \ q$

## SKILLS DEVELOPMENT

When solving a system of equations by an algebraic method, you need to eliminate one of the variables to get one equation in one variable. If the coefficients of one of the variables are opposites, simply add the equations to eliminate one of the variables. This is the **addition** method.

**CHECK UNDERSTANDING**

In Example 1, if $-2x$ had been $2x$, what would you have had to multiply the second equation by to eliminate $x$?

### Example 1

Solve. $2x + 3y = 6$
$-2x + y = 2$

### Solution

$$\begin{aligned} 2x + 3y &= 6 \\ -2x + y &= 2 \\ \hline 0 + 4y &= 8 \\ \frac{4y}{4} &= \frac{8}{4} \\ y &= 2 \end{aligned}$$

The $x$-coefficients are opposites.

Add. Solve for $y$.

$$\begin{aligned} 2x + 3y &= 6 \\ 2x + 3(2) &= 6 \\ 2x + 6 &= 6 \\ 2x + 6 - 6 &= 6 - 6 \\ \frac{2x}{2} &= \frac{0}{2} \\ x &= 0 \end{aligned}$$

Choose one of the original equations.

Substitute 2 for $y$.

Solve for $x$.

Check $x = 0$ and $y = 2$ in each original equation.

$$\begin{aligned} 2x + 3y &= 6 \\ 2(0) + 3(2) &= 6 \\ 0 + 6 &= 6 \\ 6 &= 6 \checkmark \end{aligned} \qquad \begin{aligned} -2x + y &= 2 \\ -2(0) + 2 &= 2 \\ 0 + 2 &= 2 \\ 2 &= 2 \checkmark \end{aligned}$$

The solution is (0, 2). ◄

If the coefficients of one of the variables are the same, you can use the **subtraction method** to eliminate that variable.

## Example 2

Solve. $3x + 4y = 14$
$x + 4y = 10$

**Solution**

$$\begin{array}{r} 3x + 4y = 14 \\ -\underline{(x + 4y = 10)} \end{array} \rightarrow \begin{array}{r} 3x + 4y = 14 \\ \underline{-\ x - 4y = -10} \\ 2x + 0 = 4 \\ \frac{2x}{2} = \frac{4}{2} \\ x = 2 \end{array}$$

The $y$-coefficients are the same.

Subtract.

Solve for $x$.

Choose one of the original equations.

$$\begin{aligned} x + 4y &= 10 \\ 2 + 4y &= 10 \\ 2 - 2 + 4y &= 10 - 2 \\ \frac{4y}{4} &= \frac{8}{4} \\ y &= 2 \end{aligned}$$

Substitute 2 for $x$.

Solve for $y$.

The check is left to you.

The solution is (2, 2). ◄

Sometimes you will need to multiply one or both of the equations by a number to get coefficients of one of the variables to be opposites. This is the **multiplication and addition method.**

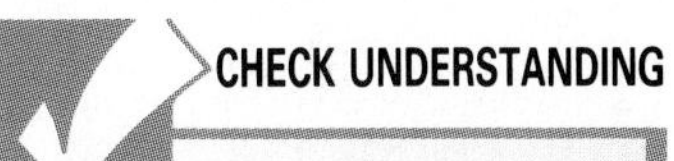

**CHECK UNDERSTANDING**

In Example 3, what would you have multiplied the second equation by to eliminate $y$?

## Example 3

Solve. $2x + 3y = 8$
$x + y = -3$

**Solution**

$$\begin{array}{r} 2x + 3y = 8 \\ \underline{-2x - 2y = 6} \\ 0 + y = 14 \\ y = 14 \end{array}$$

Multiply the second equation by $-2$.

Add.

Choose one of the original equations.

$$\begin{aligned} x + y &= -3 \\ x + 14 &= -3 \\ x + 14 - 14 &= -3 - 14 \\ x &= -17 \end{aligned}$$

Substitute for $y$.

Solve for $x$.

The check is left to you.

The solution is $(-17, 14)$. ◄

**CALCULATOR**

The *zoom* feature found on most graphing calculators allows you to magnify a section of a graph to help estimate the solution of a system of equations.

The equations $x + 3y = 2$ and $2x + 3y = 5$ are graphed below. The red box shows the section of the graph that will be magnified.

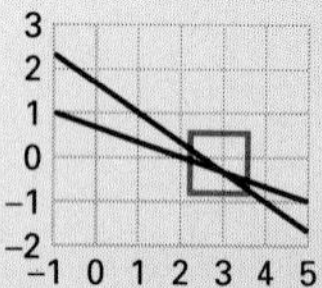


Follow these steps to zoom in on the solution.

1. Put the equations into $y = mx + b$ form. Enter and graph them on your calculator.
2. Estimate the solution to the nearest integer. Then use the ZOOM key and draw a box around the solution.
3. Repeat Step 2 until you think you have made the best possible estimate of each coordinate of the solution.

What is your best estimate for the coordinates of the solution of this system of equations?

## Example 4

Kate has nickels and dimes in her pocket. There are 11 coins. The value of the coins is $0.95. How many of each kind of coin does Kate have?

### Solution

Make a chart for the number and value of the coins. Write and solve a system of equations that represents the situation.

| | Nickels | Dimes | Total |
|---|---|---|---|
| Number | $n$ | $d$ | $n + d$ |
| Value | $5n$ | $10d$ | $5n + 10d$ |

The number of coins is 11. $\quad n + d = 11$

The value of the coins is 95¢. $\quad 5n + 10d = 95$

$$\begin{aligned} -10n - 10d &= -110 && \text{Multiply the first equation by } -10. \\ 5n + 10d &= \phantom{-}95 && \text{Add.} \\ -5n \phantom{+ 10d} &= -15 && \text{Solve for } n. \\ \frac{-5n}{-5} &= \frac{-15}{-5} \\ n &= 3 \end{aligned}$$

Choose one of the original equations.

$$\begin{aligned} n + d &= 11 && \text{Substitute for } n. \\ 3 + d &= 11 && \text{Solve for } d. \\ 3 - 3 + d &= 11 - 3 \\ d &= 8 \end{aligned}$$

Kate had 3 nickels and 8 dimes. ◄

## TRY THESE

Solve each system of equations. Check the solutions.

1. $3x - y = 15$
   $x + y = 1$
2. $-5a + 4b = -1$
   $7a + 4b = 11$
3. $4x + 3y = 2$
   $x - y = -10$
4. $r + s = 4$
   $2r + 3s = 8$
5. Juan has a jar with nickels and dimes. There are 30 coins with a total value of $1.85. How many of each coin does Juan have?

# EXERCISES

**PRACTICE/ SOLVE PROBLEMS**

Solve each system of equations. Check the solutions.

1. $3x + y = 9$
   $-3x + y = -3$
2. $-7x - 8y = 8$
   $7x - 8y = 8$
3. $7a + 2b = 16$
   $8a - 2b = 14$
4. $2r + 3s = 1$
   $9r - 3s = 54$

5. $5x - y = -23$
   $3x - y = -15$

6. $j + 3k = 10$
   $j + 2k = 7$

7. $7m - 5n = -2$
   $-8m - n = 9$

8. $3c - d = 1$
   $c + 5d = 11$

9. Steve has 35 coins made up of nickels and dimes. The value of the coins is $2.75. How many of each coin does he have?

10. Jennifer made a bank deposit of $238. Her deposit consisted of 50 bills, some one-dollar bills and the rest five-dollar bills. How many bills of each kind did she deposit?

## EXTEND/ SOLVE PROBLEMS

Solve each system of equations. Check the solutions.

11. $5x + 3y = 11$
    $-3x + 2y = 1$

12. $2r - 5s = 7$
    $3r - 2s = -17$

13. $3a + 8b = 1$
    $2a + 7b = 4$

14. $5c - 4d = 14$
    $3c + 3d = 3$

15. Joan has $2.30 in dimes and quarters. The total number of coins is 2 less than twice the number of dimes. How many of each type coin does Joan have?

16. A landscaping firm is designing a flower bed to border a rectangular pool. The perimeter of the pool is 32 m. Three times the width is the same dimension as five times the length. What are the dimensions of the pool?

17. Ricardo and Emilia are cousins. Twice Ricardo's age together with three times Emilia's age totals 34. Twice Emilia's age is one more than three times Ricardo's age. How old is Emilia?

## THINK CRITICALLY/ SOLVE PROBLEMS

Solve the following systems of equations.

18. $-5x + 2y = 12$
    $10x - 4y = -24$

19. $6x - 9y = 36$
    $-2x + 3y = -12$

20. $8x - 2y = -10$
    $-16x + 4y = 20$

Use the results of Exercises 18–20 for Exercises 21–24.

21. What do you notice about the relationship between the two equations in each of the systems?

22. What do you notice about each solution?

23. Solve the systems graphically. What do you notice?

24. Write a statement that generalizes your observations.

# 9-10 Systems of Inequalities

## EXPLORE/ WORKING TOGETHER

The $x$-axis and $y$-axis divide the coordinate plane into four quadrants. The axes act as boundary lines for these regions. Thus points on an axis are not in a quadrant and vice versa.

Working with a partner, determine which quadrant(s) satisfy the following systems of inequalities.

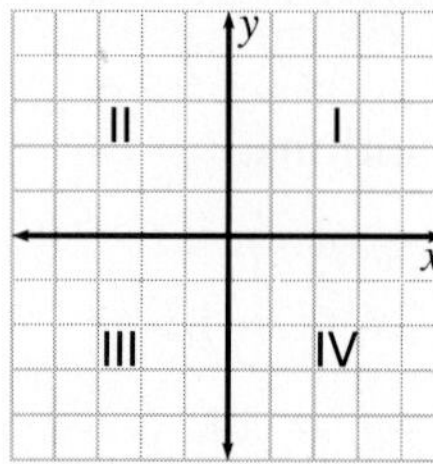


**a.** $x > 0$
$y < 0$

**b.** $x < 0$
$y < 0$

**c.** $x < 0$
$y > 0$

**d.** $x > 0$
$y > 0$

## SKILLS DEVELOPMENT

The slope-intercept form of an inequality is useful both for writing a system of inequalities for a given graph and for graphing a system of inequalities. The table you used in Section 8-8 will be useful again in graphing systems of inequalities.

| Form of Inequality | Boundary | Shaded Region |
|---|---|---|
| $y > mx + b$ | dashed | above the line |
| $y \geq mx + b$ | solid | above the line |
| $y < mx + b$ | dashed | below the line |
| $y \leq mx + b$ | solid | below the line |

### Example 1

Write a system of linear inequalities for the graph at the right.

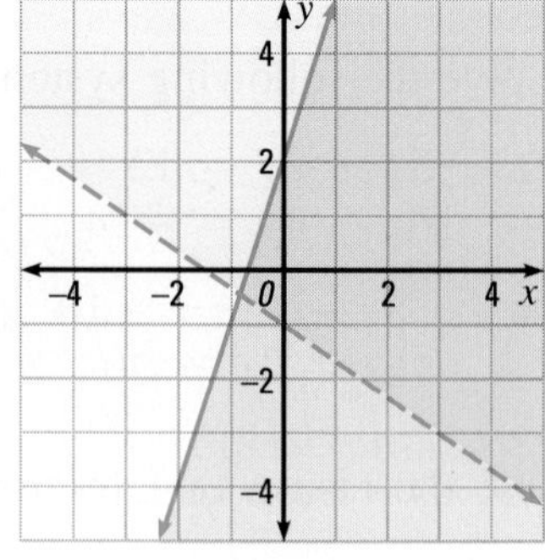


**Solution**

| | Blue | Yellow |
|---|---|---|
| **Determine *m*.** | $\frac{\text{rise}}{\text{run}} = \frac{3}{1} = 3$ | $\frac{\text{rise}}{\text{run}} = \frac{-2}{3} = -\frac{2}{3}$ |
| **Determine *b*.** | $b = 2$ | $b = -1$ |
| **Determine the inequality symbol.** | Below $\leq$<br>Solid | Above $>$<br>Dashed |

The system of linear inequalities for the graph is as follows.

$$y \leq 3x + 2$$
$$y > -\frac{2}{3}x - 1$$ ◄

The solution set of a system of linear inequalities is the intersection of the graphs of the inequalities. For the system of inequalities graphed in Example 1, the solution set includes all the points in the doubly shaded (green) region as well as those points on the solid boundary line that are not below the dashed boundary line.

## Example 2

Graph the solution set of this system of linear inequalities.

$$2x + y > 1$$
$$3x - y \geq 2$$

**Solution**

First write each inequality in slope-intercept form. Then make a chart to use for graphing.

$$2x + y > 1$$
$$-2x + 2x + y > -2x + 1$$
$$y > -2x + 1$$

$$3x - y \geq 2$$
$$-3x + 3x - y \geq -3x + 2$$
$$\frac{-y}{-1} \leq \frac{-3x}{-1} + \frac{2}{-1}$$
$$y \leq 3x - 2$$

| | $y > -2x + 1$ | $y \leq 3x - 2$ |
|---|---|---|
| Boundary | $y = -2x + 1$ | $y = 3x - 2$ |
| Shading | Above | Below |
| Line | Dashed | Solid |

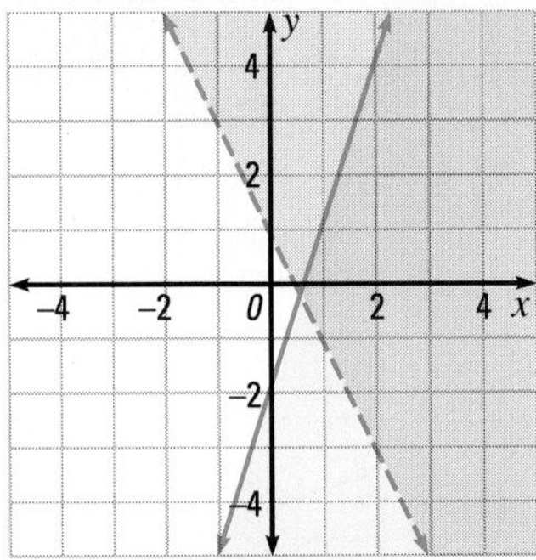


The solution set consists of all the points in the doubly shaded (green) region as well as all points on the solid boundary $y = 3x - 2$ that are not below the dashed boundary. ◄

## Try These

Write a system of linear inequalities for each graph.

1. 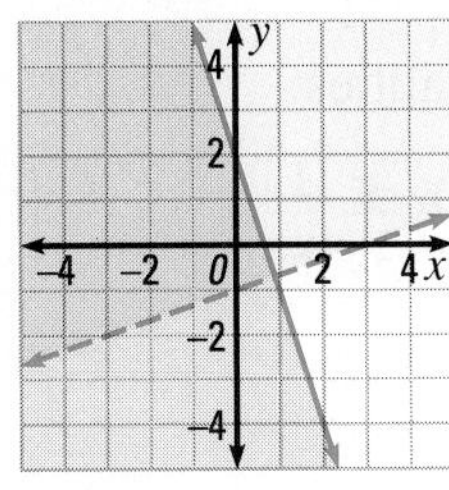


2. 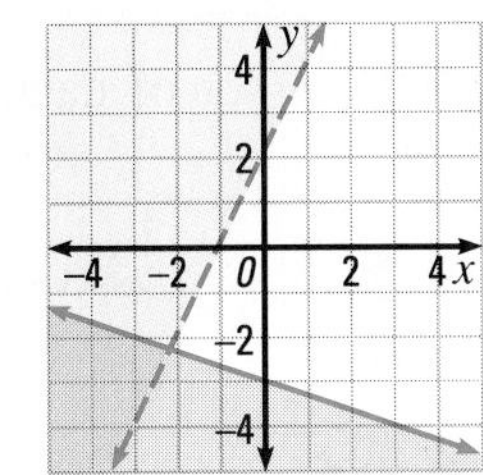

**PROBLEM SOLVING TIP**

When graphing, a quick way to check your shading is to check the point (0, 0) in the original form of the inequality. If it is a solution, it should be in the shaded section.

**3.** 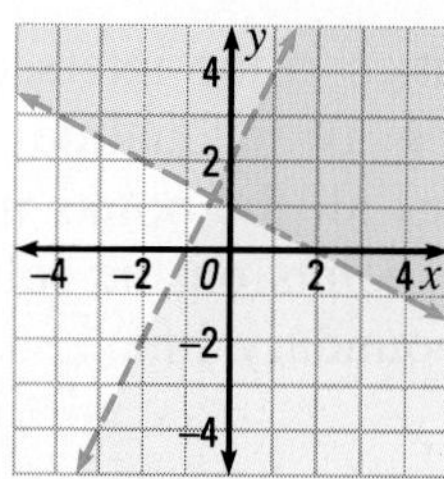

**4.** 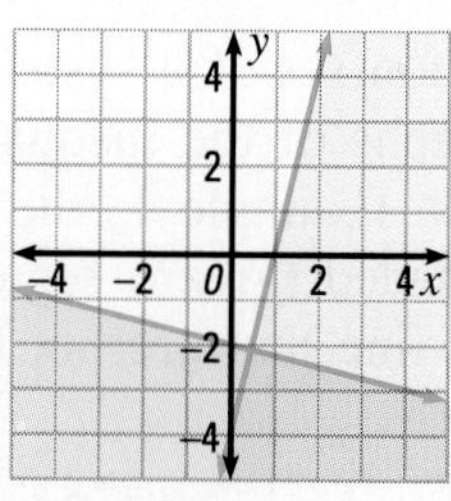

Graph the solution set of each system of inequalities.

**5.** $3x - y < -1$
$4x - y < 2$

**6.** $2x + y > 3$
$3x - y \geq -2$

**7.** $-2x + y > 3$
$5x - y \leq -10$

**8.** $-x + y < -4$
$2x - y \geq 1$

# EXERCISES

## PRACTICE/ SOLVE PROBLEMS

Write a system of linear inequalities for each graph.

**1.** 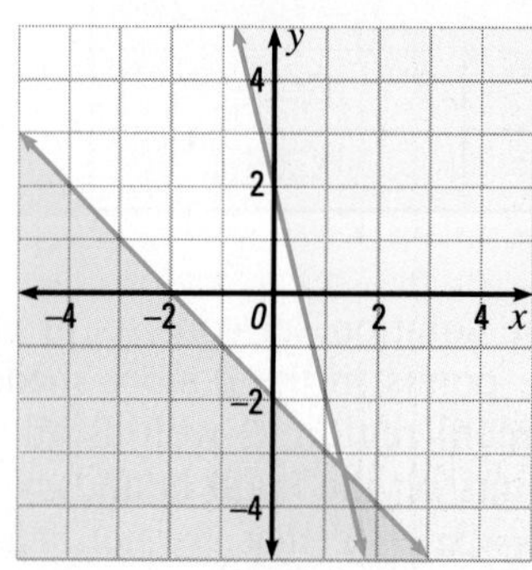

**2.** 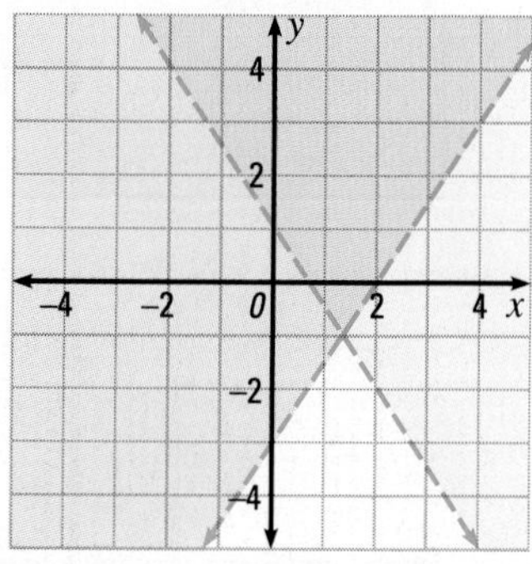

**3.** 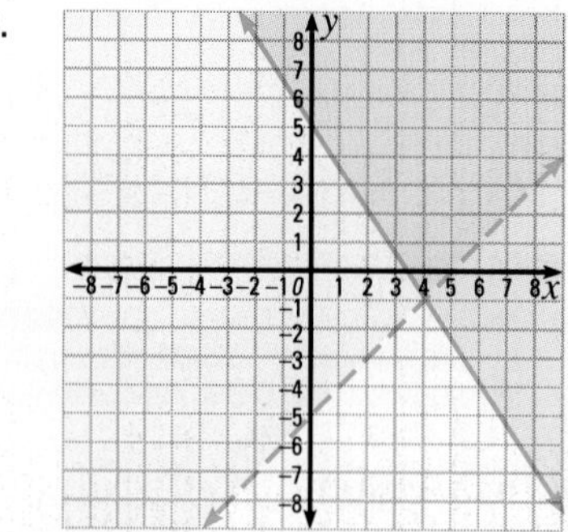

**4.** 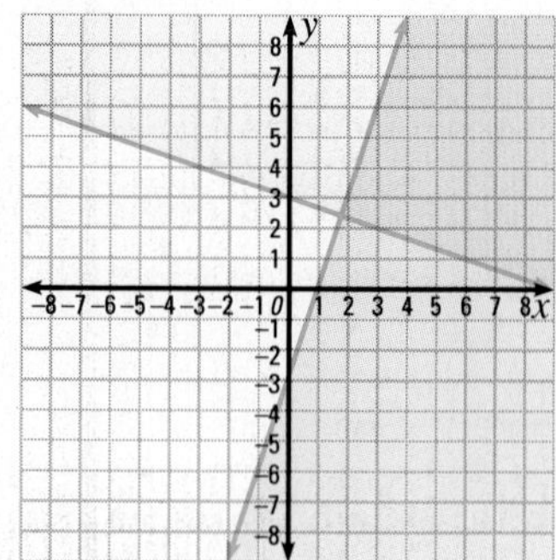

Graph the solution set of each system of inequalities.

**5.** $5x - y > -1$
$-2x + y \leq 3$

**6.** $-x + y \leq 4$
$3x - y < -2$

**7.** $-3x + y \geq 2$
$-2x - y > -1$

**8.** $4x - y > 1$
$-2x - y \leq -3$

Write a system of linear inequalities for each graph.

## EXTEND/ SOLVE PROBLEMS

**9.**

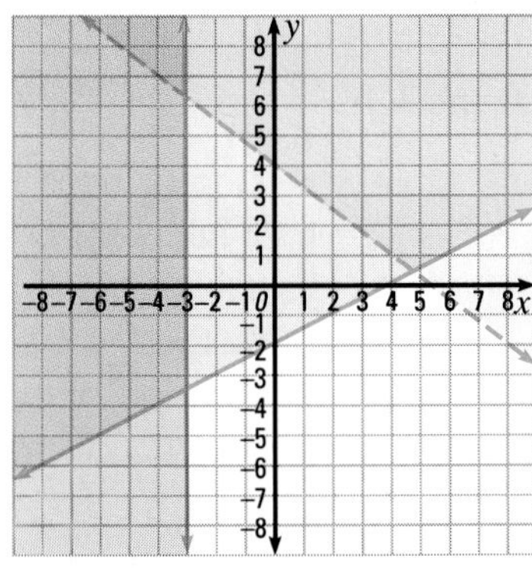

**10.**

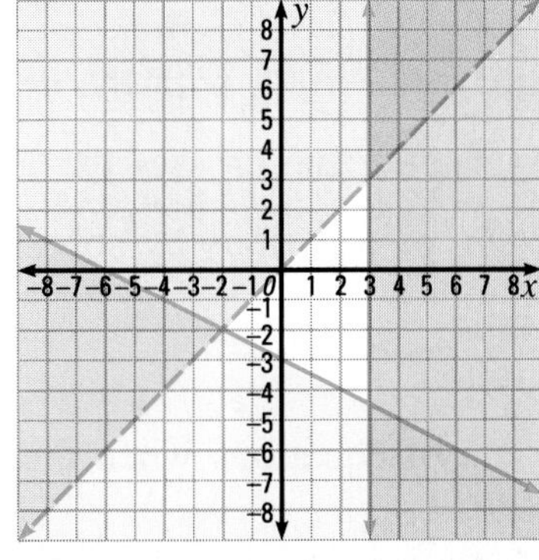

**11.**

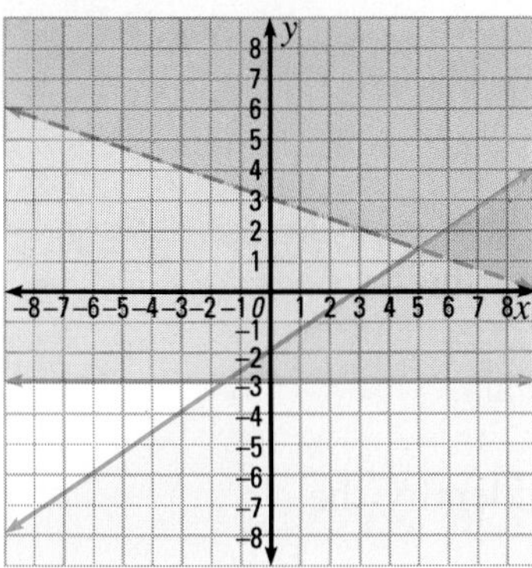

**12.**

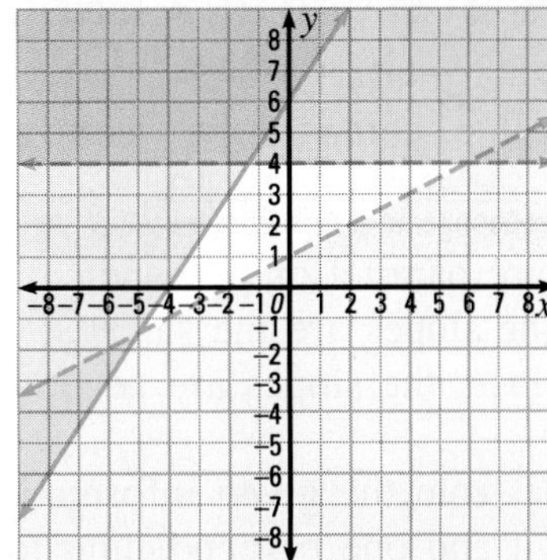

Graph the solution of each system of inequalities.

**13.** $-x - 2y > 4$
$3x + 4y \leq -12$
$y \leq -2$

**14.** $2x - 3y \geq -6$
$-x + 2y < -4$
$-x \geq -1$

## THINK CRITICALLY/ SOLVE PROBLEMS

The graph of each system of inequalities defines a figure. Graph each system and identify the figure.

**15.** $-x \geq -2$
$y \leq -3$
$x \geq 5$
$-y \leq -4$

**16.** $y \geq 2$
$3x - 2y \leq -12$
$y \leq -2$
$-4x - 5y \leq -20$

**PROBLEM SOLVING TIP**

Ask yourself these questions. What properties must the figure have to be a square? What properties must it have to be a parallelogram?

Write a system of inequalities whose graph would form the indicated region.

**17.** a square-shaped region

**18.** a region shaped like a parallelogram

## CHAPTER 9 ● REVIEW

Match the letter of the word at the right with the description at the left.

1. the point where the graph of a line crosses the $y$-axis
2. a way of solving a system of equations by solving for one variable in terms of the other
3. comparison of rise to run
4. the way to solve a system of equations when the coefficients of one variable are opposites

a. slope
b. $y$-intercept
c. substitution method
d. addition or subtraction method

### SECTIONS 9–1 AND 9–2 SLOPE OF A LINE (pages 348–355)

► To find the slope of a line, compare the rise to the run.

$$\text{slope} = \frac{\text{rise}}{\text{run}} = \frac{\text{change in } y\text{-coordinates}}{\text{change in } x\text{-coordinates}}$$

► To determine if lines are parallel or perpendicular, compare their slopes.
a. For parallel lines, the slopes are the same.
b. For perpendicular lines, the slopes are negative reciprocals.

Find the slope of each line from the given information. Then identify the slope of a line parallel to it and one perpendicular to it.

5. $A(-3, -1)$ and $B(-5, 7)$
6. $-3x + 4y = 24$

Graph the line that passes through the given point and has the given slope.

7. $P(-2, -3)$ slope $= \frac{2}{5}$
8. $P(1, 3)$ slope $= \frac{3}{4}$

### SECTION 9–3 SLOPE-INTERCEPT FORM OF AN EQUATION (pages 356–359)

► When an equation is written in the form $y = mx + b$ (slope-intercept form), the slope is the coefficient of $x$ and the $y$-intercept is the constant.

Identify the slope and $y$-intercept for the line with the given equation. Then graph the equation.

9. $y = \frac{1}{2}x + 1$
10. $y = -3x - 2$
11. $2x - 3y = 3$

### SECTION 9–5 WRITING EQUATIONS FOR LINES (pages 362–365)

► To write an equation of a line, you need the slope $m$ and the $y$-intercept $b$. Substitute $m$ and $b$ into the equation $y = mx + b$.

Write an equation of the line for the given information.

12. $m = -\frac{2}{5}$ $b = 3$
13. $m = -3$ $P(-1, 2)$
14. $A(-6, 1)$ $B(-3, 4)$

## SECTION 9–6 SYSTEMS OF EQUATIONS (pages 366–369)

► To solve a system of equations using graphs, graph both equations on the same axes. The coordinates of all points of intersection are solutions of the system.

Solve and check each system of equations.

15. $x + y = 4$
$x - y = 6$

16. $2x - y = 3$
$x + y = 6$

17. $x + 3y = 6$
$x - y = 2$

## SECTION 9–7 PROBLEM SOLVING (pages 370–371)

► To solve a problem using a system of equations, define each variable. Write a system of equations to represent the situations.

Define variables and write a system of equations that could be used to solve the problem.

18. One number is 8 more than another. Their sum is 26. What is the smaller number?

## SECTIONS 9–8 AND 9–9 SYSTEMS OF EQUATIONS (pages 372–379)

► A system of equations can be solved algebraically. To eliminate one variable, you can choose one of these three methods: substitution; addition or subtraction; multiplication and addition.

Solve.

19. $x + y = -4$
$2x - y = 1$

20. $3x - 2y = 4$
$-3x + 4y = -2$

21. $4x - 2y = -10$
$3x + 5y = 25$

## SECTION 9–10 SYSTEMS OF INEQUALITIES (pages 380–383)

► The slope-intercept form of an inequality indicates the boundary line, whether the boundary line should be solid or dashed, and where the shading should go.

Graph the solution set of each system of inequalities.

22. $-x + y < -2$
$-2x - 3y \leq -6$

23. $3x - 2y \geq -2$
$x + y > 3$

24. $2x - y > -1$
$-x - 2y < -4$

***USING DATA*** Refer to the calculations chart on page 347 to answer the following.

25. Find the number of risers and treads in an 8 ft 9 in. stairway.

# CHAPTER 9 ● TEST

Find the slope of each line from the given information.

1. 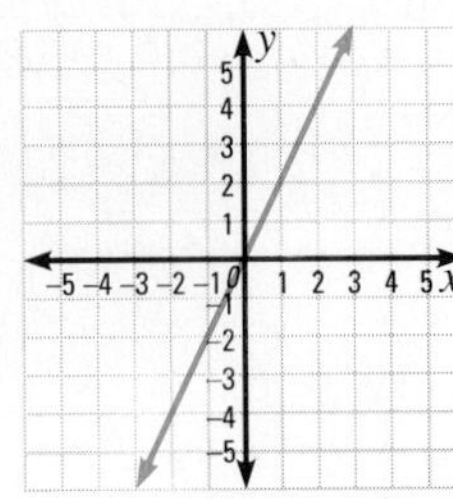

2. $R(2, -5)$ and $S(-2, 3)$

3. $-4x - 5y = 20$

4. $y = -\frac{3}{4}x + 2$

Identify the slope and $y$-intercept for the line with the given equation. Then identify the slope of a line parallel to it and the slope of a line perpendicular to it.

5. $y = -4x + 5$

6. $y = \frac{2}{3}x - 1$

7. $5x + 4y = 2$

Graph each equation.

8. $y = 2x + 1$

9. $3x - y = -2$

10. $x + 2y = 4$

Write an equation of each line from the given information.

11. 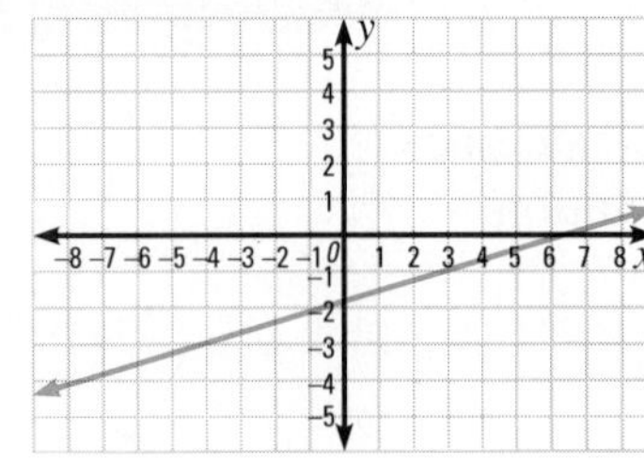

12. $m = -6 \quad b = 1$

13. $m = 3 \quad P(-2, 4)$

14. $M(3, -4) \quad N(8, -3)$

Use a graph to solve each system.

15. $3x - y = 2$
    $-x + 2y = -4$

16. $x + 2y = 6$
    $4x + 8y = -8$

17. $2x + 3y \geq -3$
    $-x - 2y < -6$

Solve.

18. $x - y = 7$
    $2x - 3y = 11$

19. $3x - 4y = 5$
    $-x + 4y = -7$

20. $2x - 3y = 4$
    $-3x + 4y = -6$

21. Jack's and Matt's ages together total 19. Matt is five years older than Jack. How old is each boy?

22. Martha has 16 dimes and quarters. There are 5 fewer quarters than twice the number of dimes. How many of each coin does she have?

Complete.

1. $4(74 + 47 + 24) = 4(74) + ■(47) + 4(24)$
2. $11.5 \times (-4.8) = ■ \times 11.5$

Find each percent to the nearest tenth.

3. 48% of 84
4. 110% of 90

Add or subtract.

5. $19 + (-11)$
6. $-18 - (-4)$

Multiply or divide.

7. $-16(-8)(-9)$
8. $-484 \div 22$

Write in scientific notation.

9. 16,800,000
10. 0.00041

11. Write the symbol for the figure.

12. Find the measure of $\angle 2$.

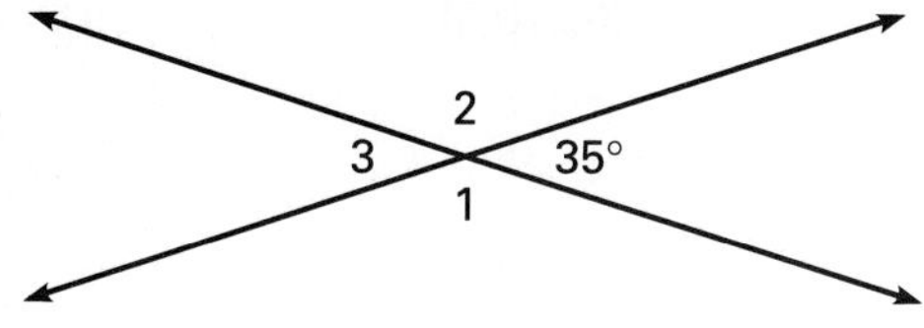


A group of randomly selected students were asked whether they collected cards. The results of this survey are shown below. Another student is selected at random. Estimate each probability. If necessary, write *not enough information.*

**TYPES OF CARDS COLLECTED**

| Celebrity | Baseball | Football |
|---|---|---|
| 28 | 9 | 7 |

13. The student collects sports cards.
14. The student is a girl who collects baseball cards.

Solve each equation.

15. $x^2 = 1.44$
16. $y^2 + 4 = 20$

For Exercises 17 and 18, refer to the figure. Use 3.14 for $\pi$.

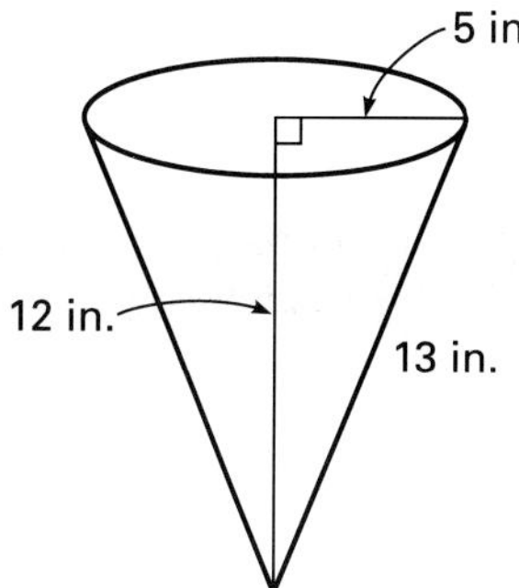


17. Find the volume.
18. Find the surface area.

Simplify.

19. $\dfrac{54a^3b^4c^5}{9a^2b^3c^5}$

Factor each polynomial.

20. $y^2 - 100$
21. $28a^2b^3 - 12ab + 20a^3b$

Tell whether the ordered pair is a solution of the inequality.

22. $(2, 3)$; $y > x - 1$
23. $(2, 0)$; $y \geq 2x - 1$

Write an equation of each line from the given information.

24. $A(-2, 3)$; $B(4, -3)$
25. $m = \frac{3}{4}$; $P(2, 2)$

1. What is the function rule for this function table?

| y | ? |
|---|---|
| −2 | 1 |
| −1 | 2 |
| 0 | 3 |
| 1 | 4 |
| 2 | 5 |

A. $f(y) = -y + 3$
B. $f(y) = 2y + 3$
C. $f(y) = y - 3$
D. $f(y) = y + 3$

2. 440 is what percent of 1,100?
A. 400% B. 44%
C. 40% D. 4%

3. Simplify. $x^8 \div x^{-2}$
A. $x^6$ B. $x^4$
C. $x^{10}$ D. $x^{16}$

4. A parallelogram has
A. one pair of congruent sides
B. four congruent angles
C. two pairs of congruent sides
D. one pair of parallel sides

In a new housing subdivision, seven different styles of houses are available. The prices are \$91,000, \$110,000, \$127,000, \$89,000, \$110,000, \$95,000, and \$92,000. The developer advertised the average price of a home as \$95,000.

5. Which measure of central tendency was used in the advertisement?
A. mean B. median
C. mode D. range

6. Which measure would have been the least misleading to potential buyers?
A. mean B. median
C. mode D. none of these

7. Solve. $\frac{4}{5}x - 3 = 1$
A. 5 B. 4 C. 20 D. 16

8. Solve. $r^2 - 18 = 7$
A. $+\sqrt{11}$ B. $\pm 5$ C. $\pm\sqrt{18}$ D. 25

Refer to the figure for Exercises 9 and 10. Use 3.14 for $\pi$. Round answers to the nearest tenth.

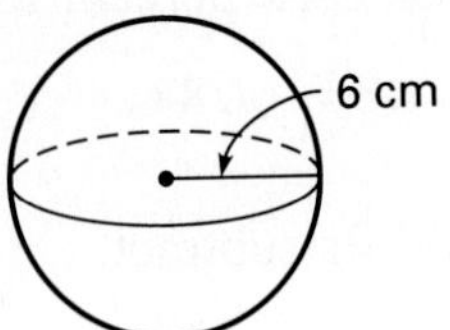


9. Find the volume.
A. 2,713.0 $cm^3$ B. 904.3 $cm^3$
C. 452.2 $cm^3$ D. 150.7 $cm^3$

10. Find the surface area.
A. 678.2 $cm^2$ B. 150.7 $cm^2$
C. 452.2 $cm^2$ D. 226.1 $cm^2$

11. Find the product. $(y - 4)(y + 5)$
A. $y^2 + y + 20$
B. $y^2 - 9y - 20$
C. $y^2 + 9y + 1$
D. $y^2 + y - 20$

12. Factor the polynomial. $x^2 - 2x - 15$
A. $(x + 5)(x - 3)$
B. $(x - 5)(x + 3)$
C. $(x - 5)(x - 3)$
D. $(x + 5)(x + 3)$

13. Which of the ordered pairs is *not* a solution of $y = 3x - 1$?
A. (1, 2) B. (3, 8)
C. (−1, 2) D. (−3, −10)

14. Identify the midpoint $P$ of the line segment whose endpoints are $A(1, 2)$ and $B(-3, -6)$.
A. $P(1, -2)$ B. $P(-1, -2)$
C. $P(2, -2)$ D. $P(-2, -4)$

15. Solve. $2x + y = 1$
$x - y = 2$
A. (1, −1) B. (1, 1)
C. (3, 1) D. (1, −3)

# CHAPTER 10 SKILLS PREVIEW

Write each ratio in lowest terms.

1. 45 m to 20 m
2. $\frac{63 \text{ L}}{18 \text{ L}}$
3. 64 h:56 h
4. 72 in.:4 ft

Find the better buy.

5. 12 lb apples for $4.95 or 10 lb apples for $4.00
6. 1.2 kg cheese for $6.52 or 0.8 kg cheese for $4.65

Solve each proportion.

7. $\frac{24}{n} = \frac{8}{7}$
8. $\frac{48}{21} = \frac{n}{14}$
9. $\frac{n}{15} = \frac{2}{25}$
10. $\frac{36}{66} = \frac{42}{n}$

Solve.

11. A machine produces 160 gaskets in 8 minutes. How many minutes does it take to produce 500 gaskets?
12. The height of a smokestack is 144 ft. What will the smokestack measure in a scale drawing with a scale of 1 in.:72 in.?
13. A company sent 8% of its employees to a training seminar. If it sent 84 people, how many employees does the company have?
14. The cost of a new car was $9,502.90. This included sales tax of 6%. What was the price of the car before sales tax was added?
15. Of 1,250 appliances a company shipped, 25 were defective. What percent of the shipped appliances were defective?

Find the unknown measure in each pair of similar polygons.

16.

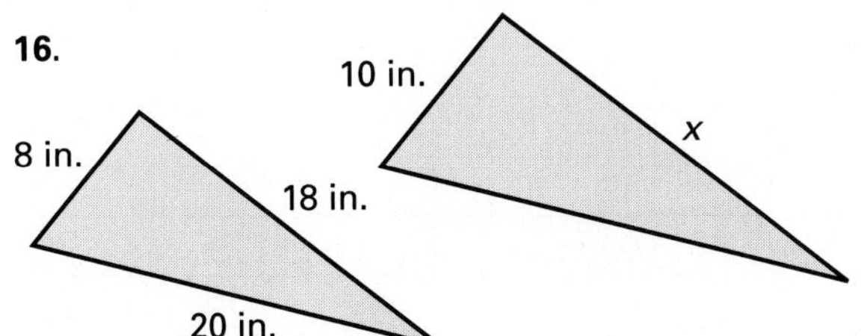


17.

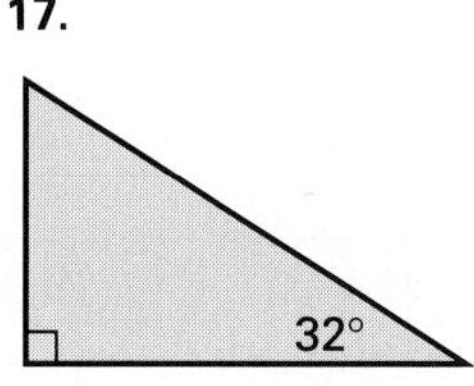


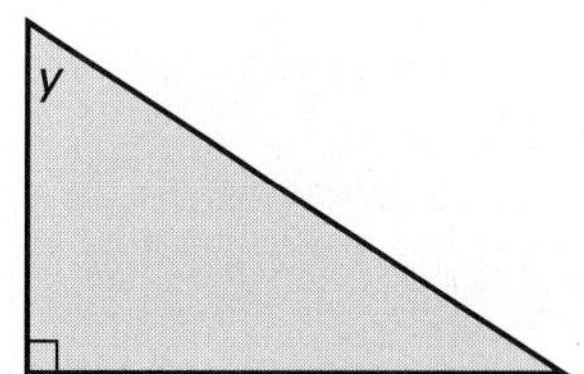


18. A tree casts a shadow that is 28.5 m long. At the same time, a boy who is 1.8 m tall casts a shadow 5.4 m long. How tall is the tree?
19. The cost of a load of sand varies directly with the volume of the sand. A load of 28 $ft^3$ of sand costs $36.40. How much would a load of 48 $ft^3$ of sand cost?
20. The time taken to complete a cross-country run varies inversely with the rate of the runner. At a rate of 8 mi/h, the run takes $1\frac{1}{2}$ h. How long would it take at a rate of 6 mi/h?

# RATIO, PROPORTION, AND VARIATION

**THEME** **Agriculture and Industry**

When you shop in a supermarket and want to find the best buy, what do you do? When you build a scale model, how do you decide on the dimensions of your model? From shopping to building, many real-world relationships can be analyzed or described by comparing numbers, measurements, or quantities. Ratios are comparisons of numbers and rates are comparisons of different kinds of quantities. Ratios and rates are especially useful in industry and agriculture. In this chapter, you will solve a wide variety of problems by using ratios, rates, and proportions.

The more information you have to study the changes that have occurred over a period of time, the greater your understanding of the data will be. The table below shows the number of people working in the United States during this century.

**THE U.S. CIVILIAN LABOR FORCE**

| Year | Agricultural (millions) | Nonagricultural (millions) | Total Population (millions) |
|---|---|---|---|
| 1900 | 11 | 16 | 76 |
| 1910 | 11 | 23 | 92 |
| 1920 | 10 | 29 | 106 |
| 1930 | 10 | 34 | 123 |
| 1940 | 10 | 38 | 132 |
| 1950 | 7 | 52 | 151 |
| 1960 | 5 | 60 | 179 |
| 1970 | 3 | 75 | 203 |
| 1980 | 3 | 96 | 227 |

# DECISION MAKING

## Using Data

Use the information in the table to answer the following questions.

1. As the population of the United States increased, what happened to the number of agricultural workers? What happened to the number of nonagricultural workers?
2. As the population of the United States increased, what change occurred in the total number of workers?
3. During the period from 1900 to 1980, did the percent of the total population who are workers change very much?
4. In which year was the greatest fraction of the total population employed?
5. Why do you think the majority of the population of the United States was not employed during the years from 1900 to 1980?

## Working Together

Use almanacs or reference books from the library to find information about the number of agricultural and nonagricultural workers for the year 1990. How do the figures compare with the trends shown in the table?

Find out about other employment trends during this century. You may wish to investigate the number of scientists, for example, or to compare the number of doctors and lawyers. Each member of your group should research one or two categories of employment. Compile your information in a table or a series of tables. Include columns to show fractions or percents to illustrate trends.

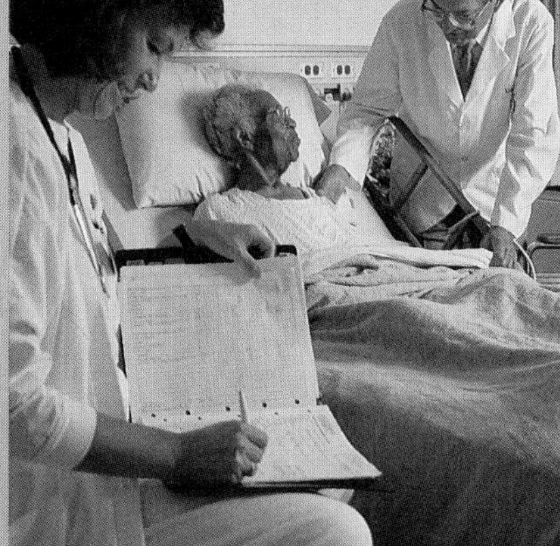

# 10-1 Ratios and Rates

**EXPLORE/ WORKING TOGETHER**

Work with a partner. Copy the table below. How many breaths have each of you taken during your lifetime? Make a quick guess, then write your answer in your table. Then, use a watch or clock to count the number of breaths each of you takes during a four-minute period. Calculate the average number of breaths you take each minute, and then complete your table.

**NUMBER OF BREATHS**

| | 1 minute | 1 hour | 1 day | 1 year | During your lifetime | |
|---|---|---|---|---|---|---|
| | | | | | Actual | Guess |
| Partner 1 | | | | | | |
| Partner 2 | | | | | | |

**SKILLS DEVELOPMENT**

A **ratio** is a quotient of two numbers. It is used to compare one number with the other. There are three ways to write a ratio. The order of the numbers is important in a ratio.

| *analogy form* | *fraction form* | *word form* |
|---|---|---|
| 9:14 | $\frac{9}{14}$ | 9 to 14 |

← Read each form as "nine to fourteen."

TALK IT OVER

In Example 1b, suppose that inches had been converted to feet. Would the ratio be different? Give reasons for your answer.

### Example 1

Write each ratio of measurement in lowest terms.

**a.** 8 in. to 20 in. **b.** 2 ft:9 in.

### Solution

| **a.** | | **b.** |
|---|---|---|
| $\frac{8 \text{ in.}}{20 \text{ in.}}$ | Write the ratio as a fraction. | $\frac{2 \text{ ft}}{9 \text{ in.}}$ |
| $\frac{8 \text{ in.}}{20 \text{ in.}}$ | If necessary, rename the measurements using the same units. | $\frac{2 \text{ ft}}{9 \text{ in.}} = \frac{24 \text{ in.}}{9 \text{ in.}}$ |
| $\frac{8 \div 4}{20 \div 4} = \frac{2}{5}$ | Write the fraction in lowest terms. | $\frac{24 \div 3}{9 \div 3} = \frac{8}{3}$ ← Do not write as a mixed number. |

**a.** The ratio of measurements is 2 to 5.
**b.** The ratio of measurements is 8:3. ◄

Sometimes you may know the total number of items in a group and the ratio of two different parts of the group, but not the number of items in each pair. Then you can write an equation to solve a ratio problem.

## Example 2

There are 630 employees at the Freemont Canning Company. The ratio of female to male employees is 4 to 3. How many male employees work at the Freemont Canning Company?

### Solution

Let the number of female employees be $4x$.
Then the number of male employees will be $3x$.

**Think: The ratio is 4 to 3.** $\frac{4}{3} = \frac{4x}{3x}$

Write an equation for the total number of employees.

$$4x + 3x = 630$$
$$7x = 630$$
$$x = 90$$

Since $x = 90$, $3x = 270$. There are 270 male employees at the Freemont Canning Company. ◄

**PROBLEM SOLVING TIP**

When you find the value 90 for $x$ in Example 2, you must check if $x$ represents the male employees. Since $3x$ represents the male employees, the answer to the problem is $3 \times 90 = 270$. Always check what quantity a variable represents and be sure you have solved the problem correctly.

A **rate** is a ratio that compares two different kinds of quantities. Many problems require you to compare the first quantity to *one unit* of the second quantity. This type of comparison is called a **unit rate.**

## Example 3

Find the unit rate.

a. 135 lb for 18 bricks

b. 248 beats in 4 min

### Solution

a. $\frac{\text{pounds} \rightarrow 135}{\text{bricks} \rightarrow 18} = \frac{135 \div 18}{18 \div 18} = \frac{7.5}{1}$

The unit rate is 7.5 lb per brick.

b. $\frac{\text{beats} \rightarrow 248}{\text{min} \rightarrow 4} = \frac{248 \div 4}{4 \div 4} = \frac{62}{1}$

The unit rate is 62 beats per min. ◄

A unit rate that identifies the cost of an item per unit is called the **unit price** of the item. Unit prices can help a consumer determine which of two items is the *better buy.*

## Example 4

A 12-oz package of Hilltop cereal costs \$2.52. A 15-oz package of Granes cereal costs \$3.40. Which brand is the better buy?

### Solution

Find the unit price for each brand and compare.

Hilltop $\frac{\text{cost in dollars} \rightarrow 2.52}{\text{weight in ounces} \rightarrow 12} = \frac{2.52 \div 12}{12 \div 12} = \frac{0.21}{1}$

Granes $\frac{\text{cost in dollars} \rightarrow 3.40}{\text{weight in ounces} \rightarrow 15} = \frac{3.40 \div 15}{15 \div 15} = \frac{0.226}{1}$ ← **Round to \$0.23.**

Hilltop costs \$0.21 per ounce and Granes costs \$0.23 per ounce. Hilltop cereal is the better buy. ◄

**CHECK UNDERSTANDING**

In Example 3, why do you keep the 1 in the denominator of each fraction?

## TRY THESE

Write each ratio in lowest terms.

1. 14 cm:91 cm  2. 64 m to 12 m  3. $\frac{125 \text{ L}}{35 \text{ L}}$  4. 18 yd:15 ft

5. A factory produced 1,419 cars. The ratio of convertibles to sedans was 2 to 9. How many convertibles were produced?

Find each unit rate.

6. 63 mi in 14 h  7. 540 $\text{ft}^2$ for 900 trees  8. $38.28 for 17.4 lb

## EXERCISES

**PRACTICE/ SOLVE PROBLEMS**

Write each ratio in lowest terms.

1. 55 m:220 m  2. 45 L to 18 L  3. 28 h:105 h  4. $\frac{24 \text{ min}}{80 \text{ sec}}$

5. A farmer uses four times as much area for wheat as for corn. If the total area used is 65 acres, how many acres are used for wheat?

Find each unit rate.

6. 15 ft in 12 min  7. 70 min for 50 pages  8. 258 lb in 43 boxes

To solve Exercise 11, use the distance formula for the length of a segment on the coordinate plane.

9. Which is the better buy: 3 apples for $0.87 or 5 apples for $1.15?

10. Water flows into the pool shown at the right at a rate of 80 $\text{ft}^3$ per min. How long will it take to fill the pool?

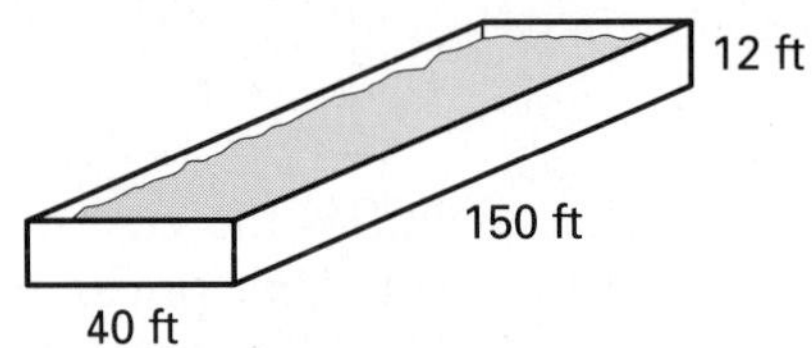


**EXTEND/ SOLVE PROBLEMS**

11. The vertices of $\triangle ABC$ are $A(-3, 7)$, $B(3, -1)$, and $C(-8, -5)$. Find the ratio of the length of $\overline{AB}$ to the length of $\overline{AC}$.

12. The legs of a right triangle measure 21 cm and 28 cm. Find the ratio of the length of each leg to the length of the hypotenuse.

13. The speed of sound is about 1,088 ft/s. What is the speed of sound in miles per hour, to the nearest whole number?

**THINK CRITICALLY/ SOLVE PROBLEMS**

14. Allen drove 300 mi to a friend's house at an average speed of 40 mi/h. On the drive back, he averaged 50 mi/h. Was his average speed for the entire trip less than 45 mi/h, exactly 45 mi/h, or greater than 45 mi/h? Give reasons for your answer.

- READ
- PLAN
- SOLVE
- ANSWER
- CHECK

# Problem Solving Applications:

## WHAT ARE THE ODDS?

When people use the term **the odds in favor** of a particular event in the case of equally likely outcomes, they are comparing the total number of favorable outcomes of an event to the total number of unfavorable outcomes. Remember that the number of favorable outcomes *plus* the number of unfavorable outcomes must be equal to the total number of possible outcomes.

**TALK IT OVER**

How do odds differ from probability? What do you think is meant by *even odds*?

Suppose 2,000 raffle tickets were sold and there are 12 prizes. If you bought one raffle ticket, what are the odds in favor of your winning a prize? What are the odds against your winning a prize?

Express the odds as a ratio.

odds in favor = favorable outcomes : unfavorable outcomes
= 12:1,988 ← Since there are 12 favorable outcomes, there are 2,000 − 12 = 1,988 unfavorable outcomes.
= 3:497 ← lowest terms

The odds in favor of your winning a prize are 3:497.

odds against = unfavorable outcomes : favorable outcomes
= 1,988:12
= 497:3 ← lowest terms

The odds against your winning a prize are 497:3.

1. A card is chosen at random from a deck of alphabet cards.
   a. What are the odds in favor of choosing a vowel?
   b. What are the odds against choosing a vowel?
2. Cal has 8 nickels, 3 quarters, and 4 dimes in his pocket. If he takes out one coin, what are the odds in favor of it being a quarter?
3. Dolores is one of 52 members of the computer club. The names of all members will be put in a hat and four will be chosen at random to attend a meeting in Washington, D.C. What are the odds in favor of Dolores attending the meeting?
4. Dolores' friend is the first member picked to attend the meeting. Now what are the odds in favor of Dolores getting picked?
5. Which, if either, are more favorable odds: 60:40 or 65:45?

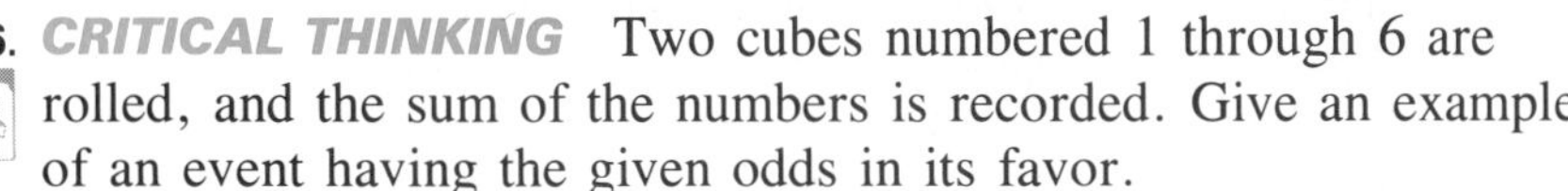

6. ***CRITICAL THINKING*** Two cubes numbered 1 through 6 are rolled, and the sum of the numbers is recorded. Give an example of an event having the given odds in its favor.
   a. 1:1 b. 5:1 c. 1:35 d. 4:5

# 10-2 Proportions

**EXPLORE**

For over 2,000 years, people have used "golden rectangles" in art and architecture. A golden rectangle is a rectangle in which the ratio of the length to the width is about 1.62:1.

| DIMENSIONS OF FRAMES (inches) | | | | | | |
|---|---|---|---|---|---|---|
| width $w$ | 4 | 5 | 8 | $8\frac{1}{2}$ | 12 | 14 |
| length $l$ | 6 | 7 | 12 | 11 | 20 | 17 |
| ratio $l{:}w$ | | | | | | |
| "golden" length | | | | | | |

Copy and complete the chart of most popular sizes of picture frames to find which come close to being golden rectangles. For each width, use guess-and-check to find the length (to the nearest half inch) that would most closely form a golden rectangle.

**SKILLS DEVELOPMENT**

Two ratios that represent the same comparison are called **equivalent ratios.** To determine if two ratios are equivalent, you can write each ratio as a fraction in lowest terms.

### Example 1

Is each pair of ratios equivalent?

**a.** 18 to 12, 16 to 10 **b.** 4:10, 6:15

**Solution**

**a.** Rewrite 18 to 12 as $\frac{18}{12}$. Rewrite 16 to 10 as $\frac{16}{10}$.

$\frac{18}{12} = \frac{18 \div 6}{12 \div 6} = \frac{3}{2}$ $\frac{16}{10} = \frac{16 \div 2}{10 \div 2} = \frac{8}{5}$

No, 18 to 12 and 16 to 10 are not equivalent ratios.

**b.** Rewrite 4:10 as $\frac{4}{10}$. Rewrite 6:15 as $\frac{6}{15}$.

$\frac{4}{10} = \frac{4 \div 2}{10 \div 2} = \frac{2}{5}$ $\frac{6}{15} = \frac{6 \div 3}{15 \div 3} = \frac{2}{5}$

Yes, 4:10 and 6:15 are equivalent ratios. ◄

A **proportion** is an equation stating that two ratios are equivalent. Since the ratios in Example 1b are equivalent, you can use them to write a proportion. The numbers 4, 10, 6, and 15 are called the **terms** of the proportion.

$4{:}10 = 6{:}15$ ← Read, "four is to ten as six is to fifteen." → $\frac{4}{10} = \frac{6}{15}$

The first and fourth terms of a proportion are called the **extremes.** The second and third terms are called the **means.**

means
$a{:}b = c{:}d$
extremes

$\frac{a}{b} = \frac{c}{d}$

In a proportion, the **cross-products** of the terms are equal.

$\frac{4}{10} = \frac{6}{15}$ $10 \times 6 = 60$ ← cross-products; $4 \times 15 = 60$ ← cross-products

$\frac{a}{b} = \frac{c}{d}$, so $ad = bc$

↑ product of extremes (ad) ↑ product of means (bc)

Sometimes one term of a proportion is unknown and is represented by a variable. Use cross-products to solve this type of proportion.

## Example 2

Solve the proportion. $\frac{9}{12} = \frac{21}{n}$

**Solution**

Find the cross-products. Solve for $n$.

$$9 \times n = 21 \times 12$$
$$9n = 252$$
$$\left(\frac{1}{9}\right)9n = \left(\frac{1}{9}\right)252$$
$$n = 28$$

Check. Substitute 28 for $n$. Determine if the two ratios are equivalent.

$\frac{9}{12} = \frac{9 \div 3}{12 \div 3} = \frac{3}{4}$ $\frac{21}{28} = \frac{21 \div 7}{28 \div 7} = \frac{3}{4}$ So, $n = 28$. ◄

**MENTAL MATH TIP**

Proportions in which one term is a multiple of another can sometimes be solved using mental math.

$\frac{7}{5} = \frac{28}{n}$ **28 is a multiple of 7**

Since $28 = 7 \times 4$,
$n = 5 \times 4 = 20$.

$\frac{12}{n} = \frac{54}{9}$ **54 is a multiple of 9**

Since $9 = 54 \div 6$,
$n = 12 \div 6 = 2$.

You can solve many problems that involve equivalent ratios by writing and solving a proportion.

## Example 3

If 6 apples cost \$1.78, what is the cost of 7 apples?

**Solution**

Write and solve a proportion.

$\frac{6}{1.78} = \frac{7}{n}$ ← number of apples; ← cost

$$6n = 7 \times 1.78$$
$$6n = 12.46$$
$$n = 2.077 \approx 2.08$$

← Since you need the cost, round to the nearest cent.

The cost of 7 apples is \$2.08. ◄

## TRY THESE

Is each pair of ratios equivalent? Write *yes* or *no*.

1. 21:35, 15:25
2. $\frac{99}{18}, \frac{44}{8}$
3. 12 to 88, 20 to 164

Solve each proportion. Check your answers.

4. $\frac{16}{n} = \frac{36}{27}$
5. $\frac{x}{9} = \frac{50}{15}$
6. $11:6 = m:18$
7. A machine produces 45 hinges in 6 min. How many minutes would it take to produce 75 hinges?

# EXERCISES

## PRACTICE/ SOLVE PROBLEMS

Is each pair of ratios equivalent? Write *yes* or *no*.

1. $\frac{96}{69}, \frac{32}{23}$
2. 12:25, 18:40
3. 16 to 44, 24 to 66

Solve each proportion. Check your answers.

4. $\frac{18}{48} = \frac{12}{n}$
5. $\frac{21}{z} = \frac{77}{33}$
6. $85:100 = c:20$

7. A load of 18 bags of lime weighs 441 lb. How many pounds will a load of 12 bags of lime weigh?
8. A water tank is 10 ft high. When the water is 4 ft deep, the volume of water is 75 $ft^3$. What volume of water is in the tank when it is full?
9. The Trux Company said that 3 out of every 25 machines they produced last month were sold outside the United States. If they made 400 machines, how many did they sell outside the United States?
10. Kerry's car used 6 gal of gasoline to travel 168 mi. She calculated that it would travel 448 mi on a full tank of gas. How many gallons of gas does the tank hold?

## EXTEND/ SOLVE PROBLEMS

Solve each proportion.

11. $\frac{36}{n + 2} = \frac{4}{3}$
12. $\frac{20}{4} = \frac{n}{n - 4}$
13. $\frac{3n + 3}{6} = \frac{2n - 1}{2}$
14. A factory has a goal of producing 10,960 lawnmowers next year. If the factory produces the same number of lawnmowers each month, how many lawnmowers should they produce by September 30?
15. In the 1800s, some farmhands plowed $7\frac{1}{2}$ acres of land in one day. If there had been 3 more workers, they could have plowed 10 acres. How many farmhands were there?

## THINK CRITICALLY/ SOLVE PROBLEMS

When the means of a proportion are the same number, that number is called the **mean proportional** between the two other numbers. In the proportion $2:4 = 4:8$, 4 is the mean proportional. Use cross multiplication to find the mean proportional between each pair of numbers.

16. 4 and 9
17. 8 and 18
18. *a* and *b*

- READ
- PLAN
- SOLVE
- ANSWER
- CHECK

# Problem Solving Applications:

## SCALE DRAWINGS

A **scale drawing** is a drawing that represents a real object. The **scale** of the drawing is the ratio of the size of the drawing to the actual size of the object. You can use a scale drawing to find the actual dimensions of an object by writing and solving a proportion.

What is the actual length of the bus shown in the scale drawing?

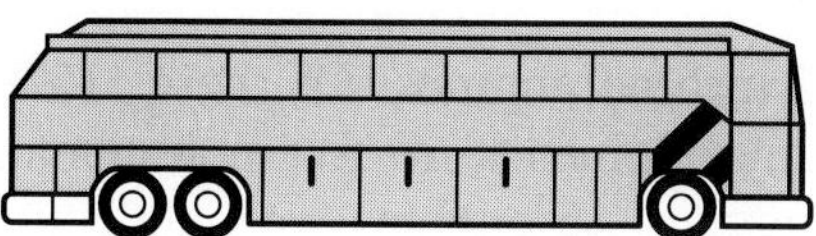

Scale: 1 cm : 2 m

Write a proportion using the scale as one of the equivalent ratios.

$$\frac{1 \text{ cm}}{2 \text{ m}} = \frac{5.5 \text{ cm}}{n} \quad \begin{array}{l} \leftarrow \text{length in drawing} \\ \leftarrow \text{actual length in meters} \end{array}$$

$$n = 5.5 \times 2$$
$$= 11 \text{ m}$$

The actual length of the bus is 11 m.

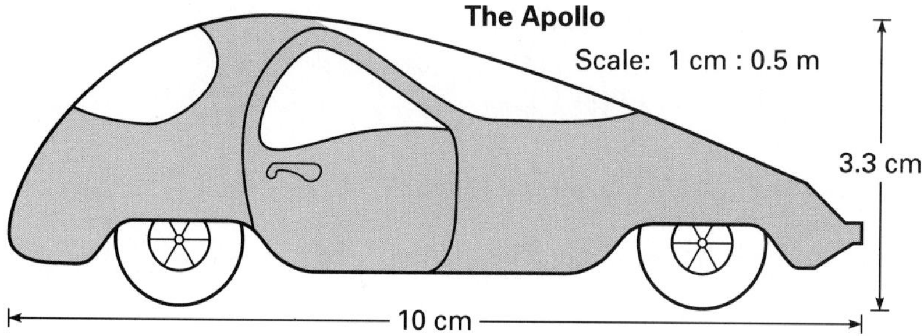


Use the scale drawing of the Apollo to solve Problems 1–3.

1. What is the height of the Apollo automobile?
2. In the drawing, the diameter of the wheel and tire is 1.5 cm. What is the actual diameter?
3. How many Apollos could park along one side of a street that is 78 m long?

For each object, you are to make a scale drawing that will fit on an $8\frac{1}{2}$-in. by 11-in. sheet of paper. Tell what scale you will use.

4. a TV set
5. a football field
6. your desk
7. Choose an object whose dimensions you know or can measure and create a scale drawing of it.

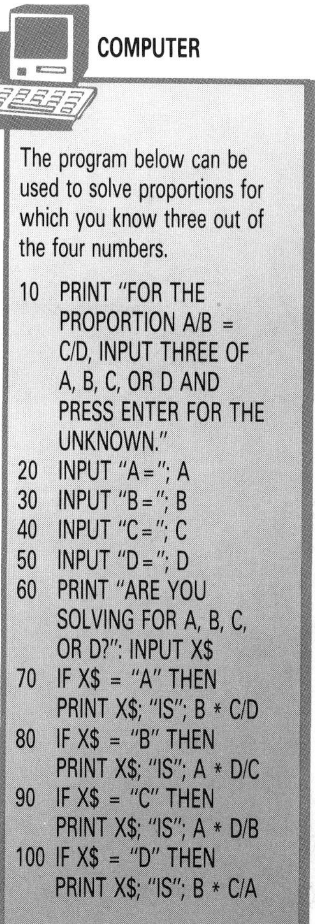

**COMPUTER**

The program below can be used to solve proportions for which you know three out of the four numbers.

```
10  PRINT "FOR THE
    PROPORTION A/B =
    C/D, INPUT THREE OF
    A, B, C, OR D AND
    PRESS ENTER FOR THE
    UNKNOWN."
20  INPUT "A="; A
30  INPUT "B="; B
40  INPUT "C="; C
50  INPUT "D="; D
60  PRINT "ARE YOU
    SOLVING FOR A, B, C,
    OR D?": INPUT X$
70  IF X$ = "A" THEN
    PRINT X$; "IS"; B * C/D
80  IF X$ = "B" THEN
    PRINT X$; "IS"; A * D/C
90  IF X$ = "C" THEN
    PRINT X$; "IS"; A * D/B
100 IF X$ = "D" THEN
    PRINT X$; "IS"; B * C/A
```

# 10-3 Problem Solving Skills:

## USE A PROPORTION TO SOLVE A PERCENT PROBLEM

**MIXED REVIEW**

Evaluate each expression when $a = -3$ and $b = 4$.

1. $b^{-3}$
2. $a^2b$
3. $a^5 \div a^4$
4. $a^{-3}b^2$

Find the median, mode, and range for each set of data.

5. 27, 28, 25, 29, 28, 27, 28
6. 7.6, 8.8, 8.3, 7.5, 8.3, 8.2, 8.1, 7.4, 7.9, 8.0

Find the surface area and volume of each rectangular prism.

7. length = 10 cm
   width = 4 cm
   height = 6 cm
8. length = 6 in.
   width = 2 in.
   height = 5.6 in.

In Section 1–6 you learned how to write equations to solve percent problems. Many problems that involve percents can also be solved by writing and solving a proportion.

## PROBLEM

**a.** A company produced 260 turbine engines. Of these, 15% were for motor boats. How many engines were for motor boats?

**b.** In a shipment of 1,150 engines, 46 engines were defective. What percent of the engines were defective?

**c.** There are 196 women working at Power, Inc. They make up 56% of the employees. How many employees are there at Power, Inc.?

## SOLUTION

**a.** You want to find the percent of a number. The ratio represented by 15% is 15 to 100. So, the proportion can be stated "$n$ is to 260 as 15 is to 100."

$$\begin{aligned} \text{part} \rightarrow \ \ & \frac{n}{260} = \frac{15}{100} \quad \leftarrow \text{whole} \\ 100 \times n &= 15 \times 260 \quad \leftarrow \text{Cross multiply.} \\ 100n &= 3{,}900 \\ n &= 39 \end{aligned}$$

There were 39 engines for motor boats.

**b.** You want to find the percent. The percent is represented by the ratio $n$ to 100. So, the proportion can be stated "46 is to 1,150 as $n$ is to 100."

$$\begin{aligned} \frac{46}{1{,}150} &= \frac{n}{100} \\ 46 \times 100 &= n \times 1{,}150 \quad \leftarrow \text{Cross multiply.} \\ 4{,}600 &= 1{,}150n \\ 4 &= n \end{aligned}$$

Four percent of the engines were defective.

**c.** You know the percent of a number and you want to find the number. The ratio represented by 56% is 56 to 100. So, the proportion can be stated "196 is to $n$ as 56 is to 100."

$$\begin{aligned} \frac{196}{n} &= \frac{56}{100} \\ 196 \times 100 &= 56 \times n \quad \leftarrow \text{Cross multiply.} \\ 19{,}600 &= 56n \\ 350 &= n \end{aligned}$$

There are 350 employees at Power, Inc.

# PROBLEMS

Solve each problem using a proportion.

1. A company shipped 1,716 cork gaskets to a client that ordered 2,640 gaskets. What percent of the order did the company send?
2. A manufacturer deducts 5% of the total charges for any order over $5,000. If the deduction is $875, what was the original charge?
3. The weight of the crate accounts for 8% of the total weight of a crate of oranges. If an empty crate weighs 3 lb, how much does a full crate of oranges weigh?
4. A company's income in 1992 was $820,650. The company's expenses for the year were $754,998. What percent of the company's income was profit? (*Hint:* $income − $expenses = $profit)
5. An automobile manufacturer employs 21,550 people. Of these, 68% are production workers. How many production workers does the company employ?

Use the data in the table to answer Exercises 6–9. Round each answer to the nearest whole number or nearest whole percent.

**AVERAGE WEEKLY CIRCULATION OF THE *LOS NOBLES COUNTY POST***

| | 1990 | 1991 | 1992 |
|---|---|---|---|
| San Marcos | 4,312 | 6,006 | 6,820 |
| Davisville | 2,490 | 4,315 | 3,264 |
| High Rocks | 853 | 1,020 | 1,115 |
| Clementon | 1,688 | 1,950 | 1,309 |
| Total | 9,343 | 13,291 | 12,508 |

6. In 1992, what percent of the *Post*'s readers lived in Davisville?
7. In 1991, the *Post* estimated that 12% of their Clementon readers were teenagers. How many teenagers in Clementon read the *Post*?
8. In 1990, the *Post* sold to 20% of the population of High Rocks. How many people lived in High Rocks in 1990?
9. In 1991, what percent of the *Post*'s circulation was new readers?
10. USING DATA Use the Data Index on pages 556–557 to find the sizes of the continents. Which continent has an area that is approximately 46% the area of Africa?

# 10-4 Similar Polygons

## EXPLORE

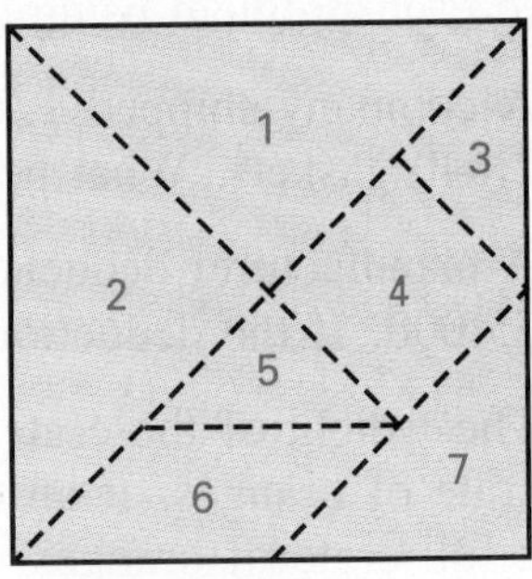


The ancient Chinese invented a puzzle called a tangram. It consists of seven pieces that can be combined to form different shapes.

Use tangram pieces or make your own tangram by tracing and cutting the diagram. Make a set of three squares using as many of the tangram pieces as necessary. Your squares should fit the following conditions.

- ► One square should have one half the area of the original tangram.
- ► One square should have one fourth the area of the original tangram.
- ► One square should have one eighth the area of the original tangram.

Which tangram piece did you not use?

## SKILLS DEVELOPMENT

**WRITING ABOUT MATH**

How would you explain the terms *similar* and *congruent*?

Write a paragraph comparing the properties of congruent and similar figures. Try to include several examples that will help demonstrate the differences.

**Similar figures** have the same shape, but not necessarily the same size. The symbol ~ means *is similar to.*

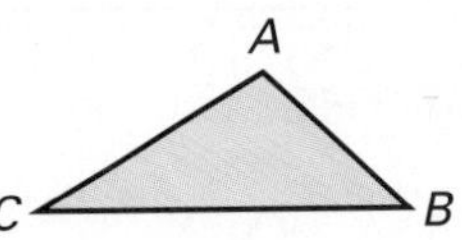


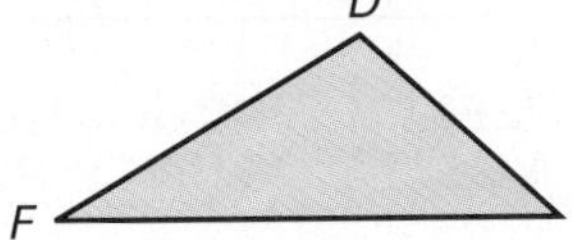


$\triangle ABC \sim \triangle DEF$

If two polygons are similar, you can match up their vertices so that all pairs of corresponding angles are congruent and all pairs of corresponding sides are in proportion.

$$\triangle ABC \sim \triangle DEF \qquad \angle A \cong \angle D \quad \angle B \cong \angle E \quad \angle C \cong \angle F$$

$$\frac{AB}{DE} = \frac{BC}{EF} = \frac{AC}{DF}$$

When you identify two polygons as similar, you must name their corresponding vertices in the same order.

- ► You can show that two triangles are similar if the corresponding angles are congruent *or* the corresponding sides are in proportion.

### Example 1

Is each pair of triangles similar?

**a.**

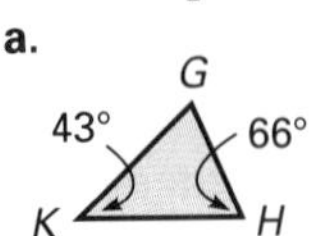


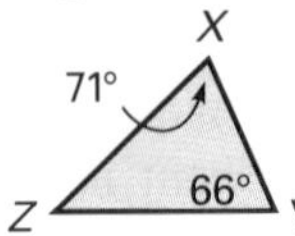


**b.**

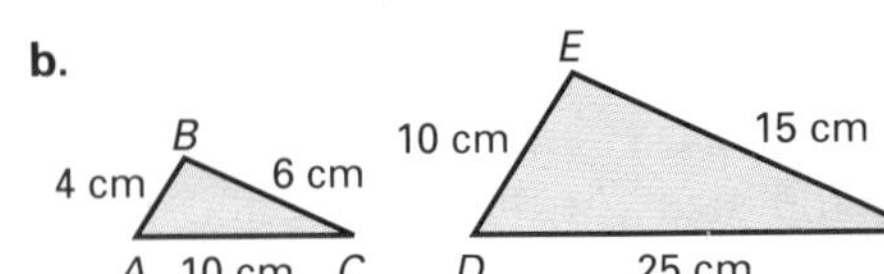

**Solution**

a. Find the missing measures.

$m\angle G = 180° - (66° + 43°) = 71°$  $\angle G$ is congruent to $\angle X$.

$m\angle Z = 180° - (66° + 71°) = 43°$  $\angle Z$ is congruent to $\angle K$.

Corresponding pairs of angles are congruent, so $\triangle GHK \sim \triangle XYZ$.

b. Find the ratios of lengths of corresponding pairs of sides.

$\frac{AB}{DE} = \frac{4}{10} = \frac{2}{5}$  $\frac{BC}{EF} = \frac{6}{15} = \frac{2}{5}$  $\frac{AC}{DF} = \frac{10}{25} = \frac{2}{5}$

Each ratio is equal to the same number. So, corresponding pairs of sides are in proportion and $\triangle ABC \sim \triangle DEF$. ◄

**CHECK UNDERSTANDING**

If $\triangle ABC \sim \triangle DEF$ and $\triangle DEF \sim \triangle GHK$, what is the relationship between $\triangle ABC$ and $\triangle GHK$?

To determine if two polygons that are *not* triangles are similar, you must compare *both* their angles and the lengths of their sides.

## Example 2

Is $ABCD \sim PQRS$?

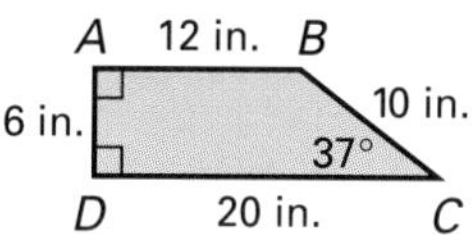


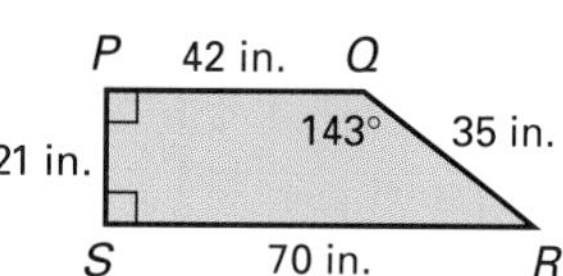


**Solution**

Find the missing measures of the angles.

$m\angle B = 360° - (90° + 90° + 37°) = 143°$

$m\angle R = 360° - (90° + 90° + 143°) = 37°$

**Remember, the sum of the angle measures of any quadrilateral is 360°.**

Find the ratios of lengths of corresponding pairs of sides.

$\frac{AB}{PQ} = \frac{12}{42} = \frac{2}{7}$  $\frac{BC}{QR} = \frac{10}{35} = \frac{2}{7}$  $\frac{CD}{RS} = \frac{20}{70} = \frac{2}{7}$  $\frac{DA}{SP} = \frac{6}{21} = \frac{2}{7}$

Pairs of corresponding angles are congruent and pairs of corresponding sides are in proportion, so $ABCD \sim PQRS$. ◄

**TALK IT OVER**

Why must you compare the measures of angles *and* sides of polygons that are not triangles to determine if the polygons are similar?

Sketch a pair of quadrilaterals that are not the same shape but whose corresponding pairs of angles are congruent.

Sketch a pair of quadrilaterals that are not the same shape but whose corresponding pairs of sides are congruent.

When you know that two polygons are similar, you can use what you know about *similar figures* to find unknown measures.

## Example 3

$\triangle MNO \sim \triangle XYZ$

What is the length of $\overline{XZ}$?

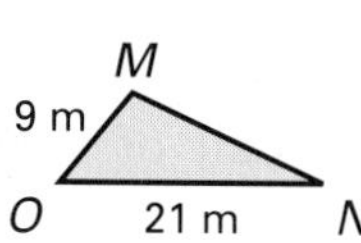


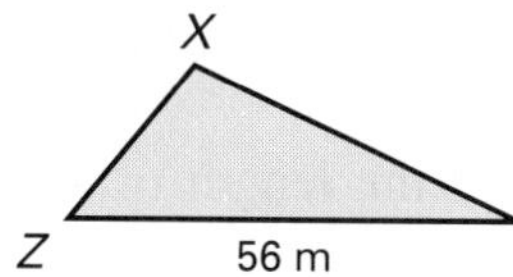


**Solution**

Since $\triangle MNO \sim \triangle XYZ$, you can write the proportion $\frac{NO}{YZ} = \frac{MO}{XZ}$.

Let $l$ = length of $\overline{XZ}$.

$$\frac{21}{56} = \frac{9}{l}$$

$$21 \times l = 56 \times 9$$ **Cross multiply.**

$$21l = 504$$

$$l = 24$$

So, the length of $\overline{XZ}$ is 24 m. ◄

## TRY THESE

Determine if each pair of polygons is similar. Write *yes* or *no*.

1.

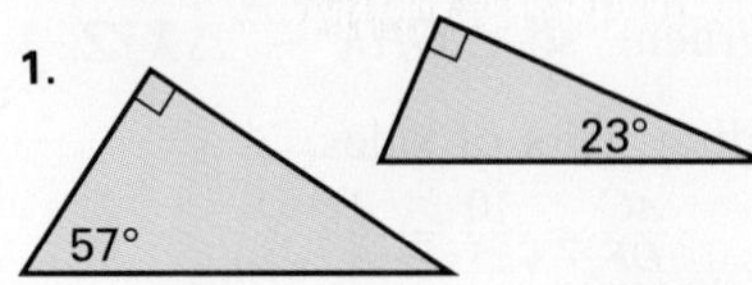


2.

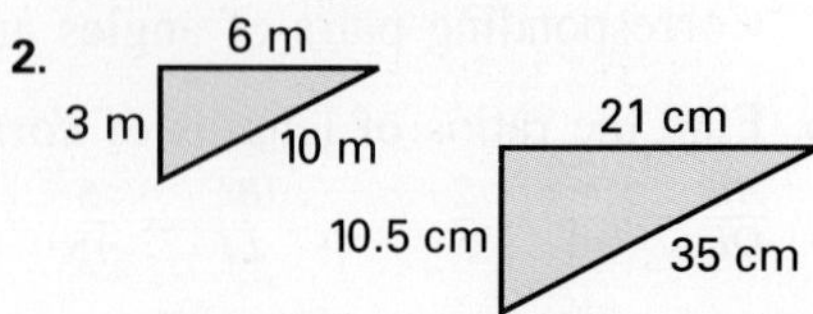


3. Both figures are parallelograms.

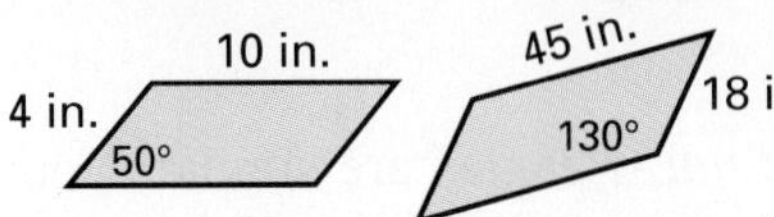


4.

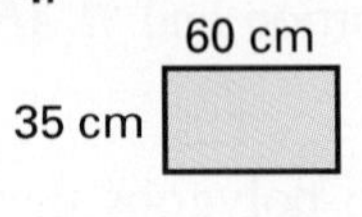


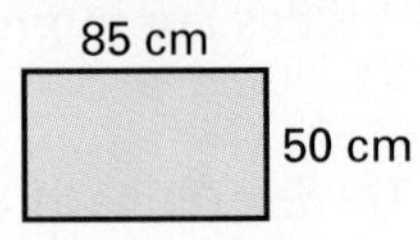


Find the length of $\overline{AB}$ in each pair of similar polygons.

5.

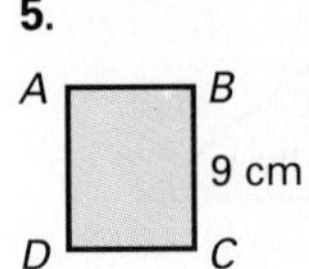


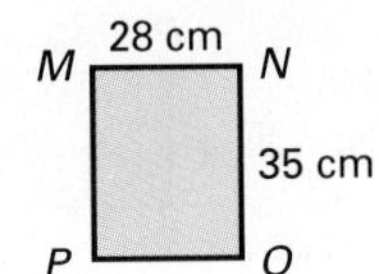


6.

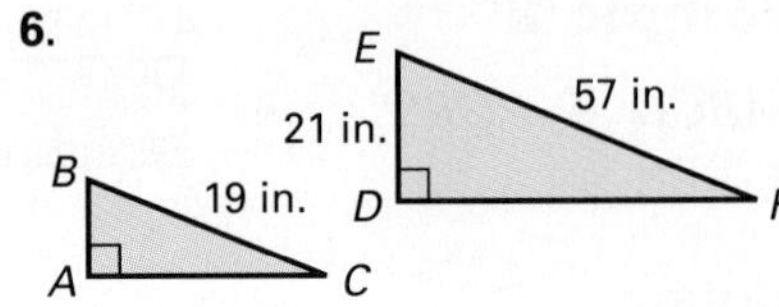


# EXERCISES

### PRACTICE/ SOLVE PROBLEMS

Determine if each pair of polygons is similar. Write *yes* or *no*.

1.

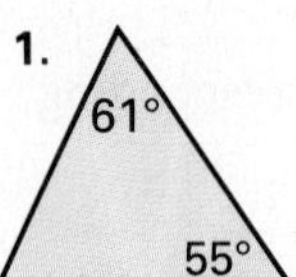


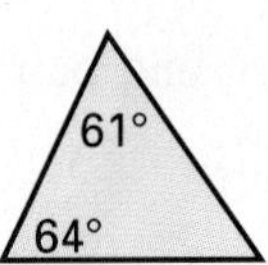


2.

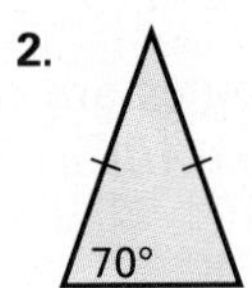


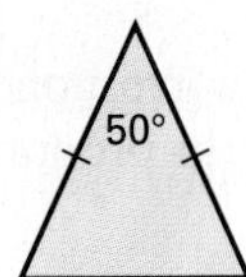


3.

55 cm

33 cm

20 cm

12 cm

4. Both figures are parallelograms.

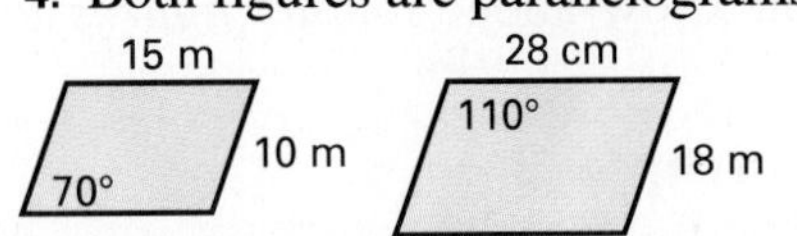


Find the length of $\overline{AB}$ in each pair of similar polygons.

5.

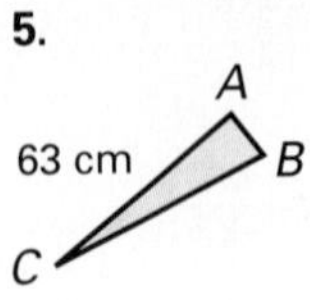


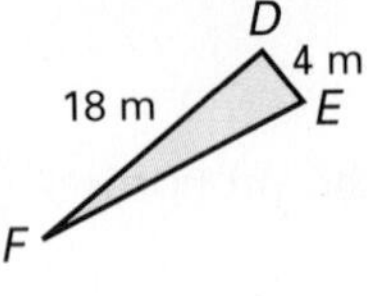


6.

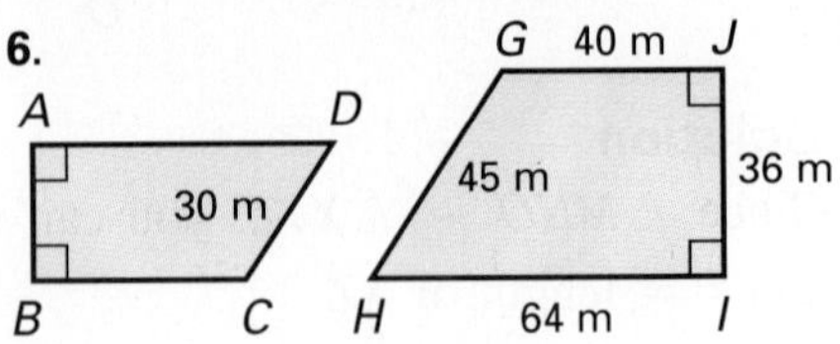


Find $m\angle A$ in each pair of similar polygons.

7.

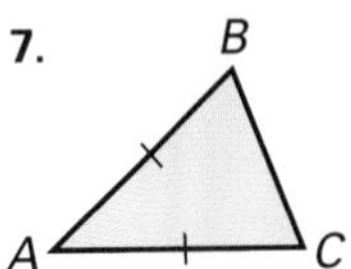


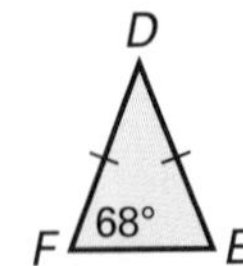


8.

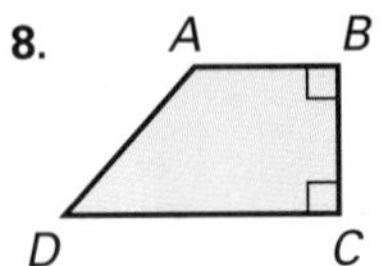


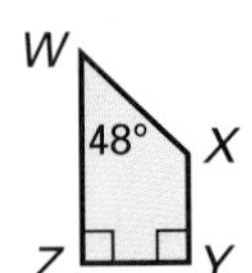

9. *ABCD* and *EFGH* are similar rectangles. The measure of $\overline{AB}$ is 12 in. The measure of $\overline{BC}$ is 16 in. If $\overline{EF}$ has a measure of 21 in., what is the length of $\overline{FG}$?

10. *ABC* and *DEF* are similar triangles. List three equivalent ratios for the sides of the triangles.

11. *ABC* and *DEF* are similar triangles. If $m\angle A = 32°$ and $m\angle F = 46°$, what are the measures of $\angle D$ and $\angle E$?

Identify each pair of similar polygons, using the correct order for the vertices.

12.

13.

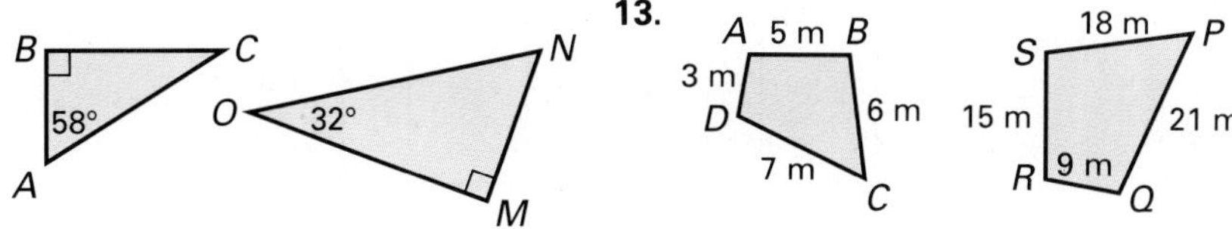


## EXTEND/ SOLVE PROBLEMS

Find the length of $\overline{AB}$ in each pair of similar polygons.

14.

D
30 m
F
72 m
E
B
12 m
A
C

15.

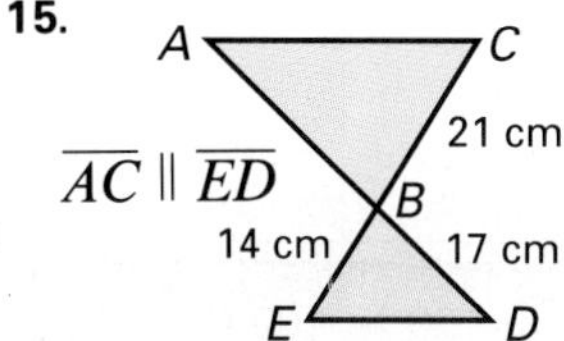


16.

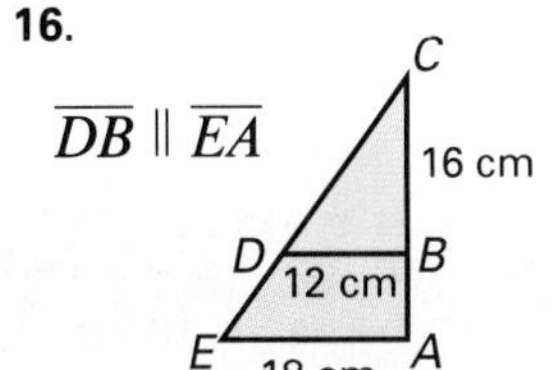

17. Rosa wishes to enlarge a rectangular photograph that is $4\frac{1}{2}$ in. long and $2\frac{3}{4}$ in. wide so that it will be $11\frac{1}{4}$ in. long. What will be the width of the enlargement?

18. *PQRS* and *WXYZ* are similar parallelograms. If $m\angle P = 72°$, what are the measures of $\angle Y$ and $\angle Z$?

19. *ABCD* and *MNOP* are similar rectangles. The measure of $\overline{AB}$ is 4 cm. The measure of $\overline{MN}$ is 15 cm. The area of *ABCD* is 48 cm². What is the area of *MNOP*?

20. *LMN* and *RST* are similar triangles and $m\angle M = 90°$. The length of $\overline{LM}$ is 8 cm and the length of $\overline{MN}$ is 15 cm. The length of $\overline{RT}$ is 34 cm. What is the area of $\triangle RST$?

## THINK CRITICALLY/ SOLVE PROBLEMS

Write *always, sometimes,* or *never* for each statement.

21. A rhombus is similar to another rhombus.

22. Congruent figures are similar.

23. Similar figures are congruent.

24. Circles are similar to each other.

25. Figures with different numbers of sides can be similar.

# 10-5 Indirect Measurement

## EXPLORE

Carla and Ray use walkie-talkies that have a range of 500 ft. Carla is at the corner of Ave. B and 4th St. Can Ray talk to her from the corner of Ave. F and 2nd St.? Can he talk to her from the corner of Ave. E and 1st St.? How did you find your answers?

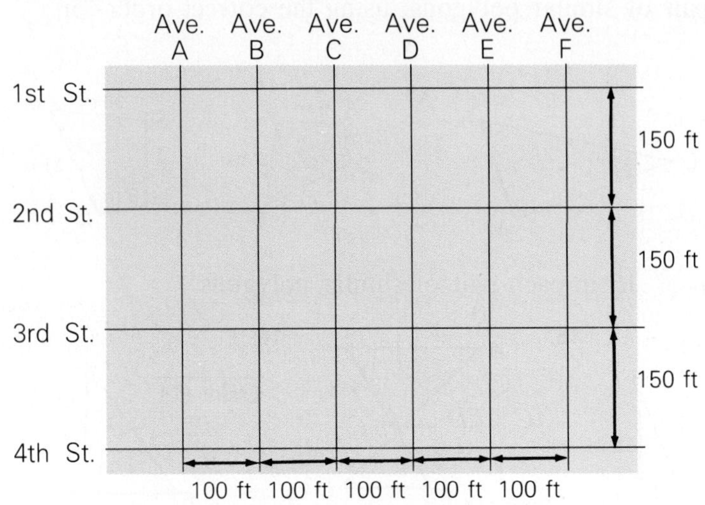


## SKILLS DEVELOPMENT

**PROBLEM SOLVING TIP**

When you write a proportion, you do not always have to convert measurements to the same units. But you must make sure that the corresponding units of the ratios are the same.

You can tell which units to use for the solution by comparing units used in the equation.

170(cm) × 12.0(m)
= *n*(?) × 300(cm)

Note that 170 and 300 are both in centimeters. Since 12.0 is in meters, the value of *n* must also be in meters.

In some situations, it is impossible or impractical to find a length by measuring the actual distance. An **indirect measurement** is one in which you take other measurements that allow you to calculate the required measurement.

### Example 1

A tree casts a shadow 12.0 m long. At the same time, a girl 170 cm tall casts a shadow 300 cm long. What is the height of the tree?

**Solution**

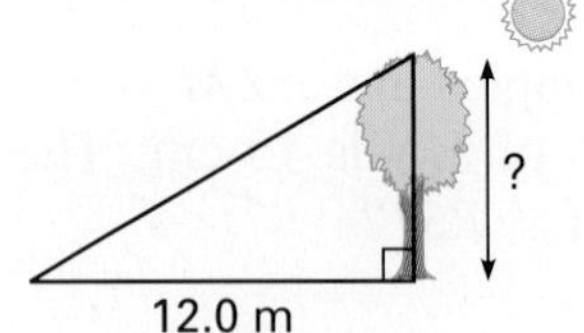


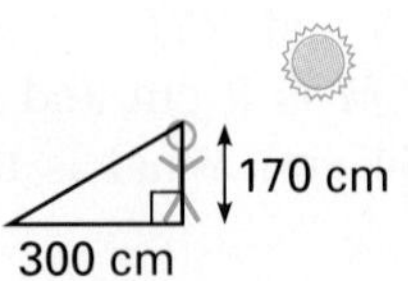


The angle at which the sun's rays meet the ground is the same in both right triangles. Since the triangles have corresponding pairs of congruent angles, they are similar. Write and solve a proportion.

height of girl → $\frac{170}{h} = \frac{300}{12.0}$ ← length of girl's shadow
height of tree → ← length of tree's shadow

$$170 \times 12.0 = 300 \times h \quad \text{Cross multiply.}$$
$$2{,}040 = 300h$$
$$6.8 = h$$

The height of the tree is 6.8 m. ◄

## Example 2

The wire supporting a radio transmitting tower touches the top of a 3.6 m pole as shown at the right. What is the height of the tower?

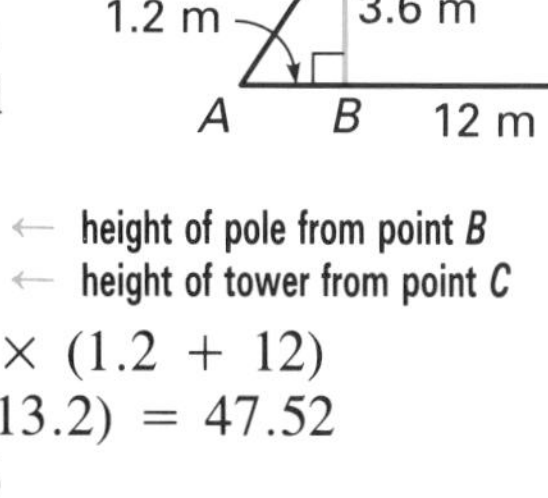


### Solution

Each right triangle shares a common angle, so they are similar. Write and solve a proportion.

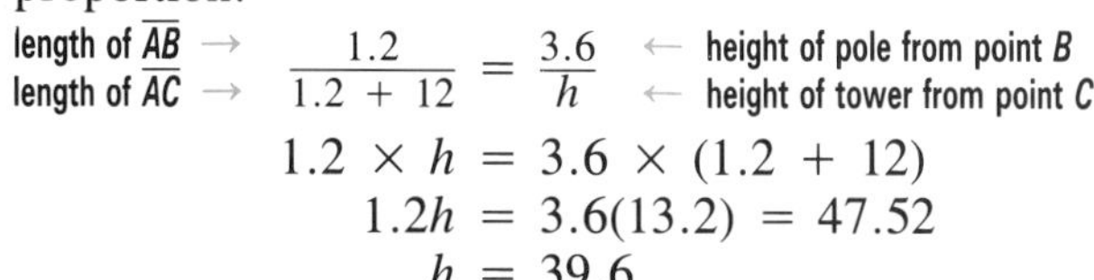

length of $\overline{AB}$ → $\dfrac{1.2}{1.2 + 12} = \dfrac{3.6}{h}$ ← height of pole from point $B$

length of $\overline{AC}$ → ← height of tower from point $C$

$$1.2 \times h = 3.6 \times (1.2 + 12)$$
$$1.2h = 3.6(13.2) = 47.52$$
$$h = 39.6$$

The height of the tower is 39.6 m. ◄

## Example 3

A surveyor took measurements along the bank of a river and made a sketch. Find the width, $w$, of the river.

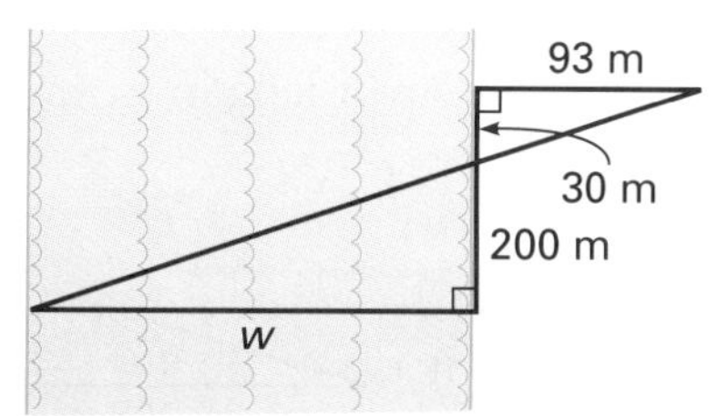


### Solution

The vertical angles formed at the common vertex of the two triangles are congruent and each triangle is a right triangle, so the two triangles are similar. Write and solve a proportion.

$\dfrac{93}{w} = \dfrac{30}{200}$ ← Be sure to match the corresponding sides correctly.

$$30 \times w = 93 \times 200$$
$$30w = 18{,}600$$
$$w = 620$$

The width of the river is 620 ft. ◄

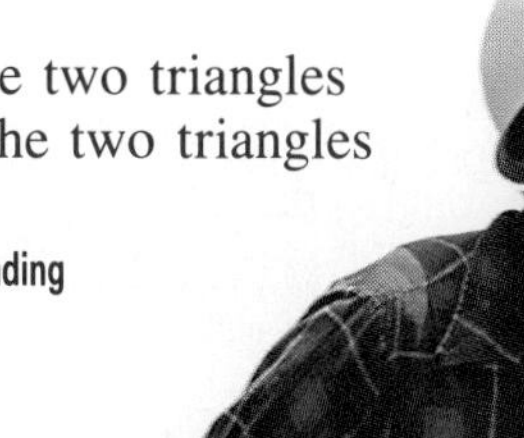

**CONNECTIONS**

Surveyors often take indirect measurements that can be used with the Pythagorean Theorem.

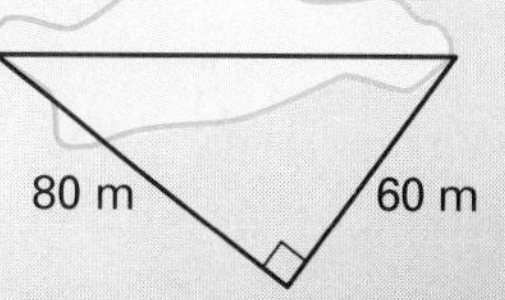


What is the length of the lake shown in the diagram?

To solve some of the exercises you will have to choose between using the Pythagorean Theorem and using similar triangles as a method of indirect measurement.

## TRY THESE

Find the unknown length, $x$, in each diagram.

1.

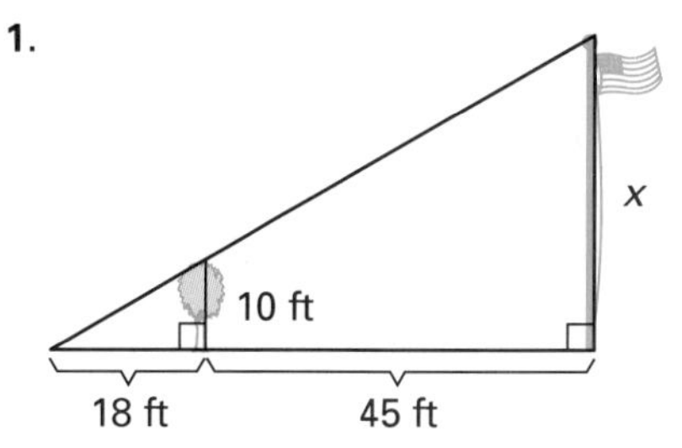


2.

150 cm

225 cm

x

18 m

**WRITING ABOUT MATH**

Write a paragraph describing the difference between direct measurement and indirect measurement. Include examples of each kind of measurement.

3.

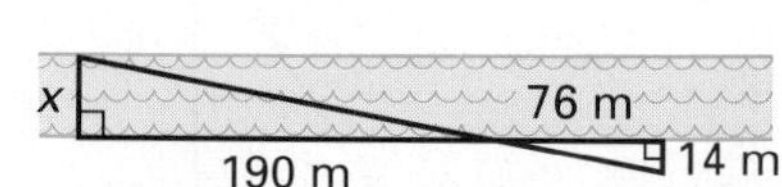


4.

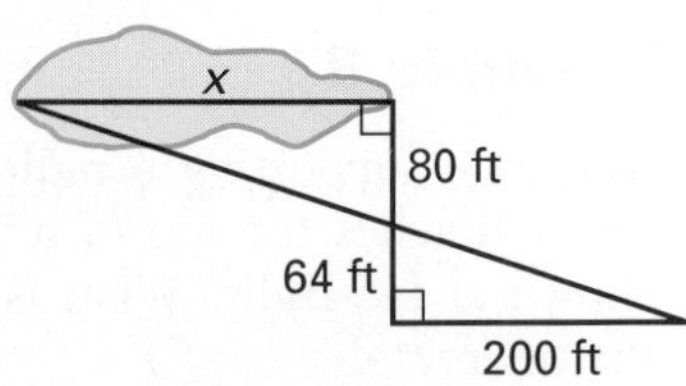


5. A flagpole casts a shadow 72 ft long. At the same time, a 5-ft-tall boy casts a shadow 12 ft long. What is the height of the flagpole?

6. A rocket traveling in a straight line reaches a height of 5.6 km above the ground after it has traveled a distance of 7 km. How many kilometers will the rocket have traveled when it reaches a height of 20 km above the ground?

# EXERCISES

## PRACTICE/ SOLVE PROBLEMS

Find the unknown length, $x$, in each diagram.

1.

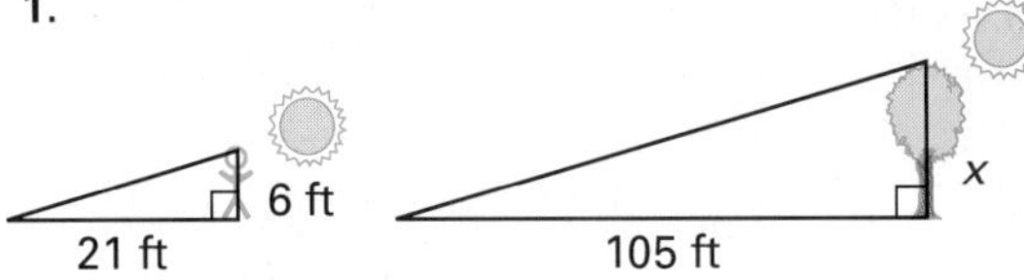


2.

x
9.5 m
20 m
80 m

3.

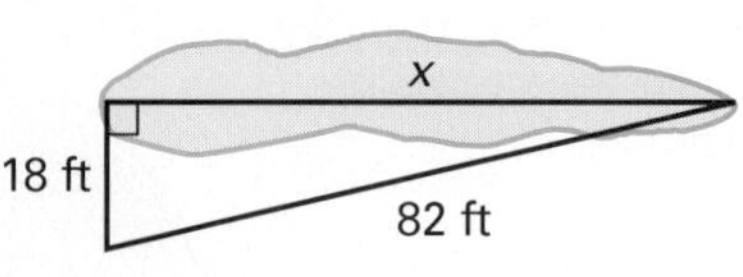


4.

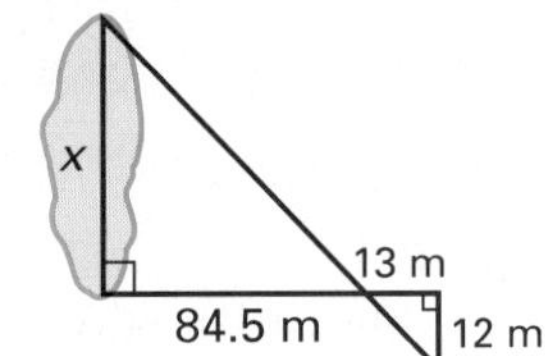


5.

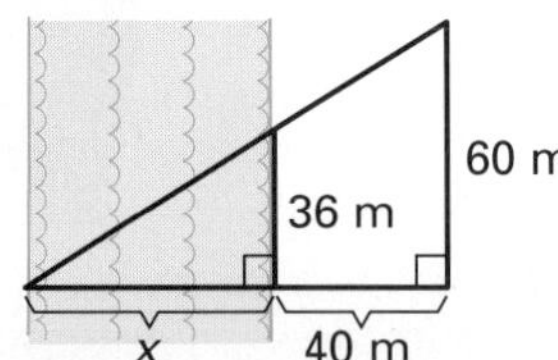


6.

x
84 ft
35 ft

7. A third-floor windowsill is 35 ft above the level of the ground. The foot of a ladder resting on the sill is 12 ft from the base of the wall. How long is the ladder?

8. A tree casts a shadow 26 m long. At the same time, a flagpole 25 m high casts a shadow 32.5 m long. How tall is the tree?

9. A 45-ft-tall tree casts a shadow 10 yd long. At the same time, Simon casts a shadow that is 3 ft long. How tall is Simon?

10. A flagpole is 96 ft tall and a tree next to it is 84 ft tall. What is the ratio of the lengths of their shadows?

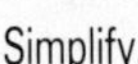

### MIXED REVIEW

Simplify.

1. $4n - 8n + 2n - 11n$
2. $5y^2 + 9z^2 + 3y^2 - 4z^2$
3. $-7ab + 10a + 3ab - 16ab$
4. $5d(1 + 2d + d^2)$
5. $8(m - 2n) - 3(m - 2n)$
6. $3(x + 2)(x + 4) + 5(x + 1)(x + 3)$
7. $(t + 6)^2 + (t + 2)(t - 2)$
8. $(z - 4)(z + 8) + (z - 9)(z + 3)$

Solve each system of equations.

9. $-7x - 8z = 8$
   $7x - 8z = 8$
10. $2a + 3b = 1$
    $3a - b = 18$

11. Sven is using wires to stabilize a flagpole that is 30 m tall. He uses three wires of equal length. Each is attached to a point one third of the way up the flagpole. If each wire makes an angle of 45° with the flagpole, what is the total length of wire, to the nearest meter, that Sven uses?

12. A 25-ft ladder is leaning against a wall. A rung that is 5 ft from the bottom of the ladder is 4 ft above the ground. How far above the ground does the top of the ladder touch the wall?

## EXTEND/ SOLVE PROBLEMS

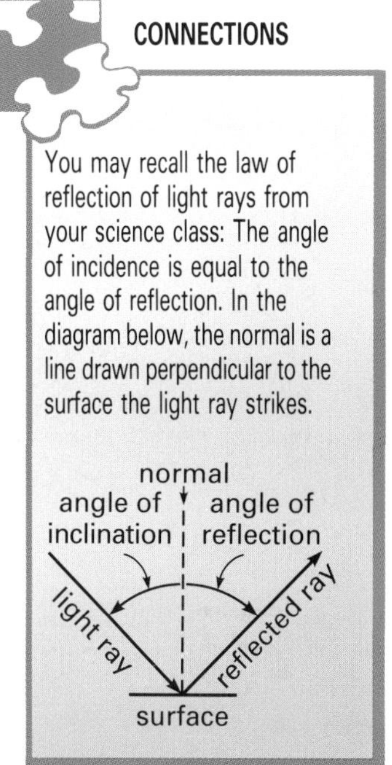


One method of indirect measurement involves placing a mirror on the ground and taking various measurements. Use the diagram at the right for Exercises 13–15.

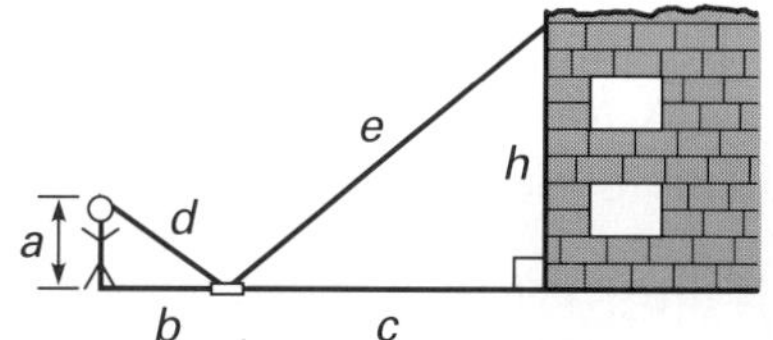


13. How do you know that the triangles shown are similar?

14. Which measurements would you take to find the height of the building using similar triangles?

15. Naima is 5 ft tall. To measure a flagpole that is 60 ft tall, she places a mirror on the ground 4 ft away from her. How far from the flagpole is Naima standing?

## THINK CRITICALLY/ SOLVE PROBLEMS

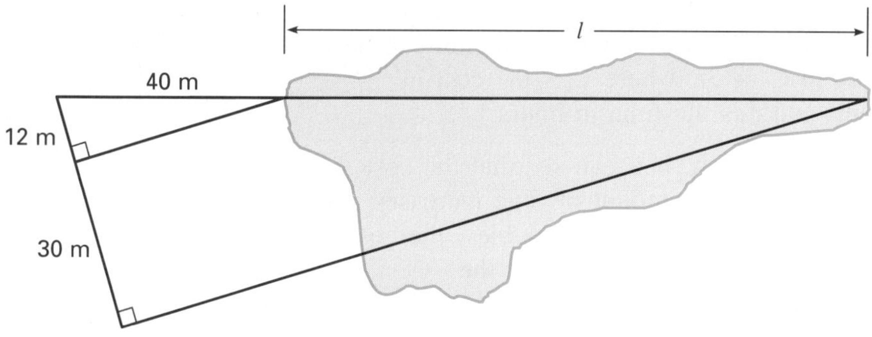


16. Write a proportion that could be used to find the length $l$ of the lake shown in the diagram above.

# 10-6 Direct Variation

## EXPLORE

Use the table at the right.

| Number of Muffins | Cost |
|---|---|
| 5 | \$3.45 |
| 10 | \$6.90 |
| 12 | \$8.28 |
| 20 | \$13.80 |
| 24 | \$16.56 |

a. Let $x$ = the number of muffins and let $y$ = cost. Find the ratio $\frac{y}{x}$ for each pair of entries. What did you discover?

b. As the number of muffins increases, what happens to the cost? When the number of muffins doubles, what happens to the cost?

c. Find the cost of 7 muffins using a proportion.

d. Write a rule for the cost as a function of the number of muffins. Then graph the function and describe the graph. What is its slope?

## SKILLS DEVELOPMENT

The table at the right shows the costs of operating a refrigerator.

| REFRIGERATOR OPERATING COSTS | | | | | |
|---|---|---|---|---|---|
| Cost (¢) | 3 | 6 | 9 | 12 | 15 |
| Time (h) | 1 | 2 | 3 | 4 | 5 |

You can see from the table that the cost of operating the refrigerator is a function of the amount of time the refrigerator is in operation. The rule for this function can be written $C = 3t$, where $C$ is the cost in cents and $t$ is the time in hours.

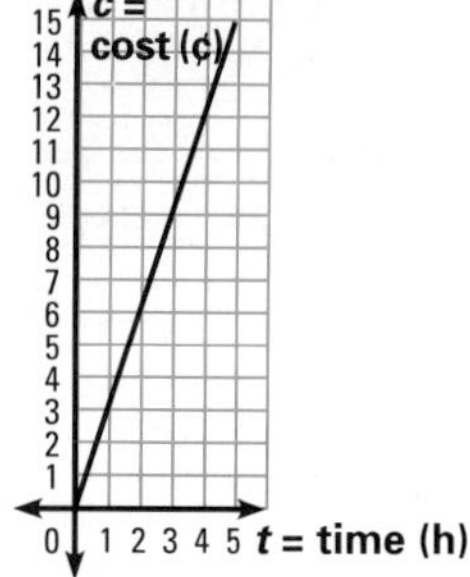


From the graph, you can see that the cost increases as the amount of time increases. When the amount of time doubles, the cost doubles. You can also see that the relationship between cost and time is linear.

Such a function is an example of a **direct variation.** A direct variation is a function that can be represented by a rule of the form $y = kx$, where $k$ is a nonzero constant and $x$ does not equal 0. You say, "$y$ varies directly as $x$." The constant $k$ is called the **constant of variation.** In the refrigerator example above, the constant of variation is 3.

### Example 1

The quantity $y$ varies directly as $x$. When $x = 36$, $y = 54$. Find $y$ when $x = 14$.

**Solution**

Substitute the known values of $x$ and $y$ in the equation $y = kx$.

$$y = kx$$
$$54 = k \times 36$$
$$\frac{54}{36} = k \quad \leftarrow \text{Solve for } k.$$
$$\frac{3}{2} = k$$
$$y = \frac{3}{2}x \quad \leftarrow \text{Write the function rule.}$$
$$y = \frac{3}{2}(14) = 21 \quad \leftarrow \text{Substitute } x = 14.$$

When $x = 14$, $y = 21$. ◄

**CONNECTIONS**

Direct variation is a special case of a linear function $y = mx + b$, where $b = 0$. When a direct variation is graphed, how does the value of the constant of variation relate to the slope of the graph? Through what point on the coordinate plane will the graph of a direct variation always pass?

Take another look at Example 1.

When $x = 36$, $y = 54$. $\quad \frac{y}{x} = \frac{54}{36} = \frac{3}{2}$

When $x = 14$, $y = 21$. $\quad \frac{y}{x} = \frac{21}{14} = \frac{3}{2}$

The constant of variation, $k$, was $\frac{3}{2}$, or the *ratio* of $y$ to $x$. Therefore, $k$ is sometimes called the **constant of proportionality.** So, you can also solve direct variation problems by writing a proportion.

## Example 2

The distance a spring stretches varies directly as the weight attached to it. A 12-kg weight stretches the spring 15 cm. How many centimeters will a 20-kg weight stretch the spring?

**CHECK UNDERSTANDING**

How could you solve Example 2 using $y = kx$?

### Solution

Write a proportion. $\quad \frac{15}{12} = \frac{y}{20}$ ← length stretched / ← weight attached

$$15 \times 20 = y \times 12 \quad \leftarrow \text{Cross multiply.}$$
$$300 = 12y$$
$$25 = y$$

A 20-kg weight will stretch the spring 25 cm. ◄

Here are data about the area and radius of a circle.

| AREA OF A CIRCLE | | | | | |
|---|---|---|---|---|---|
| **Radius** | 1 | 2 | 3 | 4 | 5 |
| **Area** | 3.14 | 12.56 | 28.26 | 50.24 | 78.5 |

**AREA OF CIRCLES**

Area: 0, 10, 20, 30, 40, 50, 60, 70, 80
Radius: 1 2 3 4 5

You can see that the area does *not* vary directly as the radius. For example, the area does not double when the radius doubles. Instead, the area varies directly as the *square* of the radius. From the graph, you can see that the relationship between the area and the radius is a quadratic function. This is an example of a **direct square variation,** a function that can be written in the form $y = kx^2$, where $k$ is a nonzero constant. You say, "$y$ varies directly as $x^2$," or "$y$ is directly proportional to $x^2$."

## Example 3

The distance a car rolls down a hill from a parked position varies directly as the square of the amount of time it is rolling. If the car rolls 65 m downhill in 4 s, how far will it roll in 10 s?

**CALCULATOR**

When you are given a table of ordered pairs, you can use the memory keys [M+] and [MR] on a calculator to decide if the function is an example of direct variation.

For example, given

| x | y |
|---|---|
| 1 | 2.5 |
| 2 | 5 |
| 3 | 7.5 |
| 4 | 10 |

Find $k = \frac{y}{x}$ for the first ordered pair and store the value.

2.5 [÷] 1 [=] [M+] 2.5

Then check each ordered pair by multiplying by the next x-value.

2 [×] [MR] [=] 5

If each ordered pair checks, the function represents a direct variation.

### Solution

Substitute known values for $x$ and $y$ in the equation $y = kx^2$.

$$y = kx^2$$
$$65 = k(4^2)$$
$$65 = 16k$$
$$\frac{65}{16} = k$$
$$4.0625 = k$$
$$y = 4.0625x^2 \quad \leftarrow \text{Write the function rule.}$$
$$y = 4.0625 \times 10^2 = 406.25 \quad \leftarrow \text{Substitute 10 for } x.$$

The car will roll 406.25 m in 10 s. ◄

## TRY THESE

1. Assume that $y$ varies directly as $x$. When $x = 24$, $y = 16$. Find $y$ when $x = 42$.
2. Assume that $y$ varies directly as $x$. When $x = 9$, $y = 33$. Find $y$ when $x = 15$.
3. Assume that $y$ varies directly as $x^2$. When $x = 3$, $y = 63$. Find $y$ when $x = 5$.
4. The annual simple interest on a loan varies directly as the amount of the loan. The interest on a loan of \$400 is \$56. Find the interest on a loan of \$650.

# EXERCISES

### PRACTICE/ SOLVE PROBLEMS

1. Assume that $y$ varies directly as $x$. When $x = 9$, $y = 30$. Find $y$ when $x = 15$.
2. Assume that $y$ varies directly as $x$. When $x = 21$, $y = 12$. Find $y$ when $x = 35$.
3. Assume that $y$ varies directly as $x^2$. When $x = 6$, $y = 198$. Find $y$ when $x = 2$.
4. Assume that $y$ varies directly as $x^2$. When $x = 10$, $y = 10$. Find $y$ when $x = 7$.
5. A length of 50 ft is equivalent to a length of 1,524 cm. How many centimeters are equivalent to 125 ft?

## EXTEND/ SOLVE PROBLEMS

Find the value of the constant of variation for each.

6. $y$ = days, $x$ = weeks
7. $y$ = days, $x$ = hours
8. $y$ = yards, $x$ = miles
9. $y$ = in.$^2$, $x$ = ft$^2$
10. The number of bacteria in a culture varies directly as the square of the time the bacteria have been growing. A culture growing for 20 min has 800 bacteria. How long does it take to grow 1,800 bacteria?
11. The distance an object falls from a given height varies directly as the square of the time the object falls. A ball falls 312 m in 8 s. How far did it fall during the first 4 s?
12. An acre is a measure of area equivalent to 4,840 yd$^2$. A chain is a special measure of length equal to 22 yd. How many square chains are equivalent to one acre?
13. For every 12 lb of apples she picks, Martha gets paid $1.80. Write a function rule relating her earnings in dollars, $E$, to the number of pounds she picks, $p$. Draw a graph showing the relationship, and use it to find how many pounds of apples Martha must pick to earn $6.

**USING DATA** The exchange rate between two currencies is a direct variation. Refer to the Data Index on pages 556–557 to find the table of exchange rates.

14. If you traded 1,000 Singapore dollars on Thursday, how many United States dollars would you get?
15. How many more Austrian schillings would $100(U.S.) purchase on Thursday than on Wednesday?
16. On Wednesday, how many Indian rupees could you buy with 10 German marks? Give your answer to the nearest whole number.

## THINK CRITICALLY/ SOLVE PROBLEMS

Assume that $y$ varies directly as $x^2$.

17. How is $y$ affected when $x$ is doubled? When $x$ is tripled?
18. How is $y$ affected when $x$ is multiplied by $k$?

Determine whether each pair of quantities is an example of a *direct variation*, a *direct square variation*, or *neither*.

19. the perimeter of a square and the measure of its side
20. the area of a square and the measure of its side
21. the volume of a cube and the measure of its side
22. the surface area of a cube and the measure of its side

# 10-7 Inverse Variation

## EXPLORE

a. Find five pairs of whole numbers whose product is 80. Identify the smaller number of each pair as $m$ and the larger number as $n$.

b. Copy and complete the chart. Begin with the least value of $m$. Write pairs in order of increasing values of $m$.

| FACTORS OF 80 | | | | | |
|---|---|---|---|---|---|
| $m$ | ■ | ■ | ■ | ■ | ■ |
| $n$ | ■ | ■ | ■ | ■ | ■ |

c. As $m$ increases, does $n$ increase or decrease? Does $m$ vary directly as $n$? Explain.

## SKILLS DEVELOPMENT

| TIME TAKEN TO TRAVEL 100 YD | | | | | | | | | |
|---|---|---|---|---|---|---|---|---|---|
| Time (s) | 2 | 4 | 5 | 8 | 10 | 12.5 | 20 | 25 | 50 |
| Rate (yd/s) | 50 | 25 | 20 | 12.5 | 10 | 8 | 5 | 4 | 2 |

The table and the graph show the time taken to travel 100 yd at various speeds. You can see from the table that the time taken is a function of the speed. If $t$ represents the time in seconds and $r$ represents the rate in yards per second, the function can be represented by the rule $t = \frac{100}{r}$.

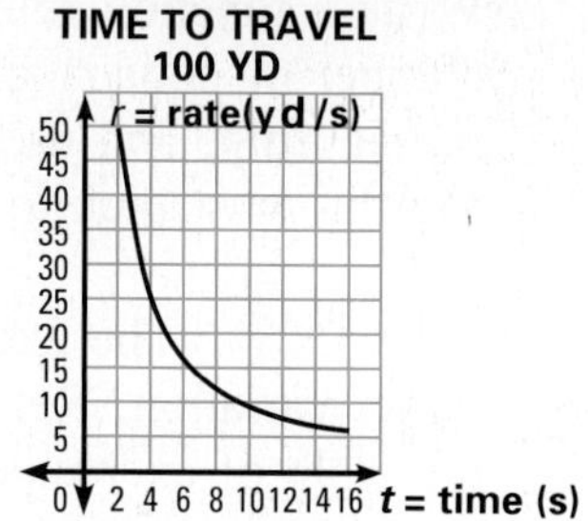


From the graph, you can see that the amount of time decreases as the rate of speed increases. The function is an example of an **inverse variation.**

An inverse variation is a function that can be represented by an equation of the form $y = \frac{k}{x}$, where $k$ is a nonzero constant and $x$ does not equal 0. You say, "$y$ varies inversely as $x$," or "$y$ is inversely proportional to $x$." So, the time taken to travel 100 yd varies inversely as the rate of speed. The constant of variation, $k$, is 100.

### Example 1

Suppose $y$ varies inversely as $x$. When $x = 80$, $y = 2$. Find $y$ when $x = 16$.

**Solution**

Substitute the known values of $x$ and $y$ in the equation $y = \frac{k}{x}$.

$$y = \frac{k}{x}$$

$$2 = \frac{k}{80}$$

$$160 = k$$

$$y = \frac{160}{x} \quad \leftarrow \text{Write the function rule.}$$

$$y = \frac{160}{16} = 10 \quad \leftarrow \text{Substitute 16 for } x.$$

When $x = 16$, $y = 10$. ◄

Take another look at Example 1.

When $x = 80$, $y = 2$. $\quad$ When $x = 16$, $y = 10$.

$xy = 160$ $\quad$ $xy = 160$

In Example 1, the constant of variation was 160. Notice that, in each case, the product $xy$ is equal to the constant of variation. So, if $y$ varies inversely as $x$, the relationship can be expressed as either $y = \frac{k}{x}$ or $xy = k$.

Suppose that $y$ varies inversely as $x$.
When $x = n$, $y = m$.
Find $y$ when $x = m$.

## Example 2

If the tension in a guitar string is constant, the frequency of a note varies inversely as the length of the string. When the length of a string is 75 cm, the frequency is 512 Hz (hertz). Find the length of the string that produces a note as a frequency of 640 Hz.

### Solution

Substitute known values of $x$ and $y$ in the equation $xy = k$.

$$xy = k$$
$$(75)512 = k$$
$$38{,}400 = k$$
$$xy = 38{,}400 \quad \leftarrow \text{Write the function rule.}$$
$$x(640) = 38{,}400 \quad \leftarrow \text{Substitute 640 for } y.$$
$$x = \frac{38{,}400}{640} = 60$$

The length of the string is 60 cm. ◄

Suppose that you are standing a certain distance from a light source. If you move twice as far away, the intensity of the light is *one fourth* as great. If you move three times as far away, the intensity is *one ninth* as great. This is an example of **inverse square variation.**

An inverse square variation is a function that can be described in the form $y = \frac{k}{x^2}$, or $x^2y = k$, where $k$ is a nonzero constant and $x$ does not equal 0. You say, "$y$ varies inversely as $x^2$," or "$y$ is inversely proportional to $x^2$."

## Example 3

The intensity of light varies inversely as the square of the distance from the source. At a distance of 8 ft from the source, a light meter measures an intensity of 24 units. How many units will the meter measure at a distance of 32 ft from the source?

### Solution

Substitute known values of $x$ and $y$ in the equation $x^2y = k$.

$$x^2y = k$$
$$(8^2)24 = k$$
$$1{,}536 = k$$
$$x^2y = 1{,}536 \quad \leftarrow \text{Write the function rule.}$$
$$(32^2)y = 1{,}536 \quad \leftarrow \text{Substitute 32 for } x.$$
$$1{,}024y = 1{,}536$$
$$y = 1.5$$

The meter will measure 1.5 units. ◄

## TRY THESE

1. Assume that $y$ varies inversely as $x$. When $x = 1$, $y = 1$. Find $y$ when $x = 5$.
2. Assume that $y$ varies inversely as $x$. When $x = 3$, $y = 8$. Find $y$ when $x = 4$.
3. Assume that $y$ varies inversely as $x^2$. When $x = 9$, $y = 12$. Find $y$ when $x = 6$.
4. The frequency of a note varies inversely as the length of a guitar string. When a string is 45 cm long, it produces a frequency of 826 Hz. What frequency will a string produce when it is 70 cm long?
5. The force of attraction between two magnets varies inversely as the square of the distance between them. When the magnets are 3 cm apart, the force is 49 newtons. How great will the force be when the magnets are 21 cm apart?

# EXERCISES

### PRACTICE/ SOLVE PROBLEMS

1. Assume that $y$ varies inversely as $x$. When $x = 10$, $y = 64$. Find $y$ when $x = 2$.
2. Assume that $y$ varies inversely as $x$. When $x = 2.5$, $y = 6$. Find $y$ when $x = 20$.
3. Assume that $y$ varies inversely as $x^2$. When $x = 10$, $y = 1$. Find $y$ when $x = 4$.
4. Assume that $y$ varies inversely as $x^2$. When $x = 4$, $y = 18$. Find $y$ when $x = 3$.
5. The time taken to fill a water tank varies inversely as the square of the diameter of the pipe used to fill it. A pipe with a diameter of 2 cm takes 10 min to fill the tank. How long would it take to fill the tank with a pipe having a diameter of 5 cm?
6. The amount paid by each member of a group chartering a bus varies inversely as the number of people in the group. When there are 30 people in the group, the cost per person is $20. What is the cost per person when there are 25 people in the group?
7. The temperature of the air varies inversely as the height above sea level. At a height of 200 m above sea level, the temperature is 15°C. What is the temperature at a height of 500 m above sea level?

Identify each relationship as a *direct variation* or an *inverse variation.*

**8.** the area of a circle and its radius

**9.** the speed of a vehicle and the time it takes to travel one mile

**10.** the capacity of a paint can and the number of cans needed to paint a wall

**11.** the weight of an object and the force needed to lift it

**12.** the loudness of a sound and the distance from the sound source

**13.** the length of a rope and its weight

## EXTEND/ SOLVE PROBLEMS

**14.** A triangle has an area of 48 $cm^2$. Write an equation that describes how the length of the base varies in relation to the height of the triangle.

**15.** A cone has a volume of 44 $cm^3$. Write an equation that describes how the height varies in relation to the radius of the cone. Use $\pi = \frac{22}{7}$.

When $x = 4$, $y = 20$. Find $y$ when $x = 100$ for each variation.

**16.** $y$ varies directly as $x$

**17.** $y$ varies directly as $x^2$

**18.** $y$ varies inversely as $x$

**19.** $y$ varies inversely as $x^2$

## THINK CRITICALLY/ SOLVE PROBLEMS

**20.** The volume of a gas varies inversely as the pressure applied to it. Air pressure at sea level is 1 atmosphere. If a balloon rises to a point where the air pressure is 0.8 atmosphere, by what percent will its volume increase?

**21.** The amount of heat varies inversely as the square of the distance from a heat source. What is the ratio of the heat at a spot 2 m from a heater to the heat at a spot 5 m from the heater?

**22.** Suppose that $y$ varies directly as $x$ and $x$ varies inversely as $z$. What is the relationship between $y$ and $z$?

**23.** Suppose that $y$ varies inversely as $x$ and $x$ varies inversely as $z^2$. What is the relationship between $y$ and $z$?

# 10-8

READ
PLAN
SOLVE
ANSWER
CHECK

# Problem Solving/ Decision Making:
## CHOOSE A STRATEGY

**PROBLEM SOLVING TIP**

Here is a checklist of the problem solving strategies you have studied so far in this book.

Use the 5-step plan
Make an algebraic model
Make a circle graph
Make an organized list
Use an equation
Choose a formula
Work backwards
Use a graph
Use a system of equations

In this book you have studied a variety of problem solving strategies. Experience in applying these strategies will help you decide which will be the most appropriate for solving a particular problem. Sometimes only one strategy will work. In other cases, any one of several strategies will lead to a solution. There may be times when you will want to use two different approaches to a problem in order to be sure that the solution you found is correct. For certain problems, you will need to use more than one strategy.

## PROBLEMS

Solve. Name the strategy you used to solve each problem.

1. Sharla is making arrangements for her company's sales meeting. The special per-person room rate at Sunny Pines Resort is \$56.50 per night. There are 47 people who will attend the meeting. The budget for rooms is \$2,800. Can Sharla make reservations at Sunny Pines?

2. Jolanda and Kerry stand next to each other. Jolanda's shadow is 17 ft long and Kerry's shadow is 14 ft long. How tall is Jolanda if she is 1 ft taller than Kerry?

3. The train to Denton travels at an average speed of 45 mi/h, and the bus to Denton travels at an average speed of 30 mi/h. They both leave Avon at the same time. If the train arrives two hours earlier than the bus, how far is it from Avon to Denton?

4. A circle graph shows the three categories of employees of a corporation. The central angles of the sectors of a circle graph are in the ratio 1:2:3 in order of administrative workers to salespeople to production workers.
   a. What are the central angles for each sector?
   b. If there is a total of 648 employees, how many of each category are there?

5. For every 5 sacks of barley a farmer buys, she buys 6 sacks of bran. For every 3 sacks of oats she buys, she buys 2 sacks of barley. For every sack of corn she buys, she buys 3 sacks of bran. One day she phones the grain store and asks for a delivery of 123 sacks. How many sacks of each grain should she receive?

6. The distance a truck needs to reach a complete stop varies directly as the square of its speed. If the truck is traveling at 20 mi/h, it needs a distance of 48 ft. How many feet will it need to stop if it is traveling at a speed of 30 mi/h?

7. Of the buses a company makes, 20% are school buses. The company sells 75% of the school buses in Texas. If the company sold 165 school buses in Texas last year, what was the total number of buses the company made?

8. The owner of a clothing store ordered 6 shirts and 8 pairs of jeans for $280. One week later, at the same prices, the owner ordered 9 shirts and 6 pairs of jeans for $264. Find the price of a shirt and the price of a pair of jeans.

9. A clock loses 5 minutes every hour. It was set to the correct time at 4:50 A.M. What time will it be when the clock next displays the correct time?

10. The same positive integer is added to both the denominator and the numerator of the fraction $\frac{1}{4}$. The value of the fraction that is formed is $\frac{2}{3}$. What integer was added?

11. Apex Car Rentals charges $28.40 per day plus $0.26 per mile. Happy Car Rentals charges $22.00 per day plus $0.34 per mile. Carlos is planning to rent a car for 2 days. Under what circumstances would it be better for him to rent from Apex? When would it be better to rent from Happy?

12. The loudness of the sound you hear from a radio varies inversely as the square of the distance you are from the radio. If you are on the couch, the sound from the radio is 16 times as loud as it is when you are in the hammock. How much farther from the radio is the hammock than the couch?

13. Complete each equation.

a. $\frac{1}{1 \times 2} = \frac{\blacksquare}{\blacksquare}$

b. $\frac{1}{1 \times 2} + \frac{1}{2 \times 3} = \frac{\blacksquare}{\blacksquare}$

c. $\frac{1}{1 \times 2} + \frac{1}{2 \times 3} + \frac{1}{3 \times 4} = \frac{\blacksquare}{\blacksquare}$

d. $\frac{1}{1 \times 2} + \frac{1}{2 \times 3} + \frac{1}{3 \times 4} + \ldots + \frac{1}{a \times b} = \frac{\blacksquare}{\blacksquare}$

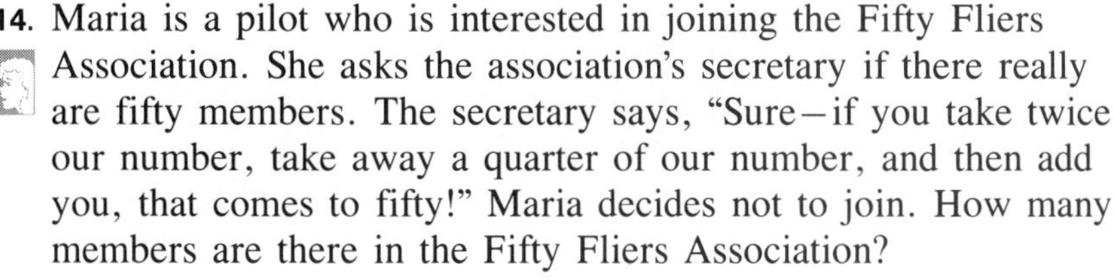

14. Maria is a pilot who is interested in joining the Fifty Fliers Association. She asks the association's secretary if there really are fifty members. The secretary says, "Sure – if you take twice our number, take away a quarter of our number, and then add you, that comes to fifty!" Maria decides not to join. How many members are there in the Fifty Fliers Association?

# CHAPTER 10 • REVIEW

Choose the word from the list that completes each statement.

a. unit
b. inverse
c. similar
d. direct
e. equivalent

1. A function in which one variable increases as the other decreases shows ___?___ variation.
2. A proportion shows that two ratios are ___?___.
3. A ratio comparing a number to a single unit is a ___?___ rate.
4. In ___?___ variation, one variable increases as the other variable increases.
5. When two figures are ___?___, corresponding pairs of angles are congruent.

## SECTION 10–1 RATIOS AND RATES

(pages 392–395)

► A **ratio** is a quotient of two numbers that is used to compare one number with the other.
► A **rate** is a ratio that compares two different kinds of quantities.
► A **unit rate** is a ratio whose second term is one unit.

6. The perimeter of a rectangle is 112 m and the ratio of its sides is 4:3. What are the lengths of its sides?
7. Which car is going faster, one that travels 46 mi in 50 min or one that travels 14 mi in 15 min?

## SECTION 10–2 PROPORTIONS

(pages 396–399)

► A **proportion** is an equation stating that two ratios are equivalent.
► The **cross-products** of the terms of a proportion are equal.

8. A stack of 144 bricks has a mass of 244.8 kg. What would be the mass of a stack of 372 bricks?

## SECTION 10–4 SIMILAR POLYGONS

(pages 402–405)

► Two polygons are **similar** if corresponding pairs of angles are congruent and corresponding pairs of sides are in proportion.

Determine whether or not the pairs of polygons are similar.

9.

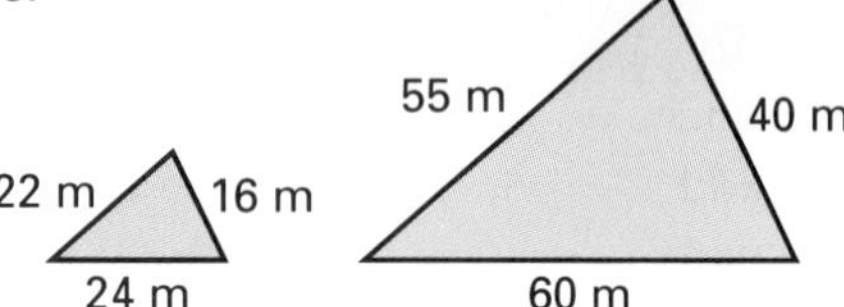


10.

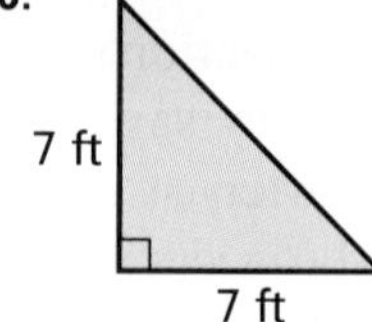

## SECTION 10–5 INDIRECT MEASUREMENT (pages 406–409)

► Use similar triangles to find a measurement that cannot be found directly.

Find the width, $w$, of each river.

**11.**

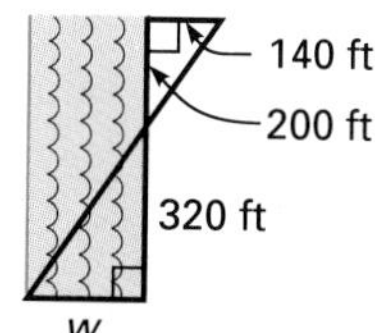


**12.**

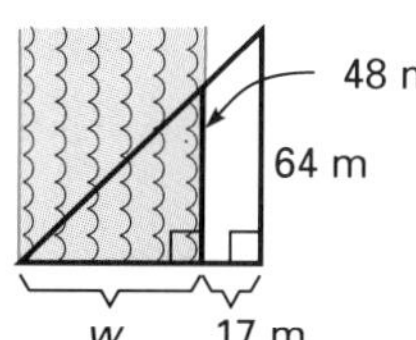


## SECTION 10–6 DIRECT VARIATION (pages 410–413)

► A **direct variation** is a function that can be described in the form $y = kx$ or $\frac{y}{x} = k$, where $k$ is a nonzero constant and $x \neq 0$.

► A **direct square variation** can be described $y = kx^2$ or $\frac{y}{x^2} = k$.

**13.** The distance a spring stretches varies directly as the weight attached to it. A weight of 4.2 kg stretches the spring 21 cm. How many centimeters will a weight of 3 kg stretch the spring?

**14.** The distance an object rolls down a slope varies directly as the square of the time it rolls. A ball rolls 80 ft down the slope in 5 s. How far will it roll in 10 s?

## SECTION 10–7 INVERSE VARIATION (pages 414–417)

► An **inverse variation** is a function that can be described in the form $y = \frac{k}{x}$ or $xy = k$, where $k$ is a nonzero constant and $x \neq 0$.

► An **inverse square variation** can be described $y = \frac{k}{x^2}$ or $x^2y = k$.

**15.** The time taken to dig a ditch varies inversely as the number of workers. If it takes 15 workers 8 h, how long would it take 16 workers?

**16.** The height of a 1-L cylindrical can varies inversely as the square of its radius. When the radius measures 5 cm, the height measures 12.7 cm. Find the height to the nearest tenth when the radius measures 4 cm.

## SECTIONS 10–3 AND 10–8 PROBLEM SOLVING (pages 400–401, 418–419)

► Percent problems can be solved by writing and solving a proportion.

► Before attempting to solve a problem, determine the strategy you will use.

**17.** A company spends four times as much on production costs as it does on advertising. Last year, it spent \$936 on commercials, which was 18% of its total advertising budget. What were the company's production costs last year?

***USING DATA*** Use the table on page 390 to solve.

**18.** About what percent of the population worked at agricultural jobs in 1910?

# CHAPTER 10 ● TEST

Write each ratio in lowest terms.

1. 44 cm:120 cm
2. $\frac{108 \text{ lb}}{45 \text{ lb}}$
3. 35 kg:119 kg
4. 14 m:28 cm

Find the better buy.

5. 15 padlocks for $49.05 or 12 padlocks for $40.08
6. $3\frac{1}{2}$ lb of fish for $16.38 or $5\frac{1}{2}$ lb of fish for $26.40

Solve each proportion.

7. $\frac{56}{n} = \frac{49}{21}$
8. $\frac{124}{24} = \frac{n}{6}$
9. $\frac{n}{18} = \frac{10}{144}$
10. $\frac{8.4}{3.6} = \frac{n}{3.9}$
11. The cost of the electricity to run a machine for 24 hours is $2.28. How much would it cost to run the machine for 14 hours?
12. A tractor is 5.2 m long. In a scale drawing, the tractor is 13 cm long. What is the scale of the drawing?
13. A manufacturer calculated that 38% of the total cost of producing an item was for materials. If the manufacturer spent $55.48 for materials, what was the total cost of producing the item?
14. A crate of apples weighs 21.4 lb. The crate itself accounts for 15% of the weight. What is the weight of the apples?
15. A corporation employs 2,864 women and 2,074 men. To the nearest whole number, what percent of the employees are women?

Find the unknown measure in each pair of similar polygons.

16.

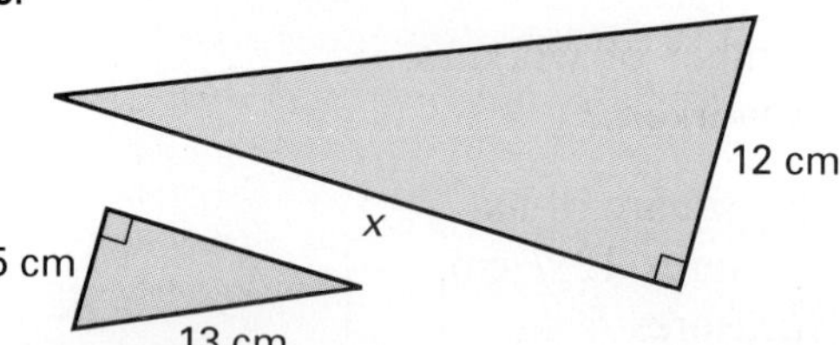


17. Both figures are parallelograms.

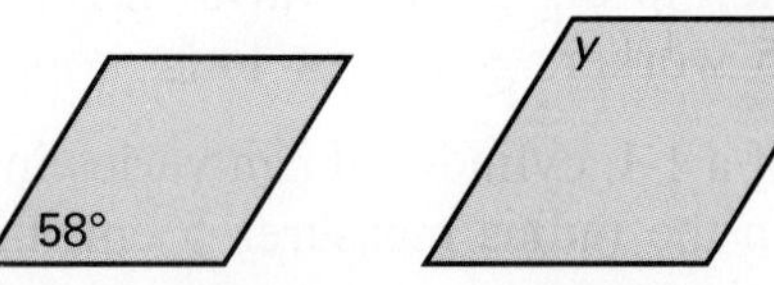


18. A flagpole casts a shadow 49.2 m long, while Hans, who is 1.8 m tall, casts a shadow 2.4 m long. How tall is the flagpole?
19. The distance a truck needs to reach a full stop varies directly as the square of its speed. From a speed of 40 mi/h, the truck needs a distance of 86 ft to stop. What distance will it need to stop from a speed of 30 mi/h?
20. The time it takes to build a wall varies inversely as the number of people doing the job. If it takes 12 people 28 h, how long would it take 16 people?

Complete.

1. $3(22 \cdot 5) = (3 \cdot ■)5$
2. $6(■ + 50) = 6(32) + 6(50)$

Perform the indicated operation.

3. $-23 + (-18) + 2$
4. $-18 - 9$
5. $-127 - (-110)$
6. $-3(-5)(-6)$
7. $108 \div (-36)$
8. $-484 \div (-121)$

Refer to the figure to identify the following.

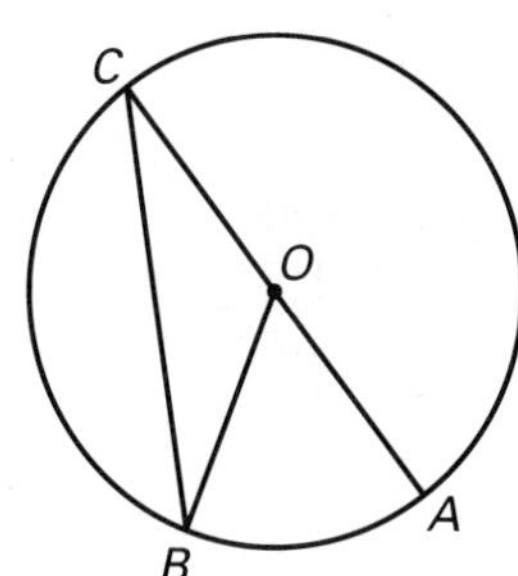


9. a minor arc
10. a major arc

Use the data to answer the questions.

| Bowler's Scores | | | | | |
|---|---|---|---|---|---|
| 156 | 145 | 148 | 145 | 184 | 176 |

11. Find the mean.
12. Find the mode.

Find each square root.

13. $\sqrt{144}$

14. $\sqrt{1.96}$

Find the area of the figure.

15.

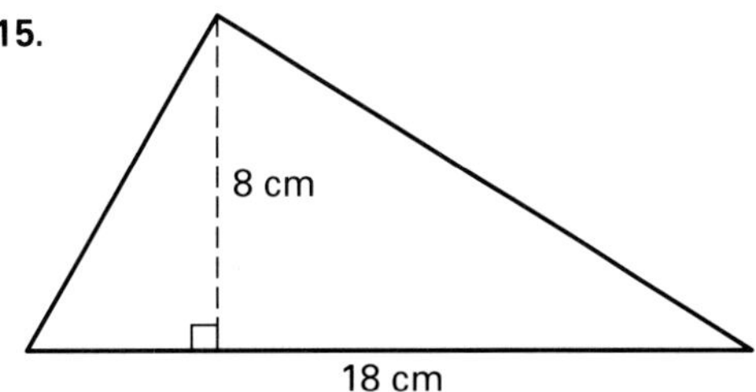


16.

16 in.

8 in.

40 in.

Simplify.

17. $3ab(-4a^3b)$

18. $b^2(2b - b^3 + 3)$

Tell whether the ordered pair is a solution of the inequality.

19. $(1, 3);\ y < x + 3$
20. $(3, 1);\ y \geq 2x - 2$

Write an equation of each line from the given information.

21. $A(-2, 3);\ B(2, 1)$
22. $m = \frac{3}{2};\ P(0, 2)$

Write each ratio of measurements in lowest terms.

23. 10 in. to 24 in.
24. 2 yd:4 ft

Solve each proportion.

25. $\frac{m}{10} = \frac{6}{5}$

26. $\frac{2.5}{5} = \frac{x}{12}$

27. $\frac{c}{5} = \frac{9}{3}$

28. $\frac{d}{120} = \frac{1.2}{3.6}$

1. Perform the indicated operation. $-3(-6)$
   A. 9 B. 18
   C. $-9$ D. $-18$

2. Estimate. 32% of 1,600
   A. 40 B. 50
   C. 400 D. 500

3. Identify the symbol for the figure.

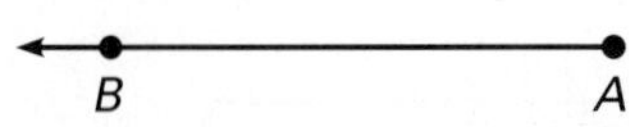


   A. $\overleftrightarrow{AB}$ B. $\overrightarrow{BA}$
   C. $\overrightarrow{AB}$ D. none of these

To learn more about television viewing habits, all the students in two classes chosen at random in a school were asked how many hours of television they watched each night.

4. What kind of sampling method is represented by this survey?
   A. random sampling
   B. cluster sampling
   C. convenience sampling
   D. systematic sampling

5. Solve. $x^2 + 12 = 133$
   A. $\pm 12$ B. $\pm 11$
   C. $\pm\sqrt{145}$ D. $\pm 121$

6. Solve. $3(y - 1) + 9 = -12$
   A. $-6$ B. $-4$
   C. $-2$ D. $-18$

7. Find the area of a circle with radius 6 cm. Use 3.14 for $\pi$.
   A. 11.304 cm$^2$ B. 113.04 cm$^2$
   C. 27.68 cm$^2$ D. 376.8 cm$^2$

8. Find the length of the side of a cube if its surface area is 61.44 in.$^2$
   A. 3.2 in. B. 10.24 in.
   C. 2.3 in. D. 5.12 in.

9. Find the GCF. $27a^2$ and $54a^3$
   A. $6a^2$ B. $27a^2$
   C. $9a^2$ D. $54a^2$

10. Find the LCM. $18x^2y^2$ and $24xy$
   A. $6xy$ B. $36x^2y^2$
   C. $24xy$ D. $72x^2y^2$

11. Which ordered pair is a solution of the inequality? $y > x + 3$
   A. (2, 4) B. (2, 6)
   C. (2, 1) D. (1, 4)

12. Identify the midpoint of the line segment whose endpoints are $A(1, 4)$ and $B(-3, -8)$.
   A. $P(-1, 2)$ B. $P(1, -20)$
   C. $P(-1, -2)$ D. $P(2, -6)$

13. Identify the slope of the line that passes through $P(-2, 8)$ and $Q(2, -4)$.
   A. 3 B. $-3$ C. 2 D. $-2$

14. Identify the equation of the line that passes through $P(1, 4)$ and has a slope of 4.
   A. $y = 4x$ B. $y = -4x$
   C. $y = 4x - 1$ D. $y = -4x - 1$

15. What is the unit price of shampoo that costs \$3.22 for a 14-oz bottle?
   A. \$45.08 B. \$0.23
   C. \$1.00 D. \$0.43

16. Roy has 495 baseball cards. The ratio of National League players to American League players is 11 to 4. How many cards with National League players does Roy have?
   A. 15 B. 132 C. 1,980 D. 363

17. A model of a train car is being built with a scale of 2 in.:5 ft. The actual length of the train car is 12 ft. What is the length of the model?
   A. 30 in. B. $2\frac{2}{5}$ in. C. $4\frac{4}{5}$ in. D. 6 ft

# CHAPTER 11 SKILLS PREVIEW

Use the figure at the right for Exercises 1–3.

1. Graph the image of $\triangle PQR$ under a translation 3 units to the right and 2 units up.
2. Graph the image of $\triangle PQR$ under a reflection across the $y$-axis.
3. Graph the rotation image of $\triangle PQR$ after a 180° turn clockwise about the origin.

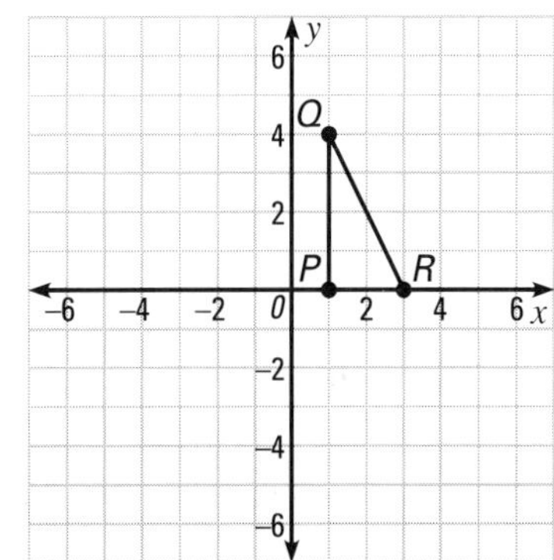

4. $\triangle F'G'H'$ is the reflected image of $\triangle FGH$. Graph the line of reflection.

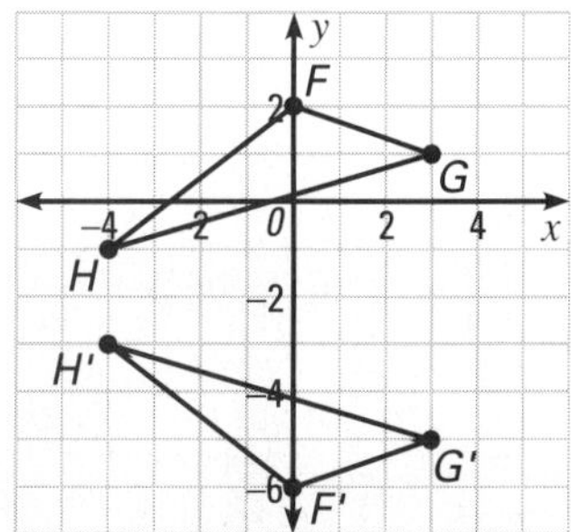

5. Parallelogram $K'L'M'N'$ is the rotation image of parallelogram $KLMN$. Identify the center of rotation, $T$, the angle of rotation, and the direction of rotation.

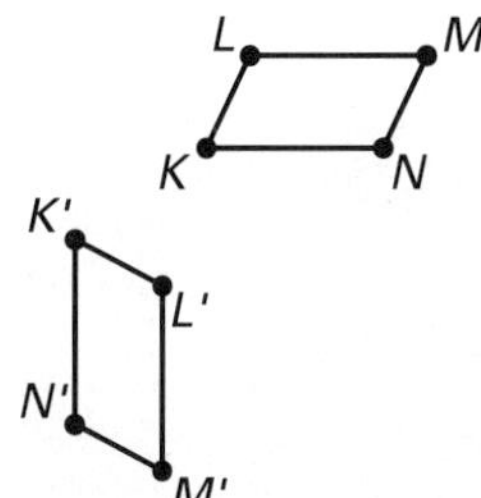


6. Describe two transformations that together could have been used to create the image shown in blue.

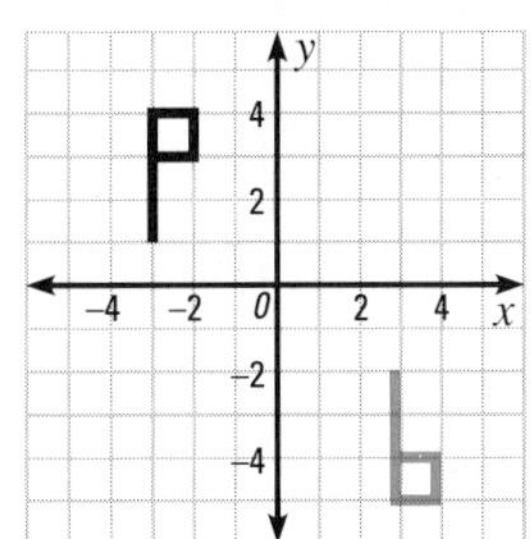

7. How many lines of symmetry does the figure below have?

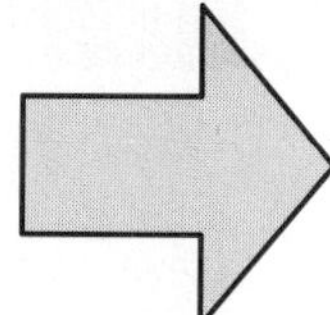

8. What is the order of rotational symmetry of the figure below?

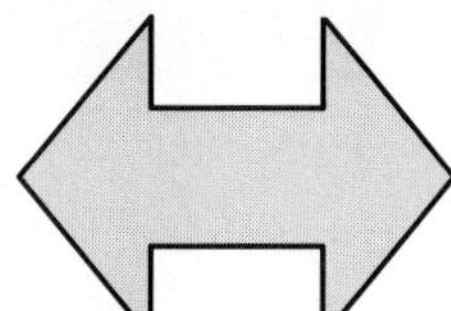

9. Draw the dilation image of $\triangle RST$ with the center of dilation at the origin and a scale factor of 3.

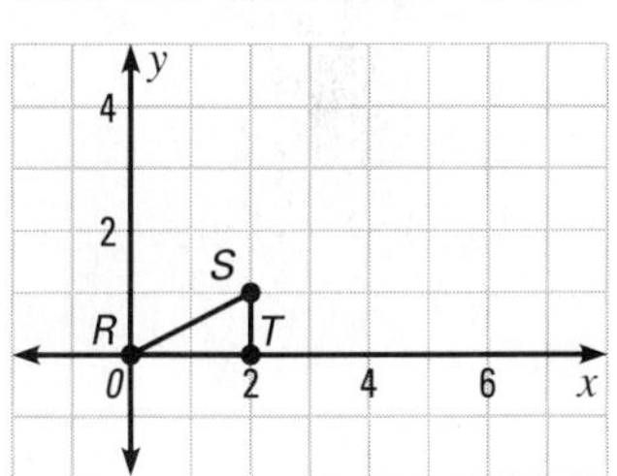

# TRANSFORMATIONS

**THEME** **Art and Design**

One feature of many designs and pieces of art that people find appealing is the repeated use of the same shape. In some designs, the repeated shapes are the same size, but they have been slid, flipped, or turned. In other designs, the shapes are the same but are of different sizes.

In this chapter, you will learn to recognize and create different types of **transformations.** Some transformations are moves that produce new figures that are the same shape and size as the original figures. Other transformations change the size, but not the shape, of the original figure.

## DECISION MAKING

### Using Data

Use the information in the diagram of the pinwheel quilt and the enlarged quilt block at the right.

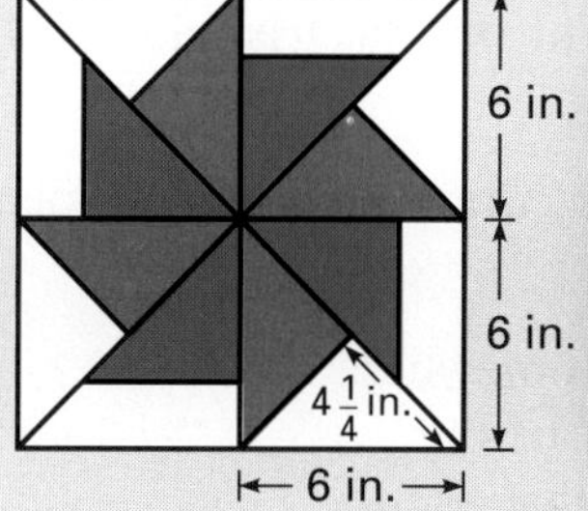

1. Name the different shapes used in this quilt.
2. How many different-sized triangles are used in this quilt?
3. What are the dimensions of each pinwheel block?
4. What are the dimensions of the entire quilt?

### Working Together

Find examples of art that includes repeated shapes that appear to have been slid, flipped, turned, enlarged, or reduced. Look in art books in your school library, check with your art teacher, or visit an art museum or gallery. Select one example that is your group's favorite. Write a short description of the piece of art including the title, artist's name, year it was completed, how shapes were used, and why your group selected it.

# 11-1 Translations

**EXPLORE/ WORKING TOGETHER**

Work with a partner. You will need a piece of graph paper, a ruler, and scissors.

**a.** Use the ruler to help you draw an isosceles triangle near one bottom corner of your paper. Make the base 4 units long and the height 5 units. Cut out the triangle and label its vertices *A*, *B*, and *C*.

**b.** On your graph paper, draw a coordinate plane. Label each axis from $-12$ to 12.

**c.** Place your triangle somewhere in the third quadrant of the coordinate plane so that each vertex is at the intersection of a horizontal and a vertical line. Trace the triangle and label each vertex to match your original triangle. Label your figure "Triangle 1."

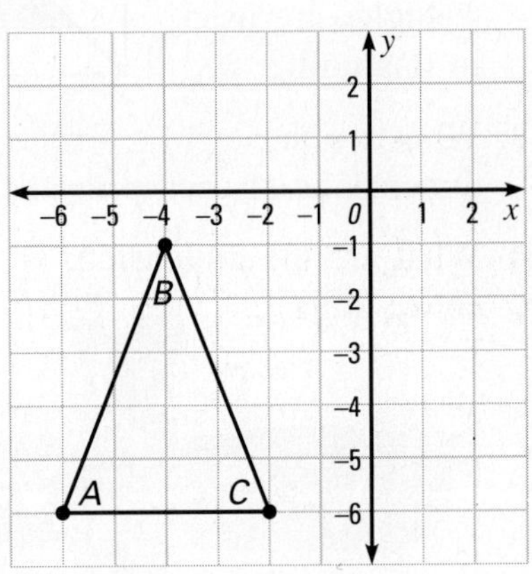


**d.** Carefully slide your triangle straight up the distance of 7 units on your coordinate plane. Trace the triangle in this new position and again label each vertex. Label this figure "Triangle 2."

**e.** From its position as Triangle 2, carefully slide the triangle 6 units straight to the right. Once more, trace the triangle and label each vertex. Call it "Triangle 3."

**f.** Record the ordered pair that describes the position of each vertex of the three triangles you traced in a table like this.

| | Triangle 1 | Triangle 2 | Triangle 3 |
|---|---|---|---|
| A | | | |
| B | | | |
| C | | | |

**g.** Compare the *x*-coordinates of the vertices of Triangles 1 and 2, Triangles 2 and 3, and Triangles 1 and 3. Then compare the *y*-coordinates of the vertices for each pair of triangles. What do you notice?

A **translation,** or *slide,* of a figure produces a new figure exactly like the original. The new figure is the **image** of the original figure, and the original is the **preimage** of the new one. A move like a translation is called a **transformation** of a figure.

## SKILLS DEVELOPMENT

As a figure is translated, you can imagine all its points sliding along a plane at once in the same direction and for the same distance. Therefore, the sides and angles of an image are equal in measure to the sides and angles of its preimage. Also, each side of an image is parallel to the corresponding side of its preimage.

### Example 1

Graph the image of parallelogram $ABCD$ with vertices $A(1, -4)$, $B(2, -2)$, $C(5, -2)$, and $D(4, -4)$ under each translation.

**a.** 6 units left  **b.** 4 units up

#### Solution

**a.** To move the image 6 units to the left, subtract 6 from the $x$-coordinate of each vertex.

Read $A'$ as "A prime."

$A(1, -4) \rightarrow A'(1 - 6, -4)$
$\rightarrow A'(-5, -4)$
$B(2, -2) \rightarrow B'(2 - 6, -2)$
$\rightarrow B'(-4, -2)$
$C(5, -2) \rightarrow C'(5 - 6, -2)$
$\rightarrow C'(-1, -2)$
$D(4, -4) \rightarrow D'(4 - 6, - 4)$
$\rightarrow D'(-2, -4)$

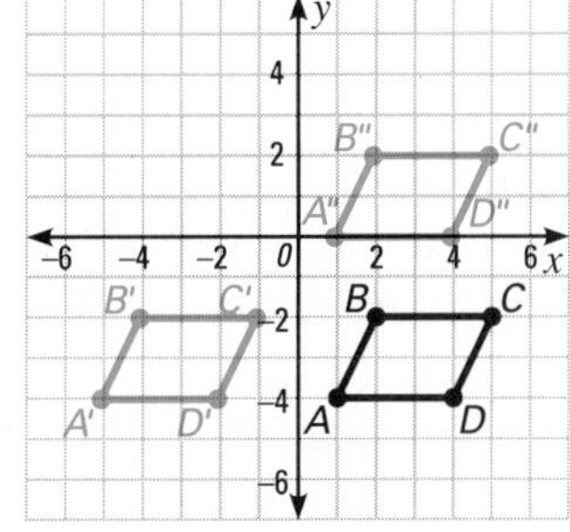


**b.** Add 4 to the $y$-coordinate of each vertex.

Read $A''$ as "A double prime."

$A(1, -4) \rightarrow A''(1, -4 + 4)$
$\rightarrow A''(1, 0)$
$B(2, -2) \rightarrow B''(2, -2 + 4)$
$\rightarrow B''(2, 2)$
$C(5, -2) \rightarrow C''(5, -2 + 4)$
$\rightarrow C''(5, 2)$
$D(4, -4) \rightarrow D''(4, -4 + 4)$
$\rightarrow D''(4, 0)$ ◀

### Example 2

Graph the image of $\triangle PQR$ with vertices $P(-5, 3)$, $Q(-2, 4)$, and $R(-2, 1)$ under a translation 7 units to the right and 5 units down.

#### Solution

Add 7 to the $x$-coordinate of each vertex. Subtract 5 from the $y$-coordinate of each.

$P(-5, 3) \rightarrow P'(-5 + 7, 3 - 5)$
$\rightarrow P'(2, -2)$
$Q(-2, 4) \rightarrow Q'(-2 + 7, 4 - 5)$
$\rightarrow Q'(5, -1)$
$R(-2, 1) \rightarrow R'(-2 + 7, 1 - 5)$
$\rightarrow R'(5, -4)$ ◀

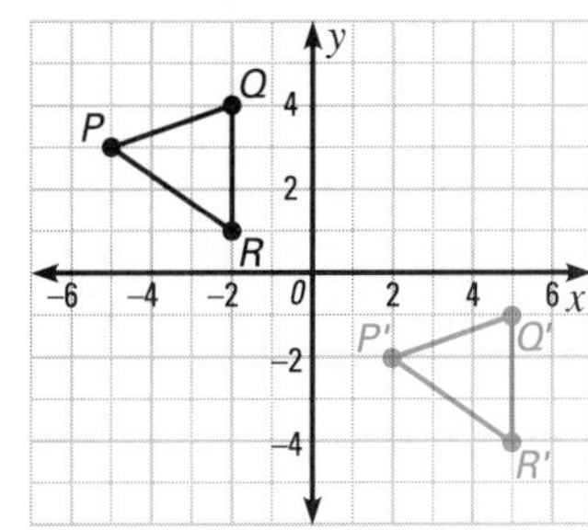

Write the rules for the translation of Triangle 1 into Triangle 2 and Triangle 1 into Triangle 3 in the Explore activity.

## Example 3

Write the rule that describes the translation for $\triangle RST$ and $\triangle R'S'T'$.

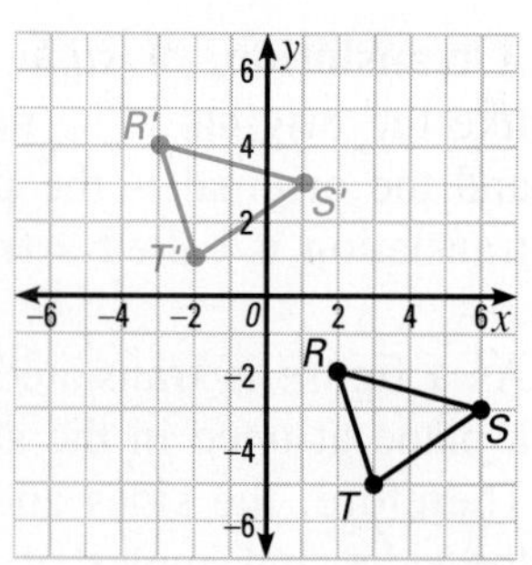


### Solution

Look for a pattern between the $x$-coordinates and $y$-coordinates of each vertex of the preimage ($\triangle RST$) and the $x$-coordinates and $y$-coordinates of each vertex of the image ($\triangle R'S'T'$).

$R(2, -2)$ $S(6, -3)$ $T(3, -5)$
$R'(-3, 4)$ $S'(1, 3)$ $T'(-2, 1)$ ◄

Suppose a figure is in the third quadrant of the coordinate plane. Describe a translation under which its image would appear in the first quadrant.

To get the $x$-coordinate of each vertex of the image, you must subtract 5 from the $x$-coordinate of each vertex of the preimage. To get the $y$-coordinate of each vertex of the image, you must add 6 to the $y$-coordinate of each vertex of the preimage. This can be stated using the rule $(x, y) \rightarrow (x - 5, y + 6)$. This rule describes the translation of a figure 5 units to the left and 6 units up.

## TRY THESE

Copy parallelogram $DEFG$ at the right with vertices $D(-3, 5)$, $E(-1, 4)$, $F(-3, 1)$, and $G(-5, 2)$ on a coordinate plane. Then graph its image under each translation from the original position.

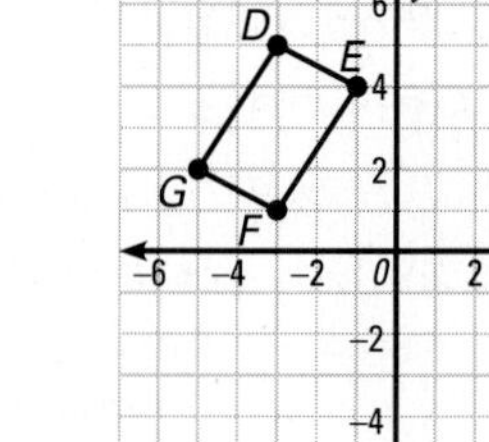


1. 4 units right
2. 7 units down
3. Copy $\triangle LMN$ at the right with vertices $L(1, -2)$, $M(3, 1)$, and $N(5, -1)$ on a coordinate plane. Then graph its image under a translation 6 units to the left and 3 units down.

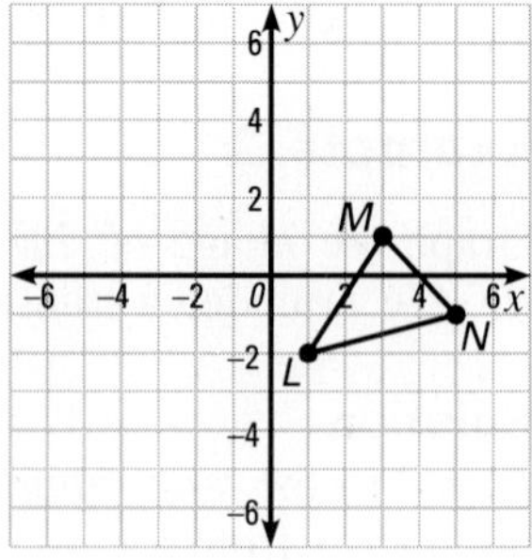


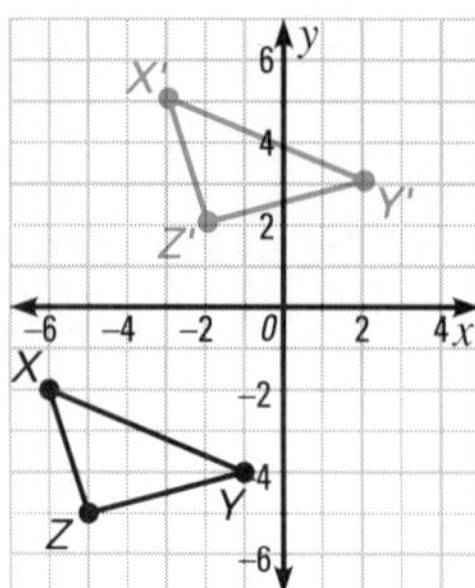


4. Write the rule that describes the translation for $\triangle XYZ$ and $\triangle X'Y'Z'$.

# EXERCISES

## PRACTICE/ SOLVE PROBLEMS

On a coordinate plane, graph pentagon *HIJKL* with vertices $H(-1, 5)$, $I(2, 6)$, $J(5, 5)$, $K(3, 2)$, and $L(1, 2)$. Then graph its image under each translation from the original position.

1. 4 units left
2. 8 units down
3. 7 units left and 6 units down

Write the rule that describes each translation.

4.

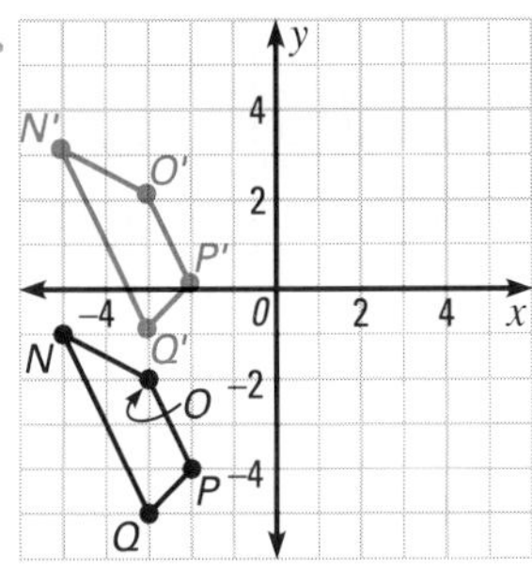

5.

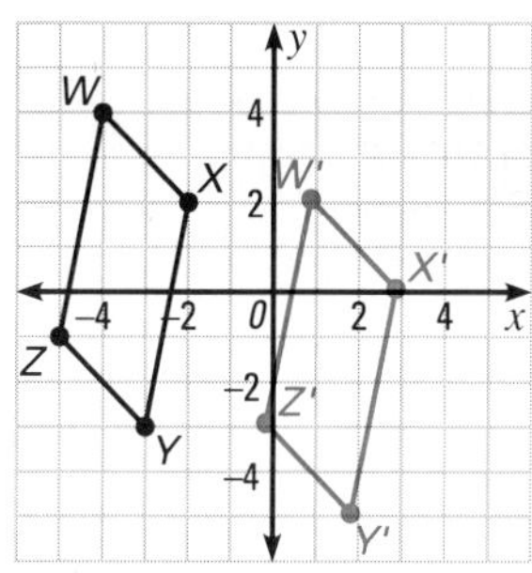

## EXTEND/ SOLVE PROBLEMS

6. Which of the lettered figures are *not* translations of the shaded figure? Explain.

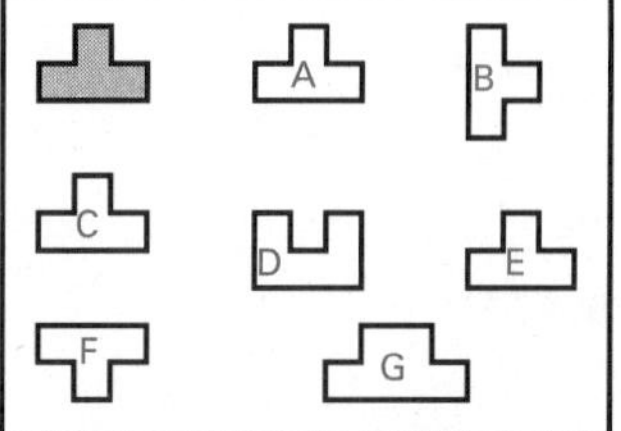


7. Quadrilateral $D'E'F'G'$ with vertices $D'(2, 3)$, $E'(5, 3)$, $F'(5, 1)$, and $G'(1, 1)$ is the image of a figure that was translated under the rule $(x, y) \rightarrow (x + 2, y - 3)$. What are the vertices of the preimage of quadrilateral $D'E'F'G'$?

## THINK CRITICALLY/ SOLVE PROBLEMS

8. Triangle *QRS* with vertices $Q(-4, 7)$, $R(-1, 3)$, and $S(-5, 2)$ was translated using the rule $(x, y) \rightarrow (x + 7, y - 1)$ to create $\triangle Q'R'S'$. Triangle $Q'R'S'$ was then translated using the rule $(x, y) \rightarrow (x - 4, y - 7)$ to create $\triangle Q''R''S''$. In words, describe the position of $\triangle Q''R''S''$ in relation to $\triangle QRS$ on the plane.

9. A **tessellation** is a pattern in which one or more shapes are repeated in a way that leaves no gaps and creates no overlaps. The brick pattern at the right shows tessellations. Tell which of these polygons can be used to create tessellations—square, isosceles triangle, equilateral triangle, parallelogram, regular pentagon, regular hexagon, regular octagon.

# 11-2 Problem Solving Strategies: SOLVE A SIMPLER PROBLEM

- ► READ
- ► PLAN
- ► SOLVE
- ► ANSWER
- ► CHECK

Sometimes you can find the answer to a problem by solving a similar problem that has simpler numbers.

## PROBLEM

Find the length of the hypotenuse of $\triangle CDE$.

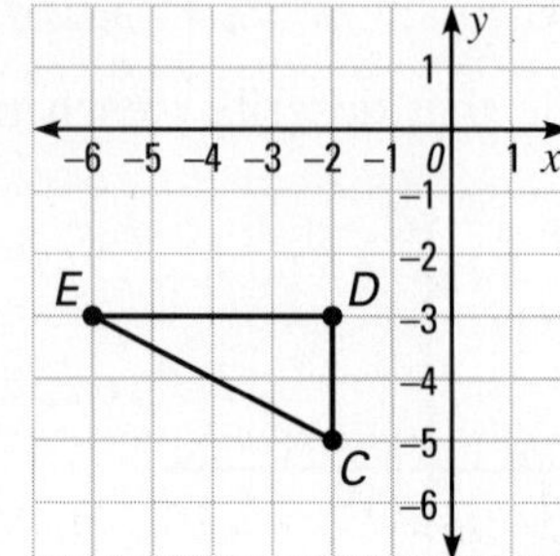


## SOLUTION

Solve a simpler problem. Translate the triangle so that one vertex of the triangle is at the origin.

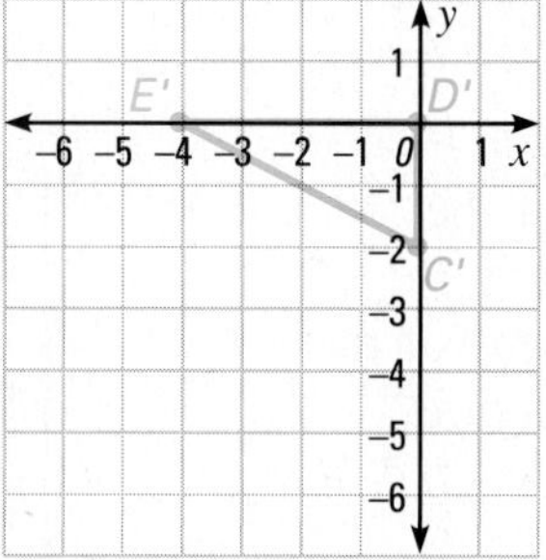


**Simpler Problem**

Translate $\triangle CDE$ using the rule $(x, y) \rightarrow (x + 2, y + 3)$.

To find the length of $\overline{C'E'}$, use the distance formula to find the distance from $C'(0, -2)$ to $E'(-4, 0)$.

$$\text{length of } \overline{C'E'} = \sqrt{(x_2 - x_1)^2 + (y_2 - y_1)^2}$$
$$\text{length of } \overline{C'E'} = \sqrt{(-4 - 0)^2 + [0 - (-2)]^2}$$
$$= \sqrt{(-4)^2 + (2)^2} = \sqrt{16 + 4}$$
$$= \sqrt{20} = 4.472$$

The length of $\overline{C'E'}$ to the nearest tenth is 4.5 units.

**Original Problem**

To find the length of $\overline{CE}$, use the distance formula to find the distance from $C(-2, -5)$ to $E(-6, -3)$.

$$\text{length of } \overline{CE} = \sqrt{(x_2 - x_1)^2 + (y_2 - y_1)^2}$$
$$\text{length of } \overline{CE} = \sqrt{[-6 - (-2)]^2 + [-3 - (-5)]^2}$$
$$= \sqrt{(-4)^2 + (-2)^2} = \sqrt{16 + 4}$$
$$= \sqrt{20} = 4.472$$

The length of $\overline{CE}$ to the nearest tenth is 4.5 units.

$\overline{CE}$ and $\overline{C'E'}$ are the same length. In the simpler problem, one $x$-coordinate and one $y$-coordinate are at *0*, so the subtraction step is easier.

# PROBLEMS

Graph the image of each figure under a translation that will give you a simpler problem to work. Then find each length to the nearest tenth.

1. Find the length of $\overline{UV}$.

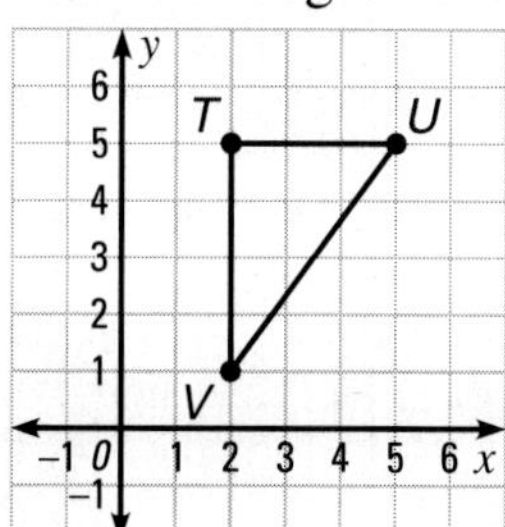


2. Find the length of $\overline{QR}$.

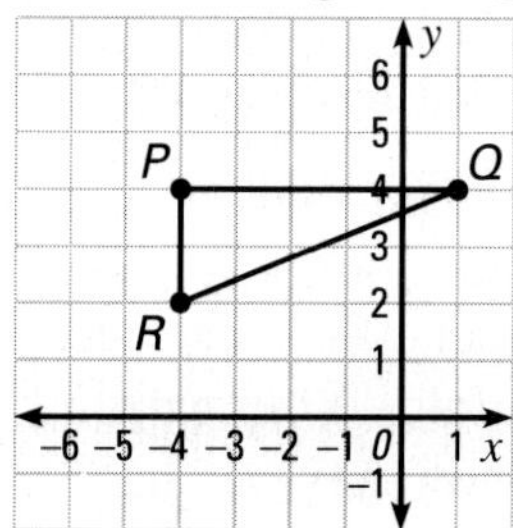


3. Find the length of $\overline{BD}$.

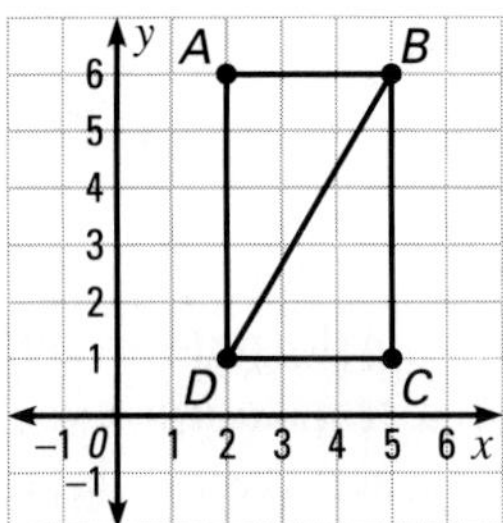


4. Find the length of $\overline{JN}$.

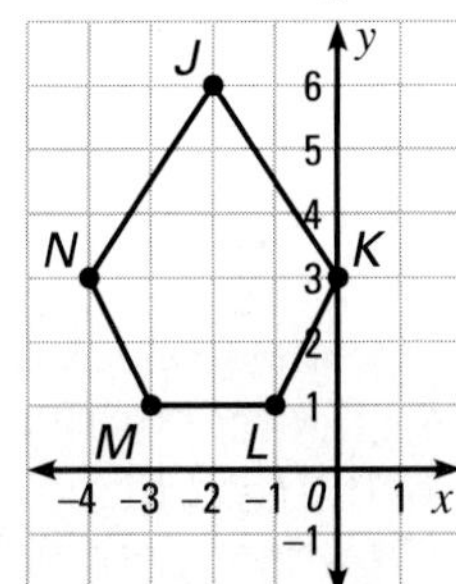

**PROBLEM SOLVING TIP**

In Problem 3, think of $\overline{BD}$ as the hypotenuse of $\triangle ABD$ or $\triangle BCD$.

In Problem 4, think of what right triangle you could draw that would have $\overline{JN}$ as its hypotenuse.

5. Find the length of $\overline{NQ}$.

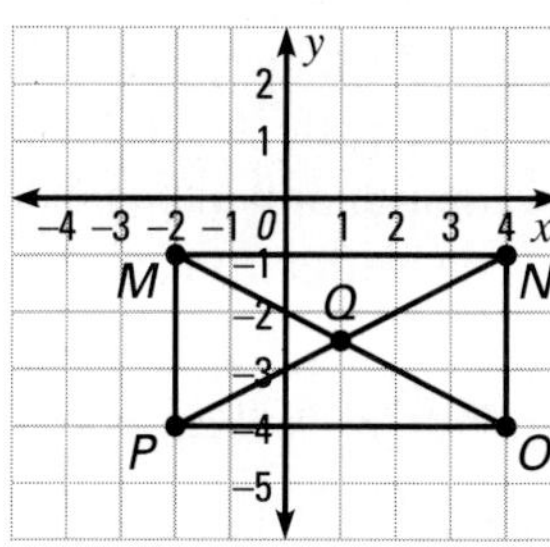


6. Find the perimeter of rhombus *STUV*.

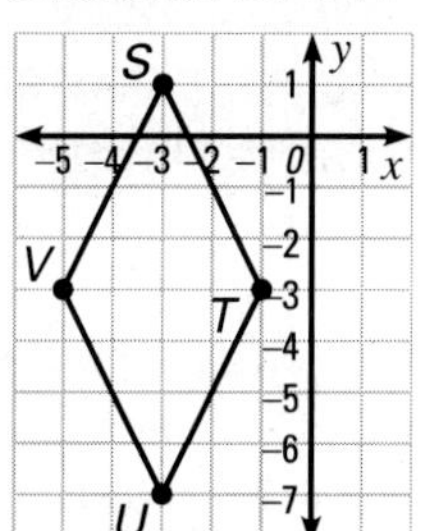


7. Find the perimeter of hexagon *BCDEFG*.

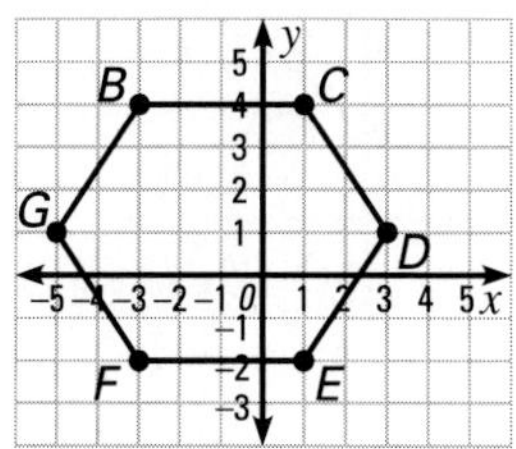


8. Find the circumference of the circle. $\overline{GH}$ is the diameter.

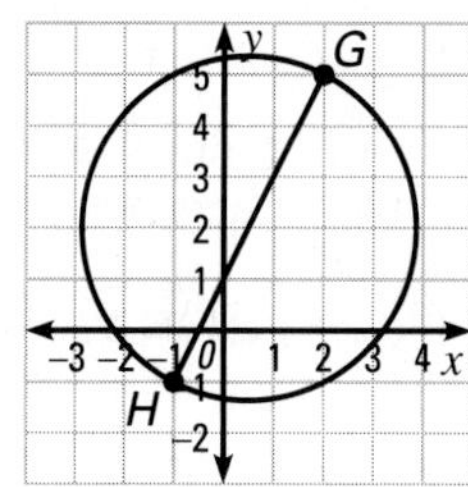

# 11-3 Reflections

## EXPLORE/ WORKING TOGETHER

You will need a piece of graph paper, a ruler, and food coloring, ink, or paint.

**a.** On graph paper, draw a coordinate plane. You do not need to label each axis.

**b.** Place a small drop of food coloring, ink, or paint on one side of the $y$-axis. Carefully fold your paper along the $y$-axis. Then unfold. Your paper should now have a shape on both sides of the $y$-axis.

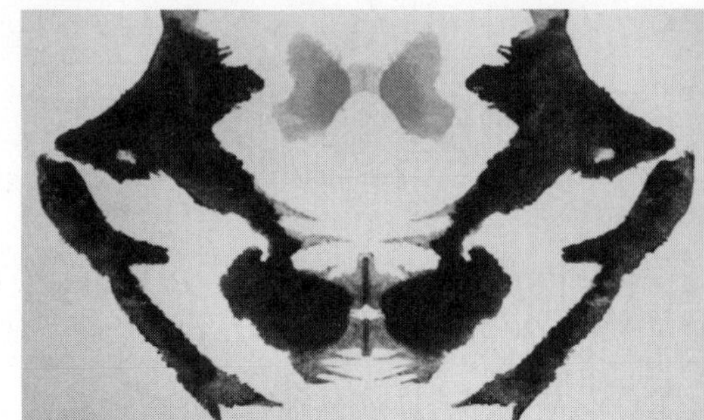

**c.** What is the relationship of corresponding points to the $y$-axis?

## SKILLS DEVELOPMENT

In Section 11–1, you learned about one kind of transformation that produces a new figure exactly like the original—a translation. Another kind of transformation that yields a congruent figure is a **reflection,** or *flip*. Under a reflection, a figure is *reflected*, or *flipped*, over a **line of reflection.**

In doing the Explore activity, you created a reflection of your original drop of fluid across the $y$-axis. When you reflect a point across the $y$-axis, the $y$-coordinate remains the same, but the $x$-coordinate is transformed into its opposite. The reflection of the point $(x, y)$ across the $y$-axis is the point $(-x, y)$. When you reflect a point across the $x$-axis, the $x$-coordinate remains the same, but the $y$-coordinate is transformed into its opposite. The reflection of the point $(x, y)$ across the $x$-axis is the point $(x, -y)$.

### Example 1

Graph the image of $\triangle DEF$ with vertices $D(4, 5)$, $E(5, 1)$, and $F(1, 3)$ across the $x$-axis.

**Solution**

Multiply the $y$-coordinate of each vertex by $-1$, using the rule $(x, y) \rightarrow (x, -y)$.

$D(4, 5) \rightarrow D'(4, -5)$
$E(5, 1) \rightarrow E'(5, -1)$
$F(1, 3) \rightarrow F'(1, -3)$ ◄

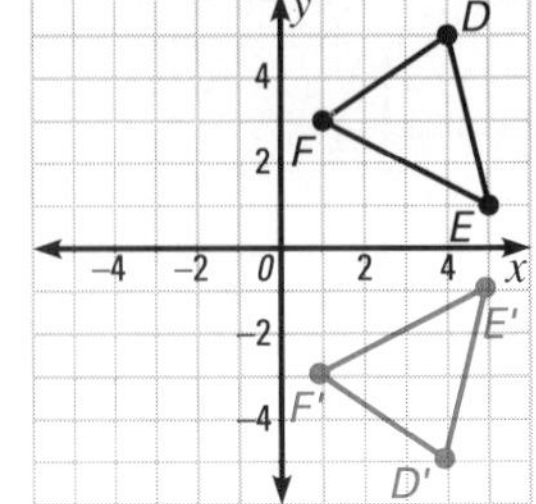

Figures can be reflected over lines other than the $x$-axis or $y$-axis. When a point $(x, y)$ is reflected across the line $y = x$, the image is the point $(y, x)$. When a point is reflected across the line $y = -x$, the image is the point $(-y, -x)$.

## Example 2

Graph the image of $\triangle JKL$ with vertices $J(-4, 2)$, $K(-2, -1)$, and $L(-5, -3)$ under a reflection across $y = x$.

**Solution**

Graph the line $y = x$. Transpose the $x$-coordinate and $y$-coordinate of each vertex, using the rule $(x, y) \rightarrow (y, x)$.

$J(-4, 2) \rightarrow J'(2, -4)$
$K(-2, -1) \rightarrow K'(-1, -2)$
$L(-5, -3) \rightarrow L'(-3, -5)$ ◄

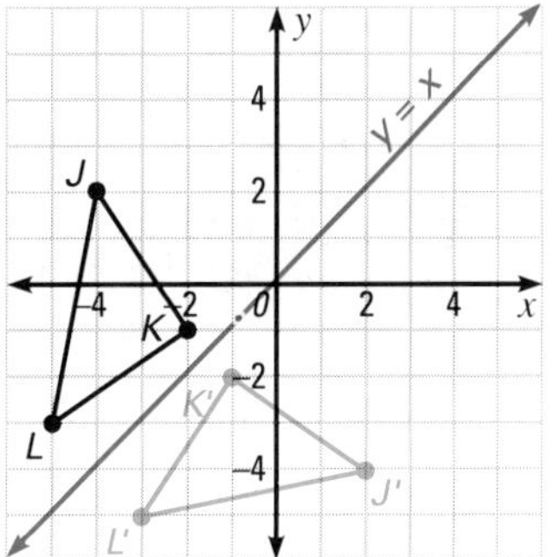


Each point of a reflected image is the same distance from the line of reflection as the corresponding point of its preimage. In other words, the line of reflection lies directly in the middle between the image and its preimage. You can use this fact to help find reflection lines, especially when they are neither on the $x$-axis nor on the $y$-axis.

**TALK IT OVER**

Is it necessary to find the midpoint of each line segment that connects corresponding vertices of a figure and its reflected image? If not, how many midpoints is it necessary to find? Why?

## Example 3

Triangle $Q'R'S'$ with vertices $Q'(5, 1)$, $R'(1, -2)$, and $S'(2, -4)$ is the reflected image of $\triangle QRS$ with vertices $Q(-7, 1)$, $R(-3, -2)$, and $S(-4, 4)$. Graph and identify the line of reflections.

**Solution**

Imagine line segments connecting each pair of corresponding vertices. Graph the midpoint of each line segment using the rule $\left(\frac{x_1 + x_2}{2}, \frac{y_1 + y_2}{2}\right) \rightarrow (x, y)$.

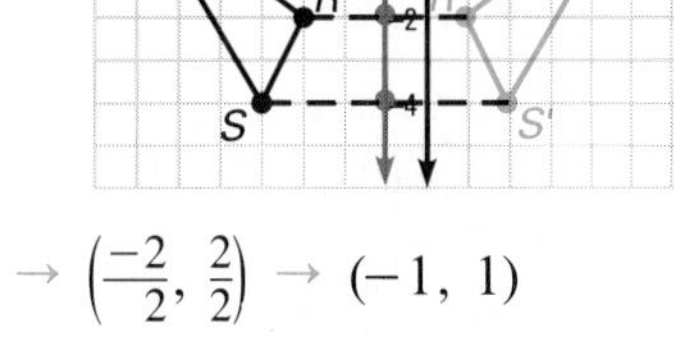


$$\text{midpoint of } \overline{QQ'}: \left(\frac{-7 + 5}{2}, \frac{1 + 1}{2}\right) \rightarrow \left(\frac{-2}{2}, \frac{2}{2}\right) \rightarrow (-1, 1)$$

$$\text{midpoint of } \overline{RR'}: \left(\frac{-3 + 1}{2}, \frac{-2 + (-2)}{2}\right) \rightarrow \left(\frac{-2}{2}, \frac{-4}{2}\right) \rightarrow (-1, -2)$$

$$\text{midpoint of } \overline{SS'}: \left(\frac{-4 + 2}{2}, \frac{-4 + (-4)}{2}\right) \rightarrow \left(\frac{-2}{2}, \frac{-8}{2}\right) \rightarrow (-1, -4)$$

Draw a line through the midpoints. The equation of the line is $x = -1$. ◄

## CONNECTIONS

A periscope uses reflections in two mirrors to make it possible for a viewer to see above the line of sight.

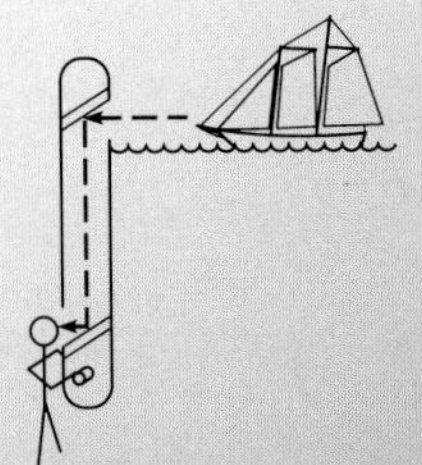

The two mirrors in a periscope, as shown in the diagram, are placed parallel to each other and at 45°-angles with the horizon. Horizontal light rays enter at the top, are reflected down to the mirror at the bottom, and are then reflected to the viewer's eye.

## TRY THESE

For each exercise, copy $\triangle XYZ$ with vertices $X(-5, 0)$, $Y(-2, -1)$, and $Z(-4, -3)$ on a coordinate plane. Then graph its image under a reflection across each given line.

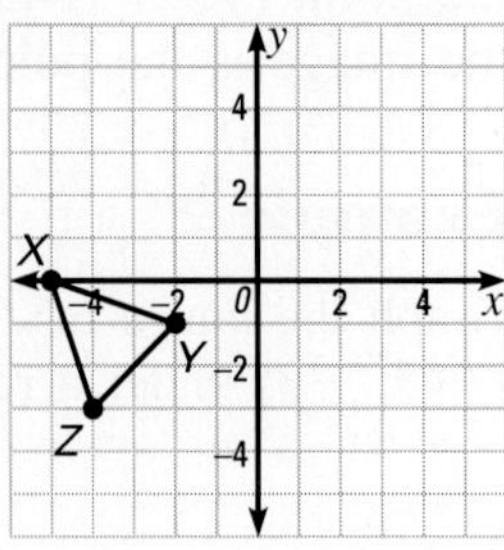


1. $y$-axis
2. $x$-axis
3. $y = x$
4. $y = -x$

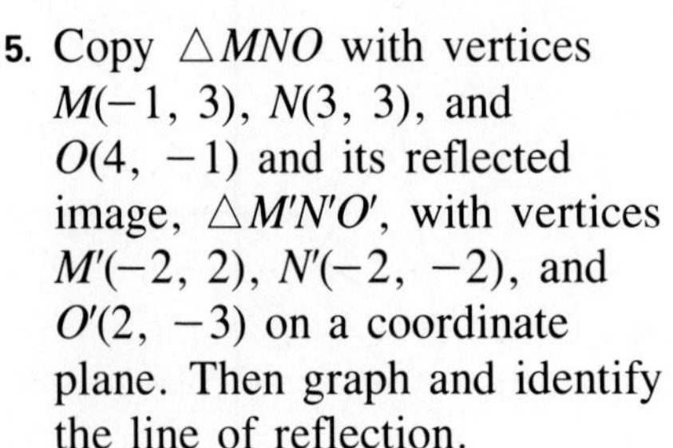

5. Copy $\triangle MNO$ with vertices $M(-1, 3)$, $N(3, 3)$, and $O(4, -1)$ and its reflected image, $\triangle M'N'O'$, with vertices $M'(-2, 2)$, $N'(-2, -2)$, and $O'(2, -3)$ on a coordinate plane. Then graph and identify the line of reflection.

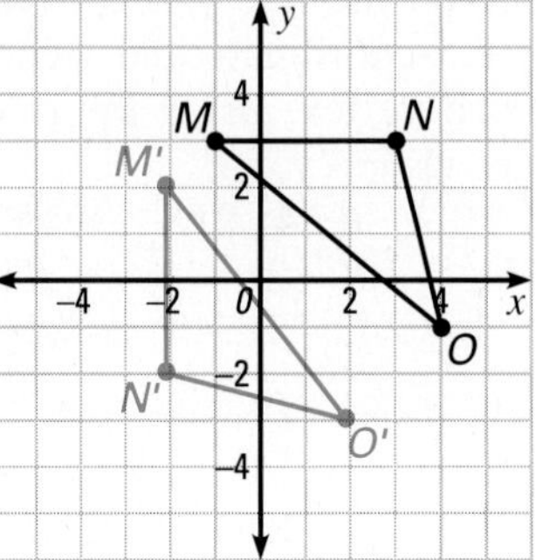


# EXERCISES

### PRACTICE/ SOLVE PROBLEMS

Give the coordinates of the image of each point under a reflection across the given line.

1. $(5, -3)$; $x$-axis
2. $(4, 0)$; $y = -x$
3. $(-2, -7)$; $y$-axis
4. $(-6, 8)$; $y = x$
5. $(0, -6)$; $x$-axis
6. $(3, -3)$; $y = -x$

Copy each figure on a coordinate plane. Then graph its image under a reflection across the given line.

7. $y = -x$

8. $y$-axis

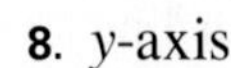

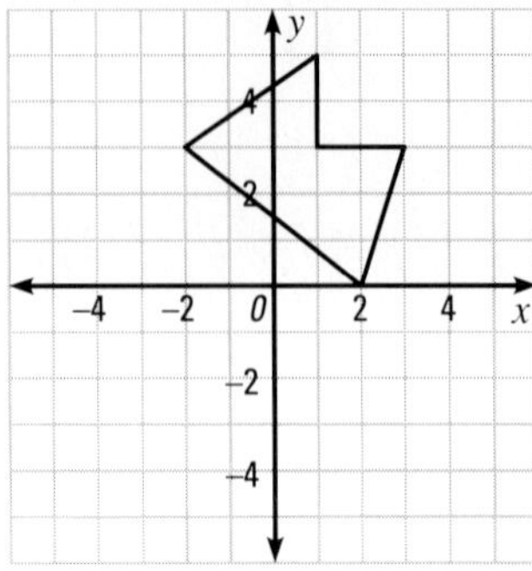


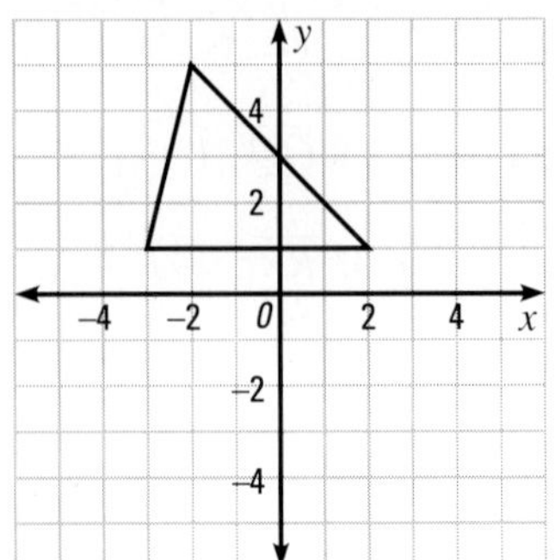

Copy each figure and its image on a coordinate plane. Then graph and identify its line of reflection.

**9.**

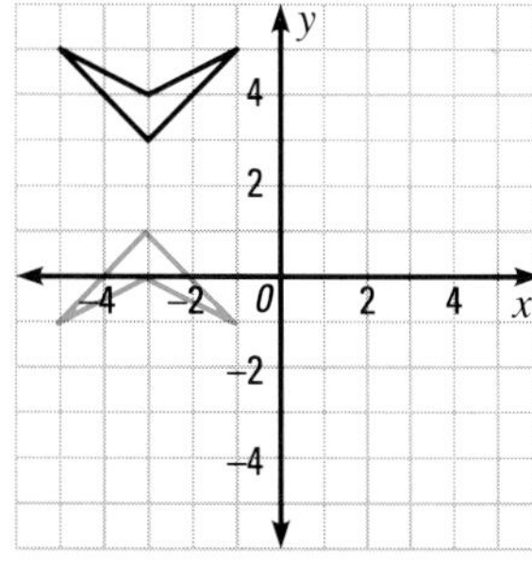


**10.**

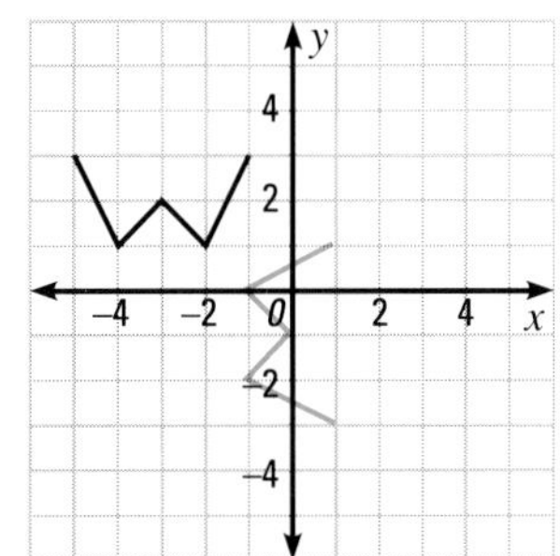


The pentagon at the right is a reflection of itself across the line shown. The rectangle is a reflection of itself across both lines shown. Copy each polygon and draw all lines across which the polygon is a reflection of itself. If there are no such lines, write *none.*

**11.**

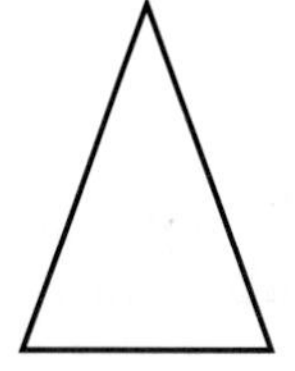

**12.**

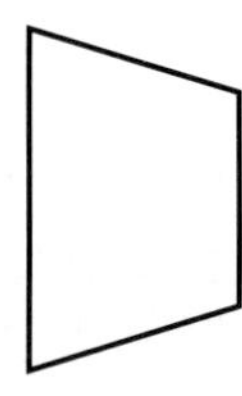

**13.**

**14.**

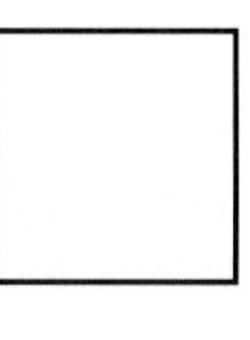

## EXTEND/ SOLVE PROBLEMS

Copy each figure and line of reflection on a coordinate plane. For each, graph the figure's image under a reflection across the line. Then write a rule that describes the reflection.

**15.**

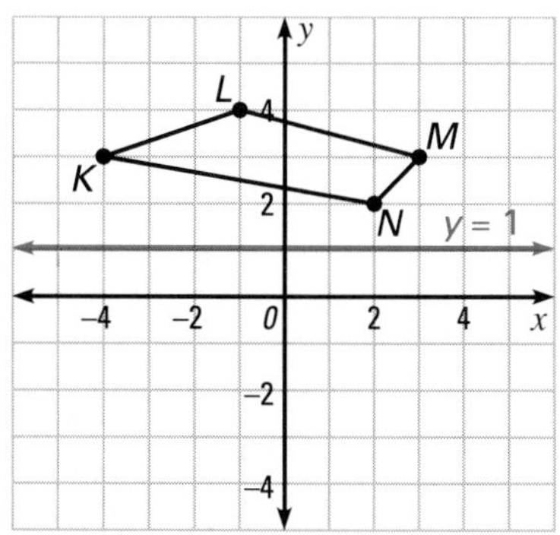


**16.**

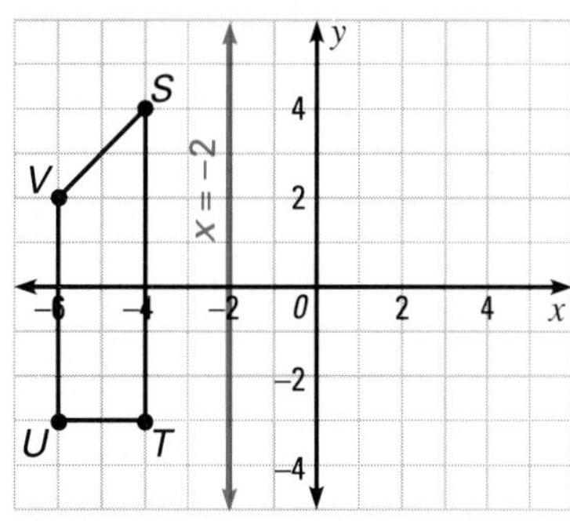


**17.** Graph the lines $y = -3$ and $x = 4$, each on a separate coordinate plane. Draw a figure on each plane and reflect it across the line you drew. Write a rule for each reflection.

**18.** Generalize from the results of Exercises 15–17. What directions could you give for finding the vertices of an image if a figure is reflected across any line parallel to the *x*-axis or *y*-axis?

## THINK CRITICALLY/ SOLVE PROBLEMS

**PROBLEM SOLVING TIP**

For Exercises 15 and 16, think about the relationship of a point and its reflected image to the line of reflection.

# 11-4 Rotations

**EXPLORE**

Think about the last time you rode on a Ferris wheel or saw one turning.

a. From the side at which you can see the Ferris wheel in the photo, would you say that this Ferris wheel turns clockwise or counterclockwise around its center?

b. What fractional part of a turn does it take for a rider to get from the bottom to the top of a Ferris wheel?

c. How many degrees does a Ferris wheel turn while a rider goes completely around one time?

**SKILLS DEVELOPMENT**

You have now learned about two types of transformations—translations and reflections. The third transformation you will learn about that produces a new figure exactly like the original is a **rotation,** or *turn*. Under a rotation, a figure is *rotated*, or *turned*, about a point.

The description of a rotation includes three pieces of information:

- ► the **center of rotation,** or point about which the figure is rotated
- ► the amount of turn expressed as a fractional part of a whole turn or as the **angle of rotation** in degrees
- ► the direction of rotation—*clockwise* or *counterclockwise*

When you rotate a point 180° clockwise about the origin, both the $x$-coordinate and the $y$-coordinate are transformed into their opposites.

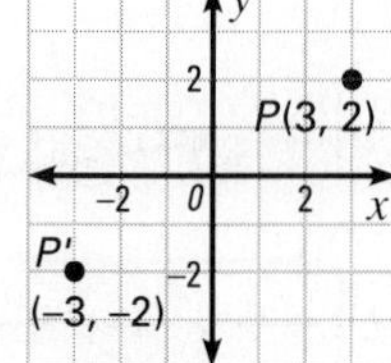


### Example 1

Draw the rotation image of $\triangle QRS$ with vertices $Q(2, 4)$, $R(1, 1)$, and $S(4, 1)$ after a turn of 180° clockwise about the origin.

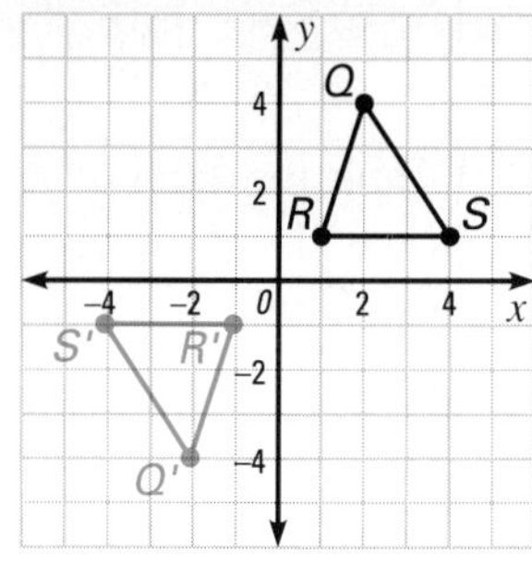


**Solution**

Multiply the $x$-coordinate and $y$-coordinate of each vertex by $-1$, using the rule $(x, y) \rightarrow (-x, -y)$.

$Q(2, 4) \rightarrow Q'(-2, -4)$
$R(1, 1) \rightarrow R'(-1, -1)$
$S(4, 1) \rightarrow S'(-4, -1)$ ◄

## Example 2

Draw the rotation image of the flag containing points $A(0, 0)$, $B(-2, 2)$, $C(-4, 4)$, and $D(-4, 2)$ after a rotation 90° counterclockwise about the origin.

**Solution—*Tracing Paper Method***

Copy the flag onto a coordinate plane. Then trace the flag and the center of rotation, the origin, onto a sheet of tracing paper.

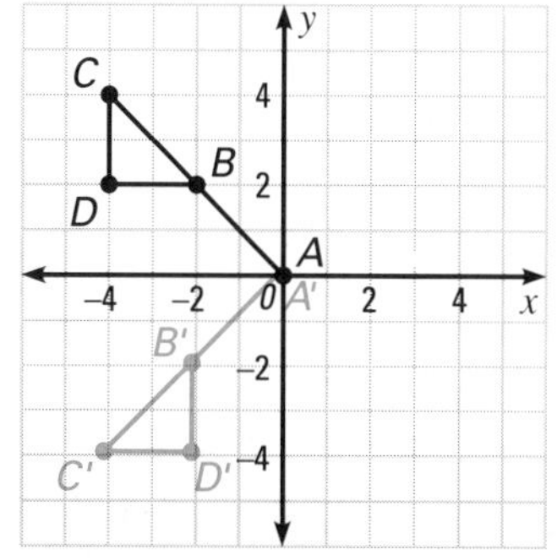


Hold your pencil point on the tracing at the center of rotation. Turn the tracing one-quarter turn, or 90°, counterclockwise. Remove the tracing paper and copy the image onto the coordinate plane.

**Solution—*Coordinates Method***

Multiply the $y$-coordinate of each point by $-1$. Then transpose the $x$-coordinate and $y$-coordinate. Use the rule $(x, y) \rightarrow (-y, x)$.

$A(0, 0) \rightarrow A'(0, 0)$ $\quad$ $B(-2, 2) \rightarrow B'(-2, -2)$

$C(-4, 4) \rightarrow C'(-4, -4)$ $\quad$ $D(-4, 2) \rightarrow D'(-2, -4)$ ◄

**CHECK UNDERSTANDING**

How many degrees are in a three-quarter turn? How many degrees are in a full turn?

**TALK IT OVER**

In Example 1, how else could you have rotated $\triangle QRS$ about the origin to form the same rotation image? In Example 2, how else could you have rotated the flag to form the same rotation image?

## Example 3

$\triangle X'Y'Z'$ is the rotation image of $\triangle XYZ$. Identify the center of rotation, $T$, the angle of rotation, and the direction of rotation.

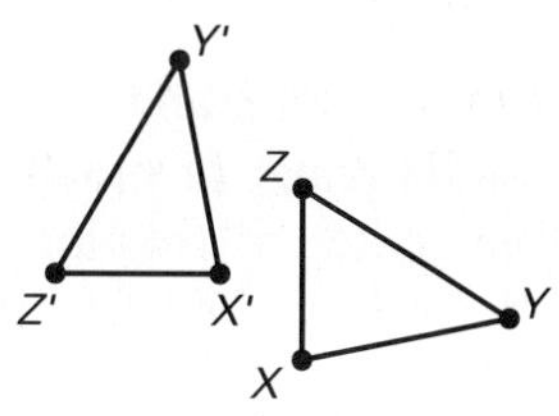


**Solution**

To identify the center of rotation, $T$, you can use these steps.

Draw a line segment connecting each pair of corresponding vertices.

Use your compass and straightedge to construct the perpendicular bisectors of line segments $XX'$, $YY'$, and $ZZ'$.

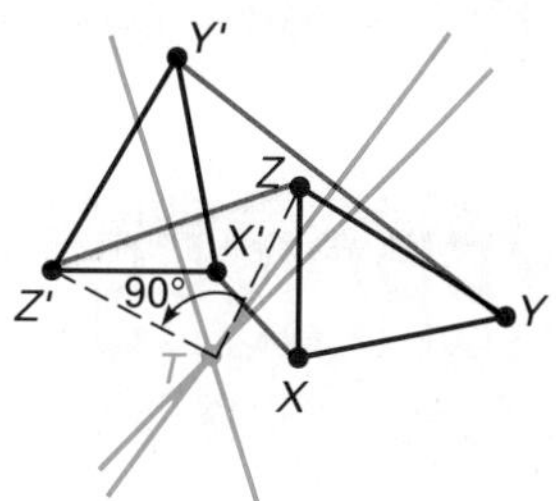


Label the point where the perpendicular bisectors intersect "$T$".

To identify the angle and direction of rotation, draw line segments connecting two corresponding vertices to the center of rotation, $T$. Then measure the angle formed by these line segments.

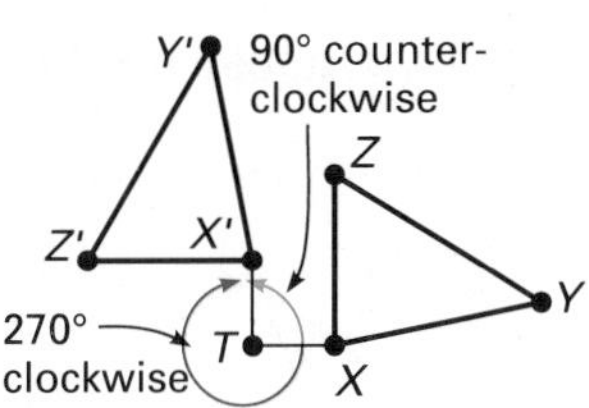


$\triangle X'Y'Z'$ is the rotation image of $\triangle XYZ$ after a turn of 90° counterclockwise about $T$ or 270° clockwise about $T$. ◄

**CHECK UNDERSTANDING**

In Example 3, what rotation would result in $\triangle XYZ$ fitting back on itself?

## TRY THESE

**WRITING ABOUT MATH**

In your journal, describe each kind of transformation so that someone who is not in your mathematics class could determine whether an image was produced by a preimage being translated, reflected, rotated, or none of these.

1. Copy the flag containing the points $J(0, 0)$, $K(-4, -2)$, $L(-6, -3)$, and $M(-3, -5)$ on a coordinate plane. Then graph its rotation image after a turn of 180° counterclockwise about the origin by finding the coordinates of points $J'$, $K'$, $L'$, and $M'$. Check your work by using tracing paper.

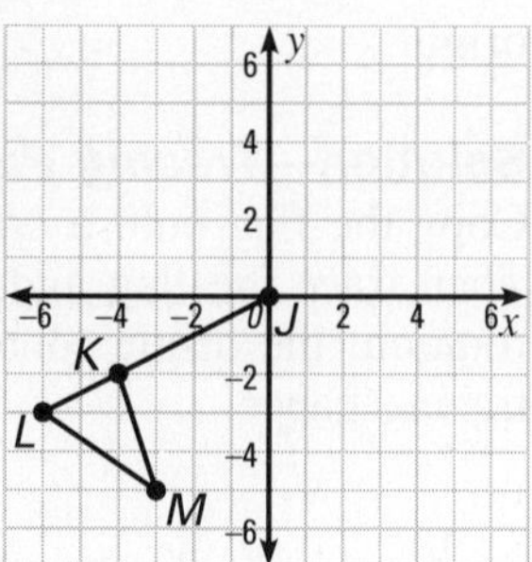


Copy $\triangle PQR$ with vertices $P(1, -2)$, $Q(5, -1)$, and $R(3, -5)$ on a coordinate plane. Then graph its image after each rotation about the origin by finding the coordinates of the vertices. Check your work by using tracing paper.

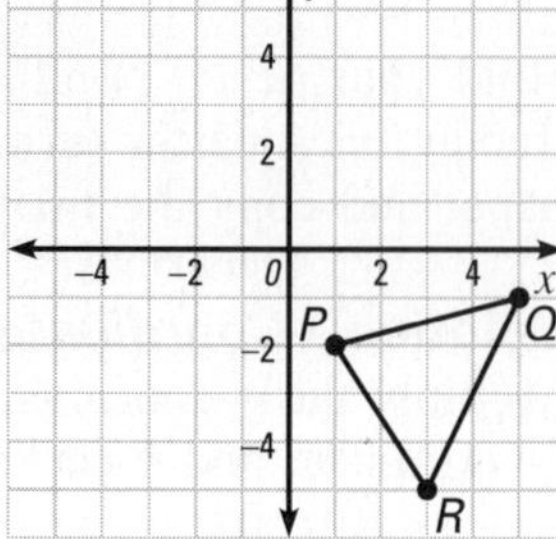

2. 90° clockwise
3. 90° counterclockwise

4. Copy $\triangle XYZ$ and its rotation image on dot paper or graph paper. Identify the center of rotation, $T$, the angle of rotation, and the direction of rotation.

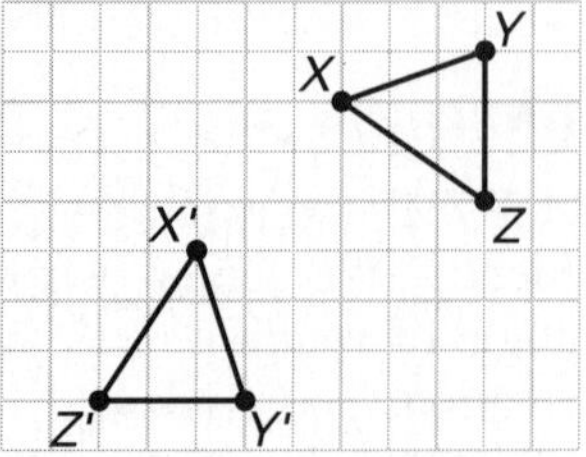


# EXERCISES

**PRACTICE/ SOLVE PROBLEMS**

Copy each figure on a coordinate plane. Then graph its image after each rotation about the origin.

1. 90° clockwise

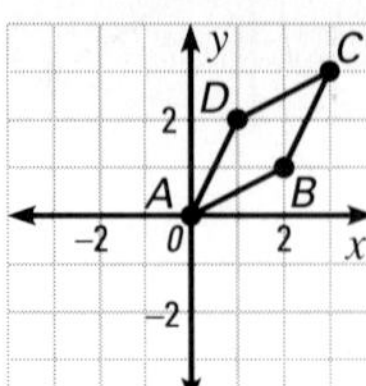


2. 180° clockwise

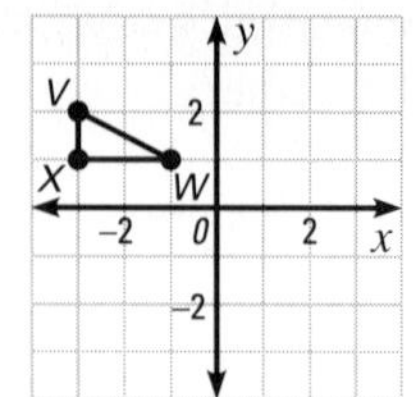


3. 90° counterclockwise

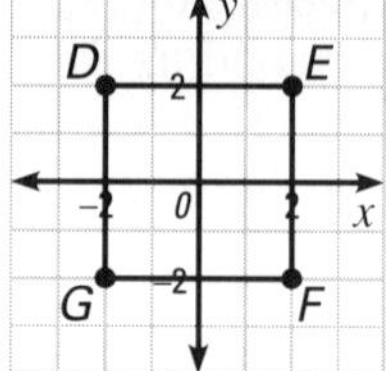

Trace each figure and its rotation image. Identify the center of rotation, *T*, the angle of rotation, and the direction of rotation.

**4.**

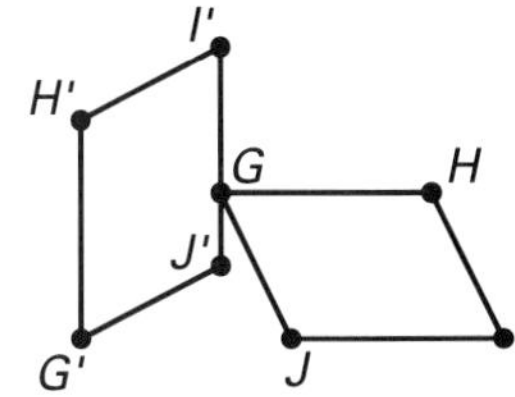


**5.**

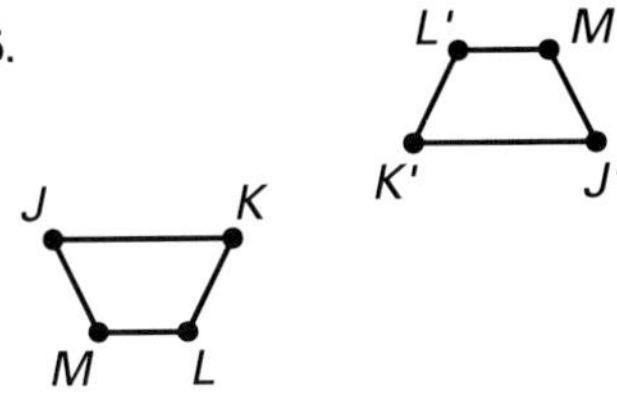


## EXTEND/ SOLVE PROBLEMS

Use the figure at the right for Exercises 6–11.

**6.** Which triangle is the rotation image of Triangle 1 about the point (−2, 0)?

**7.** Which triangle is the translation image of Triangle 1?

**8.** Which triangle is the reflection image of Triangle 1 across the *x*-axis?

**9.** Which triangle is the reflection image of Triangle 1 across the *y*-axis?

**10.** Which triangle is the rotation image of Triangle 1 after a turn of 180° clockwise about the origin?

**11.** Which triangle is the reflection image of Triangle 1 across the line $x = -2$?

## THINK CRITICALLY/ SOLVE PROBLEMS

**12.** Follow these steps for $\triangle CDE$.

**a.** Copy $\triangle CDE$ on a coordinate plane. Graph its image ($\triangle C'D'E'$) under a reflection across the *y*-axis. Then graph the reflection image of $\triangle C'D'E'$ across the *x*-axis.

**b.** Copy $\triangle CDE$ on a coordinate plane. Graph its image ($\triangle C'D'E'$) under a reflection across the *y*-axis. Then graph the line $x = 5$ and the reflection image of $\triangle C'D'E'$ across this line.

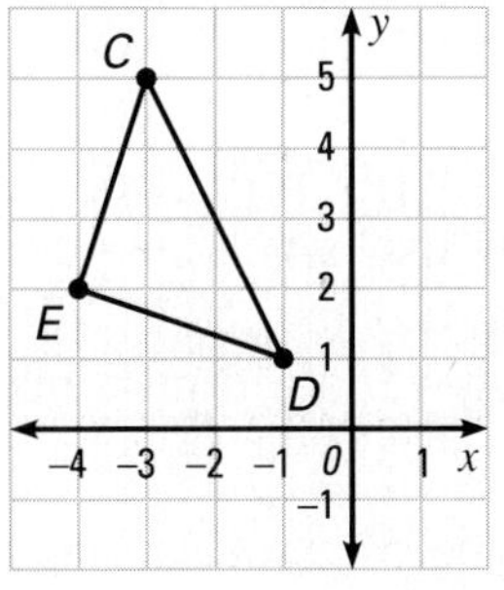


**13.** From your results in Exercise 12, make a generalization about which one transformation would produce the same image as two reflections across two parallel lines.

**14.** ***USING DATA*** Use the Data Index on pages 556–557 to find examples of African decorative patterns. Copy each pattern. Identify and describe examples of translations, reflections, and rotations.

# 11-5 Problem Solving Skills:

## USE TWO TRANSFORMATIONS

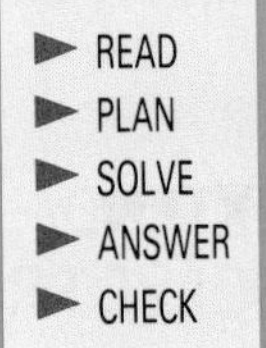


To produce the image of a figure in a certain place, you may need to use two transformations. Or, you may need to describe two transformations to explain how a given image was created.

### PROBLEM

These two footprints represent a figure and its image after two transformations. Describe these transformations.

### SOLUTION

Consider the left footprint the preimage and the right footprint the image. Use what you have learned about transformations to identify the two transformations represented by the footprints.

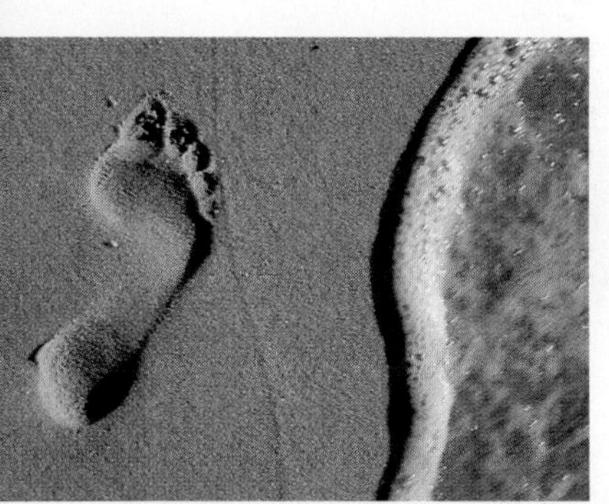

It appears that the left footprint has been reflected, or flipped, across an imaginary line running midway between the footprints. However, the right print is not directly across this line from the left print. It is moved up from where it would have been if it was simply reflected. You know that when a figure is slid along a plane in one direction, the move is a translation.

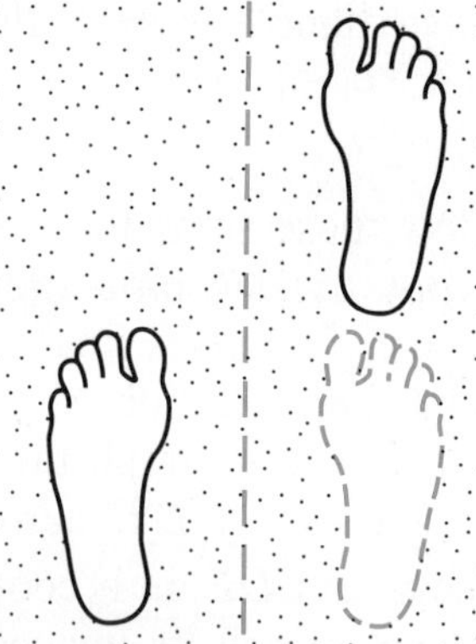

So, the transformations represented by these footprints are a reflection across a line running between the prints and a translation of a short distance up. The combination of a reflection followed by a translation, or vice versa, is called a **glide reflection.** A glide reflection is sometimes known informally as a *walk* because a good example of it is the pattern of a person's walking footprints.

To check your work, you could make a tracing of the left footprint. Then you could reflect, or flip, the tracing and slide it up to see if it matches the right footprint.

# PROBLEMS

For each problem, describe the two transformations used to create the image shown in blue. Check your work with tracings. For some problems, there is more than one possible answer.

1.

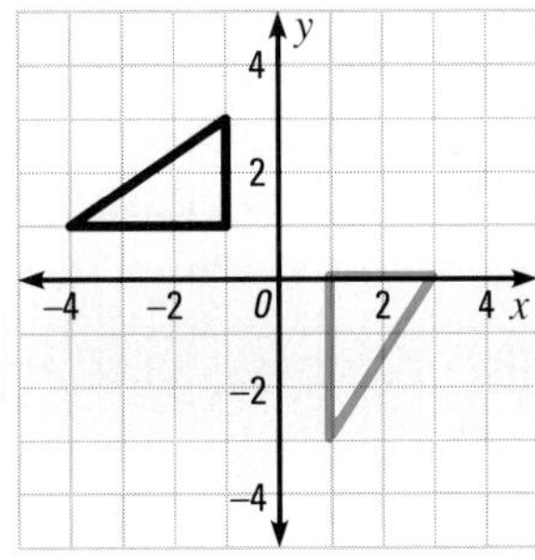

2.

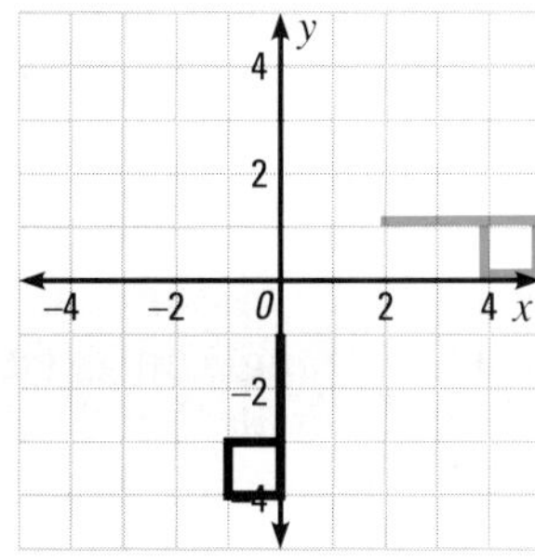

3.

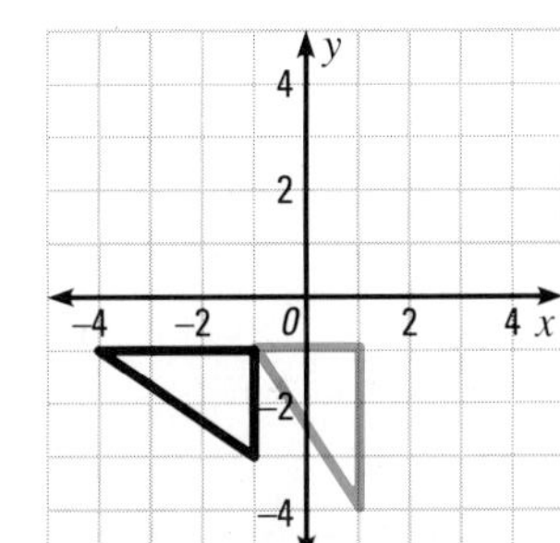

4.

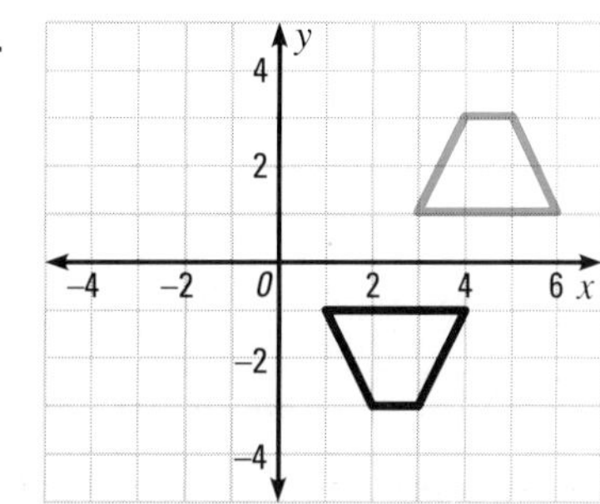

5.

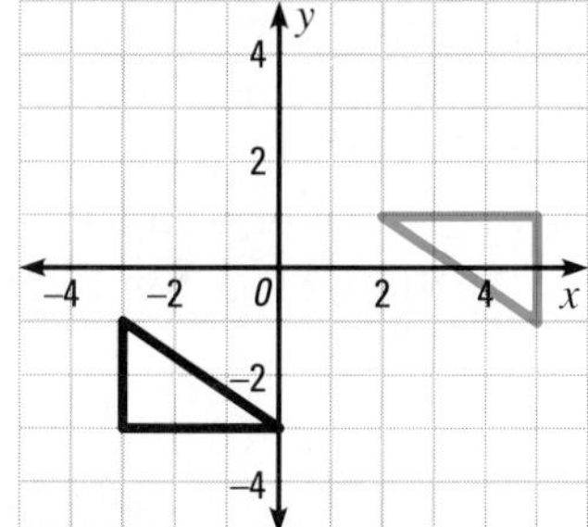

6.

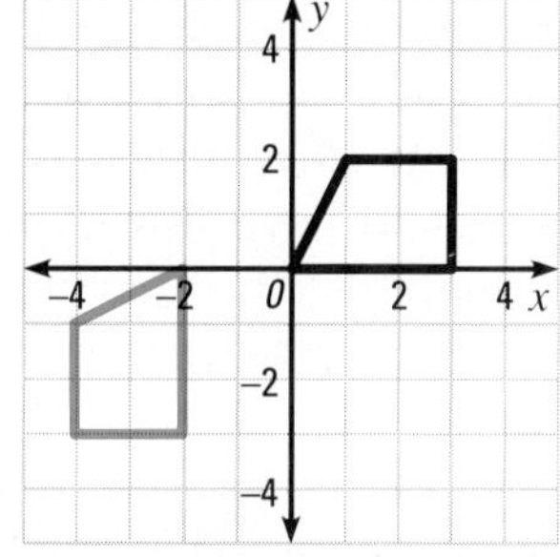

Work with a partner. Tell whether the order in which you perform each pair of transformations affects the image produced. If it does affect the image, sketch an example.

7. a translation followed by another translation
8. a translation followed by a reflection
9. a translation followed by a rotation
10. a reflection followed by another reflection
11. a reflection followed by a rotation
12. a rotation followed by another rotation

**PROBLEM SOLVING TIP**

For Problems 7–12, cut out a triangle and draw a coordinate plane. Test each pair of transformations by moving the triangle on the coordinate plane. First try the given order, then reverse the order.

# 11-6 Symmetry

## EXPLORE/ WORKING TOGETHER

Work with a partner. You will need scissors and a mirror.

a. Draw and cut out a scalene triangle, an isosceles triangle, and an equilateral triangle.

b. Place a mirror on each triangle in as many ways as possible so that the part of the triangle in front of the mirror along with the reflection in the mirror forms the original figure. For each, you may find one way, more than one way, or no way.

c. For each way you find in part **b** of forming the original triangle, have your partner hold the mirror in place while you draw a line on the triangle along the bottom edge of the mirror.

d. How many lines did you draw on each triangle?

## SKILLS DEVELOPMENT

The lines you drew in the Explore activity divided each figure into two identical parts. If you folded any of the triangles along one of the lines you drew on it, you would find that each side of the triangle exactly fits over the other side. Each of the lines you drew is called a **line of symmetry,** and each triangle on which you drew a line is said to have **line symmetry.** Lines of symmetry are the same as the lines of reflection you drew in Section 11–3 for figures that were reflections of themselves.

**CHECK UNDERSTANDING**

What types of triangles have line symmetry?

### Example 1

Trace each figure and draw all its lines of symmetry. If a figure has no lines of symmetry, write *none.*

a.

b.

c.

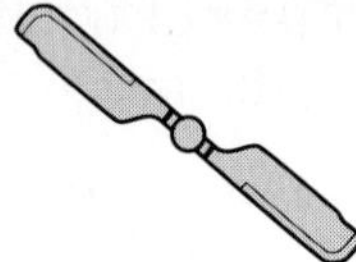

### Solution

a.

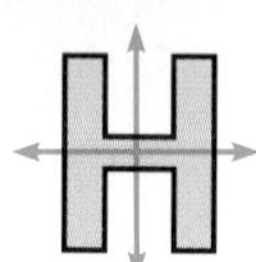

2 lines of symmetry

b.

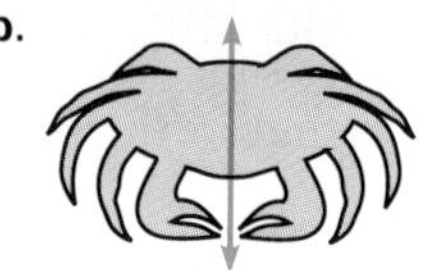

1 line of symmetry

c.

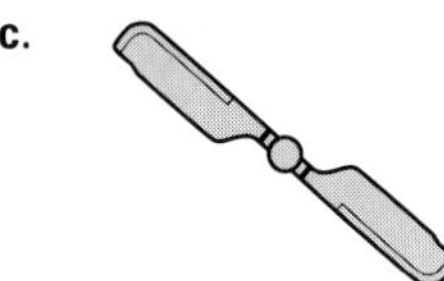

none

## Example 2

At the right is part of a figure along with that figure's line of symmetry. Complete the figure.

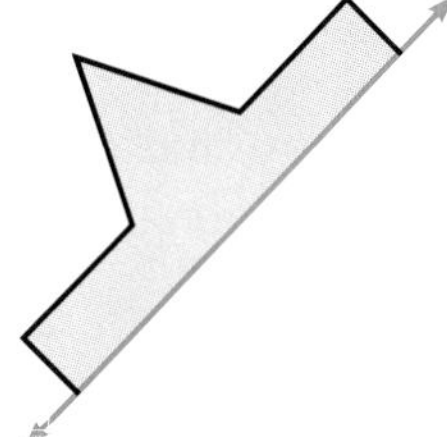

### Solution

Draw a reflection of the part of the figure shown across the line of symmetry.

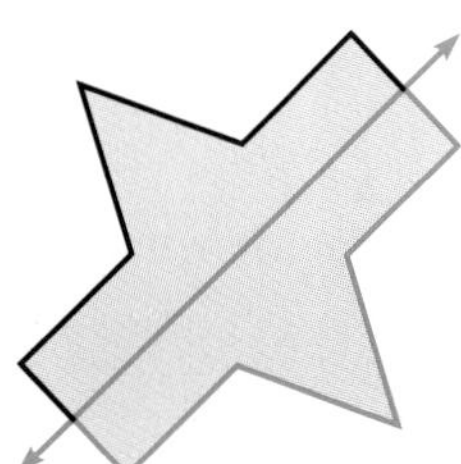

Sometimes a figure does not have line symmetry. However, the figure may have a point about which you can rotate it so that it fits exactly over the original position of the figure before you have turned it completely.

If you rotated the figure at the right about point $T$ one complete turn, it would fit over the original position of the figure 3 times. Therefore, this figure is said to have **rotational symmetry,** and its **order of rotational symmetry** is 3.

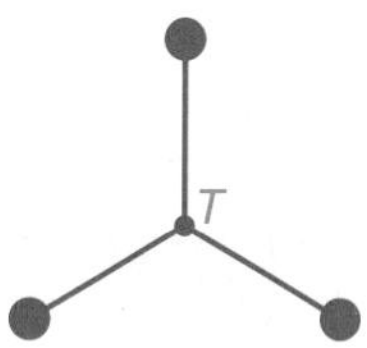


## Example 3

Give the order of rotational symmetry for each figure.

**a.**

**b.**

**c.**

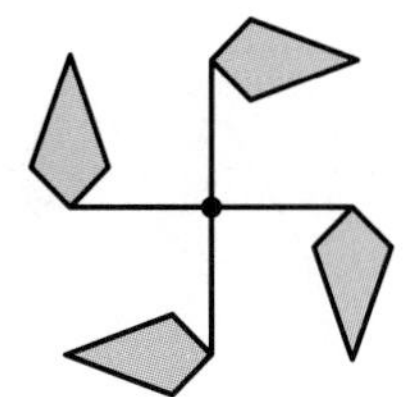

**TALK IT OVER**

Is it possible for a figure to have both line symmetry and rotational symmetry? If so, give an example.

### Solution

**a.** This figure fits over its original position 2 times during a complete turn, so its order of rotational symmetry is 2.

**b.** This figure fits over its original position 5 times during a complete turn, so its order of rotational symmetry is 5.

**c.** This figure fits over its original position 4 times during a complete turn, so its order of rotational symmetry is 4.

## TRY THESE

Trace each figure and draw all its lines of symmetry. If a figure has no lines of symmetry, write *none*.

1. 

2. 

3. 

Give the order of rotational symmetry for each figure.

4. 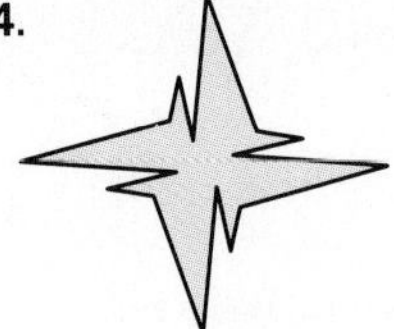

5. 

6. 

7. Copy the part of the figure at the right on graph paper along with its line of symmetry. Then complete the figure.

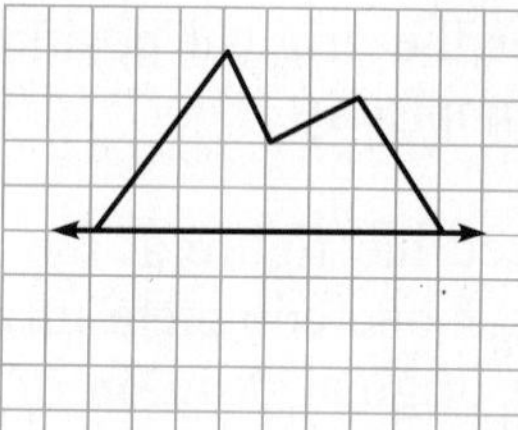

# EXERCISES

**PRACTICE/ SOLVE PROBLEMS**

Tell whether each dashed line is a line of symmetry. If it is not, trace the line and either part of the figure on one side of the line. Complete your drawing so the figure has line symmetry.

1. 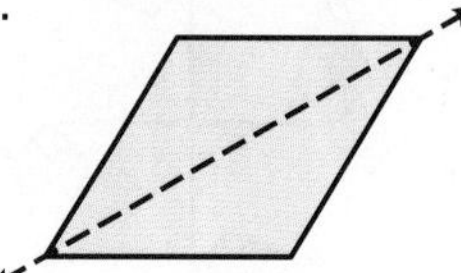

2. 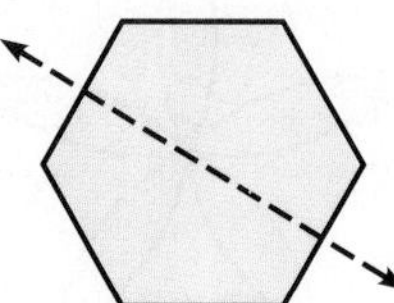

3. 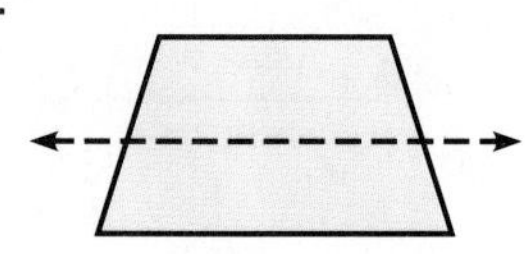

Give the order of rotational symmetry for each figure.

4. 

5. 

6. 

Try to draw each of the following figures on a piece of graph paper or dot paper. Show the line or lines of symmetry. If it is not possible to draw such a figure, write *not possible.*

## EXTEND/ SOLVE PROBLEMS

7. a quadrilateral with exactly
   a. 1 line of symmetry
   b. 2 lines of symmetry
   c. 3 lines of symmetry
   d. 4 lines of symmetry

8. a quadrilateral with an order of rotation
   a. of 2
   b. of 3
   c. of 4

## THINK CRITICALLY/ SOLVE PROBLEMS

9. How many lines of symmetry does a regular polygon with $n$ sides have?

10. What is the order of rotation of a regular polygon with $n$ sides?

11. How many lines of symmetry does a circle have?

12. What is the order of rotation of a circle?

A **game strategy** is an organized method of playing based on an analysis of the rules of the game. Here is a game for two players that has a winning strategy based on symmetry. Begin by placing a penny or counter on each vertex of a regular polygon. For example, you can use a regular 9-sided figure called a *nonagon*. The two players take turns removing one coin or two coins that lie on consecutive vertices. The player who removes the last coin wins.

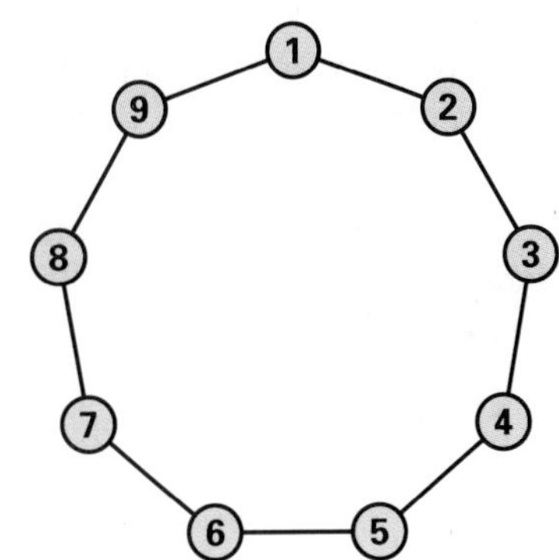


13. Play the game a few times with a partner. Take turns going first. What suggestions can you make for playing strategically?

14. Suppose the first player removes coins 5 and 6. Which coin(s) should you remove to make the arrangement of coins symmetrical? Draw a picture.

15. Continue to play symmetrically. Suppose your opponent removes coin 8 next. What should you do?

16. Explain how the game ends.

17. Suppose your opponent's first move was to take coin 9. What should your first move have been?

18. Try playing the game with a 15-sided polygon. How many coins should be in each group after the second player's first move?

# 11-7 Dilations

## EXPLORE

Study the two designs at the right.

a. How are the two designs the same?

b. How are the two designs different?

## SKILLS DEVELOPMENT

You have learned to use different types of transformations to produce new figures exactly like the original figures. A transformation that produces an image that is the same shape as the original figure but a different size is a **dilation.** A dilation used to create an image that is larger than the original figure is an **enlargement,** while one that is used to create a smaller figure is a **reduction.** A figure and its dilation image are similar.

The description of a dilation includes the **scale factor** and the **center of dilation.** The length of each side of the image is equal to the length of the corresponding side of the original figure multiplied by the scale factor. The distance from the center of the dilation to each point of the image is equal to the distance from the center of dilation to each corresponding point of the original figure times the scale factor.

### Example 1

Draw the dilation image of $\triangle ABC$ with vertices $A(-1, -1)$, $B(0, 2)$, and $C(1, -2)$ with the center of dilation at the origin and a scale factor of 3.

**Solution**

Multiply the $x$-coordinate and $y$-coordinate of each vertex by the scale factor of 3, using the rule $(x, y) \rightarrow (3x, 3y)$.

$A(-1, -1) \rightarrow A'(3 \times -1, 3 \times -1)$
$\rightarrow A'(-3, -3)$
$B(0, 2) \rightarrow B'(3 \times 0, 3 \times 2)$
$\rightarrow B'(0, 6)$
$C(1, -2) \rightarrow C'(3 \times 1, 3 \times -2)$
$\rightarrow C'(3, -6)$ ◄

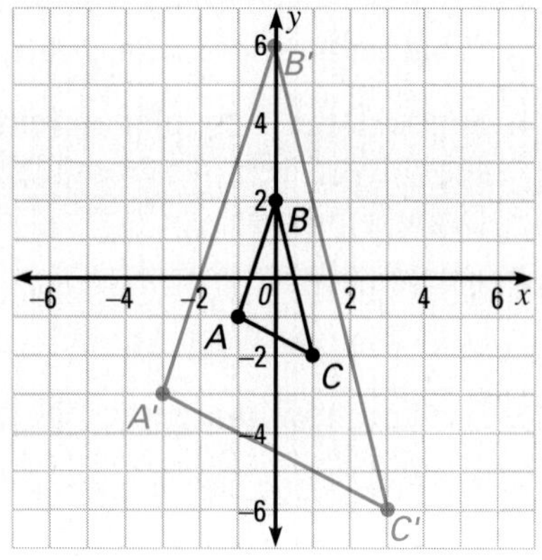


When the center of dilation is a vertex of the original figure, the corresponding vertex of the dilation image is the same point.

### Example 2

Draw the dilation image of $\triangle EFG$ with the center of dilation at $E$ and a scale factor of $\frac{1}{2}$.

What positive scale factors indicate an enlargement? What positive scale factors indicate a reduction?

**Solution**

The distance from the center of dilation, $E$, to $F$ is 6 units. So, the distance from $E$ to $F'$ is $\frac{1}{2} \times 6$, or 3 units. The distance from $E$ to $G$ is 8 units. So, the distance from $E$ to $G'$ is $\frac{1}{2} \times 8$, or 4 units. Points $E$ and $E'$ coincide. ◄

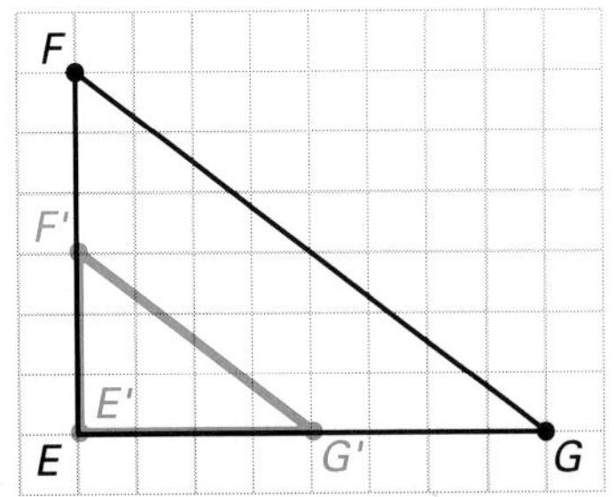


## TRY THESE

1. Copy rectangle $WXYZ$ on graph paper. Then draw its dilation image using $W$ as the center of dilation and a scale factor of 2.

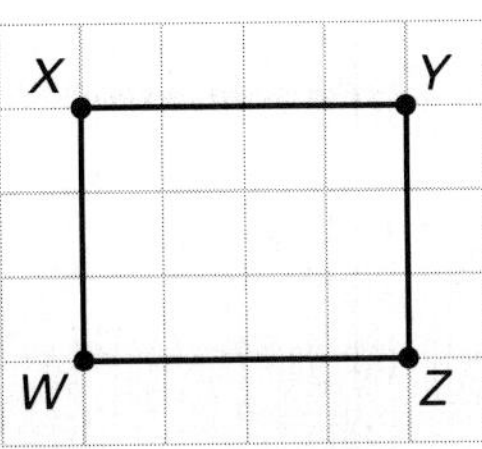


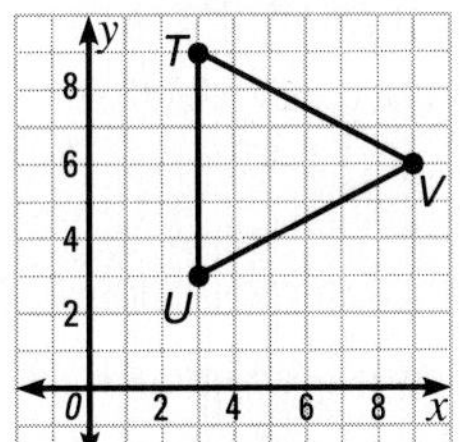


2. Copy $\triangle TUV$ on a coordinate plane. Then draw its dilation image with the origin as the center of dilation and a scale factor of $\frac{2}{3}$.

# EXERCISES

## PRACTICE/ SOLVE PROBLEMS

Copy parallelogram $PQRS$ on a coordinate plane. Draw its dilation image with the center of dilation at the origin and the given scale factor.

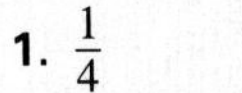

1. $\frac{1}{4}$ 2. 3

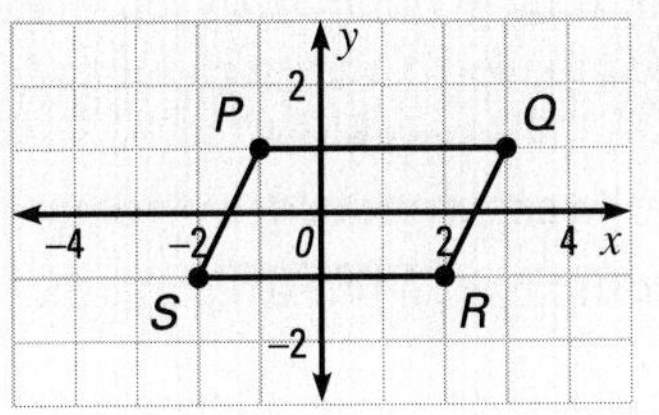


Copy trapezoid $HIJK$ on graph paper. Draw its dilation image with the center of dilation at $J$ and the given scale factor.

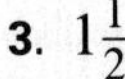

3. $1\frac{1}{2}$ 4. $\frac{3}{4}$

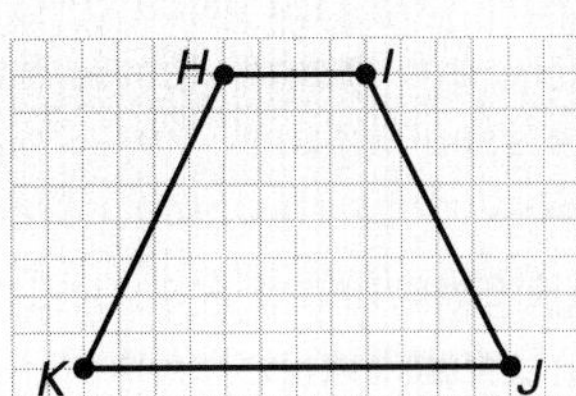

## EXTEND/ SOLVE PROBLEMS

The following sets of points are the vertices of figures and their dilation images. For every two sets of points, give the scale factor of the dilation.

5. $C(0, 0)$, $D(0, 4)$, $E(6, 0)$
   $C'(0, 0)$, $D'(0, 10)$, $E'(15, 0)$

6. $S(0, 0)$, $T(0, -3)$, $U(-3, -6)$
   $S'(0, 0)$, $T'(0, -1)$, $U'(-1, -2)$

For each pair of figures, describe the two transformations used to create the image.

7.

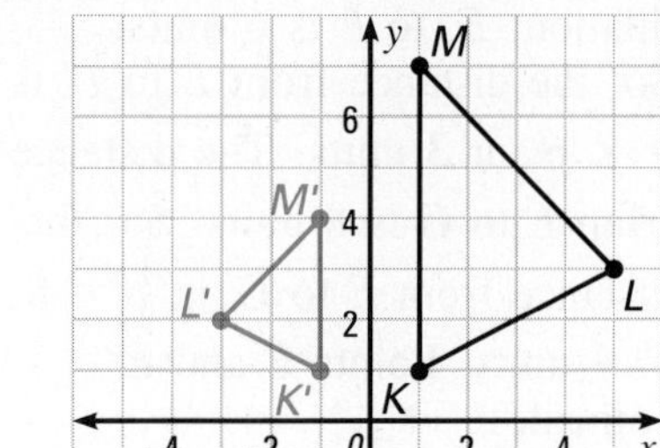


8.

9.

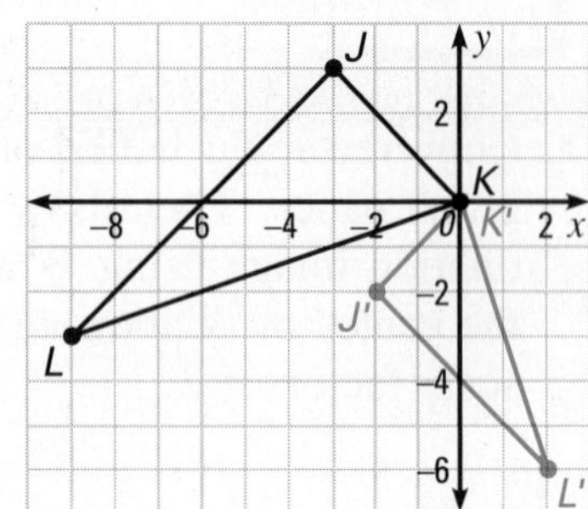


## THINK CRITICALLY/ SOLVE PROBLEMS

10. Draw $\triangle GHI$ with vertices $G(0, 0)$, $H(0, 6)$, and $I(3, 0)$ on a coordinate plane. Draw the dilation images of $\triangle GHI$ with the center of dilation at the origin and scale factors of 2 and $\frac{1}{2}$. Find the area of $\triangle GHI$, its enlargement, and its reduction.

11. Draw rectangle $TUVW$ with vertices $T(1, 1)$, $U(1, 4)$, $V(7, 4)$, and $W(7, 1)$ on a coordinate plane. Draw the dilation images of rectangle $TUVW$ with the center of dilation at $T$ and scale factors of 3 and $\frac{1}{3}$. Find the area of rectangle $TUVW$, its enlargement, and its reduction.

12. Generalize from the results of Exercises 10 and 11.
    a. How does the area of an image that was produced by enlarging a figure using a scale factor of $n$ compare to the area of the original figure?
    b. How does the area of an image that was produced by reducing a figure using a scale factor of $\frac{1}{n}$ compare to the area of the original figure?

► READ
► PLAN
► SOLVE
► ANSWER
► CHECK

# Problem Solving Applications:

## SELECTING APPROPRIATE SCALE FACTORS

Dilations can be used to enlarge or reduce designs. Many teenagers and adults enjoy using fabric paints to decorate items of clothing. Some people draw their own designs freehand. Other people find it easier to sketch a design on graph paper or to use a printed pattern such as the one at the right. When you use a pattern like the one at the right, however, you often need to enlarge or reduce the design to fit the space you want to decorate.

Suppose you want to put this design on a T-shirt. You decide that the greatest area you want to cover with the design is 6 inches square. You need to figure out the scale factor by which to enlarge the design.

1. Since the pattern is on $\frac{1}{4}$-inch graph paper, count the squares. Multiply by $\frac{1}{4}$ to find the width and length of the design.
2. Divide the available width and length by the width and length of the design to find possible scale factors.
3. You should have found a different possible scale factor for the width than for the length. But you need to use the same scale factor for all dimensions, or the design will be distorted. Which scale factor should you use? Why?

Select the most appropriate scale factor to produce a dilation of the design to fit an area of the given width and height.

4. width: 8 in.
   length: 15 in.
5. width: 3 in.
   length: 7 in.
6. width: $1\frac{1}{2}$ in.
   length: $2\frac{1}{2}$ in.

Follow these steps to prepare your own design.

7. Sketch a design that is no larger than 2 in. by 4 in. on $\frac{1}{4}$-inch graph paper.
8. Choose an item of your own clothing on which you would like to paint your design. Measure the area you want to fill.
9. Select the most appropriate scale factor to produce a dilation of your design to fit the area you want to fill.
10. Make the dilation of your design.
11. If you are able to get permission to do so, use fabric paint to actually paint your design on your clothing.

# CHAPTER 11 • REVIEW

Choose the word from the list that completes each statement.

1. An image produced by a flip across a line is a __?__.
2. A figure that can be turned part way to fit over its original position has __?__ symmetry.
3. The __?__ image of a figure has the same shape but is larger or smaller than the original figure.
4. When a figure is a reflection of itself, that figure has __?__ symmetry.
5. An image produced by a slide is a __?__.

a. dilation
b. reflection
c. rotational
d. line
e. translation

## SECTION 11–1 TRANSLATIONS (pages 428–431)

► Under a **translation,** an image is produced by sliding every point of the original figure the same distance in the same direction.

6. Graph the image of $\triangle STU$ under a translation 2 units left and 4 units up.

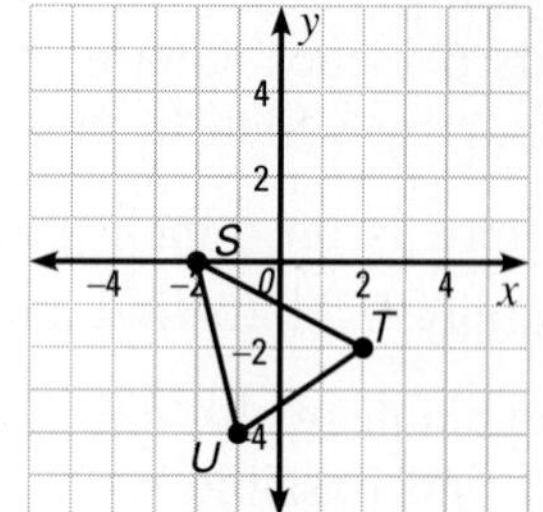


## SECTIONS 11–2 AND 11–5 PROBLEM SOLVING (pages 432–433, 442–443)

► You can sometimes solve a simpler problem by translating a figure on the coordinate plane so that one vertex is at the origin.

► Sometimes two transformations are used to produce the image of a figure.

7. Graph the image of the figure below under a translation that will give you a simpler problem to solve. Then use the distance formula to find the length of $\overline{EF}$ to the nearest tenth.

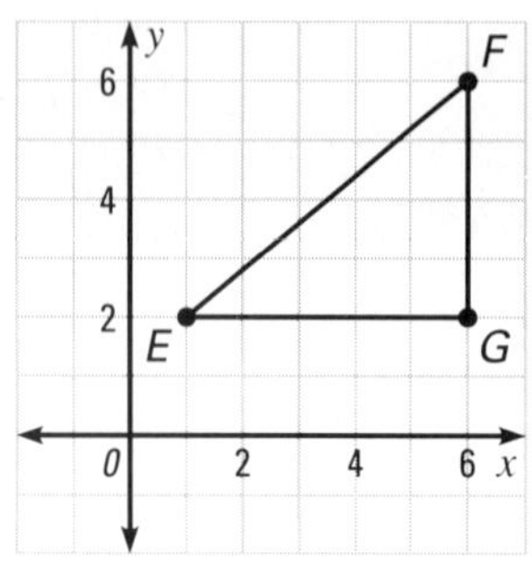


8. Describe two transformations that together could have been used to create the image shown in blue.

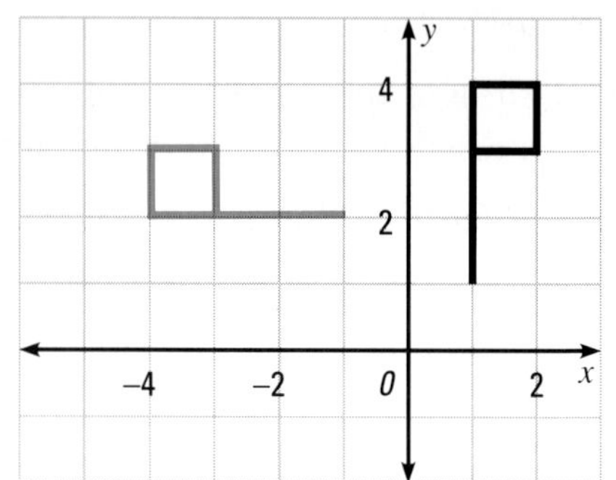

## SECTION 11–3 REFLECTIONS (pages 434–437)

► Under a **reflection,** a figure is flipped over a line of reflection.

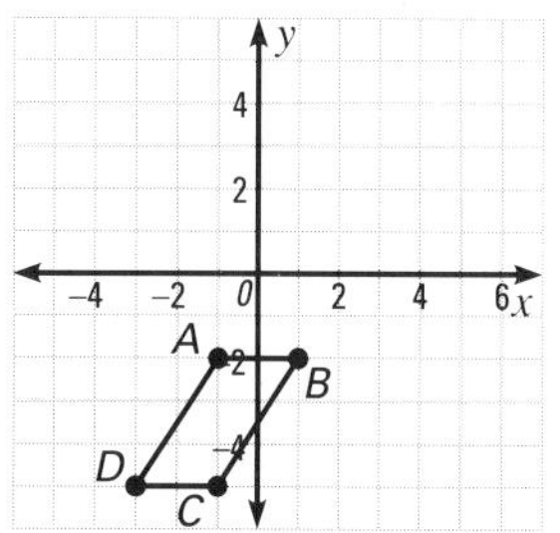


**9.** Graph the image of parallelogram $ABCD$ under a reflection across the line $y = -x$.

## SECTION 11–4 ROTATIONS (pages 438–441)

► Under a **rotation,** a figure is turned about a point.

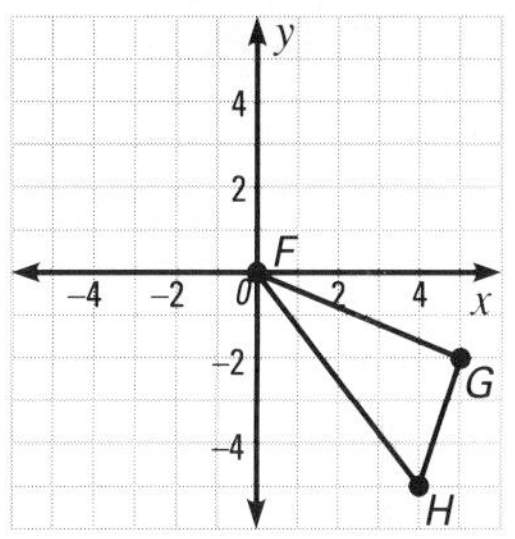


**10.** Graph the rotation image of $\triangle FGH$ after a turn of 90° clockwise about the origin.

## SECTION 11–6 SYMMETRY (pages 444–447)

► A figure has **line symmetry** if, when you fold it along a line, one side fits exactly over the other side.

► A figure has **rotational symmetry** if, when you turn it about a point, the figure fits exactly over its original position before the turn is completed.

**11.** Trace the figure at the left and draw all the lines of symmetry.

**12.** What is the order of rotational symmetry of the figure at the right?

11.

12.

## SECTION 11–7 DILATIONS (pages 448–451)

► A **dilation** is a transformation that produces the same shape but a different size image of a figure.

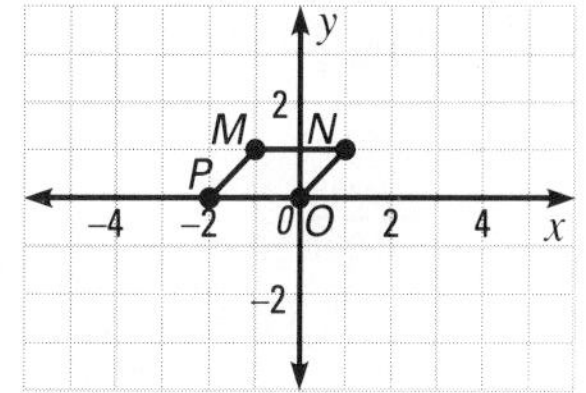


**13.** Draw the dilation image of trapezoid $MNOP$ with the center of dilation at the origin and a scale factor of 2.

**14.** ***USING DATA*** Refer to the diagram of the quilt on page 427. What scale factor was used to design the corner blocks of the border from the large blocks of the main part of the quilt?

# CHAPTER 11 ● TEST

Use the figure at the right for Exercises 1–3.

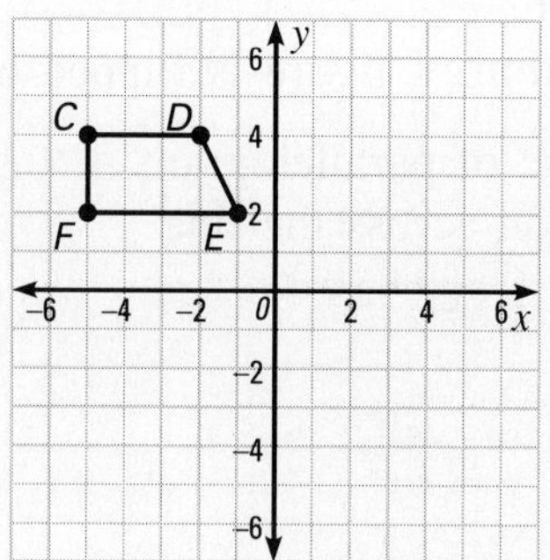

1. Graph the image of trapezoid $CDEF$ under a translation 3 units to the right and 4 units down.

2. Graph the image of trapezoid $CDEF$ under a reflection across the line $y = x$.

3. Graph the rotation image of trapezoid $CDEF$ after a 180° turn clockwise about the origin.

4. $\triangle T'U'V'$ is the reflected image of $\triangle TUV$. Graph the line of reflection.

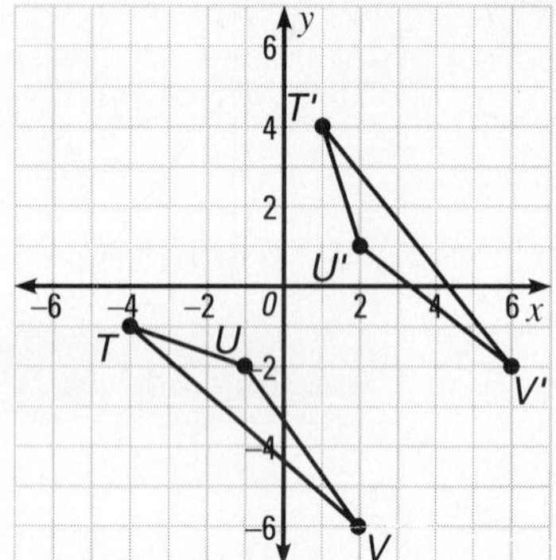

5. $\triangle A'B'C'$ is the rotation image of triangle $ABC$. Identify the center of rotation, $T$, the angle of rotation, and the direction of rotation.

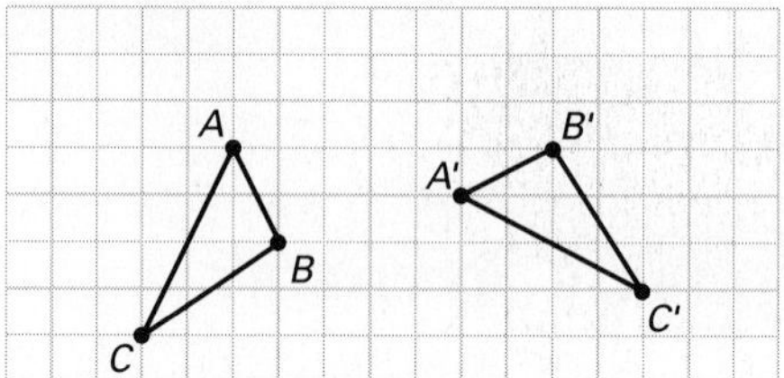


6. Describe two transformations that together could have been used to create the image shown in blue.

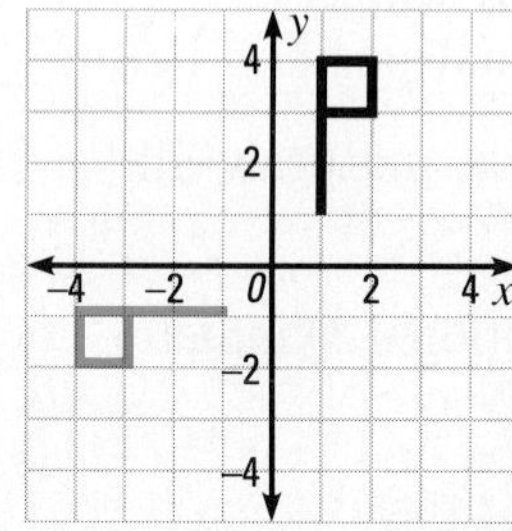

7. How many lines of symmetry does the figure below have?

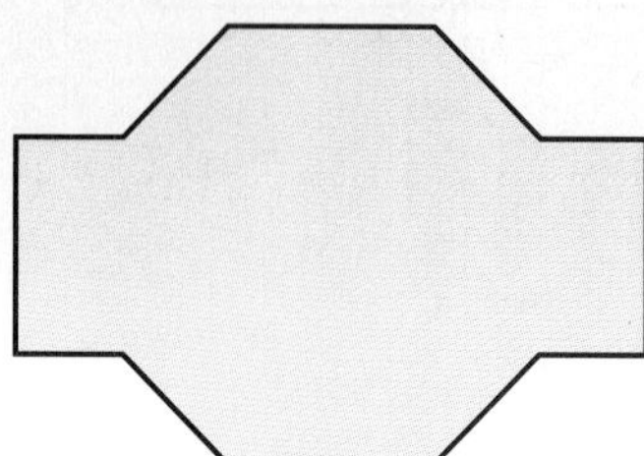

8. What is the order of rotational symmetry of the figure below?

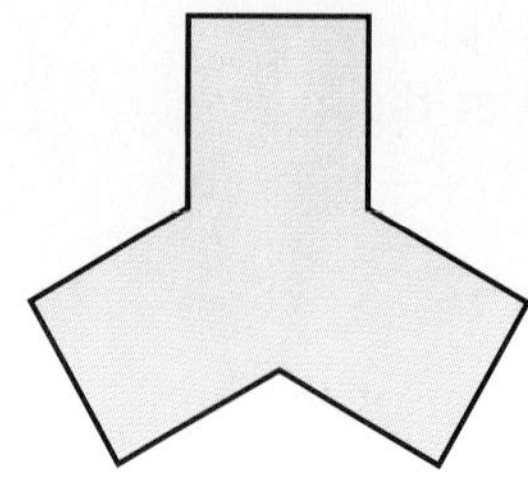

9. Draw the dilation image of $\triangle WXY$ with the center of dilation at the origin and a scale factor of $\frac{1}{3}$.

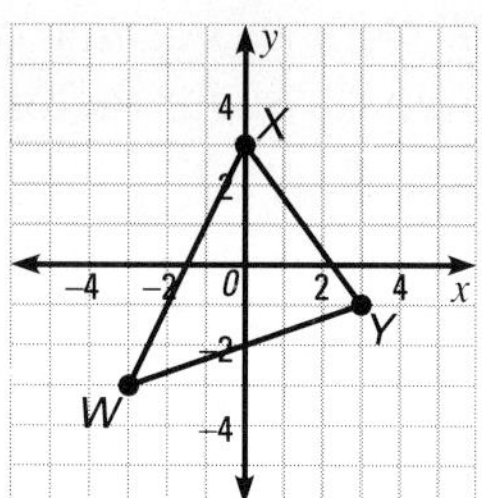

For each polygon, find the sum of the interior angles.

1. pentagon
2. nonagon

Use the graph for Exercises 3 and 4.

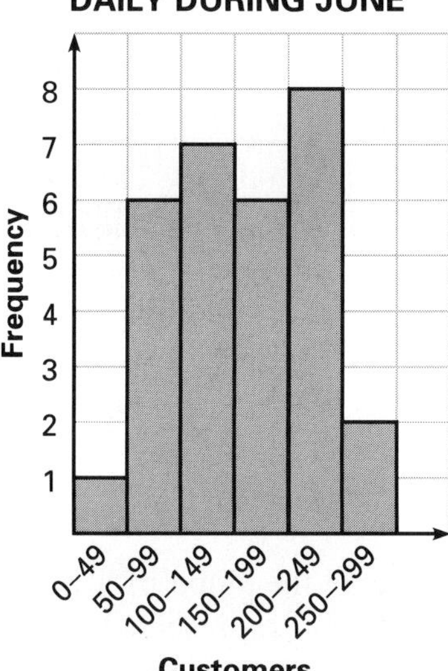


3. On how many days were there fewer than 150 customers?
4. On approximately what percent of the days were there at least 200 customers?

5. Perform the indicated operation.

$-3\frac{1}{2}\left(-4\frac{1}{3}\right)$

Solve the equation.

6. $x^2 = 169$

Find the unknown measure for each figure.

7. 

8. 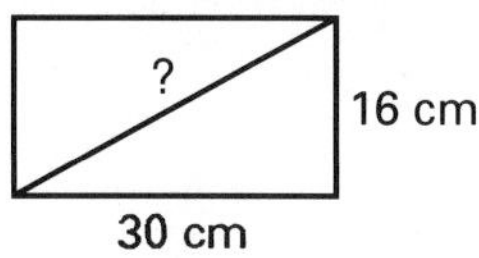


Find the product.

9. $(x + 2)(x - 1)$
10. $(b + 3)(b - 9)$

Find the LCM.

11. 15 and 60
12. $8x^3y^2$ and $12x^2y$

Use the distance formula to calculate the distance between each pair of points.

13. $A(9, 1)$ and $B(5, -2)$
14. $G(0, 1)$ and $H(0, 11)$

Write an equation of each line from the given information.

15. $P(-2, 3)$; $B(2, 1)$
16. $m = \frac{3}{2}$; $P(0, 2)$

17. On a scale drawing of a house, 2 in.:15 ft. The length of the house on the drawing is $4\frac{2}{3}$ in. Find the actual length of the house.

18. Triangles *ABC* and *DEF* are similar. Find the length of $\overline{BC}$.

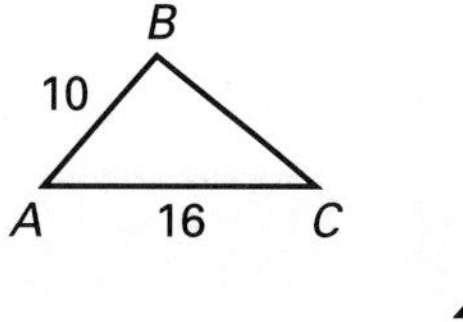


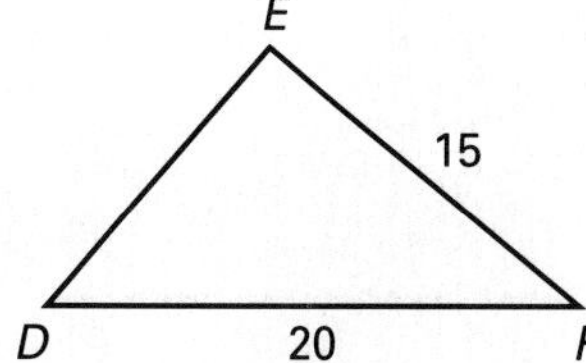


Give the coordinates of the image of $A(0, -5)$ under the translation.

19. 6 units left
20. 3 units up

1. Find the measure of $\angle 1$.

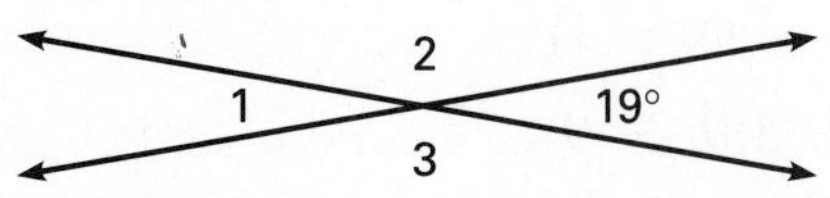


A. 161° B. 19°
C. 71° D. none of these

Use the table for Exercises 2 and 3.

2. How many vehicles get 30–34 mi/gal of gas?
A. 11 B. 6
C. 5 D. 10

3. About what percent of the vehicles get more than 24 mi/gal of gas?
A. 100% B. 21%
C. 28% D. 58%

MILES PER GALLON OF GAS

| Number of Mi per Gal | Frequency |
|---|---|
| 10–14 | 2 |
| 15–19 | 5 |
| 20–24 | 8 |
| 25–29 | 10 |
| 30–34 | 6 |
| 35–39 | 2 |
| 40–44 | 3 |

4. Perform the indicated operation. $\frac{1}{3}\left(-4\frac{1}{2}\right)$
A. $-1\frac{1}{2}$ B. $1\frac{1}{2}$ C. $-4\frac{1}{6}$ D. $4\frac{5}{6}$

5. Solve. $y^2 - 22 = 122$
A. $\pm\sqrt{122}$ B. $\pm\sqrt{110}$
C. $\pm 12$ D. $\pm 112$

6. Find the volume of the figure shown.

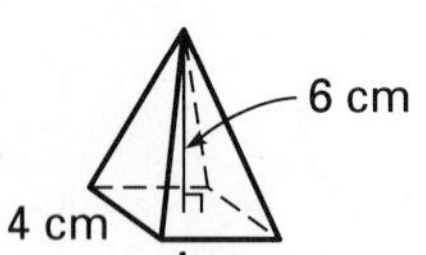


A. 48 cm³
B. 96 cm³
C. 32 cm³
D. 24 cm³

7. Find the height of the tent shown, if its volume is 756 ft³.

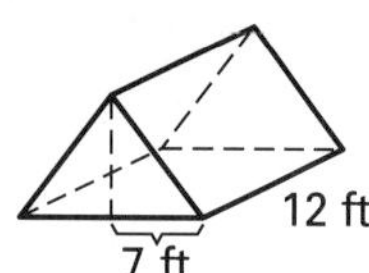


A. 9 ft B. 19 ft C. 8 ft D. 10 ft

8. Find the product. $(x - 3)(x - 4)$
A. $x^2 - x + 12$ B. $x^2 - 7x + 12$
C. $x^2 - 12x + 1$ D. $x^2 + 7x + 12$

9. Factor the polynomial. $y^2 - 16y + 64$
A. $(y - 8)^2$ B. $(y - 16)(y + 4)$
C. $(y - 2)^2$ D. $(y + 16)(y - 4)$

10. Which ordered pair is *not* a solution of the equation $4x = 2y - 2$?
A. (1, 3) B. (−1, −2)
C. (0, 1) D. (2, 5)

11. What is the distance between $P(8, -6)$ and $Q(2, 2)$?
A. 8 B. 10 C. $\sqrt{8}$ D. $\sqrt{52}$

12. Find the slope of the line.
$$2x - 3y = 6$$
A. $\frac{3}{2}$ B. $\frac{2}{3}$ C. $-\frac{3}{2}$ D. $-\frac{2}{3}$

13. If $m$ represents the slope of a line, which expression represents the slope of a line parallel to that line?
A. $-m$ B. $\frac{1}{m}$ C. $m$ D. $-\frac{1}{m}$

14. A machine seals 32 bottles in 3 minutes. How many minutes would it take to seal 80 bottles?
A. $7\frac{1}{2}$ B. $13\frac{1}{3}$ C. 15 D. 48

15. During May, a music store sold 8,000 CDs. Jazz made up 37% of the total. Which proportion could be used to find the number of jazz CDs that were sold?
A. $\frac{x}{37} = \frac{100}{8,000}$ B. $\frac{x}{8,000} = \frac{37}{100}$
C. $\frac{x}{100} = \frac{37}{8,000}$ D. $\frac{63}{100} = \frac{37}{x}$

16. Identify the coordinates of the image of (−2, 3) under a reflection across the $x$-axis.
A. (2, −3) B. (3, −2)
C. (−2, −3) D. (−3, 2)

# CHAPTER 12 SKILLS PREVIEW

1. In a random survey of 600 voters, 276 said that they would vote for Brown. What is the probability that Brown will be elected?

2. Suppose you roll a number cube with faces marked 1 through 6 and pick a letter from A, B, or C. How many possible outcomes are there?

3. In a random sample of 50 of the 1,650 students at Gaines High School, 9 said that they would join a school recycling association. How many students might be expected to join the association?

4. A store sells three brands of sneakers. Each brand is available in ten sizes and four colors. How many different pairs of sneakers does the store sell?

5. A wildlife organization sells T-shirts with a picture of a panda, an eagle, or a whale on the front. The shirts come in white, blue, black, and red. If a shirt is selected at random from a box that contains one of each type, find the probability that it is a shirt with a panda on the front.

The spinners shown here are both spun for each of Exercises 6–9.

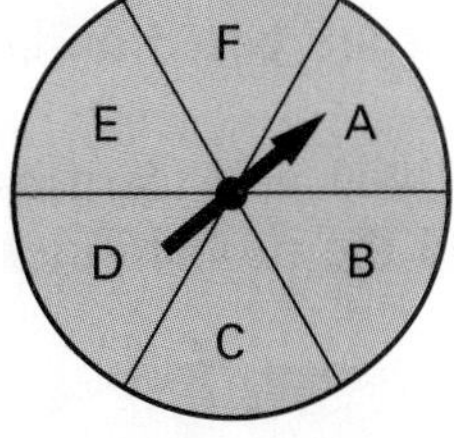

6. Find *P*(B, 5).

7. Find *P*(E or F, 3 or 4).

8. Find *P*(not A, 1).

9. Find *P*(vowel, even number).

10. Two number cubes with faces marked 1 through 6 are rolled. Find the probability that the numbers rolled are doubles or have a sum that is an odd number.

11. Find the probability that the sum of the numbers rolled on two number cubes is greater than 7 or a multiple of 4.

A bag contains two green marbles, four red marbles, and six yellow marbles.

12. Find *P*(red, then yellow) if you take one marble, replace it, and then take another marble.

13. Find *P*(red, then green) if you take one marble and then take another marble without replacing the first.

14. In how many ways could a class president and a class vice president be chosen from a class of 15 students?

15. How many volleyball teams of three players could be chosen from a group of twelve players?

# CHAPTER 12 A CLOSER LOOK AT PROBABILITY

**THEME** **Out-of-School Teen Activities**

You often hear such statements as "It will probably snow today" or "The Jays have a good chance of winning the championship." Intuitively, you know what these statements mean, even if you cannot give a mathematical explanation for them. One reason to study probability is to learn how to make the meanings of such statements precise by giving them numerical values.

Although many people think about probability in relation to games of chance, the concepts of probability can be applied to a wide variety of real-world situations that involve an element of uncertainty. For example, you may use probability to estimate how often a drug will be effective or to make business decisions such as where to locate a new movie theater and what types of films to show. Moreover, predictions that are the result of opinion polls or quality-control sampling are based on concepts of probability.

In this chapter, you will learn how to use data to make predictions and how to find the probabilities of chance events occurring.

A random group of high school students was asked how they spent a typical Saturday afternoon. Their responses are recorded in the table at the right.

**SATURDAY AFTERNOON ACTIVITIES OF HIGH SCHOOL STUDENTS**

| Playing Sports | Playing with Friends | Taking Classes | Part-Time Work/Chores | Pursuing Hobbies | Going to the Movies | Other |
|---|---|---|---|---|---|---|
| 36 | 58 | 16 | 28 | 23 | 11 | 28 |

## DECISION MAKING

### Using Data

Use the information in the table on page 458 to answer the following.

1. How many students were questioned in the survey?
2. What fraction of the students in the survey spent Saturday afternoons playing sports? What fraction spent Saturday doing part-time work or chores?
3. Assuming that the students surveyed are representative of students in your school, how many students in your school are likely to spend Saturday afternoons playing sports? How many are likely to do part-time work or chores?
4. How accurate do you think predictions based on the survey might be?
5. Do you think that a survey of students in your school would yield similar results to this survey? What might account for any differences between the results of a survey held at your school and this survey?

### Working Together

How similar are the preferences of the members of your class to those of the average American?

Find the results of a current national preference survey. The survey should focus on such topics as top-selling music albums, most-attended movies, or election outcomes. Then question the members of your class about *their* preferences in the survey category and compile your results in the same form as the data you have researched. As a class, create a chart or poster that compares the preferences of your classmates with those reflected in the survey. Discuss reasons for any significant differences.

# 12-1 Experiments and Probabilities

**EXPLORE/ WORKING TOGETHER**

There are two possible ways a thumbtack will land when it is dropped: with the point *up* or with the point *down*.

a. Prepare a table to tally the number of times a thumbtack lands with its point down and with its point up.

b. Drop a thumbtack onto the floor 20 times. Record the results.

| Event | Tally | Results |
|---|---|---|
| Point Up | ■ | ■ |
| Point Down | ■ | ■ |

c. Find the ratio of the number of times the tack landed point down to the total number of times it was dropped. How does your ratio compare with the ratios of other groups?

d. Add to find the total number of times a tack landed point down for your entire class, and divide that by the total number of times a tack was dropped. What is the ratio? Is it more likely that a dropped tack will land point down or point up?

**SKILLS DEVELOPMENT**

Most real-world situations involve more than one possible outcome. Often, these outcomes are not equally likely. While you cannot be sure of the outcome in advance, an experiment can help you find the likelihood of one particular outcome occurring. An **experiment** is an activity that is used to produce data that can be observed and recorded. The **relative frequency** of an outcome compares the number of times the outcome occurs to the total number of observations.

| | Number of Positive Responses | Total Number Questioned | Relative Frequency of Positive Responses |
|---|---|---|---|
| **Week 1** | 384 | 2,000 | $\frac{384}{2{,}000} = 0.192$ |
| **Week 2** | 420 | 2,000 | $\frac{420}{2{,}000} = 0.21$ |
| **Week 3** | 396 | 2,000 | $\frac{396}{2{,}000} = 0.198$ |
| **Week 4** | 412 | 2,000 | $\frac{412}{2{,}000} = 0.206$ |

Suppose you wanted to find whether a music department would be a successful addition to a store. If you asked 10 of the store's customers if they would shop at the proposed music department, you would not have enough data to make a reliable prediction. But if you asked 2,000 customers each week for a month, your prediction would be more accurate.

The **experimental probability** of an event, $E$, can be estimated by this formula.

$$P(E) = \frac{\text{number of observations favorable to } E}{\text{total number of observations}}$$

From the results in the table, the experimental probability that a customer would shop at the music department is about 0.20, or $\frac{1}{5}$.

## Example 1

The table shows the number of customers and skate rentals at a roller-skating rink during a week of summer vacation.

| | Customers | Pairs of Skates Rented |
|---|---|---|
| **Monday** | 192 | 130 |
| **Tuesday** | 328 | 212 |
| **Wednesday** | 296 | 222 |
| **Thursday** | 325 | 195 |
| **Friday** | 456 | 292 |

**a.** What is the probability that a customer will rent skates on Wednesday?

**b.** What is the probability that a customer will rent skates on Thursday?

### Solution

Use the experimental probability formula.

$$P(E) = \frac{\text{number of observations favorable to } E}{\text{total number of observations}}$$

$$P(\text{customers renting skates}) = \frac{\text{pairs of skates rented}}{\text{number of customers}}$$

**a.** $P(\text{customers renting skates on Wednesday}) = \frac{222}{296}$
$= 0.75$

The probability of a customer renting skates on Wednesday is 0.75.

**b.** $P(\text{customers renting skates on Thursday}) = \frac{195}{325}$
$= 0.6$

The probability of a customer renting skates on Thursday is 0.6. ◄

**TALK IT OVER**

Relative frequencies allow us to make statements that estimate probabilities.

What could you conclude if the relative frequency of an outcome was 1?

What could you conclude if the relative frequency of an outcome was 0?

## Example 2

The table shows the customers at a sporting goods store during the summer.

| CUSTOMERS BY AGE | | | | | |
|---|---|---|---|---|---|
| | 12-18 | 19-25 | 26-40 | 41-55 | Over 55 |
| **June** | 961 | 930 | 749 | 711 | 220 |
| **July** | 812 | 748 | 819 | 507 | 164 |
| **August** | 645 | 702 | 736 | 499 | 217 |

What is the probability that a July customer is over 40 years old?

### Solution

To find the probability, you need to know the total number of July customers and the number of July customers over 40 years old.

$$P(\text{July customer is over 40}) = \frac{\text{July customers over 40 years old}}{\text{Total number of July customers}}$$

Add the numbers for the age groups 41–55 and over 55. →

$$= \frac{507 + 164}{812 + 748 + 819 + 507 + 164}$$

$$= \frac{671}{3{,}050}$$

$$= 0.22$$

The probability that a customer is over 40 years old in July is 0.22. ◄

## TRY THESE

1. At a fork in a hiking trail, 82 people took the lake trail while 43 people took the mountain trail. What is the probability of a hiker taking the lake trail?

2. In a basketball tournament, 156 players were under 6 ft tall, 63 players were between 6 ft and 6 ft 6 in., and 21 players were over 6 ft 6 in. tall. What is the probability that a player is 6 ft tall or over?

# EXERCISES

### PRACTICE/ SOLVE PROBLEMS

**MENTAL MATH TIP**

How could you use your solution to Exercise 2a to solve Exercise 2b?

Use the table to answer Exercises 1–3.

**PEOPLE USING COLLEGE LIBRARY**

| | Undergraduate Students | Graduate Students | Professors |
|---|---|---|---|
| **Friday** | 737 | 105 | 33 |
| **Saturday** | 588 | 132 | 30 |
| **Sunday** | 448 | 91 | 36 |

1. On Friday, what is the probability that a person using the library is a graduate student?

2. On Saturday, what is the probability that a person using the library is
   a. a professor? b. a student?

### EXTEND/ SOLVE PROBLEMS

3. For the three-day period of library use, what is the probability that a person using the library is a professor?

4. Of the first 1,200 people to buy tickets to a concert, 840 did not want the most expensive seats. What is the probability that the next ticket buyer wants one of the most expensive seats?

5. At an automobile dealership, the ratio of red cars to cars of other colors that have been sold this year is 4 to 21. What is the probability that the next car the dealer sells will be red?

### THINK CRITICALLY/ SOLVE PROBLEMS

6. From a survey of the students in two grades, the probability of Ella being elected student-body president was found to be 0.55. In ninth grade, 280 of the 640 students asked said they would vote for Ella. In tenth grade, 500 students were asked. How many of the tenth graders said they would vote for Ella?

► READ
► PLAN
► SOLVE
► ANSWER
► CHECK

# Problem Solving Applications:

## DESIGNING A PROBABILITY EXPERIMENT

Most large companies conduct market research before they make a new product or open a new outlet in which to sell their products. One of the methods they use is to make a random survey of potential customers to determine the probability that their new product or location will receive a favorable response.

If your school were to open a store for its students, what items should it stock? Work with a partner to conduct your own market research. Copy the table below to record the results of your survey.

1. Choose four items that you think would sell well in the store. You will ask a random sample of students whether or not they would purchase each of these items regularly at the store.
2. Decide how many students from each grade you will question and what method you will use to keep your sample random.
3. For each item, find the relative frequency of positive responses by dividing the number of positive responses by the number of students surveyed.
4. Compare results with your classmates and compile an ordered list of the ten most popular items the store might carry.

| Students questioned in: | Grade ____ | Grade ____ | Grade ____ | Grade ____ | Grade ____ |
|---|---|---|---|---|---|
| ITEM A ________ Positive responses | | | | | |
| Number surveyed | | | | | |
| Relative frequency | | | | | |
| Probable customers | | | | | |
| ITEM B ________ Positive responses | | | | | |
| Number surveyed | | | | | |
| Relative frequency | | | | | |
| Probable customers | | | | | |
| ITEM C ________ Positive responses | | | | | |
| Number surveyed | | | | | |
| Relative frequency | | | | | |
| Probable customers | | | | | |
| ITEM D ________ Positive responses | | | | | |
| Number surveyed | | | | | |
| Relative frequency | | | | | |
| Probable customers | | | | | |

Total number of probable customers for each item:
Item A ________ Item B ________ Item C ________ Item D ________

**CONNECTIONS**

A Frenchman, Comte de Buffon (1707–1788), developed an interesting way of approximating $\pi$ by using probability. You may wish to try this experiment. Find a floor with parallel boards of the same width or draw parallel lines on a large sheet of paper. Cut a piece of wire of length one-half the distance between the equally spaced lines. Drop the wire onto the surface and keep a record of the number of times the wire touches the line and the number of times the wire falls between the lines. Perform at least 100 trials. The ratio of the total number of tosses to the number of times the wire touches a line should approximate $\pi$.

# 12-2 Problem Solving Strategies:
## EXPLORING WITH SIMULATIONS

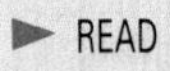

► READ
► PLAN
► SOLVE
► ANSWER
► CHECK

One problem solving strategy you have learned is to solve a simpler problem. This strategy can often be applied to problems that require you to determine experimental probability. In many cases, you can relate a complex probability problem to a simpler problem—a **simulation** of the complex problem.

Simulations can be done using coins, number cubes, or any method that involves random outcomes. When you need to conduct an experiment a great many times, it may be impractical to use such methods. A computer can be useful in generating and recording random outcomes in a very short time.

**TALK IT OVER**

How could you use a standard deck of 52 playing cards to simulate the following situations?

1. Cereal boxes include a free ring that is either red, blue, green, or yellow. If you bought 15 boxes of the cereal, how many red rings might you get?
2. Fortune cookies have 13 different fortunes. How many cookies would you have to open to read all 13 fortunes?

### PROBLEM

Experts believe that a safe new operation for arthritis will be successful about 70% of the times it is performed. If a hospital plans to perform the operation on 65 patients, how many operations might be successful?

### SOLUTION

Obviously, you cannot perform the 65 operations to find the experimental results. You can write a BASIC program to simulate the problem. Use the computer's random number generator, RND(1), to give a random decimal between 0 and 1.

In this problem, there are two outcomes: success and failure. The program should indicate that the operation is a success 70% of the time. The number generated by RND(1) is random, so there is a 70% probability that it will be less than 0.7.

The following BASIC program simulates the 65 operations.

```
10 S = 0                         ← The experiment begins with no successes.
20 FOR I = 1 TO 65               ← 65 operations
30 X = RND(1)                    ← Generate a random decimal.
40 IF X < .7 THEN S = S + 1      ← If the decimal < 0.7, increase S by 1.
50 NEXT I                        ← Simulate the next operation.
60 PRINT S                       ← total number of successes
70 END
```

Use the program to conduct a simulation of the problem.

# PROBLEMS

1. How would you change line 40 in the example if the operation had an equal probability of success or failure?

2. How would you change line 60 in the example to show the experimental probability that the operation was successful?

3. A BASIC program simulates flipping a coin to show the experimental probability of the coin showing heads. Which of the following values for I would you expect to result in an experimental probability closest to 0.5?
   **a.** 15  **b.** 1,500  **c.** 150  **d.** 15,000

4. At a farm, one-half of all the horses are brown and one-sixth of all the horses have braided manes. Design a simulation using a coin and a number cube to find the probability that the first horse a visitor to the farm sees is brown and has a braided mane.

5. The traffic light at the crossroads between Jet's house and her office shines red for 40 sec, then green for 55 sec, then yellow for 5 sec, and then repeats the sequence. Write a BASIC program that simulates the number of times Jet might have to stop at a red light the next hundred times she drives to work.

The following line in a BASIC program causes the computer to generate the integers 1, 2, 3, 4, 5, or 6 at random.

```
X = INT(6*RND(1)) + 1
```

6. Write a BASIC program that would simulate rolling a number cube 1,000 times and print the probability of rolling a 6.

7. How would you change the program so that it printed the experimental probability of rolling a 1 or a 2?

8. How would you program instructions for a computer to randomly generate the integers 3 or 4?

**MIXED REVIEW**

Give the number of significant digits for each measurement.

**1.** 18.00 m **2.** 3,000 ft
**3.** 0.005 L **4.** 1,0$\underline{0}$0 in.

Give the unit of precision for each measurement.

**5.** 9.0 m **6.** 2,9$\underline{0}$0 ft
**7.** 630 g **8.** 0.003 km

Use the given values to find four solutions of each equation. Write the solutions as ordered pairs.

**9.** $y = 3x - 4$ for $x = -5, -2, 0, 4$

**10.** $y = x^2 + 2$ for $x = -2, -1, 1, 2$

Solve each proportion.

**11.** $\frac{35}{n} = \frac{7}{11}$ **12.** $\frac{x}{20} = \frac{9}{75}$

# 12-3 Problem Solving Skills:
## MAKING PREDICTIONS

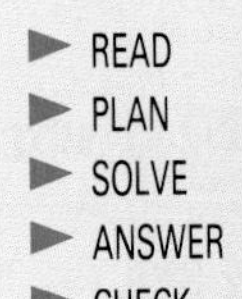


One important use of probability and statistics is to make predictions about present or future events. In order to make such predictions, people design and carry out a survey of a **random sample** of a larger group. A sample is random if each member of the larger group has an equal chance of being selected. The results of a random sample can be used to predict the probability of an outcome for much larger groups. Samples and probability are used to draw conclusions about large groups—such as the population of a country—when it is either impossible or impractical to survey each member of the group.

**CHECK UNDERSTANDING**

A school newspaper wants to predict how many students would be in favor of holding a school concert at the end of the semester.

Which of the following methods surveys a random sample of students?

1. asking every tenth person in line at the cafeteria
2. asking every student in the school band
3. asking every member of a senior class
4. asking every student whose first name begins with the letter J

### PROBLEM

A quality-control inspector tested a random sample of 2,000 light bulbs from a recent production run. The bulbs were gathered individually from different batches of the production run. Of the 2,000 bulbs selected, 3 bulbs were considered defective because they burned out before 150 hours of use. If the total number of light bulbs in the production run was 100,000, how many could be expected to be defective?

### SOLUTION

Find the experimental probability that a bulb is defective.

$$\begin{aligned} P(\text{defective}) &= \frac{\text{number of defective bulbs}}{\text{total number of observations}} \\ &= \frac{3}{2{,}000} \quad \leftarrow \textbf{number in random sample} \\ &= 0.0015 \end{aligned}$$

Substitute known values.

Let $n$ be the number of defective bulbs in the production run.

$$\begin{aligned} P(\text{defective}) &= \frac{n}{\text{total number of bulbs in production run}} \\ 0.0015 &= \frac{n}{100{,}000} \\ 100{,}000(0.0015) &= 100{,}000\left(\frac{n}{100{,}000}\right) \\ 150 &= n \end{aligned}$$

For a production run of 100,000 bulbs, 150 bulbs could be expected to be defective.

**TALK IT OVER**

Why didn't the quality-control inspector simply select the 2,000 bulbs from a single part of the production run?

# PROBLEMS

1. A manufacturer placed coupons at random in its packages of detergent. The coupons could be redeemed for cash at the store where the detergent was purchased. A store that sold 3,250 packages of the detergent redeemed 13 coupons. How many coupons would you expect a store that sold 8,500 packages of detergent to redeem?

2. An amusement park sold 27,000 admission tickets one week. During that week, one of the vendors at the park sold 810 bumper stickers. If the park sells a total of 300,000 admission tickets during the summer, how many bumper stickers would you expect the vendor to sell during the summer?

3. A survey of a random sample of 100,000 people in the United States showed that 48,700 of the people were male. If the total population of the country is 250,000,000, how many of the people would you expect to be male?

4. A company wants to advertise on the local cable television show *Teen Forum*. A random survey of 5,000 of the cable subscribers showed that 475 people watch *Teen Forum*. If there are 126,400 cable subscribers, how many people would the company expect to see their advertisement?

Reporters from the *Newtown Daily Register* asked a random sample of the city's 950,000 registered voters which of the three candidates for mayor they supported. The results are shown in the table.

| Candidate | Number of votes |
|---|---|
| Caprio | 905 |
| Mendez | 1,370 |
| Higgins | 1,087 |
| Undecided | 638 |

5. How many voters were in the sample?

6. How many voters in Newtown would you expect to be undecided?

7. If 50% of the undecided voters choose to vote for Caprio, how many votes would you expect Caprio to receive in the election if all registered voters go to the polls?

**USING DATA** Refer to the Data Index on pages 556–557 to locate information about the capacity of baseball stadiums.

8. A Seattle Mariners game at the Kingdome is sold out. Of the first 1,000 ticketholders to enter the stadium, 80 purchase Mariners baseball caps. How many people at the game would you expect to purchase Mariners caps?

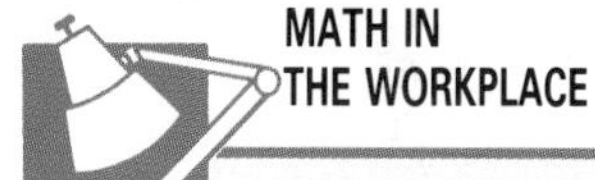

**MATH IN THE WORKPLACE**

Predicting probable outcomes is an important part of running many businesses. For one kind of business, it is the most important part.

Insurance companies collect data to find probabilities and analyze the probabilities to determine risk. They base their insurance rates on the probability that an event will occur and the amount it will cost them if it does occur. This is known as *actuarial science.*

# 12-4 Sample Spaces

**EXPLORE**

A game has cards marked with either a circle or a square.

a. If the shapes can be colored red, blue, or green, how many different kinds of cards can there be?
b. If the shapes can be colored red, blue, green, or yellow, how many different kinds of cards can there be?
c. How can you be sure you have counted each kind of card?

**SKILLS DEVELOPMENT**

The **sample space** for a probability experiment is the set of all possible outcomes of the experiment.

### Example 1

In an experiment, a coin is tossed and a number cube is rolled. How many possible outcomes are there?

**Solution**

The sample space can be shown as a set of ordered pairs. For the coin, let $H$ represent heads and $T$ represent tails. For the number cube, use the number for each face: 1, 2, 3, 4, 5, 6. The possible outcomes are:

$(H, 1)$ $(H, 2)$ $(H, 3)$ $(H, 4)$ $(H, 5)$ $(H, 6)$
$(T, 1)$ $(T, 2)$ $(T, 3)$ $(T, 4)$ $(T, 5)$ $(T, 6)$

There are 12 possible outcomes. ◄

A **tree diagram** can be used to show the sample space.

### Example 2

A pizza parlor offers three sizes of pizza: large (L), medium (M), and small (S). It also offers three toppings: cheese (C), peppers (P), and onions (O). How many different pizzas with one topping are there?

**Solution**

Use a tree diagram to show the sample space.

| SIZE | TOPPING | | OUTCOMES | | |
|---|---|---|---|---|---|
| S | C | → | SC | 3 outcomes | |
| | P | → | SP | | |
| | O | → | SO | | |
| | | | | + | |
| M | C | → | MC | 3 outcomes | 9 outcomes |
| | P | → | MP | | |
| | O | → | MO | | |
| | | | | + | |
| L | C | → | LC | 3 outcomes | |
| | P | → | LP | | |
| | O | → | LO | | |

The pizza parlor sells 9 different pizzas. ◄

In many problems, the sample space is too large to show as a set of ordered pairs or a tree diagram. The **fundamental counting principle** can help you find the number of outcomes in the sample space.

> If there are two or more stages of an activity, the total number of possible outcomes is the product of the number of possible outcomes for each stage of the activity.

## Example 3

A clothing store sells shirts in eight different sizes. For each size, there is a choice of five different colors, and for each color, there is a choice of six different patterns. If the store has one of each available type of shirt, how many different shirts does the store have?

### Solution

You must consider size, color, and pattern.

Multiply the number of possible outcomes for each.

| size | | color | | pattern | | |
|---|---|---|---|---|---|---|
| 8 | × | 5 | × | 6 | = | 240 |

The store has 240 different shirts. ◄

Probability is not always determined from the observed outcomes of an experiment. An **event** is an outcome or combination of outcomes. The **theoretical probability** of an event, $P(E)$, can be assigned using the number of favorable outcomes and the number of possible outcomes in the sample space.

$$P(E) = \frac{\text{number of favorable outcomes}}{\text{number of possible outcomes}}$$

You may think of theoretical probabilities of events in a sample space as representing the results of an *ideal* experiment.

**CHECK UNDERSTANDING**

What is the theoretical probability of getting heads on one toss of a coin?

Use the table below. In this case, what is the experimental probability of getting heads on one toss of a coin?

| Outcome | No. of Tosses |
|---|---|
| Heads | 60 |
| Tails | 40 |

## Example 4

A card is picked at random from a set marked with the numbers 1 through 12. Find $P$(odd number, greater than 5).

### Solution

There are twelve possible outcomes: 1, 2, 3, 4, 5, 6, 7, 8, 9, 10, 11, 12.
There are six odd numbers: 1, 3, 5, 7, 9, 11.
Only three odd numbers are greater than 5: 7, 9, and 11.
So, there are three favorable outcomes.

$$P(\text{odd number, greater than 5}) = \frac{3}{12} \begin{matrix} \leftarrow \text{favorable outcomes} \\ \leftarrow \text{possible outcomes} \end{matrix}$$
$$= \frac{1}{4}$$

So, $P(\text{odd number, greater than 5}) = \frac{1}{4}$. ◄

### Example 5

The faces of a cube are numbered 1 through 6, and the eight equal sections of a spinner are marked $A$ through $H$. If the number cube is rolled and the spinner is spun, find $P(1 \text{ or } 6, \text{vowel})$.

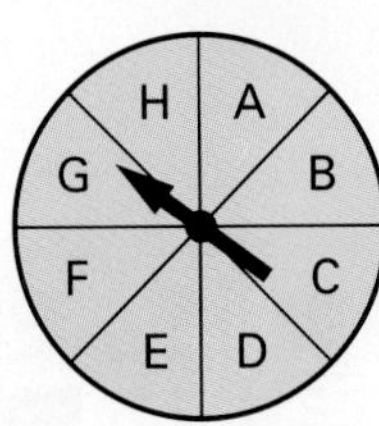


**Solution**

There are two numbers: 1 and 6. There are two vowels: $A$ and $E$.

Use the fundamental counting principle.

| cube | | spinner | | favorable outcomes | | cube | | spinner | | possible outcomes |
|---|---|---|---|---|---|---|---|---|---|---|
| 2 | × | 2 | = | 4 | | 6 | × | 8 | = | 48 |

$P(1 \text{ or } 6, \text{vowel}) = \frac{4}{48} = \frac{1}{12}$ ◄

## TRY THESE

1. Use a set of ordered pairs to show the sample space when a nickel and a dime are tossed.
2. A store sells baseball bats in three different weights and four different lengths. The bats can be either wooden or aluminum. How many different types of bats does the store sell?

What is the difference between theoretical probability and experimental probability?

Write a paragraph in your journal explaining the difference. Give examples of situations in which you might use each kind of probability.

For Exercises 3–6, a spinner with equal sections labeled $A$ through $D$ is spun and a number cube is rolled.

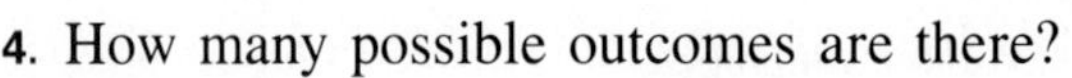

3. Use a tree diagram to show the sample space.
4. How many possible outcomes are there?
5. Find $P(A, 1)$ 6. Find $P(D, \text{odd number})$

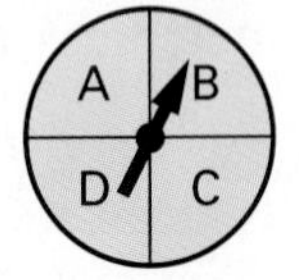


Suppose you spin the spinner at the right.

7. Find $P(\text{multiple of } 6)$
8. Find $P(\text{multiple of } 10 \text{ and less than } 50)$

## EXERCISES

**PRACTICE/ SOLVE PROBLEMS**

1. A spinner with four equal sections labeled 1 through 4 is spun and a coin is tossed. Show the sample space in each way.
   a. using ordered pairs b. using a tree diagram
2. At an audition for a three-piece band, there are eight guitarists, four bass players, and five drummers. How many different ways can a guitarist, bass player, and drummer be chosen?

The spinners shown here are both spun for each of Exercises 3–6.

3. How many possible outcomes are there?

4. Find $P(3, D)$ 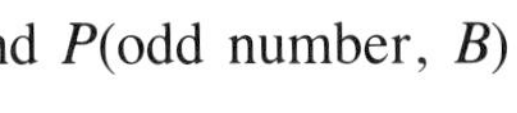 5. Find $P(\text{odd number}, B)$

6. Find $P(\text{number less than 4, vowel})$

7. A menu has a choice of two soups, five main dishes, and three desserts. How many different three-course meals can there be?

8. A combination lock has three dials, each numbered from 1 through 8. How many different ways can the lock be set?

9. A set of 26 cards marked *A* through *Z* is shuffled. The top card is turned over and a number cube is rolled. What is the probability of turning over a vowel and rolling a number greater than 2?

10. A store sells both oil-based and latex house paint. Each type of paint is available in twelve colors and in three different can sizes. How many different choices of paint does the store carry?

**PROBLEM SOLVING TIP**

To find the probabilities in Exercises 15 and 16, draw a tree diagram to show the sample space and the outcomes.

11. A jewelry company has designed five different ring styles. Each style can be made of either gold or platinum and set with either diamonds, rubies, or sapphires. If a sample of one of each kind of ring is made, how many sample rings are there?

12. Suppose one of the rings in Exercise 11 is chosen at random to be photographed for an ad. What is the probability that the ring is gold and set with rubies?

## EXTEND/ SOLVE PROBLEMS

13. A number cube is rolled three times. Find each probability.
    **a.** rolling all 6's **b.** rolling all even numbers **c.** not rolling any 6's

14. How many five-digit numbers can be made using the digits 1, 2, 3, 4, and 5? (A digit can be used more than once.)

15. If two nickels are tossed at the same time, is it more likely that they will show the same side or different sides?

16. If three nickels are tossed at the same time, what is the probability that they all show the same side when they land?

## THINK CRITICALLY/ SOLVE PROBLEMS

17. The ratio of the central angles of a spinner is 2:3:4. The sections of the spinner are marked, in the same order, *A*, *B*, *C*. If the spinner is spun, find $P(A)$.

18. A number cube is tossed twice and the outcomes are written down, in order, as a two-digit number. What is the probability that the two-digit number is a perfect square?

19. Darshana saw four puppies in the window of a pet store. She decided that since the chances were the same that each puppy was a male or a female, it was most likely that two of these puppies were male and two were female. Was she right? Explain.

# 12-5 Probability of Compound Events

## EXPLORE

The Venn diagram at the right shows {even numbers} and {multiples of 5} for the numbers 1 through 20.

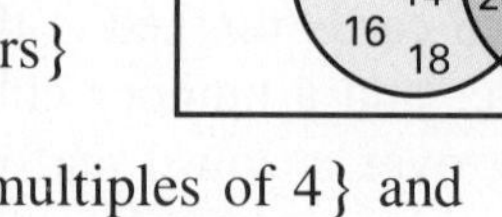


a. Draw a Venn diagram showing {numbers > 12} and {prime numbers} for the numbers 1 through 20.

b. Draw a Venn diagram showing {multiples of 4} and {prime numbers} for the numbers 1 through 20.

c. What is the main difference between your two Venn diagrams?

## SKILLS DEVELOPMENT

A **compound event** is one made up of two or more simpler events. Problems dealing with compound events may ask for the probability of one event *and* another event occurring. The probability of event $A$ and event $B$ occurring is written $P(A \text{ and } B)$. Other problems may ask for the probability of one event *or* another event occurring. The probability of event $A$ or event $B$ occurring is written $P(A \text{ or } B)$.

Sometimes two events are mutually exclusive. Events that are **mutually exclusive** are events that cannot occur at the same time.

When two events $A$ and $B$ are mutually exclusive, the probability of $A$ or $B$ can be found using the formula $P(A \text{ or } B) = P(A) + P(B)$.

**PROBLEM SOLVING TIP**

When you list the outcomes in a sample space, it is a good idea to double-check that you have listed all of them. You can often use the fundamental counting principle to find the total number of possible outcomes.

How would you use the fundamental counting principle to find the total number of possible outcomes when you roll two number cubes?

In this section, all number cubes are labeled 1 through 6.

### Example 1

Two number cubes are rolled.

a. Find the probability that the sum of the numbers rolled is even *and* greater than 9.

b. Find the probability that the sum of the numbers rolled is either 7 *or* 11.

### Solution

List the sample space for the rolls of two number cubes as ordered pairs.

There are 36 possible outcomes.

(1, 1) (1, 2) (1, 3) (1, 4) (1, 5) (1, 6)
(2, 1) (2, 2) (2, 3) (2, 4) (2, 5) (2, 6)
(3, 1) (3, 2) (3, 3) (3, 4) (3, 5) (3, 6)
(4, 1) (4, 2) (4, 3) (4, 4) (4, 5) (4, 6)
(5, 1) (5, 2) (5, 3) (5, 4) (5, 5) (5, 6)
(6, 1) (6, 2) (6, 3) (6, 4) (6, 5) (6, 6)

a. The even sums greater than 9 are 10 and 12. There are 4 outcomes with these sums: (6, 4), (5, 5), (4, 6), (6, 6).

So, $P(\text{even and greater than } 9) = \frac{4}{36}$ or $\frac{1}{9}$.

b. Six outcomes have a sum of 7: (1, 6), (2, 5), (3, 4), (4, 3), (5, 2), (6, 1). So, $P(7) = \frac{6}{36}$.

Two outcomes have a sum of 11: (5, 6), (6, 5). So, $P(11) = \frac{2}{36}$.

The sum of the numbers cannot be 7 and 11 at the same time, so the events are mutually exclusive.

$$\begin{aligned} P(7 \text{ or } 11) &= P(7) + P(11) \\ &= \frac{6}{36} + \frac{2}{36} \\ &= \frac{8}{36} = \frac{2}{9} \end{aligned}$$

The probability that the sum is 7 or 11 is $\frac{2}{9}$. ◄

**CHECK UNDERSTANDING**

Classify each of the following pairs of events as *mutually exclusive* or *not mutually exclusive.*

1. a card drawn at random from a standard deck is the 2 of hearts or the 2 of clubs
2. a card drawn at random from a standard deck is a king or a diamond
3. two number cubes show different numbers or have a sum of 10
4. two coins show two heads or a head and a tail

Events that are *not* mutually exclusive are events that can happen at the same time. The next example shows how to find the probability of two events that are not mutually exclusive.

## Example 2

Two number cubes are rolled. Find the probability that the sum of the numbers rolled is either an even number or a multiple of 3.

### Solution

Refer to the sample space for the rolls of the two number cubes shown in Example 1. The events are not mutually exclusive because a sum can be both even and a multiple of 3. Of the 36 outcomes, 18 are even sums, so $P(\text{even}) = \frac{18}{36} = \frac{1}{2}$.

Sums of 3, 6, 9, and 12 are multiples of 3. There are 12 sums that are multiples of 3. So, $P(\text{multiple of } 3) = \frac{12}{36} = \frac{1}{3}$.

However, sums that are even *and* a multiple of 3 have been counted twice. The sums that are even *and* a multiple of 3 are 6 and 12. There are 6 ordered pairs with these sums.

So, $P(\text{even and a multiple of } 3) = \frac{6}{36} = \frac{1}{6}$.

Subtract the probability of the sums that were counted twice.

$$P(\text{even or multiple of } 3) = \frac{1}{2} + \frac{1}{3} - \frac{1}{6} = \frac{2}{3}$$

You can check the answer by counting the ordered pairs in the same sample space with an even sum or a sum that is a multiple of 3. ◄

Example 2 illustrates that when two events $A$ and $B$ are not mutually exclusive, the probability of $A$ or $B$ can be found using the formula

$P(A \text{ or } B) = P(A) + P(B) - P(A \text{ and } B)$.

### Example 3

A card is drawn at random from a standard deck of 52 playing cards. Find the probability that the card is a heart or an ace.

**Solution**

The events are not mutually exclusive because a card can be both a heart and an ace.

$$P(\text{heart or ace}) = P(\text{heart}) + P(\text{ace}) - P(\text{heart and ace})$$

Of the 52 cards, there are 13 hearts, so $P(\text{heart}) = \frac{13}{52}$.

Of the 52 cards, there are 4 aces, so $P(\text{ace}) = \frac{4}{52}$.

There is 1 heart that is an ace, so $P(\text{heart and ace}) = \frac{1}{52}$.

$$P(\text{heart or ace}) = \frac{13}{52} + \frac{4}{52} - \frac{1}{52}$$
$$= \frac{16}{52} = \frac{4}{13}$$

The probability that the card is a heart or an ace is $\frac{4}{13}$. ◄

## TRY THESE

1. Two coins are tossed. Find the probability that the coins show two tails or two heads.
2. Two number cubes are rolled. Find the probability that the numbers rolled are doubles and have a sum that is a multiple of 4.
3. Two number cubes are rolled. Find the probability that the numbers rolled are doubles or have a sum of 4.
4. A card is drawn at random from a standard deck of 52 cards. Find the probability that it is a red card or a jack.

## EXERCISES

**PRACTICE/ SOLVE PROBLEMS**

1. Two number cubes are rolled. Find the probability that the sum of the numbers is 5 or 6.
2. A card is drawn at random from a standard deck of 52 cards. Find the probability that the card is a 10 or a jack.
3. Two number cubes are rolled. Find the probability that the sum of the numbers is odd and is less than 6.
4. A card is drawn at random from a standard deck of 52 cards. Find the probability that the card is a diamond or a face card (jack, queen, or king).

The spinner shown at the right is spun once. Use the diagram of the spinner to solve Exercises 5–8.

5. Find $P$(red or 7).

6. Find $P$(red and an even number).

7. Find $P$(green or less than 6).

8. Find $P$(yellow or greater than 5).

9. A card is drawn at random from a standard deck of 52 cards. Find the probability that the card is a club or an even number.

10. Three coins are tossed. Draw a tree diagram of the outcomes and find the probability that the coins show three heads or three tails.

## EXTEND/ SOLVE PROBLEMS

11. Six coins are tossed. Use the fundamental counting principle to find $P$(all heads or all tails).

12. A box contains nine ballpoint pens and nine felt-tip pens. Of the ballpoint pens, three are blue, three are red, and three are black. Of the felt-tip pens, three are blue, three are red, and three are black. If a pen is picked at random from the box, what is the probability that it is a ballpoint pen or it is blue?

A photo album contains photos of Joe, Karen, or both Joe and Karen. Joe is in 25 of the photos and Karen is in 30 of them. Joe and Karen are together in 15 of the photos.

13. How many photos are in the album?

14. If a photo is picked at random, what is the probability that it shows only Joe or only Karen?

15. One card is drawn from each of two standard decks of playing cards. Find the probability that one of the cards is a club and the other card is a 5.

## THINK CRITICALLY/ SOLVE PROBLEMS

16. Three coins are tossed. Let $A$ be "at least two heads" and let $B$ be "at most two heads." Find $P(A \text{ or } B)$.

The *complement* of the event $A$ is the event *not A*. For example, if you roll two number cubes, the event *the sum is 3* and *the sum is not 3* are complementary.

17. Does $P(A$ or complement of $A) = P(A) + P($complement of $A)$? Explain.

18. What does $P(A) + P($complement of $A)$ equal? Explain.

19. Two number cubes are rolled. Find the probability that the sum of the numbers is neither a multiple of 3 nor a multiple of 4.

# 12-6 Independent and Dependent Events

## EXPLORE

Refer to the set of cards at the right. Assume the cards are placed face down.

a. A card is selected at random, then replaced. Then another card is selected. Make a tree diagram to show the sample space. Use the tree diagram to find $P(\text{B, then L})$.

b. A card is selected at random, but it is not replaced. Then another card is selected. Make a tree diagram to show the sample space. Use the tree diagram to find $P(\text{B, then L})$.

## SKILLS DEVELOPMENT

Two events are said to be **independent** if the result of the second event is not affected by the result of the first event.

If $A$ and $B$ are independent events, the probability of both events occurring is the product of the probabilities of the individual events.

$$P(A \text{ and } B) = P(A) \times P(B)$$

Sometimes $P(A \text{ and } B)$ is written $P(A, \text{ then } B)$ to emphasize that $A$ and $B$ *do not* characterize a single event.

### Example 1

A bag contains three red marbles, four green marbles, and five blue marbles. One marble is taken at random from the bag and replaced. Then another marble is taken at random. Find the probability that the first marble is red and the second marble is blue.

**Solution**

Because the first marble is replaced, the sample space of 12 marbles does not change for each event. The two events are independent.

$$\begin{aligned} P(\text{red, then blue}) &= P(\text{red}) \times P(\text{blue}) \\ &= \frac{\text{number of red marbles}}{\text{total number of marbles}} \times \frac{\text{number of blue marbles}}{\text{total number of marbles}} \\ &= \frac{3}{12} \times \frac{5}{12} \\ &= \frac{15}{144} = \frac{5}{48} \end{aligned}$$

The probability of picking red, then blue, is $\frac{5}{48}$. ◀

If the result of one event *is* affected by the result of another event, the events are said to be **dependent.** If two events are dependent, the probability of them both occurring is the product of $P(\text{first event})$ and $P(\text{second event, given that the first event has occurred})$.

### Example 2

A bag contains three red marbles, four green marbles, and five blue marbles. One marble is taken at random from the bag. It is not replaced. Then another marble is taken at random. Find the probability that the first marble is red and the second marble is blue.

**Solution**

Because the first marble is not replaced, the second event is dependent on the first event. Find the probability of the first event.

$$P(\text{red}) = \frac{\text{number of red marbles}}{\text{total number of marbles}} = \frac{3}{12} = \frac{1}{4}$$

On the next selection, there are only eleven marbles. Assuming that a red marble was removed, there are still five blue marbles in the bag.

$$P(\text{blue after red}) = \frac{\text{number of blue marbles}}{\text{total number of marbles}} = \frac{5}{11}$$

Multiply the probabilities.

$$P(\text{red, then blue}) = \frac{1}{4} \times \frac{5}{11}$$
$$= \frac{5}{44}$$

The probability of picking red, then blue, is $\frac{5}{44}$. ◄

**TALK IT OVER**

In Example 2, how would the solution change if you were asked to find *P*(blue, then red)? Why?

### Example 3

A bag contains three red marbles, four green marbles, and five blue marbles. Two marbles are taken at random from the bag. Find the probability that both marbles are blue.

**Solution**

To solve the problem, consider that one of the marbles *must* be taken before the other. Since the first marble is not replaced, the second event is dependent on the first event.

$$P(\text{first blue marble}) = \frac{\text{number of blue marbles}}{\text{total number of marbles}} = \frac{5}{12}$$

$$P(\text{second blue marble}) = \frac{\text{number of blue marbles}}{\text{total number of marbles}} = \frac{5-1}{12-1} = \frac{4}{11}$$

Multiply the probabilities.

$$P(\text{blue, then blue}) = \frac{5}{12} \times \frac{4}{11}$$
$$= \frac{20}{132} = \frac{5}{33}$$

The probability of picking blue, then blue is $\frac{5}{33}$. ◄

**CHECK UNDERSTANDING**

Are the two events independent or dependent?

1. rolling a number cube twice
2. picking two cards without replacing the first one
3. picking a class president and then picking a class vice president
4. tossing three coins

## TRY THESE

A box contains ten 1989 pennies, four 1990 pennies, and six 1991 pennies. Pennies are taken at random from the box, one at a time, and then put back. Find each probability.

1. *P*(1989, then 1990)
2. *P*(1991, then 1990)
3. *P*(1989, then 1989)

A bag contains five blue, three red, and two green marbles. Marbles are taken at random and are not replaced. Find each probability.

4. $P$(red, then blue) 5. $P$(red, then red) 6. $P$(blue, then green)

7. The mail carrier is delivering six bills and three letters to the Jensen family. The wind blows two of the envelopes away. What is the probability that both of the envelopes are bills?

# EXERCISES

## PRACTICE/ SOLVE PROBLEMS

A group of numbered cards contains one 1, two 2s, three 3s, and four 4s. Cards are picked at random, one at a time, and then replaced. Find each probability.

1. $P$(1, then 2) 2. $P$(3, then 4) 3. $P$(4, then 4)

A box contains seven green counters, three red counters, and five blue counters. Counters are taken at random from the box, one at a time. They are not put back. Find each probability.

4. $P$(green, then red) 5. $P$(blue, then green) 6. $P$(red, then blue)

7. A farmer buys individual letters to make a sign saying NO TRESPASSING. On the way home, the farmer loses three of the letters. What is the probability that he lost *all* *S*s?

## EXTEND/ SOLVE PROBLEMS

8. Jill has two one-dollar bills, three five-dollar bills, and four ten-dollar bills in her pocket. Without looking, she pulls two bills out of her pocket. What is the probability that she pulls $15?

9. Five chicken, six cheese, and four egg sandwiches are packed in identical bags. If the bags are picked at random, find the probability that the first three picks include one of each kind.

## THINK CRITICALLY/ SOLVE PROBLEMS

The Bears and the Pumas are playing a "three out of five" series to determine a league champion. The Bears are the better team, with an estimated probability of $\frac{2}{3}$ of winning any game.

10. What is the probability that the Bears win the series in three straight games?

11. The Bears can win the series in four games, losing one and winning three. In how many ways could this happen? What is the probability that the Bears win the series in four games?

12. The series could go to five games. In how many ways could the Bears win a five-game series? What is the probability of the Bears winning a five-game series?

- READ
- PLAN
- SOLVE
- ANSWER
- CHECK

# Problem Solving Applications:

## FAIR AND UNFAIR GAMES

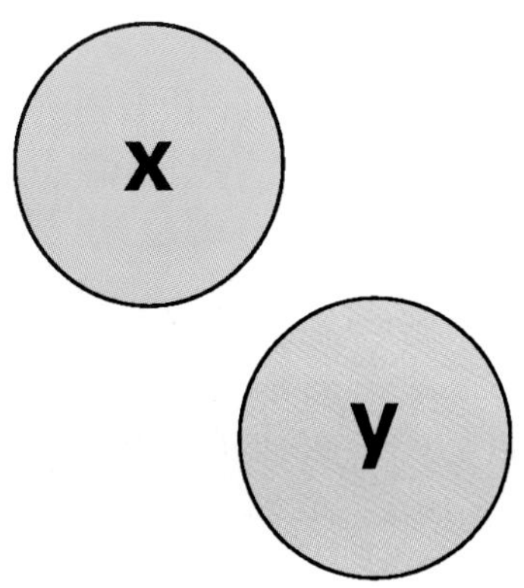


In a **fair** game, each player has the same probability of winning. In an **unfair** game, the probabilities of winning are not equal. Experimental results or theoretical probability can be used to determine whether the game is fair or unfair.

Mix and Match is played with two counters. One is marked with an *X* on both sides, the other with an *X* on one side and a *Y* on the other side. Two players take turns flipping the counters. If both counters match, Player A scores one point. If the counters do not match, Player B scores one point. The first player to score 20 points wins.

1. Play a game of Mix and Match with a partner. Record your scores.

Use a tree diagram or ordered pairs to find out the possible outcomes for each flip. (Remember to include both sides of each counter.)

2. How many outcomes show matching letters? How many show mixed letters?
3. Is Mix and Match a fair game? How do your class's experimental results compare with the theoretical probabilities?

Rolling Thunder is a game for two players who take turns rolling two number cubes. If the sum of the numbers rolled is a multiple of 3, Player A scores one point. If the sum rolled is a multiple of 4, Player B scores one point.

4. Play a game of Rolling Thunder with a partner.
5. Does a player score for each roll of the number cubes?
6. Can both players score on the same roll? Explain.

List the possible outcomes when two number cubes are rolled.

7. How many outcomes are favorable to Player A? How many are favorable to Player B?
8. Is Rolling Thunder a fair game? How do your class's experimental results compare with the theoretical probabilities?
9. Would the game be fair if Player A scored only when the sum of the numbers was a multiple of 5?
10. If Player A scored 3 points for a multiple of 3 and Player B scored 4 points for a multiple of 4, would the game be fair?
11. Devise some fair games and some unfair games using coins or number cubes. Have others tell which of your games are fair.

# 12-7 Permutations and Combinations

## EXPLORE/ WORKING TOGETHER

Work with a partner. Use coins to help you solve the problem.

a. Suppose you have a penny, a nickel, a dime, and a quarter in your pocket. If you take out two coins, how many different amounts of money might you have in your hand? How did you find your answer?

b. You have the same coins as in question **a.** In how many different ways could you take out two coins and give one each to two friends?

## SKILLS DEVELOPMENT

A word is a set of letters arranged in a definite order. For example, *charm* and *march* are different words, even though they contain the same letters. An arrangement of items in a particular order is called a **permutation.**

You can use the **fundamental counting principle** to find the number of possible permutations of a set of items. Each time you pick an item for a position in a permutation, there will be one less item to pick from for the next position.

The number of permutations of $n$ different items is $n(n - 1)(n - 2) \ldots (2)(1)$. This is written $\boldsymbol{n!}$, which is read as $\boldsymbol{n}$ **factorial.**

### CHECK UNDERSTANDING

Tell whether each question involves a *permutation* or a *combination.* Then solve.

1. How many different selections of four flowers can be made by a florist who has twelve varieties of flowers?
2. In how many different ways can six people line up at a movie ticket window?
3. How many different signals can be formed by running up three flags, one above the other, on a pole if seven different flags are available?
4. In how many different ways can a team of five be selected from a group of ten tennis players?

### Example 1

How many different "words" can be made with the letters $a$, $b$, $c$, $d$, and $e$ if all the letters are used? (The words do not have to make sense in English or any other language.)

**Solution**

Find the number of permutations of five letters.

$$\begin{aligned}\text{number of permutations of five letters} &= 5! \\ &= 5 \times 4 \times 3 \times 2 \times 1 \\ &= 120\end{aligned}$$

There are 120 different "words" that can be made using five letters. ◄

Sometimes you may want to use only part of a set. The number of permutations of $n$ different items, taken $r$ items at a time and with no repetitions, is written ${}_nP_r$. You can use the formula below to find the number of possible permutations when only part of the set is used.

$${}_nP_r = \frac{n!}{(n - r)!}$$

By definition, $n!$ is the number of permutations of $n$ items taken $n$ at a time, so ${}_nP_n = n!$ By the formula above,

$${}_nP_n = \frac{n!}{(n-n)!} = \frac{n!}{0!}$$

This means that $0! = 1$.

**PROBLEM SOLVING TIP**

If you use the formula for permutations and cancel common factors, you will find

${}_5P_2 = 5 \times 4$

${}_7P_4 = 7 \times 6 \times 5 \times 4$

${}_6P_3 = 6 \times 5 \times 4$

Look for a pattern. How many terms of $n!$ are used to find ${}_nP_r$?

Which terms would you use to find ${}_{84}P_5$?

## Example 2

How many different "words" can be made with the letters $a$, $b$, $c$, $d$, and $e$ if only three different letters are used in each word?

### Solution

There are five letters to be taken three at a time. So, $n = 5$ and $r = 3$.

Use the formula ${}_nP_r = \frac{n!}{(n-r)!}$

$${}_5P_3 = \frac{5!}{(5-3)!}$$

$$= \frac{5 \times 4 \times 3 \times 2 \times 1}{2 \times 1} = \frac{120}{2} = 60$$

There are 60 "words" that can be made. ◄

Sometimes the order of items is not important. A set of items in which order is not important is called a **combination.** For example, *acb* and *bac* are both combinations of the letters $a$, $b$, and $c$. The number of combinations of $n$ items taken $r$ items at a time is written ${}_nC_r$. You can use the formula below to find the number of possible combinations when only part of a set is used.

$${}_nC_r = \frac{n!}{(n-r)!r!}$$

## Example 3

A random drawing is held to determine which two of the six members of the math club will be sent to a regional math contest.

**a.** How many different pairs of two could be sent to the contest?

**b.** If the members of the club are Alesh, Ben, Colin, Daphne, Elsa, and Fay, what is the probability that Alesh and Fay will be picked?

**TALK IT OVER**

The value of ${}_nC_n$ is always the same number, whatever the value of $n$.

What is the value of ${}_nC_n$? Why is it always the same? Does ${}_nP_n$ always have the same value? Why?

### Solution

**a.** There are six people to be picked two at a time. So, $n = 6$ and $r = 2$.

Use the formula ${}_nC_r = \frac{n!}{(n-r)!r!}$

$${}_6C_2 = \frac{6!}{(6-2)!2!}$$

$$= \frac{6 \times 5 \times \overset{1}{\cancel{4}} \times \overset{1}{\cancel{3}} \times \overset{1}{\cancel{2}} \times \overset{1}{\cancel{1}}}{(\cancel{4} \times \cancel{3} \times \cancel{2} \times \cancel{1}) \times (2 \times 1)}$$

← You can cancel common factors to simplify the multiplication.

$$= \frac{30}{2} = 15$$

There are 15 different groups of two members.

**b.** The group of Alesh and Fay is one of the 15 possible combinations.

$$P(\text{Alesh and Fay}) = \frac{\text{number of favorable outcomes}}{\text{number of possible outcomes}} = \frac{1}{15}$$

The probability that Alesh and Fay will be picked is $\frac{1}{15}$. ◄

## CALCULATOR TIP

Some calculators have a factorial key, marked x!

To find 17!, for example, enter 1, then 7, then [x!].

On many graphing calculators, you can choose the factorial operation from a displayed mathematical menu. This menu may also include operations for finding permutations or combinations.

## TRY THESE

1. In how many different ways can you arrange the letters *m, n, o, p, q,* and *r*?
2. There are seven finalists in a swimming race. Medals are awarded for first place, second place, and third place. In how many different ways can the medals be awarded?
3. A five-piece band consists of a guitarist, a bass player, a trumpeter, a saxophone player, and a pianist. If three of the instrumentalists play an introduction to a song together, how many different groups of three instrumentalists could be used?
4. A doubles team is to be picked at random from the eight members of a tennis team. If Jo and Leah are two of the members, what is the probability that they will both be picked?

# EXERCISES

### PRACTICE/ SOLVE PROBLEMS

1. A professor is grading eight term papers in order of merit. If none of the papers are of equal merit, in how many different orders can the papers be graded?
2. A radio disc jockey has a set of nine songs to play. In how many different ways can the disc jockey play the next three songs?
3. There are six one-quart cans each containing a different fruit juice. How many different types of fruit punch could be obtained by mixing three full containers together?
4. A deck of 26 alphabet cards, marked *A* through *Z*, is shuffled and two cards are dealt. What is the probability that the cards are an *X* and a *Y*?

5. How many four-digit numbers can be formed from the digits 4, 5, 6, and 7, if each digit can be used only once?
6. A pizza parlor offers seven different toppings for its pizzas. If you were to choose two different toppings, how many different pizzas could you order?
7. A committee of three people is to be selected at random from a group of 24 people that includes Sam, Tina, and Vic. Find the probability that the committee will consist of Sam, Tina, and Vic.
8. In how many different orders could a poet recite five poems from a collection of nine poems?

## EXTEND/ SOLVE PROBLEMS

9. There are twelve people at a party. If each person shakes hands with every other person, how many handshakes are exchanged?

10. Stan and Luke are two of the twelve actors auditioning for a show. The order in which the actors will perform is picked at random. What is the probability that Stan will perform first and Luke will perform second?

11. In how many ways can the five volumes of an encyclopedia be placed next to each other on a shelf if the first volume is always on the extreme left?

12. A standard deck of 52 cards is shuffled and four cards are turned over. What is the probability that the four all have the same value (all have the same number or all have the same face)?

13. How many triangles are determined by nine points if no three of the points lie on a straight line?

**USING DATA** Refer to the Data Index on pages 556–557 to find information about the number of representatives from each state. Use the information to solve Exercises 14 and 15.

14. An environmental group wants to pick four representatives from Louisiana to join a committee. If each of the representatives will have an equal position on the committee, in how many different ways can the four members of the committee be chosen?

15. The Georgia state government votes to send one of their representatives to a conference in Cairo and another to a conference in Detroit on the same day. How many different outcomes are possible?

## THINK CRITICALLY/ SOLVE PROBLEMS

16. How many even numbers can be formed from the digits 1, 2, 3, 4, and 5 if each digit is used once in each number? How did you find the solution?

17. How many two-digit numbers can be formed using the digits 1, 2, 3, 4, and 5 if the numbers can use the same digit twice? How did you find the solution?

18. a. Find ${}_{10}C_8$ and ${}_{10}C_2$. What do you notice?
    b. Find ${}_{15}C_9$ and ${}_{15}C_6$. What do you notice?
    c. Think about how $n$ and $r$ are related in each pair of combinations. State a general rule.

19. A four-person committee is to be formed at random from six Democrats and seven Republicans. Find each probability.
    a. the committee consists of all Democrats
    b. the committee consists of all Republicans
    c. the committee consists of two Democrats and two Republicans

# CHAPTER 12 ● REVIEW

a. relative frequency
b. combination
c. simulation
d. permutation
e. sample space

1. A __?__ is an experiment that imitates a more complex problem.
2. If the arrangement of a set of items is important, the set is a __?__.
3. A __?__ contains all the possible outcomes of an experiment.
4. The __?__ of an outcome compares the number of times the outcome occurs to the total number of outcomes.
5. A set of items in which order is not important is a __?__.

## SECTION 12–1 EXPERIMENTS AND PROBABILITIES (pages 460–463)

- The **relative frequency** of a particular event compares the number of times it occurred to the total number of outcomes.
- The **experimental probability** of an event, *E*, can be estimated with this formula.

$$P(E) = \frac{\text{number of observations favorable to } E}{\text{total number of observations}}$$

6. Last month, a music store sold 1,178 records, 2,574 tapes, and 1,968 CDs. What is the relative frequency of tape sales?
7. In a random survey of 240 students, 168 said they would vote for Maria for class president. What is the probability that a student will vote for Maria?

## SECTIONS 12–2 AND 12–3 PROBLEM SOLVING (pages 464–467)

- A complex experimental probability can be found using a simulation.

8. There is a free gift in one of every six boxes of detergent. How would you use a number cube to find how many boxes you might have to buy to get three free gifts?

## SECTION 12–4 SAMPLE SPACES (pages 468–471)

- A set of ordered pairs or a tree diagram can be used to show the **sample space** of an experiment.
- The **fundamental counting principle** states that to find the number of possible outcomes for an activity, multiply the number of choices for each stage of the activity.

9. A company has printed eleven new books about American artists. Each book is available in hardcover or paperback and in regular type or large type. How many different books were printed?
10. A spinner with six equal sections marked *A* through *F* is spun and a number cube is rolled. Find *P*(vowel, number $>$ 3).

## SECTION 12–5 PROBABILITY OF COMPOUND EVENTS (pages 472–475)

- If $A$ and $B$ are mutually exclusive events, they cannot occur at the same time. $P(A \text{ or } B) = P(A) + P(B)$.
- If $A$ and $B$ are not mutually exclusive events, they can occur at the same time. $P(A \text{ or } B) = P(A) + P(B) - P(A \text{ and } B)$.

**11.** Two number cubes are rolled. Find the probability that the sum of the numbers rolled is 7 or greater than 10.

**12.** A card is drawn at random from a standard deck of 52 cards. Find the probability that it is a red card or a king.

## SECTION 12–6 INDEPENDENT AND DEPENDENT EVENTS (pages 476–479)

- Two events are independent if the outcome of one does not affect the outcome of the other. If $A$ and $B$ are independent events, $P(A \text{ and } B) = P(A) \times P(B)$.
- Two events are dependent if the outcome of one affects the outcome of the other. If $A$ and $B$ are dependent events, $P(A \text{ and } B) = P(A) \times P(B, \text{ given } A)$.

**13.** A number cube is rolled three times. Find $P(6, \text{number} > 4)$.

**14.** A box contains two green cards, three red cards, and five blue cards. Two cards are picked. Find $P(\text{blue, then red})$ if the first card is not replaced.

## SECTION 12–7 PERMUTATIONS AND COMBINATIONS (pages 480–483)

- A permutation is a set of items arranged in a particular order. The number of different ways a set of $n$ items can be arranged is $n!$ To find the number of permutations of $n$ items taken $r$ at a time, use this formula.

$$_nP_r = \frac{n!}{(n - r)!}$$

- A combination is a set of items without consideration of order. You can use this formula to find the number of combinations of a set of $n$ items taken $r$ at a time.

$$_nC_r = \frac{n!}{(n - r)!r!}$$

**15.** In how many different ways can eight paintings be awarded first, second, and third prizes?

**16.** How many different pairs of students can be chosen from a class of 28?

**17.** ***USING DATA*** Use the table on page 458 to find the probability of a high school student spending a Saturday afternoon taking classes, pursuing hobbies, or playing sports.

# CHAPTER 12 ● TEST

1. In a random survey of students, 152 students could swim and 48 could not. What is the probability that a student can swim?

2. Suppose you spin two spinners, one marked 1 through 4 and the other marked A through E. How many possible outcomes are there?

3. In a random sample of 1,000 voters in Kerr City, 525 said they would vote for Higgins for mayor. If there are 122,680 voters in the city, how many might be expected to vote for Higgins?

4. An ice cream store sells 24 different flavors and offers a choice of three sizes of cones and five types of sprinkles. How many choices of a cone with sprinkles are there?

5. Each card in a set is marked with a letter, a number, and a shape. The letters that can be used are A, B, C, D; the numbers that can be used are 1 to 5; and the shapes are a square, circle, or triangle. A card is selected at random from a box that contains one of each possible card. Find the probability that the card shows an even number and a circle.

The spinners shown here are both spun for each of Exercises 6–9.

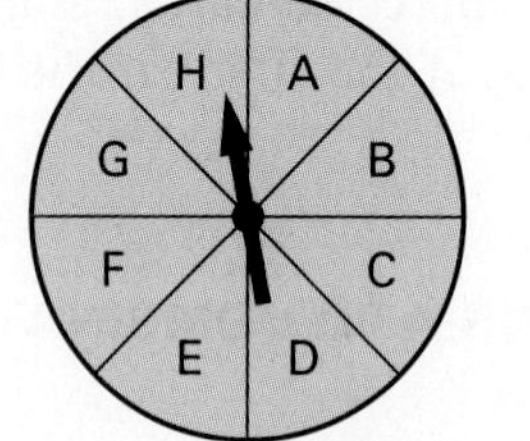

6. Find $P(\text{D}, 1)$.

7. Find $P(\text{not E}, 4 \text{ or } 5)$.

8. Find $P(\text{D}, \text{odd number})$.

9. Find $P(\text{vowel}, \text{number} > 2)$.

10. Two number cubes with faces marked 1 through 6 are rolled. Find the probability that the sum of the numbers rolled is 6 or 10.

11. Find the probability that the sum of the numbers rolled on two number cubes is less than 6 or a multiple of 3.

A bag contains seven green marbles, three red marbles, and five yellow marbles.

12. Find $P(\text{yellow, then green})$ if you take one marble, replace it, and then take another marble.

13. Find $P(\text{red, then green})$ if you take one marble and then take another marble without replacing the first.

14. In how many ways can a group of five books be lined up on a shelf if there are nine books to choose from?

15. How many ways could five runners be chosen from a track team that has ten members?

1. What percent is 14.4 of 96?

2. Simplify. $\left(\frac{1}{6} - \frac{1}{2}\right) - 3$

3. Find the measure of each interior angle of an octagon.

Last week, the salespeople at Bettina's Boutique earned commissions of \$61, \$52, \$55, \$93, \$55, \$56, and \$60, in addition to their salaries. When Laura applied for a job at the boutique, the manager told her that the average commission paid last week was \$62.

4. What measure of central tendency had the manager used to describe the average commission?

5. Was the manager's statement misleading? If so, what measure would have been a better indicator of the average commission?

Solve the equation.

6. $18 + 7y = 3y + 6$

Refer to the figure for Exercises 7 and 8. Use 3.14 for $\pi$.

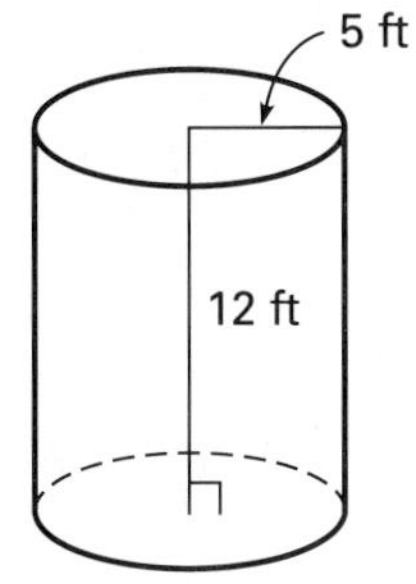


7. Find the volume of the figure.

8. Find the surface area of the figure in Exercise 7. Round to the nearest tenth.

Refer to the figure for Exercises 9 and 10.

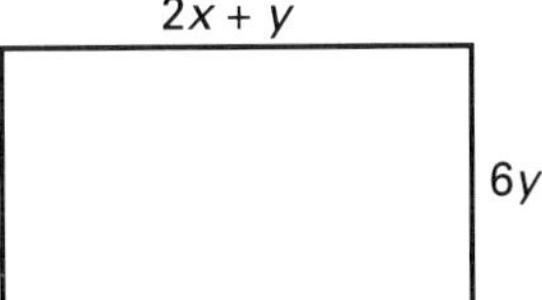


9. Write an expression for the perimeter.

10. Write an expression for the area.

Find an equation for a circle with the radius indicated and centered at the origin.

11. 7

12. 9

13. Write an equation of a line passing through $A(-2, 3)$ and $B(2, 1)$.

Solve.

14. $x + y = 12$
$-3x + y = -4$

15. $6x - 3y = -18$
$3x - y = 8$

16. The scale of a building model is 3 in.:20 ft. The actual height of the building is 250 ft. Find the height of the model.

17. Quadrilateral *GHIJ* is similar to *LMNK*. Find $m\angle G$.

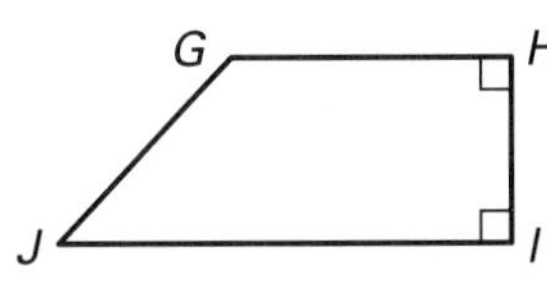


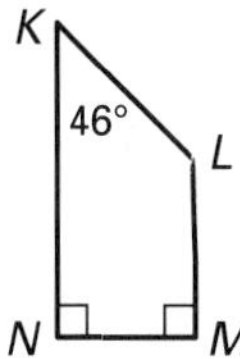


Write the coordinates of the image of each point under a reflection across the given line.

18. (2, 3); $x$-axis

19. (1, −2); $y$-axis

20. In a random survey of 60 out of the 1,800 students at Rupert High School, 12 said they would like to form a chess club. How many students might be expected to join the club?

# CHAPTER 12 • CUMULATIVE TEST

1. What percent of 142 is 28.4?
   A. 2% B. 5% C. 20% D. 22%

2. Solve. $-5\left(2\frac{1}{2}\right)\left(-\frac{1}{3}\right)$
   A. $-\frac{5}{6}$ B. $4\frac{1}{6}$ C. 5 D. $12\frac{1}{2}$

3. Classify the triangle.
   A. right scalene
   B. right isosceles
   C. acute equilateral
   D. obtuse scalene

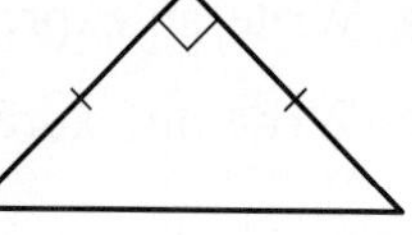

Use these data for Exercises 4 and 5.

| Rents for 2-Bedroom Apartments | | | | | |
|---|---|---|---|---|---|
| \$480 | \$560 | \$450 | \$675 | \$640 | \$520 |

4. Identify the range of the data.
   A. \$40 B. \$70 C. \$225 D. \$540

5. Identify the median of the data.
   A. \$540.00 B. \$562.50
   C. \$577.50 D. none of these

6. Solve. $\frac{3}{8}x - 4 = 44$
   A. 48 B. 128 C. 384 D. 96

7. Find the area of the figure.

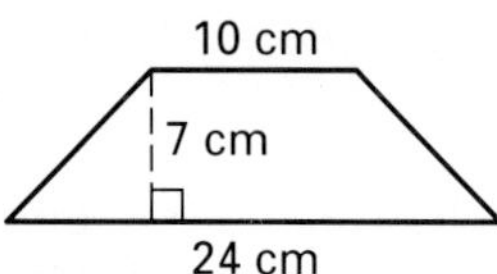


   A. 119 $cm^2$ B. 203 $cm^2$
   C. 154 $cm^2$ D. none of these

8. Find the volume of a cone with a radius of 6 cm and a height of 10 cm.
   A. 1,130.4 $cm^3$ B. 113.04 $cm^3$
   C. 376.8 $cm^3$ D. 188.4 $cm^3$

9. Simplify. $\frac{22x^2y^2z^3}{33xyz}$
   A. $11xyz$ B. $\frac{2}{3xyz}$
   C. $\frac{2xyz^2}{3}$ D. $\frac{2xy}{3}$

10. Use the distance formula to find the distance between $D(-4, -3)$ and $E(5, 9)$.
    A. 15 B. $\sqrt{37}$ C. $\sqrt{63}$ D. 12

11. Which of the ordered pairs is *not* a solution of $y \geq 3x + 2$?
    A. (0, 2) B. (1, 4)
    C. (−2, −3) D. (−1, 0)

12. Solve. $x + y = 5$
    $2x + 3y = 4$
    A. $x = -1, y = 6$
    B. $x = 19, y = -14$
    C. $x = 3, y = 2$
    D. $x = 11, y = -6$

13. Identify an equation of the line with $m = 6$ and $b = -2$.
    A. $y = -2x + 6$ B. $y = 2x - 6$
    C. $y = 6x - 2$ D. $y = -6x + 2$

14. Triangles $QRS$ and $TUV$ are similar. Find $m\angle Q$.

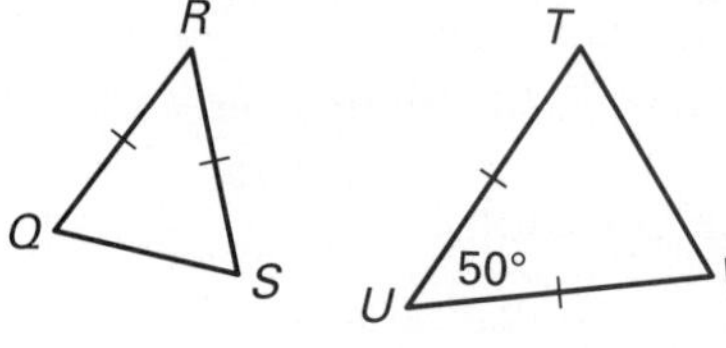


    A. 65° B. 55° C. 50° D. 40°

15. Identify the coordinates of the image of (3, −3) under the transformation.
    $(x, y) \rightarrow (x + 2, y + 6)$
    A. (5, 6) B. (5, 3)
    C. (2, 6) D. (1, −3)

16. A coin is tossed and a number cube is rolled. Identify the total number of outcomes in the sample space.
    A. 6 B. 8 C. 12 D. 24

17. Two number cubes are rolled. What is the probability that the sum of the numbers rolled is a 5 or an 11?
    A. $\frac{2}{9}$ B. $\frac{5}{6}$ C. $\frac{1}{6}$ D. $\frac{1}{3}$

# CHAPTER 13 SKILLS PREVIEW

1. Determine whether the sets are equal or equivalent.
   $P = \{2, 4, 6, 8\}$ $Q = \{4, 8, 1, 5\}$ $R = \{8, 1, 4, 5\}$

Let $U = \{e, x, t, r, a, s\}$, $A = \{t, a, x, e, s\}$, and $B = \{r, a, t, s\}$. Find each union or intersection.

2. $A \cup B$
3. $A \cap B$
4. $A' \cap B$
5. Write the inverse of the statement and tell whether the inverse is *true* or *false*. If false, give a counterexample.
   If you are studying biology, then you are studying science.

Write these statements in *if–then* form.

6. I water my plants whenever the topsoil gets dry.
7. I swim in Lake Michigan only if the water temperature is above 70°F.
8. Identify the necessary and the sufficient condition in the statement.
   If Uncle Jim brings his guitar, then we can have a singalong at the picnic.
9. Which of the four statements is a good definition of congruent triangles?
   a. Congruent triangles have equal parts.
   b. Congruent triangles have parts that can be matched.
   c. Congruent triangles are triangles whose corresponding angles and corresponding sides are equal.
   d. Congruent triangles have the same shape.

Tell whether the reasoning used is *inductive* or *deductive.*

10. Violet put three different light bulbs into the socket of her bedlamp. Each time the bulb blew out, Violet concluded that the socket and switch were defective.
11. Phil used the definitions of opposite rays and of a straight angle to conclude that opposite rays form a straight angle.

Tell whether the argument is *valid* or *invalid.*

12. If the animal is a giraffe, then it has a long neck.
    The animal does not have a long neck.
    ______________________________
    The animal is not a giraffe.
13. If the figure is an octagon, then the figure is a polygon.
    The figure is a polygon.
    ______________________________
    The figure is an octagon.

CHAPTER 13

# LOGIC AND SETS

**THEME** **Music**

Almost any group or collection of things can be called a **set.** In your kitchen you may find a set of water glasses, a set of eating utensils, or a set of cookware. Sets offer a convenient way to categorize things in order to think logically about them. Logic itself provides a way to state the relationships between classes of things.

In a symphony orchestra, the groups of related instruments form sets. There are four basic sets of instruments in an orchestra.

**Stringed Instruments**
violin, viola, cello, contrabass

**Woodwind Instruments**
clarinet, saxophone, oboe, English horn, flute, piccolo

**Brass Instruments**
trumpet, cornet, trombone, baritone, French horn, tuba

**Percussion Instruments**
xylophone, marimba, bass drum, snare drums, kettle drums, triangle, cymbals, bells

This diagram shows where the members of each section of a modern symphony orchestra are seated.

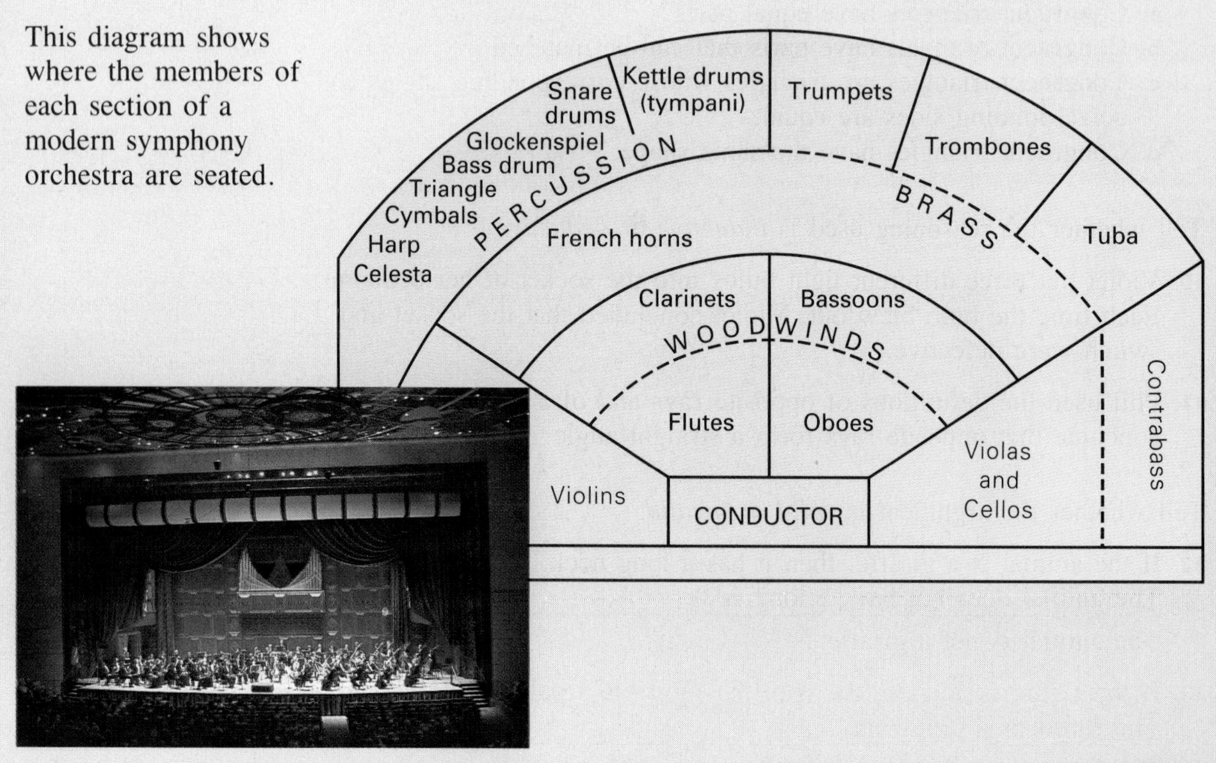

Sound is measured in **frequencies,** the number of wave vibrations per second required to produce the sound. The following table shows the range of sound frequencies, measured in units called **hertz** (Hz), for certain instruments and types of human singing voices.

**RANGE OF FREQUENCIES FOR SELECTED INSTRUMENTS AND VOICES**

| Instrument | Frequency Range (Hz) | Voice | Frequency Range (Hz) |
|---|---|---|---|
| piccolo | 512+ | soprano | 240–1,365 |
| flute | 256–2,304 | alto | 171–683 |
| clarinet | 160–1,536 | tenor | 128–683 |
| French horn | 106–853 | baritone | 96–384 |
| trombone | 80–840 | bass | 80–341 |
| tuba | 43–341 | | |
| violin | 192–3,072 | | |
| cello | 64–683 | | |
| double bass | 40–240 | | |

## DECISION MAKING

### Using Data

Use the information in the table to answer each question.

1. Which instruments or ranges of voice are able to produce sounds with a frequency of 950 hertz?
2. Which instrument or voice has the widest frequency range? Which has the narrowest range?
3. Which instruments and voices have the same highest frequency? Which instruments and voices have the same lowest frequency?
4. How could you use frequencies to create new groupings of instruments?

### Working Together

Find the frequency range of other musical instruments, either in the same family as those given (other woodwind or brass instruments) or from another family, such as percussion instruments. Each person should choose a different instrument to research. Use encyclopedias, books about musical instruments or orchestration, or textbooks. Have your group compile a table of the frequencies and instruments. Did any instrument have a range of frequencies greater than those listed? Which instruments have similar ranges of frequency?

# 13-1 Sets

**EXPLORE/ WORKING TOGETHER**

Work in a group of four or five students. Have each member of the group write the names of two classroom objects—such as "eraser" and "pen"—on slips of paper. Put the slips of paper in a bag and mix them well. Then have each member draw two slips from the bag and name a possible set to which both items might belong. As each set is named and described, discuss the description.

After you have finished, put all the slips back into the bag and begin again. This time, have each student draw three slips.

**SKILLS DEVELOPMENT**

A **set** is a well-defined collection of items. Each item is called an **element,** or a **member,** of the set. A set is usually named with a capital letter and may be defined in three ways:

1. **Description notation** (describes the set)
   $W$ = the set of whole numbers

   $S$ = the set of even whole numbers less than 20

2. **Roster notation** (lists the elements of the set)
   $W = \{0, 1, 2, 3, \ldots\}$ ← Braces are used to enclose the elements of a set. The three dots, or *ellipsis,* indicates that the pattern goes on without end.

   $S = \{0, 2, 4, 6, \ldots, 18\}$ ← This ellipsis indicates that the pattern continues through 18, but the numbers between 6 and 18 are not listed.

3. **Set-builder notation** (gives the property that defines each element)
   $W = \{x|x \text{ is a whole number}\}$ ← Read as "the set of all $x$ such that $x$ is a whole number."

   $S = \{x|x \text{ is an even whole number less than } 20\}$

A set whose elements cannot be counted or listed is called an **infinite set.** If all of the elements of a set can be counted or listed, the set is called a **finite set.**

### Example 1

Define each set in roster notation and in set-builder notation. Then determine whether the set is *finite* or *infinite.*

a. $W$, the set of whole numbers
b. $O$, the set of odd whole numbers less than 10.

### Solution

**a.** $W = \{0, 1, 2, 3, \ldots\}$; $W = \{x|x \text{ is a whole number}\}$
The ellipsis indicates that all the elements cannot be listed. So, $W$ is an infinite set.

**b.** $O = \{1, 3, 5, 7, 9\}$; $O = \{x|x \text{ is an odd whole number less than } 10\}$
All the elements in the set are listed. So, $O$ is a finite set. ◄

To show that 5 is an element of the set $\{1, 3, 5\}$, write

$5 \in \{1, 3, 5\}$ ← Read as "5 is an element of the set {1, 3, 5}."

To show that 7 is not an element of $\{1, 3, 5\}$, write

$7 \notin \{1, 3, 5\}$ ← Read as "7 is not an element of the set {1, 3, 5}."

## Example 2

Use set notation to write the following.

**a.** 5 is an element of $\{0, 1, 2, 3, 4, 5\}$.

**b.** The letter $e$ is not an element of the letters in *banana*.

### Solution

**a.** $5 \in \{1, 2, 3, 4, 5\}$ **b.** $e \notin \{b, a, n\}$ ◄

Two sets $A$ and $B$ are **equal sets** (written $A = B$) if they contain the same members (not necessarily in the same order). Two sets are **equivalent sets** if they contain the same number of elements.

## Example 3

Determine whether the following sets are equal or equivalent.

$S = \{E, G, B, D, F\}$ $T = \{F, B, D, G, E\}$ $Q = \{3, 7, 11, 5, 9\}$

### Solution

Sets $S$, $T$, and $Q$ each contain five elements. So, they are equivalent sets. Since $S$ and $T$ have exactly the same elements, $S = T$. ◄

If every element of set $A$ is also an element of set $B$, then $A$ is called a **subset** of $B$.

$A \subseteq B$ ← Read as "A is a subset of B."

Consider these sets.

$X = \{1, 2\}$
$Y = \{1, 2, 3\}$

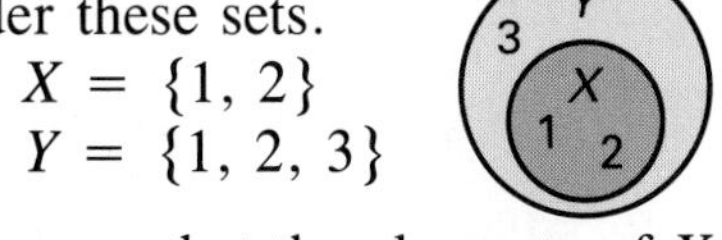


You can see that the elements of $X$ are also elements of $Y$. So, $X$ is a subset of $Y$ ($X \subseteq Y$). However, every element of $Y$ is not an element of $X$. So, $Y$ is not a subset of $X$.

$Y \nsubseteq X$ ← Read as "Y is not a subset of X."

*Any set is equal to itself.* For set $Y$,

$\{1, 2, 3\} = \{1, 2, 3\}$ or $Y = Y$

For this reason, *every set is a subset of itself.* So, $Y \subseteq Y$.

**MATH: WHO, WHERE, WHEN**

Georg Cantor (1845–1918) is credited with being the first to develop set theory as a separate branch of mathematical logic. At first, his ideas were not taken seriously. For example, you might think that there would be fewer even numbers than there would be counting numbers. Cantor proved that the set of counting numbers $\{1, 2, 3, 4, \ldots\}$ and the set of even numbers, $\{2, 4, 6, 8, \ldots\}$ both have the same infinite number of elements. In fact, Cantor showed that you could take *any* subset of the counting numbers and put it into one-to-one correspondence with the whole set. Thus, Cantor defined a **countable infinite set** as a set that can be put into one-to-one correspondence with a proper subset of itself. (A *proper subset* is a subset if $A$ is included in $B$ and there is at least one element of $B$ that is not in $A$.) Cantor symbolized the number of elements in a countable infinite set as $\aleph_0$, read "aleph-zero" or "aleph-null." ($\aleph$ is the first letter of the Hebrew alphabet.) Surprisingly, Cantor demonstrated that there were other infinite sets, uncountable ones, whose size was greater than $\aleph_0$.

Consider the set $W$, weeks containing 8 days. Since no week contains 8 days, $W$ has no elements. A set having no elements is called an **empty set** or the **null set.** To indicate that $W$ is an empty set, write either $W = \{\ \}$ or $W = \phi$.

The null set is a subset of every set.

Why is the null set a subset of any set?

### Example 4

Determine all the possible subsets of the set $\{2, 4\}$.

**Solution**

Each single element of $\{2, 4\}$ is a subset: $\{2\}$, $\{4\}$.
The set itself is a subset: $\{2, 4\}$.
The null set is a subset: $\phi$.
So, $\{2, 4\}$ has four subsets: $\{2\}$, $\{4\}$, $\{2, 4\}$, $\phi$. ◄

## TRY THESE

Define each set in roster notation and in set-builder notation. Then determine whether the set is *finite* or *infinite.*

1. $L$, the set of whole numbers less than 7.
2. $W$, the set of whole numbers

Use set notation to write the following.

3. 3 is an element of $\{3, 6, 9, 12, \ldots\}$.
4. The letter $m$ is not an element of $\{a, e, i, o, u\}$.
5. Determine whether the sets are *equal* or *equivalent.*
   $P = \{1, 3, 5, 7\}$ $Q = \{3, 7, 1, 4\}$ $R = \{1, 7, 3, 4\}$
6. Determine all the possible subsets of $\{a, c, t\}$.

## EXERCISES

### PRACTICE/ SOLVE PROBLEMS

Define each set in roster notation.

1. whole numbers less than 15
2. Great Lakes
3. vowels in the name *Figueroa*
4. months having 32 days
5. integers greater than $-4$
6. whole numbers greater than 3

Write a set equivalent to each set.

7. $\{a, b, c\}$
8. $\{1, 2, 3, 4\}$

Determine if the following sets are *equal* or *not equal.*

9. $\{r, o, v, e\}, \{o, v, e, r\}$

10. $\{a, b, c, d, e\}, \{c, d, e, 3, 1\}$

11. $\{t, i, m, e\}, \{i, t, e, m\}$

Determine if each statement is *true* or *false.*

12. $2 \in \{x|x \text{ is a whole number}\}$

13. $8 \in \{1, 3, 5, 7, \ldots\}$

14. $\{8, 12, 16\} \subseteq \{4, 8, 12, 16\}$

15. $\{x|x \text{ is a square}\} \subseteq \{x|x \text{ is a rectangle}\}$

16. $\{x|x \text{ is an isosceles triangle}\} \subseteq \{x|x \text{ is an equilateral triangle}\}$

Write all the subsets of each set.

17. $\{1\}$ 18. $\{8, 9\}$ 19. $\{m, a, t\}$

## EXTEND/ SOLVE PROBLEMS

Determine whether the statement is *true* or *false.*

20. If two sets are equal, then they are equivalent.

21. If two sets are equivalent, then they are equal.

22. $\phi$ is a subset of itself.

23. $\{1\}$ is a subset of $\phi$.

Rewrite each statement so that it is correct.

24. $\{a\} \in \{a, b\}$ 25. $\{\phi\} \subseteq \{a, c\}$ 26. $a \subseteq \{a\}$

Define each set in roster notation.

27. $\{x|x \text{ is an integer and } x < 2\}$

28. $\{x|x \text{ is a whole number and } 4x - 8 = 24\}$

29. $\{x|x \text{ is a whole number and } x < 0\}$

**JUST FOR FUN**

Suppose that you have four bills in your wallet, a one-dollar bill, a five-dollar bill, a ten-dollar bill, and a twenty-dollar bill. How many different sums of money can be made from different groupings of any one or more of these bills? What is the smallest sum of money you can make? What is the greatest sum of money you can make?

## THINK CRITICALLY/ SOLVE PROBLEMS

Determine the number of subsets for each set.

30. $\{1\}$ 31. $\{1, 2\}$ 32. $\{1, 2, 3\}$ 33. $\{1, 2, 3, 4\}$

34. How many subsets do you think a set of 5 elements has? a set of 6 elements? the null set?

35. Write a rule or definition that expresses the relationship between the number of elements in a set and the number of subsets it has.

# 13-2 Union and Intersection of Sets

## EXPLORE

Suppose that there are 32 students in your homeroom. Of those students, 8 play in the band, 12 sing in the choir, and 4 are in both the band and the choir. How many of the 32 students are in neither group? Show your answer by copying and completing the Venn diagram shown at the right.

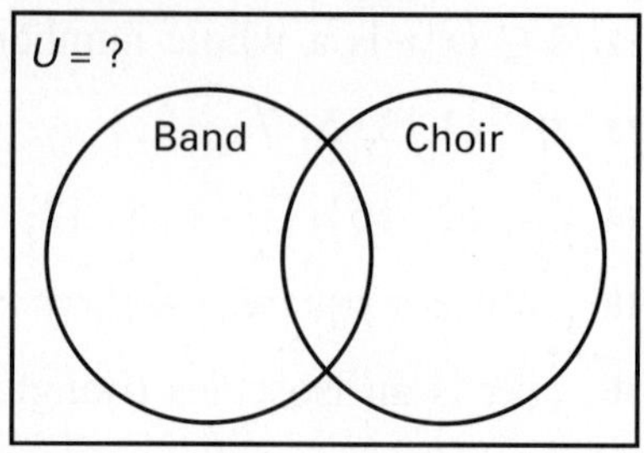


## SKILLS DEVELOPMENT

In working with sets, it is important to define the general set of elements being considered in a discussion. The set of all such elements is called the **universe** or the **universal set ($U$).** The universal set could be an infinite set such as the set of whole numbers or the set of real numbers or a finite set such as $\{a, e, i, o, u\}$.

From the universal set, a number of subsets can be formed. Suppose that $U = \{0, 1, 2, 3, 4, 5, 6, 7, 8, 9\}$. One possible subset is $A = \{2, 4, 6, 8\}$.

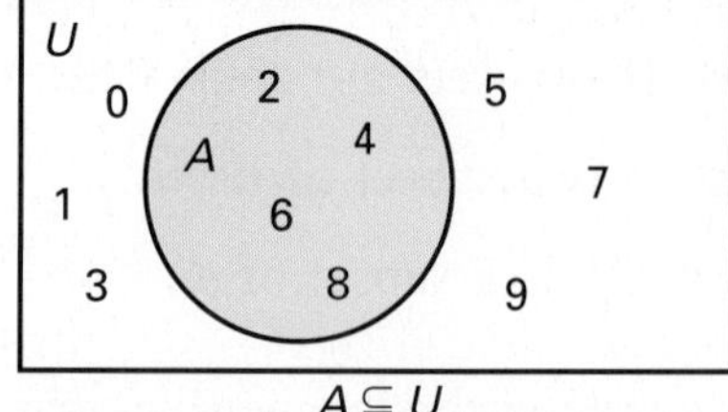


$A \subseteq U$

Another subset is the subset of all elements of $U$ that are *not* elements of $A$. This subset is called the **complement** of $A$, symbolized as $A'$. $A' = \{0, 1, 3, 5, 7, 9\}$. In set-builder notation,

$$A' = \{x | x \in U \text{ and } x \notin A\}$$

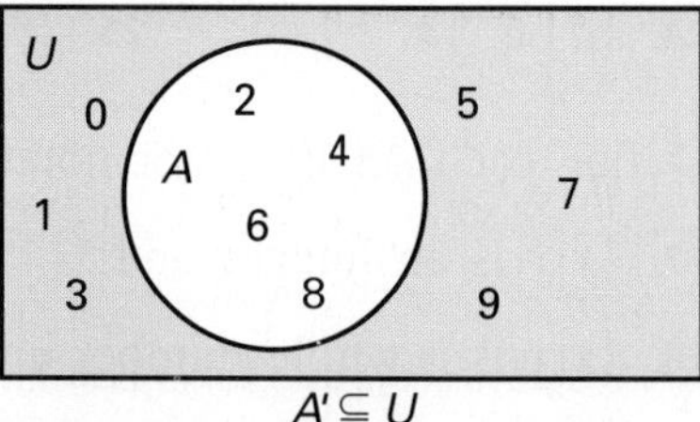


$A' \subseteq U$

### Example 1

Use roster notation to represent the sets named, given that

$$U = \{0, 1, 2, 3, 4, 5, 6, 7, 8\}$$
$$A = \{0, 2, 4, 6, 8\}$$
$$B = \{1, 3, 5\}$$

**a.** $A'$ **b.** $B'$

### Solution

**a.** $A'$ is the set of those elements in $U$ not in $A$.
So, $A' = \{1, 3, 5, 7\}$

**b.** $B' = \{0, 2, 4, 6, 7, 8\}$ ◄

Two or more sets can be combined to form new sets. Two of these new sets are the union and intersection of the sets. The **union** of any two sets $A$ and $B$ is symbolized as $A \cup B$ (read as "$A$ union $B$").

$$A \cup B = \{x | x \in A \text{ or } x \in B\}$$

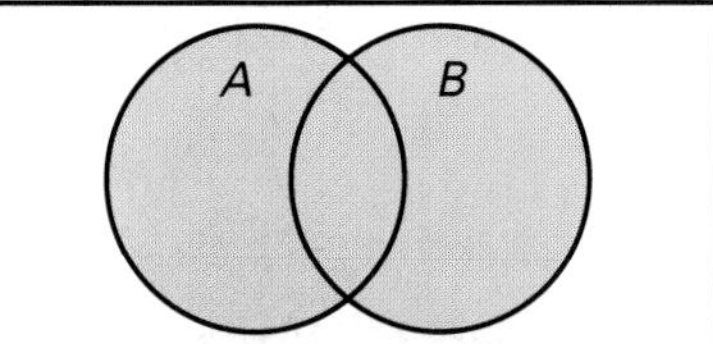


The set $A \cup B$ contains all the elements that are in $A$, in $B$, or in both. At the right is a Venn diagram of $A \cup B$.

## Example 2

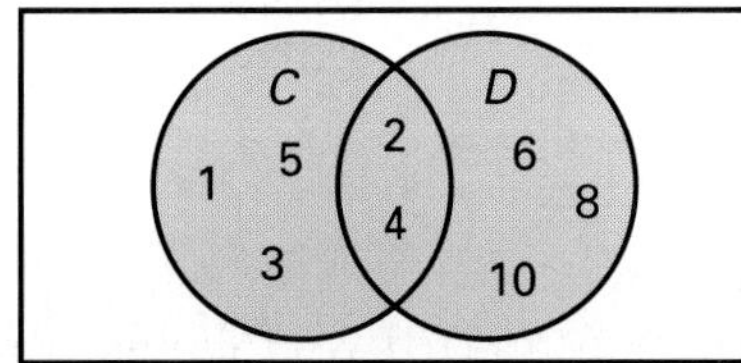


Refer to the diagram at the right. Find $C \cup D$ by listing the members.

**Solution**

$C \cup D$ is the set of those elements that are in $C$, in $D$, or in both.

$$\{1, 2, 3, 4, 5\} \cup \{2, 4, 6, 8, 10\} = \{1, 2, 3, 4, 5, 6, 8, 10\}$$
$$C \cup D = \{1, 2, 3, 4, 5, 6, 8, 10\}$$ ◄

**MATH: WHO, WHERE, WHEN**

The English mathematician John Venn (1834–1923) is credited with inventing diagrams for showing relationships between sets and classes in logic. Venn drew a rectangular region to represent the universal set $U$. He used circular regions within the rectangle to represent subsets of $U$. The diagrams used in this section to illustrate $U$, $A$, $A'$, $A \cup B$, and $A \cap B$ are Venn diagrams.

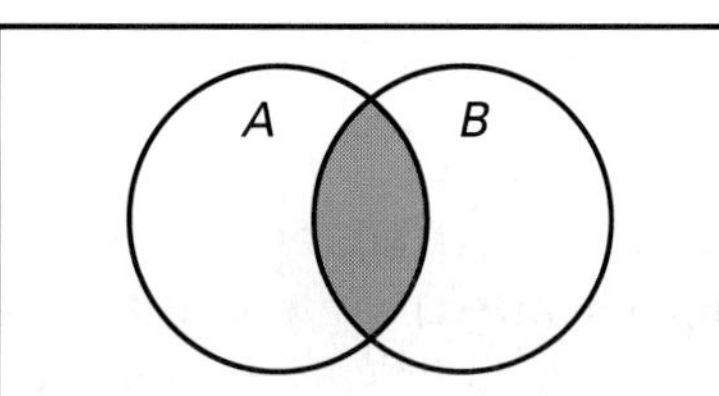


The **intersection** of two sets $A$ and $B$ is symbolized by $A \cap B$, read as "$A$ intersect $B$."

$$A \cap B = \{x | x \in A \text{ and } x \in B\}$$

The set $A \cap B$ contains the elements that are common to both $A$ and $B$. A diagram of $A \cap B$ is shown above.

## Example 3

Refer to the diagram in Example 2. Find $C \cap D$ by listing the members.

**Solution**

$C \cap D$ is the set of elements common to both $C$ and $D$. The only elements common to both $C$ and $D$ are 2 and 4.

$$\{1, 2, 3, 4, 5\} \cap \{2, 4, 6, 8, 10\} = \{2, 4\}$$

So, $C \cap D = \{2, 4\}$. ◄

If two sets have no elements in common, their intersection will be the empty set, $\phi$. Two sets whose intersection is the empty set are called **disjoint sets.**

## Example 4

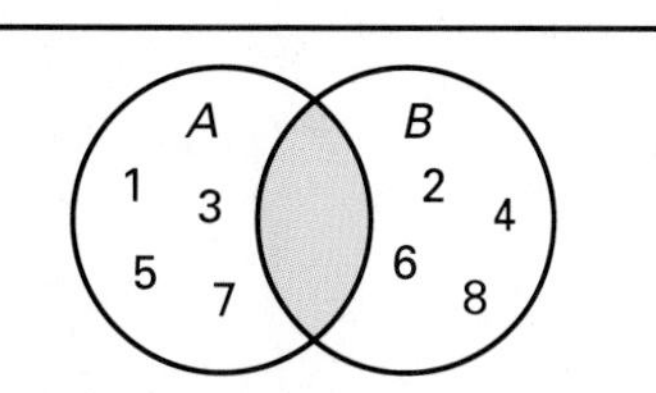


Let $A = \{1, 3, 5, 7\}$ and $B = \{2, 4, 6, 8\}$. Find $A \cap B$.

**Solution**

The two sets have no elements in common. So, $A \cap B = \phi$. ◄

## TRY THESE

Use roster notation to represent the sets named, given that
$U = \{1, 2, 4, 6, 8, 9\}$ $A = \{1, 2, 4\}$ $B = \{4, 6, 9\}$ $C = \{1, 2\}$

1. $A'$
2. $B'$
3. $C'$
4. Let $A = \{1, 3, 5, 7\}$ and $B = \{2, 4, 6, 8\}$. Find $A \cup B$.

Refer to the diagram. Find the sets named by listing the members.

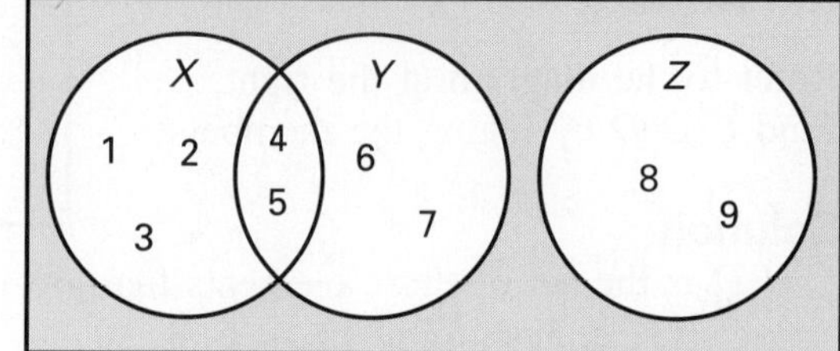


5. $X \cup Y$
6. $Y \cup Z$
7. $X \cap Y$
8. $Y \cap Z$
9. $X \cap Z$

# EXERCISES

**PRACTICE/ SOLVE PROBLEMS**

Refer to the diagram. Find the set named by listing the members.

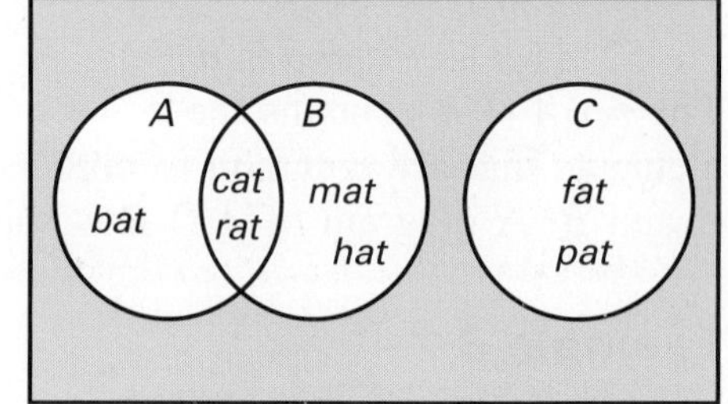


1. $A \cup B$
2. $A \cap B$
3. $A \cup C$
4. $A \cap C$

Let $U = \{p, l, a, c, e\}$, $P = \{l, a, c, e\}$, and $Q = \{l, a, p\}$. Find each union.

5. $P' \cup Q'$
6. $(P \cup Q)'$

Let $U = \{a, e, i, o, u\}$, $A = \{a, i, u\}$, and $B = \{a, e, u\}$. Find each union.

7. $A' \cup B'$
8. $(A \cup B)'$
9. Let $A = \{f, l, o, a, t\}$ and $B = \{b, r, o, t, h\}$. Find $A \cap B$.
10. Let $C = \{2, 4, 8, 16\}$ and $D = \{3, 9, 27\}$. Find $C \cap D$.

Let $U = \{1, 2, 3, 4, 5, 6, 7, 8, 9\}$, $E = \{3, 5, 7\}$, $F = \{4, 6\}$, and $G = \{2, 4, 6, 8\}$. Use roster notation to represent the following sets.

11. $E' \cap F'$
12. $E \cap F$

13. $E' \cap F$

14. $E \cup F'$

15. $G \cap E'$

## EXTEND/ SOLVE PROBLEMS

Let $U = \{1, 2, 3, \ldots, 19, 20\}$, $A = \{2, 3, 5, 7, 11, 13, 17, 19\}$, $B = \{2, 4, 6, 8, \ldots, 18\}$, $C = \{1, 2, 3, 4, 5\}$, and $D = \phi$. Find the following.

16. $A'$

17. $C'$

18. $B'$

19. $D'$

20. $A \cap B$

21. $C \cup D$

22. $B \cap C$

23. $A \cup C$

24. $B \cup C$

25. $A \cup B$

For Exercises 26–31, use the sets given for Exercises 16–25.

26. $B \cap (C \cap D)$

27. $B \cup (C \cap D)$

28. $A \cap (B \cup C)$

29. $(A \cap B) \cap C$

30. $C \cap (A \cup B)$

31. $C \cup (B \cap A)$

Let $A = \{x|x \text{ is a real number and } x > -1\}$
$B = \{x|x \text{ is a real number and } x < 3\}$

The graphs of $A$ and $B$ are as follows.

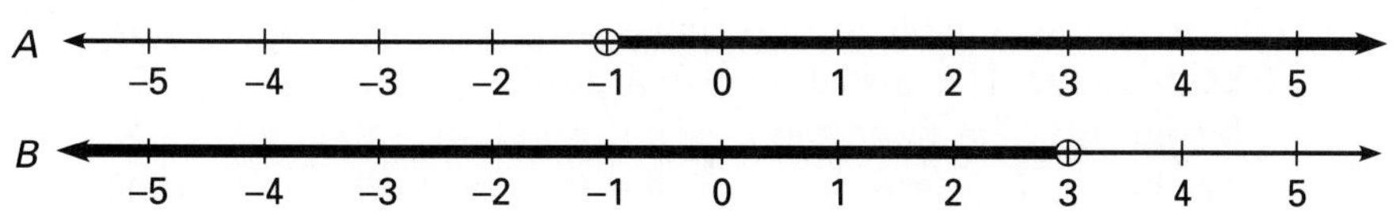


32. Describe $A \cup B$.

33. Describe $A \cap B$.

34. Write a brief description of the graph of $A \cup B$.

35. Write a brief description of the graph of $A \cap B$.

36. Graph $A \cap B$ on a number line, given that
$A = \{x|x \text{ is a real number and } x \geq -2\}$
$B = \{x|x \text{ is a real number and } x \leq 4\}$

37. Graph $C \cap D$ on a number line, given that
$C = \{x|x \text{ is a real number and } x \geq 0\}$
$D = \{x|x \text{ is a real number and } x \leq 0\}$

### MIXED REVIEW

1. Find 20% of 150.
2. Evaluate $3mn + m^2$, when $m = -7$ and $n = 12$.

Solve.

3. $5(x + 3) = 2(2x - 1)$
4. $\frac{x}{12} = \frac{15}{36}$
5. $-4x - 2 \geq 10$

## THINK CRITICALLY/ SOLVE PROBLEMS

Determine whether each statement is *true* or *false* and, if false, provide the correct answer given that
$U = \{\text{whole numbers}\}$ $S = \{0, 2, 4, 6, 8, \ldots\}$
$R = \{1, 3, 5, 7, \ldots\}$ $T = \{1, 2, 3, 4, \ldots\}$

38. $R' \cup S = S$

39. $R \cap S = \phi$

40. $S \cup T = T$

41. $S \cap U = S$

42. $R \cup S = S$

43. $R \cap T' = \phi$

# 13-3 Converse, Inverse, and Contrapositive

## EXPLORE

Read these three pairs of statements.

**Pair 1**
If the instrument is a guitar, then the instrument has strings.
If the instrument has strings, then the instrument is a guitar.

**Pair 2**
If the instrument is a guitar, then it has strings.
If the instrument is not a guitar, then it does not have strings.

**Pair 3**
If the instrument is a guitar, then it has strings.
If the instrument does not have strings, then it is not a guitar.

a. In which pair(s) are both statements true?
b. In which pair(s) is only one statement true?

## SKILLS DEVELOPMENT

Any statement can be negated. To write the **negation** of a statement, add "not" to the statement or add the words "It is not the case that."

**Statement:** The instrument is a guitar.
**Negation:** The instrument is *not* a guitar.
**Negation:** *It is not the case that* the instrument is a guitar.

**CHECK UNDERSTANDING**

In what two ways could the negation of the statement "I am not ill" be written?

Let $p$ stand for the statement. The negation can then be symbolized as $\sim p$. If a statement is true, its negation is false. If the statement is false, then its negation is true.

### Example 1

Write the negation of this statement in two ways:
**Statement:** The instrument has strings.

### Solution

**Negation:** The instrument does <u>not</u> have strings.
**Negation:** <u>It is not the case that</u> the instrument has strings. ◄

As you observed in the Explore activity, from the statement "If the instrument is a guitar, then the instrument has strings," three related conditional statements can be formed. Each one has a special name.

Let $p$ stand for the hypothesis, "the instrument is a guitar."
Let $q$ stand for the conclusion, "the instrument has strings."
Use the symbol $\rightarrow$ to represent the *if–then* relationship.

Using these symbols, you can see that each statement has a special form.

|  | | |
|---|---|---|
| **Statement** If the instrument is a guitar, then the instrument has strings. | If $p$, then $q$. | or $p \rightarrow q$ |
| **Converse** If the instrument has strings, then the instrument is a guitar. | If $q$, then $p$. | or $q \rightarrow p$ |
| **Inverse** If the instrument is not a guitar, then the instrument does not have strings. | If $\sim p$, then $\sim q$. | or $\sim p \rightarrow \sim q$ |
| **Contrapositive** If the instrument does not have strings, then the instrument is not a guitar. | If $\sim q$, then $\sim p$. | or $\sim q \rightarrow \sim p$ |

In Chapter 3 you learned that the converse of an *if–then* statement is not necessarily true.

**Statement:** If the instrument is a piano, then it has strings.
**Converse:** If the instrument has strings, then it is a piano.

The statement is true. Its converse is false. An instrument such as a violin or a banjo is not a piano, although it has strings.

## Example 2

Write the inverse of the statement and tell whether the inverse is true or false. If false, give a counterexample.

If two angles are right angles, then the angles are congruent.

### Solution

**Inverse:** If two angles are not right angles, then the angles are not congruent.

The inverse is false. Counterexample: Two angles each having a measure of 30° are not right angles, but they are congruent. ◄

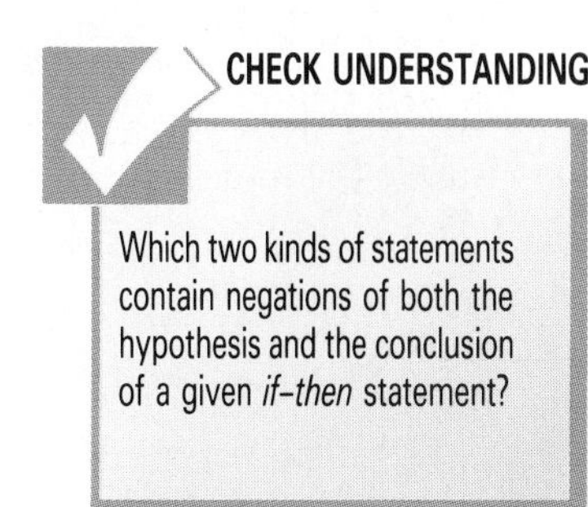


If a conditional is true, its contrapositive is necessarily true. If a conditional is false, its contrapositive is also false.

## Example 3

Write the contrapositive of each statement and tell whether the two statements are true or false. Give a reason for your answer.

**a.** If a figure is a rectangle, then the figure is a polygon.
**b.** If a figure is a rectangle, then it is a square.

**Solution**

a. **Contrapositive:** If a figure is not a polygon, then the figure is not a rectangle.
The statement is true. A rectangle is a type of polygon. The contrapositive is also true. If a figure is not a polygon, then it could not possibly be a special polygon such as a rectangle.

b. **Contrapositive:** If the figure is not a square, then the figure is not a rectangle.
The statement is clearly false. A figure may be a rectangle but not necessarily a square. Its contrapositive is also false. A figure may not be a square, but it might be a rectangle. ◄

## TRY THESE

Write the negation of each statement in two ways.

1. The trumpet is a brass instrument.
2. The number 14 is divisible by 3.

Write the inverse of each statement and tell whether the inverse is *true* or *false*. If false, give a counterexample.

3. If the plant is a tree, then it has leaves.
4. If a number is greater than 25, then that number is greater than 15.

Write the contrapositive of each statement and tell whether the two statements are *true* or *false*. Give a reason for your answer.

5. If you play the violin, then you play a stringed instrument.
6. If you live in Ohio, then you live in Cincinnati.

# EXERCISES

**PRACTICE/ SOLVE PROBLEMS**

In Exercises 1–4, write the converse, inverse, and contrapositive of each conditional statement.

1. If the object is an airplane, then it has wings.
2. If you are studying geometry, then you are studying mathematics.
3. If two lines are perpendicular, then they form four right angles.
4. If a triangle is acute, then all its angles are less than 90°.

5–8. Tell whether the converse and inverse in Exercises 1–4 are *true* or *false*. If false, give a counterexample.

## EXTEND/ SOLVE PROBLEMS

The negation of a negative statement is a positive statement. The negation of "It is not raining," for example, is "It is raining."

Write the negation of each statement.

9. My car is not a convertible.
10. There is no snow on the ground.
11. The bananas have not spoiled.
12. The tent is not torn.

Write the inverse and the contrapositive of each statement.

13. If the bill was not paid, then the telephone service will be turned off.
14. If the train arrives on time, then we will not miss our plane connection.
15. If Rainie does not study, then she will not pass the test.
16. If she cannot read French, then she cannot understand the words of this poem.
17. ***USING DATA*** Use the Data Index on pages 556–557 to find the chart on fuel economy and carbon dioxide. Using the chart as a basis, write an *if–then* statement. Then write its converse, inverse, and contrapositive.

## THINK CRITICALLY/ SOLVE PROBLEMS

Complete each statement with the word *converse, inverse,* or *contrapositive.*

18. The inverse of the converse of a conditional statement is the __?__.
19. The contrapositive of the converse of a conditional statement is the __?__.
20. The converse of the inverse of a conditional statement is the __?__.
21. The converse of the contrapositive of a conditional statement is the __?__.

# 13-4 Necessary and Sufficient Conditions

## EXPLORE

Read the following statements.

a. Which do you think are true?
b. Which are false?

1. It always rains when the sky is cloudy.
2. The sky is cloudy whenever it rains.
3. It always rains if the sky is cloudy.
4. If it rains, then the sky is cloudy.
5. If it does not rain, then the sky is not cloudy.
6. If the sky is not cloudy, then it does not rain.

## SKILLS DEVELOPMENT

An *if–then* statement says, in effect, "If you have this condition, then this result will follow." Consider the following statement.

If rain occurs, then the sky is cloudy.

What condition will guarantee that the sky will be cloudy? The answer is, the occurrence of rain. The occurrence of rain is *sufficient* to guarantee cloudy skies.

Let $p$ represent the hypothesis and $q$ represent the conclusion of an *if–then* statement. To say "If $p$, then $q$" means this.

The hypothesis $p$ is a *sufficient condition* for the conclusion $q$ to happen.

Cloudy skies do not guarantee that rain will occur. Rain is not a necessary result of cloudy skies. Rather, the opposite is true. Cloudy skies are *necessary* for rain to occur. To say "If $p$, then $q$" also means this.

The conclusion $q$ is a *necessary condition* for the hypothesis $p$ to happen.

A **necessary condition** is one in whose absence an event cannot take place. (It cannot rain unless skies are cloudy.) A **sufficient condition** is one in whose presence an event must occur. (Whenever it rains, skies are cloudy.)

### Example 1

For each statement, identify the necessary and the sufficient conditions.

a. If you are a member of this club, you must pay dues every year.
b. If fire occurs, then oxygen must be present.

**Solution**

a. **Necessary condition:** You must pay dues every year.
**Sufficient condition:** You are a member of this club.

b. **Necessary condition:** Oxygen must be present.
**Sufficient condition:** Fire occurs. ◄

The conditional statement "If $p$, then $q$" can be expressed in English in several different ways. Each of the forms below are equivalent to "If $p$ then $q$."

| | |
|---|---|
| If $p$, then $q$. | $p$ only if $q$. |
| $p$ implies $q$. | $q$ whenever $p$. |
| $q$ is implied by $p$. | $p$ is sufficient for $q$. |
| $q$, if $p$. | $q$ is necessary for $p$. |
| If $p$, $q$. | |

## Example 2

Write these statements in *if–then* form.

a. Our yard floods whenever there is a thunderstorm.

b. You get a discount at Mel's Music Corner only if you are a regular customer.

c. A temperature of 32°F is sufficient for ice to melt.

**Solution**

Each sentence is equivalent to a statement of the form "If $p$, then $q$."

a. The sentence has the form "$q$ whenever $p$."
So, the *if–then* form is "If there is a thunderstorm, then our yard floods."

b. The sentence has the form "$p$ only if $q$."
The *if–then* form is "If you get a discount at Mel's Music Corner, then you are a regular customer."

c. The *if–then* form of the sentence is "If the temperature is 32°F, then ice will melt." ◄

Two statements are said to be **logically equivalent** statements when they *always* have the same truth value; that is, both must be true or both must be false. A statement and its converse are not logically equivalent because it is *possible* for the original statement to be true and its converse false.

**Statement:** If a triangle is equilateral, then the triangle has three congruent sides.

**Converse:** If a triangle has three congruent sides, then the triangle is equilateral.

When both the original statement and its converse are true, as in the examples above, they can be combined using the phrase *if and only if*. Thus the resulting statement is "A triangle is equilateral if and only if the triangle has three congruent sides."

TALK IT OVER

In the statement "*p* if and only if *q*," why is *p* both a necessary and a sufficient condition for *q*? Is *q* a necessary and sufficient condition for *p*? Explain.

Let $p$ stand for "a triangle is an equilateral triangle" and $q$ stand for "the triangle has three congruent sides." The statement above can be symbolized as follows.

$p$ if and only if $q$.

An *if-and-only-if* statement is called a **biconditional statement.** In the biconditional "$p$ if and only if $q$," $p$ is both necessary and sufficient for $q$.

### Example 3

Write the pair of statements as a biconditional.

If $x$ is an even number, then $x$ is divisible by 2.
If $x$ is divisible by 2, then $x$ is an even number.

**Solution**

The biconditional can be stated in either of two ways.

$X$ is an even number if and only if $x$ is divisible by 2.
$X$ is divisible by 2 if and only if $x$ is an even number. ◄

## TRY THESE

For each statement, identify the necessary and the sufficient conditions.

1. If today is Wednesday, then tomorrow is Thursday.
2. If you are a citizen of Dallas, then you are a citizen of Texas.

Write each statement in *if–then* form.

3. We make fresh bread only if my cousin Robin is coming to dinner.
4. I fall asleep whenever I hear rain on the roof.
5. Sunlight is necessary for photosynthesis to occur.

Write each pair of statements as a biconditional.

6. If a number is positive, then that number is greater than zero.
   If a number is greater than zero, then that number is positive.
7. If winds are hurricane winds, then those winds blow at a speed of more than 74 mi/h.
   If winds blow at a speed of more than 74 mi/h, then those winds are considered hurricane winds.

# EXERCISES

## PRACTICE/ SOLVE PROBLEMS

Identify which condition is necessary and which condition is sufficient in each of the following statements.

1. If the bottle contains water, then the bottle is not empty.
2. If Marcy has a job, then she receives a paycheck.
3. A number is odd if the number is not divisible by 2.
4. If you write a research paper, then you receive 25 extra points.
5. If the figure is a square, then the figure is a rhombus.

## EXTEND/ SOLVE PROBLEMS

Write the *if–then* statement that is equivalent to each statement.

6. The presence of water is necessary for the presence of life on Earth.
7. Passing Algebra I is both necessary and sufficient for enrollment in Geometry I.
8. Winning the final game in the tournament is sufficient for the team to win the division championship.
9. Writing original music is a necessary condition for being a composer.
10. To be divisible by 6, it is both necessary and sufficient that a number be divisible by 2 and by 3.

Write these statements in *if–then* form.

11. I sneeze whenever the east wind blows.
12. I get good grades whenever I'm really interested in a course.
13. You will be admitted only if you have a ticket.
14. A 4-in. rainfall is sufficient to flood the banks of the river.

## THINK CRITICALLY/ SOLVE PROBLEMS

Rewrite each sentence as two different *if–then* statements.

15. I never drive my car unless my brakes are working well.
16. You will be seated, provided that you are not late.
17. Only registered voters are allowed to vote.
18. I will not pass the test tomorrow unless I have slept well.

# 13-5 Problem Solving Skills:
## WORK WITH DEFINITIONS

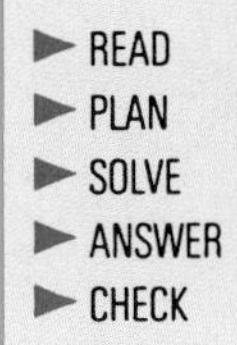


Clear, well-stated definitions are important in every field of study. In general, a good definition should be neither too narrow nor too broad. The language should be clear and precise. The definition should not be stated negatively. In mathematics, definitions have a special form.

- ► A definition must name the term being defined.
- ► A definition must name the set or classification to which that term belongs.
- ► A definition must name the characteristics that distinguish the defined term from other things that belong to the same classification.

### PROBLEM

Which of the four statements that follow is a good definition of *rectangle*?

1. A rectangle is not a square.
2. A rectangle is a quadrilateral with opposite sides parallel.
3. A rectangle is a quadrilateral with opposite sides parallel and congruent.
4. A rectangle is a quadrilateral with four right angles and opposite sides that are congruent and parallel.

### SOLUTION

Definition 1 is stated negatively and is not helpful. A description of what a rectangle is not does not tell you what it is.

Definition 2 is too broad. It names the set to which the term belongs (quadrilaterals) but gives only one characteristic. Thus, it does not distinguish a rectangle from other types of quadrilaterals such as parallelograms or rhombuses.

Definition 3 is also too broad. Its wording does not distinguish a rectangle from a parallelogram or a rhombus.

Definition 4 is the only good definition. It is neither too narrow nor too broad. It clearly defines only a rectangle and no other quadrilateral.

A mathematical definition can be stated as a biconditional statement.

## PROBLEM

Write the definition as a biconditional statement:

An isosceles triangle is a triangle with two equal sides.

## SOLUTION

Write the definition as an *if–then* statement and write its converse.

If a triangle is isosceles, then it has two equal sides.

If a triangle has two equal sides, then it is isosceles.

Both statements are true. Combine them into a biconditional statement.

A triangle is isosceles if and only if it has two equal sides.

# PROBLEMS

Tell whether each definition is a good one. If not, explain.

1. A violin is a musical instrument with strings.
2. A transversal is a line that intersects two or more coplanar lines in different points.
3. Thanksgiving Day is a national holiday.
4. A bisector of an angle divides an angle into two equal parts.
5. A triangle has three segments joining three points.
6. A power of 2 is a number that is divisible by 4.
7. A quadrilateral is when four sides make four angles.
8. A symphony orchestra is a group of musicians who play stringed, brass, woodwind, and percussion instruments.

Write your own definition for each term.

9. timepiece
10. metric ruler

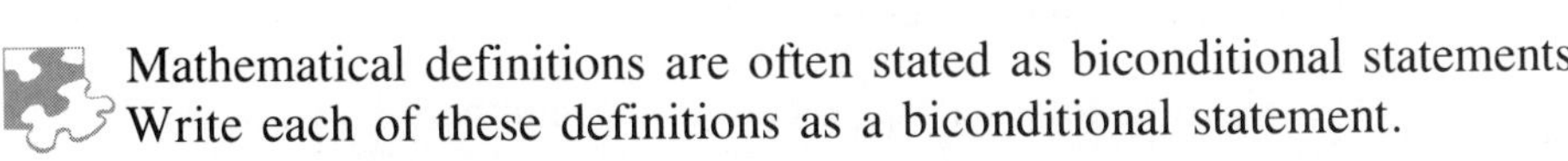

Mathematical definitions are often stated as biconditional statements. Write each of these definitions as a biconditional statement.

11. Collinear points are points that lie on the same line.
12. Angles that have equal measures are congruent angles.
13. The point that divides a segment into two equal segments is the midpoint of the segment.
14. Congruent triangles have corresponding parts that are congruent.

# 13-6 Inductive and Deductive Reasoning

**EXPLORE**

You are asked to draw some disks from a bag, one at a time, without returning them to the bag. You have been told that there are exactly 15 disks in the bag. The first disk you draw is green, as are the next four disks.

a. Suppose that you have been told that the bag contains an equal number of red, green, and white disks. What can you conclude with certainty about the color of the next disk that you will draw from the bag?

b. Suppose that you have been told nothing about the color of the disks in the bag. You draw five more disks from the bag and they are also green. What can you conclude about the color of the next disk that you will draw from the bag?

**SKILLS DEVELOPMENT**

When you examine a number of individual instances of an event and then make a conjecture that is supposed to apply to all such events, you are reasoning **inductively.** The greater the number of instances that support a conclusion, the greater the probability that the conclusion may be correct. You cannot be sure, however. You cannot test all possible instances. So, you cannot know that an inductive conclusion is true beyond a doubt.

### Example 1

Find a pattern for the sum of the first $n$ odd integers.

$1 = 1 = 1^2$ $\quad$ $1 + 3 + 5 + 7 = 16 = 4^2$

$1 + 3 = 4 = 2^2$ $\quad$ $1 + 3 + 5 + 7 + 9 = 25 = 5^2$

$1 + 3 + 5 = 9 = 3^2$ $\quad$ $1 + 3 + 5 + 7 + 9 + 11 = 36 = 6^2$

**Solution**

Let $n$ = the number of odd numbers added together. The series of statements suggests this conjecture.

The sum of the first $n$ odd whole numbers equals $n^2$.

Test the conjecture for successive odd whole numbers beyond 11.

$36 + 13 = 49 = 7^2$ $\quad$ $49 + 15 = 64 = 8^2$

The conjecture seems to be true, but we cannot know without testing all possible instances of $n$. ◄

Inductive reasoning leads to conclusions that may be true but can never be proved to be so. **Deductive reasoning** begins with a set of statements, called **premises,** that are accepted as true. The **conclusion** of a deductive argument follows from the premises and is implied by them. The premises and conclusion make up an **argument.**

In a deductive argument, the premises themselves lead logically to the conclusion. In other words, given the premises, you can readily see what the conclusion is.

## Example 2

Complete this deductive argument by writing the conclusion that follows from the premises.

**Premise 1:** If a student is a member of the school band, then the student plays a musical instrument.

**Premise 2:** Francine is a member of the school band.

**Conclusion:**

### Solution

Francine is a member of the school band. So, she fits the condition stated in Premise 1. Given these premises, only one conclusion follows.

**Conclusion:** Francine plays a musical instrument. ◄

## Example 3

Tell whether the reasoning used is *inductive* or *deductive.*

a. Frank finds that on five different Friday evenings before the basketball game, a coffee and doughnut cart is outside the gymnasium. He concludes that the cart is outside the gymnasium every night before a game.

b. Karen uses the definition of a square (a rectangle with four congruent sides) and of perimeter (the sum of the lengths of the sides of a figure) to conclude that the perimeter of any square is equal to four times the length of a side.

### Solution

a. Frank draws a conclusion about the presence of the cart on every night of a game from evidence about five Friday evenings. He reasons from evidence of a few instances to a conclusion about every instance. The reasoning is inductive.

b. Karen reasons from the congruence of the sides of a square and the definition of perimeter to a conclusion about the perimeter of a square. The conclusion follows logically from the premises. The reasoning is deductive. ◄

## TRY THESE

1. Look at the following statements and make a conjecture based on them.

| | | | |
|---|---|---|---|
| $3 = 2 + 1$ | $5 = 2 + 3$ | $7 = 2 + 5$ | $9 = 7 + 2$ |
| $4 = 3 + 1$ | $6 = 3 + 3$ | $8 = 3 + 5$ | $10 = 7 + 3$ |

In Exercises 2 and 3, complete the deductive argument by writing the conclusion that follows from the premises.

2. **Premise 1:** If a person is 16 years of age or older, then that person may apply for a driver's license.
   **Premise 2:** Ramon is 17 years old.
3. **Premise 1:** If the animal is a spaniel, then it is a canine.
   **Premise 2:** Bowzer is a spaniel.

Tell whether the reasoning used is inductive or deductive.

4. Whenever Magda enters a room with a vase of roses, her eyes water and she begins to sneeze. Based on these experiences, Magda reasons that she must be allergic to roses.
5. Marcia notices that macaroni and cheese has been on the school cafeteria menu each Tuesday for the past six weeks. She reasons that the cafeteria always serves macaroni and cheese on Tuesdays.
6. Sam knows that to be a member of the school choir, you must be able to read music. Sam's friend Mark is a member of the school choir. Sam concludes that Mark can read music.

## EXERCISES

### PRACTICE/ SOLVE PROBLEMS

1. Ron noticed a pattern for some whole numbers. He made a conjecture about the pattern. State the conjecture. Then test it by looking for a counterexample that shows the conjecture to be false.

| | |
|---|---|
| $5 = 2 + 3$ | $51 = 16 + 17 + 18$ |
| $15 = 1 + 2 + 3 + 4 + 5$ | $42 = 9 + 10 + 11 + 12$ |

Look for a pattern and predict the next number in each sequence.

2. 3, 6, 12, 24, __?__
3. 2, 4, 7, 11, 16, __?__
4. 10, 1, 0.1, 0.01, __?__
5. 12, 3, $\frac{3}{4}$, __?__

Complete each argument by writing the conclusion that follows by deductive reasoning from the premises.

6. If Lila finishes her homework, then she will join us for dinner. Lila has finished her homework.

7. All parallelograms have two pairs of parallel sides. A rectangle is a parallelogram.

Tell whether the reasoning used is deductive or inductive.

8. Albert reasoned as follows: Earth has nitrogen, oxygen, and water, all necessary for life. Evidence of nitrogen, oxygen, and water on any other planet would mean that life must be present or have appeared on such a planet.

## EXTEND/ SOLVE PROBLEMS

For each set of premises, write the conclusion that follows by deductive reasoning from the premises. If no conclusion is possible, write *none*.

9. $\overline{AB}$ is longer than $\overline{BC}$.
$\overline{BC}$ is longer than $\overline{CD}$.

10. If the instrument is a clarinet, then it has valves.
This instrument has no valves.

11. Ned is older than Sandra.
Sandra is older than Megan.

12. Mark is taller than Merv.
Doug is taller than Merv.

13. Polygon $M$ has more than 5 sides.
Polygon $N$ has more than 5 sides.

14. All dancers have flexibility and strength.
Petra is neither flexible nor strong.

## THINK CRITICALLY/ SOLVE PROBLEMS

15. Is a conclusion possible from the given premises? Explain.
If Cleo is taking French III, she will be able to translate this letter for me.
Cleo is able to translate this letter for me.

16. Suppose that you are taking a quiz. You know three things.
(1) If the first answer is true, then the next answer is false.
(2) The last answer is the same as the first answer.
(3) The second answer is true.

How could you show that the last answer on the quiz is false? (*Hint:* Assume that the last answer on the quiz is true.)

# 13-7 Patterns of Deductive Reasoning

## EXPLORE/ WORKING TOGETHER

With a partner, read and discuss each of the four arguments below.

**Argument 1**
If you live in San Francisco, then you live in California.
You live in San Francisco.
Therefore, you live in California.

**Argument 2**
If you live in San Francisco, then you live in California.
You do not live in California.
Therefore, you do not live in San Francisco.

**Argument 3**
If you live in San Francisco, then you live in California.
You live in California.
Therefore, you live in San Francisco.

**Argument 4**
If you live in San Francisco, then you live in California.
You do not live in San Francisco.
Therefore, you do not live in California.

a. In which arguments does the conclusion follow logically from the premises?
b. In which arguments does the conclusion not follow logically from the premises?

## SKILLS DEVELOPMENT

Depending on just how the premises of a deductive argument are related, the argument may be valid or invalid. In a **valid** argument, if the premises are true, then the conclusion must be true.

Arguments 1 and 2 in the Explore activity are valid. The validity of these arguments becomes apparent if you look at their form. To do this, let $p$ stand for "you live in San Francisco." Let $q$ stand for "you live in California." The first premise of each of these arguments can be symbolized as $p \rightarrow q$ (read "If $p$ then $q$" or "$p$ implies $q$").

Argument 1

Argument 2

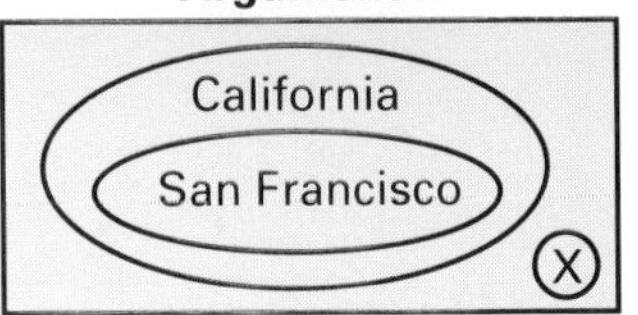


**Valid Argument Forms**

**Argument 1**
Premise 1: $p \rightarrow q$
Premise 2: $\underline{p}$
Conclusion: $q$

**Argument 2**
Premise 1: $p \rightarrow q$
Premise 2: $\underline{\sim q}$
Conclusion: $\sim p$

Thus, given the premise $p \rightarrow q$, a valid argument results from asserting $p$ or denying $q$. Argument 1 illustrates the Law of Detachment, an argument form known as *modus ponens*. Argument 2 illustrates the

Law of the Contrapositive, an argument form known as *modus tollens*. The conclusion of each argument becomes apparent as the premises are diagrammed.

Arguments 3 and 4 have invalid argument forms. As the diagrams at the right show, the stated conclusions do not result from diagramming the premises.

**Argument 3**

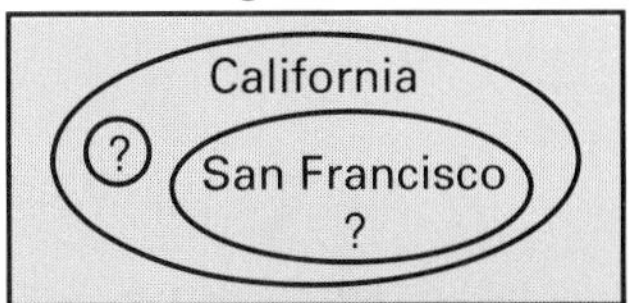


**Argument 4**

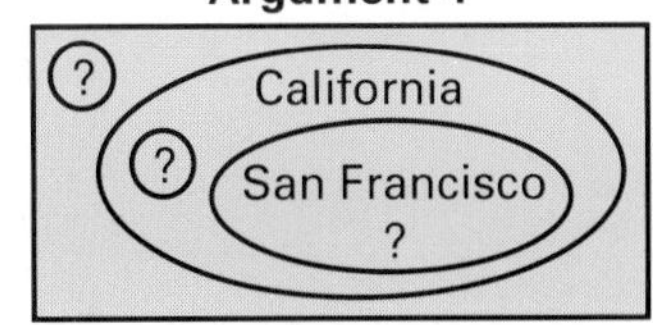


**Invalid Argument Forms**

**Argument 3**
Premise 1: $p \rightarrow q$
Premise 2: $q$
Conclusion: $p$

**Argument 4**
Premise 1: $p \rightarrow q$
Premise 2: $\sim p$
Conclusion: $\sim q$

### Example

Determine by form whether the following arguments are valid or invalid.

**a.** If $2x = 4$, then $x = 2$
$2x = 4$
$x = 2$

**b.** If this polygon is a square, then the measures of its angles total 360°.
This polygon is not a square.
The measures of its angles do not total 360°.

### Solution

**a.** Let $p$ = "$2x = 4$" and $q$ = "$x = 2$." The argument asserts $p \rightarrow q$ and $p$. You can conclude $q$, or $x = 2$. The argument is valid. It illustrates the Law of Detachment.

**b.** Let $p$ = "this polygon is a square." Let $q$ = "the measures of its angles total 360°." This argument asserts $p \rightarrow q$ and $\sim p$. This argument form is invalid. Both premises are true, but the conclusion is not necessarily true. Rectangles are polygons that are not squares but their angles total 360°. ◀

It is important to stress that the validity of an argument does not depend on the truth or falsity of the statements in it. Validity is concerned only with the form of an argument. A valid argument form guarantees that *if* the premises are true, then the conclusion *must* be true. A deductive argument whose form is valid and whose premises are true is called a **sound argument,** whereas a deductive argument whose form is valid but contains at least one false premise is said to be **unsound.** The argument below is valid because of its form, but it is unsound because its first premise is false.

If an animal has four legs, then it is a dog.
Tabby has four legs.
Therefore, Tabby is a dog.

## TRY THESE

Determine whether the following arguments are *valid* or *invalid*. Give a reason for each valid argument.

1. If $3b = 12$, then $b = 4$.
   $3b = 12$
   ______________
   $b = 4$

2. If it rains, then our picnic will be canceled.
   Our picnic was canceled.
   Therefore, it must have rained.

# EXERCISES

**PRACTICE/ SOLVE PROBLEMS**

Determine whether the following arguments are *valid* or *invalid*. Give a reason for each valid argument.

1. If the animal is a whale, then it is a mammal.
   The animal is a whale.
   ______________
   The animal is a mammal.

2. If the bulb is burned out, then the lamp will not light.
   The lamp does not light.
   ______________
   The bulb is burned out.

**EXTEND/ SOLVE PROBLEMS**

Determine whether the following arguments are *valid* or *invalid*. Then tell whether the argument is *sound* or *unsound*.

3. If roses were not flowers, then weeds would bear seeds.
   Weeds bear seeds.
   ______________
   Roses are not flowers.

4. If a number is not a whole number, then it is not greater than zero.
   1.2 is greater than zero.
   ______________
   1.2 is a whole number.

**THINK CRITICALLY/ SOLVE PROBLEMS**

5. Prove the following statement: *If the square of a number is odd, then the number is odd.*
   (*Hint:* Show that the contrapositive is true: *If the number is not odd, then the square of that number is not odd.*
   Let $n$ be an even number: $2x$.)

- READ
- PLAN
- SOLVE
- ANSWER
- CHECK

# Problem Solving Applications:

## INVALID ARGUMENTS IN ADVERTISING

Many advertisements are actually arguments in which only one premise and the conclusion are stated. When the other premise is hidden, the argument is often invalid. No conclusion is possible when the argument form is invalid. Read this example:

Really popular people chew Tasty Gum.

In this slogan the key premise is never stated. The full argument would look like this.

If you chew Tasty Gum, then you will be popular.
You want to be popular.
Therefore, you should chew Tasty Gum.

The truth of the key premise is questionable. In addition, the second premise is not necessarily true. However, even if the premises were true, the conclusion does not follow from the premises, $p \rightarrow q$ and $q$.

**WRITING ABOUT MATH**

Write one or two examples of your own of an invalid argument. Use your examples as a basis for a paragraph in which you explain why no certain conclusion can be drawn from such an argument.

Determine the unstated key premise in each advertisement. Then write the argument underlying the slogan.

1. Not getting fiber in your diet?
   Eat Granny Granola bars.
2. Feeling tired? Eat Pep Crackles for breakfast.
3. Don't be a fool.
   Buy the best—Springsole running shoes!
4. Without life insurance, you can't feel secure about your future.
5. Don't like rich coffee flavor?
   Then don't buy Golden Bean coffee.
6. Get rid of tangled hair for good. Use Smoothie Shampoo regularly.
7. Wowicin gets rid of even the worst headache in almost no time at all.
8. Haven't joined Stretchgym yet? Then you're not getting the exercise you need.
9. Are you unable to fight germs that cause a sore, raspy throat? Don't despair—use Clean Green Mouthwash.
10. Tired of feeling fat? Try Speedy Slim today.

# 13-8 Problem Solving Strategies: USE LOGICAL REASONING

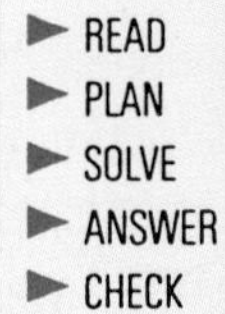


In the study of mathematics, deductive reasoning is important in solving problems. In geometry, for example, you reason from definitions and postulates that form true premises to reach true conclusions. In algebra, you use properties of addition and multiplication and the laws of exponents in order to simplify algebraic expressions and solve equations. Knowing which facts to use and how to organize your argument logically is an important part of solving problems successfully.

## PROBLEM

Use geometric facts to show that the conclusion stated about the isosceles triangle shown at the right is true. Present your reasons in a logical order.

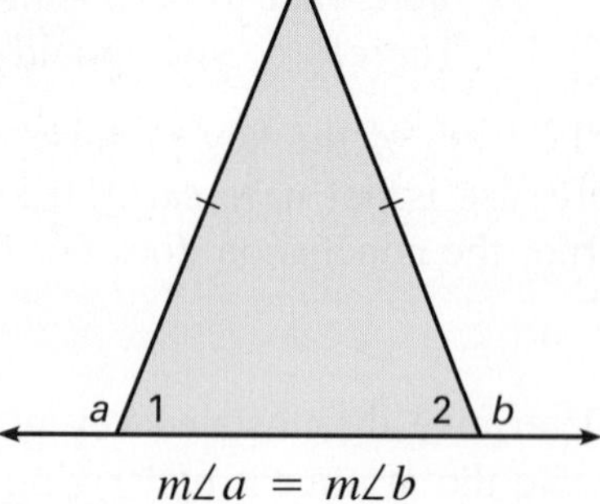


$m\angle a = m\angle b$

## SOLUTION

The triangle is an isosceles triangle.

$\angle 1 \cong \angle 2$ ← Base angles of an isosceles triangle are congruent.
$m\angle 1 = m\angle 2$

$m\angle 1 + m\angle a = 180°$ ← $\angle 1$ and $\angle a$ are supplementary angles.
$m\angle 2 + m\angle b = 180°$ $\angle 2$ and $\angle b$ are supplementary angles.

$m\angle 1 = 180° - m\angle a$
$m\angle 2 = 180° - m\angle b$
$180° - m\angle a = 180° - m\angle b$
$m\angle a = m\angle b$

## PROBLEM

Prove that the following statement is true: *The sum of an odd number and an even number is an odd number.*

## SOLUTION

To prove this statement inductively would require testing every possible combination of even and odd counting numbers. Use a deductive argument to show that the statement is true.

### MIXED REVIEW

Solve.

1. $3x + 4 = 43$
2. $x^2 - 12 = 69$

Find the midpoint, $P$, of the line segment whose endpoints are given.

3. $A(4, 6)$ and $B(-2, -2)$
4. $C(-5, 5)$ and $D(5, -5)$
5. Solve. $x + 2y = 12$
   $2x - y = 9$

Write the coordinates of the image of each point under a reflection across the given axis.

6. $(-3, 4)$; $x$-axis
7. $(-2, 1)$; $y$-axis
8. A store sells 8 kinds of pizza with 8 different toppings. How many different combinations are available?

Let $n$ be any whole number. Then, any odd number can be represented by the expression $2n + 1$.
Let $m$ be another whole number, such that $m \neq n$. Then any even number can be represented by the expression $2m$.

Use these expressions to write an algebraic expression for the sum of an odd number and an even number.

$$(2n + 1) + 2m = (2n + 2m) + 1 \quad \leftarrow \text{commutative and associative laws}$$
$$= 2(n + m) + 1 \quad \leftarrow \text{distributive law}$$

The expression $2(n + m) + 1$ is the form of an odd number. So, the statement is true for any odd number and any even number.

# PROBLEMS

For Exercises 1 and 2, refer to the figures. Use geometric facts to show that the conclusion given is true. State your reasons in a logical order.

**1.** $a + b = 180°$

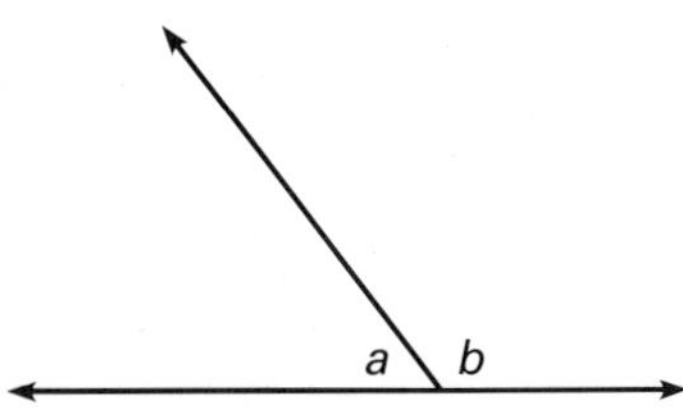


**2.** $a = b = 45°$

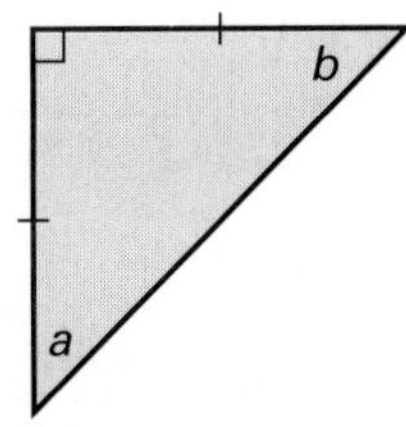


**3.** Use algebraic facts to show that the sum of two even numbers is an even number. (*Hint:* Let $2m$ be an even number. Let $2n$ be another even number.)

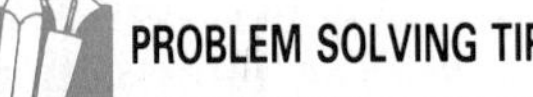


In order to solve the problem, look at the triangles and determine whether they are similar or congruent.

**4.** Use geometric facts and algebraic processes to find the measure of each angle in the figure below.

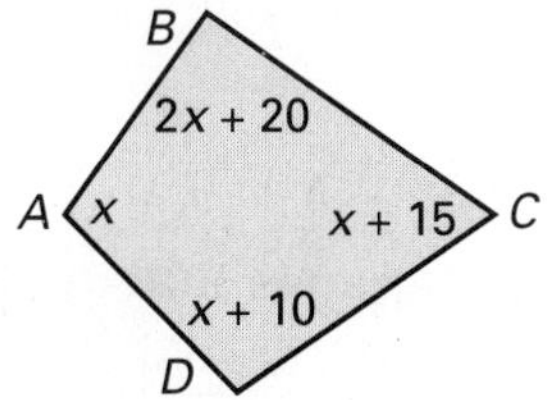


**5.** Use the diagram below of a river and its banks to find the width of the river. Present reasons to support your answer.

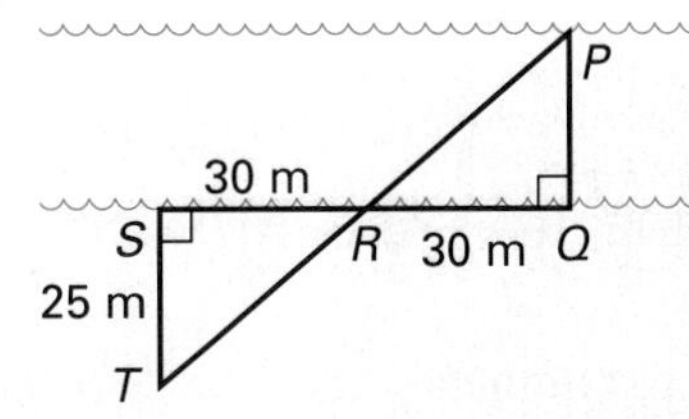


**6.** Use algebraic facts to show that the sum of two odd numbers is an even number. (*Hint:* Let $2n + 1$ be an odd number. Let $2m + 1$ be another odd number.)

## CHAPTER 13 ● REVIEW

Write the letter of the word at the right that matches each description.

1. sets that contain the same number of elements
2. sets that contain the same members
3. reasoning from a number of instances to a generalization that is probably true about all such instances
4. reasoning from a set of premises to a conclusion implied by those premises
5. a statement of the form "$p$ if and only if $q$."

a. biconditional statement
b. deductive reasoning
c. inductive reasoning
d. equivalent sets
e. equal sets

### SECTION 13–1 SETS (pages 492–495)

► Each item in a **set** is called an **element,** or **member,** of the set.
► One set is a **subset** of another if every element of the first set is a member of the second.

Determine all the possible subsets of each set.

6. $\{1, 5\}$
7. $\{2, 4, 5\}$

### SECTION 13–2 UNION AND INTERSECTION OF SETS (pages 496–499)

► The **union** of two sets $A$ and $B$ is the set of all elements that are elements of $A$ *or* of $B$ *or* of both.
► The **intersection** of two sets $A$ and $B$ is the set of all elements that are elements of $A$ *and* elements of $B$. Two sets whose intersection is the null set are **disjoint.**

Let $U = \{1, 2, 3, 4, \ldots, 12\}$, $A = \{1, 2, 3, 4\}$, and $B = \{3, 6, 9, 12\}$. Find the following.

8. $A'$
9. $A \cup B$
10. $A \cap B$

### SECTION 13–3 CONVERSE, INVERSE, AND CONTRAPOSITIVE (pages 500–503)

► To write the **negation** of a statement, add "not" to the statement.
► The **converse** and **inverse** of a true statement are not necessarily true. A statement and its **contrapositive** are either both true or both false.

11. Write the inverse and the contrapositive of the statement.
    If you are studying Russian, then you are studying a foreign language.

## SECTION 13–4 NECESSARY AND SUFFICIENT CONDITIONS (pages 504–507)

- In the statement "If $p$, then $q$," $q$ is a **necessary condition** for $p$ to happen. $p$ is a **sufficient condition** for $q$ to happen.
- If a statement and its converse are both true, then the two statements may be written as a **biconditional statement.**

12. Write the pair of statements as a biconditional statement.
    If 2 is an even number, then 4 is an even number.
    If 4 is an even number, then 2 is an even number.

## SECTIONS 13–5 AND 13–8 PROBLEM SOLVING (pages 508–509, 518–519)

- Reasoning with mathematical facts and organizing them is important in solving problems.

13. Find the measure of each angle in the triangle shown.

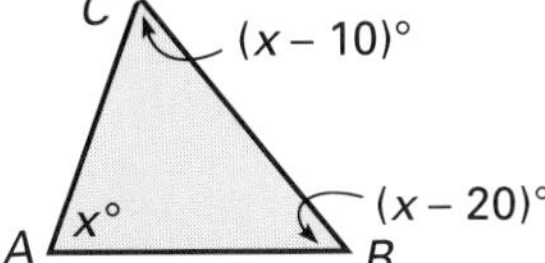


## SECTION 13–6 INDUCTIVE AND DEDUCTIVE REASONING (pages 510–513)

- In **inductive reasoning,** a limited number of individual instances provide evidence for a conjecture about all instances of something. In **deductive reasoning** a conclusion is drawn from premises that are accepted as true.

14. Complete this deductive argument by writing the conclusion that follows from the premises.
    If the animal is a lion, then it has a mane.
    Elsa is a lion.

## SECTION 13–7 PATTERNS OF DEDUCTIVE REASONING (pages 514–517)

- In a valid deductive argument, the conclusion will be true if the premises are true.

15. Tell whether the argument is *valid* or *invalid*. Give a reason.
    If the bird is a canary, then the bird is not a macaw.
    My bird Chatter is a macaw.
    ___
    My bird Chatter is not a canary.

***USING DATA*** Use the table of frequency ranges for instruments and human voices on page 491.

16. Use set-builder notation or roster notation to describe one of the sets of instruments or voices in the table.

# CHAPTER 13 ● TEST

1. Determine which sets are equal and which are equivalent.

$A = \{9, 3, 1, 7, 11\}$ $B = \{11, 7, 1, 9, 3\}$ $C = \{7, 1, 5, 9, 4\}$

Let $U = \{t, i, m, e, s\}$, $P = \{t, i, m, e\}$, and $Q = \{s, i, t, e\}$.
Find each union or intersection.

2. $P' \cup Q'$
3. $(P \cup Q)'$
4. $P' \cap Q'$
5. Write the inverse of the statement and tell whether the inverse is *true* or *false*. If false, give a counterexample.

If a number is less than 10, then that number is less than 15.

Write these statements in *if–then* form.

6. Our collie Max barks whenever he hears a knock at the door.
7. You receive a ticket to the dress rehearsal only if you are a contributor to the Opera Guild Fund.
8. Identify the necessary and the sufficient condition in the statement.

If a substance burns, then oxygen is present in the atmosphere.

9. Which of the four statements is a good definition of an inscribed angle?
   a. An inscribed angle is the opposite of a central angle.
   b. An inscribed angle has its vertex on the circle.
   c. An inscribed angle intercepts an arc on the circle.
   d. An inscribed angle is an angle whose vertex lies on a circle and each side of which contains a chord of the circle.

Tell whether the reasoning used is *inductive* or *deductive*.

10. Ravi measured 12 pairs of alternate interior angles. He then concluded that alternate interior angles have the same measure.
11. Harry used the definitions of a square and of the area of a rectangle to conclude that the area of a square with sides of length $x$ is equal to $x^2$.

Tell whether the argument is *valid* or *invalid*.

12. If a number is odd, then the square of the number is odd.
    The number 7 is odd.
    ---
    The square of 7 is odd.
13. If the figure is a pentagon, then the figure is a polygon.
    The figure is not a pentagon.
    ---
    The figure is not a polygon.

# CHAPTER 13 • CUMULATIVE REVIEW

1. Find what percent 14.4 is of 80.

2. Evaluate $-|r|$ when $r = -3$.

Identify the figure.

3.

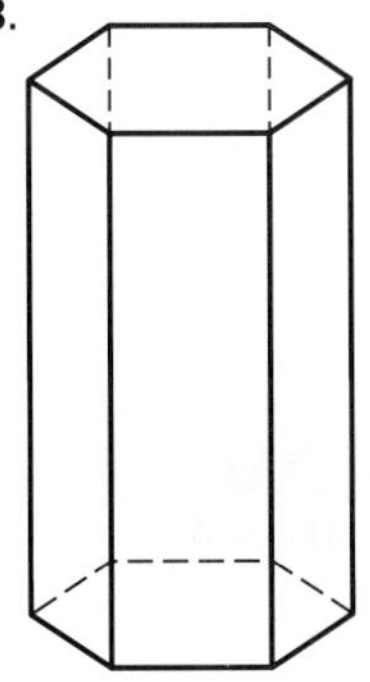

4.

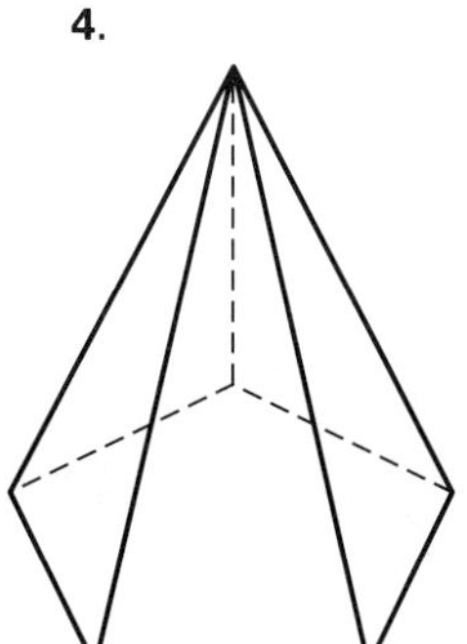

5. Find the first quartile, median, and third quartile for the set of data.

| HEIGHTS OF VOLLEYBALL TEAM MEMBERS (in cm) | | | | | | | |
|---|---|---|---|---|---|---|---|
| 179 | 168 | 171 | 185 | 175 | 165 | 182 | 168 |
| 177 | 175 | 180 | 181 | 170 | 177 | 173 | 180 |

6. Solve. $r + 6.2 = 4.8$

Find the volume of the figure. Use 3.14 for $\pi$. Round answers to the nearest tenth.

7. sphere with radius of 8 in.

8. cone with radius of 7 cm and height of 10 cm

Factor the polynomial.

9. $4mn - 16\ m^3n^4$

Find an equation for each circle described.

10. center at origin, radius 14

11. center at origin, radius 4

Solve.

12. $3x + 4y = 12$
$2x - y = 8$

13. $3x + 3y = 15$
$-2x - y = -14$

14. A flagpole casts a shadow 64 ft long. At the same time, a girl who is 5 ft tall casts a shadow 8 ft long. Find the height of the flagpole.

Write the coordinates of the image of each point under a reflection across the given line.

15. $(-4, 5)$; $y = x$

16. $(-4, -2)$; $y = -x$

17. A store sells 4 brands of sweatshirts. Each brand is available in 4 sizes and 5 colors. How many different sweatshirts does the store sell?

18. Determine which of the given sets, if any, are equal.

$P = \{1, 14, 10, 8\}$
$Q = \{8, 1, 12, 6\}$
$R = \{8, 10, 1, 14\}$

# CHAPTER 13 ● CUMULATIVE TEST

1. Find what percent 11.4 is of 76.
A. $6\frac{2}{3}\%$ B. 10%
C. 15% D. 20%

2. Evaluate $-|-n|$ when $n = -4$.
A. $|-4|$ B. $-4$
C. 4 D. $|4|$

3. Find the measure of arc $CAB$.

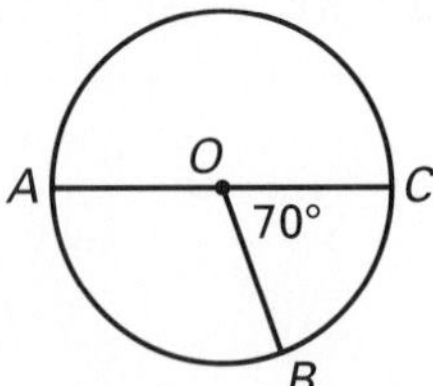


A. 110°
B. 20°
C. 290°
D. 300°

Use the table for Exercises 4 and 5.

**DISTANCES TRAVELED BY HIKING CLUB MEMBERS**

| Number of Miles | Frequency |
|---|---|
| 5 | 4 |
| 8 | 6 |
| 10 | 3 |
| 15 | 6 |
| 20 | 1 |

4. Find the mean.
A. 11.5 B. 11.6
C. 15 D. none of these

5. Find the mode.
A. 8 and 15 B. 10
C. 5 and 20 D. none of these

6. Solve. $y^2 + 5 = 69$
A. $\pm\sqrt{74}$ B. $\pm 64$
C. $\pm 8$ D. $\pm 4$

7. Find the area of the rectangle.

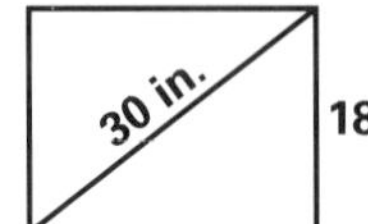


A. 576 in.$^2$
B. 324 in.$^2$
C. 432 in.$^2$
D. 540 in.$^2$

8. Find the product. $(a + 2)(a - 3)$
A. $a^2 - a + 6$ B. $a^2 - 5a + 6$
C. $a^2 - a - 6$ D. $a^2 - a - 5$

9. Which of the ordered pairs is *not* a solution of $5x = y + 2$?
A. (1, 3) B. (0, 2)
C. (3, 13) D. (−1, −7)

10. Solve. $x + 3y = -12$
$2x - y = 4$
A. (4, 0) B. (0, −4)
C. (24, −44) D. (4, −4)

11. A tree 9 ft tall casts a shadow that is $4\frac{1}{2}$ ft long. At the same time of day, Dana casts a shadow $2\frac{1}{2}$ ft long. How tall is Dana?
A. 3 ft B. 5 ft
C. $5\frac{1}{2}$ ft D. $5\frac{3}{4}$ ft

12. Identify the coordinates of the image of (4, −3) under a reflection across the line whose equation is $y = x$.
A. (3, 4) B. (−3, −4)
C. (−4, 3) D. (−3, 4)

13. Two number cubes are rolled. What is the probability that the sum of the numbers rolled will be odd and greater than 8?
A. $\frac{5}{36}$ B. $\frac{1}{18}$
C. $\frac{1}{6}$ D. $\frac{1}{9}$

14. Find $X \cup Y$.
$X = \{a, l, m, s\}$
$Y = \{m, a, s, t\}$
A. {l, a, m, s, t} B. {l, t}
C. {a, m, s} D. $\phi$

# CHAPTER 14 SKILLS PREVIEW

Write each ratio in fraction form.

1. sin $P$
2. cos $P$
3. tan $T$

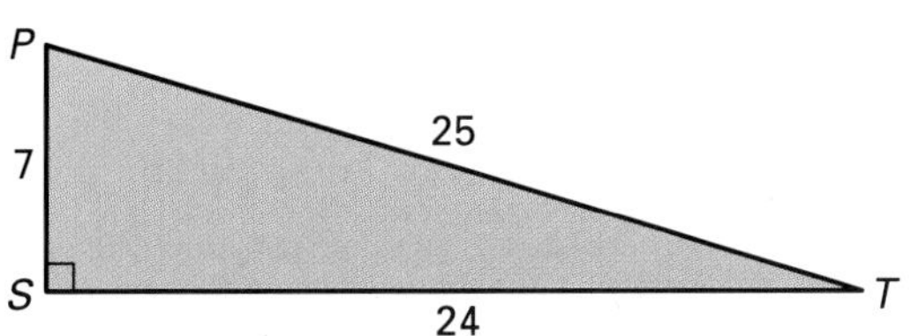


Use a calculator or the Table of Trigonometric Ratios on page 581 to find each value.

4. sin 53°
5. $m\angle A$, if tan $A$ = 1.6
6. $AB$ to the nearest tenth

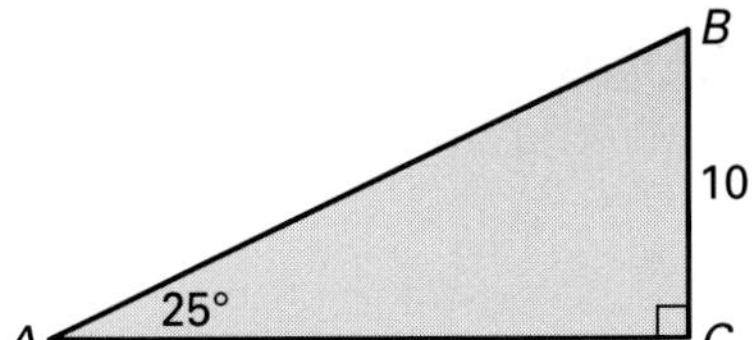


7. $m\angle F$ to the nearest degree

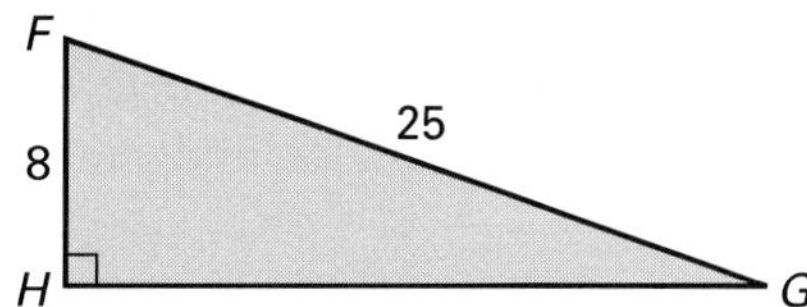


8. Solve $\triangle UVW$. Round each length to the nearest tenth and each angle measure to the nearest degree.

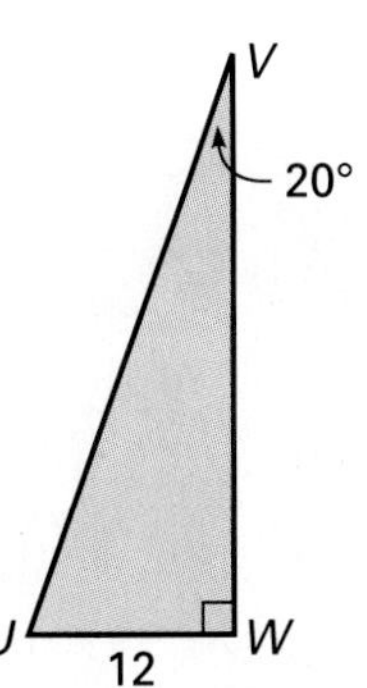


Find each length in $\triangle DEF$. If necessary, express your answers using square roots.

9. $DE$
10. $DF$

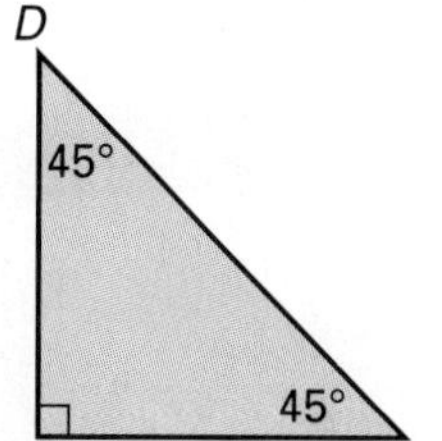

Solve.

11. A 12-ft ladder leaning against a wall makes a 70° angle with the ground. To the nearest tenth of a foot, how far from the base of the wall is the foot of the ladder?

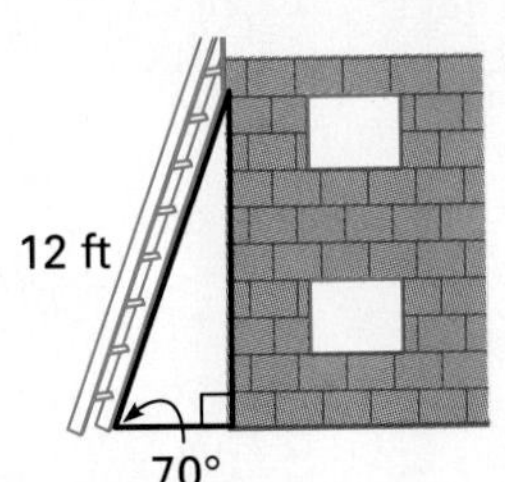


12. When the sun shines on a tree 20 m tall, the tree casts a shadow 15 m long. To the nearest degree, what angle do the sun's rays make with the ground?

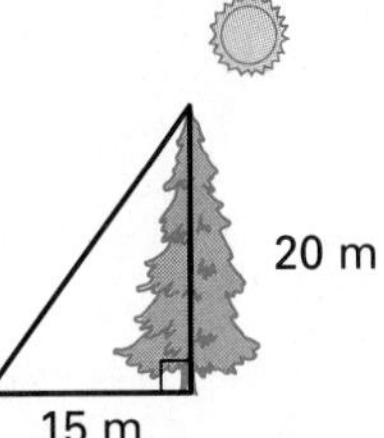


13. The tailgate of a truck is 1.14 m from the ground. How long should a ramp be so that the incline from the ground up to the tailgate is 9.5°? Round your answer to the nearest tenth.

# CHAPTER 14 TRIGONOMETRY

**THEME Crafts**

About 2,000 years ago, the Greek astronomer Hipparchus discovered that he could use measurements of certain parts of a right triangle to find the measurements of the remaining parts of the triangle. He called his method **trigonometry,** which means "triangle measure." Trigonometry is used in many fields, such as surveying, navigation, and astronomy. In this chapter, you will learn how to use Hipparchus's method to find missing parts of a right triangle.

Craftspeople who work with materials such as glass, wood, metal, lucite, or fabric often must find angles and lengths of sides of triangles and other figures.

## DECISION MAKING

### Using Data

Use the table and the pattern for the front of one type of birdhouse to answer the following questions.

1. If the house for each species is set at the maximum recommended height above the ground, which house's entrance would be highest above the ground?
2. The houses for which species of birds require the most wood?
3. For the woodpecker and wren houses, which measure(s) – a, b, c, or d – would be different?

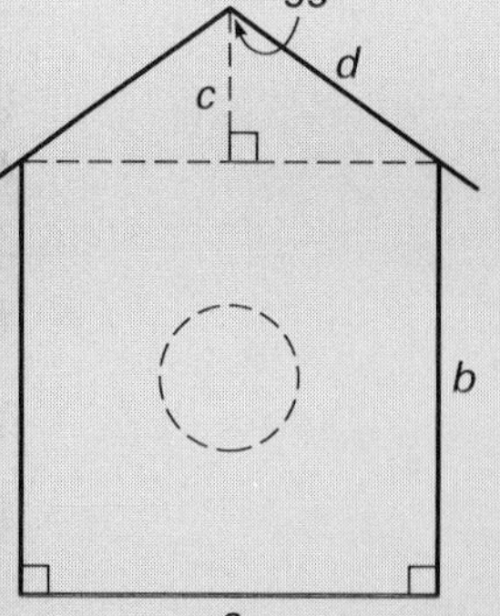


**BIRDHOUSE SPECIFICATIONS**

| Bird Species | Floor (in., $a \times a$) | Height (in., $b$) | Entrance Above Floor (in.) | Diameter of Entrance (in.) | Height Above Ground (ft) |
|---|---|---|---|---|---|
| Bluebird | $5 \times 5$ | 8 | 6 | $1\frac{1}{2}$ | 5–10 |
| Chickadee | $4 \times 4$ | 10 | 6–8 | $1\frac{1}{8}$ | 6–15 |
| Finch | $6 \times 6$ | 6 | 4 | 2 | 8–12 |
| Martin | $6 \times 6$ | 6 | 1 | $2\frac{1}{2}$ | 15–20 |
| Nuthatch | $4 \times 4$ | 10 | 6 | $1\frac{1}{4}$ | 12–20 |
| Woodpecker | $4 \times 4$ | 10 | 6–8 | $1\frac{1}{4}$ | 6–20 |
| Wren | $4 \times 4$ | 8 | 1–6 | $1\frac{1}{4}$ | 6–10 |

### Working Together

Your group's task is to find out about a craft that involves working with right triangles. Use your own knowledge, library reference materials, or interviews with family members or craftspeople in your area. Present your findings to the class in the form of a poster, display, or demonstration. Include a short description of the craft, an actual item or picture of an item made with this craft, and an explanation of where triangles are used in the craft. Tell whether the craftsperson needs to find the measures of the sides or angles of the triangles. If so, how is this done? If not, why isn't it necessary?

# 14-1 The Sine, Cosine, and Tangent Ratios

## EXPLORE

Juanita uses a stencil to create designs. Her stencil includes three similar right triangles *ABC, DEF,* and *GHI.*

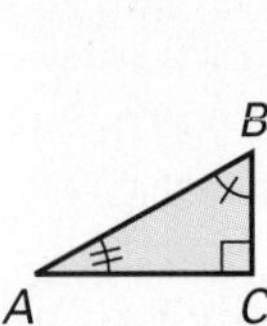

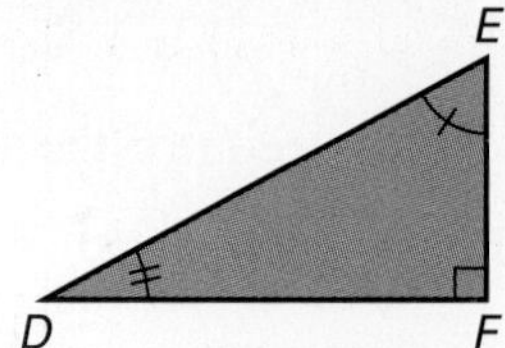

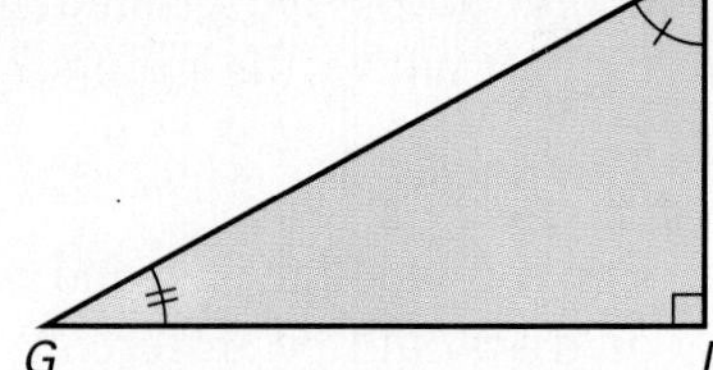


a. Measure the sides of each triangle to the nearest millimeter.

Find each ratio to the nearest hundredth.

b. $\frac{BC}{AB}$, $\frac{EF}{DE}$, and $\frac{HI}{GH}$ c. $\frac{AC}{AB}$, $\frac{DF}{DE}$, and $\frac{GI}{GH}$ d. $\frac{BC}{AC}$, $\frac{EF}{DF}$, and $\frac{HI}{GI}$

e. What can you conclude about the ratio of the length of one leg of a right triangle to its hypotenuse compared to the ratio of the length of the corresponding leg and hypotenuse of a similar triangle?

f. What can you conclude about the ratio of the lengths of the legs of one right triangle compared to the ratio of the lengths of the corresponding legs of a similar right triangle?

## SKILLS DEVELOPMENT

**CHECK UNDERSTANDING**

In $\triangle RST$ at the right, which leg is opposite $\angle S$? Which leg is adjacent to $\angle S$?

The legs of a right triangle are often described by relating them to one of the acute angles of the triangle.

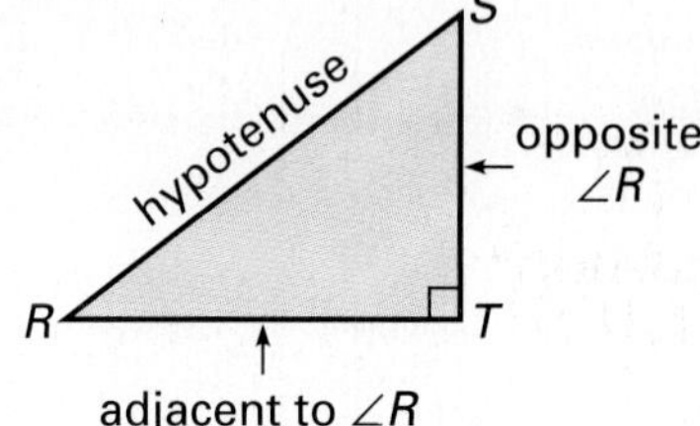


In relation to $\angle R$: $\overline{RT}$ is the side *adjacent* to $\angle R$.
$\overline{ST}$ is the side *opposite* $\angle R$.

For any right triangle, there are three **trigonometric ratios** of the lengths of the sides of the triangle. These ratios are the same for all congruent angles in right triangles even though the lengths of the sides of these triangles may be different. The trigonometric ratios for triangle *ABC* at the right are summarized in the following chart.

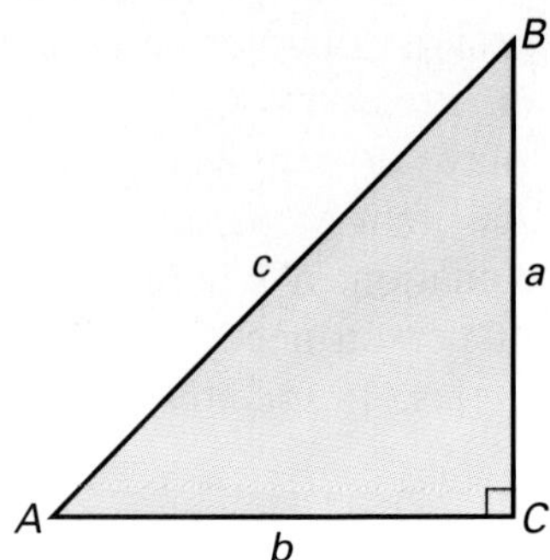

| TRIGONOMETRIC RATIOS | | |
|---|---|---|
| Name of Ratio | Abbreviation | Ratio |
| **sine** of $\angle A$ | $\sin A$ | $\frac{\text{length of side opposite } \angle A}{\text{hypotenuse}}$ or $\frac{a}{c}$ |
| **cosine** of $\angle A$ | $\cos A$ | $\frac{\text{length of side adjacent to } \angle A}{\text{hypotenuse}}$ or $\frac{b}{c}$ |
| **tangent** of $\angle A$ | $\tan A$ | $\frac{\text{length of side opposite } \angle A}{\text{length of side adjacent to } \angle A}$ or $\frac{a}{b}$ |

**READING MATH**

You may find it easier to remember these shortened forms of the trigonometric ratios.

$\sin A = \frac{\text{opposite}}{\text{hypotenuse}}$

$\cos A = \frac{\text{adjacent}}{\text{hypotenuse}}$

$\tan A = \frac{\text{opposite}}{\text{adjacent}}$

A memory device that some people use is
SOH CAH TOA
**SOH**: **S**ine is **O**pposite over **H**ypotenuse.
**CAH**: **C**osine is **A**djacent over **H**ypotenuse.
**TOA**: **T**angent is **O**pposite over **A**djacent.

## Example 1

Use the triangle at the right.
Find sin $D$, cos $D$, and tan $D$.

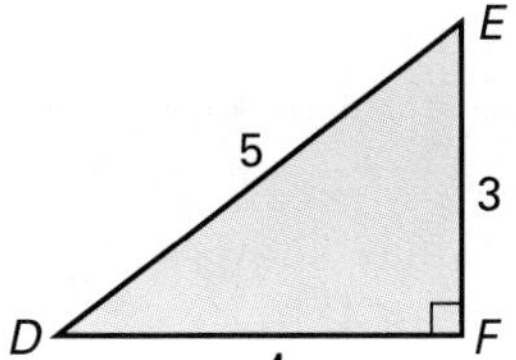


### Solution

$$\sin D = \frac{\text{length of side opposite } \angle D}{\text{hypotenuse}} = \frac{3}{5}$$

$$\cos D = \frac{\text{length of side adjacent to } \angle D}{\text{hypotenuse}} = \frac{4}{5}$$

$$\tan D = \frac{\text{length of side opposite } \angle D}{\text{length of side adjacent to } \angle D} = \frac{3}{4}$$ ◄

## Example 2

In right triangle $PQR$, $\angle R$ is the right angle. If $\sin P = \frac{12}{13}$, find cos $P$ and tan $P$.

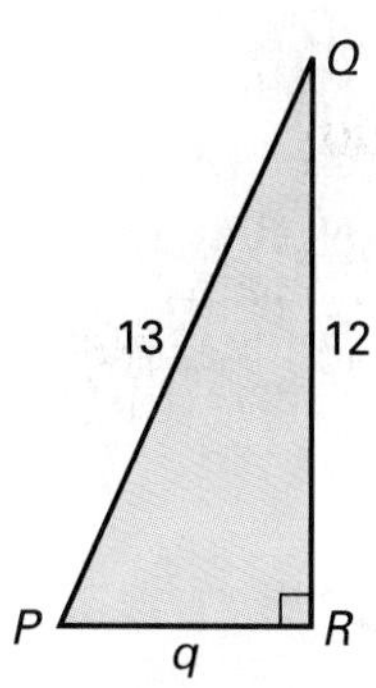


### Solution

Use the information given in the problem to make a diagram of $\triangle PQR$.

$$\sin P = \frac{\text{length of side opposite } \angle P}{\text{hypotenuse}} = \frac{12}{13}$$

Use the Pythagorean Theorem to find the length of $q$, the side adjacent to $\angle P$.

$$12^2 + q^2 = 13^2$$
$$144 + q^2 = 169$$
$$q^2 = 169 - 144$$
$$q^2 = 25$$
$$q = \sqrt{25} = 5$$

So, the length of the side adjacent to $\angle P$ is 5.

$$\cos P = \frac{\text{length of side adjacent to } \angle P}{\text{hypotenuse}} = \frac{5}{13}$$

$$\tan P = \frac{\text{length of side opposite } \angle P}{\text{length of side adjacent to } \angle P} = \frac{12}{5}$$ ◄

When right triangles are similar, corresponding angles are congruent and the ratios of the lengths of corresponding sides are equal. This means that the trigonometric ratios for corresponding angles are the same. For example, in any right triangle with an angle that measures 60°, sin 60° is always the same number. Because these ratios are the same, every scientific calculator has sine, cosine, and tangent keys. You can also find these ratios by using a table of trigonometric ratios like the one on page 581.

**CALCULATOR TIP**

When you use a calculator to solve trigonometry problems, you must specify the type (mode) of angle measure you are using. For the problems in this book, you should always set the calculator for the degree (Deg) mode. If the calculator you are using does not have a special memory feature, you may need to set the mode each time you turn the calculator on.

## Example 3

Find sin 35°, cos 35°, and tan 35°.

**Solution 1**

Use a scientific calculator.

Enter this key sequence: 35 [sin]

The display shows 0.573576436. Rounded to four decimal places, sin 35° ≈ 0.5736.

Enter this key sequence: 35 [cos]

The display shows 0.819152044. Rounded to four decimal places, cos 35° ≈ 0.8192.

Enter this key sequence: 35 [tan]

The display shows 0.700207538. Rounded to four decimal places, tan 35° ≈ 0.7002.

**Solution 2**

Use a table. Find 35° in the Angle column of the Table of Trigonometric Ratios on page 581. Read the numbers across from 35° in the Sine, Cosine, and Tangent columns.

sin 35° ≈ 0.5736    cos 35° ≈ 0.8192    tan 35° ≈ 0.7002 ◄

## TRY THESE

Use △*DEF* in Example 1. Find each ratio.

**1.** sin *E*    **2.** cos *E*    **3.** tan *E*

In right triangle *XYZ*, ∠*Z* is the right angle. If cos $Y = \frac{8}{10}$, find these ratios.

**4.** sin *Y*    **5.** tan *Y*

Use a calculator or the Table of Trigonometric Ratios on page 581 to find these ratios. Round answers to four decimal places.

**6.** sin 60°    **7.** cos 42°    **8.** tan 65°

# EXERCISES

## PRACTICE/ SOLVE PROBLEMS

Use $\triangle TVW$ for Exercises 1–6. Find each ratio in lowest terms.

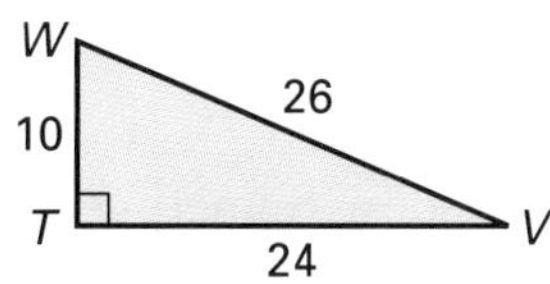


1. sin $W$
2. tan $V$
3. cos $W$
4. tan $W$
5. sin $V$
6. cos $V$

For each triangle, use the information given to find the trigonometric ratio indicated. Write the answer in lowest terms.

7. In $\triangle ABC$, $\angle B = 90°$, $AB = 6$, and $BC = 8$. Find cos $C$.
8. In $\triangle DEF$, $\angle F = 90°$, $EF = 15$, and $DE = 25$. Find sin $E$.
9. In $\triangle TAM$, $\angle A = 90°$, $TM = 10$, and $AT = 8$. Find tan $T$.

***USING DATA*** Use a calculator or the Table of Trigonometric Ratios on page 581 to find each ratio. Round the answer to four decimal places.

10. cos 52°
11. tan 17°
12. cos 33°
13. sin 81°
14. tan 58°
15. sin 19°
16. cos 5°
17. tan 85°
18. sin 45°

## EXTEND/ SOLVE PROBLEMS

***USING DATA*** Refer to the Table of Trigonometric Ratios on page 581.

19. Tell whether each ratio increases or decreases as the angle increases from 0° to 90°.
    a. sine b. cosine c. tangent
20. Which angle has a cosine equal to sin 31°? Which angle has a cosine equal to sin 48°? What do you notice?
21. Which angle has a sine equal to cos 78°? Which angle has a sine equal to cos 25°? What do you notice?
22. For which angle are the sine and cosine ratios equal?
23. What is tan 90°? Explain why this occurs.

**PROBLEM SOLVING TIP**

For Exercise 23, draw right triangles with angles of 80°, 83°, 85°, and 88°. Think about how a right triangle with a second 90° angle would look.

## THINK CRITICALLY/ SOLVE PROBLEMS

Tell whether each statement is *true* or *false*. Refer to the triangle at the right to justify your answers.

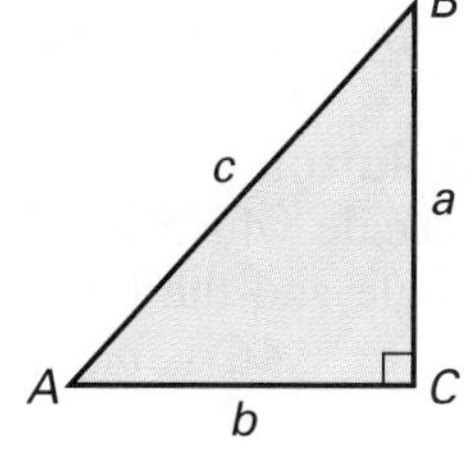


24. $\sin A = \cos B$
25. $\cos A = \frac{1}{\sin A}$
26. $\tan A = (\sin A)(\cos A)$
27. $\tan A = \frac{\sin A}{\cos A}$

# 14-2 Finding Lengths of Sides in Right Triangles

## EXPLORE

Terry makes stained glass windows and lamp shades. Sometimes he knows the measure of one angle and the length of one side of a piece of glass he needs to cut. An example is the right triangle indicated by the dotted lines in the figure at the right.

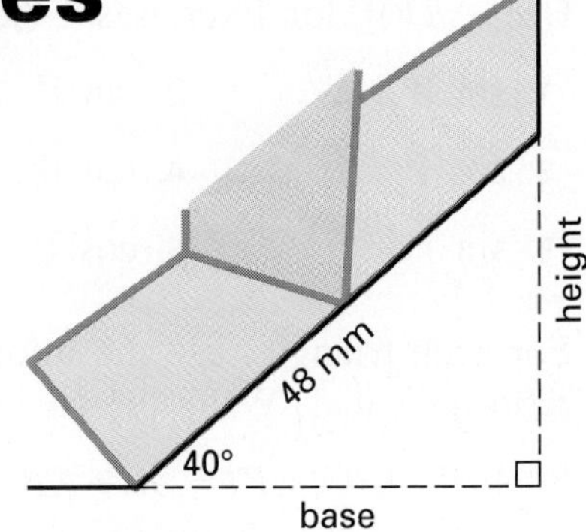

For each equation, decide if you can substitute two values using the information in the figure above and the Table of Trigonometric Ratios on page 581. If you can, show the substitution; otherwise, write *not enough information.*

a. $\sin 40° = \frac{\text{opposite}}{\text{hypotenuse}}$

b. $\cos 40° = \frac{\text{adjacent}}{\text{hypotenuse}}$

c. $\tan 40° = \frac{\text{opposite}}{\text{adjacent}}$

d. Which of the equations in questions **a–c** could you solve to find the length of the base of the piece Terry needs to cut?

e. Which of the equations in questions **a–c** could you solve to find the height of the piece Terry needs to cut?

f. Solve the equations you identified in questions **d** and **e** to find the base and height to the nearst millimeter of the piece Terry needs.

g. Measure the base and height of the piece to be cut in the figure above. Compare your measurements to your results in question **f.**

## SKILLS DEVELOPMENT

In Section 14-1, you learned that the trigonometric ratios for any given angle are the same for any right triangle. If you know the measures of one acute angle and one side of a right triangle, you can use them to find the lengths of the other sides of the triangle.

### Example 1

In $\triangle ABC$, find $AB$ to the nearest tenth.

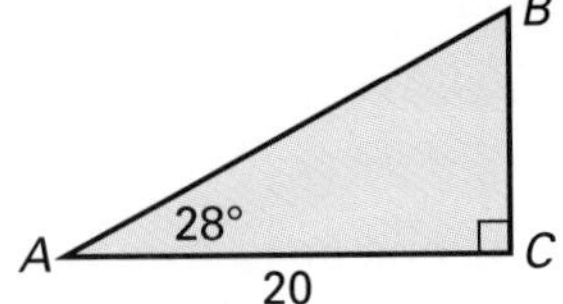


**Solution**

Decide which trigonometric ratio relates the unknown side to the known angle and the known side.

$\overline{AC}$ is *adjacent* to $\angle A$.
$\overline{AB}$ is the *hypotenuse*.

Think: $\frac{\text{adjacent}}{\text{hypotenuse}} = \text{cosine}$

Write and solve an equation involving the trigonometric ratio and the values you know. Use your calculator or the Table of Trigonometric Ratios on page 581 to find cos 28°.

$$\cos 28^\circ = \frac{AC}{AB}$$

$$0.8829 \approx \frac{20}{AB}$$

$$AB \approx \frac{20}{0.8829} \approx 22.7$$ ◄

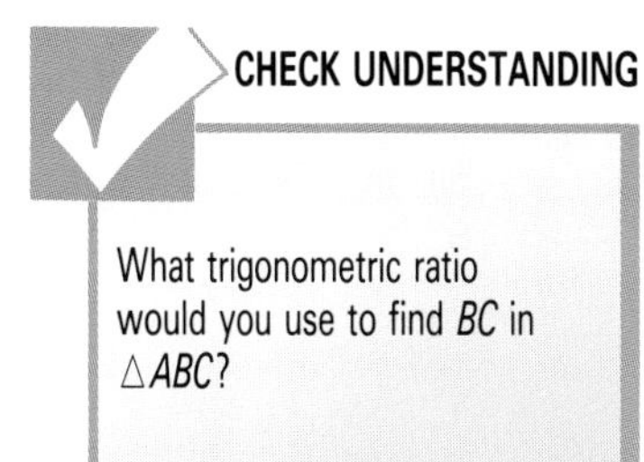

CHECK UNDERSTANDING

What trigonometric ratio would you use to find $BC$ in $\triangle ABC$?

## Example 2

From a point at eye level with the base, an observer looks up at a 60° angle to the top of a monument. The distance between the observer and the monument is 26.5 m. How tall is the monument?

### Solution

Think of the monument, the horizontal, and the line of sight forming right triangle $XYZ$. Find $YZ$.

Decide which trigonometric ratio to use.

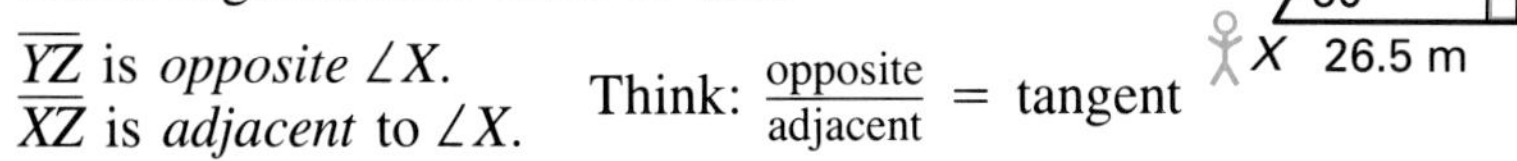

$\overline{YZ}$ is *opposite* $\angle X$. $\overline{XZ}$ is *adjacent* to $\angle X$. Think: $\frac{\text{opposite}}{\text{adjacent}}$ = tangent

Y
60°
X 26.5 m Z

Then write and solve an equation involving the trigonometric ratio and the values you know.

$$\tan 60^\circ = \frac{YZ}{26.5}$$

$$1.7321 \approx \frac{YZ}{26.5}$$

$$YZ \approx 26.5(1.7321) \approx 45.90065$$

Round to the nearest tenth. The monument is approximately 45.9 m tall. ◄

## Example 3

Find $x$ in the figure at the right.

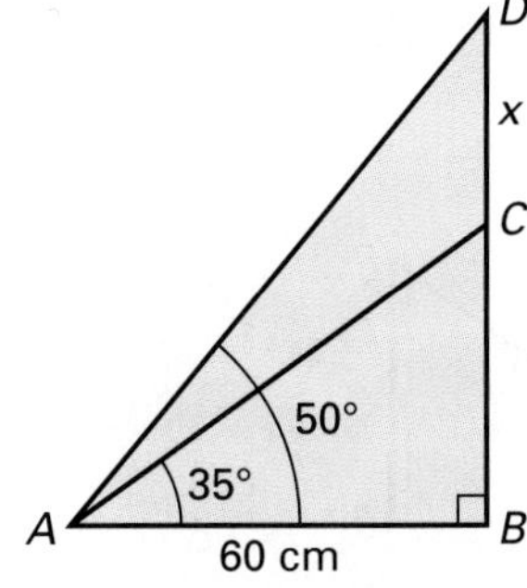


### Solution

Use trigonometric ratios to find $DB$, the side opposite $\angle DAB$ in right triangle $ADB$, and $CB$, the side opposite $\angle CAB$ in right triangle $ACB$.

$$\tan 50^\circ = \frac{DB}{60} \qquad \tan 35^\circ = \frac{CB}{60}$$

$$1.1918 \approx \frac{DB}{60} \qquad 0.7002 \approx \frac{CB}{60}$$

$$DB \approx 60(1.1918) \approx 71.508 \qquad CB \approx 60(0.7002) \approx 42.012$$

Subtract $CB$ from $DB$ to find $x$, length $DC$. Round to the nearest tenth.

$$71.508 - 42.012 = 29.496$$

So, $x$ is approximately 29.5 cm. ◄

**MIXED REVIEW**

Find the slope and y-intercept of each line.

1. $y = \frac{3}{2}x + 2$
2. $2x + y = 5$
3. $3y = 6x + 9$
4. $4x - 3y = 0$
5. Assume that $y$ varies directly with $x^2$. When $x = 3$, $y = 15$. Find the value of $y$ when $x = 6$.
6. Triangle $ABC$ with vertices $A(2, 3)$, $B(6, -1)$, and $C(1, -3)$ is reflected across the y-axis. Give the coordinates of the vertices of triangle $A'B'C'$.

Find the value of each expression.

7. $6!$
8. ${}_4P_4$
9. ${}_8P_3$
10. ${}_{10}C_5$

## TRY THESE

Use your calculator or the Table of Trigonometric Ratios.

In $\triangle DEF$, find each length. Round your answer to the nearest tenth.

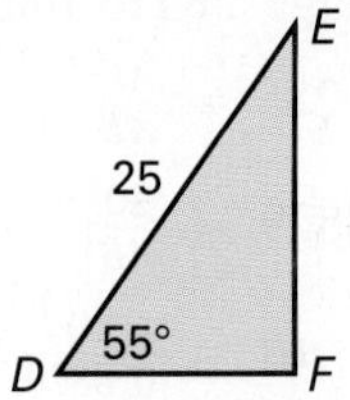

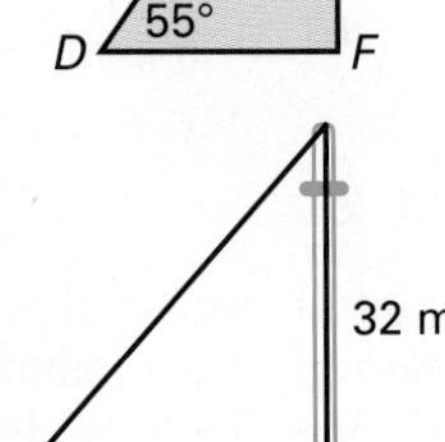


1. *DF*
2. *EF*
3. A guy wire is secured near the top of a television transmitting tower. The guy wire meets the ground at an angle of 48°. If the height of the tower is 32 m, how far from the base of the tower is the guy wire secured?

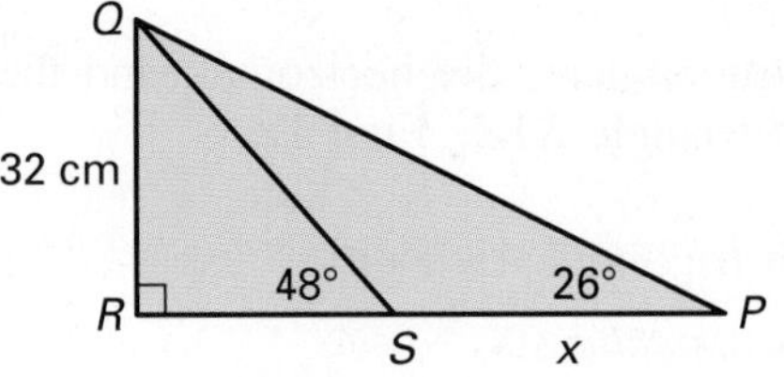


4. Find $x$ in the figure at the right.

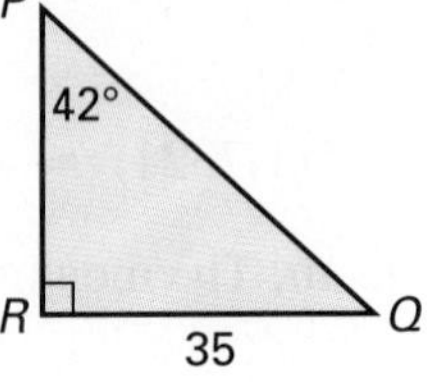


# EXERCISES

**PRACTICE/ SOLVE PROBLEMS**

Solve. Round answers to the nearest tenth.

Use the triangle shown at the right for Exercises 1 and 2.

1. Find *PQ*.
2. Find *PR*.
3. Find the height of roof beam *RB*.

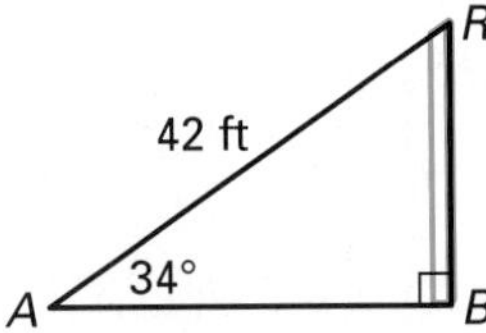


4. Find the width of the river, *w*.

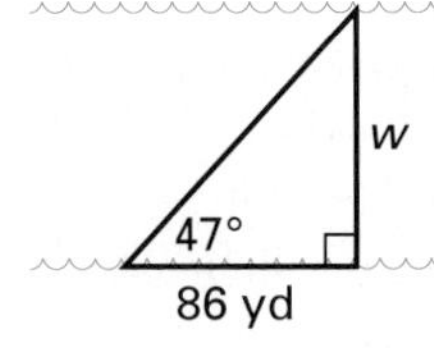


5. Find $x$ in the figure at the right.

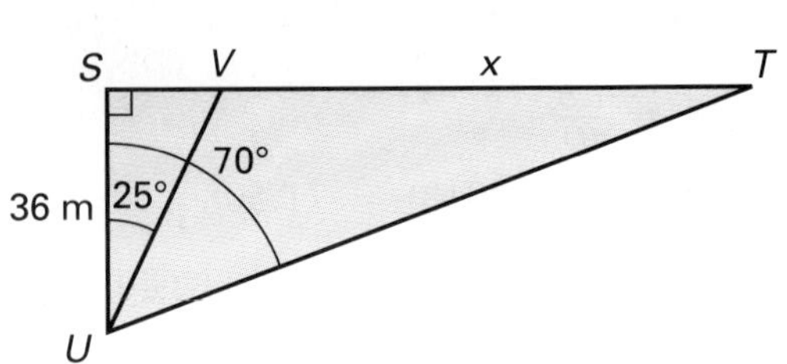

## EXTEND/ SOLVE PROBLEMS

Draw a diagram for each problem. Then solve. Round the answer to the nearest tenth.

6. A 15-ft ladder leaning against a wall makes a 53° angle with the wall. How far up the wall does the ladder reach?

7. The sun is shining on a tree 12 m tall. The sun's rays form a 30° angle with the ground. How long is the shadow cast by this tree?

8. An airplane is flying at 8,000 ft. To avoid the mountain top, the pilot flies the route shown. What is the difference, to the nearest whole mile, between the route actually taken and the direct route (if the plane were flying at a higher elevation)?

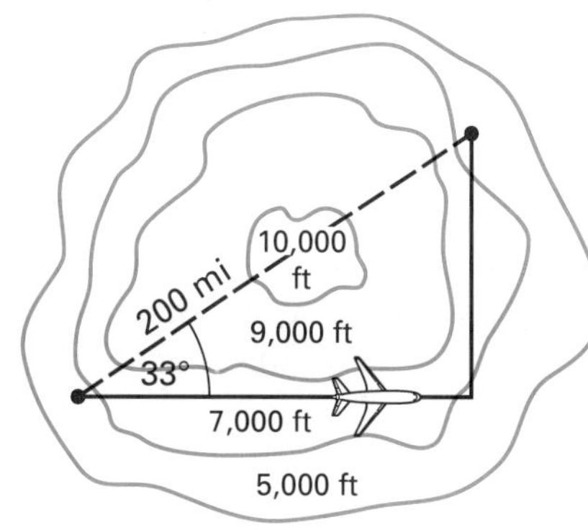

**PROBLEM SOLVING TIP**

Which trigonometric ratio could you use to find the length of the rectangle in Exercise 10? To find the width?

Find the area of each figure. Round the answer to the nearest tenth.

9.

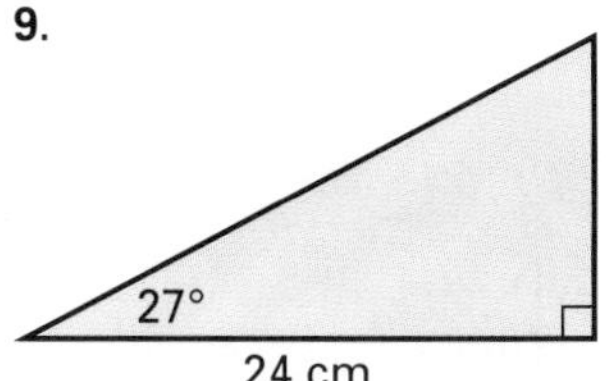


10.

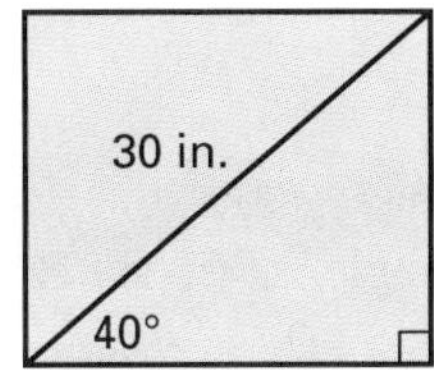


## THINK CRITICALLY/ SOLVE PROBLEMS

11. Find $KL$ in the figure at the right to the nearest tenth.

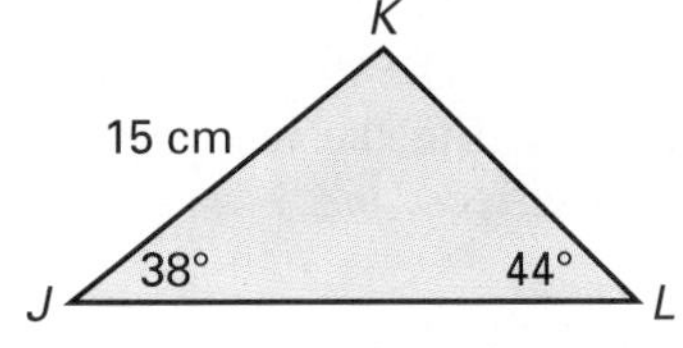

**PROBLEM SOLVING TIP**

How could you form a right triangle with $\overline{KL}$ as one of its sides?

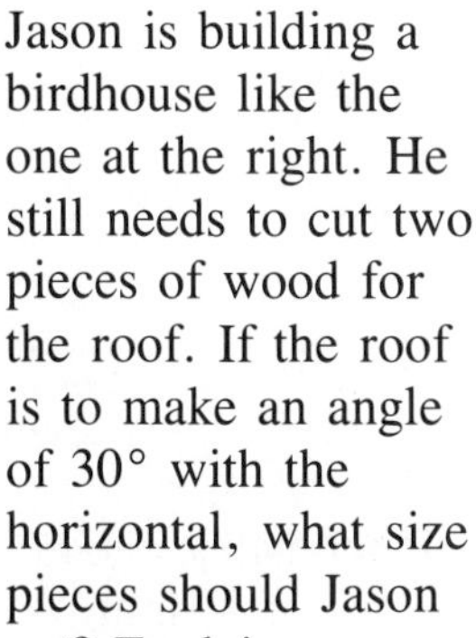

12. Jason is building a birdhouse like the one at the right. He still needs to cut two pieces of wood for the roof. If the roof is to make an angle of 30° with the horizontal, what size pieces should Jason cut? Explain.

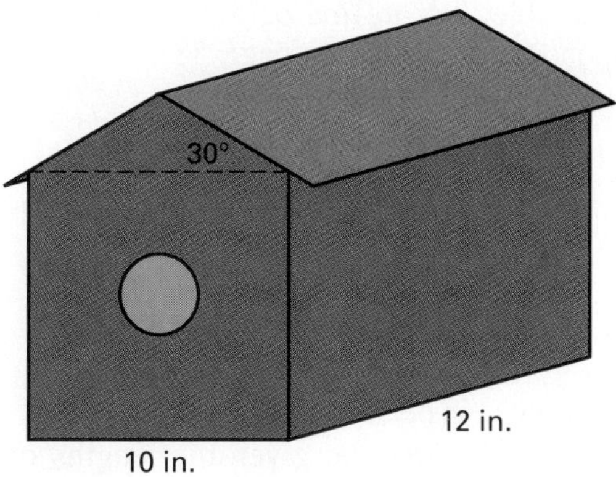

# 14-3 Finding Measures of Angles in Right Triangles

## EXPLORE

Explain how you could use each fact to find the measure of $\angle E$ in the triangle shown at the right.

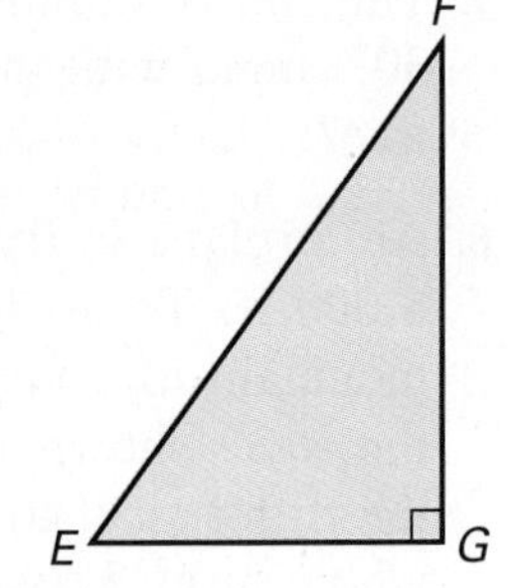


a. The measure of $\angle F$ is 36°.
b. The tangent of $\angle E$ is 1.3764.
c. Find the measure of $\angle E$ two ways. First, use fact **a,** then use fact **b.**
d. Check your findings in question **c** by using a protractor.

## SKILLS DEVELOPMENT

If you know the sine, cosine, or tangent of an angle, you can use that information to find the measure of the angle.

**READING MATH**

Read 0.75 [sin⁻¹] as "the angle whose sine is 0.75."

### Example 1

In right triangle $ABC$, $\sin A = 0.75$. Find $m\angle A$.

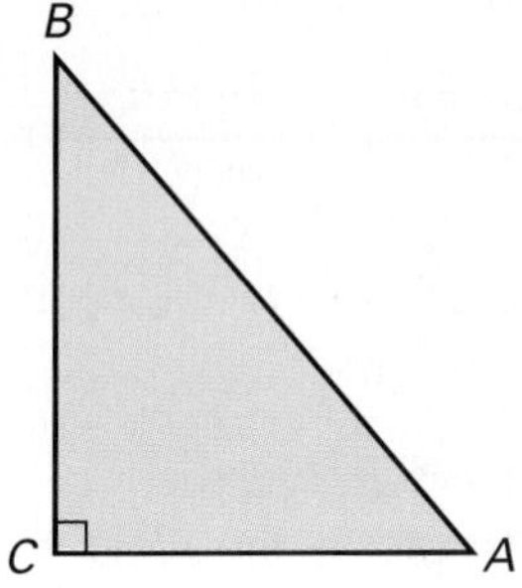


**Solution 1**
Use a scientific calculator. Enter this key sequence: 0.75 [$\sin^{-1}$]. The display shows 48.59037789. Rounded to the nearest whole number of degrees: $m\angle A \approx 49°$.

**CHECK UNDERSTANDING**

How could you use a calculator to find the measure of an angle if you were given its cosine? If you were given its tangent?

**Solution 2**
Use a table. Try to find 0.75 in the Sine column of the Table of Trigonometric Ratios on page 581.

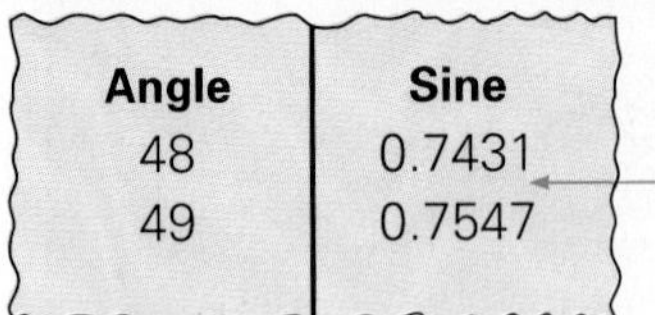

| Angle | Sine |
|---|---|
| 48 | 0.7431 |
| 49 | 0.7547 |

**0.75 is between sin 48° and sin 49°. Since 0.75 is closer to 0.7547 than to 0.7431, $m\angle A \approx 49°$.** ◄

Sometimes, instead of being given one of the trigonometric ratios of an angle, you may be given the lengths of two sides of a right triangle. You can use this information to write a trigonometric ratio and, in turn, use the ratio to find the measure of the angle.

## Example 2

Find $m\angle M$ in the triangle shown at the right.

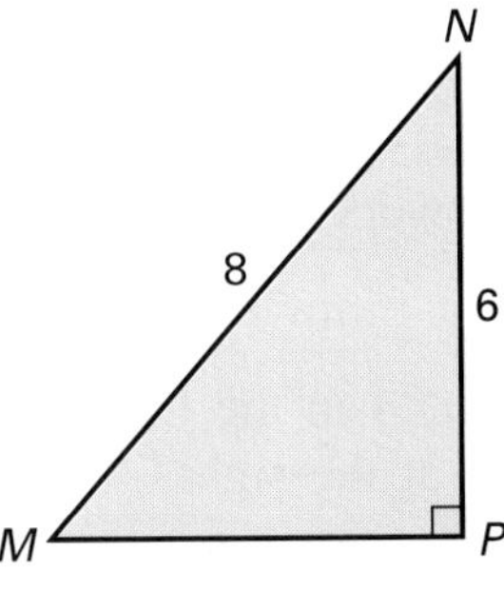

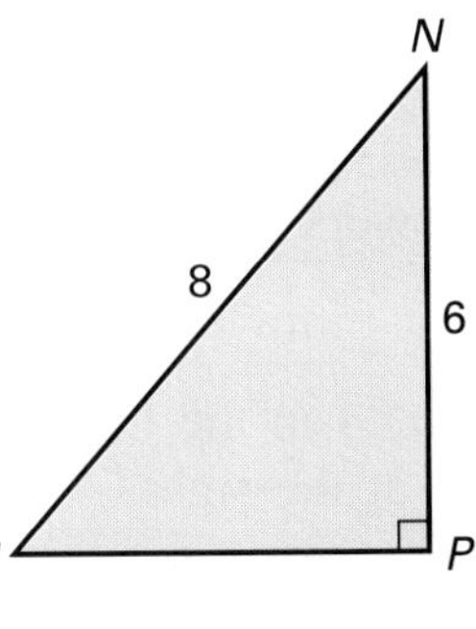

### Solution

Decide which trigonometric ratio relates the angle whose measure you want to find and the sides whose lengths are known.

$\overline{NP}$ is *opposite* $\angle M$.
$\overline{MN}$ is the *hypotenuse*.  Think: $\frac{\text{opposite}}{\text{hypotenuse}} = \text{sine}$

Write and solve an equation involving the trigonometric ratio and the values you know.

$$\sin M = \frac{NP}{MN}$$

$$\sin M = \frac{6}{8}$$

Divide to find the decimal value of sin $M$. $\sin M = 0.75$

In Example 1, you found that the measure of an angle whose sine is 0.75 is approximately 49°. So, $m\angle M \approx 49°$. ◄

**CHECK UNDERSTANDING**

You would need to know the length of which sides in order to use the cosine to find $m\angle M$? In order to use the tangent to find $m\angle N$?

Sometimes you need to combine your knowledge of trigonometric ratios with other geometric concepts to find the measure of an angle.

## Example 3

In the figure at the right, $\overleftrightarrow{AB} \parallel \overleftrightarrow{CD}$. Find $m\angle BAC$.

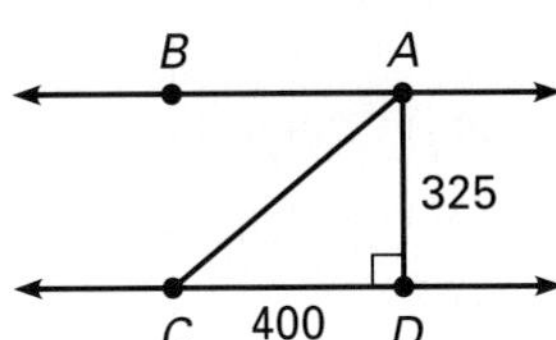


### Solution

Note that $\angle BAC$ and $\angle ACD$ are alternate interior angles. You can use the lengths of the two sides given for $\triangle ADC$ to find $m\angle ACD$. Since $\overleftrightarrow{AB} \parallel \overleftrightarrow{CD}$, $m\angle ACD = m\angle BAC$.

To find $m\angle ACD$, decide which trigonometric ratio relates $AD$ and $CD$ to $\angle ACD$.

$\overline{AD}$ is *opposite* $\angle ACD$.
$\overline{CD}$ is *adjacent* to $\angle ACD$.  Think: $\frac{\text{opposite}}{\text{adjacent}} = \text{tangent}$

Write and solve an equation.

$$\tan\angle ACD = \frac{AD}{CD}$$

$$\tan\angle ACD = \frac{325}{400}$$

$$\tan\angle ACD = 0.8125$$

Use a calculator or table to find the measure of the angle whose tangent is 0.8125. You will find $m\angle ACD \approx 39°$. So, $m\angle BAC \approx 39°$. ◄

**TALK IT OVER**

How could you solve the problem in Example 3 by finding the measure of a different angle in $\triangle ADC$?

PROBLEM SOLVING TIP

When you have found the measure of an angle in a triangle, check your answer by working backwards. See if the appropriate trigonometric ratio—sine, cosine, or tangent—of the angle is the quotient of the relevant side lengths in the triangle.

## TRY THESE

Round answers for these exercises to the nearest whole degree.

1. In right triangle $QRS$, $\cos S = 0.4848$. Find $m\angle S$.
2. Find $m\angle A$ in the triangle shown at the right.
3. In the figure at the right, $\overleftrightarrow{PQ} \parallel \overleftrightarrow{SR}$. Find $m\angle QPR$.

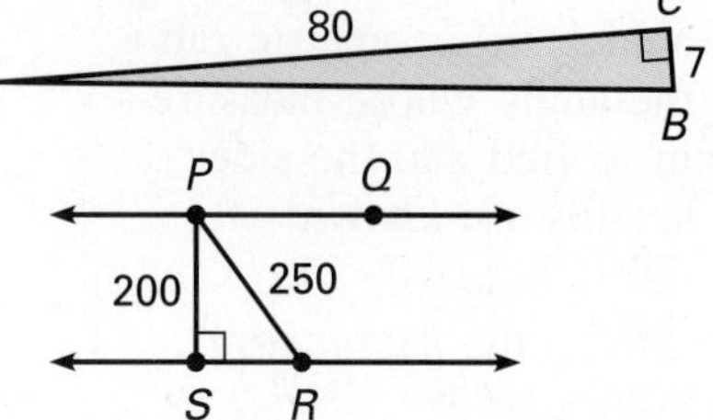


# EXERCISES

**PRACTICE/ SOLVE PROBLEMS**

Find each measure to the nearest whole degree.

1. $m\angle B$.

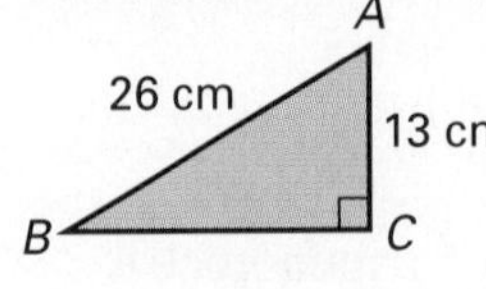


2. $m\angle S$.

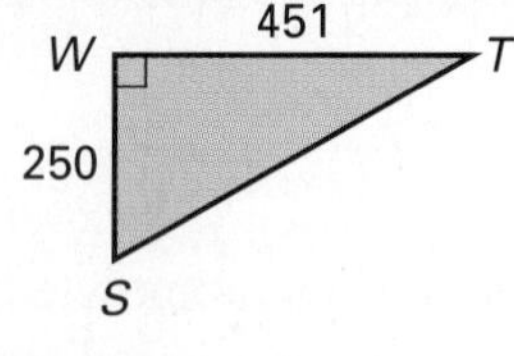


3. $m\angle G$.

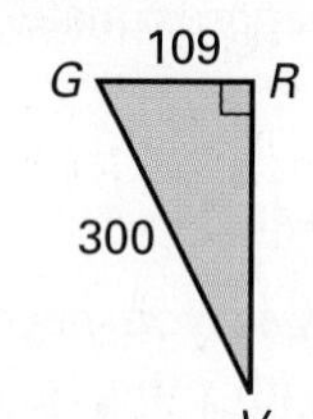


WRITING ABOUT MATH

In your journal, describe two situations in which you use trigonometric ratios to find the measure of a side or an angle of a right triangle.

4. When a certain airplane has reached a distance of 1,500 m from its starting point, it is 970 m above the ground. What angle does the plane's path make with the ground?

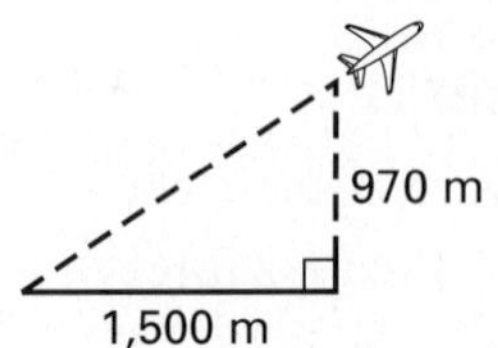


**EXTEND/ SOLVE PROBLEMS**

Solve each problem. Round the answer to the nearest degree.

5. San Francisco's Filbert Street, the steepest street in the world, rises 1 ft for every 3.17 ft of horizontal distance. Find the angle at which Filbert Street rises.
6. The Chamonix Line of the French National Railroad is the steepest climbing train track in the world. It ascends 1 ft for every 11 ft of horizontal distance. Find the angle at which the Chamonix Line ascends.

**THINK CRITICALLY/ SOLVE PROBLEMS**

***USING DATA*** Refer to the Table of Trigonometric Ratios on page 581 for Exercise 7.

7. One leg of a right triangle is twice as long as the other leg. To the nearest whole degree, what are the measures of the acute angles of this triangle?

► READ
► PLAN
► SOLVE
► ANSWER
► CHECK

# Problem Solving Applications:

## TRIGONOMETRY AND SAFETY

If you lean a ladder against a wall, you should know that at certain angles the ladder may slip and crash to the ground. To avoid accidents, some construction companies follow the rule that the measure of the angle that an unsecured ladder makes with the ground should not be less than 75°.

The diagrams show ladders placed against walls. Which ladders are safe?

**1.**

8 m

1.5 m

**2.**

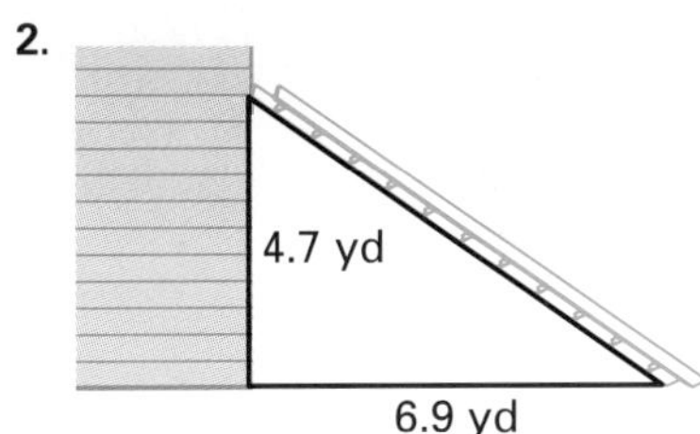


**3.**

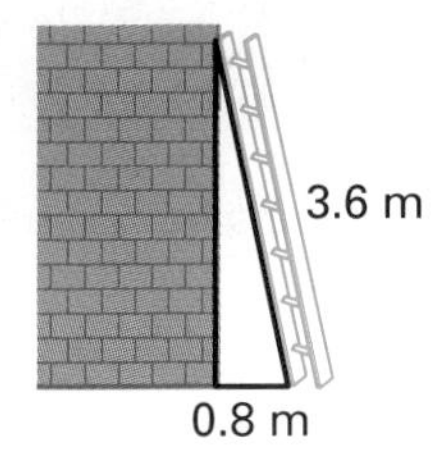


Solve. Round angle measures to the nearest degree and lengths to the nearest tenth.

**4.** A ladder 6.7 m in length is placed against a tree. At what distance should the base of the ladder be placed from the trunk of the tree so that the ladder is safe to climb?

**5.** A painter places the base of a ladder 3.2 m from a building. The ladder is 9 m in length. How far up the wall does the ladder reach? Is this a safe ladder to be on?

A rough guide for placing a ladder safely is shown in the diagram.

$$\frac{\text{height up the wall}}{\text{distance from the wall}} = \frac{4}{1}$$

**6.** Calculate the angle at which the ladder shown in the diagram meets the ground.

**7.** A ladder is placed against a wall so that it reaches 14 ft up the wall. According to the rough guide, how far away from the wall should the base of the ladder be placed?

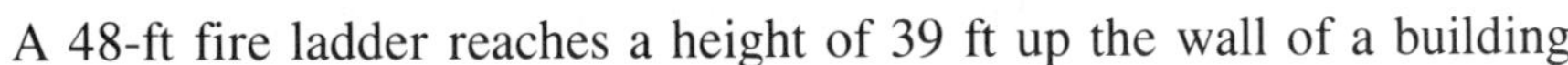

A 48-ft fire ladder reaches a height of 39 ft up the wall of a building.

**8.** How far is the base of the ladder from the building?

**9.** What angle does the ladder make with the ground? Is the ladder in a safe position?

# 14-4 Problem Solving Skills:

## SOLVE A RIGHT TRIANGLE

- ► READ
- ► PLAN
- ► SOLVE
- ► ANSWER
- ► CHECK

Sometimes when you are working with right triangles, it may seem that you are given very little information about a triangle. However, with your knowledge of trigonometric ratios, the information is often enough to find the measures of all of the triangle's angles and sides. To **solve a right triangle,** find the measures of all angles and sides.

### PROBLEM

Solve $\triangle ABC$ shown at the right.

### SOLUTION

Find $AB$ to the nearest tenth using the Pythagorean Theorem.

$$\begin{aligned} AB^2 &= AC^2 + BC^2 \\ AB^2 &= 11^2 + 8^2 \\ AB^2 &= 121 + 64 \\ AB^2 &= 185 \\ AB &= \sqrt{185} \approx 13.6 \end{aligned}$$

Find $m\angle A$ to the nearest degree using the tangent ratio.

$$\begin{aligned} \tan A &= \frac{BC}{AC} \\ \tan A &= \frac{8}{11} \\ \tan A &\approx 0.7273 \end{aligned}$$

So, $m\angle A \approx 36°$.

**TALK IT OVER**

How else could you have used your knowledge of trigonometric ratios and the information you were given to solve $\triangle ABC$?

Find $m\angle B$ to the nearest degree using the sum of the measures of the angles of a triangle.

$$\begin{aligned} m\angle A + m\angle B + m\angle C &= 180° \\ 36° + m\angle B + 90° &\approx 180° \\ m\angle B &\approx 180° - 36° - 90° \\ m\angle B &\approx 54° \end{aligned}$$

Summarize the results.

| | |
|---|---|
| $AB \approx 13.6$ | $m\angle A \approx 36°$ |
| $BC = 8$ | $m\angle B \approx 54°$ |
| $AC = 11$ | $m\angle C = 90°$ |

# PROBLEMS

Solve each right triangle. Round angle measures to the nearest degree and lengths of sides to the nearest tenth. If there is not enough information, tell the least amount of additional information needed.

**1.**

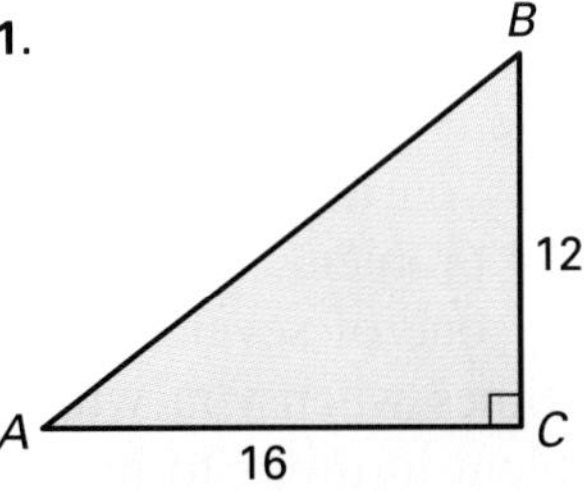


**2.**

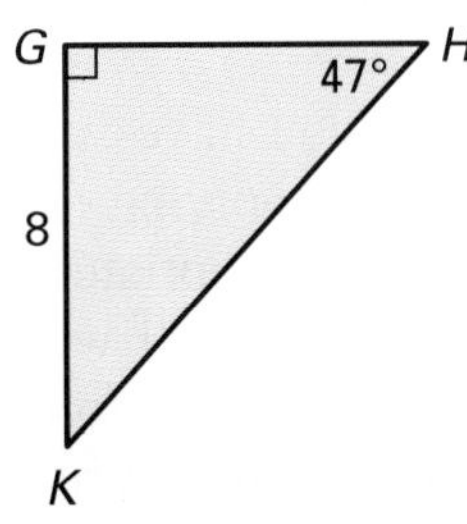


**3.**

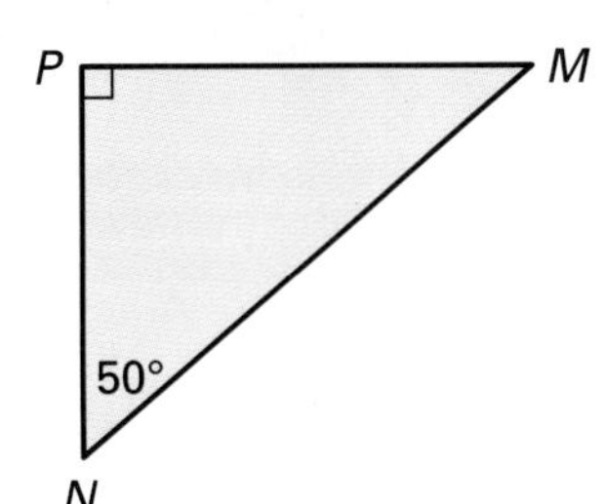


**4.**

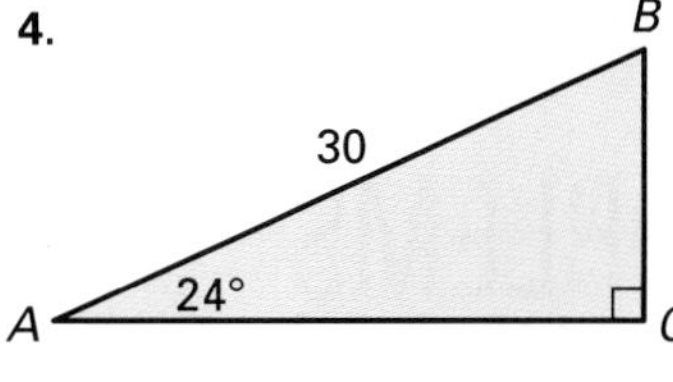


**5.**

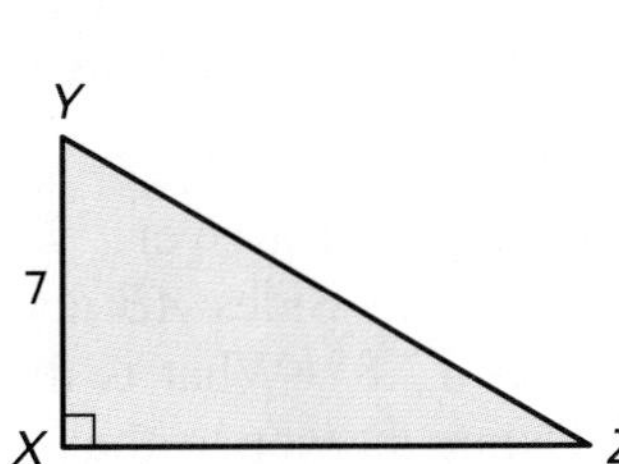


**6.**

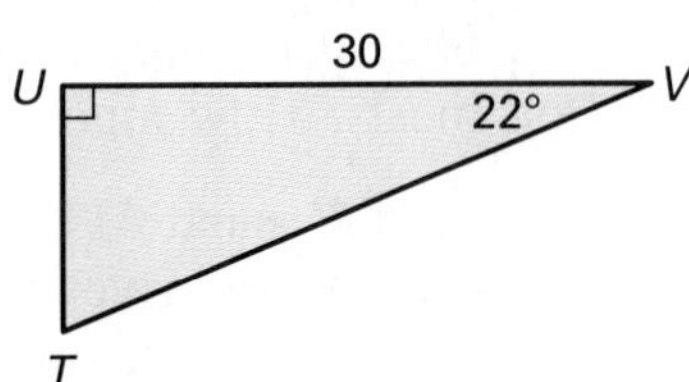


Ana plans to build a bookcase like the one at the right to hold craft books. There will be four equally spaced shelves and two diagonal braces. For each set of information given below, find the following measures to the nearest tenth.

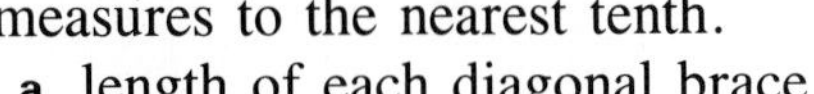

**a.** length of each diagonal brace
**b.** space between shelves

B

A

C

If there is not enough information, tell what other measure is needed.

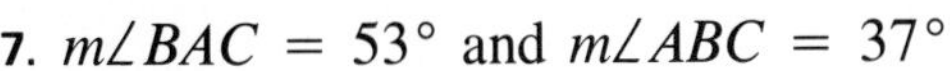

**7.** $m\angle BAC = 53°$ and $m\angle ABC = 37°$

**8.** The width of the bookcase will be 36 in. and $m\angle BAC = 53°$.

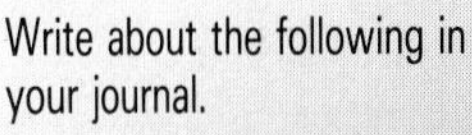

**WRITING ABOUT MATH**

Write about the following in your journal.

What is the least information you can have about the measures of the angles and sides of a right triangle and still be able to solve the triangle? How can you use this information to solve the triangle?

# 14-5 Problem Solving/ Decision Making:

## CHOOSE A STRATEGY

► READ
► PLAN
► SOLVE
► ANSWER
► CHECK

In this book you have studied a variety of problem solving strategies. Experience in applying these strategies will help you decide which will be most appropriate for solving a particular problem. Sometimes only one strategy will work. In other cases, any one of several strategies will offer a solution. There may be times when you will want to use two different approaches to a problem in order to be sure that the solution you found is correct. For certain problems, you will need to use more than one strategy in order to find the solution.

**PROBLEM SOLVING TIP**

Here is a checklist of the problem solving strategies you have studied so far in this book.

Use the 5-Step Plan
Make an Algebraic Model
Make a Circle Graph
Make an Organized List
Write and Solve an Equation
Choose a Length, Area, or Volume Formula
Work Backwards
Use a Graph
Use a System of Equations
Solve a Simpler Problem
Explore with Simulations
Use Logical Reasoning

## PROBLEMS

Solve. Name the strategy you used to solve each problem.

1. Carmen spent a total of \$86 last week on clothing and restaurant meals. She spent three times as much money on clothing as she did on meals. How much money did she spend on each?

2. Posts $A$ and $B$ stand 411.1 ft apart on one side of a river. Directly across the river from Post $A$ is Post $C$. Sides $AB$ and $AC$ form a right angle and $\angle ABC$ measures 17.8°. What is the distance from Post $A$ to Post $C$? Round your answer to the nearest tenth of a foot.

3. Suppose that the equator was a perfect circle and that you tied a steel cable around it. If you lengthened the cable by 15 m and supported it evenly above the ground, about how far above the ground would it be? Use 3.14 for $\pi$ and round your answer to the nearest tenth of a meter.

4. a. How many different ways can the digits 1 through 9 be arranged in the figure below?
   b. How many different arrangements are there with 1, 2, and 3 in the corners?
   c. There are two distinct ways of arranging the digits 1 through 9 in the figure so that the sum of the numbers on each side is 17. Can you find both of them?

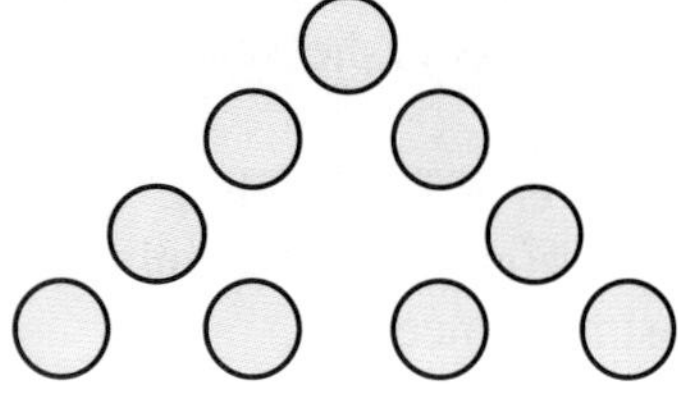

5. Ellen was born on a Friday, exactly 2,000 days before her brother Allen was born. On what day of the week was Allen born?

6. Which quadrant on a coordinate plane contains the points that are solutions to each of the following inequalities: $y < -2$, $x + y < 1$, $x > 5$?

7. What is the greatest whole number that is a factor of the sum of any three consecutive whole numbers? (Remember that 0 is a whole number.)

8. Lena, Mina, and Nina always have either pizza or hamburgers for lunch. If Lena has pizza, Mina has hamburgers. Either Lena or Nina, but not both, has pizza. Mina and Nina do not both have hamburgers. Who had pizza yesterday and hamburgers today?

Amanda started at Centerville, drove due south to Baker, and then drove 20.2 km east to Diamond. On a map, a line connecting Centerville and Diamond makes a 37° angle with a line connecting Baker and Diamond.

9. If Amanda could drive in a straight line from Centerville to Diamond, what is the distance she would drive?

10. How does the distance you found in Exercise 9 compare to the actual distance Amanda drove?

11. A health food store mixes its own granola. The mixture contains 40 oz of rolled oats. Twice as much raisins as walnuts is used. The amount of sunflower seeds is one-fourth the amount of oats and one-third the amount of raisins. Half as much coconut as sunflower seeds is used. Make a circle graph to show the amount of each ingredient in the granola.

12. A bus travels between Youngsville and Zenia, a distance of 200 mi. If the bus breaks down, what is the probability that it has broken down within 10 mi of either city?

13. Five people chose seats in one empty row of a movie theater. There were 40,320 different ways they could arrange themselves. How many empty seats were in the row?

14. Jorge drove to the library at an average speed of 40 mi/h and he returned home at an average speed of 50 mi/h. What was his average speed if the trip took a total of 2 h 15 min? Round your answer to the nearest tenth of a mile.

# 14-6 Special Right Triangles

**EXPLORE/ WORKING TOGETHER**

Work with a partner. You will need a ruler and a protractor.

a. Draw two equilateral triangles, one with 2-in. sides and one with 4-in. sides.
b. Fold each triangle in half. Then unfold and cut along the fold line. Use only one of the two triangles formed by each cut for the rest of your work.
c. Measure each angle, the shorter leg, and the hypotenuse of each triangle. Record these measures on the triangles.
d. Next, draw a 1-in. square and a 2-in. square and their diagonals.
e. Cut each square along its diagonal. Use only one of the two triangles formed by each cut for the rest of your work.
f. Measure each angle and the legs of each triangle. Record these measures on the triangles.
g. Use the Pythagorean Theorem to find the measure of the third side of each triangle. Express this measure as a square root and record it on the triangle.

Use the measures on your triangles to answer the following questions.

h. How do the measures of the legs of a right triangle with a 30° angle compare? How do the measures of the legs compare in the right triangle with a 45° angle?
i. How does the measure of the hypotenuse of a right triangle with a 30° angle compare to the measure of its shorter leg. How does the measure of the hypotenuse of a right triangle with a 45° angle compare to the measures of its legs?

**SKILLS DEVELOPMENT**

Many engineers, draftspeople, architects, and designers use special instruments for drawing. These include two special right triangles. The triangles are a **30°–60°–90° right triangle** and a **45°–45°–90° right triangle.**

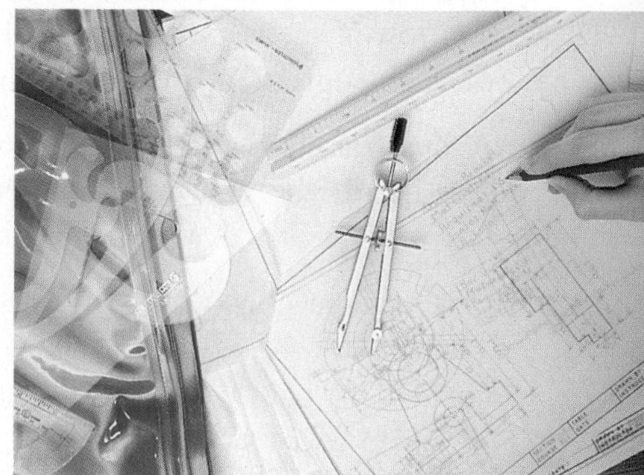

In earlier sections, you learned to use trigonometric ratios to solve problems involving right triangles. The special geometric properties of 30°–60°–90° right triangles and 45°–45°–90° right triangles are summarized in the following table. These properties can also be used to solve certain problems.

| 30°–60°–90° right triangles | 45°–45°–90° right triangles |
|---|---|
| length of side opposite 30° angle: $x$<br>length of side opposite 60° angle: $x\sqrt{3}$<br>length of hypotenuse: $2x$ | length of each leg: $x$<br>length of hypotenuse: $x\sqrt{2}$ |

## Example 1

Find $MP$ and $NP$.

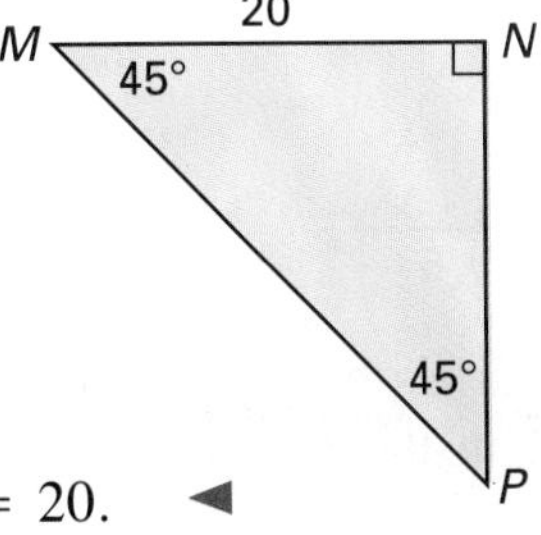


**Solution**

$\triangle MNP$ is a 45°–45°–90° triangle, and $\overline{MP}$ is the hypotenuse.

$$MP = MN\sqrt{2}$$
$$MP = 20\sqrt{2}$$

Since the legs are congruent, $NP = 20$. ◄

## Example 2

Find $BC$ and $AB$.

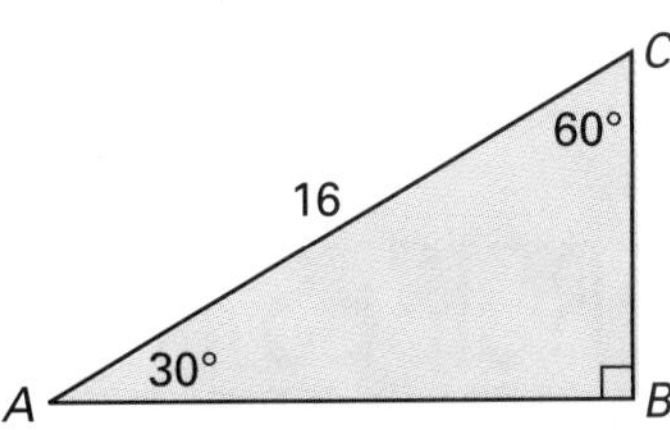
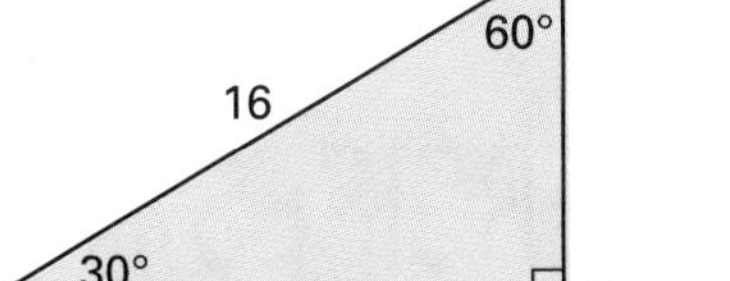


**Solution**

$\triangle ABC$ is a 30°–60°–90° triangle, and $\overline{AC}$ is the hypotenuse.

$\overline{BC}$ is the side opposite the 30° angle.

$$AC = 2BC$$
$$16 = 2BC$$
$$\frac{16}{2} = BC$$
$$8 = BC$$

$\overline{AB}$ is the side opposite the 60° angle.

$$AB = BC\sqrt{3}$$
$$AB = 8\sqrt{3}$$

◄

## Example 3

A ladder leaning against a wall makes a 60° angle with the ground. For safety, the ladder is tied to a post at its base. If the ladder is extended to 32 ft, how high up the wall does it reach? Find a decimal value to the nearest tenth.

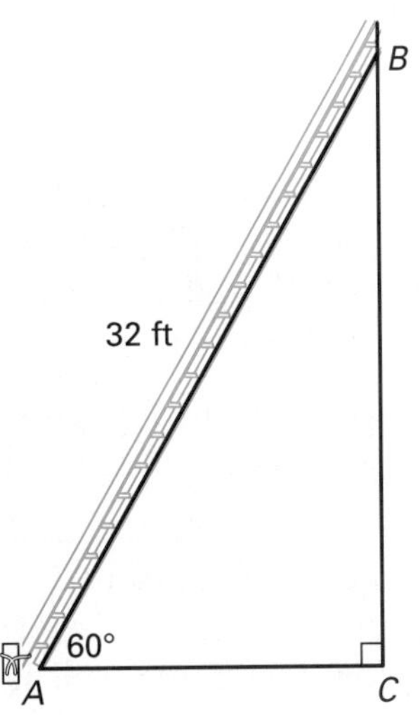


**Solution**

$\triangle ABC$ is a 30°–60°–90° triangle. Since $\overline{AB}$ is the hypotenuse, 32 ft $= 2x$, and $x = \frac{1}{2}(32)$, or 16. $\overline{BC}$ is the side opposite the 60° angle.

$$BC = x\sqrt{3}$$
$$BC = 16\sqrt{3}$$
$$BC \approx 16(1.7321)$$
$$BC \approx 27.7136$$

So, the ladder reaches 27.7 ft up the side of the wall. ◄

**PROBLEM SOLVING TIP**

To remember which special triangle uses $\sqrt{2}$ and which uses $\sqrt{3}$, think:

- There are two different side lengths in the 45°–45°–90° right triangle, so it uses $\sqrt{2}$.
- There are three different side lengths in the 30°–60°–90° right triangle, so it uses $\sqrt{3}$.

**TALK IT OVER**

How could you solve Example 2 using trigonometric ratios? Do you think you will prefer using trigonometric ratios or geometric properties to solve problems involving 30°–60°–90° and 45°–45°–90° triangles? Why?

## TRY THESE

1. Find $QS$ and $QR$ in $\triangle QRS$ shown below. Write answers using square roots.

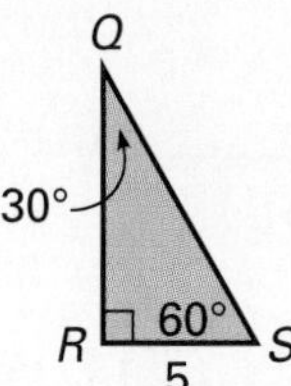


2. Find $GF$ and $GH$ in $\triangle FGH$ shown below. Find decimal values to the nearest tenth.

3. Find $h$, the kite's height above the ground, in the figure at the right.

# EXERCISES

**PRACTICE/ SOLVE PROBLEMS**

Find each length. Write answers using square roots.

1. $BC$
2. $AC$

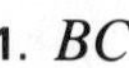

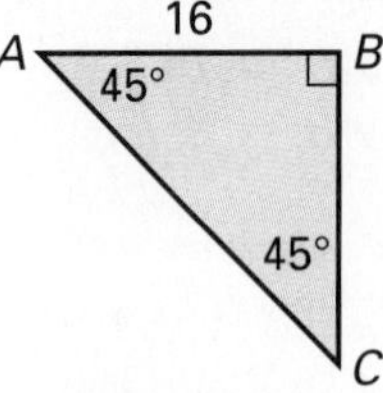


3. $NP$
4. $MN$

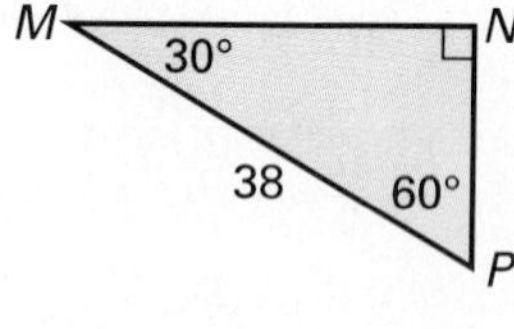


5. $GH$
6. $HK$

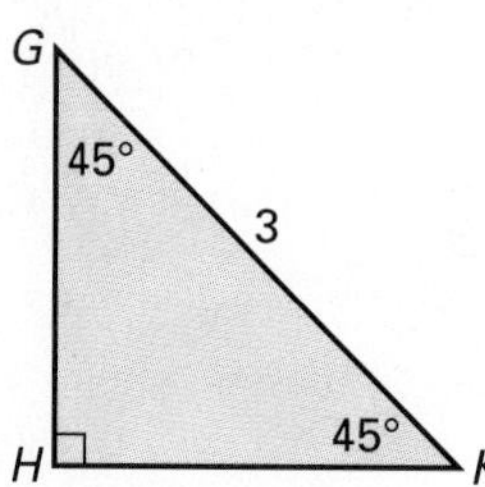


7. $VW$
8. $SW$

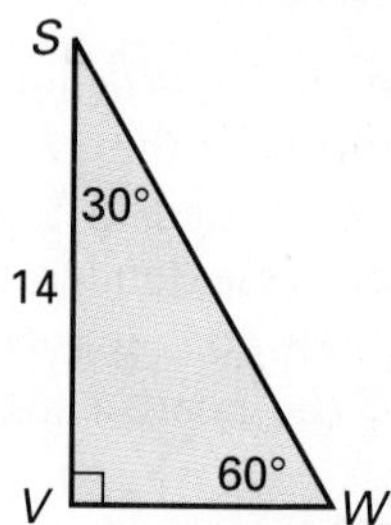


A 4.2-m ladder leaning against a wall makes a 60° angle with the ground.

9. How far from the wall is the foot of the ladder?

10. How far up the wall is the top of the ladder? Find a decimal value to the nearest tenth.

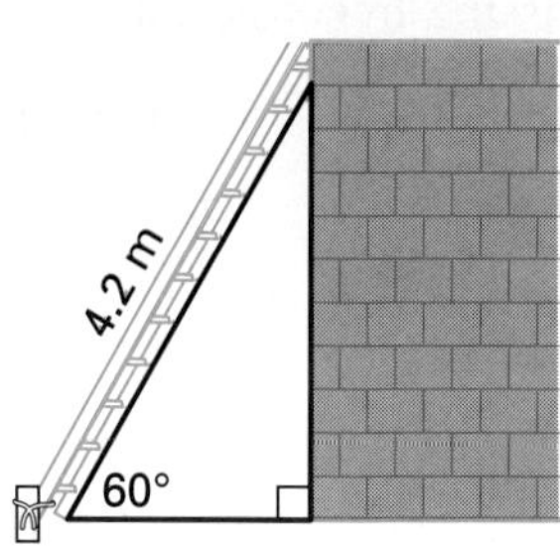

11. When the sun is at an angle of 30°, a tree casts a 90-ft shadow. Find a decimal value for the height of the tree. Round to the nearest tenth.

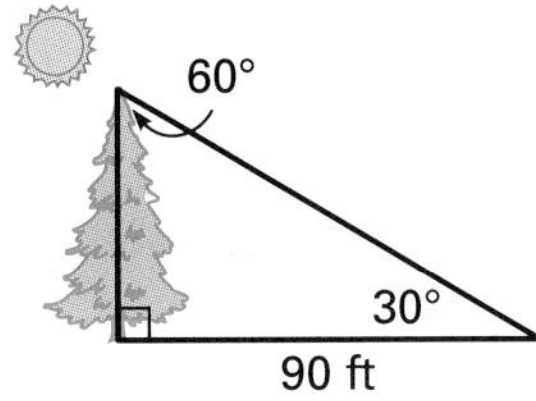


12. A swimmer tries to swim across the river from point $K$ to point $M$. Because of the current, the swimmer reaches point $L$ instead. How far does the swimmer travel? Find a decimal value to the nearest tenth.

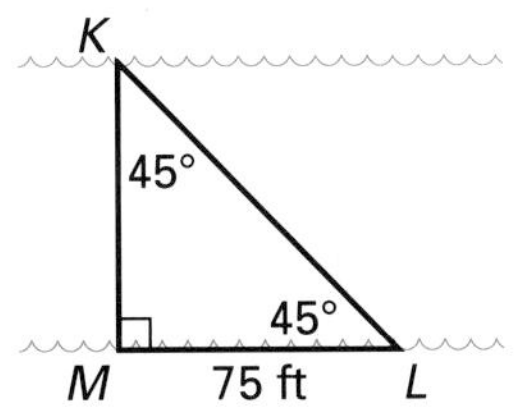


## EXTEND/ SOLVE PROBLEMS

13. Find the measures of the other sides of the triangles shown at the right. Use these measures to complete the table. Write answers using square roots.

| | sine | cosine | tangent |
|---|---|---|---|
| 30° | | | |
| 45° | | | |
| 60° | | | |

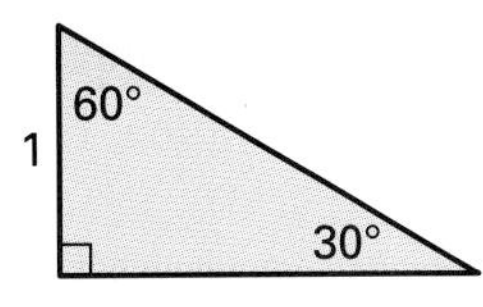


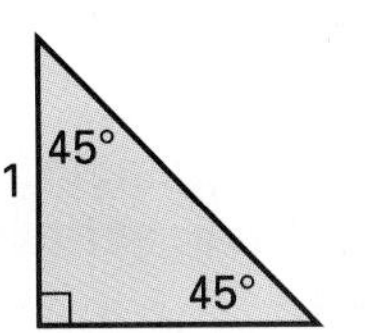


14. Use a calculator to rewrite the table in Exercise 13 with decimal values rounded to four decimal places.

15. Compare the values in the table in Exercise 13 with those in the Table of Trigonometric Ratios on page 581. What did you find?

16. Find the length and width of the rectangle at the right. Write answers using square roots.

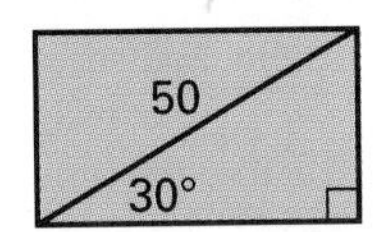


17. Each side of a square measures 9 inches. Find the length of the diagonal. Write the answer using square roots.

**PROBLEM SOLVING TIP**

To solve Exercises 16 and 17, think of what figures are formed by the bottom, side, and diagonal of the rectangle and the two sides and diagonal of the square.

## THINK CRITICALLY/ SOLVE PROBLEMS

Write an expression that represents each of the following.

18. area of a right triangle with a 30° angle whose shortest side measures $x$

19. area of a right triangle with a 45° angle whose shorter sides measure $x$

20. area of a rectangle whose diagonal forms two triangles each having a 60° angle and whose shorter legs measure $x$

# CHAPTER 14 ● REVIEW

Choose the word from the list that completes each statement.

1. If you know the measure of an acute angle in a right triangle and the length of the side adjacent to it, you can use the __?__ to find the length of the triangle's hypotenuse.
2. A __?__ ratio is formed by the measures of any two sides of a right triangle.
3. If you know the measures of the sides opposite and adjacent to an acute angle in a right triangle, you can use the __?__ to find the measure of the angle.
4. Finding the measures of all of the triangle's sides and angles is called __?__.
5. The ratio of the length of the side opposite an acute angle in a right triangle to the length of the triangle's hypotenuse is the __?__.

**a.** cosine
**b.** trigonometric
**c.** sine
**d.** solving a right triangle
**e.** tangent

## SECTION 14–1 THE SINE, COSINE, AND TANGENT RATIOS (pages 528–531)

► For each acute angle in a right triangle, the lengths of the sides can be used to form these **trigonometric ratios: sine, cosine,** and **tangent.**

Write each ratio in fraction form.

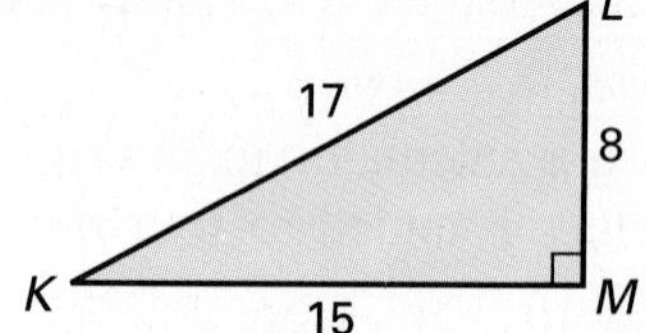


**6.** sin *K* **7.** cos *L* **8.** tan *L*

**9.** cos *K* **10.** sin *L* **11.** tan *K*

Use a calculator or a table of trigonometric ratios to find each ratio.

**12.** tan 17° **13.** sin 9° **14.** cos 64°

## SECTION 14–2 LENGTHS OF SIDES IN RIGHT TRIANGLES (pages 532–535)

► If you know the measure of one acute angle and the measure of one side of a right triangle, you can use trigonometric ratios to find the lengths of the other two sides.

In $\triangle STU$, find these lengths to the nearest tenth.

**15.** *ST*

**16.** *SU*

**17.** An airplane took off on a path forming a 29° angle with the ground. After the plane had traveled 1,200 m, how far above the ground was it?

## SECTION 14–3 MEASURES OF ANGLES IN RIGHT TRIANGLES (pages 536–539)

► If you know the measures of two sides of a right triangle, you can use them to write a trigonometric ratio for each acute angle in the triangle. Then you can use these ratios to find the measures of the angles.

Round your answers to the nearest whole degree.

**18.** In right triangle *DEF*, $\cos D = 0.25$. Find $m\angle D$.

**19.** Find $m\angle I$ in the triangle at the right.

**20.** Joel attached 750 ft of string to a kite. If he would like the kite to reach a height of 500 ft using all the string, what angle must the kite make with the ground?

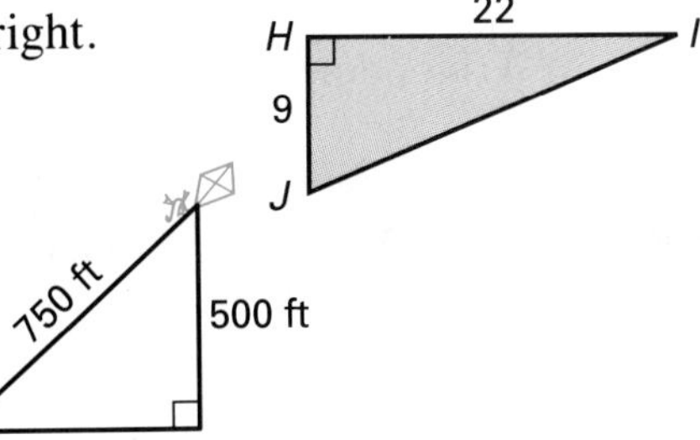


## SECTIONS 14–4 AND 14–5 PROBLEM SOLVING (pages 540–543)

► Finding the measures of all the sides and all the angles of a right triangle is known as **solving a right triangle.**

► When a problem seems difficult because it has complex numbers, substitute simpler numbers and look for a method of solution. Then apply that method to the original problem.

**21.** Solve $\triangle JEN$. Round angle measures to the nearest degree and lengths of sides to the nearest tenth.

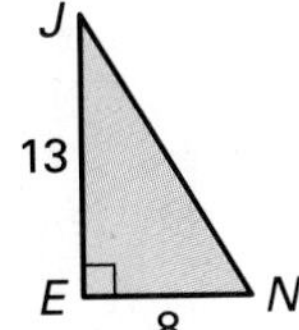


**22.** If Leta could have swum directly across a river, it would have been a distance of 124.8 m. However, the current carried her on a path that formed a 40.7° angle with the riverbank. To the nearest tenth of a meter, how far did Leta actually swim?

## SECTION 14–6 SPECIAL RIGHT TRIANGLES (pages 544–547)

► Sometimes it is easier to use the special geometric properties of a **30°–60°–90° right triangle** or a **45°–45°–90° right triangle** to solve problems than to use trigonometric ratios.

Find each length. Express answers using square roots.

**23.** *MP*

**24.** *MN*

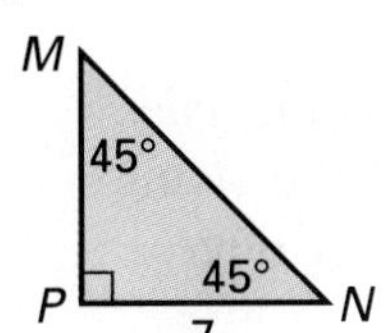


**25.** *EF*

**26.** *DF*

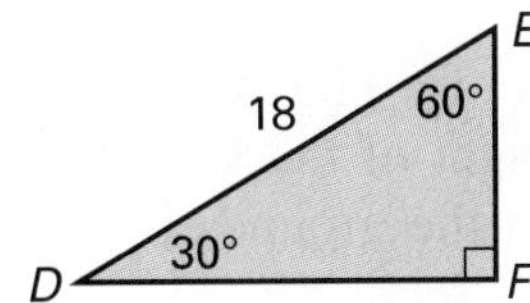


**27.** **USING DATA** Use the table and pattern on page 527 to find the total height of a wren house. Round your answer to the nearest tenth.

# CHAPTER 14 ● TEST

Write each ratio in fraction form.

1. $\cos S$
2. $\tan Q$
3. $\sin Q$

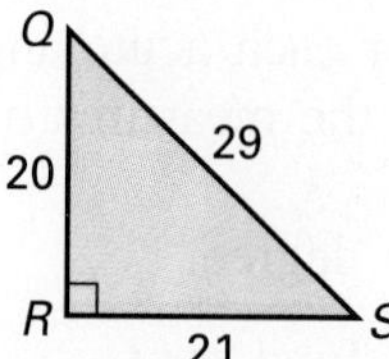


Use a calculator or the Table of Trigonometric Ratios on page 581 to find each value.

4. $\cos 28°$
5. $m\angle Z$, if $\sin Z = 0.65$
6. $KL$ to the nearest tenth
7. $m\angle C$ to the nearest degree

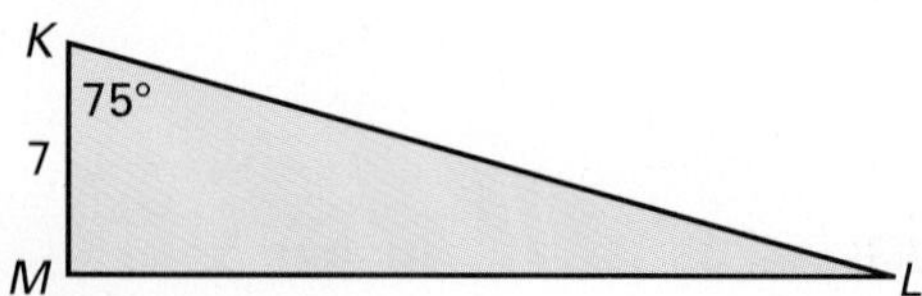


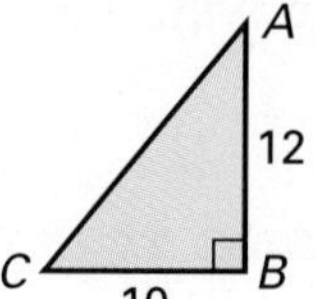


8. Solve $\triangle FGH$. Round each length to the nearest tenth and each angle measure to the nearest degree.

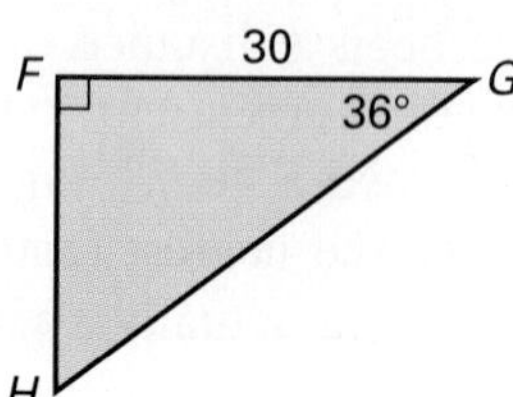


Find each length in $\triangle XYZ$. Express your answers using square roots.

9. $YZ$
10. $XZ$

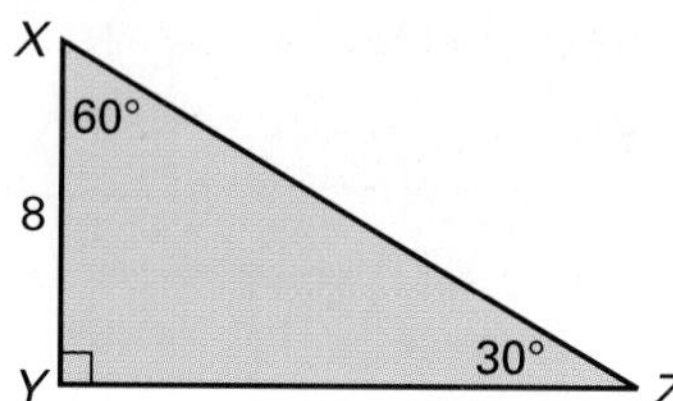


Solve each problem.

11. When the sun's rays make an angle of 40° with the ground, a certain tree casts a shadow 30 ft long. To the nearest tenth of a foot, how tall is the tree?

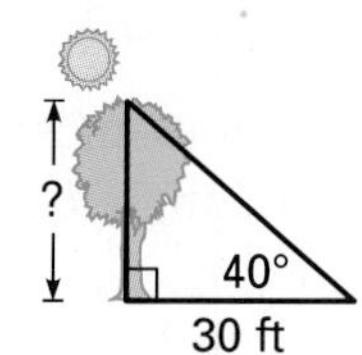


12. The length of a kite string fastened to the ground is 86 m. The vertical height of the kite is 52 m. Find the angle that the string makes with the ground. Round the answer to the nearest degree.

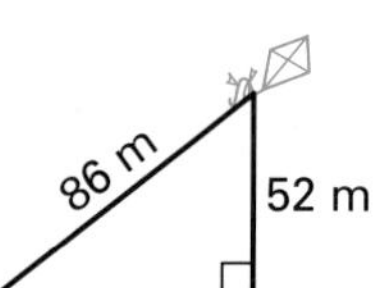


13. A guy wire is anchored to the ground 95.6 m from the base of a transmitting tower. If it forms a 62.5° angle with the ground, how long is the wire? Round your answer to the nearest tenth.

Find the next three terms in each pattern.

1. 49, 46, 43, 40, ■, ■, ■
2. 4500, 450, 45, 4.5, ■, ■, ■

Write in scientific notation.

3. 45,000,000
4. 0.00082

Find the unknown angle measure.

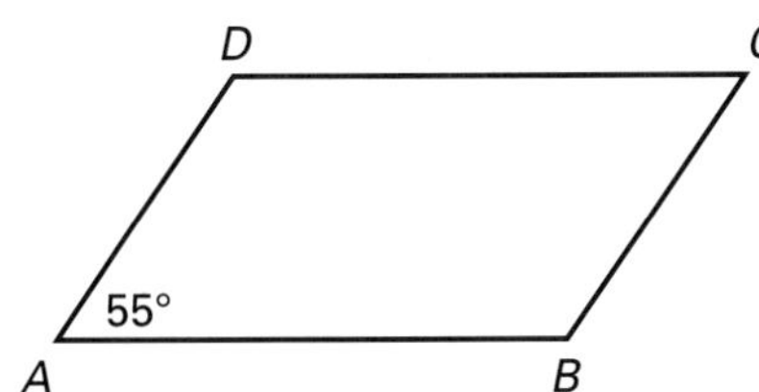


5. $m\angle B$
6. $m\angle C$

Use the scatter plot for Exercises 7 and 8.

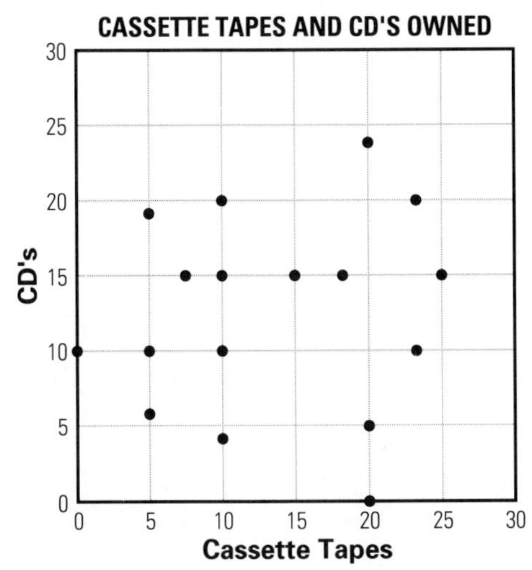


7. How many students owned exactly 20 cassette tapes?
8. Find the mode of the data.

Solve each equation.

9. $6a - 4 + 4a = 56$
10. $r^2 = 8.41$

Find the volume of each figure. Use 3.14 for $\pi$.

11.

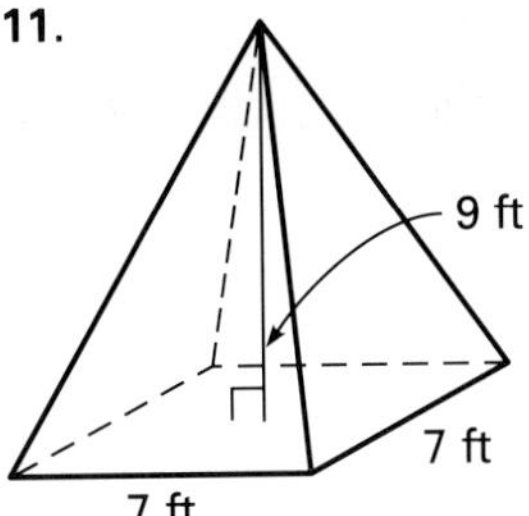


12.

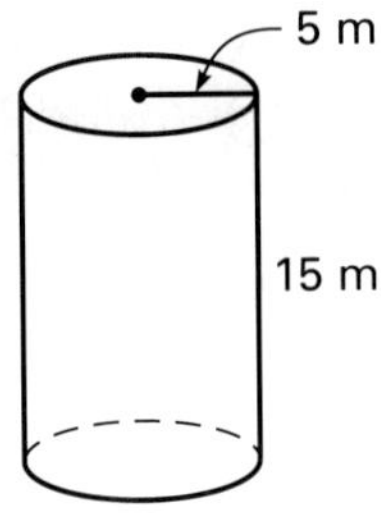


13. Find the CGF. $18a^4$ and $45a^3$
14. Find the LCM. $7x^2$ and $8xy^2$

Factor each polynomial.

15. $x^2 - 12x + 36$
16. $m^2 - 121$

Tell whether the ordered pair is a solution of the inequality.

17. $(1, 4)$; $y > x + 4$
18. $(-1, 2)$; $y \geq 2x + 1$

Find the midpoint, $P$, of the line segment whose endpoints are given.

19. $A(3, 5)$ and $B(-3, -3)$
20. $C(5, 8)$ and $D(11, 12)$

For Exercises 21 and 22, write an equation of each line from the information given.

21. $m = 3$; $P(1, -2)$

22. $A(0, -3)$; $B(2, 1)$

23. Marsha's and Al's ages together total 24. Marsha is 4 years older than Al. How old is each person?

Solve each proportion.

24. $\frac{x}{14} = \frac{9}{6}$

25. $\frac{12}{3.6} = \frac{c}{9}$

26. In the figure below, the triangles are similar. Find $h$.

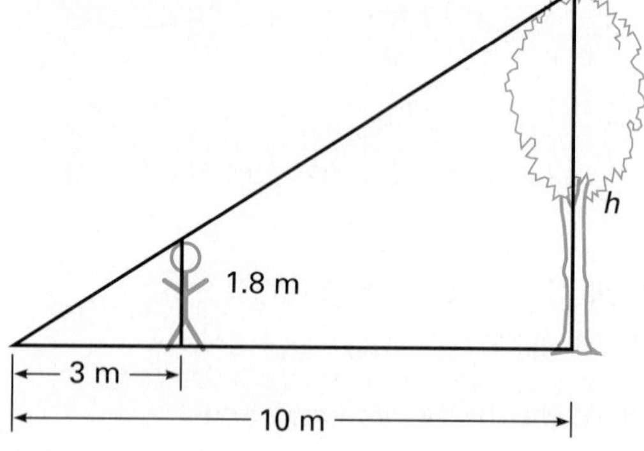


Give the coordinates of the image of $A(4, -1)$ under each translation.

27. 4 units left

28. 4 units down

Write the coordinates of the image of each point under a reflection across the given line.

29. $(-2, 3)$; $x$-axis

30. $(4, -1)$; $y$-axis

31. In how many ways could a doubles tennis team be chosen from a group of 12 players?

32. How many different five-digit numbers can be formed from 2, 3, 5, 8, 9 if each number includes each digit once?

For Exercises 33 and 34, let $A = \{2, 4, 6\}$ and $B = \{1, 3, 5\}$.

33. Find $A \cup B$.

34. Find $A \cup B$.

35. Identify the necessary and sufficient conditions in the statement. If $n$ is even, then $3n$ is even.

Refer to the figure below for Exercises 36 and 37.

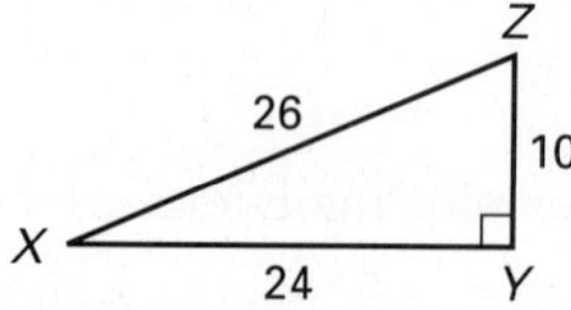


36. Find tan $Z$.

37. Find tan $X$. Round to the nearest hundredth.

38. Find the height of the tree.

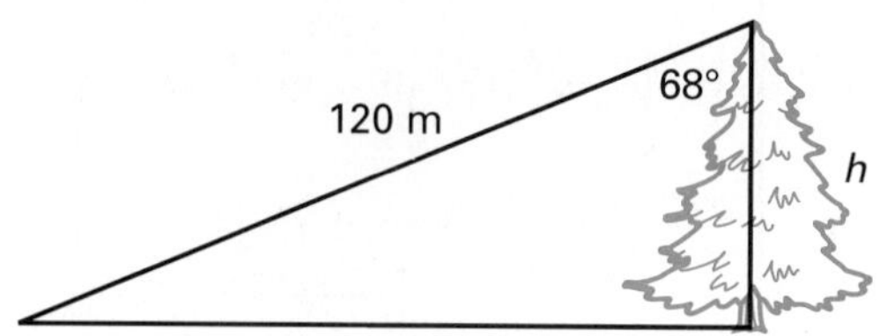

# CHAPTER 14 ● CUMULATIVE TEST

1. Find 55% of 1,800.
   A. 99 B. 990 C. 900 D. 90

2. What percent of 2,250 is 450?
   A. 2% B. 22% C. 20% D. 25%

3. Multiply. $-3(-5)$
   A. $-8$ B. $-15$ C. 8 D. 15

4. Divide. $-140 \div (-0.5)$
   A. 28 B. 280 C. 2.8 D. 2,800

5. Perform the indicated operations.
   $-\left(\frac{1}{4} + \frac{1}{6}\right) \div (-5)$
   A. $-\frac{1}{6}$ B. $\frac{1}{6}$ C. $\frac{1}{12}$ D. $-\frac{1}{12}$

6. A rhombus has
   A. two congruent sides.
   B. four congruent sides.
   C. four congruent angles.
   D. one pair of parallel sides.

7. Find $m\angle ABC$.

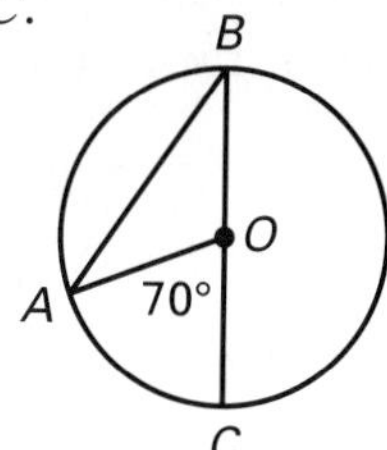


   A. 70° B. 35°
   C. 105° D. none of these

Use this stem-and-leaf plot for Exercises 8 and 9.

**Weekly Cost of Groceries for a Family of Four (in dollars)**

| Stem | Leaves |
|---|---|
| 8 | 7 |
| 9 | 3 7 7 8 |
| 10 | 2 4 5 6 7 7 9 |
| 11 | 3 5 5 5 8 |
| 12 | 4 7 |

8. What is the mode of the data?
   A. 7 B. 10 C. 107 D. 115

9. Where is there a cluster in the data?
   A. between 87 and 93
   B. around 105
   C. around 110
   D. between 124 and 127

10. On a test, Bruna had the sixth-highest score. If 28 students took the test, what was Bruna's percentile rank?
    A. 21st B. 82nd
    C. 72nd D. none of these

11. Solve. $13.8 = 2x + 9.4 - 6x$
    A. 4.4 B. $-5.6$ C. 0.55 D. $-1.1$

12. Solve. $16 - y^2 = -9$
    A. $\pm 4$ B. $\pm 5$ C. $\pm\sqrt{7}$ D. $\pm\sqrt{5}$

13. Find the area of the rectangle.

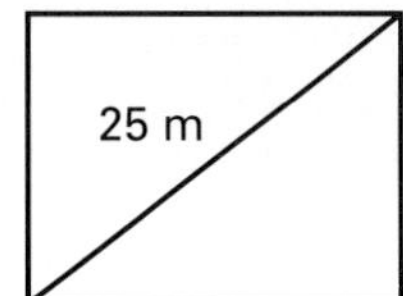


    A. 150 m$^2$
    B. 300 m$^2$
    C. 400 m$^2$
    D. 375 m$^2$

14. Find the surface area of the figure. Use 3.14 for $\pi$.
    A. 401.92 cm$^2$
    B. 1,004.8 cm$^2$
    C. 1,406.72 cm$^2$
    D. none of these

8 cm

20 cm

15. Simplify. $\frac{45x^3y^3z^4}{9x^2y^3z^5}$

A. $5xyz$ B. $54x^5y^6z$
C. $\frac{5xy}{z}$ D. $\frac{5x}{z}$

16. Factor the polynomial. $z^2 - 14z + 49$

A. $(z + 7)^2$ B. $(z - 7)^2$
C. $(z + 7)(z - 7)$ D. none of these

17. Factor the polynomial. $x^2 - 100$

A. $(x + 10)^2$
B. $(x - 10)^2$
C. $(x + 10)(x - 10)$
D. none of these

18. Which of the ordered pairs is a solution of $4x = y - 2$?

A. $(0, -2)$ B. $(2, 10)$
C. $(1, 2)$ D. $(-1, -6)$

19. What is the distance between points $A(6, 3)$ and $B(-6, -6)$?

A. 15 B. 9 C. $\sqrt{63}$ D. 3

20. Find the midpoint, $P$, of the line segment whose endpoints are $D(-4, -2)$ and $E(4, 6)$.

A. $P(4, 4)$ B. $P(0, 2)$
C. $P(4, 2)$ D. $P(0, 4)$

21. Write an equation of the line with $m = -4$ and $P(0, -4)$.

A. $y = 4x - 4$ B. $y = x - 4$
C. $y = -4x + 4$ D. $y = -4x - 4$

22. Identify the slope of the line that is perpendicular to the line whose equation is $y = -3x - 4$.

A. 3 B. $-3$ C. $\frac{1}{3}$ D. $-\frac{1}{3}$

23. Marina has 30 dimes and quarters in all. The number of dimes is two more than three times the number of quarters. How many of each coin does Marina have?

A. 15 dimes, 8 quarters
B. 20 dimes, 10 quarters
C. 23 dimes, 7 quarters
D. none of these

24. What is the unit price of cereal that costs \$3.10 for a 24-oz box?

A. \$7.40 B. \$0.13
C. \$7.74 D. none of these

25. A machine assembles 18 ball-point pens in 1.5 min. How many minutes would it take to assemble 144 pens?

A. 12 min B. 18 min
C. 45 min D. 27 min

26. The triangles in the figure are similar. Find $h$.

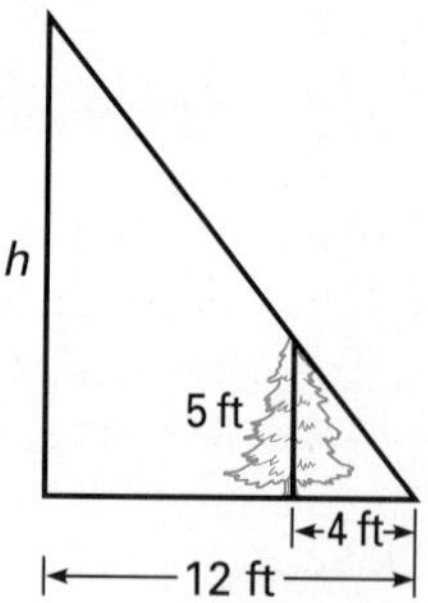


A. $1\frac{2}{3}$ ft B. 60 ft
C. 13 ft D. 15 ft

27. Write the coordinates of the image of $P(2, 4)$ under a reflection across the $y$-axis.

A. $(-2, 4)$ B. $(-2, -4)$
C. $(2, -4)$ D. $(2, 4)$

28. Write the coordinates of the image of $P(-1, 5)$ under a reflection across the line whose equation is $y = -x$.

A. $(5, -1)$ B. $(-1, -5)$
C. $(1, -5)$ D. none of these

29. A bag contains 4 green marbles, 6 red marbles, and 6 blue marbles. Find $P$(green, then blue) if you take one marble, replace it, and then take another marble.

A. $\frac{3}{32}$ B. $\frac{3}{16}$ C. $\frac{5}{8}$ D. $\frac{3}{4}$

30. Two number cubes are tossed. Find the probability that the sum is even and greater than 7.

A. $\frac{7}{36}$ B. $\frac{1}{4}$ C. $\frac{1}{6}$ D. $\frac{4}{9}$

**31.** Let $G = \{m, u, b\}$ and $H = \{m, u, p\}$. Find $G \cup H$.

A. $\{m, u\}$ B. $\{b, u, m, p\}$
C. $\{b, p\}$ D. $\phi$

**32.** Let $A = \{1, 3, 5, 7, \ldots\}$ and $B = \{1, 2, 4, 8, \ldots\}$. Find $A \cap B$.

A. $\{x|x \text{ is a whole number}\}$
B. $\phi$
C. $\{0\}$
D. $\{1\}$

**33.** Given this statement, which of the following is its inverse?

*If the figure is a pentagon, then the figure is a polygon.*

A. If the figure is not a polygon, then the figure is not a pentagon.
B. If the figure is not a pentagon, then the figure is not a polygon.
C. If the figure is a polygon, then the figure is a pentagon.
D. none of these

**34.** Given this statement, which of the following is its converse?

*If two angles are right angles, then they are congruent.*

A. If two angles are not right angles, then they are not congruent.
B. If two angles are congruent, then they are right angles.
C. If two angles are not congruent, then they are not right angles.
D. none of these

**35.** In a 30°–60°–90° triangle,

A. the side opposite the 60° angle is $\frac{\sqrt{3}}{2}$ times the length of the side opposite the 30° angle.
B. the side opposite the 60° angle is twice the length of the side opposite the 30° angle.
C. the side opposite the 60° angle is $\sqrt{3}$ times the length of the side opposite the 30° angle.
D. none of these

Refer to the figure for Exercises 36 and 37.

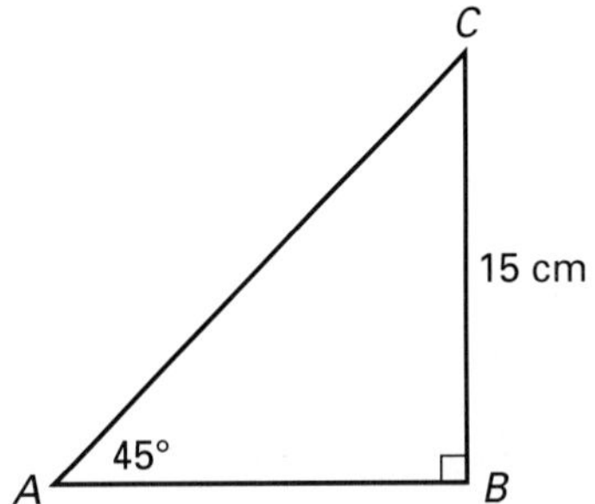


**36.** Find $AB$.

A. $\frac{15\sqrt{2}}{2}$ cm B. 15 cm
C. $15\sqrt{2}$ cm D. none of these

**37.** Find $AC$.

A. $\sqrt{2}$ cm B. $\frac{15\sqrt{2}}{2}$ cm
C. $15\sqrt{2}$ cm D. none of these

# How to Use the Data File

The *Data File* contains interesting information presented in tables and graphs for your use in solving problems. The data is organized into the following categories: animals, architecture, the arts, earth science, economics, entertainment, health and sports, science, and United States. You will need to refer to this material often as you work through this book. Whenever you come upon the words "USING DATA" at the beginning of an exercise, refer to the *Data Index* to help you locate the information you will need to complete the exercise. For a quick review of commonly used symbols, formulas, and other information, you may wish to refer to the last part of the *Data File*, Useful Mathematical Data.

## DATA INDEX

## Entertainment 568–569



## Health and Sports 570–571



## Science 572–573



## United States 574–575



## Useful Mathematical Data 576–581

# Animals

## LONGEST RECORDED LIFE SPANS OF SOME ANIMALS

| Animal | Years |
|---|---|
| Marion's tortoise | 152 |
| Deep-sea clam | 100 |
| Killer whale | 90 |
| Blue whale | 90 |
| Fin whale | 90 |
| Freshwater oyster | 80 |
| Cockatoo | 70 |
| Condor | 70 |
| Indian elephant | 70 |
| Ostrich | 62 |
| Horse | 62 |
| Chimpanzee | 50 |
| Termite | 50 |
| Lobster | 50 |
| Cow | 40 |
| Domestic pigeon | 35 |
| Domestic cat | 34 |
| Dog (Labrador) | 29 |
| Budgerigar | 28 |
| Sheep | 20 |
| Goat | 18 |
| Rabbit | 18 |
| Golden hamster | 10 |
| House mouse | 6 |
| Housefly | 0.2 |

## MAJOR U.S. PUBLIC ZOOLOGICAL PARKS

Source: World Almanac questionnaire, 1991; budget and attendance in millions.
(*) Park has not provided up-to-date data.

| Zoo | Budget | Attendance | Species | Major Attractions |
|---|---|---|---|---|
| **Arizona-Sonora Desert Museum** (Tucson) | $12.0 | 0.6 | 600 | "Living" museum, 90% outdoors |
| **Audubon** (New Orleans) | 9.0 | 1.0 | 433 | Asian Domain, Louisiana Swamp |
| **Bronx** (N.Y.C.) | 24.6 | 2.1 | 674 | Himalayan Highlands, African Plains, Wild Asia, MouseHouse, endangered species |
| **Buffalo** | 3.0 | 0.5 | 215 | Habicat, Gorilla Habitat, Children's Zoo |
| **Chicago** (Brookfield) | 21.0 | 1.9 | 385 | 7 Seas Seascape, Tropic World |
| **Cincinnati** | 10.0 | 1.3 | 730 | Gorilla World, Insect World, white bengal tigers |
| **Cleveland** | 6.0 | 0.9 | 470 | African Plains, rhino/cheetah, Animals of China |
| **Dallas** | 7.1 | 0.7 | 330 | 25-acre Wilds of Africa with monorail, nature trail, Gorilla Conservation Center |
| **Denver** | 6.6 | 1.2 | 312 | Bear Mountain, Wolf Pack Woods, Feline House |
| **Detroit*** | 11.4 | 1.0 | 300 | Penguinarium, Chimps of Harambee |
| **Houston** | 3.7 | 1.2 | 724 | Children's Zoo, white tigers, white rhinoceros |
| **Lincoln Park** (Chicago) | 8.8 | 4.0 | 312 | Great Ape House, bird house, Children's Zoo |
| **Los Angeles** | 7.2 | 2.0 | 500 | Adventure Island, World of Birds |
| **Louisville*** | 2.5 | 0.4 | 245 | African Panorama, polar bear, Siberian tiger |
| **Memphis*** | 2.5 | 0.8 | 403 | Cat Country, Primate World, The Forest |
| **Miami Metrozoo** | 7.2 | 0.8 | 300 | Koalas, Aviary, cageless exhibits, Asian River Life |
| **Milwaukee** | 12.0 | 1.5 | 350 | Sea Lion Exhibit, Predator Prey, new Aviary |
| **Minnesota*** | 10.0 | 1.0 | 311 | Tropics Trail, Minnesota Trail, koalas, dolphins |
| **National** (Wash., D.C.)* | 13.0 | 3.0 | 509 | Giant pandas, Komodo dragon lizards, gorillas |
| **Oklahoma City** | 9.0 | 0.6 | 500 | Aquaticus dolphin & sea lion shows, 300 aquariums |
| **Philadelphia*** | 12.0 | 1.3 | 550 | World of Primates, Treehouse, Rare Animal House, African Plains, Small Mammals |
| **Phoenix*** | 7.0 | 1.0 | 300 | African Veldt, Children's Zoo, Arizona Trail |
| **Gladys Porter** (Brownsville, Tex.) | 2.0 | 0.3 | 434 | Free-flight aviary, Herpetarium, Aquatic wing |
| **Rio Grande** (Albuquerque)* | 2.2 | 0.5 | 289 | Ape Country, Free-flying Bird Show, Rainforest |
| **Riverbanks** (Columbia, S.C.) | 3.2 | 1.0 | 439 | Aquarium Reptile Complex, Riverbanks Farm |
| **St. Louis** | 14.7 | 2.7 | 657 | Living World, Bear Pits, Jungle of the Apes |
| **San Antonio** | 6.8 | 1.0 | 700 | Children's Zoo, Australian Walkabout |
| **San Diego*** | 40.0 | 3.5 | 800 | Tiger River, Southeast Asian exhibit, koalas |
| **San Diego** (Wild Animal Park) | 15.0 | 1.3 | 450 | Mixed-species enclosures; exotic species, monorail with 50-minute, narrated tour |
| **San Francisco*** | 8.0 | 1.2 | 270 | Primate Discovery Center, Koala Crossing, Gorilla World, Penguin Island |
| **Toledo** | 7.5 | 0.8 | 400 | Hippoquarium, African Savanna, Children's Zoo |
| **Washington Pk** (Portland) | 11.0 | 1.0 | 195 | Alaska Tundra, Penguinarium, Africa exhibit |
| **Woodland Pk** (Seattle) | 5.3 | 0.9 | 258 | African Savanna, gorillas, Asian Elephant Forest |

## MAXIMUM SPEEDS OF ANIMALS

The data on this topic are notoriously unreliable because of the many inherent difficulties of timing the movement of most animals—whether running, flying, or swimming—and because of the absence of any standardization of the method of timing, of the distance over which the performance is measured, or of allowance for wind conditions.

The most that can be said is that a specimen of the species below has been timed to have attained as a maximum the speed given.

| mi/h | | mi/h | |
|---|---|---|---|
| 219.5 | Spine-tailed Swift | 37 | Dolphin |
| 180* | Peregrine Falcon | 36 | Dragonfly |
| 120* | Golden Eagle | 35 | Flying Fish |
| 96.29 | Racing Pigeon | 35 | Rhinoceros |
| 88 | Spurwing Goose | 35 | Wolf |
| 70† | Cheetah | 33 | Hawk Head Moth |
| 60 | Pronghorn Antelope | 32 | Giraffe |
| 60 | Mongolian Gazelle | 32 | Guano Bat |
| 57 | Quail | 30 | Blackbird |
| 57 | Swordfish | 28 | Grey Heron |
| 53 | Partridge | 25 | California Sea Lion |
| 45 | Red Kangaroo | 24 | African Elephant |
| 45 | English Hare | 23 | Salmon |
| 40 | Red Fox | 22.8 | Blue Whale |
| 40 | Mute Swan | 22 | Leatherback Turtle |
| 38 | Swallow | 22 | Wren |
| | | 20 | Monarch Butterfly |

*Stooping
†Unable to sustain a speed of over 44 mi/h over 500 yards.

## HOW FAST DO INSECTS FLY?

| | Wingbeats per second | Flight speed (mi/h) |
|---|---|---|
| White butterfly | 8–12 | 4–9 |
| Damselfly | 16 | 2–4 |
| Dragonfly | 25–40 | 16–34 |
| Cockchafer beetle | 50 | 17 |
| Hawkmoth | 50–90 | 11–31 |
| Hoverfly | 120 | 7–9 |
| Bumblebee | 130 | 7 |
| Housefly | 200 | 4 |
| Honeybee | 225 | 4–7 |
| Mosquito | 600 | 0.6–1.2 |
| Midge | 1,000 | 0.6–1.2 |

## TOP 25 AMERICAN KENNEL CLUB REGISTRATIONS

| Breed | Rank | 1990 | Rank | 1989 |
|---|---|---|---|---|
| Cocker Spaniels | 1 | 105,642 | 1 | 111,636 |
| Labrador Retrievers | 2 | 95,768 | 2 | 91,107 |
| Poodles | 3 | 71,757 | 3 | 78,600 |
| Golden Retrievers | 4 | 64,848 | 4 | 64,269 |
| Rottweilers | 5 | 60,471 | 6 | 51,291 |
| German Shepherd Dogs | 6 | 59,556 | 5 | 58,422 |
| Chow Chows | 7 | 45,271 | 7 | 50,150 |
| Dachshunds | 8 | 44,470 | 8 | 44,305 |
| Beagles | 9 | 42,499 | 9 | 43,314 |
| Miniature Schnauzers | 10 | 39,910 | 10 | 42,175 |
| Shetland Sheepdogs | 11 | 39,870 | 11 | 39,665 |
| Shih Tzu | 12 | 39,503 | 13 | 38,131 |
| Yorkshire Terriers | 13 | 36,033 | 12 | 39,268 |
| Pomeranians | 14 | 34,475 | 14 | 32,109 |
| Chihuahuas | 15 | 24,593 | 16 | 24,917 |
| Lhasa Apsos | 16 | 24,024 | 15 | 28,810 |
| Boxers | 17 | 23,659 | 18 | 22,037 |
| Siberian Huskies | 18 | 21,944 | 19 | 21,875 |
| Dalmatians | 19 | 21,603 | 24 | 17,488 |
| English Springer Spaniels | 20 | 21,342 | 22 | 20,910 |
| Basset Hounds | 21 | 20,945 | 21 | 21,517 |
| Doberman Pinschers | 22 | 20,255 | 20 | 21,782 |
| Pekingese | 23 | 18,505 | 17 | 22,966 |
| Collies | 24 | 17,337 | 23 | 18,227 |
| Boston Terriers | 25 | 15,401 | 25 | 15,355 |

# Architecture

**THE TALLEST BUILDINGS IN THE WORLD (as of 1990)**

| Building | City | Year Built | Stories | Height (meters) | Height (feet) |
|---|---|---|---|---|---|
| Sears Tower | Chicago | 1974 | 110 | 443 | 1,454 |
| World Trade Center, North | New York | 1972 | 110 | 417 | 1,368 |
| World Trade Center, South | New York | 1973 | 110 | 415 | 1,362 |
| Empire State | New York | 1931 | 102 | 381 | 1,250 |
| Bank of China Tower | Hong Kong | 1988 | 72 | 368 | 1,209 |
| Amoco | Chicago | 1973 | 80 | 346 | 1,136 |
| John Hancock | Chicago | 1968 | 100 | 344 | 1,127 |
| Chrysler | New York | 1930 | 77 | 319 | 1,046 |

**BUILDING CODES STANDARDS FOR SLOPE AND SAFETY**

| | Maximum Slope |
|---|---|
| Ramps—wheelchair | 0.125 |
| Ramps—walking | 0.3 |
| Driveway or street parking | 0.22 |
| Stairs | 0.83 |

**DIMENSIONS OF CROSS SECTIONS OF FINISHED LUMBER**

| Name | Actual size (in.) | Name | Actual size (in.) |
|---|---|---|---|
| one-by-one | $\frac{3}{4} \times \frac{3}{4}$ | | |
| one-by-two | $\frac{3}{4} \times 1\frac{1}{2}$ | two-by-two | $1\frac{1}{2} \times 1\frac{1}{2}$ |
| one-by-four | $\frac{3}{4} \times 3\frac{1}{2}$ | two-by-four | $1\frac{1}{2} \times 3\frac{1}{2}$ |
| one-by-six | $\frac{3}{4} \times 5\frac{1}{2}$ | two-by-six | $1\frac{1}{2} \times 5\frac{1}{2}$ |
| one-by-eight | $\frac{3}{4} \times 7\frac{1}{4}$ | two-by-eight | $1\frac{1}{2} \times 7\frac{1}{4}$ |
| one-by-ten | $\frac{3}{4} \times 9\frac{1}{4}$ | two-by-ten | $1\frac{1}{2} \times 9\frac{1}{4}$ |
| one-by-twelve | $\frac{3}{4} \times 11\frac{1}{4}$ | two-by-twelve | $1\frac{1}{2} \times 11\frac{1}{4}$ |
| four-by-four | $3\frac{1}{2} \times 3\frac{1}{2}$ | six-by-six | $5\frac{1}{2} \times 5\frac{1}{2}$ |
| four-by-six | $3\frac{1}{2} \times 5\frac{1}{2}$ | eight-by-eight | $7\frac{1}{2} \times 7\frac{1}{2}$ |

Source: *Projects in Wood*, by David Field, G. P. Putnam's Sons, NY, 1985

## NOTED RECTANGULAR STRUCTURES

| Structure | Country | Length (meters) | Width (meters) |
|---|---|---|---|
| Parthenon | Greece | 69.5 | 30.9 |
| Palace of the Governors | Mexico | 96 | 11 |
| Great Pyramid of Cheops | Egypt | 230.6 | 230.6 |
| Step Pyramid of Zosar | Egypt | 125 | 109 |
| Temple of Hathor | Egypt | 290 | 280 |
| Cleopatra's Needle (base) | England* | 2.4 | 2.3 |
| Ziggurat of Ur (base) | Middle East | 62 | 43 |
| Guanyin Pavilion of Dule Monastery | China | 20 | 14 |
| Izumo Shrine | Japan | 10.9 | 10.9 |
| Kibitsu Shrine (main) | Japan | 14.5 | 17.9 |
| Kongorinjo Hondo | Japan | 21 | 20.7 |
| Bakong Temple, Roluos | Cambodia | 70 | 70 |
| Ta Keo Temple | Cambodia | 103 | 122 |
| Wat Kukut Temple, Lampun | Thailand | 23 | 23 |
| Tsukiji Hotel | Japan | 67 | 27 |

*Gift to England from Egypt

## LONGEST HIGHWAY TUNNELS IN THE UNITED STATES

| Name | Length (mi) |
|---|---|
| E. Johnson Memorial | 1.70 |
| Eisenhower Memorial | 1.69 |
| Allegheny | 1.15 |
| Liberty Tubes | 1.12 |
| Zion National Park | 1.09 |
| East River Mountain | 1.03 |
| Tuscarora | 1.02 |
| Kittatinny | 0.88 |

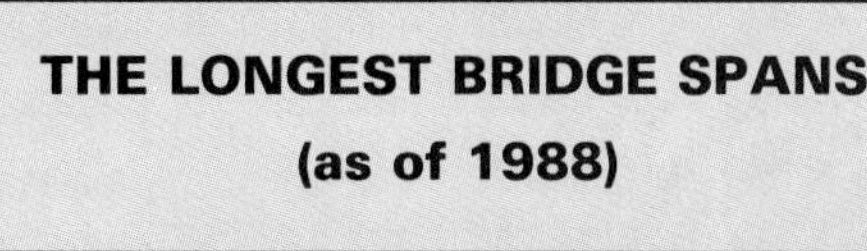

## THE LONGEST BRIDGE SPANS (as of 1988)

| Bridge | Country | Year Built | Span (meters) | Span (feet) |
|---|---|---|---|---|
| Akashi-Ohashi* | Japan | 1988 | 1,780 | 5,840 |
| Humber | England | 1981 | 1,410 | 4,626 |
| Verrazano-Narrows | United States | 1964 | 1,298 | 4,260 |
| Golden Gate | United States | 1937 | 1,280 | 4,200 |
| Mackinac | United States | 1957 | 1,158 | 3,800 |
| Bosphorus | Turkey | 1973 | 1,074 | 3,524 |
| George Washington | United States | 1931 | 1,067 | 3,500 |
| Tagus River | Portugal | 1966 | 1,013 | 3,323 |
| Forth | Scotland | 1964 | 1,006 | 3,300 |
| Severn | England/Wales | 1966 | 988 | 3,240 |

*Scheduled completion information

# The Arts

## AFRICAN TEXTILE PATTERNS

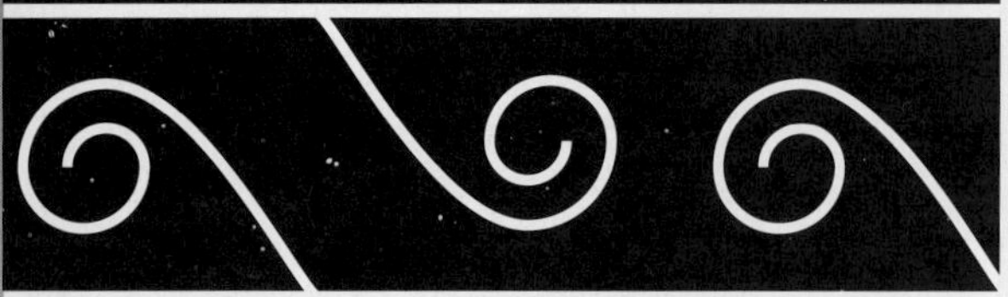
Ashanti pattern, Ghana

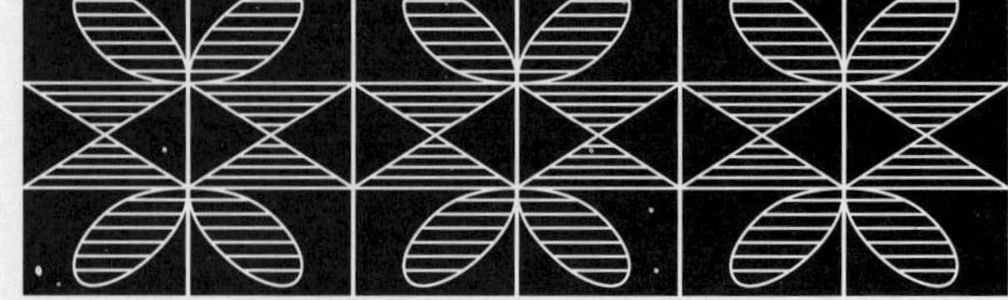
Kuba pattern, Congo-Kinshasa

Bambara pattern, Mali

### WHICH WAY IS UP?

Here are three views of the same box. Which way is "UP↑" in View 3?

View 1

View 2

View 3

### ALL-TIME BESTSELLING CHILDREN'S BOOKS

| Hardcover | |
|---|---|
| 1. *The Tale of Peter Rabbit*, Beatrix Potter, 1902 | 9,000,000 |
| 2. *Pat the Bunny*, Dorothy Kunhardt, 1940 | 4,857,417 |
| 3. *The Littlest Angel*, Charles Tazewell, 1946 | 4,665,209 |
| 4. *The Cat in the Hat*, Dr. Seuss, 1957 | 3,693,197 |
| 5. *Green Eggs and Ham*, Dr. Seuss, 1960 | 3,684,097 |
| 6. *The Children's Bible*, 1965 | 3,683,097 |
| 7. *The Real Mother Goose*, illus. Blanche F. Wright, 1916 | 3,600,000 |
| 8. *Richard Scarry's Best Word Book Ever*, 1963 | 3,303,583 |
| 9. *One Fish, Two Fish, Red Fish, Blue Fish*, Dr. Seuss, 1960 | 2,970,833 |
| 10. *Hop on Pop*, Dr. Seuss, 1963 | 2,953,324 |

| Paperback | |
|---|---|
| 1. *The Outsiders*, S. E. Hinton, 1968 | 5,855,085 |
| 2. *Are You There, God? It's Me, Margaret*, Judy Blume, 1974 | 5,278,412 |
| 3. *Charlotte's Web*, E. B. White, illus. Garth Williams, 1973 | 4,607,131 |
| 4. *Tales of a Fourth Grade Nothing*, Judy Blume, 1976 | 4,582,039 |
| 5. *Little House on the Prairie*, Laura Ingalls Wilder, illus. Garth Williams, 1971 | 3,803,209 |
| 6. *The Little Prince*, Antoine de Saint-Exupery, 1968 | 3,667,861 |
| 7. *Little House in the Big Woods*, Laura Ingalls Wilder, illus. Garth Williams, 1971 | 3,495,079 |
| 8. *That Was Then, This Is Now*, S. E. Hinton, 1972 | 3,351,194 |
| 9. *Where the Red Fern Grows*, Wilson Rawls, 1974 | 3,347,000 |
| 10. *Superfudge*, Judy Blume, 1981 | 3,243,442 |

## PITCH RANGES OF MUSICAL INSTRUMENTS

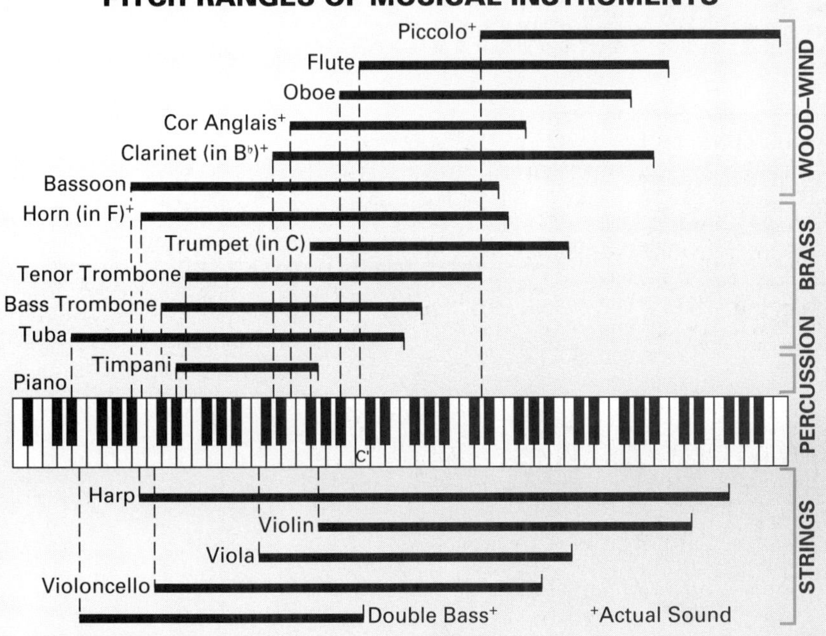

## PUBLIC LIBRARIES

Source: World Almanac questionnaire (1990)

| City | No. Bound Volumes | Circulation | Annual Acquisitions Expend. |
|---|---|---|---|
| Albuquerque, N.M. (11) | 610,502 | 2,283,512 | $1,168,447 |
| Anchorage, Alas. (5) | 395,700 | 1,099,780 | 662,790 |
| Atlanta, Ga. (33) | 1,678,750 | 2,303,681 | 3,011,362 |
| Baltimore, Md. (30) | 2,289,857 | 1,532,279 | 2,418,089 |
| Birmingham, Ala. (19) | 1,092,285 | 2,072,444 | 916,174 |
| Boston, Mass. (25) | 4,916,277 | 1,454,414 | NA |
| Buffalo, N.Y. (53) | 3,602,478 | 6,237,516 | 2,023,551 |
| Charlotte, N.C. (19) | 1,200,000 | 2,700,000 | 1,400,000 |
| Cincinnati, Oh. (39) | 3,700,000 | 7,179,389 | 4,205,140 |
| Cleveland, Oh. (29) | 2,261,043 | 4,550,994 | 4,588,427 |
| Columbus, Oh. (20) | 1,600,000 | 4,100,000 | 4,475,615 |
| Dallas, Tex. (19) | 2,409,864 | 4,364,027 | 17,201,527 |
| Denver, Col. (21) | 2,239,106 | 3,396,268 | 1,856,485 |
| Des Moines, Ia. (5) | 487,444 | 1,246,350 | 292,163 |
| Detroit, Mich. (25) | 2,746,021 | 1,743,314 | 1,900,000 |
| District of Columbia (25) | 1,507,556 | 1,843,098 | 1,530,000 |
| Fairfax, Va. (22) | 1,761,264 | 8,454,714 | 3,387,721 |
| Ft. Worth, Tex. (10) | 1,126,931 | 3,356,148 | 1,157,965 |
| Honolulu, Ha. (49) | 2,310,843 | 6,454,824 | 2,886,133 |
| Indianapolis, Ind. (21) | 1,500,000 | 5,376,388 | 2,170,167 |
| Los Angeles, Cal. (62) | 5,663,240 | 10,382,321 | 4,469,384 |
| Louisville, Ky. (14) | 903,084 | 3,084,620 | 979,836 |
| Memphis, Tenn. (23) | 1,700,000 | 2,700,000 | 3,228,007 |
| Miami, Fla. (28) | 2,386,204 | 4,200,000 | 2,200,000 |
| Milwaukee, Wis. (12) | 2,051,114 | 3,382,982 | 1,360,814 |
| Minneapolis, Minn. (14) | 1,888,934 | 3,012,111 | $1,771,400 |
| Nashville, Tenn. (16) | 608,346 | 1,945,454 | 871,515 |
| New Orleans, La. (15) | 1,039,264 | 1,159,303 | 750,063 |
| New York, N.Y. (research) | 9,189,489 | — | — |
| Branches (81) | 3,180,866 | 9,669,316 | 14,931,000 |
| Brooklyn (58) | 4,637,167 | 8,610,459 | 5,133,434 |
| Queens (61) | 5,500,000 | 11,320,000 | 5,500,000 |
| Norfolk, Va. (11) | 880,564 | 823,908 | 540,747 |
| Oklahoma City, Okla. (11) | 933,241 | 4,184,437 | 1,346,726 |
| Omaha, Neb. (10) | 588,013 | 1,860,728 | 892,109 |
| Philadelphia, Pa. (53) | 3,458,700 | 4,800,000 | 6,200,000 |
| Phoenix, Az. (10) | 1,683,458 | 4,529,337 | 2,500,000 |
| Pittsburgh, Pa. (18) | 1,880,241 | 2,841,120 | 13,764,737 |
| Portland, Ore. (15) | 4,000,000 | 2,000,000 | 1,400,000 |
| Richmond, Va. (10) | 780,000 | 858,213 | 450,000 |
| Rochester, N.Y. (11) | 500,000 | 1,587,691 | 7,200,000 |
| Sacramento, Cal. (24) | 1,663,893 | 3,977,515 | 2,045,262 |
| St. Louis, Mo. (18) | 1,903,218 | 7,915,559 | 2,700,000 |
| St. Paul, Minn. (12) | 740,171 | 2,414,263 | 949,246 |
| San Diego, Cal. (31) | 1,561,232 | 4,568,116 | 1,862,547 |
| San Francisco, Cal. (26) | 1,749,129 | 2,470,091 | NA |
| Tampa, Fla. (17) | 1,100,600 | 3,200,000 | 1,800,000 |
| Tucson, Ariz. (15) | 760,000 | 3,900,000 | NA |
| Tulsa, Okla. (20) | 825,000 | 3,236,081 | 1,419,000 |
| Wichita, Kan. (12) | 894,942 | 1,396,667 | 400,000 |

# Earth Science

## SIZE OF THE CONTINENTS

| Continent | Mi² |
|---|---|
| Asia | 17,297,000 |
| Africa | 11,708,000 |
| North America | 9,406,000 |
| South America | 6,883,000 |
| Antarctica | 5,405,000 |
| Europe | 3,835,000 |
| Australia | 2,968,000 |

## SOME PRINCIPAL RIVERS OF THE WORLD

| River | Length (miles) |
|---|---|
| Amazon | 4,000 |
| Arkansas | 1,459 |
| Columbia | 1,243 |
| Danube | 1,776 |
| Ganges | 1,560 |
| Indus | 1,800 |
| Mackenzie | 2,635 |
| Mississippi | 2,340 |
| Missouri | 2,540 |
| Nile | 4,160 |
| Ohio | 1,310 |
| Orinoco | 1,600 |
| Paraguay | 1,584 |
| Red | 1,290 |
| Rhine | 820 |
| Rio Grande | 1,900 |
| St. Lawrence | 800 |
| Snake | 1,038 |
| Thames | 236 |
| Tiber | 252 |
| Volga | 2,194 |
| Zambezi | 1,700 |

## HIGHEST AND LOWEST CONTINENTAL ALTITUDES

Source: *National Geographic Society*, Washington, D.C.

| Continent | Highest point | Feet of elevation | Lowest point | Feet below sea level |
|---|---|---|---|---|
| Asia | Mount Everest, Nepal-Tibet | 29,028 | Dead Sea, Israel-Jordan | 1,312 |
| South America | Mount Aconcagua, Argentina | 22,834 | Valdes Peninsula, Argentina | 131 |
| North America | Mount McKinley, Alaska | 20,320 | Death Valley, California | 282 |
| Africa | Kilimanjaro, Tanzania | 19,340 | Lake Assai, Djibouti | 512 |
| Europe | Mount El'brus, USSR | 18,510 | Caspian Sea, USSR | 92 |
| Antarctica | Vinson Massif | 16,864 | Unknown | . . . |
| Australia | Mount Kosciusko, New South Wales | 7,310 | Lake Eyre, South Australia | 52 |

From *The World Almanac and Book of Facts*, 1991.

## SIZE AND DEPTH OF THE OCEANS

| Ocean | Mi² | Greatest Depth (ft) |
|---|---|---|
| Pacific | 63,800,000 | 36,161 |
| Atlantic | 31,800,000 | 30,249 |
| Indian | 28,900,000 | 24,441 |
| Arctic | 5,400,000 | 17,881 |

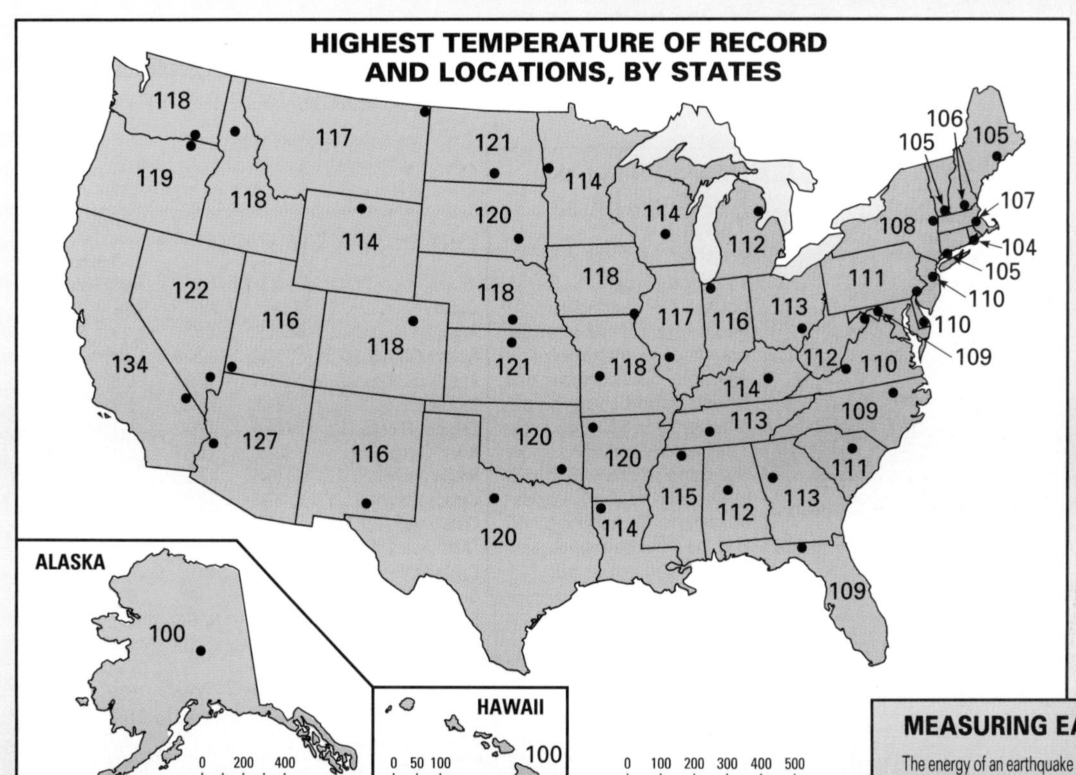


## AVERAGE DAILY TEMPERATURES (°F)

| | San Diego, CA | Milwaukee, WI |
|---|---|---|
| January | 65 | 26 |
| February | 66 | 30 |
| March | 66 | 39 |
| April | 68 | 54 |
| May | 69 | 65 |
| June | 71 | 75 |
| July | 76 | 80 |
| August | 78 | 78 |
| September | 77 | 71 |
| October | 75 | 60 |
| November | 70 | 45 |
| December | 66 | 32 |

## MEASURING EARTHQUAKES

The energy of an earthquake is generally reported using the Richter scale, a system developed by American geologist Charles Richter in 1935, based on measuring the heights of wave measurements on a seismograph.

On the Richter scale, each single-integer increase represents 10 times more ground movement and 30 times more energy released. The change in magnitude between numbers on the scale can be represented by $10^x$ and $30^x$, where *x* represents the change in the Richter scale measure. Therefore, a 3.0 earthquake has 100 times more ground movement and 900 times more energy released than a 1.0 earthquake.

| Richter Scale | |
|---|---|
| 2.5 | Generally not felt, but recorded on seismometers. |
| 3.5 | Felt by many people. |
| 4.5 | Some local damage may occur. |
| 6.0 | A destructive earthquake. |
| 7.0 | A major earthquake. About ten occur each year. |
| 8.0 and above | Great earthquakes. These occur once every five to ten years. |

From *The Universal Almanac* © 1990.

# Economics

## THE SHRINKING VALUE OF THE DOLLAR

**The average retail cost of certain foods in selected years, 1890–1975 (to the nearest cent)**

| | 5 lb flour | 1 lb round steak | 1 qt milk | 10 lb potatoes |
|---|---|---|---|---|
| 1890 | $0.15 | $0.12 | $0.07 | $0.16 |
| 1910 | 0.18 | 0.17 | 0.08 | 0.17 |
| 1930 | 0.23 | 0.43 | 0.14 | 0.36 |
| 1950 | 0.49 | 0.94 | 0.21 | 0.46 |
| 1970 | 0.59 | 1.30 | 0.33 | 0.90 |
| 1975 | 0.98 | 1.89 | 0.45 | 0.99 |

### How to Read the Exchange Rates Table

The first two columns show how many U.S. dollars are needed to equal one unit of another country's currency. For example, on Thursday, one British pound could be exchanged for $1.6045 U.S. The last two columns show the value of one U.S. dollar in another country. For example, on Thursday, $1.00 U.S. had the same value as 0.6232 British pound.

## EXCHANGE RATES

Thursday, July 11, 1991

The New York foreign exchange selling rates below apply to trading among banks in amounts of $1 million and more, as quoted at 3 p.m. Eastern time by Bankers Trust Co. and other sources. Retail transactions provide fewer units of foreign currency per dollar.

| | U.S. $ equiv. | | Currency per U.S. dollar | |
|---|---|---|---|---|
| Country | Thurs. | Wed. | Thurs. | Wed. |
| **Argentina** (Austral) | .0001010 | .0001010 | 9902.00 | 9902.00 |
| **Australia** (Dollar) | .7667 | .7670 | 1.3043 | 1.3038 |
| **Austria** (Schilling) | .07746 | .07846 | 12.91 | 12.75 |
| **Bahrain** (Dinar) | 2.6525 | 2.6525 | .3770 | .3770 |
| **Belgium** (Franc) | .02648 | .02682 | 37.76 | 37.28 |
| **Brazil** (Cruzeiro) | .00320 | .00320 | 312.34 | 312.34 |
| **Britain** (Pound) | 1.6045 | 1.6225 | .6232 | .6163 |
| **Canada** (Dollar) | .8705 | .8716 | 1.1487 | 1.1473 |
| **Chile** (Peso) | .002945 | .002944 | 339.51 | 339.71 |
| **China** (Renmimbi) | .186567 | .186567 | 5.3600 | 5.3600 |
| **Colombia** (Peso) | .001753 | .001751 | 570.38 | 571.00 |
| **Denmark** (Krone) | .1409 | .1425 | 7.0949 | 7.0163 |
| **Ecuador** (Sucre) | .000965 | .000965 | 1036.00 | 1036.00 |
| **Finland** (Markka) | .22656 | .22911 | 4.4138 | 4.3647 |
| **France** (Franc) | .16085 | .16260 | 6.2170 | 6.1500 |
| **Germany** (Mark) | .5451 | .5516 | 1.8345 | 1.8130 |
| **Greece** (Drachma) | .004995 | .005062 | 200.20 | 197.55 |
| **Hong Kong** (Dollar) | .12877 | .12882 | 7.7660 | 7.7625 |
| **India** (Rupee) | .04167 | .04167 | 24.00 | 24.00 |
| **Indonesia** (Rupiah) | .0005123 | .0005123 | 1952.00 | 1952.00 |
| **Ireland** (Punt) | 1.4579 | 1.4765 | .6859 | .6773 |
| **Israel** (Shekel) | .4276 | .4291 | 2.3386 | 2.3302 |
| **Italy** (Lira) | .0007337 | .0007419 | 1363.01 | 1347.85 |
| **Japan** (Yen) | .007211 | .007220 | 138.67 | 138.50 |
| **Jordan** (Dinar) | 1.4535 | 1.4535 | .6880 | .6880 |
| **Kuwait** (Dinar) | z | z | z | z |
| **Lebanon** (Pound) | .001110 | .001110 | 901.00 | 901.00 |
| **Malaysia** (Ringgit) | .3588 | .3588 | 2.7870 | 2.7870 |
| **Malta** (Lira) | 2.9028 | 2.9028 | .3445 | .3445 |
| **Mexico** (Peso) | .0003305 | .0003305 | 3026.00 | 3026.00 |
| **Netherlands** (Guilder) | .4842 | .4918 | 2.0654 | 2.0335 |
| **New Zealand** (Dollar) | .5620 | .5627 | 1.7794 | 1.7771 |
| **Norway** (Krone) | .1410 | .1410 | 7.0937 | 7.0937 |
| **Pakistan** (Rupee) | .0410 | .0410 | 24.40 | 24.40 |
| **Peru** (New Sol) | 1.2231 | 1.2231 | .82 | .82 |
| **Philippines** (Peso) | .03724 | .03724 | 26.85 | 26.85 |
| **Portugal** (Escudo) | .006389 | .006351 | 156.51 | 157.46 |
| **Saudi Arabia** (Riyal) | .26660 | .26660 | 3.7510 | 3.7510 |
| **Singapore** (Dollar) | .5690 | .5700 | 1.7575 | 1.7545 |
| **South Africa** (Rand) | .3454 | .3452 | 2.8950 | 2.8968 |
| **South Korea** (Won) | .0013805 | .0013805 | 724.35 | 724.35 |
| **Spain** (Peseta) | .008702 | .008787 | 114.92 | 113.80 |
| **Sweden** (Krona) | .1508 | .1525 | 6.6308 | 6.5576 |
| **Switzerland** (Franc) | .6285 | .6365 | 1.5910 | 1.5710 |
| **Taiwan** (Dollar) | .037397 | .037355 | 26.74 | 26.77 |
| **Thailand** (Baht) | .03891 | .03891 | 25.70 | 25.70 |
| **Turkey** (Lira) | .0002300 | .0002309 | 4347.01 | 4330.02 |
| **United Arab** (Dirham) | .2723 | .2723 | 3.6730 | 3.6730 |
| **Uruguay** (New Peso) | .000500 | .000500 | 2000.00 | 2000.00 |
| **Venezuela** (Bolivar) | .01803 | .01803 | 55.46 | 55.46 |

z—Not quoted.

## STATE GENERAL SALES AND USE TAXES, JULY 1990

| State | Percent Rate | State | Percent Rate | State | Percent Rate |
|---|---|---|---|---|---|
| Alabama | 4 | Kentucky | 6 | Ohio | 5 |
| Arizona | 5 | Louisiana | 4 | Oklahoma | 4.5 |
| Arkansas | 4.5 | Maine | 5 | Pennsylvania | 5 |
| California | 4.75 | Maryland | 5 | Rhode Island | 7 |
| Colorado | 3 | Massachusetts | 5 | South Carolina | 5 |
| Connecticut | 8 | Michigan | 4 | South Dakota | 4 |
| D.C. | 6 | Minnesota | 6 | Tennessee | 5.5 |
| Florida | 6 | Mississippi | 6 | Texas | 6.25 |
| Georgia | 4 | Missouri | 4.225 | Utah | 5 |
| Hawaii | 4 | Nebraska | 5 | Vermont | 5 |
| Idaho | 5 | Nevada | 5.75 | Virginia | 3.5 |
| Illinois | 6.25 | New Jersey | 7 | Washington | 6.5 |
| Indiana | 5 | New Mexico | 5 | West Virginia | 6 |
| Iowa | 4 | New York | 4 | Wisconsin | 5 |
| Kansas | 4.25 | North Carolina | 3 | Wyoming | 3 |
| | | North Dakota | 5 | | |

NOTE: Alaska, Delaware, Montana, New Hampshire, and Oregon have no statewide sales and use taxes.
Source: *Information Please Almanac* questionnaires to the states.

## UNITED STATES FOREIGN TRADE (millions of dollars)

| | United States Exports | | United States Imports | |
|---|---|---|---|---|
| | 1980 | 1989 | 1980 | 1989 |
| Norway | 843 | 1,037 | 2,632 | 1,991 |
| Turkey | 540 | 2,003 | 175 | 1,372 |
| Ireland | 836 | 2,182 | 411 | 1,566 |
| Australia | 4,093 | 8,331 | 2,509 | 3,873 |
| France | 7,485 | 10,086 | 5,247 | 13,014 |

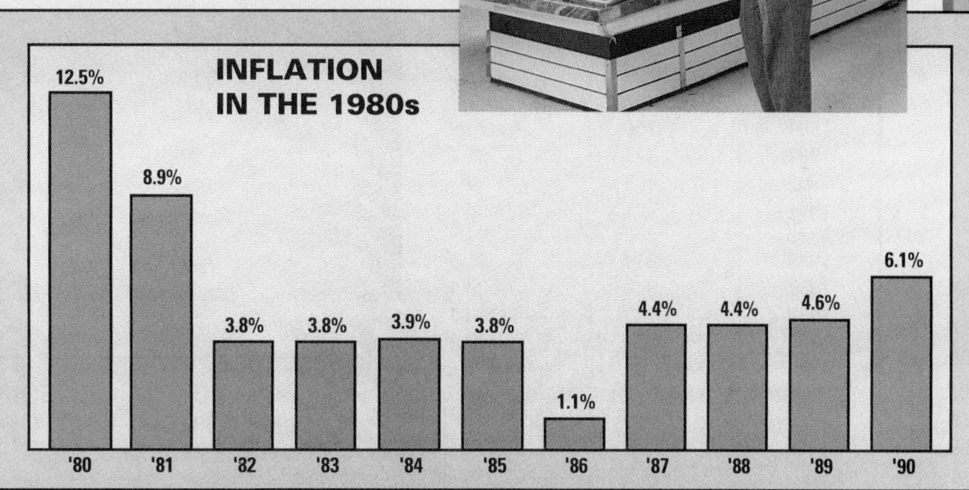


## PUBLIC DEBT OF THE UNITED STATES

Source: U.S. Treasury Department, Bureau of Public Debt

| Fiscal year | Debt (billions) | Per. cap. (dollars) | Interest paid (billions) | Pct. of federal outlays | Fiscal year | Debt (billions) | Per. cap. (dollars) | Interest paid (billions) | Pct. of federal outlays |
|---|---|---|---|---|---|---|---|---|---|
| 1870 | $ 2.4 | $61.06 | — | — | 1976 | $ 620.4 | $ 2,852 | $ 37.1 | 10.0 |
| 1880 | 2.0 | 41.60 | — | — | 1977 | 698.8 | 3,170 | 41.9 | 10.2 |
| 1890 | 1.1 | 17.80 | — | — | 1978 | 771.5 | 3,463 | 48.7 | 10.6 |
| 1900 | 1.2 | 16.60 | — | — | 1979 | 826.5 | 3,669 | 59.8 | 11.9 |
| 1910 | 1.1 | 12.41 | — | — | 1980 | 907.7 | 3,985 | 74.9 | 12.7 |
| 1920 | 24.2 | 228 | — | — | 1981 | 997.9 | 4,338 | 95.6 | 14.1 |
| 1930 | 16.1 | 131 | — | — | 1982 | 1,142.0 | 4,913 | 117.4 | 15.7 |
| 1940 | 43.0 | 325 | $ 1.0 | 10.5 | 1983 | 1,377.2 | 5,870 | 128.8 | 15.9 |
| 1945 | 258.7 | 1,849 | 3.8 | 4.1 | 1984 | 1,572.3 | 6,640 | 153.8 | 18.1 |
| 1950 | 256.1 | 1,688 | 5.7 | 13.4 | 1985 | 1,823.1 | 7,598 | 178.9 | 18.9 |
| 1955 | 272.8 | 1,651 | 6.4 | 9.4 | 1986 | 2,125.3 | 8,774 | 190.2 | 19.2 |
| 1960 | 284.1 | 1,572 | 9.2 | 10.0 | 1987 | 2,350.3 | 9,615 | 195.4 | 19.5 |
| 1965 | 313.8 | 1,613 | 11.3 | 9.6 | 1988 | 2,602.3 | 10,534 | 214.1 | 20.1 |
| 1970 | 370.1 | 1,814 | 19.3 | 9.9 | 1989 | 2,857.4 | 11,545 | 240.8 | 25.0 |
| 1975 | 533.2 | 2,475 | 32.7 | 9.8 | 1990 | 3,233.3 | 13,000 | 264.8 | 21.1 |

Note: Through 1976 the fiscal year ended June 30. From 1977 on, fiscal year ends Sept. 30.

# Entertainment

## U.S. TV STATIONS AND HOUSEHOLDS WITH TV, 1950–1989

| Year | Commercial TV Stations on Air | Households with TV (millions) |
|---|---|---|
| 1950 | 98 | 3.8 |
| 1955 | 411 | 32.0 |
| 1960 | 515 | 45.2 |
| 1965 | 569 | 53.8 |
| 1970 | 677 | 60.1 |
| 1975 | 706 | 71.5 |
| 1980 | 734 | 77.8 |
| 1981 | 756 | 79.9 |
| 1982 | 777 | 81.5 |
| 1983 | 813 | 83.3 |
| 1984 | 841 | 83.8 |
| 1985 | 883 | 84.9 |
| 1986 | 919 | 85.9 |
| 1987 | 968 | 87.4 |
| 1988 | 1,028 | 88.6 |
| 1989 | 1,064 | 90.4 |

## 20 TOP-GROSSING FEATURE FILMS OF 1989

| Title | Distributor | Box-Office Gross (millions) |
|---|---|---|
| The Abyss | 20th Century Fox | $ 54 |
| Back to the Future II | Universal | 95 |
| Batman | Warner Bros. | 251 |
| Dead Poets Society | BV/Touchstone | 95 |
| Field of Dreams | Universal | 63 |
| Ghostbusters II | Columbia | 113 |
| Harlem Nights | Paramount | 55 |
| Honey, I Shrunk the Kids | BV/Touchstone | 130 |
| Indiana Jones and the Last Crusade | Paramount | 196 |
| Lethal Weapon 2 | Warner Bros. | 147 |
| Look Who's Talking | TriStar | 115 |
| Major League | Paramount | 50 |
| National Lampoon's Christmas Vacation | Warner Bros. | 61 |
| Parenthood | Universal | 96 |
| Pet Sematary | Paramount | 57 |
| Sea of Love | Universal | 57 |
| Star Trek V | Paramount | 52 |
| Turner & Hooch | BV/Touchstone | 71 |
| Uncle Buck | Universal | 64 |
| When Harry Met Sally | Columbia | 92 |

## 25 CONTEMPORARY ENTERTAINERS

| Name | Birth Date |
|---|---|
| Carson, Johnny | 10/23/25 |
| Charles, Ray | 9/23/30 |
| Clark, Dick | 11/30/29 |
| Cole, Natalie | 2/6/50 |
| Cosby, Bill | 7/12/37 |
| DeVito, Danny | 11/17/44 |
| Donohue, Phil | 12/21/35 |
| Field, Sally | 11/6/46 |
| Fox, Michael J. | 6/9/61 |
| Franklin, Aretha | 3/25/42 |
| Griffith, Andy | 6/1/26 |
| Hopkins, Anthony | 12/31/37 |
| Jackson, Michael | 8/19/58 |
| Joel, Billy | 5/9/49 |
| John, Elton | 3/25/47 |
| King, B. B. | 9/16/25 |
| Leno, Jay | 4/28/50 |
| McCartney, Paul | 6/18/42 |
| Murphy, Eddie | 4/3/61 |
| Nelson, Willie | 4/30/33 |
| Nicholson, Jack | 4/22/37 |
| Parton, Dolly | 1/19/46 |
| Stallone, Sylvester | 7/6/46 |
| Tandy, Jessica | 6/7/09 |
| Wonder, Stevie | 5/18/50 |

Source: *Information Please Almanac*, 1989

## LONGEST BROADWAY RUNS

| Show | Performances |
|---|---|
| 1. A Chorus Line (1975–90) | 6,137 |
| 2. Oh! Calcutta (1976–89) | 5,959 |
| 3. 42nd Street (1980–89) | 3,486 |
| 4. Grease (1972–80) | 3,388 |
| 5. Cats (1982– ) | 3,269 |
| 6. Fiddler on the Roof (1964–72) | 3,242 |
| 7. Life with Father (1939–47) | 3,224 |
| 8. Tobacco Road (1933–41) | 3,182 |
| 9. Hello, Dolly! (1964–71) | 2,844 |
| 10. My Fair Lady (1956–62) | 2,717 |
| 11. Annie (1977–83) | 2,377 |
| 12. Man of La Mancha (1965–71) | 2,328 |
| 13. Abie's Irish Rose (1922–27) | 2,327 |
| 14. Oklahoma! (1943–48) | 2,212 |
| 15. Pippin (1971–77) | 1,944 |
| 16. South Pacific (1949–54) | 1,925 |
| 17. Magic Show (1974–78) | 1,920 |
| 18. Deathtrap (1978–82) | 1,792 |
| 19. Gemini (1977–81) | 1,788 |
| 20. Harvey (1944–49) | 1,775 |
| 21. Dancin' (1978–82) | 1,774 |
| 22. La Cage aux Folles (1983–87) | 1,761 |
| 23. Hair (1968–72) | 1,750 |
| 24. The Wiz (1975–79) | 1,672 |
| 25. Born Yesterday (1946–49) | 1,642 |

## THE RECORDING INDUSTRY ASSOCIATION OF AMERICA

### 1990 Sales Profile

**Released September 1991**

PERCENTAGE OF DOLLAR VALUE BY TYPE OF MUSIC

| Type of Music | 1986 | 1987 | 1988 | 1989 | 1990 |
|---|---|---|---|---|---|
| Rock | 46.8 | 47.2 | 46.2 | 42.9 | 37.4 |
| Pop | 14.2 | 12.9 | 15.2 | 14.4 | 13.6 |
| Urban Contemporary | 10.1 | 11.6 | 13.3 | 14.0 | 18.3 |
| Country | 9.7 | 9.5 | 7.4 | 6.8 | 8.8 |
| Classical | 5.9 | 2.6 | 3.5 | 4.3 | 4.1 |
| Jazz | 0.4 | 5.2 | 4.7 | 5.7 | 5.2 |
| Gospel | 3.0 | 3.9 | 2.5 | 3.1 | 2.4 |
| Other | 5.6 | 6.6 | 6.0 | 8.0 | 9.3 |

PERCENTAGE OF DOLLAR VALUE BY CONFIGURATION

| Configuration | 1986 | 1987 | 1988 | 1989 | 1990 |
|---|---|---|---|---|---|
| Cassette | 57.4 | 59.3 | 60.4 | 50.4 | 48.4 |
| Cassette Singles | 0 | 0 | 1.0 | 2.7 | 2.6 |
| LPs | 28.4 | 18.2 | 13.9 | 8.3 | 4.3 |
| 7″ Singles | 1.5 | 0.8 | 3.4 | 0.5 | 0.3 |
| 12″ Singles | 2.1 | 1.5 | 2.2 | 1.6 | 0.7 |
| CDs 3″ Singles | 0 | 0 | 0 | 0.4 | 1.1 |
| CDs 5″ Full | 10.1 | 19.8 | 18.6 | 35.9 | 42.5 |

| TOTAL U.S. SALES IN BILLIONS OF DOLLARS—1986 THROUGH 1990 | 1986 | 1987 | 1988 | 1989 | 1990 |
|---|---|---|---|---|---|
| | 4.6 | 5.5 | 6.2 | 6.4 | 7.5 |

# Health and Sports

## NUMBER OF CALORIES BURNED PER HOUR BY PEOPLE OF DIFFERENT BODY WEIGHTS

| | Calories Burned per Hour | | |
|---|---|---|---|
| Exercise | 110 lb | 154 lb | 198 lb |
| Martial arts | 620 | 790 | 960 |
| Racquetball (2 people) | 610 | 775 | 945 |
| Basketball (full-court game) | 585 | 750 | 910 |
| Skiing—cross country (5 mi/h) | 550 | 700 | 850 |
| downhill | 465 | 595 | 720 |
| Running—8-min mile | 550 | 700 | 850 |
| 12-min mile | 515 | 655 | 795 |
| Swimming—crawl, 45 yd/min | 540 | 690 | 835 |
| crawl, 30 yd/min | 330 | 420 | 510 |
| Stationary bicycle—15 mi/h | 515 | 655 | 795 |
| Aerobic dancing—intense | 515 | 655 | 795 |
| moderate | 350 | 445 | 540 |
| Walking—5 mi/h | 435 | 555 | 675 |
| 3 mi/h | 235 | 300 | 365 |
| 2 mi/h | 145 | 185 | 225 |
| Calisthenics—intense | 435 | 555 | 675 |
| moderate | 350 | 445 | 540 |
| Scuba diving | 355 | 450 | 550 |
| Hiking—20-lb pack, 4 mi/h | 355 | 450 | 550 |
| 20-lb pack, 2 mi/h | 235 | 300 | 365 |
| Tennis—singles, recreational | 335 | 425 | 520 |
| doubles, recreational | 235 | 300 | 365 |
| Ice skating | 275 | 350 | 425 |
| Roller skating | 275 | 350 | 425 |

## SIZES AND WEIGHTS OF BALLS USED IN VARIOUS SPORTS

| Type | Diameter (cm) | Average Weight (g) |
|---|---|---|
| Baseball | 7.6 | 145 |
| Basketball | 24.0 | 596 |
| Croquet ball | 8.6 | 340 |
| Field hockey ball | 7.6 | 160 |
| Golf ball | 4.3 | 46 |
| Handball | 4.8 | 65 |
| Soccer ball | 22.0 | 425 |
| Softball, large | 13.0 | 279 |
| Softball, small | 9.8 | 187 |
| Table tennis ball | 3.7 | 2 |
| Tennis ball | 6.5 | 57 |
| Volleyball | 21.9 | 256 |

## CALORIES USED PER MINUTE BY PEOPLE OF DIFFERENT BODY WEIGHTS

| Activity | 100 lb | 120 lb | 150 lb | 200 lb |
|---|---|---|---|---|
| Volleyball | 2.3 | 2.7 | 3.4 | 4.6 |
| Walking (3 mi/h) | 2.7 | 3.2 | 4.0 | 5.4 |
| Tennis | 4.5 | 5.4 | 6.8 | 9.1 |
| Swimming (crawl) | 5.8 | 6.9 | 8.7 | 11.6 |
| Skiing (cross-country) | 7.2 | 8.7 | 10.8 | 14.5 |

## LIFE EXPECTANCY

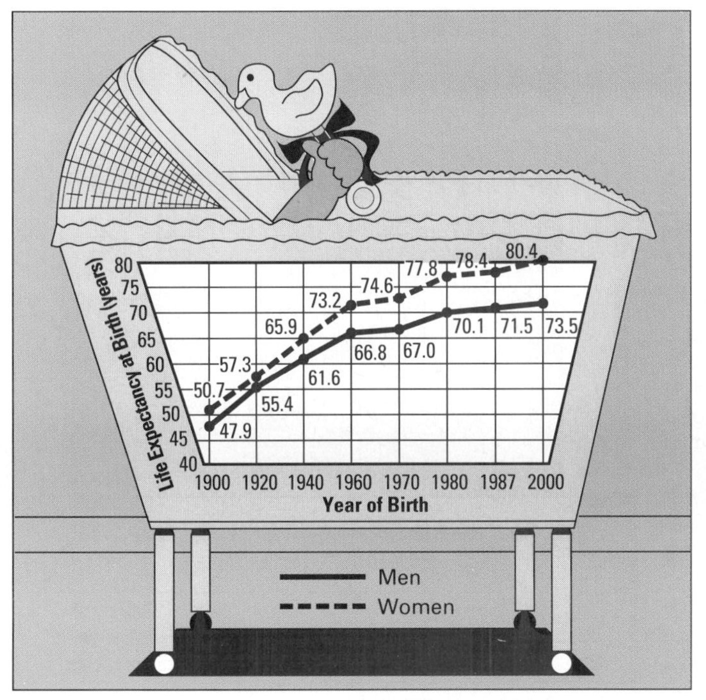


## NFL STADIUMS

| Name, Location | Capacity |
|---|---|
| Anaheim Stadium, Anaheim, Cal. | 69,007 |
| Arrowhead Stadium, Kansas City, Mo. | 78,067 |
| Astrodome, Houston, Tex. | 60,502 |
| Atlanta-Fulton County Stadium | 59,673 |
| Candlestick Park, San Francisco, Cal. | 65,729 |
| Cleveland Stadium | 80,098 |
| Foxboro Stadium, Mass. | 60,794 |
| Giants Stadium, E. Rutherford, N.J. | 76,891 |
| Hoosier Dome, Indianapolis, Ind. | 60,127 |
| Robert F. Kennedy Stadium, Wash., D.C. | 55,672 |
| Kingdome, Seattle, Wash. | 64,984 |
| Lambeau Field, Green Bay, Wis. | 59,543 |
| Los Angeles Memorial Coliseum | 92,488 |
| Louisiana Superdome, New Orleans | 69,065 |
| Metrodome, Minneapolis | 63,000 |
| Mile High Stadium, Denver, Col. | 76,273 |
| Milwaukee County Stadium | 56,051 |
| Pontiac Silverdome, Mich. | 80,500 |
| Rich Stadium, Buffalo, N.Y. | 80,290 |
| Riverfront Stadium, Cincinnati, Oh. | 59,754 |
| Joe Robbie Stadium, Miami, Fla. | 75,000 |
| San Diego Jack Murphy Stadium, San Diego | 60,750 |
| Soldier Field, Chicago, Ill. | 66,949 |
| Sun Devil Stadium, Tempe, Ariz. | 72,000 |
| Tampa Stadium, Tampa, Fla. | 74,314 |
| Texas Stadium, Irving, Tex. | 65,024 |
| Three Rivers Stadium, Pittsburgh, Pa. | 59,000 |
| Veterans Stadium, Philadelphia, Pa. | 65,356 |

## THE SPORTS WITH THE MOST INJURIES

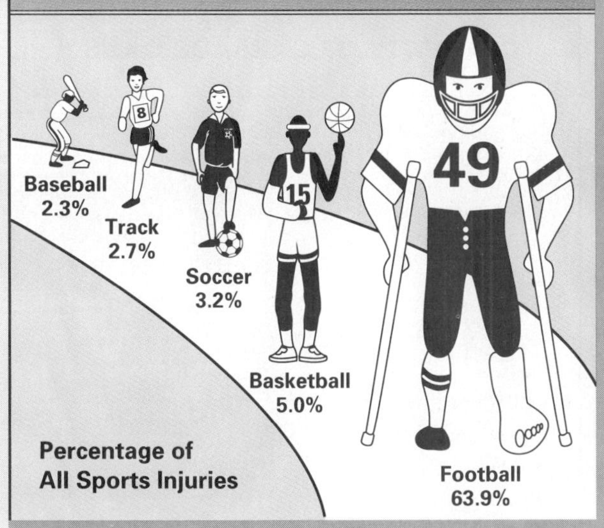

# DATA FILE Science

## FUEL ECONOMY AND CARBON DIOXIDE

The better your car's gas mileage, the less carbon dioxide your car will emit into the environment. As stated earlier, cars and light trucks emit one-fifth of all carbon dioxide in the United States, one of the main causes of the greenhouse effect.

Here are estimates of the amount of carbon dioxide emissions over a single car's lifetime, courtesy of the Energy Conservation Coalition, a project of Environmental Action:

$17\frac{29}{100}$ Tons

60 mi/gal

$25\frac{93}{100}$ Tons

45 mi/gal

$39\frac{29}{100}$ Tons

26 mi/gal

$57\frac{3}{4}$ Tons

18 mi/gal

## QUANTITIES OF HAZARDOUS WASTE BURNED IN U.S.

| Incinerator Type | Number of Facilities | Quantity of Hazardous Waste Burned, lbs/yr |
|---|---|---|
| Commercial incinerators | 17 | 1.3 billion (1) |
| Captive/on-site incinerators | 154 | 2.3 billion (1) |
| Cement kilns | 25–30 | 1.8 billion (2) |
| Aggregate kilns | 6 | 1.2 billion (2) |
| Boilers/other furnaces | 900 + | 1.0 billion (2) |
| Total | 1,100 + | 7.6 billion |

(1) Highum 1990. Data from review of states' capacity assurance plans.
(2) Holloway 1990. Data from U.S. EPA'S Office of Solid Waste and Emergency Response.

## CONTENTS OF GARBAGE CANS

Our garbage cans are filled with a diverse mix (by weight):

paper and paperboard $\frac{9}{25}$
yard wastes $\frac{19}{100}$
glass $\frac{2}{25}$
metals $\frac{9}{100}$
food $\frac{9}{100}$
plastics $\frac{2}{25}$
wood/fabric $\frac{1}{25}$
rubber and leather $\frac{3}{100}$
textiles $\frac{1}{50}$
other $\frac{4}{250}$
household hazardous waste $\frac{1}{250}$

From: *Nontoxic, Natural, & Earthwise* by Debra Lynn Dood.

## HALF-LIVES OF SELECT SUBSTANCES

| Substance | Half-Life |
|---|---|
| uranium-238 | 4.5 billion years |
| carbon-14 | 5,730 years |
| radium-226 | 1,620 years |
| strontium-90 | 28 years |
| hydrogen-3 | 12.3 years |
| polonium-210 | 138 days |
| thorium-234 | 25 days |
| iodine-131 | 8 days |
| bismuth-210 | 5 days |
| radium-222 | 4 days |
| sodium-24 | 15 hours |
| lead-212 | 10.6 hours |
| nitrogen-13 | 10 minutes |
| polonium-194 | 0.7 seconds |

## PLANETARY DIAMETERS AND DISTANCES FROM SUN

| Planet | Equatorial Diameter (km) | Average Distance from Sun (km) |
|---|---|---|
| Mercury | 4,880 | 58,000,000 |
| Venus | 12,100 | 108,000,000 |
| Earth | 12,756 | 150,000,000 |
| Mars | 6,780 | 228,000,000 |
| Jupiter | 142,800 | 778,000,000 |
| Saturn | 120,000 | 1,427,000,000 |
| Uranus | 50,800 | 2,870,000,000 |
| Neptune | 48,600 | 4,497,000,000 |
| Pluto | 2,200 | 5,900,000,000 |

## PERCENTAGE OF MAIN ELEMENTS MAKING UP EARTH'S CRUST

| | | | |
|---|---|---|---|
| Oxygen | 47% | Calcium | 4% |
| Silicon | 28% | Magnesium | 2% |
| Aluminum | 8% | Sodium | 3% |
| Iron | 5% | Potassium | 3% |

## LARGEST MOONS OF THE SOLAR SYSTEM

| Moon | Planet | Diameter (mi) |
|---|---|---|
| Titan | Saturn | 3,500 |
| Triton | Neptune | 3,300 |
| Callisto | Jupiter | 3,220 |
| Ganymede | Jupiter | 3,200 |
| Io | Jupiter | 2,310 |
| Moon | Earth | 2,160 |
| Europa | Jupiter | 1,950 |

# United States

## UNITED STATES RESIDENT POPULATION CHANGE BY REGION

(to the nearest hundred thousand)

| | 1980 | 1990 |
|---|---|---|
| Northeast | 49,100,000 | 50,800,000 |
| Midwest | 58,900,000 | 59,700,000 |
| South | 75,300,000 | 85,400,000 |
| West | 43,200,000 | 52,800,000 |

## PERCENT DISTRIBUTION OF U.S. POPULATION GROWTH FOR SELECTED AREAS, BY DECADE: 1900 TO 1990

| | 1900–1910 | 1910–1920 | 1920–1930 | 1930–1940 | 1940–1950 | 1950–1960 | 1960–1970 | 1970–1980 | 1980–1990 |
|---|---|---|---|---|---|---|---|---|---|
| Northeast and Midwest Regions | 52.3 | 57.5 | 54.4 | 34.6 | 40.8 | 44.1 | 39.0 | 10.1 | 11.2 |
| Remainder of South and West Regions | 35.4 | 27.8 | 22.9 | 40.3 | 28.7 | 23.1 | 28.8 | 48.2 | 34.5 |
| California, Florida, and Texas | 12.3 | 14.7 | 22.7 | 25.1 | 30.5 | 32.8 | 32.2 | 41.7 | 54.3 |

## U.S. CITIES, 1988

| Population Size | Number of Cities | Total Population (millions) |
|---|---|---|
| 1 million or more | 7 | 19.1 |
| 500,000–1 million | 17 | 11.5 |
| 250,000–500,000 | 37 | 13.2 |
| 100,000–250,000 | 125 | 18.4 |
| 50,000–100,000 | 300 | 20.6 |
| 25,000–50,000 | 575 | 19.9 |
| 10,000–25,000 | 1,323 | 20.7 |
| Under 10,000 | 16,868 | 28.7 |

## 1990 POPULATION AND NUMBER OF REPRESENTATIVES, BY STATE

**TOTAL POPULATION 249,632,692**

| State | Apportionment Population | Number of Representatives Based on the 1990 Census | Change from 1980 Apportionment |
|---|---|---|---|
| U.S. TOTAL* | 249,022,783 | 435 | |
| Alabama | 4,062,608 | 7 | – |
| Alaska | 551,947 | 1 | – |
| Arizona | 3,677,985 | 6 | +1 |
| Arkansas | 2,362,239 | 4 | – |
| California | 29,839,250 | 52 | +7 |
| Colorado | 3,307,912 | 6 | – |
| Connecticut | 3,295,669 | 6 | – |
| Delaware | 668,696 | 1 | – |
| Florida | 13,003,362 | 23 | +4 |
| Georgia | 6,508,419 | 11 | +1 |
| Hawaii | 1,115,274 | 2 | – |
| Idaho | 1,011,986 | 2 | – |
| Illinois | 11,466,682 | 20 | −2 |
| Indiana | 5,564,228 | 10 | – |
| Iowa | 2,787,424 | 5 | −1 |
| Kansas | 2,485,600 | 4 | −1 |
| Kentucky | 3,698,969 | 6 | −1 |
| Louisiana | 4,238,216 | 7 | −1 |
| Maine | 1,233,223 | 2 | – |
| Maryland | 4,798,622 | 8 | – |
| Massachusetts | 6,029,051 | 10 | −1 |
| Michigan | 9,328,784 | 16 | −2 |
| Minnesota | 4,387,029 | 8 | – |
| Mississippi | 2,586,443 | 5 | – |
| Missouri | 5,137,804 | 9 | – |
| Montana | 803,655 | 1 | −1 |
| Nebraska | 1,584,617 | 3 | – |
| Nevada | 1,206,152 | 2 | – |
| New Hampshire | 1,113,915 | 2 | – |
| New Jersey | 7,748,634 | 13 | −1 |
| New Mexico | 1,521,779 | 3 | – |
| New York | 18,044,505 | 31 | −3 |
| North Carolina | 6,657,630 | 12 | +1 |
| North Dakota | 641,364 | 1 | – |
| Ohio | 10,887,325 | 19 | −2 |
| Oklahoma | 3,157,604 | 6 | – |
| Oregon | 2,853,733 | 5 | – |
| Pennsylvania | 11,924,710 | 21 | −2 |
| Rhode Island | 1,005,984 | 2 | – |
| South Carolina | 3,505,707 | 6 | – |
| South Dakota | 699,999 | 1 | – |
| Tennessee | 4,896,641 | 9 | – |
| Texas | 17,059,805 | 30 | +3 |
| Utah | 1,727,784 | 3 | – |
| Vermont | 564,964 | 1 | – |
| Virginia | 6,216,568 | 11 | +1 |
| Washington | 4,887,941 | 9 | +1 |
| West Virginia | 1,801,625 | 3 | −1 |
| Wisconsin | 4,906,745 | 9 | – |
| Wyoming | 455,975 | 1 | – |

*Total population, not including the District of Columbia.

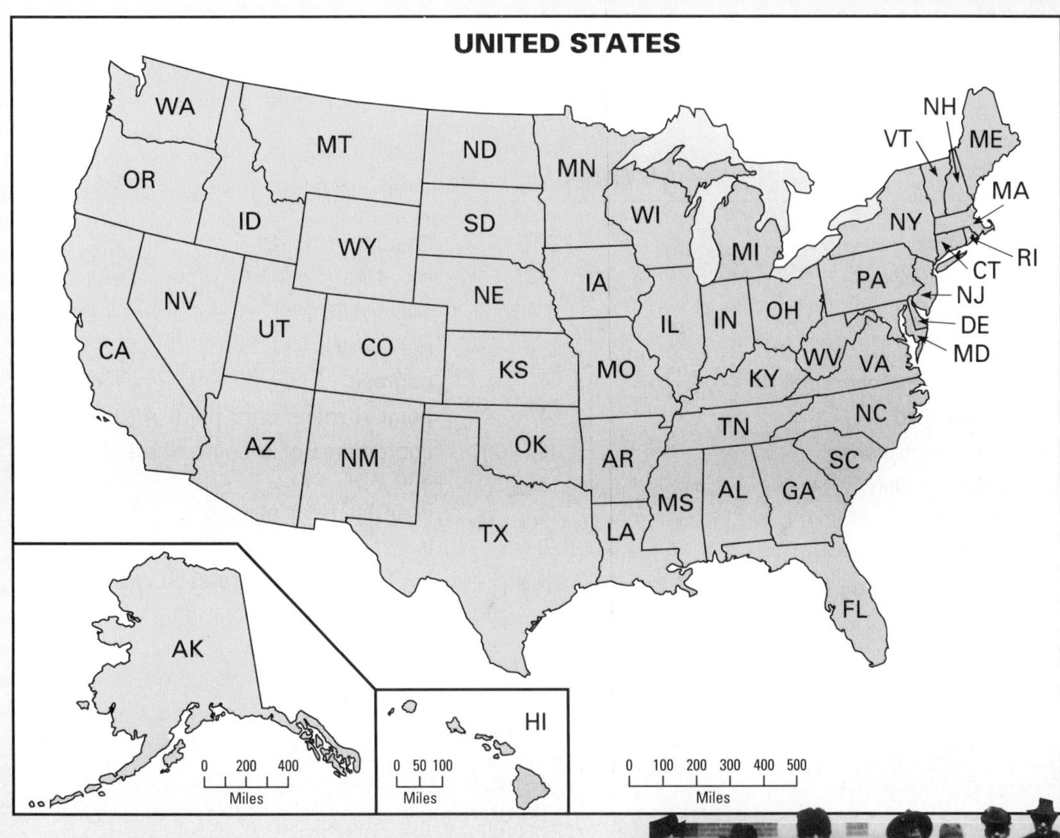


## IMMIGRATION TO THE U.S. IN THE 1980s

| Country | Legal Immigrants |
|---|---|
| Mexico | 975,657 |
| Philippines | 477,485 |
| China | 306,108 |
| South Korea | 302,782 |
| Vietnam | 266,027 |
| India | 221,977 |
| Dominican Republic | 209,899 |
| Jamaica | 184,481 |
| United Kingdom | 140,119 |
| Cuba | 135,142 |
| El Salvador | 133,938 |
| Canada | 132,296 |
| Haiti | 118,510 |
| Iran | 101,267 |

Source: U.S. Dept. of Justice *1989 Statistical Yearbook of the Immigration and Naturalization Service* (1990).

## AREAS OF STATES

| State | Total Area square miles | Land Area square miles | Forested Land square miles |
|---|---|---|---|
| Arizona | 114,000 | 113,508 | 28,900 |
| California | 158,700 | 156,300 | 62,700 |
| Colorado | 104,100 | 103,600 | 34,800 |
| Nevada | 110,600 | 109,900 | 12,000 |
| New Mexico | 121,600 | 121,300 | 28,200 |
| Utah | 84,900 | 82,100 | 24,300 |

# Useful Mathematical Data

## SYMBOLS

|  |  |
|---|---|
| $=$ | is equal to |
| $\neq$ | is not equal to |
| $\cong$ | is congruent to |
| $\sim$ | is similar to |
| $\approx$ | is approximately equal to |
| $\parallel$ | is parallel to |
| $\perp$ | is perpendicular to |
| $>$ | is greater than |
| $<$ | is less than |
| $\geq$ | is greater than or equal to |
| $\leq$ | is less than or equal to |
| ( ) | parentheses: "Do this operation first." |
| $3^2$ | exponent (2); base (3) |
| % | percent |
| $0.\overline{3}$ | repeating decimal |
| $\pi$ | pi: $\approx \frac{22}{7}$ or 3.14 |
| $\sqrt{\phantom{x}}$ | square root |
| $\lvert x \rvert$ | absolute value of $x$ |
| $\overleftrightarrow{AB}$ | line $AB$ |

|  |  |
|---|---|
| $\overline{AB}$ | line segment $AB$ |
| $\overrightarrow{AB}$ | ray $AB$ |
| $\angle ABC$ | angle $ABC$ |
| ∟ | right angle |
| $^\circ$ | degrees |
| $A \rightarrow A'$ | point $A$ maps onto point $A'$ |
| $(1, -2)$ | coordinates of a point where $x = 1$ and $y = -2$ |
| $\in$ | is an element of |
| $\notin$ | is not an element of |
| { } | set inclusion; empty set |
| $\subseteq$ | is a subset of |
| $\nsubseteq$ | is not a subset of |
| $\phi$ | empty set |
| $U$ | universal set |
| $A'$ | the complement of set $A$ |
| $A \cup B$ | the union of set $A$ and set $B$ |
| $A \cap B$ | the intersection of set $A$ and set $B$ |
| $p \rightarrow q$ | If $p$, then $q$. |
| $\sim p$ | not $p$ |

## PRIME NUMBERS TO 499

| 2 | 3 | 5 | 7 | 11 | 13 | 17 | 19 |
|---|---|---|---|---|---|---|---|
| 23 | 29 | 31 | 37 | 41 | 43 | 47 | 53 |
| 59 | 61 | 67 | 71 | 73 | 79 | 83 | 89 |
| 97 | 101 | 103 | 107 | 109 | 113 | 127 | 131 |
| 137 | 139 | 149 | 151 | 157 | 163 | 167 | 173 |
| 179 | 181 | 191 | 193 | 197 | 199 | 211 | 223 |
| 227 | 229 | 233 | 239 | 241 | 257 | 263 | 269 |
| 271 | 277 | 281 | 283 | 293 | 307 | 311 | 313 |
| 317 | 331 | 337 | 347 | 349 | 353 | 359 | 367 |
| 373 | 379 | 383 | 389 | 397 | 401 | 409 | 419 |
| 421 | 431 | 433 | 439 | 443 | 449 | 457 | 461 |
| 463 | 467 | 479 | 487 | 491 | 499 | | |

# FORMULAS/USEFUL EQUATIONS

## Geometric

$l$ = length
$b$ = base
$s$ = side
$d$ = diameter
$B$ = area of the base of a three-dimensional figure
$w$ = width
$h$ = height
$r$ = radius

## Perimeter: $P$

Rectangle: $P = 2l + 2w$
Square: $P = 4s$
Regular Polygon with $n$ sides: $P = ns$
Triangle: $P = s_1 + s_2 + s_3$
Circle (Circumference): $C = \pi d$ or $C = 2\pi r$

## Area: $A$

Rectangle: $A = lw$
Parallelogram: $A = bh$
Trapezoid: $A = \frac{1}{2}h(b_1 + b_2)$
Triangle: $A = \frac{1}{2}bh$
Circle: $A = \pi r^2$
Square: $A = s^2$

## Surface Area: $SA$

Rectangular Prism:
$SA = 2(lw + lh + wh)$
Cube: $SA = 6s^2$
Cylinder: $SA = 2\pi rh + 2\pi r^2$
Sphere: $SA = 4\pi r^2$
Cone: $SA = \pi rs + \pi r^2$

## Volume: $V$

Prism: $V = Bh$
Rectangular Prism: $V = lwh$
Cube: $V = s^3$
Cylinder: $V = \pi r^2 h$
Sphere: $V = \frac{4}{3}\pi r^3$
Pyramid: $V = \frac{1}{3}Bh$
Cone: $V = \frac{1}{3}\pi r^2 h$

## Right Triangles

Pythagorean Theorem: In a right triangle, as shown.

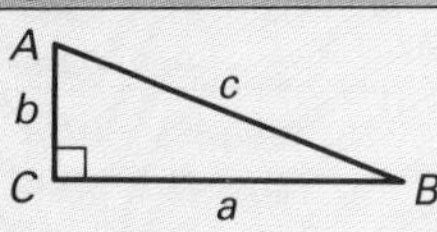


$$a^2 + b^2 = c^2 \text{ or } c = \sqrt{a^2 + b^2}$$

Trigonometric Ratios:

$\sin\angle A = \frac{a}{c}$ $\sin\angle B = \frac{b}{c}$

$\cos\angle A = \frac{b}{c}$ $\cos\angle B = \frac{a}{c}$

$\tan\angle A = \frac{a}{b}$ $\tan\angle B = \frac{b}{a}$

## Temperature

$C$: degrees Celsius $F$: degrees Fahrenheit

$C = \frac{5}{9}(F - 32)$ $F = \frac{9}{5}C + 32$

## Distance

$d$: distance $r$: rate $t$: time

$d = rt$ $r = \frac{d}{t}$ $t = \frac{d}{r}$

On a coordinate plane, the distance between points $P(x_1, y_1)$ and $Q(x_2, y_2)$ is

$$PQ = \sqrt{(x_2 - x_1)^2 + (y_2 - y_1)^2}$$

## Laws of Exponents

$a^m \times a^n = a^{m+n}$ $\frac{a^m}{a^n} = a^{m-n}$
$(a^m)^n = a^{m \times n}$

## Probability

Factorial: $n! = n(n - 1)(n - 2) \ldots (1)$
Permutations: ${}_nP_n = n!$

$${}_nP_r = \frac{n!}{(n - r)!}$$

Combination: ${}_nC_r = \frac{n!}{(n - r)!\, r!}$

## MEASUREMENT: METRIC UNITS

### Length

1 centimeter (cm) = 10 millimeters (mm)
10 centimeters or 100 millimeters = 1 decimeter (dm)
10 decimeters or 100 centimeters = 1 meter (m)
1,000 meters = 1 kilometer (km)

### Area

100 square millimeters ($mm^2$) = 1 square centimeter ($cm^2$)
10,000 square centimeters = 1 square meter ($m^2$)
100 square meters = 1 are (a)
10,000 square meters = 1 hectare (ha)

### Volume

1,000 cubic millimeters ($mm^3$) = 1 cubic centimeter ($cm^3$)
1,000 cubic centimeters = 1 cubic decimeter ($dm^3$)
1,000,000 cubic centimeters = 1 cubic meter ($m^3$)

### Capacity

1,000 milliliters = 1 liter (L)
1,000 liters = 1 kiloliter (kL)

### Mass

1,000 milligrams (mg) = 1 gram (g)
1,000 grams = 1 kilogram (kg)
1,000 kilograms = 1 metric ton (T)

### Temperature

0°C = freezing point of water
37°C = normal body temperature
100°C = boiling point of water

## MEASUREMENT: CUSTOMARY UNITS

### Length

12 inches (in.) = 1 foot (ft)
3 feet or 36 inches = 1 yard (yd)
1760 yards or 5280 feet = 1 mile (mi)
6,076 feet = 1 nautical mile

### Area

144 square inches ($in.^2$) = 1 square foot ($ft^2$)
9 square feet = 1 square yard ($yd^2$)
4,840 square yards = 1 acre (A)

### Volume

1,728 cubic inches ($in.^3$) = 1 cubic foot ($ft^3$)
27 cubic feet = 1 cubic yard ($yd^3$)

### Capacity

8 fluid ounces (fl oz) = 1 cup
2 cups = 1 pint (pt)
2 pints = 1 quart (qt)
4 quarts = 1 gallon (gal)

### Weight

16 ounces (oz) = 1 pound (lb)
2,000 pounds = 1 ton (T)

### Temperature

32°F = freezing point of water
98.6°F = normal body temperature
212°F = boiling point of water

### Metric/Customary Comparisons

5 centimeters is about the same length as 2 inches.
1 meter is slightly longer than 1 yard.
5 kilometers is about the same length as 3 miles.

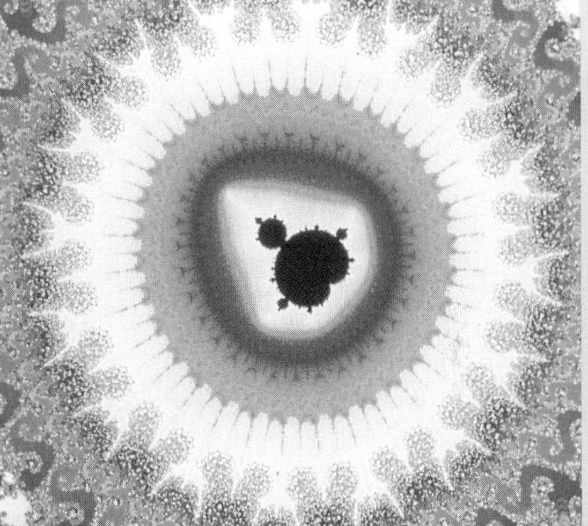

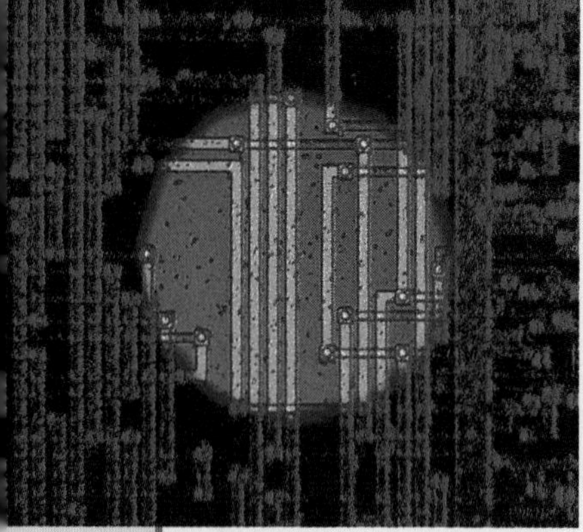

## TABLE OF SQUARES AND APPROXIMATE SQUARE ROOTS

| $n$ | $n^2$ | $\sqrt{n}$ | $n$ | $n^2$ | $\sqrt{n}$ |
|---|---|---|---|---|---|
| 1 | 1 | 1.000 | 51 | 2601 | 7.141 |
| 2 | 4 | 1.414 | 52 | 2704 | 7.211 |
| 3 | 9 | 1.732 | 53 | 2809 | 7.280 |
| 4 | 16 | 2.000 | 54 | 2916 | 7.348 |
| 5 | 25 | 2.236 | 55 | 3025 | 7.416 |
| 6 | 36 | 2.449 | 56 | 3136 | 7.483 |
| 7 | 49 | 2.646 | 57 | 3249 | 7.550 |
| 8 | 64 | 2.828 | 58 | 3364 | 7.616 |
| 9 | 81 | 3.000 | 59 | 3481 | 7.681 |
| 10 | 100 | 3.162 | 60 | 3600 | 7.746 |
| 11 | 121 | 3.317 | 61 | 3721 | 7.810 |
| 12 | 144 | 3.464 | 62 | 3844 | 7.874 |
| 13 | 169 | 3.606 | 63 | 3969 | 7.937 |
| 14 | 196 | 3.742 | 64 | 4096 | 8.000 |
| 15 | 225 | 3.873 | 65 | 4225 | 8.062 |
| 16 | 256 | 4.000 | 66 | 4356 | 8.124 |
| 17 | 289 | 4.123 | 67 | 4489 | 8.185 |
| 18 | 324 | 4.243 | 68 | 4624 | 8.246 |
| 19 | 361 | 4.359 | 69 | 4761 | 8.307 |
| 20 | 400 | 4.472 | 70 | 4900 | 8.367 |
| 21 | 441 | 4.583 | 71 | 5041 | 8.426 |
| 22 | 484 | 4.690 | 72 | 5184 | 8.485 |
| 23 | 529 | 4.796 | 73 | 5329 | 8.544 |
| 24 | 576 | 4.899 | 74 | 5476 | 8.602 |
| 25 | 625 | 5.000 | 75 | 5625 | 8.660 |
| 26 | 676 | 5.099 | 76 | 5776 | 8.718 |
| 27 | 729 | 5.196 | 77 | 5929 | 8.775 |
| 28 | 784 | 5.292 | 78 | 6084 | 8.832 |
| 29 | 841 | 5.385 | 79 | 6241 | 8.888 |
| 30 | 900 | 5.477 | 80 | 6400 | 8.944 |
| 31 | 961 | 5.568 | 81 | 6561 | 9.000 |
| 32 | 1024 | 5.657 | 82 | 6724 | 9.055 |
| 33 | 1089 | 5.745 | 83 | 6889 | 9.110 |
| 34 | 1156 | 5.831 | 84 | 7056 | 9.165 |
| 35 | 1225 | 5.916 | 85 | 7225 | 9.220 |
| 36 | 1296 | 6.000 | 86 | 7396 | 9.274 |
| 37 | 1369 | 6.083 | 87 | 7569 | 9.327 |
| 38 | 1444 | 6.164 | 88 | 7744 | 9.381 |
| 39 | 1521 | 6.245 | 89 | 7921 | 9.434 |
| 40 | 1600 | 6.325 | 90 | 8100 | 9.487 |
| 41 | 1681 | 6.403 | 91 | 8281 | 9.539 |
| 42 | 1764 | 6.481 | 92 | 8464 | 9.592 |
| 43 | 1849 | 6.557 | 93 | 8649 | 9.644 |
| 44 | 1936 | 6.633 | 94 | 8836 | 9.695 |
| 45 | 2025 | 6.708 | 95 | 9025 | 9.747 |
| 46 | 2116 | 6.782 | 96 | 9216 | 9.798 |
| 47 | 2209 | 6.856 | 97 | 9409 | 9.849 |
| 48 | 2304 | 6.928 | 98 | 9604 | 9.899 |
| 49 | 2401 | 7.000 | 99 | 9801 | 9.950 |
| 50 | 2500 | 7.071 | 100 | 10000 | 10.000 |

## TABLE OF TRIGONOMETRIC RATIOS

| Angle | Sine | Cosine | Tangent | Angle | Sine | Cosine | Tangent |
|---|---|---|---|---|---|---|---|
| 0° | 0.0000 | 1.0000 | 0.0000 | 45° | 0.7071 | 0.7071 | 1.0000 |
| 1° | 0.0175 | 0.9998 | 0.0175 | 46° | 0.7193 | 0.6947 | 1.0355 |
| 2° | 0.0349 | 0.9994 | 0.0349 | 47° | 0.7314 | 0.6820 | 1.0724 |
| 3° | 0.0523 | 0.9986 | 0.0524 | 48° | 0.7431 | 0.6691 | 1.1106 |
| 4° | 0.0698 | 0.9976 | 0.0699 | 49° | 0.7547 | 0.6561 | 1.1504 |
| 5° | 0.0872 | 0.9962 | 0.0875 | 50° | 0.7660 | 0.6428 | 1.1918 |
| 6° | 0.1045 | 0.9945 | 0.1051 | 51° | 0.7771 | 0.6293 | 1.2349 |
| 7° | 0.1219 | 0.9925 | 0.1228 | 52° | 0.7880 | 0.6157 | 1.2799 |
| 8° | 0.1392 | 0.9903 | 0.1405 | 53° | 0.7986 | 0.6018 | 1.3270 |
| 9° | 0.1564 | 0.9877 | 0.1584 | 54° | 0.8090 | 0.5878 | 1.3764 |
| 10° | 0.1736 | 0.9848 | 0.1763 | 55° | 0.8192 | 0.5736 | 1.4281 |
| 11° | 0.1908 | 0.9816 | 0.1944 | 56° | 0.8290 | 0.5592 | 1.4826 |
| 12° | 0.2079 | 0.9781 | 0.2126 | 57° | 0.8387 | 0.5446 | 1.5399 |
| 13° | 0.2250 | 0.9744 | 0.2309 | 58° | 0.8480 | 0.5299 | 1.6003 |
| 14° | 0.2419 | 0.9703 | 0.2493 | 59° | 0.8572 | 0.5150 | 1.6643 |
| 15° | 0.2588 | 0.9659 | 0.2679 | 60° | 0.8660 | 0.5000 | 1.7321 |
| 16° | 0.2756 | 0.9613 | 0.2867 | 61° | 0.8746 | 0.4848 | 1.8040 |
| 17° | 0.2924 | 0.9563 | 0.3057 | 62° | 0.8829 | 0.4695 | 1.8807 |
| 18° | 0.3090 | 0.9511 | 0.3249 | 63° | 0.8910 | 0.4540 | 1.9626 |
| 19° | 0.3256 | 0.9455 | 0.3443 | 64° | 0.8988 | 0.4384 | 2.0503 |
| 20° | 0.3420 | 0.9397 | 0.3640 | 65° | 0.9063 | 0.4226 | 2.1445 |
| 21° | 0.3584 | 0.9336 | 0.3839 | 66° | 0.9135 | 0.4067 | 2.2460 |
| 22° | 0.3746 | 0.9272 | 0.4040 | 67° | 0.9205 | 0.3907 | 2.3559 |
| 23° | 0.3907 | 0.9205 | 0.4245 | 68° | 0.9272 | 0.3746 | 2.4751 |
| 24° | 0.4067 | 0.9135 | 0.4452 | 69° | 0.9336 | 0.3584 | 2.6051 |
| 25° | 0.4226 | 0.9063 | 0.4663 | 70° | 0.9397 | 0.3420 | 2.7475 |
| 26° | 0.4384 | 0.8988 | 0.4877 | 71° | 0.9455 | 0.3256 | 2.9042 |
| 27° | 0.4540 | 0.8910 | 0.5095 | 72° | 0.9511 | 0.3090 | 3.0777 |
| 28° | 0.4695 | 0.8829 | 0.5317 | 73° | 0.9563 | 0.2924 | 3.2709 |
| 29° | 0.4848 | 0.8746 | 0.5543 | 74° | 0.9613 | 0.2756 | 3.4874 |
| 30° | 0.5000 | 0.8660 | 0.5774 | 75° | 0.9659 | 0.2588 | 3.7321 |
| 31° | 0.5150 | 0.8572 | 0.6009 | 76° | 0.9703 | 0.2419 | 4.0108 |
| 32° | 0.5299 | 0.8480 | 0.6249 | 77° | 0.9744 | 0.2250 | 4.3315 |
| 33° | 0.5446 | 0.8387 | 0.6494 | 78° | 0.9781 | 0.2079 | 4.7046 |
| 34° | 0.5592 | 0.8290 | 0.6745 | 79° | 0.9816 | 0.1908 | 5.1446 |
| 35° | 0.5736 | 0.8192 | 0.7002 | 80° | 0.9848 | 0.1736 | 5.6713 |
| 36° | 0.5878 | 0.8090 | 0.7265 | 81° | 0.9877 | 0.1564 | 6.3138 |
| 37° | 0.6018 | 0.7986 | 0.7536 | 82° | 0.9903 | 0.1392 | 7.1154 |
| 38° | 0.6157 | 0.7880 | 0.7813 | 83° | 0.9925 | 0.1219 | 8.1443 |
| 39° | 0.6293 | 0.7771 | 0.8098 | 84° | 0.9945 | 0.1045 | 9.5144 |
| 40° | 0.6428 | 0.7660 | 0.8391 | 85° | 0.9962 | 0.0872 | 11.4301 |
| 41° | 0.6561 | 0.7547 | 0.8693 | 86° | 0.9976 | 0.0698 | 14.3007 |
| 42° | 0.6691 | 0.7431 | 0.9004 | 87° | 0.9986 | 0.0523 | 19.0811 |
| 43° | 0.6820 | 0.7314 | 0.9325 | 88° | 0.9994 | 0.0349 | 28.6363 |
| 44° | 0.6947 | 0.7193 | 0.9657 | 89° | 0.9998 | 0.0175 | 57.2900 |
| 45° | 0.7071 | 0.7071 | 1.0000 | 90° | 1.0000 | 0.0000 | — |

# Credits

## PHOTO ACKNOWLEDGMENTS

### CONTENTS

p. v: National Park Service, Ellis Island Immigration Museum; p. vi: Dale E. Boyer/Photo Researchers, Inc.; p. vii: J.M. Labat/National Audubon Society Collection, Photo Researchers, Inc.; p. viii: John Carter/Photo Researchers, Inc.; p. ix: Robert Isear/Science Source, Photo Researchers, Inc.; p. x: Photri; p. xi: Douglas Faulkner/Photo Researchers, Inc.; p. xii: Photri

### CHAPTER 1

p. 2: National Park Service, Ellis Island Immigration Museum (all); p. 6: David Grossman/Photo Researchers, Inc.; p. 7: Mark Duran/Greyhound Lines, Inc.; p. 8: Syd Greenberg/Photo Researchers, Inc.; p. 11: Junebug Clark/Photo Researchers, Inc.; p. 13: Jackson & Perkins; p. 15: J. Love/Photo Researchers, Inc. (top); Lee F. Snyder/Photo Researchers, Inc. (bottom); p. 16: David Weintraub/San Francisco Convention & Visitors Bureau; p. 18: Oldsmobile; p. 19: Dr. P. James, University of Toledo and NASA; p. 20: Larry Mulvehill/Photo Researchers, Inc.; p. 21: Spencer Grant/Photo Researchers, Inc.; p. 25: Y. Arthus Bertrand/Photo Researchers, Inc.; p. 26: Blair Seitz/Photo Researchers, Inc.; p. 27: Eunice Harris/Photo Researchers, Inc.; p. 29: J.M. Barrey/Photo Researchers, Inc.; p. 31: Jackson & Perkins; p. 32: Joanne Savio/Falletta Associates; p. 35: Michael P. Gadomski/Photo Researchers, Inc.

### CHAPTER 2

p. 42: Art Twomey/Photo Researchers, Inc. (top); Andrew J. Martinez/Photo Researchers, Inc. (bottom); p. 46: Margot Granitsas/Photo Researchers, Inc.; p. 49: Dale E. Boyer/Photo Researchers, Inc.; p. 51: Francois Gohier/Photo Researchers, Inc.; p. 55: Mike James/Photo Researchers, Inc.; p. 56: Art Twoney/Photo Researchers, Inc.; p. 57: NASA (top); Courtesy of Evelyn B. Granville (bottom); p. 60: Renee Lynn/Photo Researchers, Inc.; p.: 61: Shenandoah Valley Travel Association; p. 62: Bonnie Rauch/Photo Researchers, Inc.; p. 63: Oldsmobile (top); David R. Frazier/Photo Researchers, Inc. (bottom); p. 66: Art Stein/Photo Researchers, Inc.; p. 67: Richard Hutchings/Photo Researchers, Inc.; p. 71: Andy Levin/Photo Researchers, Inc.; p. 75: David Parker/Photo Researchers, Inc.

### CHAPTER 3

p. 82: The registered trademark TRIVIAL PURSUIT® and related proprietary rights are owned by Horn Abbot Ltd. Used with permission. CLUE® is a registered trademark of Waddington Games Ltd., used with permission of Parker Brothers, the exclusive licensee. MONOPOLY® is a registered trademark of Tonka Corporation. © 1992 Parker Brothers, a division of Tonka Corporation. Used with permission. Edward Lettau/Photo Researchers, Inc. (top right); Dale E. Boyer/Photo Researchers, Inc. (bottom right); p. 84: Rappahannock Area Development Commission; p. 89: Greater Houston Partnership; p. 91: Greater Houston Partnership; p. 93: Carol Simowitz/San Francisco Convention & Visitors Bureau; p. 97: David Weintraub/San Francisco Convention & Visitors Bureau; p. 101: Kerrick James/San Francisco Convention & Visitors Bureau; p. 109: Eunice Harris/Photo Researchers, Inc.; p. 123: George E. Jones III/Photo Researchers, Inc.

### CHAPTER 4

p. 130: Dennis Purse/Photo Researchers, Inc. (top); Laima Drukis/Photo Researchers, Inc. (bottom); p. 131: Jeff Isaac Greenberg/Photo Researchers, Inc.; p. 132: Eunice Harris/Photo Researchers, Inc.; p. 133: Richard Hutchings/Photo Researchers, Inc.; p. 134: Lawrence Migdale/Photo Researchers, Inc.; p. 135: Jeff Isaac Greenberg/Photo Researchers, Inc.; p. 136: David R. Frazier/Photo Researchers, Inc.; p. 138: Lawrence Migdale/Photo Researchers, Inc.; p. 141: Spencer Grant/Photo Researchers, Inc.; p. 143: Renee Lynn/Photo Researchers, Inc.; p. 145: Blair Seitz/Photo Researchers, Inc.; p. 146: Blair Seitz/Photo Researchers, Inc.; p. 148: Jeff Isaac Greenberg/Photo Researchers, Inc.; p. 149: Richard Hutchings/Photo Researchers, Inc.; p. 152: Rafael Macia/Photo Researchers, Inc.; p. 154: Tim Davis/Photo Researchers, Inc.; p. 156: Borrfdon/ Explorer/Photo Researchers, Inc.; p. 159: Laima Druskis/Photo Researchers, Inc.; p. 160: David R. Frazier/Photo Researchers, Inc.; p. 163: Rappahannock Area Development Commission; p. 164: Kerrick James/San Francisco Convention & Visitors Bureau; p. 166: Mark Burnett/Photo Researchers, Inc.

### CHAPTER 5

p. 174: Roy Whitehead/Photo Researchers, Inc. (left top); Jessie Cohen/National Zoological Park, Smithsonian Institution (left bottom); Leonard Lee Rue III, The National Audubon Society

Collection/Photo Researchers, Inc. (bottom center); Jessie Cohen/Office of Graphics and Exhibits, National Zoological Park (bottom right); p. 175: National Wildlife Federation (top); Patrice Thomas/Photo Researchers, Inc. (bottom); p. 179: Stephen J. Kraseman, The National Audubon Society Collection/Photo Researchers, Inc. (top left); Bill Dyer, The National Audubon Society Collection/Photo Researchers, Inc. (left bottom); William H. Mullins, The National Audubon Society Collection/Photo Researchers, Inc. (right); p. 182: Alan Carey, The National Audubon Society Collection/Photo Researchers, Inc.; p. 183: James Foote/Photo Researchers, Inc.; p. 185: Jessie Cohen/National Zoological Park, Smithsonian Institution; p. 187: Lawrence Migdale/Photo Researchers, Inc.; p. 188: Tim Davis/Photo Researchers, Inc.; p. 190: Jessie Cohen/National Zoological Park, Smithsonian Institution (both); p. 191: Bill Bachman, The National Audubon Society Collection/Photo Researchers, Inc.; p. 194: Renee Lynn/Photo Researchers, Inc.; p. 195: International Business Machines Corporation; p. 197: Leonard Lee Rue III, The National Audubon Society Collection/Photo Researchers, Inc.; p. 199: Blair Seitz/Photo Researchers, Inc.; p. 200: Garry D. McMichael/Photo Researchers, Inc.; p. 203: Blair Seitz/Photo Researchers, Inc.; p. 207: Ken Eward/Photo Researchers, Inc.

**CHAPTER 6**

p. 214: Michael P. Gadomski/Photo Researchers, Inc. (top); Lawrence Migdale/Photo Researchers, Inc. (bottom); Vinnie Fish/Photo Researchers, Inc. (bottom center); G. Buttner/Naturbild/OKAPIA/ Photo Researchers, Inc. (bottom right); p. 215: Andy Levin/Photo Researchers, Inc. (top); Joyce Photographics/Photo Researchers, Inc. (bottom); p. 217: P. Plisson/Photo Researchers, Inc.; p. 220: Michael P. Gadomski/Photo Researchers, Inc.; p. 223: Ellan Young/Photo Researchers, Inc.; p. 224: Jerry Cooke/Photo Researchers, Inc.; p. 225: Joseph Nettis/Photo Researchers, Inc.; p. 229: Rappahannock Area Development Commission; p. 235: Shenandoah Valley Travel Association; p. 237: C. Seghers, Photo Researchers, Inc. (top); Jerome Wexler/Photo Researchers, Inc. (bottom); p. 239: Sylvain Grandadam/Photo Researchers, Inc.; p. 240: Bonnie Sue Rauch/Photo Researchers, Inc.; p. 241: Lawrence Migdale/Photo Researchers, Inc.; p. 245: Anthony Mercieca/ Photo Researchers, Inc.; p. 249: Joseph Nettis/Photo Researchers, Inc.; p. 252: Brian Brake/Photo Researchers, Inc.; p. 254: Blair Seitz/Photo Researchers, Inc.; p. 255: Tom Hollyman/Photo Researchers, Inc.; p. 256: George Whiteley/Photo Researchers, Inc.; p. 258: Kent & Donna Dannen/Photo Researchers, Inc.; p. 259: Charlie Ott/Photo Researchers, Inc.

**CHAPTER 7**

p. 266: Oldsmobile (all); p. 267: Oldsmobile (both); p. 271: Oldsmobile; p. 275: Greater Houston Partnership; p. 279: Gregory K. Scott/Photo Researchers, Inc.; p. 281: International Business Machines Corporation; p. 285: Blair Seitz/Photo Researchers, Inc; p. 286: Mark Burnett/Photo Researchers, Inc.; p. 287: Michael P. Gadomski/Photo Researchers, Inc.; p. 289: Will McIntyre/Photo Researchers, Inc.; p. 290: Andy Levin/Photo Researchers, Inc.; p. 293: Craig Buchanan/San Francisco Convention & Visitors Bureau; p. 297: John G. Ross/Photo Researchers, Inc.; p. 299: Blair Seitz/Photo Researchers, Inc.; p. 301: Will McIntyre/Photo Researchers, Inc.; p. 303: Allen Green/Photo Researchers, Inc.

**CHAPTER 8**

p. 310: Michael Abbey/Science Source/Photo Researchers, Inc.; p. 311: International Business Machines Corporation; p. 313: Nicholas De Sciose/Photo Researchers, Inc.; p. 315: International Business Machines Corporation; p. 317: International Business Machines Corporation; p. 319: Oldsmobile; p. 320: Louis Goldman/Photo Researchers, Inc.; p. 323: P. Plisson/Photo Researchers, Inc.; p. 325: A. Fabiani/Photo Researchers, Inc.; p. 326: Exroy-Explorer/Photo Researchers, Inc.; p. 327: Rafael Macia/Photo Researchers, Inc.; p. 328: Blair Seitz/Photo Researchers, Inc.; p. 330: Will McIntyre/Photo Researchers, Inc.; p. 335: Sylvain Grandadam/Photo Researchers, Inc.; p. 336: International Business Machines Corporation; p. 338: International Business Machines Corporation

**CHAPTER 9**

p. 346: D & I MacDonald/Photri (top); David R. Frazier/Photo Researchers, Inc. (bottom); p. 347: Michael & Elvan Habicht/Photri; p. 351: Larry Mulvehill/Photo Researchers, Inc.; p. 355: B. Grunzweig/Photo Researchers, Inc.; p. 357: Photri; p. 359: Kees Van Ben Berg/Photo Researchers, Inc.; p. 361: Blair Seitz/Photo Researchers, Inc.; p. 365: Photri; p. 366: D & I MacDonald/ Photri; p. 370: Robert J. Bennett/Photri; p. 371: Joanne Savio/Falletta Associates; p. 373: Dennis MacDonald/Photri; p. 374: Joseph Nettis/Photo Researchers, Inc.; p. 379: Joanne Savio/Falletta Associates (both)

**CHAPTER 10**
p. 390: International Business Machines Corporation (right); p. 390: Robert J. Bennett/Photri (left); p. 391: Ken Biggs/Photo Researchers, Inc. (center); Louis Goldman/Photo Researchers, Inc. (top); Photri (bottom left); Stevie Grand/Photo Researchers, Inc. (bottom right); p. 393: Larry Mulvehill/ Photo Researchers, Inc.; p. 395: International Business Machines Corporation; p. 398: Richard Hutchings/Photo Researchers, Inc.; p. 401: Robert W. Ginn/Photri; p: 405: William E. Townsend, Jr./ Photo Researchers, Inc.; p. 407: Bruce Roberts/Photo Researchers, Inc.; p. 408: NASA; p. 409: Alan Carey/Photo Researchers, Inc.; p. 410: Photri; p. 413: Catherine Ursillo/Photo Researchers, Inc.; p. 415: Ellsworth/Photri; p. 417: Photri; p. 419: International Business Machines Corporation

**CHAPTER 11**
p. 426: Photri (both); p. 427: Blair Seitz/Photo Researchers, Inc. (left); Earl Roberge/Photo Researchers, Inc.(right); p. 429: Scott Berner/Photri; p. 431: Lani/Photri; p. 434: Stan Goldblatt/Photo Researchers, Inc.; p. 437: Photri; p. 438: Photri; p. 442: Photri; p. 444: Charles Philip/Photri

**CHAPTER 12**
p. 458: Skjold/Photri (far left); Dennis MacDonald/Photri (center left); Skjold/Photri (far right); Vic Bider/Photri (center right); p. 459: Photri (left); Skjold/Photri (right); p. 461: International Business Machines Corporation; p. 465: International Business Machines Corporation; p. 467: MacDonald Photography/Photri; p. 473: Joanne Savio/Falletta Associates; p. 469: International Business Machines Corporation; p. 476: Joanne Savio/Falletta Associates; p. 482: Photri; p. 483 Atlanta Chamber of Commerce

**CHAPTER 13**
p. 490: Blair Seitz/Photo Researchers, Inc.; p. 496: C. Seghers/Photo Researchers, Inc.; p. 501: Photri; p. 503: Chevrolet; p. 504: Jim Dixon/Photo Researchers, Inc.; p. 507: Joanne Savio/Falletta Associates; p. 509: Will McIntyre/Photo Researchers, Inc.; p. 511: Lawrence Migdale/Photo Researchers, Inc.; p. 513: Lawrence Migdale/Photo Researchers, Inc.; p. 514: Kerrick James/San Francisco Convention & Visitors Bureau; p. 517: C. Seghers/Photo Researchers, Inc.; p. 518: Greater Houston Partnership

**CHAPTER 14**
p. 526: Blair Seitz/Photo Researchers, Inc. (bottom right) Richard Hutchings/Photo Researchers, Inc. (top right); Will McIntyre/Photo Researchers, Inc. (left); p. 527: Maslowski/Photo Researchers, Inc.; p. 532: Melissa Grimes-Guy/Photo Researchers, Inc.; p. 535: Christian Pinson/Photo Researchers, Inc.; p. 539: Photri; p. 543: Oldsmobile; p. 544: Will & Deni McIntyre/Photo Researchers, Inc.

**DATA FILES**
p. 558: Jessie Cohen/National Zoological Park, Smithsonian Institution (right); Greater Houston Partnership (left); p. 559: William H. Mullins/Photo Researchers, Inc.; p. 560: Bruce Roberts/Photo Researchers, Inc. (center); Michael & Elvan Habicht/Photri (bottom); Greater Houston Partnership (top); p. 561: Borje Svensson/San Francisco Convention & Visitors Bureau (bottom); Rappahannock Area Development Commission (top); p. 563: Joseph Nettis/Photo Researchers, Inc.; p.564: Ted Kerasote/Photo Researchers, Inc. (top); Bill Curtsinger/Photo Researchers, Inc. (bottom); p. 565: Renee Lynn/Photo Researchers, Inc.; p. 566: Bill Bachmann/Photri; p. 567: Richard Hutchings/ Photo Researchers, Inc.; p. 568: Jeff Isaac Greenberg/Photo Researchers, Inc.; p. 569: Will & Deni McIntyre/Photo Researchers, Inc.; p. 570 Michael Ponzini/Baseball Hall of Fame (bottom); Richard Raphael/Baseball Hall of Fame (top); p. 571: Tim Davis/Photo Researchers, Inc.; p. 572: Michael P. Gadomski/Photo Researchers, Inc. (bottom); Rafael Macia/Photo Researchers, Inc. (top); p. 573: NASA (both); p. 574: Greater Houston Partnership;. p. 575: National Park Service, Ellis Island Immigration Museum; p. 576: International Business Machines Corporation; p. 578: International Business Machines Corporation (both); p. 579: International Business Machines Corporation; p. 580: C. Falco/Photo Researchers, Inc.; p. 581: David Parker/Photo Researchers, Inc.

All statistical information that appears in the Data Files is in the public domain unless otherwise noted. We gratefully acknowledge the following sources:

*Games,* November, 1984, New York, NY, "Which Way Is Up?" p. 562.
*Introducing Music,* by Otto Karolyi, © 1965 Penguin Books, Middlesex, England, "Pitch Ranges of

Musical Instruments," p. 563.
*1989 Information Please Almanac,* © Houghton Mifflin Co., New York, NY, "Twenty-five Contemporary Entertainers," p. 569.
*1989 Statistical Yearbook of the Immigration and Naturalization Service,* © 1990, U.S. Dept. of Justice, "Immigration to the United States in the 1980s," p. 575.
*1990 Census Profile,* U.S. Dept. of Commerce, Bureau of the Census,"Percent Distribution of U.S. Population Growth for Selected Areas, by Decade: 1900 to 1990," p. 574.
*1992 Information Please Almanac,* © Houghton Mifflin Co., New York, NY, "State General Sales and Use Taxes," p. 566.
*Playing with Fire—Hazardous Waste Incineration,* © 1990 Greenpeace U.S.A., Washington, DC, "Quantities of Hazardous Waste Burned in U.S.," p. 572.
*Prevention's Giant Book of Health Facts,* © 1991, Rodale Press, Inc., Emmaus, PA, "Life Expectancy," p. 571; "Number of Calories Burned by People of Different Weights," p. 570; "The Sports with the Most Injuries," p. 571.
*Projects in Wood,* by David Field, © 1985, G. P. Puntam's Sons, New York, NY, "Dimensions of Cross Sections of Finished Lumber," p. 560.
*Reader's Digest Book of Facts,* © 1987, The Reader's Digest Association, Inc., Pleasantville, NY, "How Fast Do Insects Fly?" p. 559; "Longest Recorded Life Spans of Some Animals," p. 558.
The Recording Industry Association of America,"The Recording Industry Association of America 1990 Sales Profile," p. 569.
*The Universal Almanac,* © 1990 Andrews and McMeel, Kansas City, MO, "Measuring Earthquakes," p. 565.
U.S. Bureau of the Census, "1990 Population and Number of Representatives by State," p. 574.
*The World Almanac and Book of Facts,* 1991 © 1990 Pharos Books, "Highest and Lowest Continental Altitudes," p. 564.
*The World Almanac and Book of Facts,* 1992 © 1991 Pharos Books, "All-Time Bestselling Children's Books," p. 562; "Major U.S. Public Zoological Parks," p. 558; "NFL Stadiums," p. 571; "Public Debt of the U.S.," p. 567; "Public Libraries." p. 563; "Top 25 American Kennel Club Registrations," p. 559.

# Selected Answers

## Chapter 1 MATHEMATICS AND NUMBER SENSE

### Section 1–1, pages 6–7

**Exercises** **1.** 34 **3.** 6 **5.** 0.5 **7.** 14 **9.** 16 **11.** 38 **13.** 48 **15.** 12.5 **17.** 4 **19.** 17.5 **21.** 93.75 **23.** 8 + (8 – 2); $14 **25.** 1.7 **27.** 61.2 **29.** $\frac{17}{18}$ **31.** 15.6 **33.** thirty–seven and five tenths plus a number *n* **35.** the sum of two numbers *c* and *t* times 2 **37.** $(2 \times 88) \times 5 = 176 \times 5 = 880$ **39.** Answers will vary.

### Section 1–2, page 11

**Exercises** **1.** 5 **3.** $\frac{3}{4}$ **5.** 24 **7.** 0 **9.** 0 **11.** 9,300 **13.** 65 **15.** 9.7 in. **17.** 900 **19.** 105 **21.** 90 **23.** 367 **25.** 81 **27.** 220 **29.** no: $10 - 5 \neq 5 - 10$. Examples will vary. no: $25 - (12 - 2) \neq (25 - 12) - 2$ **31.** yes: $4(7 - 2) = (4 \times 7) - (4 \times 2)$ no: $3(7 \times 5) \neq (3 \times 7)(3 \times 5)$ Examples will vary.

### Section 1–3, pages 14–15

**Exercises** **1.** 3.1, 3.3, 3.5 **3.** 60, 45, 30 **5.** 16, 11, 7 **7.** 0.038; 0.0038; 0.00038 **9.** 23,328; $255,150 **11.** a 5 × 5 square, a 6 × 6 square, a 7 × 7 square; 144 **13.** Answers will vary. Samples are given. **a.** 6, 7, 8, 9 or 9, 10, 11, 12 or 16, 17, 18, 19 **b.** 30, 90, 93, 279 or 31, 93, 98, 286

**Problem Solving Applications** **1.** 1, 1, 2, 3, 5, 8, 13, 21, 34, 55, 89, 144, 233, 377, 610 **3.** 54; 1 less than tenth term **5.** 1, 1, 2, 3, 1, 0, 1, 1, 2, 3, 1, 0, 1, 1, 2 The numbers 1, 1, 2, 3, 1, 0 repeat.

### Section 1–4, pages 18–19

**Exercises** **1.** 5, 26, 40, 68, 145 **3.** 0, $\frac{1}{5}$, 1, 4, 9 **5.** $y \div 2$ **7.** 54 lb **9. a.** $492.50 **b.** $2,250.00 **11.** 23; 59; 122 **13.** $f(x) = x - 8$

### Section 1–5, page 21

**Problems** **1.** 217 dozen **3.** 674.7 mi. **5.** Dominican Republic, Cuba, Canada **7.** Bears–12 games; Cheetahs–9 games; Lions–3 games

### Section 1–6, pages 24–25

**Exercises** **1.** 79% **3.** 85% **5.** $83\frac{1}{3}\%$ **7.** 60 **9.** 48 **11.** 64% **13.** 1,500,655 **15.** $2,748; 11.4% **17.** They are lower by 1%.

**Problem Solving Applications** **1.** $1.40 **3.** $1.04 **5. a.** $31.35 **b.** $411.34 **7.** $5.71 **9.** $21.38 **11.** $23.67; $22.59

### Section 1–7, pages 28–29

**Exercises** **1. a.** 4,000 **3. b.** 102 **5. b.** 350 **7.** 1,400 **9.** 80,000 **11.** estimate; no **13.** exact; 5 **15.** less than; estimate: 7,200 – 4,800 = 2,400 **17.** chicken **19.** 8 h; underestimate the distance, but overestimate the racer's pace to make sure the photographer gets there before the racer.

### Section 1–8, page 31

**Problems** **1.** $69.77 **3.** Yes **5.** $7.05 or $7.06 **7.** 1,024 people

### Section 1–9, pages 34–35

**Exercises** **1.** Golden Gate; about 22 million **3.** about 10 million **5.** about 4.5 million **7.** $7.00 **9.** $3.50

**11.**

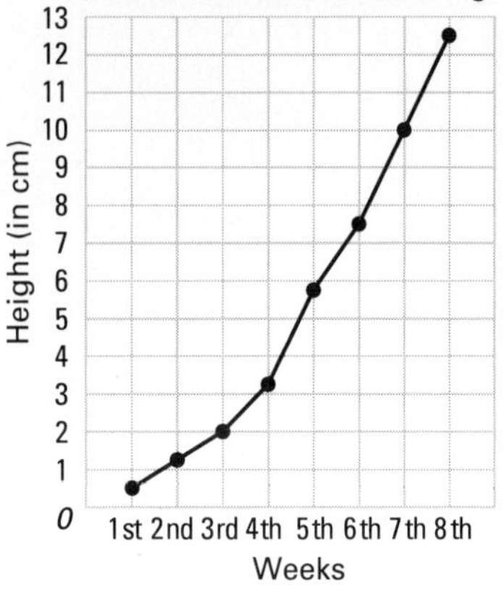

13. It has vertical bars that are divided to reflect percentages according to the three geographical parts of the U.S. **15.** The values for the first bar reflect the heavy immigration into the eastern states from Europe. The values for the last bar reflect the heavy immigration into the western states from Asia and South America.

## CHAPTER 2 REAL NUMBERS AND ALGEBRAIC EXPRESSIONS

### Section 2–1, page 47

**Exercises** **1.**

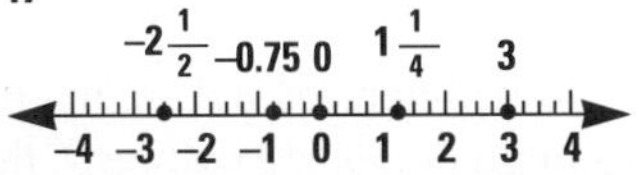


**3.**

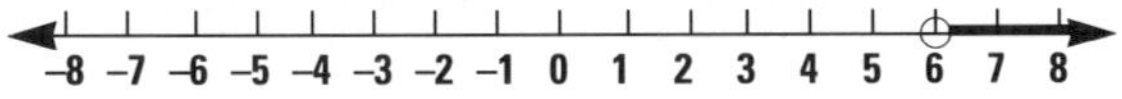


**5.** 4 **7.** −2,018 ft **9.** > **11.** −2.4 **13.** $-\frac{1}{5}$
**15.** $\{-2\frac{3}{4}, -\frac{1}{2}, 2, 2\frac{1}{4}\}$ **17.** real numbers greater than −3 **19.** false; there is no integer between 1 and 2 **21.** false; pi is a real number that is not a rational number **23.** false; −3 is less than zero, and $|-3| = 3$, so if $n = -3$, then $|n| = -n$

### Section 2–2, pages 50–51

**Exercises** **1.** −14 **3.** 0 **5.** −19 **7.** 5 **9.** 17 **11.** −50 **13.** −6 **15.** 6 **17.** = **19.** < **21.** −$954 million **23.** Australia **25.** a; −6 **27.** d; 8

### Section 2–3, page 55

**Exercises** **1.** −35 **3.** −9 **5.** 41 **7.** −9 **9.** 80 **11.** $\frac{1}{5}$ **13.** −14°F **15.** −592 **17.** −1,349 **19.** −84 **21.** −25 **23.** 12 **25.** −2°F **27.** +, + **29.** −, − **31.** false; if $a = 6$ and $b = 2$ then $-\frac{6}{2} = -3$ and $\frac{-6}{-2} = -(-3) = 3$ **33.** false; if $a = 6$ and $b = 2$ then $-\frac{6}{2} = -3$ and $-\frac{-6}{-2} = -3$

### Section 2–4, page 57

**Problems** **1.** 3 min 23 s **3.** −$27 **5.** from 1,200 to 1,300 ft; estimates will vary **7.** −$150 **9.** + $302 million

### Section 2–5, pages 59–61

**Exercises** **1.** $2\frac{1}{2}$ **3.** −28.8 **5.** 30 **7.** $-1\frac{1}{2}$ **9.** $-13\frac{4}{5}$ **11.** −3.5 **13.** 11.6 **15.** −17.5 **17.** 0 **19.** $2,850 **21.** Rebecca's balance **23.** 0.3 **25.** 2.9 **27.** 0.1 **29.** 1 h **31.** 65 calories

**Problem Solving Applications** **1.** −18°F **3.** 30 mi/h, 15°F **5.** 20 mi/h, −10°F **7.** −11°F **9.** −24°F

### Section 2–6, page 63

**Problems** **1. a.** Miles Plus is the better choice, since the cost of renting a car from Drive More is $45/day, or $13/day more than renting a car from Miles Plus.

**b.**

| Drive More | Miles Plus |
|---|---|
| $f(n) = 30n + 15n$ | $f(n) = 32n$ |
| $f(9) = 30(9) + 15(9)$ | $f(9) = 32(9)$ |
| $= 270 + 135$ | $= \$288$ |
| $= \$405$ | |

**3. a.**

| Number of days | Type of Movie | Cost at See–More | Cost at Movie Magic |
|---|---|---|---|
| 1 | new | $5.50 | $4.50 |
| | old | $4.50 | $4.50 |
| 2 | new | $8.00 | $6.50 |
| | old | $6.00 | $6.50 |
| 3 | new | $10.50 | $8.50 |
| | old | $7.50 | $8.50 |

A new movie that is 1 day overdue costs $1 more at See–More while an old movie that is 1 day overdue costs the same at both video clubs. For movies that are overdue by 2 or more days, however, any new movie is more expensive at See–More; and any old movie is more expensive at Movie Magic.

**b.**

| See–More | Movie Magic |
|---|---|
| $f(n) = 3 + 2.5n + 3 + 1.5n$ | $f(n) = 2[2.5 + 2n]$ |
| $f(5) = 3 + 2.5(5) + 3 + 1.5(5)$ | $f(5) = 2[2.5 + 2(5)]$ |
| $= 3 + 12.50 + 3 + 7.50$ | $= 2[12.5]$ |
| $= 26$ | $= 25$ |

To have a new movie and an old movie 5 days overdue would cost $26 at the See–More Video Club and $25 at the Movie Magic Video Club

### Section 2–7, pages 66–67

**Exercises** **1.** $14a$ **3.** $11rs$ **5.** $-3ab$ **7.** $2a + ab$ **9.** $3a + 2$ **11.** $7f + 14$ **13.** $13x - 3y$ **15.** $-p + 14q$ **17.** $-9r - 8s$ **19.** $5wz + 7w - 2z$ **21.** $7h + 16t$ **23.** $2 * a + 5$ **25.** $-6 * a + 3 * b$ **27.** $(5 * r)/2 + 12 * s$ **29.** $19 - 4 * a + 4 * c$ **31.** −14 **33.** $6a - 9b$ **35.** $7n + 112$

### Section 2–8, pages 70–71

**Exercises** **1.** 27 **3.** 256 **5.** −144 **7.** −8 **9.** $c^7$ **11.** $y^5$ **13.** $r$ **15.** $z^{16}$ **17.** $\frac{m^3}{64}$ **19.** $3y^{12}$ **21.** $n^3$ **23.** $a^{10}$ **25.** $x^6y^3$ **27.** $fg^2$ **29.** $t$ **31.** $x^{18}$

**33.** $\frac{16}{w^4}$ **35.** $100^4 = (10^2)^4 = 10^8$; $(1{,}000^3) = (10^3)^3 = 10^9$; $(10{,}000)^2 = (10^4)^2 = 10^8$; so $1{,}000^3$ is the greatest. **37.** false; $-5 < 1$ but $(-5)^2 > (1)^2$ **39.** 9:55 A.M.

**Section 2–9, page 75**

**Exercises** **1.** $z^{-12}$ **3.** $w^{-9}$ **5.** $x^{-3}$ **7.** $\frac{1}{1{,}024}$ **9.** 16 **11.** $4.658 \times 10^{-2}$ **13.** 0.000201 **15.** 0.0000086 **17.** $6.5 \times 10^{-7}$m **19.** $(-5)^2(-5)$, $(-4)^{-2}(-4)$, $(-5)(-5)^{-2}$ **21.** $1.67 \times 10^{-14}$ **23.** 4; $a^n$ **25.** $\frac{y^2}{x^2}$ **27.** $\frac{1}{m^7n^3}$

## CHAPTER 3 REASONING IN GEOMETRY

**Section 3–1, pages 86–87**

**Exercises** **1.** $\angle DEF$, $\angle FED$, or $\angle E$ **3.** $\overline{BC}$ or $\overline{CB}$

**5.** 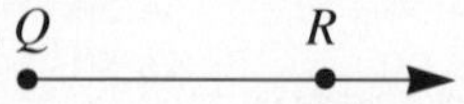

**7.** 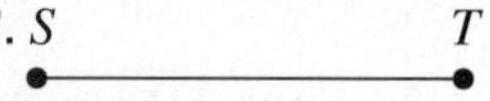


**9.** collinear, coplanar **11.** noncoplanar, noncollinear **13.** f **15.** b **17.** a **19.** Postulate 1 **21.** $\overleftrightarrow{PQ}$ **23.** The intersection is a point for each pair of figures.

**Section 3–2, pages 90–91**

**Exercises** **1.** 76°, 166° **3.** 60°, 150° **5.** $\angle TOS$, $\angle TOR$ **7.** $\angle ROQ$, $\angle TOQ$, and $\angle POS$ **9.** $\angle ROQ$ and $\angle POS$; $\angle POR$ and $\angle QOS$ **11.** $\overleftrightarrow{AD} \perp \overleftrightarrow{AB}$, $\overleftrightarrow{CD} \perp \overleftrightarrow{CB}$, and $\overleftrightarrow{AC} \perp \overleftrightarrow{DB}$ **13.** $C, X, Y$ **15.** $Y$, 2, $D$ **17.** 45° **19.** false; by definition, vertical angles are nonadjacent angles **21.** true, if the measure of each is 45°

**Section 3–3, page 95**

**Exercises** **1.** corresponding **3.** corresponding **5.** alternate exterior **7.** alternate exterior **9.** 105° **11.** 75° **13.** 75° **15.** 105° **17.** $N$ **19.** $k = 38°$; $x = 142°$; $p = 142°$ **21.** true **23.** true

**Section 3–4, pages 98–99**

**Exercises** **1.** scalene **3.** equilateral **5.** 140° **7.** 50° **9.** 140° **11.** 40° **13.** 50° **15.** p = 18; $\angle B = 54°$; $\angle C = 36°$ **17.** $y = 99°$ **19.** sometimes **21.** sometimes **23.** $m\angle 3 = 2x°$ **25.** Show that $\angle 2 \cong \angle 4$ and $\angle 3 \cong \angle 5$ because these are pairs of alternate interior angles. Then show that $m\angle 4 + m\angle 1 + m\angle 5 = 180°$ because they form a straight angle. By substitution, $m\angle 1 + m\angle 2 + m\angle 3 = 180°$.

**Section 3–5, page 101**

**Problems** **1.** Converse: If a triangle is isosceles, then it has two congruent sides. Both the statement and its converse are true. **3.** Converse: If two angles are congruent, then the angles are vertical angles. The statement is true; its converse is false. Counterexample: Two angles that are congruent need not be vertical angles. Adjacent angles are not vertical angles and yet may be congruent. **5.** Converse: If two lines are parallel, then the two lines do not intersect. The converse is true, but the statement itself is false. Counterexample: The two lines may be skew lines. Skew lines never intersect and are not parallel. **7.** If two or more points are collinear, then those points lie on the same line. **9.** If a figure is a rhombus, then it has four congruent sides. **11.** If two coplanar lines are parallel, then they do not intersect.

**Section 3–6, pages 104–105**

**Exercises** **1.** $\angle B$ **3.** $\overline{AB}$ **5.** ASA **7.** SSS **9.** ASA **11.** b **13.** a **15.** You know that two angles of one triangle are congruent to two angles of the other. You can show that the third angles are congruent, since the sum of the angles of each triangle is 180°. You can then apply ASA. **17.** $\overline{AB} \cong \overline{AC}$; $\overline{BD} \cong \overline{DC}$, and $\angle B \cong \angle C$. So, $\Delta ADB \cong \Delta ADC$ by SAS. Since corresponding parts of congruent triangles are congruent, $\angle 1 \cong \angle 2$, and therefore $\overline{AD}$ is the bisector of $\angle A$.

**Section 3–7, pages 108–109**

**Exercises** **1.** parallelogram, rectangle, rhombus, square **3.** parallelogram, rhombus, rectangle, square **5.** rectangle, square **7.** 122° **9.** 58° **11.** 61° **13.** 12 **15.** 106° **17.** 106° **19.** 10 **21.** 6 **23.** If a quadrilateral is a parallelogram, then both pairs of opposite sides of the quadrilateral are congruent. If both pairs of opposite sides of the quadrilateral are congruent, then the quadrilateral is a parallelogram. **25.** Points $A$, $B$, $C$, and $D$ form a quadrilateral. It is given that the diagonals $AC$ and $BD$ bisect each other. So quadrilateral $ABCD$ must be a parallelogram. Since by definition, opposite sides of a parallelogram are parallel, $\overline{CD}$ (representing the top surface of the ironing board) is parallel to $\overline{AB}$ (representing the surface of the floor).

**Section 3–8, page 113**

**Exercises** **1.** 360° **3.** 720° **5.** 1,440° **7.** 2,160°

**9.** 150° **11.** 172° **13.** false **15.** true **17.** false
**19.** sum of exterior angles $= n(180°) - (n - 2)180°$
$= n(180°) - n(180°) + 360°$
$= 360°$

### Section 3–9, pages 116–117

**Exercises** **1.** 60° **3.** 105° **5.** $\widehat{POQ}$, $\widehat{PNQ}$, $\widehat{NPQ}$, $\widehat{NQO}$ **7.** 110° **9.** All radii of the same circle are congruent. **11.** false **13.** false **15.** false **17.** 180° **19.** Size of angles will vary. **21.** $\widehat{MN}$ is twice the size of the inscribed angles. **23.** no, yes

### Section 3–10, page 119

**Problems** **1.** Title of circle graph: How Americans get to Work: Sectors should be formed by the following central angles: Drive alone, 230°; Carpool, 72°; Bus or train, 22°; Walk only, 22°; Work at home, 7°; Other, 7°. **3.** Title of circle graph: Monthly Budget. Sectors should be formed by the following central angles: Car, 54°; Travel, 68°; Food 97°; Clothing, 83°; Savings, 43°; Miscellaneous, 14°. 1° may be added to the miscellaneous sector to make a total of 360°.
**5.** 29%, 18%, 33%, 9%, 11%

### Section 3–11, page 122

**Exercises** **1.**

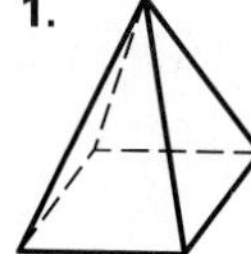

**3.**

**5.** right cylinder **7.** hexagonal prism

| | Polyhedron | Number of Faces (F) | Number of Vertices (V) | Number of Edges (E) | F + V – E |
|---|---|---|---|---|---|
| **9.** | hexagonal prism | 6 | 8 | 12 | 2 |
| **11.** | square pyramid | 5 | 5 | 8 | 2 |

**13.**

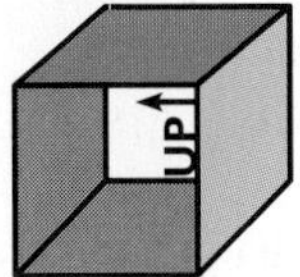


**Problem Solving Applications** **1.** triangular pyramid **3.** triangular prism **5.** circle **7.** oval **9.** rectangle

## CHAPTER 4 STATISTICS

### Section 4–1, pages 134–135

**Exercises** **1.** systematic sampling; people who attend a sporting event usually have a greater interest in that sport than another. **3.** convenience sampling; teenagers at the same party are most likely friends, and a common interest of friends can be the sport they participate in or enjoy watching **5.** People who attended the meeting probably felt more strongly in favor of or against the proposed law than citizens who did not attend. **7.** A possible method is to survey residents of fifty homes that have more trash cans set out on trash pick-up day than their neighbors. **9.** A possible method is to survey all citizens whose phone number ends with a particular digit. Because phone numbers are randomly assigned, the results of this sampling method should be less biased than any others. **11.** Answers will vary. **13.** Answers will vary. **15.** Answers will vary.

### Section 4–2, page 137

**Problems** **1.**

| | Action | Comedy | Thriller | Horror | Adventure |
|---|---|---|---|---|---|
| Matinee | 140 | 185 | 176 | 97 | 192 |
| Early evening | 242 | 350 | 325 | 276 | 148 |
| Late evening | 258 | 187 | 346 | 240 | 175 |

**3.** Comedy, early evening **5.** 24% **7.** Possible question: How many more adults than children attended the adventure movie? To answer this, reorganize the data in a matrix using the movie types as row headings and ticket types as column headings.

### Section 4–3, pages 140

**Exercises** **1.**

**Number of Games Bowled by League Members**

| Number of Games | Tally | Frequency |
|---|---|---|
| 75 | \|\| | 2 |
| 78 | \|\|\|\| | 4 |
| 81 | 𝍸 𝍸 \| | 11 |
| 84 | 𝍸 \|\| | 7 |
| 87 | \| | 1 |

19 league members bowled at least 81 games.

**3.** Birthdates of Entertainers

| Year of Birth | Tally | Frequency |
|---|---|---|
| 1900–1909 | \| | 1 |
| 1910–1919 | | 0 |
| 1920–1929 | \|\|\|\| | 4 |
| 1930–1939 | 𝍸 \| | 6 |
| 1940–1949 | 𝍸 \|\|\| | 8 |
| 1950–1959 | \|\|\|\| | 4 |
| 1960–1969 | \|\| | 2 |

**5.** 15, 3, 7 **7.** Yes; she could make a new frequency table with Marissa's score missing and see which score was left out.

**Problem Solving Applications** **1.** Answers will vary. **3.** Answers will vary. **5.** Answers will vary.

## Section 4–4, pages 143–145

**Exercises** **1.** 125.5, 120, 120, 47 **3.** 4.8, 4.5, 4, 4 **5.** The mean would be \$1.09 less, the median \$0.51 less, and the range \$5.87 less. **7. a.** Multiply to find the needed sum of the data with the new test score: $6 \times 86 = 516$. Subtract the sum of the original test scores: $516 - 417 = 99$ **b.** no; if you score 87 or higher, the median will be 85.5. **9.** always **11.** sometimes: mode does not change if 5 6 6 7 is changed to 5 6 6 8

**Problem Solving Applications** **1.** 75, 70 **3.** \$4,917 **5.** \$7,217

## Section 4–5, pages 146–147

**Problems** **1. a.** median **b.** no; the amount of rainfall during each summer month was much greater than 10 cm **c.** mode; the rainfall during each of the summer months was close to or equal to 25 cm **3. a.** The mean (\$52,875) is misleading because only three out of eight positions have salaries higher than this. The mode (\$30,000) is misleading because five out of eight positions have salaries higher than this, and some are considerably higher. The median (\$45,000) is probably misleading also because it is likely that there are more employees in the positions paying less than \$45,000 than in the positions paying more **b.** the number of employees in each position

## Section 4–6, pages 150–151

**Exercises** **1.** 5, 10 **3.** 17 **5.** 6

Percents of Free Throws Made

| Stem | Leaves |
|---|---|
| 2 | 4 |
| 3 | 9 |
| 4 | 1 3 3 5 6 6 7 |
| 5 | 3 5 6 7 9 |
| 6 | 0 4 6 |
| 7 | 2 9 |
| 8 | 6 |

2|4 represents 24%

**7.** 43% and 46% **9.** 90–99 cm and 150–159 cm; 100–109 cm and 130–139 cm

**11.** Box Office Grosses

| Stem | Leaves |
|---|---|
| 5 | 0 2 4 5 7 7 |
| 6 | 1 3 4 |
| 7 | 1 |
| 9 | 2 5 5 6 |
| 11 | 3 5 |
| 13 | 0 |
| 14 | 7 |
| 19 | 6 |
| 25 | 1 |

5|0 represents \$50 million

Students should include that the median is \$81.5 million, the modes are \$57 million and \$95 million, and the range is \$201 million. Outliers are \$196 million and \$251 million. Data clusters are in the middle \$50 millions and middle \$90 millions. The largest gaps occur between \$71 million and \$92 million and between the majority of the data and the outliers. **13.** The heights of the bars of the histogram are in the same proportion to each other as the lengths of the lists of leaves are in the stem–and–leaf plot.

## Section 4–7, pages 154–155

**Exercises** **1.** 15 **3.** 1 h **5.** Acceptable range of answers: 975–990 **7.** The scatter plot should show that there is a *negative correlation* between the sets of data in the table. 11.7 s, 11.65 s **9.** The winning time, made by Gail Devers, was 10.82 s. **11.** Answers will vary. **13.** Answers will vary.

## Section 4–8, pages 158-159

**Exercises** **1.** 73, 83, 86.5 **3.** 5.3, 6.75, 8.65 **5.** 44th **7.** 88th **9.** 87th **11.** 80th **13.** 27.5 **15.** 11 **17.** 30th **19.** 25 **21.** 75th **23.** 72nd **25.** Answers will vary.

## Section 4–9, pages 162–163

**Exercises** **1.** Hours Cleo Worked Daily

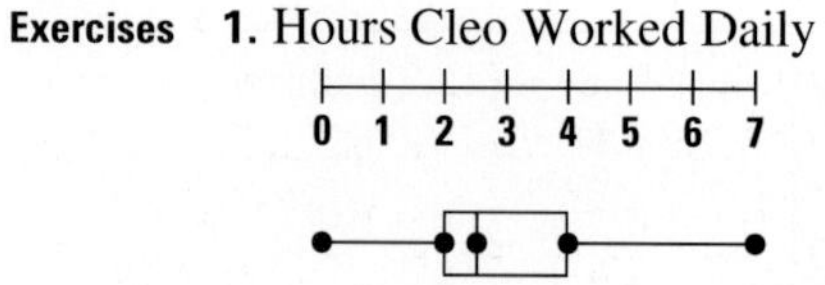


**3.** 1,900; 100 **5.** 25%

**7.** Average Daily Temperature per Month

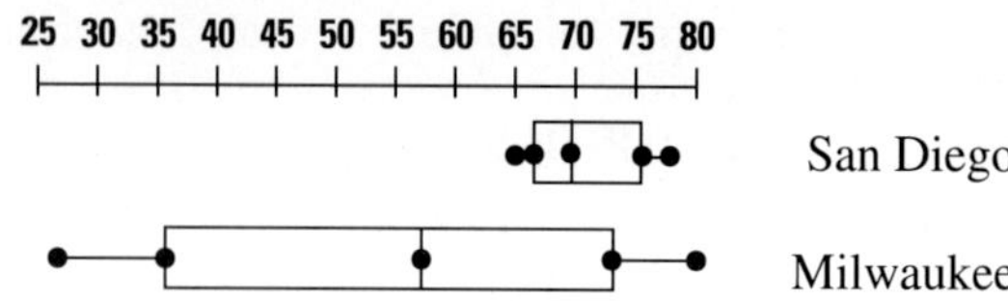

9. The set of data with a short box and long whiskers has a wide range, but the middle 50% is closely clustered together. The set of data with a long box, short whiskers, and one outlier has a wide range, but the lower 25% and the upper 25% is closely clustered except for one piece of data which is very different from the rest. **11.** Individual game scores can still be seen in the stem-and-leaf plot. Distribution and clusters of scores can be seen more easily at a glance in the box-and-whisker plot.

### Section 4–10, page 167

**Exercises** **1.** about 6% **3.** not enough information **5.** Thin crust: teenagers 9; adults 12; Pan: teenagers 3; adults 16 **7.** Answers will vary. **9.** Answers will vary.

## CHAPTER 5 SOLVING EQUATIONS AND INEQUALITIES

### Section 5–1, pages 178–179

**Exercises** **1.** –2 **3.** 7 or –7 **5.** 0.5 **7.** 48 **9.** 0.7 **11.** 4 **13.** –13 **15.** no solution **17.** 3.2, –3.2 **19.** –7.4 **21.** $1\frac{4}{5}, -1\frac{4}{5}$ **23.** $\frac{1}{6}$ **25.** –4 **27.** 3, –3 **29.** 80, –80 **31.** 48 **33.** No; the square of a number is never negative.

### Section 5–2, page 182–183

**Exercises** **1.** 12 **3.** 21 **5.** 36 **7.** –1.2 **9.** –13 **11.** 24 **13.** –4 **15.** 28 **17.** –3 **19.** 7 lb **21.** –2.5 **23.** $-6\frac{1}{2}$ **25.** $-2\frac{1}{2}$ **27.** 65 **29.** –25 **31.** 12 **33.** 5, –5 **35.** 2, –2 **37.** 52.21; $52.21 **39.** 20 **41.** 45

### Section 5–3, page 187

**Exercises** **1.** 18 **3.** 2 **5.** 8 **7.** 2.2 **9.** 24 **11.** –11 **13.** 1 **15.** 3 **17.** 7 **19.** $m\angle ADB = 39°$; $m\angle BDC = 51°$ **21.** –4 **23.** –2 **25.** $\angle J = 42.5°$; $\angle K = 85°$; $\angle L = 52.5°$ **27.** 19 **29.** In solving the equation, you arrive at a false statement, 4 = 10. The equation has no solution.

### Section 5–4, page 189

**Problems** **1.** – **3.** × **5.** c **7.** g **9.** b **11.** d **13.** $p + 40$; $p$ represents last week's price **15.** $s - 15$; $s$ represents Karen's score **17.** $(7.5)h$; $h$ represents hours **19.** Answers will vary. **21.** Answers will vary. **23.** $6(n + 3)$ or $6n + 18$ **25.** $6(\frac{n}{2})$ or $3n$ **27.** $25(3x)$ or $(\$0.25)(3x)$ or $75x$ or $(\$0.75)x$ **29.** $4n + 10$ or $(\$0.01)(4n + 10)$ or $(\$0.04)n + \$0.10$

### Section 5–5, page 191

**Problems** **1.** c. **3.** $46 **5.** 73 beats per minute **7.** panda, $28.50; monkey, $23.50

### Section 5–6, pages 194–195

**Exercises**
**1.** $n < -5$
**3.** $x \geq -4$
**5.** $z > -2.5$
**7.** $e > 2$
**9.** $c \leq 3$
**11.** $x \leq -2.5$
**13.** $t < -3$
**15.** $x \geq -2$
**17.** 14
**19.** $c > -1\frac{3}{4}$
**21.** $m \leq \frac{2}{3}$
**23.** $b < 2$
**25.** $x > -5$
**27.** 19 **29. a.** A.
B.

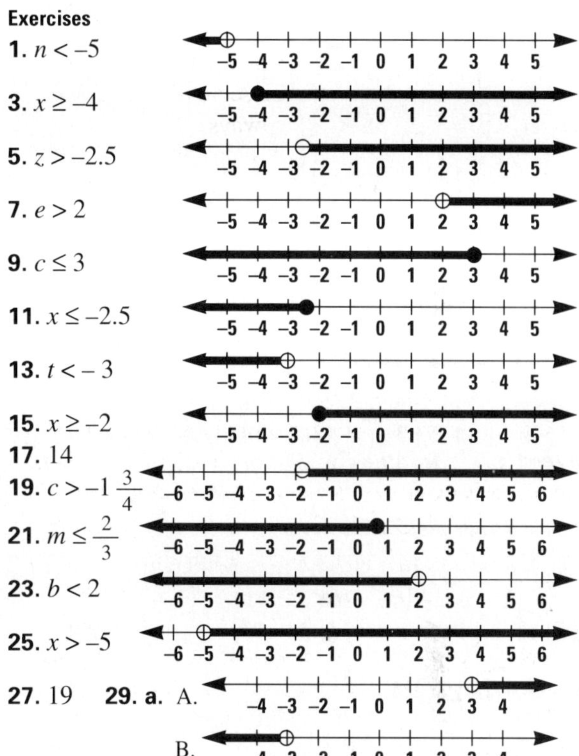

**b.** Graphs A and B together are the graph of $|x| > 3$. **31.** Sometimes. Examples will vary. True if $a = 5$, $b = 3$, $c = 2$, $d = 1$. Not true if $a = 2$, $b = 1$, $c = -10$, $d = -20$.

### Section 5–7, pages 198–199

**Exercises** **1.** 12 s **3.** 180.46 cm **5.** $h = \frac{w + 190}{5}$ **7.** 93.5 **9.** 48 **11.** 32.5 ft **13.** 82 ft **15.** 12t **17.** 4 – 4t

**Problem Solving Applications** **1.** 0.8 kWh **3.** $49.10; assume that it is operating 24 h a day **5.** $2.41

### Section 5–8, pages 202–203

**Exercises** **1.** $\frac{3}{5}$ **3.** $\frac{11}{14}$ **5.** 1.7 **7.** 0.01 **9.** 4.472

**11.** −17.088 **13.** 3 and 4 **15.** 8 and 9 **17.** 3 and 4 **19.** 9 and 10 **21.** 14 **23.** 4 **25.** 15 **27.** 196 **29.** 2.898 **31.** 8.361 **33.** 2 **35.** 5 **37.** 8 **39.** $\sqrt{7}$ **41.** 5 **43.** 60.7 in. **45.** 6; 3; 2; 6 **47.** 20; 5; 4; 20 **49.** 2; $\frac{6}{3}$; 2 **51.** $\sqrt{\frac{x}{y}} = \frac{\sqrt{x}}{\sqrt{y}}$

**Section 5–9, pages 206–207**

**Exercises** **1.** ±13 **3.** ±5 **5.** ± $\sqrt{59}$ **7.** no solution **9.** ±16 **11.** ±0.9 **13.** ±2.828 **15.** ±1.612 **17.** ±10.247 **19.** 40 mi/h **21.** ±0.07 **23.** ±0.12 **25.** ±0.5 **27.** ±9.055 **29.** ±4.583 **31.** $2(y - 7)^2 = 128$; $(y - 7)^2 = 64$; $y - 7 = \pm 8$; if $y - 7 = 8$, $y = 15$; if $y - 7 = -8$, $y = -1$; so, $y = 15$ or $y = -1$. Check: $2(15 - 7)^2 = 128$; $2(8)^2 = 128$; $2(64) = 128$; $128 = 128$; $2(-1 - 7)^2 = 128$; $2(-8)^2 = 128$; $2(64) = 128$; $128 = 128$.

## CHAPTER 6 USING FORMULAS IN GEOMETRY

**Section 6–1, pages 218–219**

**Exercises** **1.** 11.7 m **3.** 26 ft **5.** 22.4 m **7.** 41 ft **9.** 15 in. **11.** 53.3 m **13.** 5 cm **15.** 9.9 cm **17.** 127.3 ft **19.** 17 in.

**Section 6–2, pages 222–223**

**Exercises** **1.** 188 m **3.** 34.4 m **5.** 48.6 m **7.** 490 m **9.** 220 in. **11.** 22.4 m **13.** 42 cm **15.** 14 ft by 11 ft **17.** 15 m **19.** only one: 200 ft

**Section 6–3, page 225**

**Problems** **1.** 70 in. or 5 ft 10 in. **3.** 58 ft or $19\frac{1}{3}$ yd **5.** the square one **7.** 56 ft **9.** 88 in. or $7\frac{1}{3}$ ft **11.** 4.8 m or 480 cm

**Section 6–4, pages 228–229**

**Exercises** **1.** 198 ft$^2$ **3.** 71.5 cm$^2$ **5.** 182 cm$^2$ **7.** 13 m **9.** 15 m **11.** 432 ft$^2$ **13.** 84 m$^2$ **15.** 10 m$^2$ **17.** 30 m$^2$ **19.** $\frac{3}{8}$ **21.** The length of $\overline{BX}$ is $\frac{2}{3}$ the length of $\overline{BC}$.

**Section 6–5, pages 232–233**

**Exercises** **1.** about 48.4 cm **3.** about 2.24 cm **5.** 136.5 cm$^2$ **7.** 16 **9.** 7.74 ft$^2$ **11.** 882 cm$^2$

**Problem Solving Applications** **1.** 256 **3.** 24 **5.** 21 **7.** ≈ 20 **9.** 734 ft$^2$

**Section 6–6, pages 236–237**

**Exercises** **1.** 0.31 **3.** 0.16 **5.** 0.125 **7.** 0.17 **9.** $\frac{1}{6}$ **11.** 0.41 **13.** 0.32 or $\frac{1}{\pi}$ **15.** $\frac{1}{25}$ **17.** the point in Utah **19.** 0.25

**Section 6–7, pages 240–241**

**Exercises** **1.** 960 m$^2$ **3.** 2,896 ft$^2$ **5.** 10.88 cm$^2$ **7.** 620 in.$^2$ **9.** 420 m$^2$ **11.** 925,344 ft$^2$ **13.** 340 ft$^2$ **15.** $SA = b^2 + 2bl$ **17.** It would be 4 times greater.

**Section 6–8, pages 244–245**

**Exercises** **1.** 71 m$^2$ **3.** 406 m$^2$ **5.** 9,156 in.$^2$ **7.** 251 m$^2$ **9.** 691 m$^2$ **11.** 327 m$^2$ **13.** 259 cm$^2$ **15.** 730 in.$^2$ **17.** 441 m$^2$ **19.** 13 in. **21.** $A = \frac{C^2}{\pi}$ **23.** Doubling the height doubles the area of only the curved surface; doubling the radius doubles the area of the curved surface and doubles the area of the bases.

**Section 6–9, pages 248–249**

**Exercises** **1.** 658.8 m$^3$ **3.** 29,791 cm$^3$ **5.** 1,122 ft$^3$ **7.** 480 m$^3$ **9.** 6 ft **11.** 84 ft$^3$ **13.** 720 m$^3$ **15.** 594 in.$^3$ **17.** 330 in.$^3$ **19.** a 2-cm cube, 24 cm$^2$, 8 cm$^3$ **21.** 10 cm by 10 cm by 10 cm

**Section 6–10, pages 251–253**

**Exercises** **1.** 7,235 m$^3$ **3.** 1,417 m$^3$ **5.** 1,884 ft$^3$ **7.** 359 cm$^3$ **9.** 489.8 cm$^3$ **11.** $47\frac{2}{3}$% **13. a.** $h = 12.7$ cm **b.** $h = 8.8$ cm **c.** $h = 19.9$ cm **d.** $h = 35.4$ cm **e.** $h = 6.5$ cm **f.** $h = 5$ cm **15.** 5.4 cm

**Problem Solving Applications** **1.** 2,323,200 ft$^3$ **3.** 144,233,549 lb; 72,117 tons **5.** 264,135,547 gal

**Section 6–11, page 255**

**Problems** **1.** 168 ft **3.** 98.9 in.$^3$ **5.** 7 lb **7.** 1,134.3 ft$^3$ **9.** 138.8 ft$^2$ **11.** 166 ft$^3$

**Section 6–12, pages 258–259**

**Exercises** **1.** 2 **3.** 5 **5.** 1 **7.** 1 **9.** 0.1 m **11.** 0.001 L **13.** 10 g **15.** 1 mi **17.** 2,996 km **19.** about 190 ft **21.** 6,280 mi **23.** 2,600 g **25.** 20.0 L **27.** about 387,000 acres **29.** about 833,000 acres **31.** no; the unit of precision of the least precise measurement is 10 km

## CHAPTER 7 POLYNOMIALS

### Section 7–1, pages 270–271

**Exercises** **1.** $3x^3 + x^2 + 2$ **3.** $-x^5 + x^3 - x$ **5.** $x^3y + x^2y^2 + xy^4$ **7.** $-2n$ **9.** 0 **11.** $2x^3 + 11x$ **13.** $4n^2 + 8n$ **15.** $x + xy$ **17.** $8x - 2$ **19.** $k$ **21.** $4p + 5$ **23.** $4x^2 + 2x - 2$ **25.** $(15r + 46t)$ dollars **27.** $-1.9m + 11.6n - 2.3$ **29.** $4a^2b^2 + b^3 - 4$ **31.** $12x + 18$ **a.** $16x - 10$ **b.** $28x + 8$ **33.** Answers will vary; examples are $(9x^2 + 4x + 10)$ and $(6x^2 + 9x + 3)$

### Section 7–2, pages 273–275

**Exercises** **1.** $21pq$ **3.** $-12mn$ **5.** $-18a$ **7.** $-9ke$ **9.** $-20m^3$ **11.** $-6h^5$ **13.** $a^8b^6$ **15.** $-16e^4f$ **17.** $49h^2kmn$ **19.** $9x^2$ **21.** $32a^5b^5$ **23.** $m^8n^{20}$ **25.** $6a^2b^3$ **27.** $16.1r^4$ yd$^2$ **29.** $8a^{16}$ **31.** $-40x^{11}$ **33.** $-a^{18}b^9$ **35.** $128h^{13}$ **37.** $e^{11}f^{15}$ **39.** $54x^2$ **41.** $12x^5$ **43.** $32a^4b^3$ **45.** $n = 3$ **47.** $n = e^2$ **49.** $12x, x$; $12x^2, 1$; $x^2, 12$; $2, 6x^2$; $6, 2x^2$; $6x, 2x$; $3x^2, 4$; $4x^2, 3$; $3x, 4x$

### Section 7–3, page 277

**Problems** **1.** $(2x)(2y) = 4xy$ **3.** $(3m^2)(2n^3) = 6m^2n^3$; other solutions possible

**5.** $12xy$

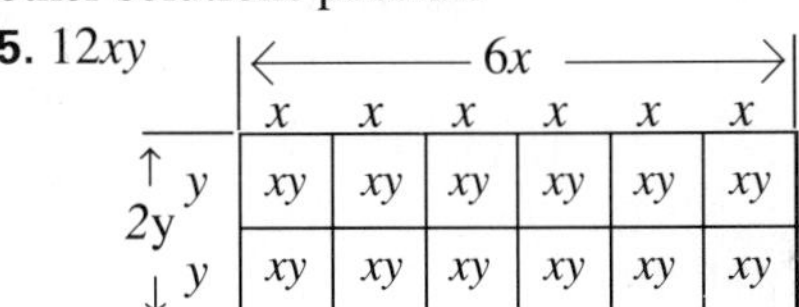


**7.** $4m^2n^4$

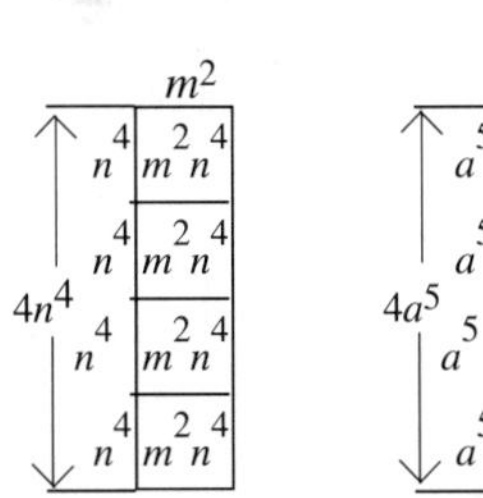


**9.** $8a^5bc$

| | ← 2bc → | |
|---|---|---|
| | $bc$ | $bc$ |
| $a^5$ | $a^5bc$ | $a^5bc$ |
| $4a^5$ $a^5$ | $a^5bc$ | $a^5bc$ |
| $a^5$ | $a^5bc$ | $a^5bc$ |
| $a^5$ | $a^5bc$ | $a^5bc$ |

**11.** $2y(3x + 2)$ **13.** $6xy + 4y$

**15.** $2x^2 + 2x$

| | ← x + 1 → | |
|---|---|---|
| | $x$ | 1 |
| $2x$ $x$ | $x^2$ | $x$ |
| $x$ | $x^2$ | $x$ |

**17.** $mn + 3m$

| | ← n + 3 → | | | |
|---|---|---|---|---|
| | $n$ | 1 | 1 | 1 |
| $m$ | $mn$ | $m$ | $m$ | $m$ |

### Section 7–4, pages 280–281

**Exercises** **1.** $6m^2 + 3mn$ **3.** $6a^2 + 8a$ **5.** $27c^2 + 9c^3 + 54c^4$ **7.** $-12e^3 - 20e^2 + 12e$ **9.** $-3y^4 + 3y^3 + 3y^2$ **11.** $5h^2 + 4$ **13.** $5w^2 - 20$ **15.** $C = 8r + 2rh$ **17.** $280 - 5m$ **19.** $12e^5 - 10e^4 + 4e^2$ **21.** $17y^4 + 16y^3 - 25y^2$ **23.** $48x^3 - 20x^2$ **25. a.** $40w$ **b.** 2,400 words **27.** $192 + 12n$ **29.** $72k^2 + 32k$

### Section 7–5, pages 284–285

**Problems** **1.** $a^2 + 5a + 6$ **3.** $p^2 - 2p - 3$ **5.** $e^2 + 4e + 4$ **7.** $36 - x^2$ **9.** $84 + 19t + t^2$ **11.** $w^2 - 17w + 72$ **13.** $2y^2 - 4y - 51$ **15.** $x^2 + 10x + 18$ **17.** $6x^2 + 13x + 6$ **19.** $25c^2 - 1$ **21.** $30x^2 + 61x + 30$ **23.** $5k^2 + 3k - 2$ **25.** $x^3 + 3x^2 + 3x + 1$ **27.** $2y^3 + 14y^2 + 24y$ **29.** $12n^3 + 2n^2 - 2n$

**Problem Solving Applications** **1.** 160 ft × 40 ft **3.** 21 cm × 7 cm

### Section 7–6, pages 287

**Problems** **1.** 8:20 A.M. **3.** \$72 **5.** 112 **7.** $x + 5$ **9.** $c^2 + 3c - 6$ **11.** $4w^2y - 5wk^2 + 1$

### Section 7–7, pages 290–291

**Exercises** **1.** $2^5$ **3.** $3^3 \times 2$ **5.** $2^3 \times 19$ **7.** 4 **9.** 12 **11.** 12p **13.** 90 **15.** 63 **17.** $24x^3y^2$ **19.** $420p^2t^2w^3$ **21.** 20 seconds after **23.** 3 **25.** 7 **27.** $7mn^2p^2$ **29.** 24 **31.** 288 **33.** $416j^3k^3l^4$ **35.** 48 yd **37.** yes **39.** no **41.** no **43.** 1; 1 is their only common factor **45.** the other number

### Section 7–8, page 294–295

**Exercises** **1.** $8m(n - p)$ **3.** $4(x - 5)$ **5.** $9p^2(p + 3)$ **7.** $6k(2k - 7)$ **9.** $w^4(w - 1)$ **11.** $3x(y^2 + 2)$ **13.** $v^3(v^2 - 6v + 3)$ **15.** $a^2(3a^4 - 5a + 2)$ **17.** $3x^2y^2(4x^2y^2 + x - 2)$ **19.** $x(x + 6y - 1)$ **21.** $25(2m^2 + 5mn + n^2)$ **23.** $8x^2(8x^4 - 6x^2 + 3)$ **25.** $P = 2(l + w)$ **27.** $D = \frac{1}{2}n(n - 3)$ **29.** $9m^2n(2mn + 5n^2 + 3m^2 - 6n)$ **31.** $4ab(12 - 10a^2b + 6ab^2 + 7ab)$ **33.** $2(8n + 3)$ **35.** $6y(6y + 1)$ **37.** $3(7x^2 + 3x + 6)$ or $21x^2 + 9x + 18$ **39.** no **41.** no **43.** no **45.** $n^2 - 49$ **47.** $k^2 - 81$ **49.** $(w + 8)(w - 8)$ **51.** $(c + 10)(c - 10)$ **53.** $(6y + 11z^3)(6y - 11z^3)$

### Section 7–9, pages 298–299

**Exercises** **1.** $4x$ **3.** $-\frac{10}{7}$ **5.** $\frac{b}{2c}$ **7.** $15xy$ **9.** $11abc^3$ **11.** 3 **13.** $a + 5$ **15.** $3a - 2$ **17.** $2 - 3x$ **19.** $6c^3 + 5c^2 - 9c$ **21.** $3x^2 - x + 8$ **23.** $3x - 4y + 2z - 1$ **25.** $4b$ **27.** $6p$ **29.** $7m$ **31.** $3x + y$ units **33.** $49x^2 + 140xy + 100y^2$ **35.** 4 **37.** 8 **39.** $p^2$ **41.** $9 + m$

**Section 7–10, pages 302–303**

**Exercises** **1.** yes **3.** no **5.** yes **7.** $(c-2)^2$ **9.** $(y+11)^2$ **11.** $(e-3)^2$ **13.** $(n+10)(n-10)$ **15.** $(h+5)(h-5)$ **17.** not factorable **19.** $(p+12)(p-12)$ **21.** not factorable **23.** $(t+20)(t-20)$ **25.** $3(m+3)(m-3)$ **27.** $2(m+8)^2$ **29.** $3a(k+10)(k-10)$ **31.** 1.3 mi **33.** $(2n+5)(2n-5)$ **35.** $(5x-4)^2$ **37.** $x=9$; $y=7$ or $x=6$; $y=2$

**Problem Solving Applications**

**1.**

| | |
|---|---|
| Think of a number. | $n$ |
| Subtract 7. | $n-7$ |
| Multiply by 3. | $3(n-7)=3n-21$ |
| Add 30. | $3n-21+30=3n+9$ |
| Divide by 3. | $\frac{3n+9}{3}=\frac{3n}{3}+\frac{9}{3}=n+3$ |
| Subtract the original number. | $n+3-n=3$ |

The result is 3.

**3.**

| | |
|---|---|
| Think of a number. | $n$ |
| Add 5. | $n+5$ |
| Multiply by 6. | $6(n+5)=6n+30$ |
| Divide by 2. | $\frac{6n+30}{2}=3n+15$ |
| Subtract 15. | $3n+15-15=3n$ |
| Divide by 3. | $\frac{3n}{3}=n$ |

## CHAPTER 8 GRAPHING FUNCTIONS

**Section 8–1, pages 314–315**

**Exercises** **1.** (2, 3) **3.** $E(5, 0)$ and $H(-2, 0)$ **5.** $C(-1, 4)$ or $F(-5, 5)$ **7.** $I(-3, -3)$ **9.–15.** See graph below.

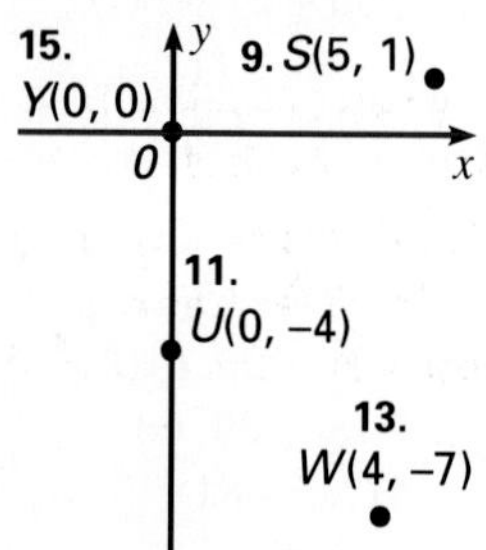


**17. a.**

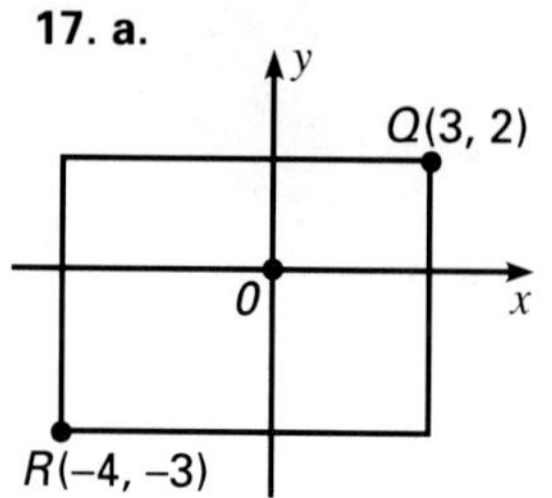


**b.** $P=24$ units **c.** $A=35$ square units

**19.** square; 36 square units

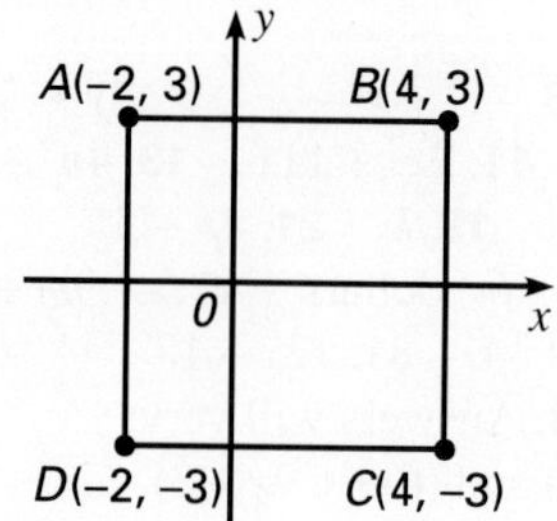


**21.** The points are part of a line. Ordered pairs will vary.

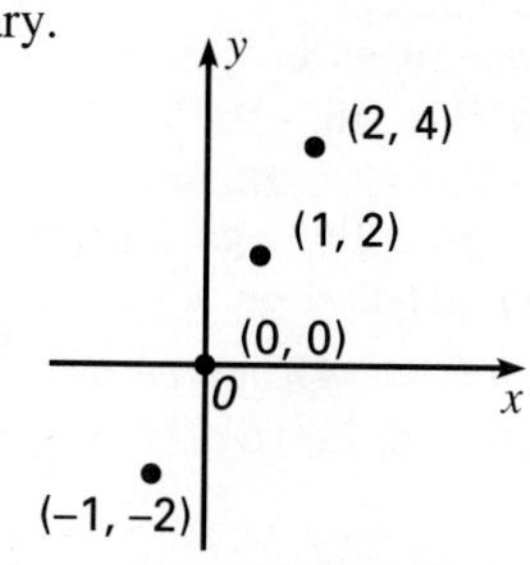


**23. a.** same $y$–coordinate: (1, 5), (0, 5), (–2, 5) **b.** same $x$–coordinate: (–4, 2), (–4, 0), (–4, –1)

**Section 8–2, pages 318–319**

**Exercises** **1.** (–3, –3), (–1, 1), (0, 3), (2, 7), (4, 11) **3.** (–6, 0), (–4, 1), (0, 3), (2, 4), (8, 7)

**5.**

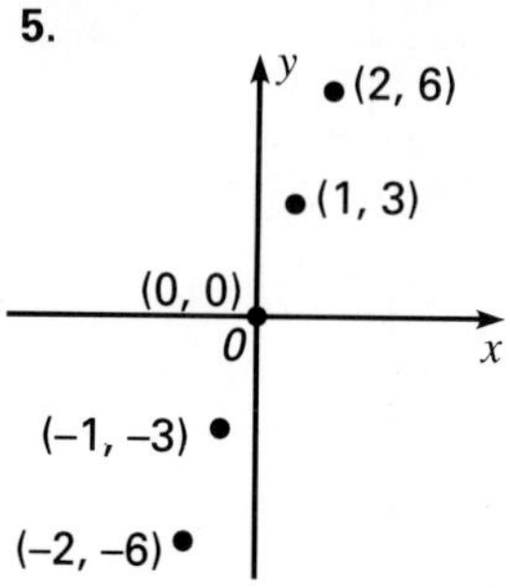

**7.**

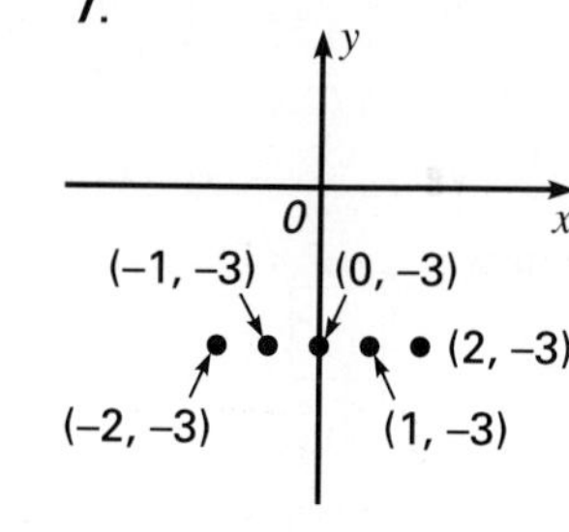


**9.**

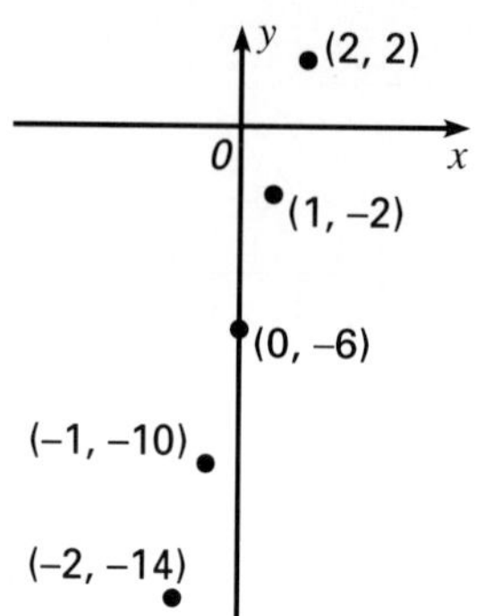

**11.**

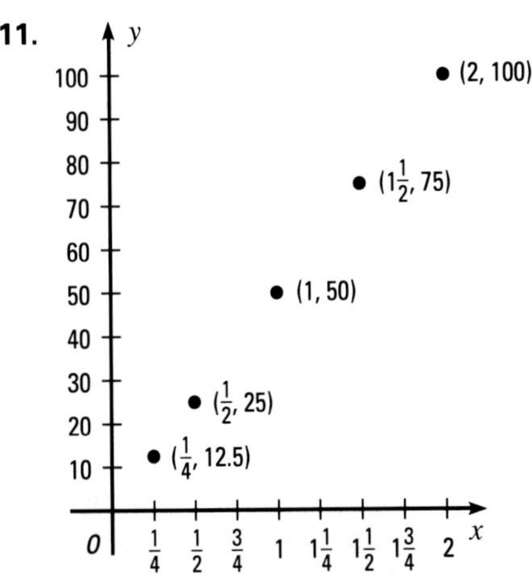


**13.** (18, 216) and (21, 294)

**15.**

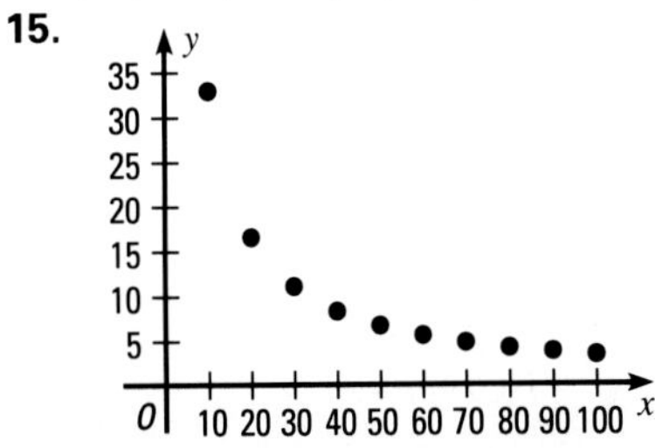


**17.** all negative real numbers and 0 **19.** all real numbers **21.** no; for each value of $x$ there are two values for the square root and, thus, two values for $y$

## Section 8–3, pages 322–323

### Exercises

**1.**

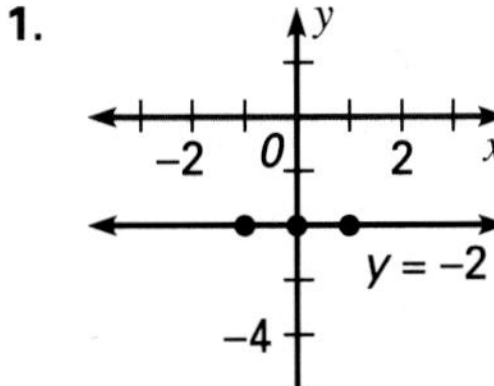


**3.**

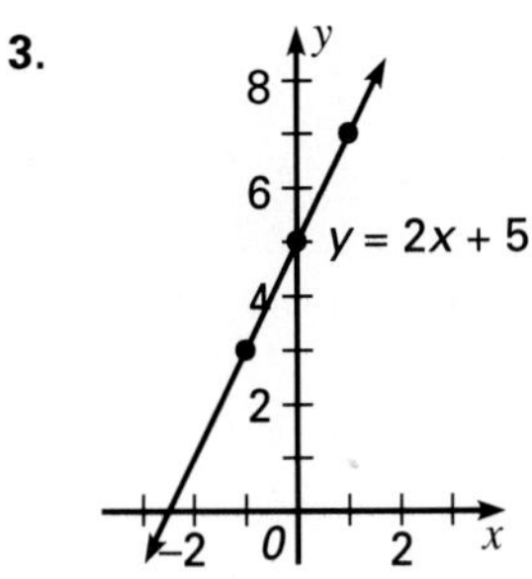


**5.**

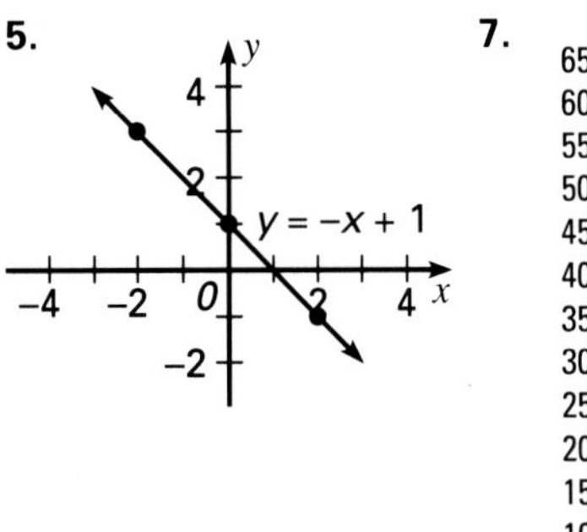


**7.**

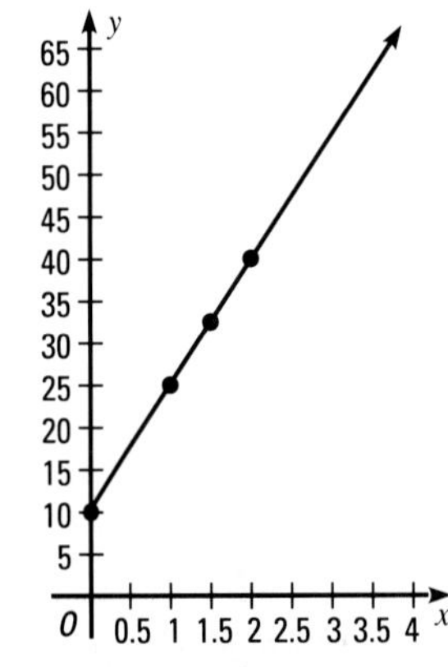


**9.** 3.5 s

**11.**

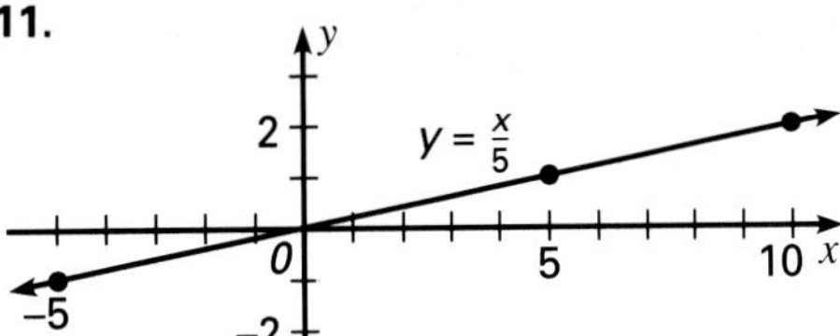


**13.**

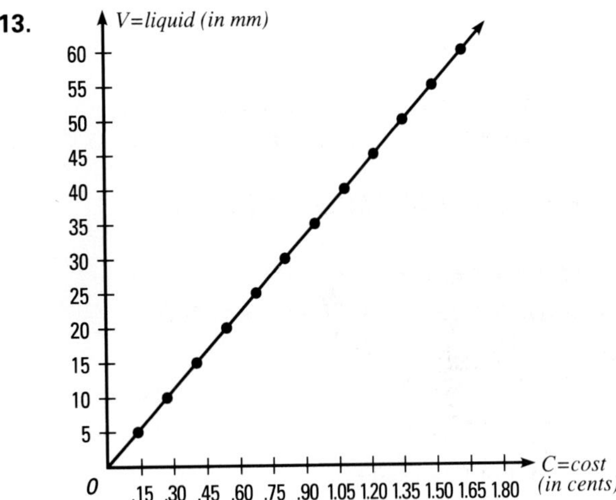


**15.** 13 mL **17.** $l = 8 - w$ **19.** As the values in the domain increase, the values in the range decrease. The domain is $0 < w \le 4$; the range is $4 \le l < 8$. **21.** 68° F

## Section 8–4, page 327

### Exercises

**1.**

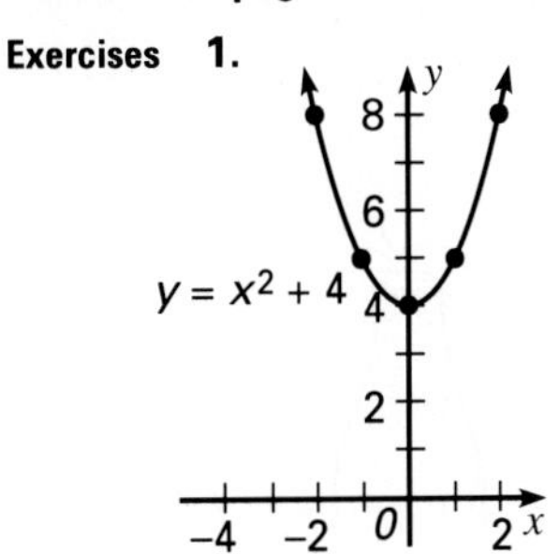


**3.**

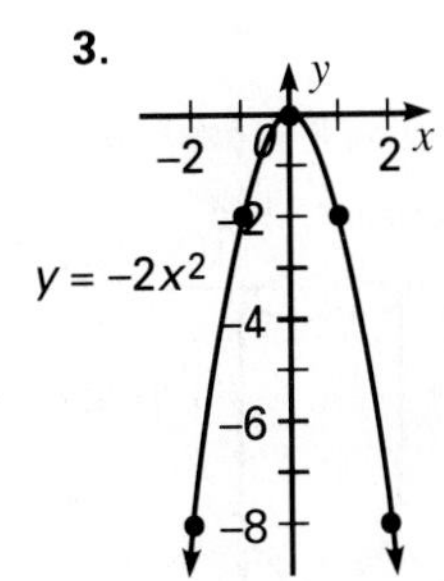

**5. a.** \$60 **b.** 5

**7. a.**

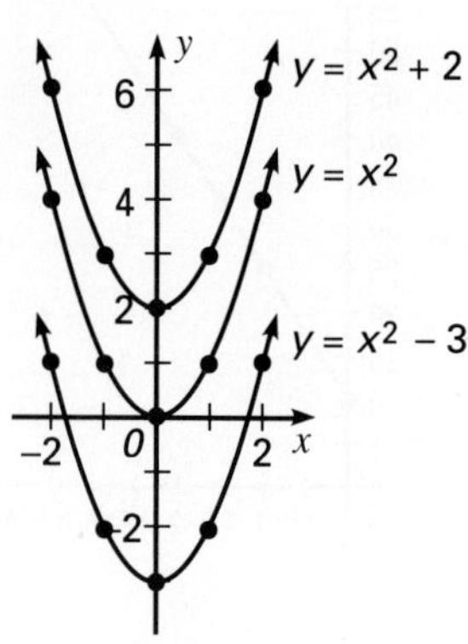


**b.** $y = x^2 + 10$

**Section 8–5, page 329**

**Problems**

**1. a.**

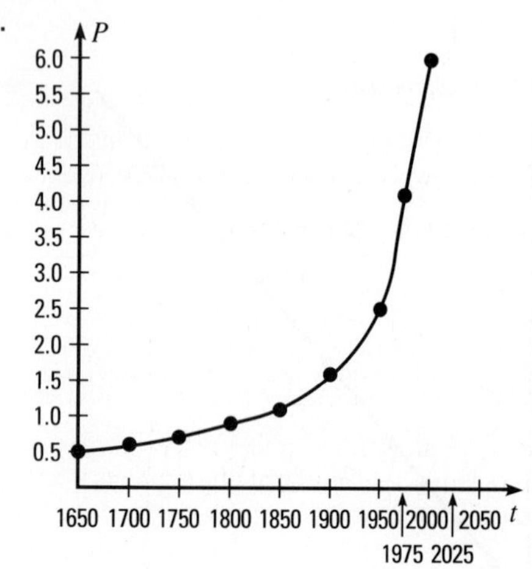


**b.** 1.3 billion, 1.9 billion **c.** about 6 billion **d.** in 2025 **3. a.** Days elapsed: 5, 10, 15, 20, 25, 30, 35, 40, 45; Amount of substance remaining (grams): 64, 32, 16, 8, 4, 2, 1, 0.5 **b.** after 60 days

**Section 8–6, page 331**

**Problems** **1.** c **3.** b

**5.**

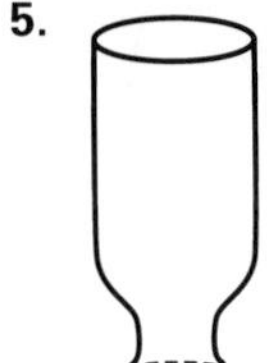

**7.**

**Section 8–7, pages 334–335**

**Exercises** **1.** 6 units **3.** 4 units **5.** 3.2 **7.** 13.0 **9.** $x^2 + y^2 = 9$ **11.** $x^2 + y^2 = 64$ **13.** $AB = 5$; $BC = 5$; $CA = 6$; isosceles **15.** $(x - 0)^2 + (y - 3)^2 = 100$; $x^2 + y^2 - 6y = 91$ **17.** outside

**Problem Solving Applications** **1.** about 8 km **3.** about 13 km **5.** about 298 km **7.** Answers may vary. about 2,000 m

**Section 8–8, page 339**

**Exercises** **1.** solid **3.** solid **5.** below **7.** above

**9.**

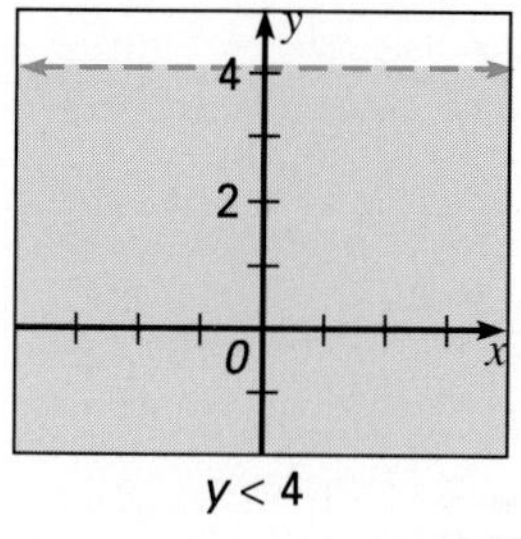


**11.**

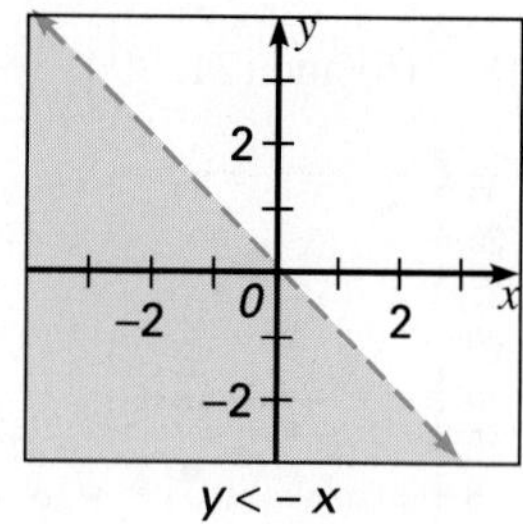


**13.**

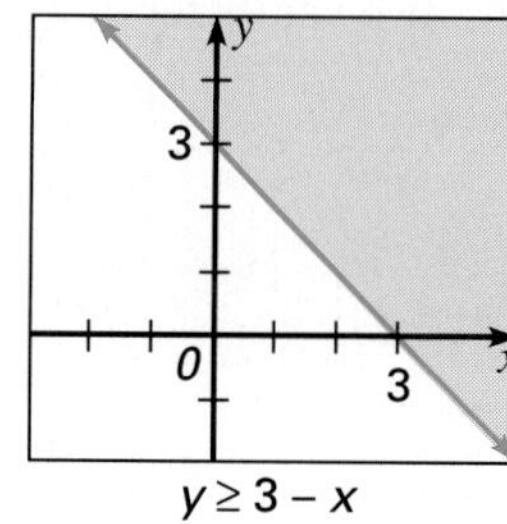


**15.**

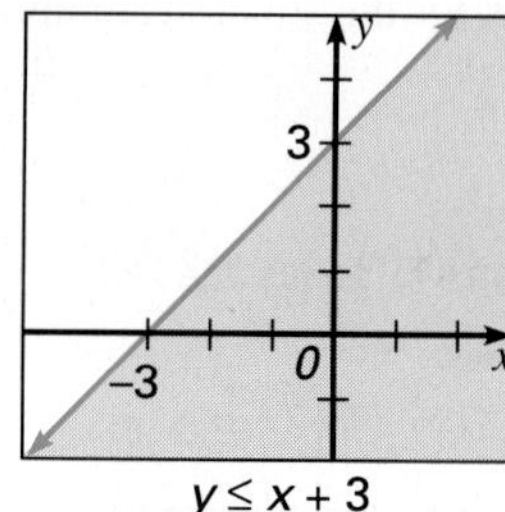


**17.**

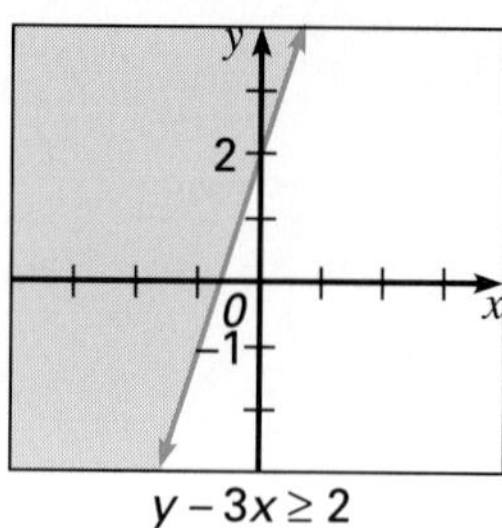


**19.**

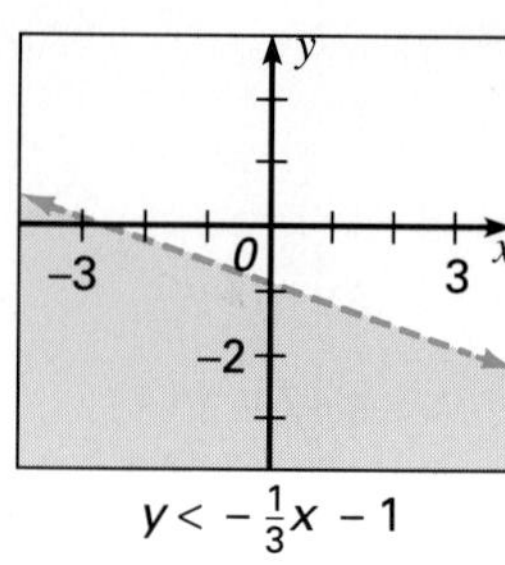


**21.** $y \geq -1$ **23.** $y > -x + 1$

**25.**

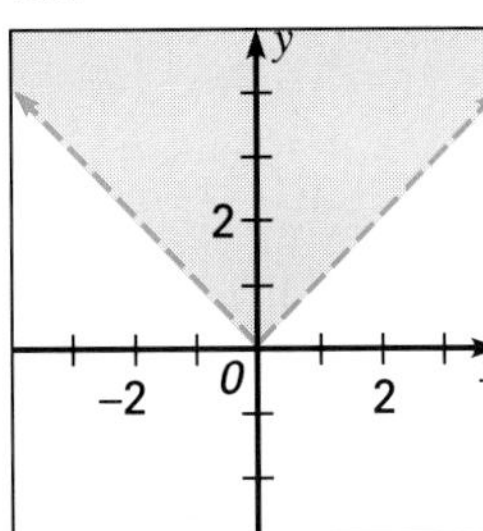


**27.**

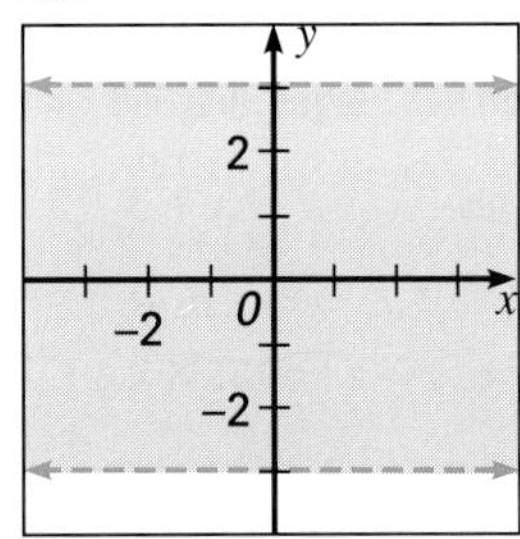


**29.**

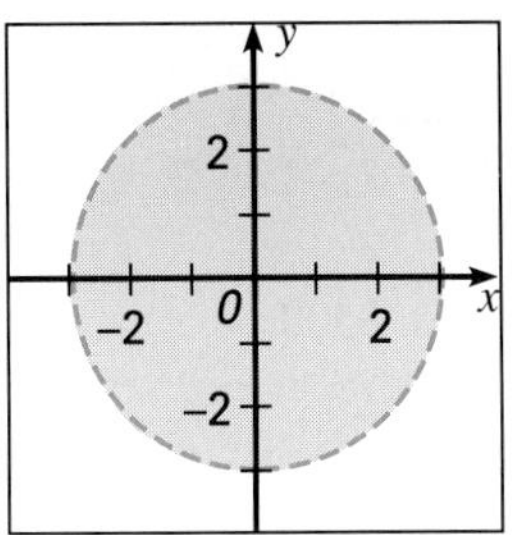


## CHAPTER 9 SLOPE AND SYSTEMS

### Section 9–1, pages 350–351

**Exercises** **1.** $\frac{2}{5}$ **3.** $\frac{2}{5}$ **5.** –2 **7.** $\frac{2}{3}$

**9.**

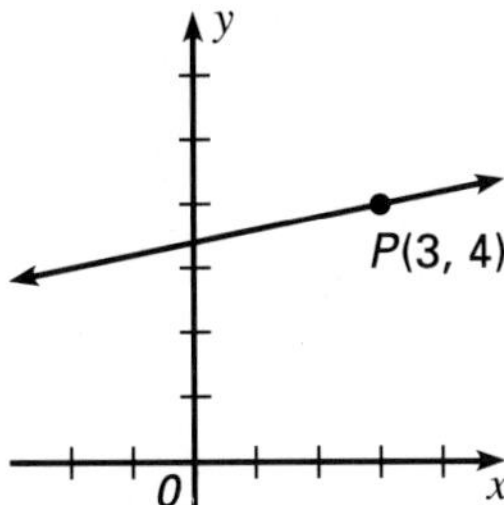


**11.**

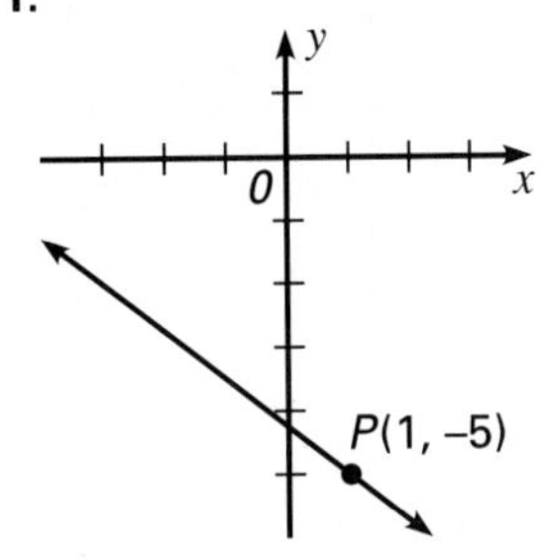


**13.**

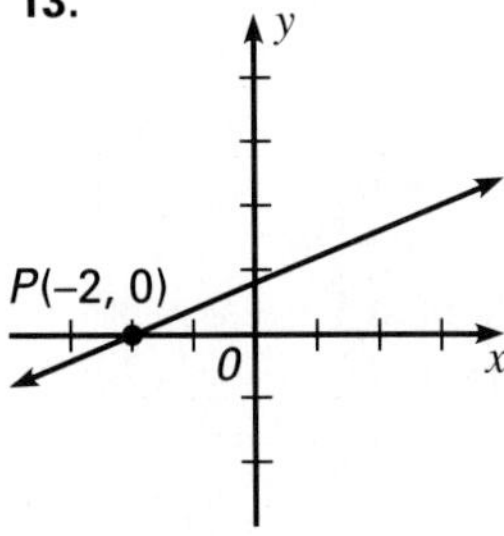


**15.** 5 **17.** –3

**19.** 0;

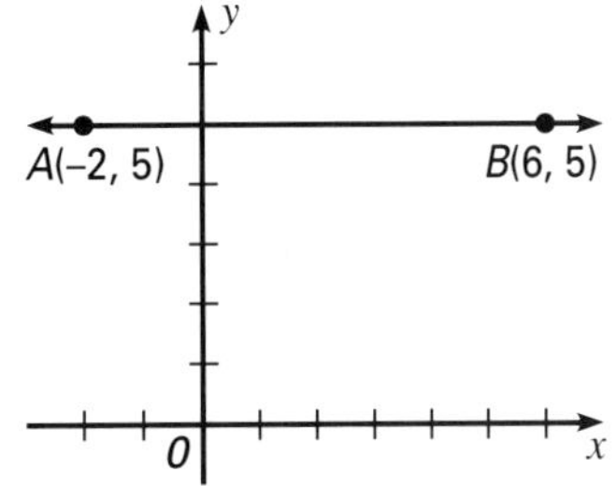


**21.** undefined;

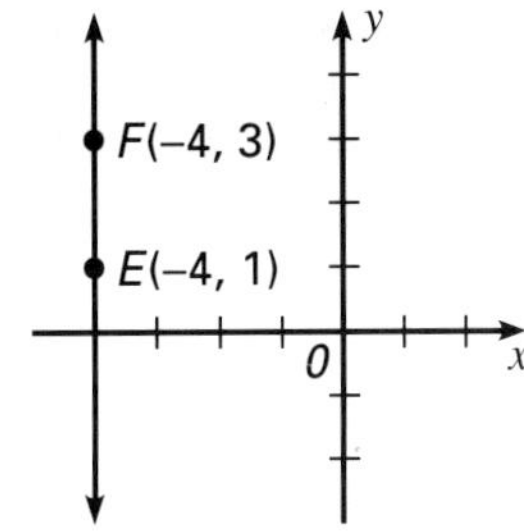


**23.** slope ≈ 0.85 and 0.85 > 0.83; no **25.** slope ≈ 0.81 and 0.81 < 0.83; yes **27.** slope = $0.\overline{3}$ and $0.\overline{3} > 0.\overline{2}$; no

**29.** $\frac{4}{3}$;

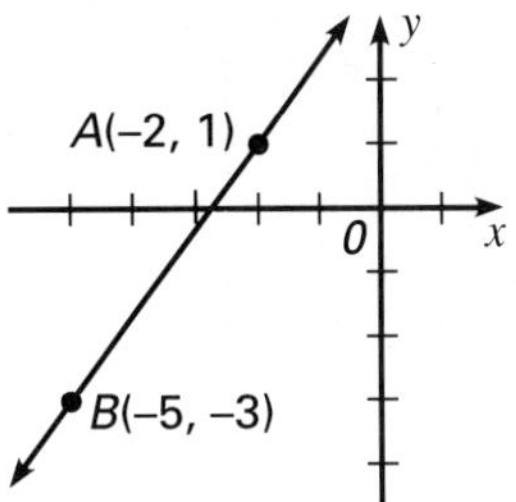


**31.** –1;

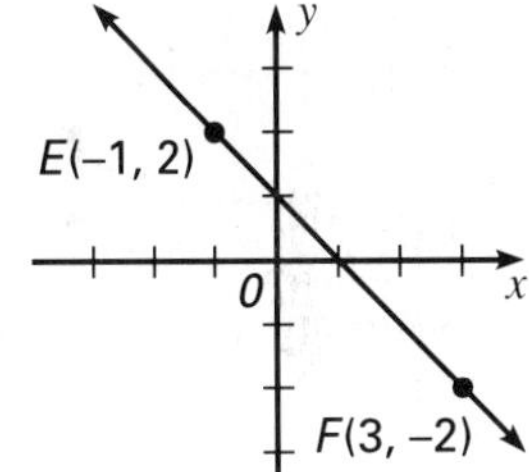


**33.** horizontal: $y$–coordinates the same; vertical: $x$–coordinates the same

### Section 9–2, pages 354–355

**Exercises** **1.** $-\frac{1}{2}$, 2 **3.** $-\frac{7}{5}$, $\frac{5}{7}$ **5.** perpendicular **7.** parallel **9.** no; sides are not parallel **11.** –4 **13.** 2 **15.** 4 **17.** 1 **19.** no; sides are not perpendicular **21.** always; slope is the same **23.** never; they are perpendicular

### Section 9–3, pages 358–359

**Exercises** **1.** $m = -4$; $b = 3$ **3.** $m = \frac{7}{3}$; $b = -2$ **5.** $m = -\frac{2}{5}$; $b = -1$

**7.**

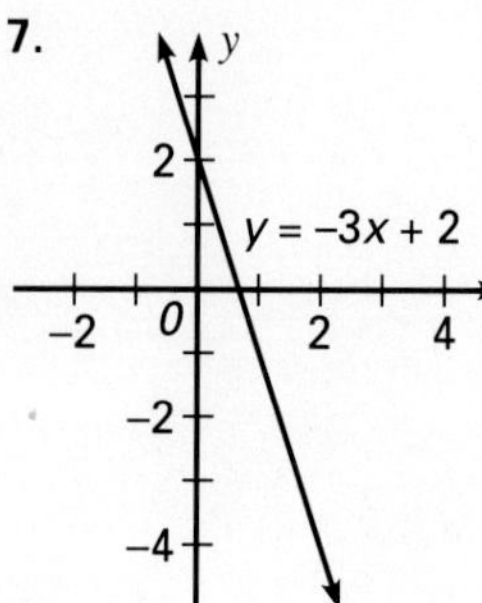


**9.**

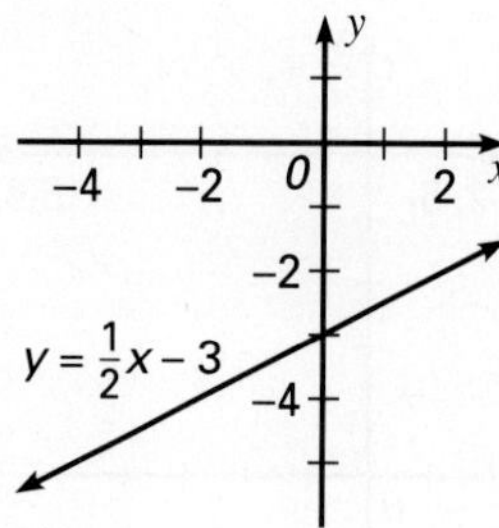


**11.**

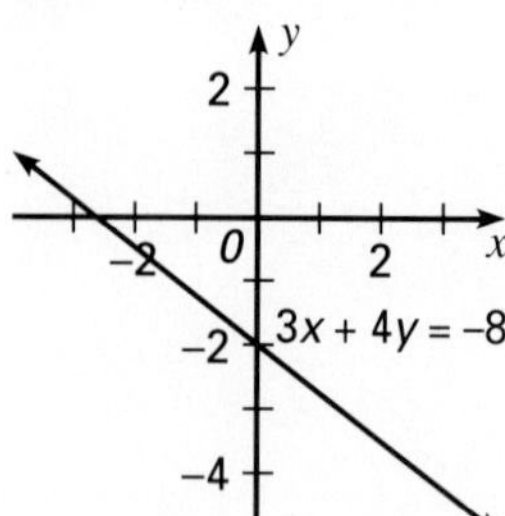


**13.**

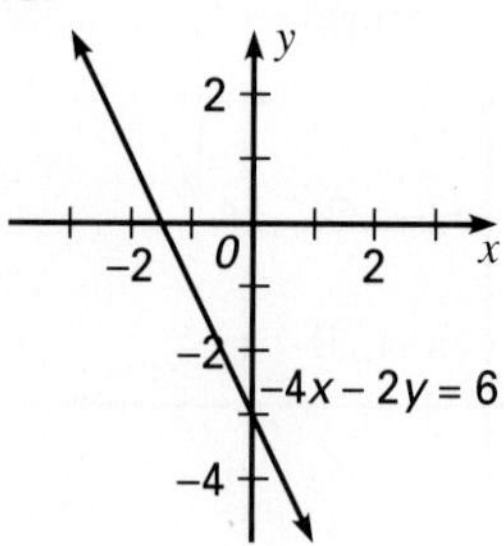


**15.** neither **17.** neither **19.** parallel **21.** no; slopes are the same so lines are parallel

**23.**

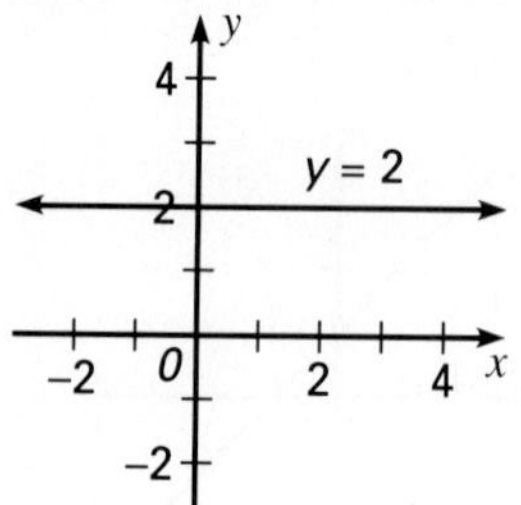


**25.**

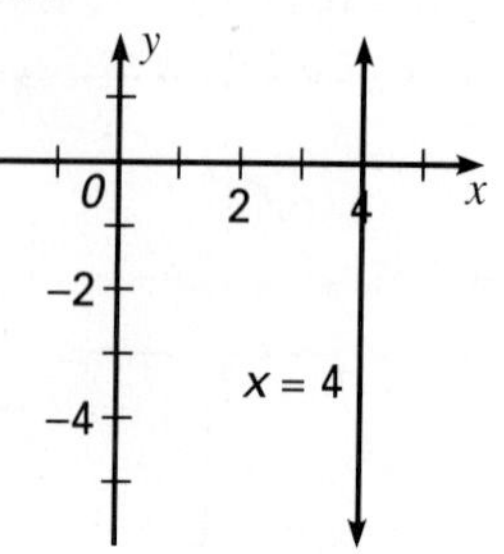


**27.** parallel **29.** perpendicular

**31.**

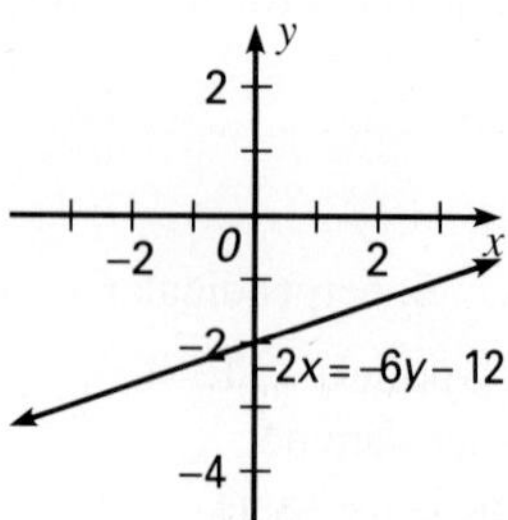


**33.** $m = \frac{1}{3}$; $b = -2$

**Section 9–4, page 361**

**Problems** **1.** 25 mi/h **3.** 65 mi/h **5.** 2 min **7.** 15 gal/min **9.** 10 gal/min **11.** Kai **13.** Marco

**Section 9–5, pages 364–365**

**Exercises** **1.** $y = -7x + 3$ **3.** $y = -\frac{3}{5}x - 4$ **5.** $y = 3x + 2$ **7.** $y = 2x - 8$ **9.** $y = 2x - 1$ **11.** $y = 5x - 8$ **13.** $y = x + 1$ **15.** $y = -\frac{1}{3}x - \frac{5}{3}$ **17.** $y = 2$ **19.** $y = -2x + 7$ **21.** $y = -\frac{1}{2}x - 3$ **23.** Two of the values can equal 0: $Ax = 0$ and $By = 0$ **25.** vertical **27.** $x$–axis **29.** $y = -\frac{A}{B}x + \frac{C}{B}$

**Section 9–6, pages 368–369**

**Exercises** **1.** no **3.** (4, 3) **5.** yes (–1, –1) **7.** 1 **9.** none

**Problem Solving Applications** **1.** 20 **3.** $500

**5.**

**GRAMATAN'S INCOME AND EXPENSES**

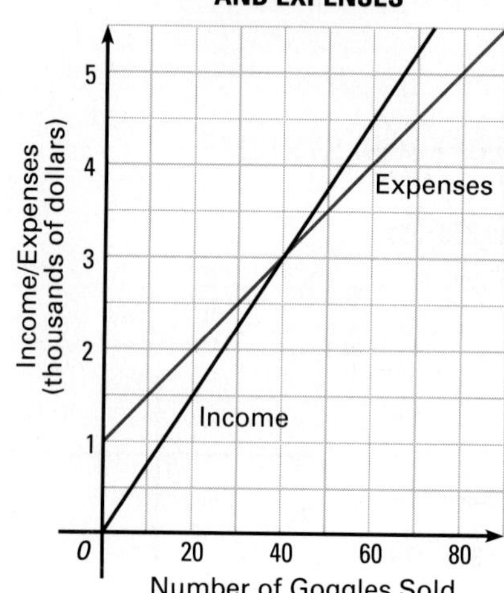


**7.** deluxe; they make a profit more quickly **9.** 25

**Section 9–7, page 371**

**Problems**

**1.** $m$ = number of men $m + w = 52$
$w$ = number of women $w = 2m - 1$

**3.** $l$ = length $w = l - 9$
$w$ = width $2l + 2w = 26$

**5.** $v$ = number of workers in 1 $v + w = 30$
$w$ = number of workers in 2 $v = 2w - 6$

**7.** $s$ = smaller number $s + l = 46$
$l$ = larger number $\frac{1}{2}l + s = 26$

**9.** $s$ = number of gal super $s + r = 12{,}074$
r = number of gal regular $s = 2r + 5$

**11.** $a$ = length of each of the two congruent sides $\quad 2a = b + 5$
$b$ = length of third side $\quad 2a + b = 35$
Choice of letters for variables will vary.

### Section 9–8, pages 374–375

**Exercises** **1.** (2, 2) **3.** no solution **5.** (3, 1) **7.** (2, 1) **9.** (3, 2) **11.** (0, 5) **13.** no solution **15.** 6 **17.** $2.36 **19.** infinite solutions **21.** no solution
**23.**

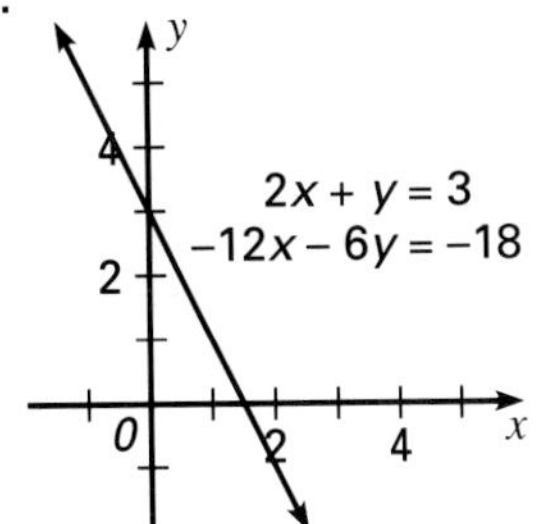


**25.**

**27.** parallel; 0 **29.** $a = 4$, $b = -2$, $c = -1$ **31.** $j = -1$, $k = 3$, $l = 5$

### Section 9–9, pages 378–379

**Exercises** **1.** (2, 3) **3.** (2, 1) **5.** (–4, 3) **7.** (–1, –1) **9.** 15 nickels, 20 dimes **11.** (1, 2) **13.** (–5, 2) **15.** 6 quarters, 8 dimes **17.** 8 years old **19.** For each there are an infinite number of solutions **21.** One has coefficients and constants that are twice or 3 times the other.
**23.** coincident lines;

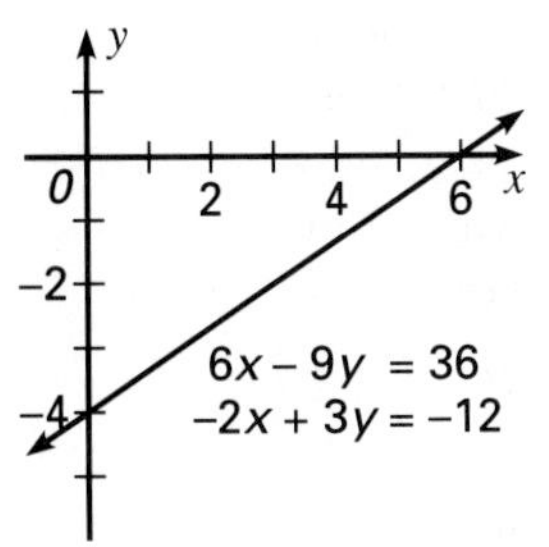


### Section 9–10, pages 382–383

**Exercises** **1.** $y \geq -4x + 2$; $y \leq -x - 2$ **3.** $y \geq -\frac{3}{2}x + 5$; $y > x - 5$ **5.**

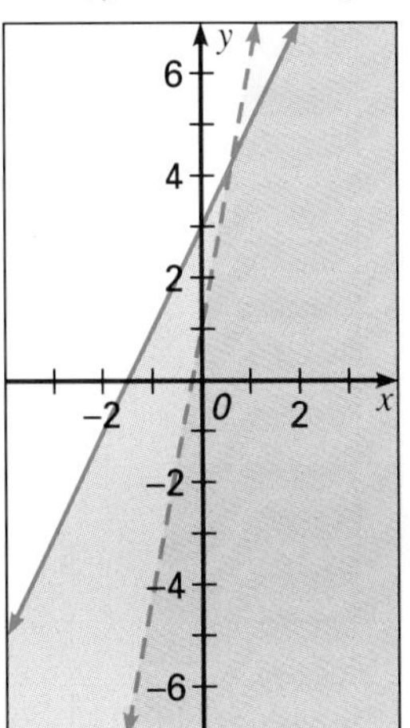

**7.**

**9.** $x \leq -3$; $y \geq \frac{1}{2}x - 2$; $y > -\frac{3}{4}x + 4$
**11.** $y \geq -3$; $y \leq \frac{2}{3}x - 2$; $y > -\frac{1}{3}x + 3$
**13.**

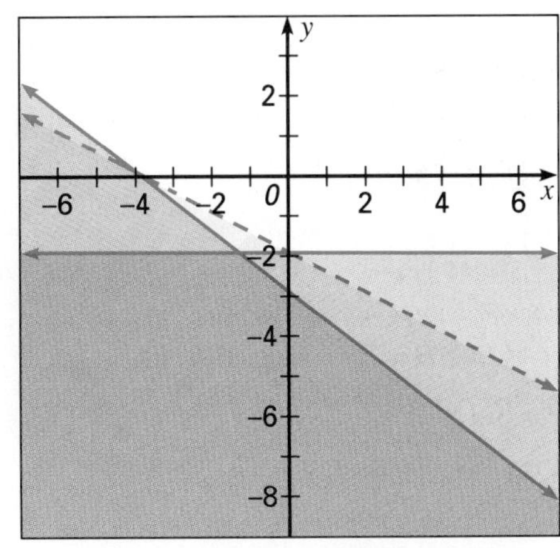

**15.** a rectangle;

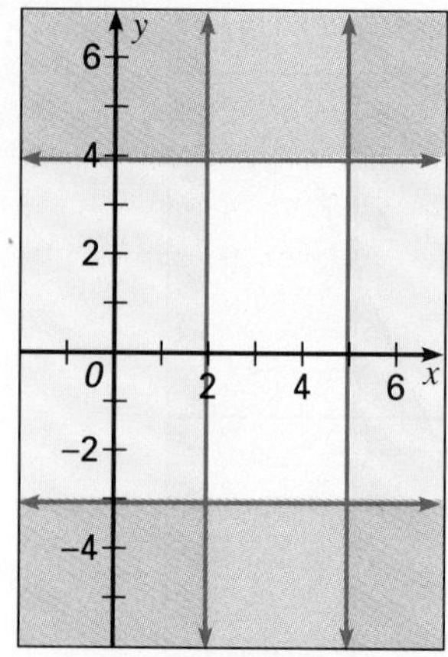


**17.** Answers will vary. Sample answer is given. $x \le -4$; $x \ge 5$; $y \ge 5$; $y \le -4$

## CHAPTER 10 RATIO, PROPORTION, AND VARIATION

### Section 10–1, pages 394–395

**Exercises** **1.** 1:4 **3.** 4:15 **5.** 52 acres **7.** 1.4 min/page **9.** 5 apples for $1.15 **11.** 10:13 **13.** 742 mi/h

**Problem Solving Applications** **1. a.** 5:21 **b.** 21:5 **3.** 1:12 **5.** 60:40 is 3:2 in lowest terms; 65:45 is 13:9 in lowest terms; $\frac{3}{2} > \frac{13}{9}$

### Section 10–2, pages 398–399

**Exercises** **1.** yes **3.** yes **5.** $z = 9$ **7.** 294 lb **9.** 48 **11.** $n = 25$ **13.** $n = 2$ **15.** 9 **17.** 12

**Problem Solving Applications** **1.** 1.65 m **3.** 15 **5.** 1 in.:10 yd **7.** Answers will vary.

### Section 10–3, page 401

**Problems** **1.** 65% **3.** $37\frac{1}{2}$ lb **5.** 14,654 workers **7.** 234 teenagers **9.** 30%

### Section 10–4, pages 404–405

**Exercises** **1.** yes **3.** yes **5.** 14 cm **7.** 44° **9.** 28 in. **11.** $m\angle D = 32°$; $m\angle E = 102°$ **13.** $ABCD \sim RSPQ$ **15.** 25.5 cm **17.** $6\frac{7}{8}$ in. **19.** 675 $cm^2$ **21.** sometimes **23.** sometimes **25.** never

### Section 10–5, pages 408–409

**Exercises** **1.** 30 ft **3.** 80 ft **5.** 60 m **7.** 37 ft **9.** $4\frac{1}{2}$ ft **11.** 42 m **13.** The corresponding angles of the two triangles are congruent. **15.** 52 ft

### Section 10–6, pages 412–413

**Exercises** **1.** 50 **3.** 22 **5.** 3,810 cm **7.** $\frac{1}{24}$ **9.** 144 **11.** 78 m **13.** $E = 0.15p$; 40 lb **15.** 16 schillings **17.** $y$ is multiplied by 4; by 9 **19.** direct variation **21.** neither

### Section 10–7, pages 416–417

**Exercises** **1.** 320 **3.** 6.25 **5.** 1.6 min **7.** 6°C **9.** inverse **11.** direct **13.** direct **15.** $h = \frac{42}{r^2}$ **17.** 12,500 **19.** 0.032 **21.** 25:4 **23.** $y$ varies directly as $z^2$

### Section 10–8, pages 418–419

**Problems** **1.** choose a method of computation; yes **3.** use a system of equations; 180 mi **5.** make an organized list; 12 corn, 36 bran, 30 barley, 45 oats **7.** work backwards; 1,100 buses **9.** find a pattern; 4:50 P.M. **11.** make an algebraic model; rent from Happy when total miles are expected to be less than 160 mi, otherwise use Apex **13.** find a pattern; **a.** $\frac{1}{2}$ **b.** $\frac{2}{3}$ **c.** $\frac{3}{4}$ **d.** The pattern is $\frac{a}{b}$.

## CHAPTER 11 TRANSFORMATIONS

### Section 11–1, page 431

**Exercises** **1. and 3.**

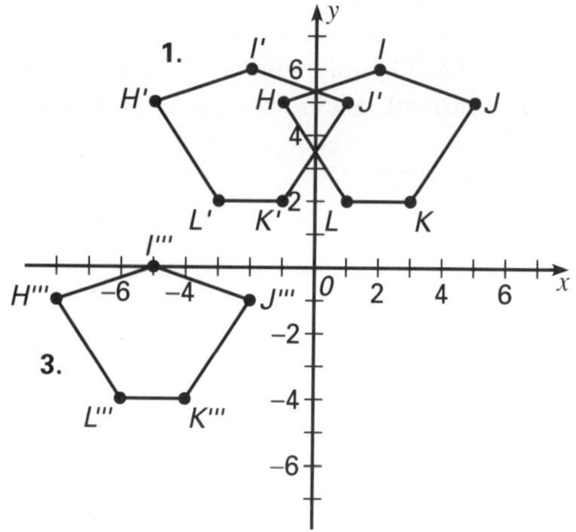

5. $(x, y) \rightarrow (x + 5, y - 2)$ **7.** $D(0, 6)$, $E(3, 6)$, $F(3, 4)$, $G(-1, 4)$ **9.** These polygons tessellate—square, isosceles triangle, equilateral triangle, parallelogram, regular hexagon.

### Section 11–2, page 433

**Problems**

**1.** 5 units;

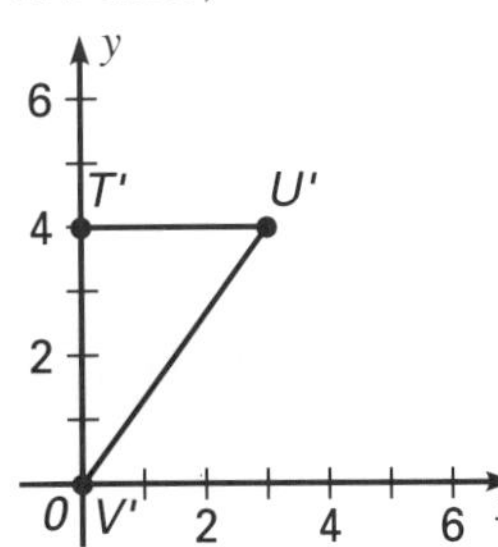


**3.** 5.8 units;

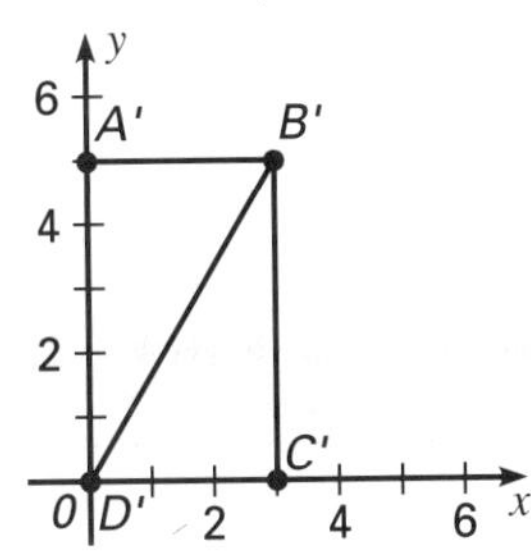


**5.** Find the length of $\overline{NP}$ and divide by 2; 3.4 units;

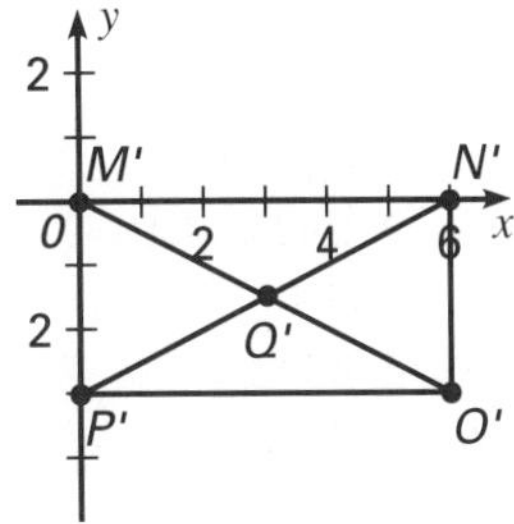


**7.** $\overline{BC}$ and $\overline{EF}$ are each 4 units; $\overline{CD}$, $\overline{DE}$, $\overline{FG}$, and $\overline{BG}$ are all the same length, 3.6 units; $(2 \times 4) + (4 \times 3.6) = 22.4$ units;

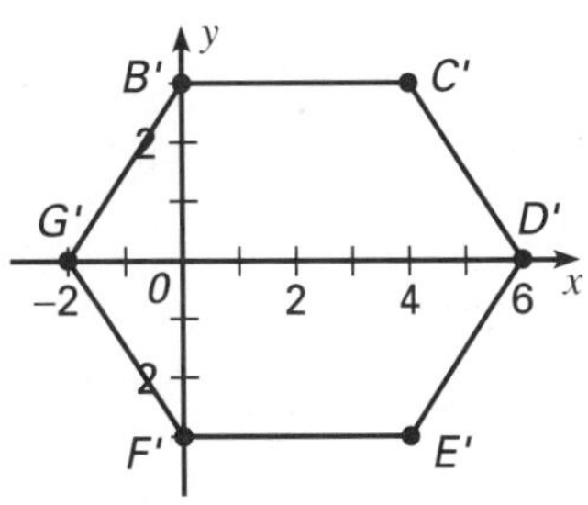


### Section 11–3, pages 436–437

**Exercises** **1.** (5, 3) **3.** (2, –7) **5.** (0,6)

**7.**

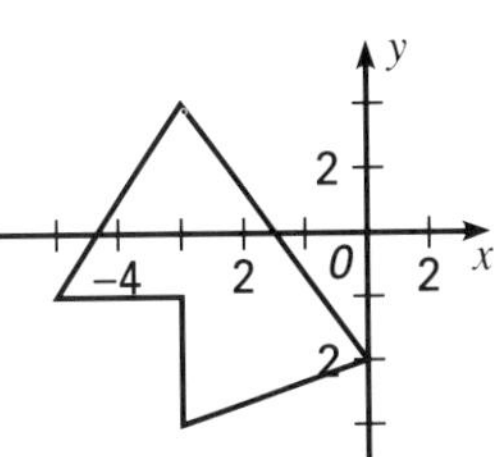


**9.**

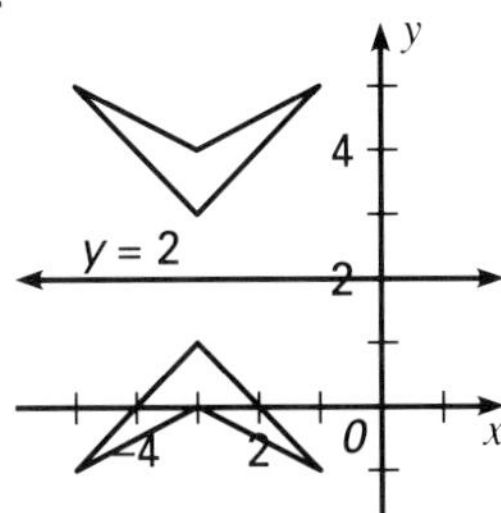


**11.**

**13.** none

**15.** $(x, y) \rightarrow (x, -y + 2)$

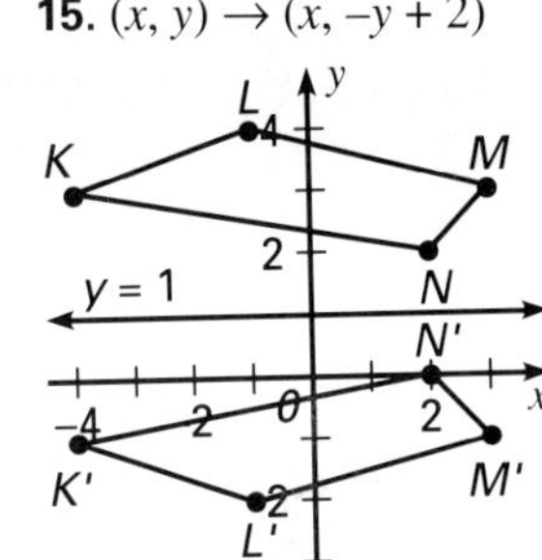


**17.** Answers will vary.

### Section 11–4, pages 440–441

**Exercises**

**1.**

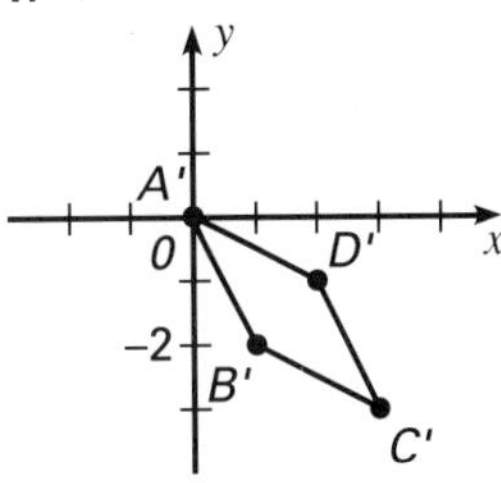


**3.**

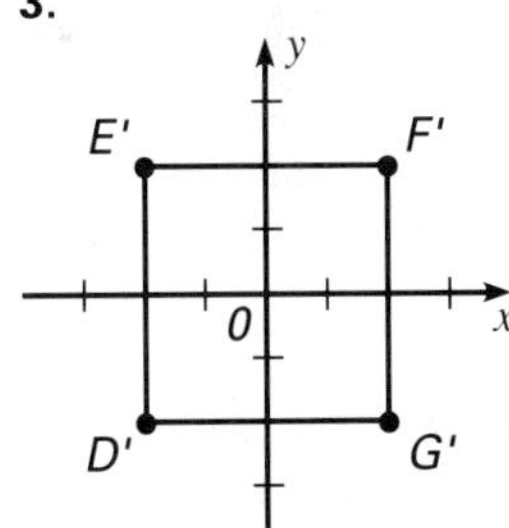


**5.** 180° clockwise or counterclockwise about T
**7.** Triangle 3 **9.** Triangle 4 **11.** Triangle 2
**13.** translation in one direction

### Section 11–5, page 443

**Problems** Possible answers are given. **1.** translation 1 unit to the right and reflection across the line $y = x$ **3.** reflection across $y$–axis and rotation 90° clockwise about the point $(1, -1)$ **5.** translation 2 units to the left and 2 units up and rotation 180° clockwise about the origin **7.** no **9.** no **11.** yes

## Section 11–6, pages 446–447

**Exercises** **1.** yes
**3.** no;

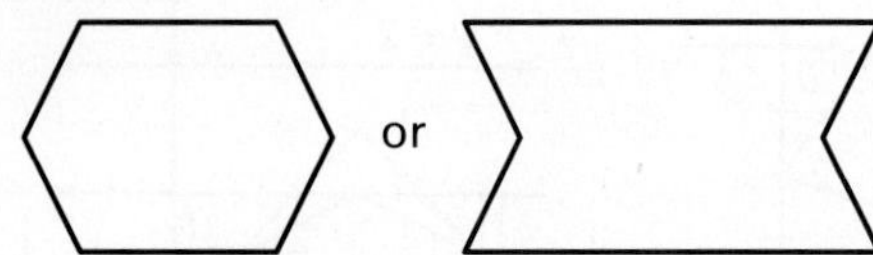


**5.** 6 **7.–13.** Answers will vary. Possible answers are given. **7. a.** isosceles trapezoid **b.** rectangle or rhombus **c.** not possible **d.** square **9.** $n$
**11.** infinite number **13.** Answers will vary.
**15.** remove coin 3 **17.** remove coins 4 and 5

## Section 11–7, pages 449–451

**Exercises**
**1.**

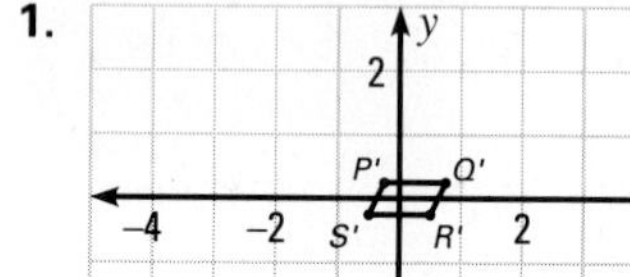


**3.**

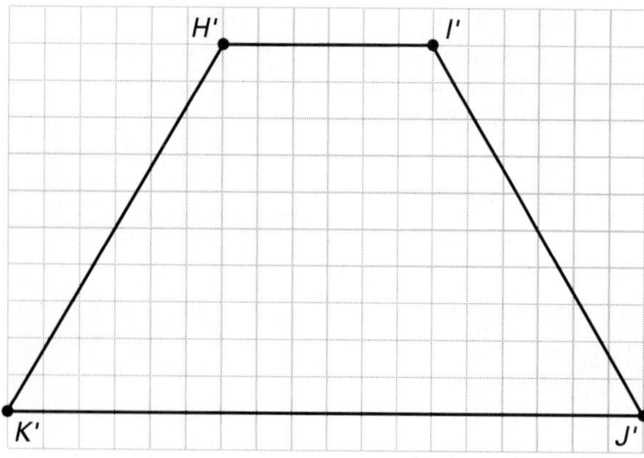


**5.** $2\frac{1}{2}$ **7.** dilation with center of dilation at $K$, scale factor of $\frac{1}{2}$, and reflection across the $y$–axis
**9.** dilation with center of dilation at the origin, scale factor of $\frac{2}{3}$, and rotation 90° counterclockwise about the origin **11.** 18 square units; 162 square units; 2 square units

**Problem Solving Applications** **1.** 2 in.; 3 in. **3.** 2; if you used the larger scale factor, the design would be too long **5.** $1\frac{1}{2}$ **7.** Answers will vary. **9.** Answers will vary. **11.** Answers will vary.

# CHAPTER 12 A CLOSER LOOK AT PROBABILITY

## Section 12–1, pages 462–463

**Exercises** **1.** 0.12 **3.** 0.045 **5.** $\frac{4}{25} = 0.16$

**Problem Solving Applications** **1.** Answers will vary.
**3.** Answers will vary.

## Section 12–2, page 465

**Problems** **1.** IF X < .5 THEN S = S + 1 **3.** d
**5.** The light shines red for $\frac{40}{40 + 55 + 5}$, or 0.4, of the total time it shines.

```
10 R = 0
20 FOR I = 1 to 100
30 X = RND(1)
40 IF X < .4 THEN R = R + 1
50 NEXT I
60 PRINT R
70 END
```

**7.**

```
10 A = 0
20 B = 0
30 FOR I = 1 to 1000
40 X = INT(6*RND(1)) + 1
50 IF X = 1 THEN A = A + 1
60 IF X = 2 THEN B = B + 1
70 NEXT I
80 PRINT (A + B)/1000
90 END
```

Alternative:

```
10 A = 0
20 FOR I = 1 to 1000
30 X = RND(1)
40 IF X < 1/3 THEN A = A + 1
50 NEXT I
60 PRINT A/1000
70 END
```

## Section 12–3, page 467

**Problems** **1.** 34 **3.** 121,750,000 **5.** 4,000
**7.** 290,700

### Section 12–4, pages 470–471

**Exercises** **1. a.** $(H, 1)$ $(H, 2)$ $(H, 3)$ $(H, 4)$ $(T, 1)$ $(T, 2)$ $(T, 3)$ $(T, 4)$ **b.**

| Coin | Spinner | Outcomes |
|---|---|---|
| H | 1 | H1 |
| | 2 | H2 |
| | 3 | H3 |
| | 4 | H4 |
| T | 1 | T1 |
| | 2 | T2 |
| | 3 | T3 |
| | 4 | T4 |

**3.** 64 **5.** $\frac{1}{16}$ **7.** 30 **9.** $\frac{5}{39}$ **11.** 30 **13. a.** $\frac{1}{216}$ **b.** $\frac{1}{8}$ **c.** $\frac{215}{216}$ **15.** equally likely; both probabilities are $\frac{1}{2}$ **17.** $\frac{2}{9}$ **19.** No; ordered pairs or a tree diagram show that the sample space consists of 16 possible outcomes. There are only 6 outcomes showing 2 male and 2 female puppies, but there are 8 outcomes showing 3 of one sex and 1 of the other sex. (There are 2 outcomes showing 4 of one sex.)

### Section 12–5, pages 474–475

**Exercises** **1.** $\frac{1}{4}$ **3.** $\frac{1}{6}$ **5.** $\frac{5}{8}$ **7.** $\frac{7}{8}$ **9.** $\frac{7}{13}$ **11.** $\frac{1}{64}$ **13.** A Venn diagram can be used to show the total of 40 photos, including 10 of Joe alone and 15 of Karen alone. **15.** $\frac{1}{26}$ **17.** yes; an event and its complement are mutually exclusive. **19.** $\frac{4}{9}$

### Section 12–6, pages 478–479

**Exercises** **1.** $\frac{1}{50}$ **3.** $\frac{4}{25}$ **5.** $\frac{1}{6}$ **7.** $\frac{1}{286}$ **9.** $\frac{24}{91}$ **11.** 3 ways; $\frac{8}{27}$

**Problem Solving Applications** **1.** Answers will vary. **3.** yes; Answers will vary. **5.** no **7.** 12; 9 **9.** No. **11.** Answers will vary.

### Section 12–7, pages 482–483

**Exercises** **1.** 40,320 **3.** 20 **5.** 24 **7.** $\frac{1}{2,024}$ **9.** 66 **11.** 24 **13.** 84 **15.** 110 **17.** 25 numbers **19. a.** $\frac{3}{143}$ **b.** $\frac{7}{143}$ **c.** $\frac{63}{143}$

## CHAPTER 13 LOGIC AND SETS

### Section 13–1, pages 494–495

**Exercises** **1.** {1, 2, 3, 4, 5, 6, 7, 8, 9, 10, 11, 12, 13, 14} **3.** {*a*, *e*, *i*, *o*, *u*} **5.** {–3, –2, –1, 0, 1, . . .} **7.** Sample: {1, 2, 3} **9.** equal **11.** equal **13.** false **15.** true **17.** {1}, $\phi$ **19.** {*m*}, {*a*}, {*t*}, {*m*, *a*}, {*a*, *t*}, {*m*, *t*}, {*m*, *a*, *t*}, { } **21.** false **23.** false **25.** $\phi \subseteq \{a, c\}$ **27.** {. . . , –2, –1, 0, 1} **29.** $\phi$ **31.** 4 **33.** 16 **35.** A set of *n* elements has $2^n$ subsets.

### Section 13–2, pages 498-499

**Exercises** **1.** {*bat*, *cat*, *rat*, *mat*, *hat*} **3.** {*bat*, *cat*, *rat*, *fat*, *pat*} **5.** {*p*, *c*, *e*} **7.** {*e*, *i*, *o*} **9.** {*o*, *t*} **11.** {1, 2, 8, 9} **13.** {4, 6} **15.** {2, 4, 6, 8} or *G* **17.** {6, 7, 8, 9, 10, 11, 12, 13, 14, 15, 16, 17, 18, 19, 20} **19.** {1, 2, 3, . . . , 20} **21.** {1, 2, 3, 4, 5} **23.** {1, 2, 3, 4, 5, 7, 11, 13, 17, 19} **25.** {2, 3, 4, 5, 6, 7, 8, 10, 11, 12, 13, 14, 16, 17, 18, 19} **27.** {2, 4, 6, 8, . . . , 18} or *B* **29.** {2} **31.** {1, 2, 3, 4, 5} or *C* **33.** $A \cap B = \{x | x$ is a real number and $-1 < x < 3\}$ **35.** The graph will be the parts of the graphs of *A* and *B* that overlap.

**37.**



**39.** true **41.** true **43.** true

### Section 13–3, pages 502–503

**Exercises** **1.** Converse: If the object has wings, then it is an airplane. Inverse: If the object is not an airplane, then it does not have wings. Contrapositive: If the object does not have wings, then it is not an airplane. **3.** Converse: If two lines form four right angles, then the lines are perpendicular. Inverse: If two lines are not perpendicular, then the lines do not form four right angles. Contrapositive: If two lines do not form four right angles, then the lines are not perpendicular. **5.** Exercise 1: false, false; a bird has wings, as does an insect; neither is an airplane. **7.** Exercise 3: true, true **9.** My car is a convertible. **11.** The bananas have spoiled. **13.** If the bill was paid, then the telephone service will not be turned off. If the telephone service is not turned off, then the bill was paid. **15.** If Rainie studies, then she will pass the test. If Rainie passed the test, then she studied. **17.** Answers will vary. Sample answer: If your car gets an average of 45 miles to the gallon, then you will produce 25.93 tons of $CO_2$ over the lifetime of your car. **19.** inverse **21.** inverse

### Section 13–4, page 507

**Exercises** **1.** Necessary: The bottle is not empty. Sufficient: The bottle contains water. **3.** Necessary: A number is odd. Sufficient: The number is not divisible by 2. **5.** Necessary: The figure is a rhombus. Sufficient: The figure is a square. **7.** You may enroll in Geometry I if and only if you have passed Algebra I. **9.** If you are a composer, then you write original music. **11.** If the east wind blows, then I sneeze. **13.** If you are admitted, then you have a ticket. **15.** If I drive my car, then my brakes are working well. If my brakes are not working well, then I never drive my car. **17.** If you are allowed to vote, then you are a registered voter. If you are not a registered voter, then you are not allowed to vote.

### Section 13–5, page 509

**Problems** **1.** too broad; definition does not distinguish characteristics that make a violin different from other stringed instruments **3.** too broad; definition does not specify characteristics that make Thanksgiving Day different from any other national holiday **5.** language is vague and too little information is given; the points must be noncollinear, for example **7.** too vague and imprecise; the definition does not name the set to which the term belongs or state the defining characteristics at all accurately **9.** Answers will vary. **11.** Two or more points are collinear if and only if those points lie on the same line. **13.** A point is a midpoint of a segment if and only if that point divides the segment into two equal parts.

### Section 13–6, pages 512–513

**Exercises** **1.** Conjecture: Every whole number can be written as the sum of consecutive whole numbers.

Counterexamples: 8, 16 **3.** 22 **5.** $\frac{3}{16}$

**7.** Conclusion: A rectangle has two pairs of parallel sides. **9.** Conclusion: $\overline{AB}$ is longer than $\overline{CD}$.
**11.** Conclusion: Ned is older than Megan.
**13.** none **15.** No conclusion is possible because Cleo may be able to translate the letter but may not be taking French III. She may be taking French I or French II or may know the language without having taken any courses.

### Section 13–7, pages 516–517

**Exercises** **1.** valid; law of detachment **3.** invalid; unsound **5.** If $n = 2x$, then $n^2 = (2x)^2 = 4x^2 = 2(2x^2)$ $2(2x^2)$ is an even number. Thus, the contrapositive is proved to be true. So, the original statement is true.

**Problem Solving Applications** **1.** If you eat Granny Granola bars, then you'll get fiber in your diet. You want fiber in your diet. You should eat Granny Granola bars. **3.** If you buy Springsole running shoes, then you won't be a fool. You don't want to be a fool. You should buy Springsole running shoes. **5.** If you don't like rich coffee flavor, then you won't buy Golden Bean coffee. You like rich coffee flavor. You buy Golden Bean coffee. **7.** If you take Wowicin, then you can get rid of your worst headache. You want to get rid of your worst headache. You should take Wowicin. **9.** If you want to fight germs that cause sore throats, then you will take Clean Green mouthwash. You want to fight germs that cause sore throats. You should take Clean Green mouthwash.

### Section 13–8, page 519

**Problems** **1.** $\angle a$ and $\angle b$ are supplementary. Therefore, the sum of their measures is 180°. **3.** $2m + 2n = 2(m + n)$; this expression has the form of an even number.

| **5.** $\angle SRT \cong \angle PRQ$ | **1.** $\angle SRT$ and $\angle PRQ$ are vertical angles. Vertical angles are congruent. |
|---|---|
| $\angle TSR \cong \angle PQR$ | **2.** $\angle TSR$ and $\angle PQR$ are right angles (given). Right angles are congruent. |
| $SR = RQ = 30$ m | **3.** Given |
| $\Delta PQR \cong \Delta TSR$ | **4.** By the ASA postulate |
| Therefore, $PQ \cong ST$ | **5.** Corresponding parts of congruent triangles are congruent. |

So, $ST = PQ = 25$ m

## CHAPTER 14 TRIGONOMETRY

### Section 14–1, page 531

**Exercises** **1.** $\frac{12}{13}$ **3.** $\frac{5}{13}$ **5.** $\frac{5}{13}$ **7.** $\frac{4}{5}$ **9.** $\frac{3}{4}$
**11.** 0.3057 **13.** 0.9877 **15.** 0.3256 **17.** 11.4301
**19. a.** increases **b.** decreases **c.** increases **21.** 12°; 65°; $\cos A = \sin(90 - A)$ **23.** There is none. Since the side adjacent to the 90° angle would be 0, $\tan 90° = \frac{\text{opposite}}{0}$. Division by 0 is undefined. **25.** False; $\frac{1}{\sin A} = \frac{1}{\frac{a}{b}} = \frac{b}{a}$; $\cos A = \frac{b}{c}$; $\frac{b}{c} \neq \frac{b}{a}$ **27.** True;
$\frac{\sin A}{\cos A} = \frac{a}{c} \div \frac{b}{c} = \frac{ac}{bc} = \frac{a}{b} = \tan A$

**Section 14–2, pages 534–535**

**Exercises** **1.** 52.3 **3.** 23.5 ft **5.** 82.1 m **7.** 20.8 m **9.** $\frac{1}{2}(12.2 \times 24) = 146.4 \text{ cm}^2$ **11.** The altitude of $\Delta JKL$ is 9.2 cm; $KL \approx 13.2$ cm

**Section 14–3, pages 538–539**

**Exercises** **1.** 30° **3.** 69° **5.** 18° **7.** 63° and 27°

**Problem Solving Applications** **1.** safe **3.** safe **5.** 8.4 m; no **7.** 3.8 ft **9.** 54°; no

**Section 14–4, page 541**

**Problems** Answers may vary by a tenth depending on the solution method used. **1.** $AB = 20$, $BC = 12$, $AC = 16$; $m\angle A \approx 37°$, $m\angle B \approx 53°$, $m\angle C = 90°$
**3.** The length of one side is also needed.
**5.** The measure of one other side or the measure of one acute angle is also needed. **7.** The height or width of the bookcase is needed.

**Section 14–5, pages 542–543**

**Problems** Strategies will vary. **1.** \$21.50 on meals, \$64.50 on clothing; use a system of equations
**3.** 2.4 m; choose a length, area, or volume formula
**5.** Wednesday; work backwards **7.** 3; make an algebraic model **9.** 25.3 km; draw a diagram
**11.** oats, 40 oz (144°); raisins, 30 oz (108°); walnuts, 15 oz (54°); sunflower seeds, 10 oz (36°); coconut, 5 oz (18°) **13.** 8 seats; guess and check, or work backwards

**Section 14–6, pages 546–547**

**Exercises** **1.** 16 **3.** 19 **5.** $\frac{3}{\sqrt{2}}$ **7.** $\frac{14}{\sqrt{3}}$ **9.** 2.1 m
**11.** 52.0 ft
**13.**

| | sine | cosine | tangent |
|---|---|---|---|
| 30° | $\frac{1}{2}$ | $\frac{\sqrt{3}}{2}$ | $\frac{1}{\sqrt{3}}$ |
| 45° | $\frac{1}{\sqrt{2}}$ | $\frac{1}{\sqrt{2}}$ | 1 |
| 60° | $\frac{\sqrt{3}}{2}$ | $\frac{1}{2}$ | $\sqrt{3}$ |

**15.** The values are the same. **17.** $9\sqrt{2}$ **19.** $\frac{x^2}{2}$

# Glossary

## ▶A

**absolute value** (p. 46) The distance of any number, $x$, from zero on the number line. Represented by $|x|$.

**acute angle** (p. 88) An angle measuring less than 90°.

**acute triangle** (p. 97) A triangle with three acute angles.

**addition property of equality** (p. 180) When two expressions are equal, you can add the same number to each expression and the resulting sums will be equal. If $a = b$, then $a + c = b + c$ and $c + a = c + b$.

**addition property of inequality** (p. 192) If $a < b$, then $a + c < b + c$ and $c + a < c + b$. If $a > b$, then $a + c > b + c$.

**addition property of opposites** (p. 49) The sum of a number and its opposite equals 0.

**additive inverse** (p. 49) The opposite of a number. The additive inverse of $a$ is $-a$, and of $-a$ is $a$.

**adjacent angles** (p. 89) Two angles that have a common vertex and a common side, but no interior points in common.

**alternate exterior angles** (p. 93) Two nonadjacent exterior angles on opposite sides of the transversal.

**alternate interior angles** (p. 93) Two nonadjacent interior angles on opposite sides of the transversal.

**angle** (p. 85) The figure formed by two rays that have a common endpoint.

**angle of rotation** (p. 438) In a rotation, the amount of turn expressed as a fractional part of a whole turn or as the angle of rotation in degrees.

**Angle-Side-Angle Postulate (ASA)** (p. 103) If two angles and the included side of one triangle are congruent to two corresponding angles and the included side of another triangle, then the triangles are congruent.

**arc** (p. 115) Part of a circle.

**associative property of addition** (p. 8) Changing the grouping of terms does not change the sum. $a + (b + c) = (a + b) + c$

**associative property of multiplication** (p. 8) Changing the grouping of terms does not change the product. $a(bc) = (ab)c$

**axes** (p. 312) The perpendicular lines used for reference in a coordinate plane.

## ▶B

**bar graph** (p. 32) A means of displaying statistical information in which horizontal or vertical bars are used to compare quantities.

**base** (p. 68) The repeating factor in a power. For example, in $3^2$, 3 is the base.

**biconditional statement** (p. 506) A statement in the *if-and-only-if* form. In the biconditional "$P$ if and only if $Q$," $P$ is both a necessary condition and a and sufficient condition for $Q$.

**binomial** (p. 268) A polynomial with two terms.

**bisector of an angle** (p. 89) A ray that divides the angle into two congruent adjacent angles.

**bisector of a segment** (p. 85) Any line, segment, ray, or plane that intersects the segment at its midpoint.

**boundary** (p. 336) The line separating two half-planes in the coordinate plane.

**box-and-whisker plot** (p. 160) A means of displaying data that shows the median of a set of data, the median of each half of data, and the least and greatest value of the data.

## ▶C

**center of rotation** (p. 438) The point about which the figure is rotated in a rotation.

**central angle (of a circle)** (p. 115) An angle with its vertex at the center of a circle. The measure of a central angle is always less than 180°.

**chord** (p. 114) A segment with both endpoints on the circle.

**circle** (p. 114) The set of all points in a plane that are a given distance from a fixed point in the plane. The fixed point is the center of the circle.

**circle graph** (p. 33) A means of displaying data where items are represented as parts of the whole circle. The parts are called *sectors.*

**circumference** (p. 230) The distance around a circle.

**closed half-plane** (p. 337) The graph of either half-plane *and* the line that separates them.

**clusters** (p. 149) See *stem-and-leaf plot.*

**cluster sampling** (p. 132) The members of the population are randomly selected from particular parts of the population and then surveyed in groups, not individually.

**coefficient** (p. 268) The numerical part of a monomial.

**collinear points** (p. 84) Points that lie on the same line.

**combination** (p. 481) A set of items chosen from a larger set without regard to order.

**combining like terms** (p. 64) Process using the distributive property to simplify an expression that contains like terms.

**commutative property of addition** (p. 8) Changing the order of two or more terms that are added does not change the sum. $a + b = b + a$

**commutative property of multiplication** (p. 8) Changing the order of two or more terms that are being multiplied does not change the product. $a \times b = b \times a$

**compatible numbers** (p. 27) Numbers easily used in mental computation.

**complement** (p. 496) If $A$ is a subset of $U$, then the subset of all elements of $U$ that are not elements of $A$ is called the complement of $A$, symbolized as $A'$.

**complementary angles** (p. 88) Two angles whose measures have a sum of 90°.

**composite number** (p. 288) A whole number greater than 1 that is not a prime number.

**compound event** (p. 472) An event made up of two or more simple events.

**concentric circles** (p. 115) Two or more circles having a common centerpoint and radii of different lengths.

**conditional statement** (p. 100) An *if-then* statement having two parts, hypothesis and conclusion. A conditional statement is symbolized as $p \rightarrow q$.

**cone** (p. 120) A three-dimensional figure with a curved surface and one circular base. The axis is a segment that joins the vertex to the center of the base.

**congruent angles** (p. 89) Angles having the same measure.

**congruent line segments** (p. 85) Line segments that have the same measure.

**congruent sides** (p. 96) Sides of a figure that have the same length.

**conjecture** (p. 510) A conclusion reached as a part of the process of inductive reasoning.

**constant** (p. 268) A monomial that contains no variables.

**constant of proportionality** (p. 411) See *constant of variation.*

**constant of variation** (p. 410) The nonzero constant $k$ in the rule $y = kx$. The constant $k$ represents the ratio of $y$ to $x$.

**contrapositive of a statement** (p. 501) For the statement "if $p$, then $q$," the contrapositive is "if $\sim q$, then $\sim p$.' A statement and its contrapositive are either both true or both false.

**convenience sampling** (p. 132) In this type of sampling method, members of a population are selected because they are readily available, and all are surveyed.

**converse** (p. 100) A conditional statement formed by interchanging the hypothesis and conclusion of the original conditional statement.

**convex polygon** (p. 110) A polygon is convex if each line containing a side has no points in the interior of the polygon. A polygon that is not convex is called concave.

**coordinate of the point** (p. 44) The number that corresponds to a point in a number line. An ordered pair of numbers associated with a point on a grid are the coordinates of the point.

**coordinate plane** (p. 312) A mathematical system in which two number lines are drawn perpendicular to each other. The horizontal number line is called the $x$-axis. The vertical number line is called the $y$-axis.

**coplanar points** (p. 84) Points that lie on the same plane.

**corresponding angles** (p. 93) Two angles in corresponding positions relative to two lines cut by a transversal. Also angles in the same position in congruent or similar polygons.

**cosine** (p. 529) See *trigonemetric ratios.*

**counterexample** (p. 100) An instance that satisfies the hypothesis, but not the conclusion of the conditional statement. A single counterexample proves that the conditional statement is false.

**cross-products** (p. 397) The cross-products of $\frac{a}{b} = \frac{c}{d}$ are $ad$ and $bc$. In a proportion, the cross-products of the terms are equal.

**cylinder** (p. 120) A three-dimensional shape made up of a curved region and two congruent circular bases that lie in parallel planes. The axis is a segment that joins the centers of the bases.

## ▶D

**deductive reasoning** (p. 510) A process of reasoning in which the truth of the conclusion necessarily follows from the truth of the premises. The premises and conclusion make up a deductive argument.

**degree** (p. 88) The unit commonly used to measure the size of an angle.

**dependent events** (p. 476) If the result of one event is affected by the result of another event, the events are said to be dependent.

**diagonal** (p. 111) A segment that joints two vertices of a polygon but is not a side.

**diameter** (p. 114) A chord that passes through the center of the circle and has endpoints on the circle. Any diameter is equal to twice the measure of any radius of a given circle.

**dilation** (p. 448) A transformation that produces an image that is the same shape as the original figure but a different size.

**direct square variation** (p. 411) A function that can be written in the form $y = kx^2$, where $k$ is a nonzero constant.

**direct variation** (p. 410) A function that can be written in the form $y = kx$, where $k$ is a nonzero constant.

**disjoint sets** (p. 497) Two sets whose intersection is the empty set ø are called disjoint sets.

**distance formula** (p. 333) For any points $P_1(x_1, y_1)$ and $P_2(x_2, y_2)$, the distance between $P_1$ and $P_2$ is given by the formula $P_1P_2 = \sqrt{(x_2 - x_1)^2 + (y_2 - y_1)^2}$.

**distributive property** (p. 9) Each factor outside parentheses can be used to multiply each term within the parentheses. $a(b + c) = ab + ac$

**domain** (p. 317) The set of all possible values of $x$ for the function $y = f(x)$.

## ▶E

**empty set** (p. 494) See *null set.*

**enlargement** (p. 448) A dilation that creates an image that is larger than the original figure.

**equation** (p. 176) A mathematical statement that two numbers or expressions are equal.

**equiangular triangle** (p. 97) A triangle having all three angles congruent.

**equilateral triangle** (p. 96) A triangle with all three sides congruent.

**equivalent sets** (p. 493) Two sets are equivalent sets if they contain the same number of elements.

**event** (p. 469) An outcome or combination of outcomes.

**experiment** (p. 460) An activity that is used to produce data that can be observed and recorded.

**experimental probability** (p. 460) The probability of an event based on the results of an experiment.

**exponent** (p. 68) A number showing how many times the base is used as a factor. For example, in $2^3$ the exponent is 3.

**exponential form** (p. 68) A number written with a base and an exponent. For example, the exponential form of $2 \times 2 \times 2 \times 2$ is $2^4$.

**extremes** (p. 396) The first and fourth terms of a proportion.

## F

**factor** (p. 288) Any number or polynomial multiplied by another to produce a product is called a factor.

**factorial** (p. 480) The product of all whole numbers from $n$ to 1. This product is written $n!$.

**factoring** (p. 292) Finding two or more factors of a number or a polynomial.

**finite set** (p. 492) See *infinite set.*

**formula** (p. 196) An equation stating a relationship between two or more quantities.

**frequency table** (p. 138) A method of recording data that shows how often an item appears in a set of data.

**front-end estimation** (p. 27) A method of estimating in which only the front-end digits are used and then the estimate is adjusted.

**function** (p. 16) A function $f$ is a correspondence or relationship that pairs each member of a given set with exactly one member of another set.

**function rule** (p. 16) The description of a function.

**function table** (p. 16) A table used to show values of the variable expression that is the rule for a given function.

**fundamental counting principle** (p. 469) If there are two or more stages of an activity, the total number of possible outcomes is the product of the number of possible outcomes for each stage of the activity.

## G

**gaps** (p. 149) See *stem-and-leaf plot.*

**geometry** (p. 84) The study of sets of points in space (from the Greek *geo,* meaning "earth" and *metria* meaning "measurement").

**graph of an inequality** (p. 337) The set of all ordered pairs that make the inequality true.

**graph of an equation** (p. 321) The set of all points whose coordinates are solutions of an equation.

**graph of a function** (p. 317) The graph of an equation that represents the function.

**graph of a number** (p. 44) The point that corresponds to a number; indicated on the number line by a solid dot.

**greatest common factor (GCF)** (p. 288) The greatest integer that is a factor of two or more integers. The GCF of two or more monomials is the common factor having the greatest numerical factor and the least power of the common factors.

## H

**histogram** (p. 148) A type of bar graph used to show frequencies. In a histogram the bars usually represent grouped intervals of numbers.

**hypotenuse** (p. 216) The side opposite the right angle in a right triangle.

## I

**identity property of addition** (p. 8) The sum of any number and zero is that number. $a + 0 = 0 + a = a$

**identity property of multiplication** (p. 8) The product of any number and 1 is that number. $a \times 1 = 1 \times a = a$

**independent events** (p. 476) If the result of one event is not affected by the result of another event, the events are said to be independent.

**indirect measurement** (p. 406) Calculating a measurement that is difficult to measure directly by using other measurements. Using similar triangles is one method of indirect measurement.

**inductive reasoning** (p. 510) Logical reasoning where the premises of an argument provide some, but not absolute, support for the conclusion. The conclusion is called a *conjecture.*

**inequality** (p. 192) A mathematical sentence that contains one of the symbols $<$, $>$, $\leq$ or $\geq$.

**infinite set** (p. 492) A set whose elements cannot be counted or listed. If all of the elements of a set can be counted or listed, the set is called a *finite* set.

**inscribed angle** (p. 116) An angle whose vertex lies on the circle and whose sides contain chords of the circle.

**integers** (p. 44) The set of whole numbers and their opposites.

**interquartile range** (p. 156) The difference between the values of the first and third quartiles.

**intersection of two figures** (p. 85) The set of points that are in both figures. Two lines intersect in a point. A plane and a line intersect in a point.

**intersection (of two sets)** (p. 497) The intersection of two sets, *A* and *B,* contains all the elements that are common to both *A* and *B.*

**inverse (of a statement)** (p. 501) For the statement, "if *p*, then *q*," the inverse is "if ~*p*, then ~*q*.' The inverse of a true statement is not necessarily true.

**inverse square variation** (p. 415) A function that can be written in the form $y = k/x^2$, or $x^2y = k$, where *k* is a nonzero constant.

**inverse variation** (p. 414) A function that can be written in the form $y = k/x$, where *k* is a nonzero constant.

**irrational numbers** (p. 44) Nonterminating and non-repeating decimals, such as π or the square root of 2.

**isosceles triangle** (p. 96) A triangle with at least two congruent sides.

## ▶L

**lateral faces** (p. 121) The faces of prisms and pyramids that are not bases.

**least common multiple (LCM)** (p. 289) The least integer that is a multiple of two or more numbers. The LCM of two or more monomials is the product when the LCM of the coefficients is multiplied by the greatest power of all variable factors.

**legs** (p. 216) In a right triangle, the two sides that are not the hypotenuse.

**line** (p. 84) A set of points that extends infinitely in opposite directions.

**linear equation** (p. 321) An equation for which the graph is a line. Linear equations may be written in the form $Ax + By = C$, where *A, B,* and *C* are real numbers and *A* and *B* are not both zero.

**linear function** (p. 321) A function that can be represented by a linear equation.

**line graph** (p. 33) A means of displaying data using points and line segments to show changes in data over periods of time.

**line of symmetry** (p. 444) A line on which a figure can be folded, so that when one part is reflected over that line it matches the other part exactly.

**line segment** (p. 85) Part of a line consisting of two endpoints and all points that lie between these two endpoints.

**line symmetry** (p. 444) A figure has line symmetry if a line drawn through it divides the figure into two matching parts.

**logically equivalent** (p. 505) Two statements are logically equivalent when they *always* have the same truth value; that is, both must be true or both must be false.

## ▶M

**major arc** (p. 115) An arc that is larger than a semicircle.

**mean** (p. 142) The sum of the data divided by the number of data. Also known as *arithmetic average.*

**means** (p. 396) The second and third terms of a proportion.

**measures of central tendency** (p. 142) Statistics used to describe a set of data. These measures are the mean, the median, and the mode.

**median** (p. 142) The middle value of the data when the data are arranged in numerical order.

**midpoint of a segment** (p. 85) The point that divides the segment into two congruent segments.

**minor arc** (p. 115) An arc that is smaller than a semicircle.

**mode** (p. 142) The number that occurs most often in a set of data.

**monomial** (p. 268) An expression that is a number, a variable, or a product of a number and one or more variables with whole number exponents.

**multiple** (p. 289) The product of any number and another whole number or the product of two or more monomials.

**multiplication property of equality** (p. 180) When two expressions are equal, you can multiply each expression by the same number and the resulting products will be equal. If $a = b$, then $ac = bc$.

**multiplication property of –1** (p. 54) Multiplying any real number by –1 results in a product that is the additive inverse of the number. $-1(a) = -a$, and $a(-1) = -a$

**multiplication property of 0** (p. 8) The product of any term and 0 is 0. $a \times 0 = 0 \times a = 0$

**mutually exclusive** (p. 472) Events that cannot occur at the same time are mutually exclusive.

## ▶N

**necessary condition** (p. 504) A condition in whose absence an event cannot take place.

**negation** (p. 500) To write the negation of a statement, add "not" to the statement or add the words "it is not the case that." The symbol for negation is ~.

**negative integer** (p. 44) An integer that is less than zero.

**negative reciprocals** (p. 353) Two numbers are negative reciprocals when their product is –1.

**noncollinear points** (p. 84) Points that do not lie on the same line.

**noncoplanar points** (p. 84) Points that do not lie on the same plane.

**nonlinear function** (p. 324) A function represented by an equation whose graph is not a line.

**null set** (p. 494) A set containing no elements. The symbol for the null set is ø.

**numerical expression** (p. 4) An expression that does not contain variables. A numerical expression names a number.

## ▶O

**obtuse angle** (p. 88) An angle whose measure is greater than 90° but less than 180°.

**obtuse triangle** (p. 97) A triangle that has one obtuse angle.

**open half-plane** (p. 336) The region on either side of a line on a coordinate plane.

**opposite of the opposite property** (p. 46) The opposite of the opposite of any real number is the number. $-(-n) = n$

**opposite rays** (p. 85) Two rays, $\overrightarrow{BA}$ and $\overrightarrow{BC}$ in which $B$ is in $\overleftrightarrow{AC}$ and between $A$ and $C$.

**ordered pair** (p. 312) Two numbers named in a specific order.

**order of operations** (p. 5) Rules followed to evaluate expressions.

**order of rotational symmetry** (p. 445) The number of times a figure fits exactly over its original position when the figure is being turned about a point during a complete rotation.

**origin** (p. 312) The point of intersection of the $x$-axis and $y$-axis in a coordinate plane.

**outliers** (p. 149) See *stem-and-leaf plot.*

## P

**parallel lines** (p. 92) Coplanar lines that do not intersect.

**parallelogram** (p. 107) A quadrilateral having two pairs of parallel sides.

**parallel planes** (p. 92) Planes that do not intersect.

**percentiles** (p. 157) A measure of a rank, or standing, within a group. A percentile divides a group of data into two parts, those at or below a certain score and those above.

**perfect square** (p. 201) An integer whose square roots are integers.

**perfect square trinomial** (p. 300) A trinomial that results from squaring a binomial.

**permutation** (p. 480) An arrangement of items in a particular order.

**perpendicular lines** (p. 89) Two lines that intersect to form adjacent right angles.

**pictograph** (p. 32) A means of displaying data that uses pictures or symbols to represent data. The key identifies the number of data items represented by each symbol.

**plane** (p. 84) A flat surface that exists without end in all directions. A plane contains an infinite set of points and has no boundaries.

**point** (p. 84) A location in space having no dimensions. A point is represented by a dot, which is named by a letter.

**polygon** (p. 110) A closed plane figure formed by joining three or more line segments at their endpoints. Each segment, or *side* of the polygon, intersects exactly two other segments, one at each endpoint. The point at which the endpoints meet is called a *vertex*.

**polyhedron** (p. 120) Plural: polyhedra. A closed, three-dimensional figure made of polygons alone. The polygonal surfaces are called *faces*. Two faces meet, or intersect, to form an *edge*. The point at which three or more edges intersect is called a *vertex*.

**polynomial** (pp. 268–269) The sum of monomials. Each monomial is called a *term* of the polynomial. A polynomial is in *standard form* when its terms are ordered from the greatest to the least powers of one of the variables.

**positive integer** (p. 44) An integer that is greater than zero.

**power of a product rule** (p. 69) To find the power of a product, find the power of each factor and multiply. $(ab)^m = a^m b^m$

**power of a quotient rule** (p. 69) To find the power of a quotient, find the power of each number and divide. $\left(\frac{a}{b}\right)^m = \frac{a^m}{b^m}$

**power rule** (p. 69) To raise an exponential number to a power, multiply exponents. $(a^m)^n = a^{mn}$

**premises** (p. 511) A set of statements accepted as true and from which a conclusion is drawn in deductive and inductive arguments.

**prime factorization** (p. 288) A whole number written as a product of prime factors.

**prime number** (p. 288) A whole number greater than 1 that has exactly two factors, 1 and the number itself.

**prism** (p. 121) A polyhedron with two identical parallel faces called *bases*. The other faces are parallelograms.

**probability** (pp. 164–165) The chance or likelihood that an event will occur. An impossible event has a probability of 0. A certain event has a probability of 1.

**product rule** (p. 69) To multiply numbers with the same base, add the exponents. $a^m \times a^n = a^{m+n}$

**property of the opposite of a sum** (p. 65) For all real numbers $a$ and $b$, the opposite of the sum of $a$ and $b$ is the sum of the opposites of $a$ and $b$. Thus, $-(a + b) = -a + (-b)$.

**proportion** (p. 396) An equation stating that two ratios are equivalent.

**pyramid** (p. 121) A polyhedron with only one base. The other faces are triangles that meet at a vertex.

**Pythagorean Theorem** (p. 216) In any right triangle, the square of the length of the hypotenuse $c$ is equal to the sum of the squares of the lengths of the legs $a$ and $b$. The Pythagorean theorem is expressed as $c^2 = a^2 + b^2$.

## ▶ Q

**quadrant** (p. 312) One of the four regions formed by the axes of the coordinate plane.

**quadratic equation** (p. 326) An equation of the form $Ax^2 + Bx + C = 0$, where $A$, $B$, and $C$ are real numbers and $A$ is not zero.

**quadratic function** (p. 326) A function that can be represented by a quadratic equation.

**quadrilateral** (p. 106) A polygon that has four sides.

**quartiles** (p. 156) The three values which divide an ordered set of data into four equal parts. The *first quartile* is the median of the lower half of the data. The *third quartile* is the median of the upper half. The *second quartile* is another name for the median of the entire set of data.

**quotient rule** (p. 69) To divide numbers with the same base, subtract the exponents. $a^m \div a^n = a^{m-n}$

## ▶ R

**radius** (p. 114) A line segment that has one endpoint at the center of the circle and the other endpoint on the circle.

**random sampling** (p. 132) Each member of the population has an equal chance of being selected.

**range** (p. 142) The difference between the greatest and least values in a set of data.

**range of a function** (p. 317) The set of all possible values of $y$ for the function $y = f(x)$.

**rate** (p. 393) A ratio that compares two different kinds of quantities.

**ratio** (p. 392) A comparison of two numbers represented in one of the following ways: 9:14, $\frac{9}{14}$, 9 to 14.

**rational number** (p. 44) A number that can be expressed in the form $\frac{a}{b}$, where $a$ is any integer and $b$ is any integer except zero.

**ray** (p. 85) Part of a line that starts at one endpoint and extends without end in one direction.

**real numbers** (p. 44) The set of rational and irrational numbers together.

**reciprocals** (p. 353) Two numbers are reciprocals when their product is 1.

**rectangle** (p. 107) A parallelogram having four right angles.

**reduction** (p. 448) A dilation that creates an image that is smaller than the original figure.

**reflection** (p. 434) A transformation in which a figure is reflected, or flipped, over a line of reflection. A reflection image is congruent to the preimage but oriented in the opposite direction.

**regular polygon** (p. 111) A polygon that has all sides congruent and all angles congruent.

**relative frequency** (p. 460) A comparison of the number of times a particular outcome occurs to the total number of observations.

**relatively prime** (p. 291) If two numbers have 1 as their greatest common factor, the numbers are relatively prime.

**repeating decimal** (p. 44) A decimal in which a digit or group of digits repeats. A bar above the group of repeating digits is used to express the repeating decimal.

**rhombus** (p. 107) A parallelogram with four congruent sides.

**right angle** (p. 88) An angle measuring exactly 90°.

**right triangle** (p. 97) A triangle having one right angle.

**rotation** (p. 438) A transformation in which a figure is rotated, or turned, about a point.

**rotational symmetry** (p. 445) A figure has rotational symmetry if it fits exactly over its original position at least once during a complete rotation about a point.

**rounding** (p. 27) Expressing a number to the nearest ten, hundred, thousand, and so on.

## ▶ S

**same-side interior angles** (p. 93) Interior angles on the same side of a transversal.

**sample** (p.132) A representative part of the population.

**sample space** (p. 468) The set of all possible outcomes of an event.

**sampling** (p. 466) See *cluster sampling, convenience sampling, random sampling,* and *systematic sampling.*

**scale drawing** (p. 399) A drawing that represents a real object. All lengths in the drawing are proportional to actual lengths in the object. The *scale* of the drawing is the ratio of the size of the drawing to the actual size of the object.

**scale factor** (p. 448) The number that is multiplied by the length of each side of a figure to create an altered image in a dilation.

**scalene triangle** (p. 96) A triangle with no congruent sides and no congruent angles.

**scatter plot** (p. 152) A method of displaying the relationship between two sets of data. The data are represented by unconnected points.

**scientific notation** (p. 73) A notation for writing a number as the product of a factor that is greater than or equal to 1 and less than 10 and a second factor that is a power of 10.

**semicircle** (p. 115) An arc of a circle with endpoints that are the endpoints of a diameter.

**sequence** (p. 12) An arrangement of numbers according to a pattern.

**set** (p. 492) A well-defined collection of items. Each item is called an *element,* or *member,* of the set.

**Side-Angle-Side Postulate (SAS)** (p. 103) If two sides and the included angle of one triangle are congruent to two corresponding sides and the included angle of another triangle, then the triangles are congruent.

**Side-Side-Side Postulate (SSS)** (p. 103) If three sides of one triangle are congruent to three corresponding sides of another triangle, then the triangles are congruent.

**significant digits** (p. 256) Digits used in a measurement that indicate the number of measured units.

**similar figures** (p. 402) Figures that have the same shape but not necessarily the same size.

**simplify** (p. 64) To simplify a variable expression is to perform as many of the indicated operations as possible.

**sine** (p. 529) See *trigonometric ratios.*

**skew lines** (p. 92) Noncoplanar lines that do not intersect and are not parallel.

**slant height** (p. 243) The length of a line drawn from the tip to the base of a cone along the side of the figure. The slant height of a pyramid is the height of a face.

**slide** (p. 429) See *translation.*

**slope** (p. 348) The ratio of the change in $y$ to the change in $x$, or the ratio of the change in the *rise* (vertical distance) compared to the *run* (horizontal distance).

**slope-intercept form** (p. 356) A linear equation in the form $y = mx + b$ where $m$ is the slope of the graph of the equation and $b$ is the $y$-intercept.

**solution** (p. 192) A replacement set for a variable that makes a mathematical sentence true.

**sound argument** (p. 515) A deductive argument whose form is valid and whose premises are true.

**sphere** (p. 120) A three-dimensional figure consisting of the set of all points that are a given distance from a given point, called the *center* of the sphere.

**square** (p. 107) A parallelogram with four right angles and four congruent sides.

**square root** (p. 200) A number, $a$, is a square root of another number, $b$, if $a^2 = b$. A square root is one of two equal factors of a number.

**statement** (p. 500) A sentence that can be said to be true or false. See also *conditional statement.*

**stem-and-leaf plot** (p. 149) A method of displaying data in which certain digits are used as *stems* and the remaining digits are used *leaves*. *Outliers* are data values that are much greater than or much less than most of the other values. *Clusters* are isolated groups of values. *Gaps* are large spaces between values.

**straight angle** (p. 88) An angle measuring 180°.

**subset** (p. 493) If every element of set A is also an element of set B, then A is called a subset of B.

**substitution method for solving systems of equations** (p. 372) One equation is used to write an expression for one variable in terms of the other variable. Then the expression is substituted into the other equation.

**sufficient condition** (p. 504) A condition in whose presence an event must occur.

**supplementary angles** (p. 89) Two angles whose measures have a sum of 180°.

**surface area** (p. 238) The sum of the areas of all the faces of a three-dimensional figure.

**survey** (p. 132) A means of collecting data for the analysis of some aspect of a group or area.

**systematic sampling** (p. 132) Members of a population that has been ordered in some way are selected according to a pattern.

**system of equations** (p. 366) Two or more equations with the same variables.

## ▶T

**tangent** (p. 529) See *trigonometric ratios.*

**terminating decimal** (p. 44) A decimal in which the the only repeating digit is 0.

**terms** (p. 64) The parts of a variable expression that are separated by addition or subtraction signs. Terms that have identical variable parts are called *like* terms. Terms that have different variable parts are called *unlike* terms.

**tessellation** (p. 431) A pattern in which one or more shapes are repeated in a way that covers the plane without gaps or overlaps.

**theoretical probability** (p. 469) The theoretical probability of an event, $P(E)$, can be assigned by determining the number of favorable outcomes and the number of possible outcomes in the sample space. $P(E) = \frac{\text{number of favorable outcomes}}{\text{number of possible outcomes}}$

**transformation** (p. 426) A way of moving or changing the size of a geometric figure in the coordinate plane. The new figure is referred to as the *image* of the original figure, and the original is referred to as the *preimage* of the new.

**translation** (p. 429) A change in position of a figure such that all the points in the plane slide exactly the same distance and in the same direction at once. A translation image is congruent to the image.

**transversal** (p. 92) A line that intersects each of two other coplanar lines in different points. The intersection of a transversal with two lines produces interior and exterior angles.

**trapezoid** (p. 107) A quadrilateral having exactly one pair of parallel sides.

**tree diagram** (p. 468) A diagram that shows all the possible outcomes in a sample space.

**trend line** (p. 153) A line that can be drawn near most of the points on a scatter plot. A trend line that slopes upward to the right indicates a *positive correlation* between the sets of data. A trend line that slopes downward to the right indicates a *negative correlation.*

**triangle** (p. 96) The polygon formed by three line segments joining three noncollinear points. Each point is called a *vertex*. Each vertex names an angle of the triangle.

**trigonometric ratios** (p. 528) Ratios of the lengths of the sides of a right triangle. The *sine* of an acute angle in a right triangle is the ratio of the length of the leg opposite the angle to the length of the hypotenuse. The *cosine* of an acute angle is the ratio of the length of the leg adjacent to the angle to the length of the hypotenuse. The *tangent* of an acute angle is the ratio of the length of the leg opposite the angle to the length of the leg adjacent to the angle.

**trinomial** (p. 268) A polynomial with three terms.

**two-step equations** (p. 184) Equations involving two operations. To solve, use the addition property of equality first, then use the multiplication property.

## ▶U

**unit of precision** (p. 256) The smallest unit on a measuring instrument.

**unit price** (p. 393) A ratio comparing the price of an item to the unit of its measure,

**unit rate** (p. 393) A rate that has a denominator of 1 unit.

**universal set** (p. 496) The general set of all elements being considered in a discussion. The universal set could be an infinite set such as the set of whole numbers or a finite set.

**unsound argument** (p. 515) A deductive argument whose form is valid but contains at least one false premise.

## ▶V

**valid argument** (p. 514) A deductive argument such that if the premises are true, then the conclusion must be true.

**values of the function** (p. 317) The set of all possible values of $y$ for a function $y = f(x)$.

**variable** (p. 4) A symbol used to represent a number.

**variable expression** (p. 4) An expression that contains one or more variables.

**vertex** (p. 85) Plural: vertices. The common endpoint of two rays that form an angle. The rays are called the *sides* of the angle.

**vertical angles** (p. 89) The angles that are not adjacent to each other when two lines intersect. Vertical angles are congruent.

**volume** (p. 246) A measure of the number of cubic units needed to fill a region of space.

## ▶X

***x*-axis** (p. 312) The horizontal number line.

***x*-coordinate** (p. 312) The first number in an ordered pair.

## ▶Y

***y*-axis** (p. 312) The vertical number line.

***y*-coordinate** (p. 312) The second number in an ordered pair.

***y*-intercept** (p. 356) The $y$-intercept of a line is the $y$-coordinate of the point where the line intersects the $y$-axis.

# Index

## ▶D



## ▶E

## ▶ J



## ▶ L



## ▶ M



## ▶ N



## ▶ O

## P



## Q

## ▶U



## ▶V



## ▶W



## ▶X



## ▶Y



## ▶Z